鱼类病毒和细菌：病理学与防控

Fish Viruses and Bacteria：Pathobiology and Protection

[加] Patrick T. K. Woo
[美] Rocco C. Cipriano 主编

王启要 主译

马　悦　刘晓红 副主译

华东理工大学出版社
EAST CHINA UNIVERSITY OF SCIENCE AND TECHNOLOGY PRESS
·上海·

图书在版编目(CIP)数据

鱼类病毒和细菌：病理学与防控/(加) 胡栋祺(Patrick T. K. Woo)，(美) 罗科·C.西普里亚诺(Rocco C. Cipriano)主编；王启要主译.—上海：华东理工大学出版社，2021.12

书名原文：Fish Viruses and Bacteria: Pathobiology and Protection

ISBN 978-7-5628-6794-4

Ⅰ.①鱼… Ⅱ.①胡…②罗…③王… Ⅲ.①鱼类病毒病-介绍 Ⅳ.①S941.41

中国版本图书馆 CIP 数据核字(2021)第 277193 号

First Published by CAB International in the year 2017.

著作权合同登记号：图字 09-2019-1057 号

项目统筹 / 吴蒙蒙
责任编辑 / 赵子艳
装帧设计 / 徐　蓉
出版发行 / 华东理工大学出版社有限公司
地址：上海市梅陇路 130 号，200237
电话：021-64250306
网址：www.ecustpress.cn
邮箱：zongbianban@ecustpress.cn
印　　刷 / 上海中华商务联合印刷有限公司
开　　本 / 720 mm×1000 mm　1/16
印　　张 / 32.75
字　　数 / 642 千字
版　　次 / 2021 年 12 月第 1 版
印　　次 / 2021 年 12 月第 1 次
定　　价 / 198.00 元

编者及编者单位

第 1 章　传染性胰腺坏死病毒

阿伦·K. 达尔(Arun K. Dhar)

BrioBiotech LLC, Glenelg, Maryland, USA

Aquaculture Pathology Laboratory, School of Animal and Comparative Biomedical Sciences, The University of Arizona, Tucson, Arizona, USA

斯科特·拉帕特(Scott LaPatra)

Clear Springs Foods, Buhl, Idaho, USA

鲁·奥里(Andrew Orry)

Molsoft, San Diego, California, USA

F. C. 托马斯·奥尔纳特(F. C. Thomas Allnutt)

BrioBiotech LLC, Glenelg, Maryland, USA

第 2 章　传染性造血器官坏死病病毒

乔安·C. 莱昂(Jo-Ann C. Leong)

Hawai'i Institute of Marine Biology, University of Hawai'i at Mānoa, Kāne'ohe, Hawai'i, USA

盖尔·库拉特(Gael Kurath)

Western Fisheries Research Center, US Geological Survey, Seattle, Washington, USA

第 3 章　病毒性出血性败血症病毒

约翰·S. 拉姆斯登(John S. Lumsden)

Department of Pathobiology, Ontario Veterinary College, University of Guelph, Guelph, Ontario, Canada and Adjunct Professor, Department of Pathobiology, St. George's University, True Blue, Grenada

第 4 章　流行性造血器官坏死病与欧鲇病毒

保罗·希克(Paul Hick)
Faculty of Veterinary Science, University of Sydney, Camden, New South Wales, Australia

埃伦·阿里尔(Ellen Ariel)
College of Public Health, Medical and Veterinary Sciences, James Cook University, Townsville, Queensland, Australia

理查德·惠廷顿(Richard Whittington)
Faculty of Veterinary Science, University of Sydney, Camden, New South Wales, Australia

第 5 章　致瘤病毒：马苏大麻哈鱼病毒和鲤疱疹病毒

吉水真守(Mamoru Yoshimizu)
Faculty of Fisheries Sciences, Hokkaido University, Minato, Hakodate, Japan

葛西久江(Hisae Kasai)
Faculty of Fisheries Sciences, Hokkaido University, Minato, Hakodate, Japan

坂田义弘(Yoshihiro Sakada)
Graduate School of Veterinary Medicine, Hokkaido University, Sapporo, Japan

佐野夏美(Natsumi Sano)
Graduate School of Bioresources, Mie University, Tsu, Japan

佐野元彦(Motohiko Sano)
Faculty of Marine Science, Tokyo University of Marine Science and Technology, Tokyo, Japan

第 6 章　传染性鲑贫血症

克努特·福尔克(Knut Falk)
The Norwegian Veterinary Institute, Oslo, Norway

玛丽亚·阿梅尔福特(Maria Aamelfot)
The Norwegian Veterinary Institute, Oslo, Norway

第 7 章　鲤春病毒血症

彼得·狄克逊(Peter Dixon)
Centre for Environment, Fisheries and Aquaculture Science (Cefas) Weymouth

Laboratory, Weymouth, UK

戴维 · 斯通(David Stone)
Centre for Environment, Fisheries and Aquaculture Science (Cefas) Weymouth Laboratory, Weymouth, UK

第 8 章　斑点叉尾鮰病毒病

拉里 · A. 汉森(Larry A. Hanson)
Department of Basic Sciences, College of Veterinary Medicine, Mississippi State University, Mississippi, USA

莱斯特 · H. 库(Lester H. Khoo)
Thad Cochran Warmwater Aquaculture Center, Stoneville, Mississippi, USA

第 9 章　大口黑鲈虹彩病毒病

罗德曼 · G.格彻尔(Rodman G. Getchell)
Veterinary Medical Center, Department of Microbiology and Immunology, Cornell University, Ithaca, New York, USA

杰弗里 · H. 格鲁科克(Geoffrey H. Groocock)
Transit Animal Hospital, Depew, New York, USA

第 10 章　锦鲤疱疹病毒病

基思 · 韦(Keith Way)
Centre for Environment, Fisheries and Aquaculture Science (Cefas) Weymouth Laboratory, Weymouth, UK

彼得 · 狄克逊(Peter Dixon)
Centre for Environment, Fisheries and Aquaculture Science (Cefas) Weymouth Laboratory, Weymouth, UK

第 11 章　病毒性脑病和视网膜病

安娜 · 托凡(Anna Toffan)
OIE Reference Centre for Viral Encephalopathy and Retinopathy, Istituto Zooprofilattico Sperimentale delle Venezie, Legnaro (Padova), Italy

第 12 章　虹彩病毒病：真鲷虹彩病毒和高首鲟虹彩病毒

安彦川户(Kawato Yasuhiko)

National Research Institute of Aquaculture, Japan Fisheries Research and Education Agency, Nakatsuhamaura, Minami-Ise, Mie, Japan

库蒂昌川・苏巴马廉(Kuttichantran Subramaniam)
Department of Infectious Diseases and Pathology, University of Florida, Gainesville, Florida, USA

中岛和弘(Kazuhiro Nakajima)
National Research Institute of Aquaculture, Japan Fisheries Research and Education Agency, Nakatsuhamaura, Minami-Ise, Mie, Japan

托马斯・沃尔泽克(Thomas Waltzek)
Department of Infectious Diseases and Pathology, University of Florida, Gainesville, Florida, USA

理查德・惠廷顿(Richard Whittington)
Faculty of Veterinary Science, University of Sydney, Camden, New South Wales, Australia

第 13 章　鲑 甲 病 毒

马柳斯・卡尔森(Marius Karlsen)
PHARMAQ AS, Oslo, Norway

雷纳特・约翰逊(Renate Johansen)
PHARMAQ Analytiq, Bergen, Norway

第 14 章　杀鲑气单胞菌和嗜水气单胞菌

比恩海杜尔・K. 古德蒙兹多蒂尔(Bjarnheidur K. Gudmundsdottir)
Faculty of Medicine, University of Iceland, Reykjavik, Iceland

布林迪斯・比约恩斯多蒂尔(Bryndis Bjornsdottir)
Matís, Reykjavik, Iceland

第 15 章　爱德华氏菌

马特・J. 格里芬(Matt J. Griffin)
Thad Cochran National Warmwater Aquaculture Center, Mississippi State University, Stoneville, Mississippi, USA

特伦斯・E. 格林韦(Terrence E. Greenway)
Thad Cochran National Warmwater Aquaculture Center, Mississippi State

University, Stoneville, Mississippi, USA

戴维·J. 怀斯(David J. Wise)

Thad Cochran National Warmwater Aquaculture Center, Mississippi State University, Stoneville, Mississippi, USA

第16章　黄 杆 菌

托马斯·P. 洛克(Thomas P. Loch)

Department of Pathobiology and Diagnostic Investigation, College of Veterinary Medicine, Michigan State University, East Lansing, Michigan, USA

穆罕默德·费萨尔(Mohamed Faisal)

Department of Pathobiology and Diagnostic Investigation, College of Veterinary Medicine, Michigan State University, East Lansing, Michigan, USA

第17章　诺神弗朗西斯菌

埃斯特班·索托(Esteban Soto)

Department of Medicine and Epidemiology, School of Veterinary Medicine, University of California Davis, Davis, California, USA

约翰·P. 霍克(John P. Hawke)

Department of Pathobiological Sciences, School of Veterinary Medicine, Louisiana State University, Baton Rouge, Louisiana, USA

第18章　分 枝 杆 菌

戴维·T. 高蒂尔(David T. Gauthier)

Department of Biological Sciences, Old Dominion University, Norfolk, Virginia, USA

玛莎·W. 罗兹(Martha W. Rhodes)

Department of Aquatic Health Sciences, Virginia Institute of Marine Science, The College of William and Mary, Gloucester Point, Virginia, USA

第19章　美人鱼发光杆菌

约翰·P. 霍克(John P. Hawke)

Department of Pathobiological Sciences, School of Veterinary Medicine,

Louisiana State University, Baton Rouge, Louisiana, USA

第 20 章　鲑立克次氏体

杰里・巴索洛缪(Jerri Bartholomew)
Department of Microbiology, Oregon State University, Corvallis, Oregon, USA

克里斯滕・D. 阿库什(Kristen D. Arkush)
formerly of Bodega Marine Laboratory, University of California-Davis, Bodega Bay, California, USA

埃斯特班・索托(Esteban Soto)
Department of Medicine and Epidemiology, School of Veterinary Medicine, University of California, Davis, California, USA

第 21 章　鲑 肾 杆 菌

戴安娜・G. 埃利奥特(Diane G. Elliott)
US Geological Survey, Western Fisheries Research Center, Seattle, Washington, USA

第 22 章　海豚链球菌和无乳链球菌

克雷格・A. 休梅克(Craig A. Shoemaker)
US Department of Agriculture-Agricultural Research Service, Aquatic Animal Health Research Unit, Auburn, Alabama, USA

徐德海(De-Hai Xu)
US Department of Agriculture-Agricultural Research Service, Aquatic Animal Health Research Unit, Auburn, Alabama, USA

埃斯特班・索托(Esteban Soto)
Department of Medicine and Epidemiology, School of Veterinary Medicine, University of California Davis, Davis, California, USA

第 23 章　弧菌病：鳗弧菌，奥氏弧菌和杀鲑别弧菌

阿莉西亚・E. 托兰索(Alicia E. Toranzo)
Departamento de Microbiología y Parasitología, Universidade de Santiago de Compostela, Santiago de Compostela, Spain

贝亚特里斯・马加里尼奥斯(Beatriz Magariños)

Departamento de Microbiología y Parasitología, Universidade de Santiago de Compostela, Santiago de Compostela, Spain

鲁本·阿文达尼奥-埃雷拉(Ruben Avendaño-Herrera)

Laboratorio de Patología de Organismos Acuáticos y Biotecnología Acuícola, Universidad Andrés Bello, Viña del Mar, Chile

第 24 章　鲸魏斯氏菌

蒂莫西·J. 韦尔奇(Timothy J. Welch)

US Department of Agriculture Agricultural Research Service, National Center for Cool and Cold Water Aquaculture, Kearneysville, West Virginia, USA

戴维·P. 马兰西克(David P. Marancik)

Department of Pathobiology, St. George's University of Veterinary Medicine, True Blue, St. George's, Grenada, West Indies

克里斯托弗·M. 古德(Christopher M. Good)

The Conservation Fund's Freshwater Institute, Shepherdstown, West Virginia, USA

第 25 章　鲁氏耶尔森菌

迈克尔·奥姆斯比(Michael Ormsby)

Institute of Infection, Immunity and Inflammation, College of Medical, Veterinary and Life Sciences, University of Glasgow, Glasgow, UK

罗伯特·戴维斯(Robert Davies)

Institute of Infection, Immunity and Inflammation, College of Medical, Veterinary and Life Sciences, University of Glasgow, Glasgow, UK

译者的话

历经三载磨砻淬砺，译著付梓成书之时，适逢十年举办一次的全球水产养殖盛会——第四届全球水产养殖大会首次在中国召开，本次大会以“面向食物供给和可持续发展的水产养殖”为主题，并发布了《促进全球水产养殖业可持续发展的上海宣言》。我们了解到，主要由饮食相关因素导致的疾病已成为全球公共健康的重要威胁。以鱼类为代表的水产食品作为一个高度多样化的食物类别，可以提供对预防营养不良和疾病至关重要的营养物质并改善公众健康，在全球食物系统中扮演着重要角色。水产食品对人类营养和粮食安全的重要性从未像现在这样显著，而全球粮食安全和营养在很大程度上依赖于水产养殖。随着世界人口到2050年将达到近100亿，水产养殖的可持续增长至关重要。水产养殖在全球粮食安全中扮演的日益重要的作用，在中国这样一个人口第一大国和全球最大渔品生产国、消费国、加工和出口国可能尤其突出。

水产养殖作为人类影响环境的重要农业活动之一，不可避免地推动了水生动物的疾病出现。事实上，疾病已被广泛认为是水产养殖可持续和绿色健康发展的最大威胁，并限制了水产养殖的快速增长及其对全球粮食安全贡献的全部潜力。随着水生动物对粮食安全的重要性日益增强，水生动物健康已成为全球关注的问题。过去50年来，随着水产养殖业的发展，各种水生动物疾病不断出现，其中许多疾病已在世界各地蔓延成为流行病。对2002年至2017年期间全球监测和记录的400余起水生动物新发疾病事件的分析显示，其中一半以上的新发疾病是由病毒引起的，20%～30%的则由细菌引起。随着水产养殖和渔品贸易的继续增长，新的病原体将出现、传播并加剧当前的疾病挑战。如果要抓住水产养殖提供的机遇并保护粮食安全，就必须加强和深入进行水生病原的生物学、病理学以及病害预防控制的基础研究，改善水生动物健康管理，以充分发挥水产养殖对全球粮食安全持续和不断增长的贡献。

《鱼类病毒和细菌：病理学与防控》的引进与翻译是适时的，正如本书副标题所示，本书除了重点介绍各种鱼类病毒和细菌病原及它们的鱼类宿主、传播方式、地理分布以及对水产养殖业的影响外，更为重要的是深入阐述了各种病害发生的病理学及防御和控制措施。本书原版中收录的作者都是世界知名的水生动物病原学和健康管理方面的专家。这些作者在应对疾病威胁和加强水生动物健康管理的研究和实践领域积累了丰富经验。他们的经验和观点对于我国水生动物卫生专业人员和决策者都有着重要参考价值。同时，随着世界水产养殖业和渔品国际贸易的不断增长，一些水产疾病的大流行表明，全球生产系统在流行病学上相互联系，水生动物疾病是一种需要全球合作研究与协作控制的共同威胁。因此本书的出版目的主要是服务于水产养殖业、高等院校、水族馆、专业机构等的科研人员、本科生/研究生、主管、兽医师以及政府监管部门的管理者等。本书也可以作为水生微生物感染性疾病相关大学课程的参考书。

我国水产病原学的研究在某些领域已经跻身国际先进水平，但相关教材的编撰相对匮乏，因此跟踪学科发展、引进国外优秀教材已成为相关课程教学内容改革和课程建设工作的重要内容之一。为了促进国内相关专业教学内容的改革，华东理工大学出版社引进了这本 Patrick T. K. Woo 和 Rocco C. Cipriano 教授主编的经典之作，我们有幸为本书的翻译推广尽自己的绵薄之力。在本书的引进过程中也感谢来自广东以色列理工学院 Ka Yin Leung 教授的推荐和介绍，得以与 Woo 博士联系并取得引进的许可。

本译著共计 25 章，涵盖了目前世界范围内主要水产养殖鱼类病原的 25 种重要疾病。其中第 1～13 章阐述鱼类的主要病毒，包括传染性胰腺坏死病毒、传染性造血器官坏死病病毒等；第 14～25 章介绍鱼类的主要细菌性病原，主要包括杀鲑气单胞菌、嗜水气单胞菌、爱德华氏菌、弧菌、黄杆菌和分枝杆菌等。本书对各种病原进行了详细的病原生物学的介绍，并且配有非常精美的病理学照片，为病原的正确诊断提供了新颖丰富的知识，在国内同类教材中并不多见。本书所涉及的内容既有一定广度，重要之处又很有深度，不失为一本优秀的参考书，可供农、林、水产、理工院校水生动物医学、水产养殖学，水产生物学、生物工程、生物技术人员学习参考。

本书涉及的鱼类病原种类繁多，既有一些传统的且研究比较深入的病原，又有一些感染多种宿主并且是新发的病原。本书涉及的疾病，一些是我们无法有效控制的已知疾病，另一些是可能成为流行病的新疾病。另外，书中不但详细介绍了各种病原的生物学特性，同时也深入阐明了鱼类疾病的病理生理学，以及相应的生物学、生理学、法规相关的防疫和防控措施等内容，而译者对这些领域的工作未必都非常熟悉，书中难免会有翻译不确切甚至错误之处，恳请同行不吝指正。

译　者
2021 年 10 月于华理校园青春河畔

前　　言

本书的编写主旨是研究对具有重要经济意义的鱼类造成疾病和死亡的主要病毒和细菌的病理生物学和防控策略。书中25章分别由对所选病原微生物具有相当专业知识的科学家撰写，其中绝大多数章节是关于世界动物卫生组织(world organization for animal health，OIE)认定的应具报病原微生物。撰稿人尽一切努力引用新进出版物的研究成果，最新可追溯到2016年的相关文献。

本书经过精心编辑筛选，纳入相关病原和撰稿人。病原微生物和疾病的选择基于许多标准。其中包括：

- 仅在2011年出版的前一卷《鱼类疾病与障碍症(第3卷)：病毒、细菌和真菌感染》(第2版，由Woo，P. T. K.和Bruno，D. W.编写)中简要地讨论相关问题(如锦鲤疱疹病毒和鲸魏斯氏菌)；
- 研究相对充分的鱼类病原(如传染性造血器官坏死病病毒、气单胞菌)，可作为其他病原的疾病模型；
- 给水产养殖业的特定部门造成相当大的经济损失(如病毒性出血性败血症、弧菌)；
- 已通过感染的鱼类(如欧洲的锦鲤疱疹病毒)被引入新的地理区域，并对当地鱼类种群构成重大威胁；
- 对特定鱼类种群具有致病性(如日本地区的马苏大麻哈鱼病毒病和鲑致瘤病毒)；
- 适应性强且不具有宿主特异性，因此在世界范围内广泛分布(如流行性造血器官坏死病病毒、链球菌)。

每一章都对所选定的病原体、宿主、传播、地理分布和对鱼类生产的影响进行了简要描述，提供了关于感染检测和诊断的最新信息，讨论了疾病的临床体征，并提供了关于外部/内部病变(宏观和微观)的更多细节。每一章都重点阐述了疾病的病理生理学，包括它对渗透调节的影响，对宿主内分泌系统、生长和繁殖的影响。

最后，提出了最新的预防和保护控制策略，包括生物、物理和立法监管等多层面的方法。

其中许多病原已被广泛研究，然而，有一些还没有得到很好的研究，包括新发病原（如甲病毒、鲸魏斯氏菌）。在这些情况和案例中，撰稿人强调了我们知识的不足之处，我们希望这些论述将促进对这些"被忽视"领域的进一步研究。

本书主要面向水产养殖业和大学的科研工作者、鱼类健康顾问、鱼类健康实验室的管理人员和监督人员以及商业水族馆的兽医专家。同时，本书也适用于鱼类健康专家的培训，以及从事鱼类疾病研究的高年级本科生/研究生和兽医专业的学生。此外，本书还可作为传染病、普通微生物学和疾病对水产养殖业的影响等高校相关课程的有用参考书。对一些希望研究微生物/寄生虫感染对鱼类健康的综合影响的病理学家，以及正在研究污染物和微生物感染的协同效应的环境毒理学家和免疫学家也具有一定参考价值。我们预计，随着鱼类健康作为评价生态系统质量的指标变得越来越明显，这类读者数量将会增加。

胡栋祺和罗科·C. 西普里亚诺
Patrick T. K. Woo and Rocco C. Cipriano

目　　录

1

传染性胰腺坏死病毒

Arun K. Dhar*, Scott LaPatra, Andrew Orry 和 F. C. Thomas Allnutt

1.1 引言

传染性胰腺坏死病毒(infectious pancreatic necrosis virus, IPNV)为传染性胰腺坏死病(infectious pancreatic necrosis, IPN)的病原体,是一种双链 RNA(dsRNA)病毒,属双 RNA 病毒科(Leong et al., 2000; ICTV, 2014)。该科包括四个属:水生双 RNA 病毒属(*Aquabirnavirius*)、禽双 RNA 病毒属(*Avibirnavirus*)、斑鳢病毒属(*Blosnavirus*)和昆虫双 RNA 病毒属(*Entomobirnavirus*)(Delmas et al., 2005),它们可感染脊椎动物和无脊椎动物。水生双 RNA 病毒属主要感染水生物种(如鱼类、软体动物和甲壳类动物),包括传染性胰腺坏死病毒(IPNV)、鰤腹水病毒(*Yellowtail ascites virus*)和樱蛤病毒(*Tellina virus*)3 个种,其中,感染鲑科鱼类的 IPNV 是其模式病毒种。

IPNV 基因组由两个 dsRNA 片段组成,即片段 A 和 B(图 1.1; Leong et al., 2000)。片段 A 的大小约为 3 100 bp(碱基对),包含两个部分重叠的可读框(open reading frames, ORFs)。长 ORF 编码一个 106 kDa① 的多聚蛋白(NH_2 - pVP2 - VP4 - VP3 - COOH),该蛋白被 VP4(病毒蛋白 4)蛋白酶(29 kDa)共翻译切割产生 pVP2(62 kDa;主要衣壳蛋白 VP2 的前体)和 31 kDa 的 VP3(Petit et al., 2000)。短 ORF 则编码 VP5,这是一种在复制早期产生的富含精氨酸的非结构蛋白,其大小为 17 kDa。VP5 是一种类似于原癌基因 Bcl-2 家族的抗凋亡蛋白。VP5 对于 IPNV 的体内复制不是必需的,它的缺失不会改变 IPNV 在宿主体内的毒力或持留(Santi et al., 2005)。片段 B 的大小约为 2 900 bp,编码多肽 VP1(94 kDa),VP1 是一种 RNA 依赖性 RNA 聚合酶。在成熟的病毒粒子内,VP1 是一种具有 RNA 依赖性 RNA 聚合酶相关活性的游离多肽,同时通过鸟酰化成为一种基因组连接蛋白 VPg(图 1.1 和表 1.1)。

---

* 通信作者邮箱:arun_dhar@hotmail.com 或 adhar@email.arizona.edu。

① 1 kDa=1 000 Da; 1 Da=1 U。

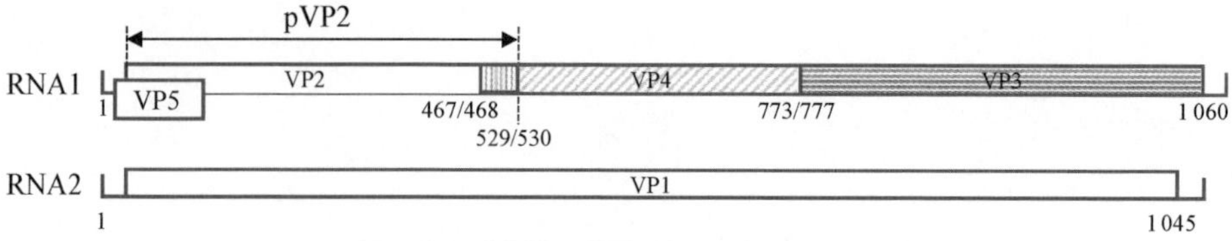

图 1.1　传染性胰腺坏死病毒(IPNV)的基因组结构。图中各段底部的数字表示氨基酸数。pVP2 是 IPNV 蛋白 VP2 的前体。RNA1 是病毒基因组 dsRNA 的 A 段,RNA2 是 B 段

表 1.1　传染性胰腺坏死病毒编码的蛋白及其功能

蛋白	相对分子质量	功　　能	参 考 文 献
VP1	94 kDa	RNA 依赖性 RNA 聚合酶;病毒复制	Calvert et al., 1991; Leong et al., 2000; Graham et al., 2011
VP2	54 kDa	主要衣壳蛋白,含大部分抗原决定簇;具有结构功能	Coulibaly et al., 2010
VP3	31 kDa	次要衣壳蛋白,在衣壳形成过程中与主要衣壳蛋白 VP2 相互作用,与 dsRNA 基因组结合,招募聚合酶进入衣壳,含有一些抗原表位	Leong et al., 2000; Bahar et al., 2013
VP4	29 kDa	参与加工编码于传染性胰腺坏死病毒 dsRNA A 段的多蛋白的蛋白酶	Feldman et al., 2006; Lee et al., 2007
VP5	17 kDa	富含精氨酸的抗凋亡蛋白,类似于原癌基因的 Bcl－2 家族	Magyar and Dobos, 1994; Santi et al., 2005

水生双 RNA 病毒的宿主范围广泛并具有不同的最佳复制温度。它们由 A、B、C 和 D 四种血清群组成(Dixon et al., 2008),但绝大多数属于 A 血清群,分为 A1～A9 血清型。A1 血清型包含大部分美国分离株(参考毒株 West Buxton),A2～A5 血清型主要是欧洲分离株(参考毒株 Ab 和 Hecht),A6～A9 血清型包括来自加拿大的分离株(参考毒株 C1、C2、C3 和 Jasper)。

1.1.1　IPNV 形态发生

IPNV 在感染期间产生两种类型的颗粒(A 和 B)。复制后,dsRNA 被组装进直径为 66 nm 的非感染性颗粒 A 中,其衣壳由成熟病毒多肽(VP2)和未成熟病毒多肽(pVP2)组成。未经处理的 pVP2 蛋白被水解加工成 VP2,其衣壳被压紧至直径为 60 nm 的感染性颗粒,称为颗粒 B(Villanueva et al., 2004)。VP2 蛋白组成外衣壳,而 VP3 蛋白形成成熟病毒颗粒的内层。此外,VP3 仍与 VP1、VP4 以及聚合酶相关基因组保持相关。

1.1.2　IPNV 三级结构

IPNV 病毒颗粒是一种无囊膜的直径约 60 nm 的 T13 晶格二十面体,在

CsCl 中的浮力密度为 1.33 g/cm^3(Delmas et al., 2005)。病毒衣壳表面含有 VP2 蛋白,在 IPNV 和 IBDV(鸡传染性法氏囊病病毒)中,VP2 的三维(3D)结构已被解析[图 1.2(A)]。IPNV 的 VP2 衣壳由 260 个三聚体刺突组成,它们呈放射状突起,携带抗原决定簇以及毒力和细胞适应的决定因子,并在病毒颗粒内部与 VP3 相连[图 1.2(B)]。不过,IPNV 与 IBDV 的刺突排列不同,这体现在控制毒力和细胞适应性的氨基酸位于 IPNV 外周,而 IBDV 位于中心区域。刺突底部包含一个位于外露沟内的整合素结合基序,这在所有属的双 RNA 病毒中都是保守的(Coulibaly et al., 2010)。

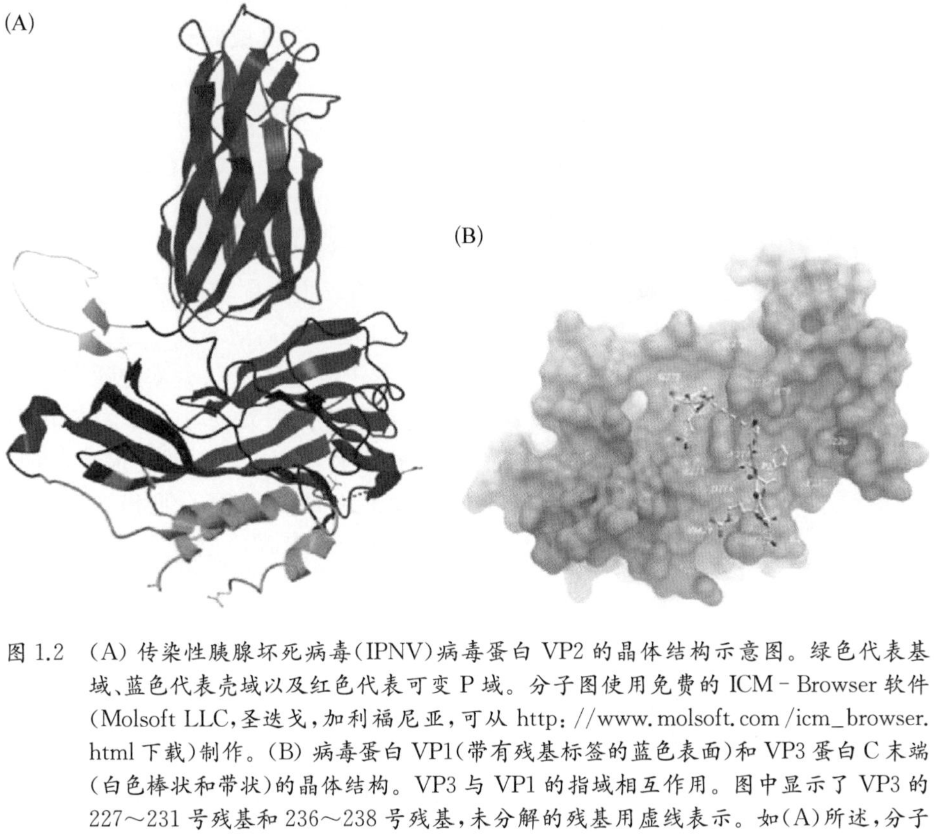

图 1.2　(A) 传染性胰腺坏死病毒(IPNV)病毒蛋白 VP2 的晶体结构示意图。绿色代表基域、蓝色代表壳域以及红色代表可变 P 域。分子图使用免费的 ICM - Browser 软件(Molsoft LLC,圣迭戈,加利福尼亚,可从 http://www.molsoft.com/icm_browser.html 下载)制作。(B) 病毒蛋白 VP1(带有残基标签的蓝色表面)和 VP3 蛋白 C 末端(白色棒状和带状)的晶体结构。VP3 与 VP1 的指域相互作用。图中显示了 VP3 的 227~231 号残基和 236~238 号残基,未分解的残基用虚线表示。如(A)所述,分子图使用 ICM - Browser 软件绘制

1.2 地理分布

传染性胰腺坏死病普遍存在于世界范围内的养殖和野生鲑科鱼类中。20 世纪 40 年代在加拿大和 20 世纪 50 年代在美国的淡水鳟中首次发现该疾病(Wood et al., 1955)。IPNV 病毒于 1960 年首次被分离出来(Wolf et al.,

1960)。20世纪70年代初欧洲也报道发现了该病毒，随后在许多从事鲑科鱼类进口或水产养殖业活跃的国家(如日本、韩国、中国、泰国、老挝、新西兰、澳大利亚、土耳其)也相继报道。IPN的暴发通常可追溯至受感染鱼卵/种鱼的引进和随后的经销。

1.3 传染性胰腺坏死病对经济的影响

从历史上看，IPN是造成鲑鱼产业损失的三大原因之一。根据2001年设得兰群岛鲑鱼养殖者协会的一项调查显示，IPN造成了平均20%～30%的经济损失，现金价值达200万英镑(Ruane et al., 2007)。根据挪威的动物流行病学研究显示，从1991年到2002年，IPN对入海后二龄幼鲑生存的影响相对密度由6.4%提升到12.0%(Munro and Midtlyng, 2011)。1998年，由IPN导致的经济损失估计超过1 200万欧元(Munro and Midtlyng, 2011)。即使在现在，IPN仍然是鲑科鱼类养殖的一个重要威胁。例如，据报道2014年挪威有48个鲑鱼养殖场发现有IPN发生，尽管数量少于往年(Norwegian Veterinary Institute, 2015)。IPN在鱼迁徙入海后的前6个月最为严重。报道称，因死亡或随后幸存鱼的衰减，渔业仍有严重损失。近期一份关于迁徙入海后前6个月累计死亡率的报告显示，与3.4%的死亡率基准相比，死亡率上升到7.2%，为基准死亡率的两倍多(Jensen and Kristoffersen, 2015)。这项研究同时发现，IPNV感染鱼群在受到其他胁迫时，累计死亡率增加。例如，在无其他因素影响的前提下，鱼群胰腺疾病(pancreas disease, PD)的死亡率增加到12.9%，而心肌和骨骼肌炎(heart muscle and skeletal muscle inflammation, HSMI)的死亡率增加到16.6%。

1.4 感染诊断

1.4.1 临床体征和病毒传播

已经从鲑科鱼类及非鲑科鱼类[如鲤(*Cyprinus carpio*)、黄金鲈(*Perca flavescens*)、欧鳊(*Abramis brama*)和梭子鱼(*Esox lucius*)]、软体动物、甲壳类动物和假体腔动物中分离出IPNV以及IPNV类双RNA病毒(McAllister, 2007)。外部临床体征包括体色变黑、眼球突出、腹胀、有黏液样假管型(“粪便铸型”)从肛门挤出、体表和鳍基部出血等。患病鱼沿纵轴旋转游动，一般在几个小时内死亡。内部临床体征包括肝和脾苍白，以及消化道内没有食物，但却充满清亮或乳白色黏液等。内脏器官可能发生出血(Munro and Midtlyng, 2011)。IPN暴发的特点是鱼苗和幼鱼的死亡率突然增加。这种疾病也可能在幼鲑初次迁移入海后的最初几周发生(Jensen and Kristoffersen, 2015)。对宿主的胁迫在增强病毒复制、突变，甚至毒力回复方面起着关键作用(Gadan et al.，2013)。在疾病暴发中幸存下来的鱼通常终生携带 IPNV 且无临床症状。这些病毒携带者作为传染源，可通过粪便和尿液排泄进行水平传播，或通过受污染的生殖产品垂直传播(Roberts and Pearson，2005)。

1.4.2 病毒检测

临床体征和病理不能用来区分 IPN 与其他病毒疾病，无相关临床体征也不能确保鱼体内没有 IPNV。IPN 的初步诊断是基于养殖场和鱼类种群的既往病史、临床症状和大体尸检结果。确诊诊断则包括在细胞培养中分离病毒及随后的免疫学和分子生物学鉴定。血清学或分子技术尤其适用于监测有无临床症状的鱼。适合进行病毒学检查的组织包括肾、肝、脾、产卵时亲鱼的卵巢液或整尾初孵鱼苗。蓝鳃太阳鱼细胞系(BF-2)、大鳞大麻哈鱼胚胎细胞系(CHSE-214)或虹鳟生殖腺细胞系(RTG-2)可用于分离培养 IPNV(OIE，2003)。细胞中培养分离出来的病毒可通过中和试验、荧光抗体试验、酶联免疫吸附试验(enzyme-linked immunosorbent assay，ELISA)、IPNV 特异性抗体免疫组织化学染色或逆转录聚合酶链反应(RT-PCR)进行鉴定(OIE，2003；USFWS and AFS-FHS，2007)。

近年来，基于 SYBR Green 和 TaqMan 的实时荧光定量 PCR 已被开发用于检测 IPNV(Bowers et al.，2008；Orpetveit et al.，2010)。实时荧光定量 PCR 比传统 PCR 灵敏 100 多倍，并且可以检测亚临床动物中的病毒(Orpetveit et al.，2010)。利用实时荧光定量 PCR 技术，发现胸鳍剪样中的 IPNV 载量与脾和头肾中的一样高(Bowers et al.，2008)。因此，非致死性组织采样与实时荧光定量 PCR 相结合，可作为监测野生和养殖鱼类的重要工具，还可减少产卵时牺牲种鱼来进行检测的需要。

1.5 病理学

IPNV 感染呈现多种病理变化。胰腺组织发生严重坏死，表现为细胞核固缩(染色质凝聚)、核碎裂(细胞核碎裂)和胞质包涵体形成(图 1.3)，幽门、幽门盲囊和前肠也发生广泛坏死，肠上皮细胞脱落并与黏液结合形成浓稠的可从肛门排出的白色渗出物，肾脏、肝脏和脾脏组织也会发生退化病变。在持续感染的鱼体内，IPNV 存在于肾脏造血组织内的巨噬细胞中，并可在携带病毒鱼分离出的黏附白细胞中繁殖(Johansen and Sommer，1995)。有迹象表明，从携带病毒的鱼中分离出的白细胞免疫应答水平降低，并且在用植物血凝素刺激静止白细胞后，病毒的体外复制增加(Knott and Munro，1986)。

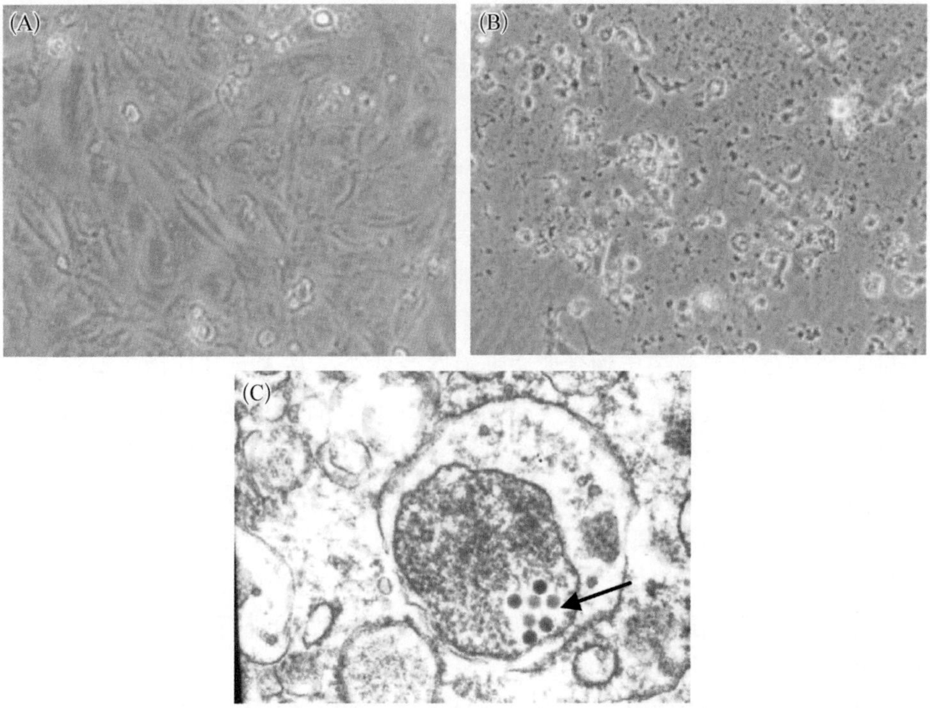

图 1.3 (A) 未受感染的大鳞大麻哈鱼胚胎细胞系(CHSE-214)；(B) 出现了溶解型致细胞病变效应的感染传染性胰腺坏死病毒(IPNV)的细胞(放大 100 倍)；(C) 利用透射电子显微镜(TEM)拍摄的 IPNV 颗粒图像(箭头所指，放大了 27 500 倍)，在细胞质小泡中呈现特征性的六角形轮廓

1.6 病理生理学

鱼类对 IPN 的易感性和死亡率取决于鱼种、鱼龄或发育阶段、宿主的生理条件、病毒株、宿主的遗传背景以及环境和养殖管理因素(Munro and Midtlyng，2011)。在养殖鲑鳟中，感染程度从较低死亡率或无死亡的亚临床感染到高死亡率的急性感染不等。虽然这种疾病在鲑鳟中主要导致严重的胰腺坏死，但也会引起肾造血组织、肠道和肝的组织学变化。肝脏是一个重要感染靶器官(Ellis et al.，2010)，而病毒也存在于胰岛和肾脏斯坦尼氏小体中(图 1.4)，这表明它也可能影响宿主新陈代谢过程。McKnight 和 Roberts(1976)报道了“黏膜损伤”的临床症状，其描述与目前所称的由粪便铸型引起的急性肠炎相符。他们推测这种损伤可能比胰腺坏死更致命。消化腺和肠黏膜上皮细胞坏死也被认为是病毒随粪便排出的原因。肠黏膜和胰腺的严重坏死也可能导致厌食症，加重诸如“针头鱼”和“银化失败”等情况，这些可以经常在动物流行病后幸存的鱼群中观

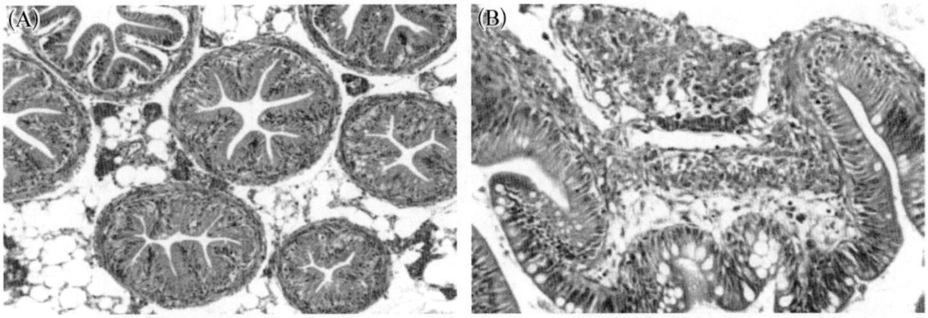

图 1.4 位于感染传染性胰腺坏死病毒(IPNV)鱼的圆柱形幽门盲囊和肠黏膜脱落之间的脂肪组织中坏死的胰腺腺泡细胞[苏木精-伊红(haematoxylin-eosin, HE)染色,400 倍放大]

察到(Smail et al., 1995)。Roberts 和 Pearson(2005)还发现,在海水阶段,IPN 造成鱼群损失 50%或更多后,许多鱼将无法生长,变得长期消瘦,并容易受到海虱的侵扰。

亚临床感染可能不会影响大西洋鲑(*Salmo salar*)入海前的幼鱼和入海后二龄鲑的生长。然而,在实验室研究中,健康二龄鲑的摄食量和比生长速率在浸泡感染 IPNV 后均降低(Damsgard et al., 1998)。在实验感染前后对肾脏和幽门盲囊中的病毒滴度进行了测定,感染组和对照组均无死亡。在感染鱼中,肾脏和幽门盲囊的病毒滴度均显著增加。感染后 16～44 d,幽门盲囊中的病毒滴度由 10^6 PFU/g (PFU 为噬斑形成单位)显著下降至 10^3～10^4 PFU/g。从感染后约 20 d 开始,感染鱼的摄食量和比生长速率明显低于未感染鱼。结果表明,感染 IPNV 的鱼在摄食量降低前,需在肾脏和幽门盲囊中获得相对较高的病毒滴度。

IPNV 可诱导细胞程序性死亡,因为在肝、肠和胰腺组织中已发现与病毒积累和病理变化相对应的细胞凋亡标志物(Imatoh et al., 2005; Santi et al., 2005),据推测,细胞凋亡可能限制而不是加剧 IPNV 感染的负面后果。

Sadasiv(1995)发现,即使存在病毒中和抗体的情形,病毒清除率也很低,提示病毒可能持续感染白细胞,从而破坏中和抗体反应。已有 IPNV 和 IHNV (infectious haematopoietic necrosis virus,传染性造血器官坏死病病毒)双重感染虹鳟(*Oncorhynchus mykiss*)的相关报道,但未说明混合感染对免疫系统的潜在影响(LaPatra et al., 1993)。与之前未接触过 IPNV 的鱼相比,预先暴露于 IPNV 并随后遭受病毒性出血性败血症病毒(viral haemorrhagic septicaemia virus, VHSV)攻毒的虹鳟对 VHSV 具有明显的抗性(de Kinkelin et al., 1992),作者将这种现象称为"干扰介导的抗性",并怀疑这是由干扰素的产生所引起的。

同样地,携带 IPNV 的大西洋鲑暴露于传染性鲑贫血症病毒(infectious salmon anaemia virus, ISAV)时,未感染 IPNV 的二龄鲑死亡率始终高于 3 周前已暴露于 IPNV 的鱼(Johansen and Sommer, 2001)。相反地，当鱼在感染 IPNV 6 周后接受 ISAV 攻毒时，携带 IPNV 与否对鱼的死亡率没有影响。这些研究人员还报道了类似的亚临床 IPNV 感染对杀鲑弧菌(*Vibrio salmonicida*)感染的短期保护作用，并将其归因于 IPNV 诱导的干扰素产生的非特异性效应。此外，在 IPNV 携带组和非携带对照组中，未观察到与使用三价油佐剂疫苗进行腹腔接种相关的显著效果。然而，携带 IPNV 的鱼最终有中度 IPN 暴发，未接种疫苗的 IPNV 携带鱼的累计死亡率为 24%，而接种疫苗的 IPNV 携带鱼的累计死亡率为 7%。在另一项研究中，免疫 IPNV 携带组和非携带组在进行疖疮病(furunculosis)或冷水性弧菌病攻毒后的死亡率以及两组对杀鲑气单胞菌(*Aeromonas salmonicida*)的体液免疫反应方面均无差异(Johansen et al., 2009)。此外，当对携带或非携带 IPNV 的大西洋鲑鱼苗(平均体重为 2～4 g)进行抗肠红嘴病(enteric redmouth disease, ERM)接种免疫后，它们对鲁氏耶尔森菌(*Yersinia ruckeri*)实验攻毒的保护作用没有差异(Bruno and Munro, 1989)。这些研究表明，即使在小鱼体内，IPNV 感染对细菌疫苗诱导的保护作用也无不利影响。

1.7 防控策略

由于目前没有 IPN 的治疗方法，预防是最好的策略。对鲑鱼养殖场中 IPNV 传播的流行病学研究表明，病毒传播是不可预测的。由于非临床感染携带者通过粪便和生殖产品中的病毒脱落而成为传染源，加强监测和生物安全可以降低病毒的流行率。每当引进新鱼时，必须从无病原体的来源获取种苗，并在无病原水供应上维持严格的生物安全保障。对孵化场的进水进行紫外消毒处理是一种合适的控制措施。此外，用甲醛(3%，5 min)、氢氧化钠(pH 为 12.5，10 min)、氯(30 ppm①，5 min)和碘化合物等消毒剂处理也能灭活病毒(OIE, 2003)。

1.7.1 IPN 抗性家系鱼选育

由于相当数量的鱼在 IPN 疫病中幸存，因此推测可以通过育种提高其抗病力。Ozaki 及其同事报道，数量性状位点(quantitative trait loci, QTLs)可能与 IPN 抗性提高有关。近来有一篇关于 DNA 标记辅助育种提高具有重要商业价值鱼类抗病能力的研究进展综述(Ozaki et al., 2012)。利用基因组性状进行更多抗性品系的标记辅助选择(MAS)是开发抗 IPNV 鲑科鱼类家系的有力工具(Moen et al., 2009)，并正拓展到其他疾病抗性选育中(Houston et al., 2008; Ozaki et al., 2012)。最近的一项研究报道，将上皮钙黏蛋白基因(*cdh1*)与 IPN

① 1 ppm=10^{-6}。

抗性联系在一起(Moen et al., 2015)。目前,挪威特隆赫姆的 Aquake 等公司正在销售抗 IPN 的大西洋鲑品种(http://aquagen.no/en/products/salmon-eggs/product-documentation/resistance-against-ipn/);这些品系鱼的 IPN 抗性都与一个数量性状位点有关,该 QTL 可能对今后利用 MAS 开发鱼类抗性品系有所帮助(Moen et al., 2009)。

1.7.2 可用生物制品

目前,已有多种 IPN 疫苗上市(表 1.2),但更具成本效益并适用于所有养殖阶段的疫苗仍有待研发。利用灭活的野生型病毒诱导免疫是鱼类病毒疫苗早期的技术手段,它仍是评价其他疫苗的可靠标准。Alpha Ject® micro 1 ISA (Pharma/Novartis)和 Alpha Ject® 1000(表 1.2)是分别针对传染性鲑贫血症(infectious salmon anaemia, ISA)和 IPN 的此类疫苗的代表性产品(http://www.pharmaq.no/products/injectable/)。病毒灭活疫苗由于保留了表面暴露的抗原和灭活的基因组组分,可引起宿主强烈的免疫反应。

表 1.2 已批准的传染性胰腺坏死病疫苗

名称	疫苗内容物/预防疾病[a]	类型	接种途径	制造商	许可地区
Alpha Ject® 1000	IPNV 全病毒	灭活	IP	挪威 Pharma AS	智利
Alpha Ject® 2.2	传染性胰腺坏死病、疖疮病		IP	挪威 Pharma AS	英国
Alpha Ject® 4-1	传染性胰腺坏死病、SRS、疖疮病、弧菌病		IP	挪威 Pharma AS	智利
Alpha Ject® 5-1	传染性胰腺坏死病、ISA、SRS、疖疮病、弧菌病		IP	挪威 Pharma AS	智利
Alpha Ject® 6-2	传染性胰腺坏死病、疖疮病、冷水弧菌病、冬季冻疮		IP	挪威 Pharma AS	挪威 法罗群岛
Alpha Ject® IPNV-Flevo 0.025	IPNV、黄杆菌病		IP	挪威 Pharmaq AS	智利
Alpha Ject® micro 2	传染性胰腺坏死病、SRS		IP	挪威 Pharmaq AS	智利
Alpha Ject® micro 3	传染性胰腺坏死病、SRS、弧菌病		IP	挪威 Pharmaq AS	智利
Alpha Ject® micro 7 ILA	传染性胰腺坏死病、弧菌病、疖疮病、冷水弧菌病、冬季冻疮、ISA		IP	挪威 Pharmaq AS	挪威 法罗群岛

续 表

名 称	疫苗内容物/预防疾病[a]	类型	接种途径	制 造 商	许可地区
AquaVac® IPN Oral	VP2 和 VP3 衣壳蛋白	亚单位	口服	美国 Merck Animal Health	加拿大、美 国
Birnagen Forte	IPNV 全病毒	灭活	IP	加拿大诺华 Aqua Health Ltd.	加拿大
IPNV Norvax® Minova 6 Norvax® Compact 6	IPNV 全病毒、VP2 衣壳蛋白	灭活 亚单位	IP IP	智利 Centrovet 荷兰 Intervet International BV Merck Animal Health	智 利 智 利
SRS/IPNV/Vibrio	VP2 蛋白	亚单位	IP	加拿大 Microtek International, Inc.	加拿大、智 利

[a]IPNV，传染性胰腺坏死病毒；ISA，传染性鲑贫血症；SRS，鲑立克次体败血症；VP2，VP3，病毒蛋白VP2 和 VP3

基于主要病毒抗原的亚单位疫苗是生产病毒疫苗的另一种选择。一些病毒结构蛋白具有自组装成病毒颗粒的固有能力，并在大小和宿主加工过程等方面模拟天然病毒，从而发展出一类称为病毒样颗粒(virus-like particles, VLPs)的亚单位疫苗(Kushnir et al., 2012)。现在已可利用细菌、酵母、转基因植物和细胞培养表达 VLP。许多人用疫苗(例如 Gardasil vaccine® 9，9 价人乳头瘤病毒重组疫苗)已经使用该技术生产(Kushnir et al., 2012)。最近也有研究利用这种技术生产 IPNV 疫苗。利用杆状病毒表达系统，已在昆虫细胞和粉纹夜蛾(*Trichoplusia ni*)幼虫中产生含有 VP2 和 VP3 蛋白的 IPNV 病毒样颗粒，其直径为 60 nm。用纯化抗原腹腔接种免疫大西洋鲑 2 龄幼鱼后浸泡攻毒，4 周后累计死亡率(56%)低于对照组(77%)(Shivappa et al., 2005)。

另一种 IPNV 疫苗仅基于 IPNV 的 VP2 蛋白(Allnutt et al., 2007)。酵母中表达的基于 VP2 的亚病毒颗粒(subviral particles, SVPs)大小为 22 nm，而天然病毒为 60 nm。SVPs 在虹鳟体内诱导了很强的抗 IPNV 抗体反应。抗原通过注射或饲喂进行递送，注射接种和口服接种后的鱼体内的 IPNV 载量分别降低为原来的$\frac{1}{22}$和$\frac{1}{12}$(Allnutt et al., 2007)。为了进一步探索利用 IPNV 亚病毒颗粒研发多价疫苗的可能性，研究人员在 SVP 上表达了一个外源表位(人癌基因*c-myc*)，嵌合表达的 SVPs 诱导宿主产生了对 IPNV 和 *c-myc* 的抗体反应(Dhar et al., 2010)。在此 IPNV 的 SVP 表面进一步成功表达出 ISAV 血凝素表位，将此嵌合 SVPs 注射免疫虹鳟后，诱导了对 IPNV 和 ISAV 的抗体反应

(Dhar et al.,未发表的数据)。另有三种基于 IPNV VP2 衣壳蛋白的疫苗也已上市销售,包括 IPNV(Centrovet,在智利获得许可)、NORVAX(荷兰 Intervet International BV)和 SRS/IPNV/*Vibrio*(在加拿大和智利获得许可,加拿大不列颠哥伦比亚省 Microtek International Inc.;自 2010 年 12 月由加拿大 Zoetis 兽药研发公司全面收购)(Gomez-Casado et al., 2011)。Centrovet 公司的疫苗可通过灵活的口服给药方式实施 IPNV 灭活疫苗和重组蛋白疫苗的接种(http://www.centrovet.com/index.php/products/aqua/vaccines99)。NORVAX 疫苗是另一种重组蛋白疫苗,通过腹腔注射给药且只针对 IPN。SRS/IPNV/*Vibrio* 疫苗是一种三价重组蛋白疫苗,也是通过腹腔注射给药,为用户同时应对三种不同的病原提供了便利。

通过注射给药一种实验性的 IPNV DNA 疫苗(表达 VP2 抗原),为 1～2 g 虹鳟鱼苗抵御传染性同源病毒攻毒提供了近 80%的免疫保护率(relative percent survival, RPS)(Cuesta et al., 2010)。另一种包埋在海藻酸盐中的 DNA 疫苗以饵料颗粒的方式进行给药,在病毒水传播攻毒后,可降低或消除虹鳟体内的 IPNV 滴度(Ballesteros et al., 2015)。在这项研究中,VP2 基因被克隆到一个 DNA 载体中,并包埋到海藻酸盐微球中,随后使用移液管将其口服递送到虹鳟体内,以确保疫苗均量递送(Ballesteros et al., 2012)。海藻酸盐包埋的 DNA 疫苗也被包含进饵料颗粒中,以诱导宿主免疫反应(Ballesteros et al., 2014)。IgM 和 IgT 均在免疫接种 15 d 后增加,但在接种 30 d 后,显著增加。此外,细胞免疫反应通过观察 T 细胞标记物 CD4 和 CD8 来监测。这两种标记物在接种第 15 天升高,但在第 30 天恢复到基础水平。用 IPNV 对鱼实施攻毒,接种后第 15 天和第 30 天的相对免疫保护率分别为 85.9%和 78.2%。最近,另一项研究报道了一种在大肠杆菌中表达 IPNV 的 VP2 - VP3 融合蛋白,通过注射给药,可诱导产生针对 IPNV 的 IgM;对虹鳟幼鱼的相对免疫保护率达到 83%(Dadar et al., 2015)。

1.7.3 亚单位疫苗

以 IPNV 的 Sp 株为出发株,采用反向遗传学方法制备了设计型全病毒疫苗(Munang'andu et al., 2012)。将无毒和有毒的基序加入 Sp 株中,然后将其灭活作为疫苗使用。通过共栖感染系统,将 IPNV 灭活病毒与 DNA 疫苗、亚单位疫苗和纳米颗粒亚单位疫苗进行了比较。全病毒灭活疫苗的抗体效价与其他疫苗相似,但免疫保护率优于其他疫苗,全病毒灭活疫苗的 RPS 为 48%～58%,而 VP2 融合蛋白疫苗、亚单位疫苗和 DNA 纳米颗粒疫苗的 RPS 分别为 25.4%～30.7%、22.8%～34.2%和 16.7%～27.2%(Munang'andu et al., 2012)。

通过干酪乳杆菌(*Lactobacillus casei*)递送 VP2 或 VP3 抗原的方法被评价为一种潜在的疫苗研发策略(Liu et al., 2012)。通过基因工程手段,VP2 和

VP3 被设计为由 *L. casei* 分泌或表面展示。当这些重组 *L. casei* 疫苗通过口服方式接种给虹鳟后，VP2 分泌疫苗株相比其他 *L. casei* 株提供了更高的血清 IgM 滴度。在 IPNV 攻毒实验中，分泌 VP2 的干酪乳杆菌在降低鱼体内病毒载量方面也更有效$\left(约下降为原来的\frac{1}{46}，VP3 分泌株减少为原来的\frac{1}{3}\right)$。

其他正在开展的研究包括改善 IPNV 抗原的口服递送效果。例如，最近对大西洋鲑的一项研究表明，海藻酸盐包埋的 IPNV 抗原与非包埋的 IPNV 抗原相比，可显著提高 IPNV 靶向抗体滴度和免疫相关基因诱导表达（Chen et al.，2014）。注射疫苗的接种成本和操作不便性限制了其在超过一个接种周期的大规模水产养殖中的应用，因此使用口服疫苗增强免疫激活更具吸引力。将海藻酸盐包埋疫苗与单独接种的抗原进行比较发现，海藻酸盐对抗原的保护提高了疫苗效力。最初的强化免疫是在疫苗注射接种一年后进行的，两次口服间隔 7 周。第一次口服免疫后，血清中 IgM 水平显著提高，但 IgT 却未上调。相反地，第二次口服免疫后，血清中 IgT 和 IgM 水平均上调（Chen et al.，2014）。表 1.3 列出了一些针对 IPNV 的实验疫苗示例。

表 1.3　针对 IPNV 的实验疫苗

疫　苗	说　明	参考文献
IPNV 病毒样颗粒（VLP）	基于 VP2 和 VP2/VP3 病毒蛋白的病毒样颗粒疫苗已被制备出来，并证实可诱导强烈免疫应答	Martinez-Alonso et al.，2012
表达 VP2 的干酪乳杆菌重组活疫苗	口服表达分泌型 VP2 的重组细菌可显著降低病毒载量	Liu et al.，2012
表达 IPNV 的 dsRNA 全部或部分 A 段的质粒	只对所有大型可读框（ORF）多聚蛋白的质粒有保护作用	Mikalsen et al.，2004
重组全病毒灭活疫苗（IWV）	反向遗传法构建的有毒和无毒 IPNV 的纳米颗粒载体疫苗	Munang'andu et al.，2012

1.8　总结与研究展望

水生双 RNA 病毒属是双 RNA 病毒科中最大和最多样化的属。IPNV 是目前海洋和淡水鱼类感染中研究最广泛和分布最广的病毒之一。自 20 世纪 40 年代首次报道 IPN 相关疾病暴发以来，已从世界范围内患病和看起来很健康的鲑科鱼类、非鲑科鱼类以及无脊椎动物中分离出大量 IPNV 和 IPNV 样病毒（ICTV，2014）。由于水生双 RNA 病毒属病毒种类繁多，很难将病毒分类到种级别（Crane and Hyatt，2011）。目前尚不清楚基于全基因组序列信息的系统发育和 VP2/VP3 蛋白结构分析是否有助于将水生双 RNA 病毒划分到种级别。

因其对生长于淡水中的鲑鱼苗和迁徙入海后的二龄幼鲑都具有致命性威胁，所以，IPN 具有重要的经济影响。生物安全措施的改进、有针对性的疫苗研发和抗性亲鱼种群等工作的开展对控制 IPN 非常有益，尽管做出了这些努力，但这种疾病仍然是全球鲑养殖业面临的严重挑战。研发有效疫苗的主要制约因素之一是缺乏可重复的感染模型来评价疫苗的有效性。然而，最近在大西洋鲑中开发了一种针对 IPNV 的共栖攻毒方法(Munang'andu et al., 2016)。结合非侵入性组织取样测定病毒滴度，进一步验证该感染模型，可能会促进对 IPN 的控制。

考虑到 IPNV 和 IPNV 类病毒的广泛多样性，也许可以采用基于结构的疫苗设计方法来提高 IPNV 疫苗的有效性。在感染人快速进化诺如病毒 GⅡ.4 中，将含有来自多个病毒株抗原决定表位的嵌合 VLP 整合到单一 VLP 中可产生广谱的阻遏性抗体反应，不仅对 GⅡ.4 VLP 家族中的 GⅡ.4 - 1987 到GⅡ.4 - 2012，也对那些不包括在嵌合 VLP 中的毒株产生抗性(Debbink et al., 2014)。类似的方法可能适用于研发 IPNV 疫苗，为鲑科鱼类养殖提供针对多种流行毒株的免疫保护。

参考文献①

[1] Allnutt, F.C., Bowers, R.M., Rowe, C.G., Vakharia, V.N., LaPatra, S.E. *et al.* (2007) Antigenicity of infectious pancreatic necrosis virus VP2 subviral particles expressed in yeast. *Vaccine* 25, 4880 - 4888.

[2] Bahar, M.W., Sarin, L.P., Graham, S.C., Pang, J., Bamford, D.H. *et al.* (2013) Structure of a VP1 - VP3 complex suggests how birnaviruses package the VP1 polymerase. *Journal of Virology* 87, 3229 - 3236.

[3] Ballesteros, N.A., Saint-Jean, S.S.R., Encinas, P.A., Perez-Prieto, S.I. and Coll, J.M. (2012) Oral immuni-zation of rainbow trout to infectious pancreatic necro-sis virus (IPNV) induces different immune gene expression profiles in head kidney and pyloric ceca. *Fish and Shellfish Immunology* 33, 174 - 185.

[4] Ballesteros, N.A., Rodriguez Saint-Jean, S. and Perez-Prieto, S.I. (2014) Food pellets as an effective delivery method for a DNA vaccine against infectious pancre-atic necrosis virus in rainbow trout (*Oncorhynchus mykiss*, Walbaum). *Fish and Shellfish Immunology* 37, 220 - 228.

[5] Ballesteros, N.A., Rodriguez Saint-Jean, S. and Perez-Prieto, S.I. (2015) Immune responses to oral pcDNA-VP2 vaccine in relation to infectious pancreatic necrosis virus carrier state in rainbow trout *Oncorhynchus mykiss*. *Veterinary Immunology and Immunopathology* 165, 127 - 137.

[6] Bowers, R.M., Lapatra, S.E. and Dhar, A.K. (2008) Detection and quantitation of infectious pancreatic necrosis virus by real-time reverse transcriptase-polymerase chain reaction using lethal and non-lethal tissue sampling. *Journal of Virological Methods*

① 为方便读者查阅，本书按原版复制参考文献。

147, 226 - 234.
[7] Bruno, D. and Munro, A. (1989) Immunity of Atlantic salmon, *Salmo salar* L., fry following vaccination against *Yersinia ruckeri*, and the influence of body weight and infectious pancreatic necrosis virus (IPNV) infection on the detection of carriers. *Aquaculture* 81, 205 - 211.
[8] Calvert, J.G., Nagy, E., Soler, M. and Dobos, P. (1991) Characterization of the VPg-dsRNA linkage of infec-tious pancreatic necrosis virus. *Journal of General Virology* 72, 2563 - 2567.
[9] Chen, L., Klaric, G., Wadsworth, S., Jayasinghe, S., Kuo, T.-Y. *et al*. (2014) Augmentation of the antibody response of Atlantic salmon by oral administration of alginate-encapsulated IPNV antigens. *PLoS ONE* 9(10): e109337.
[10] Coulibaly, F., Chevalier, C., Delmas, B. and Rey, F.A. (2010) Crystal structure of an *Aquabirnavirus* particle: insights into antigenic diversity and virulence deter-minism. *Journal of Virology* 84, 1792 - 1799.
[11] Crane, M. and Hyatt, A. (2011) Viruses of fish: an over-view of significant pathogens. *Viruses* 3, 2025 - 2046. Cuesta, A., Chaves-Pozo, E., de Las Heras, A.I., Saint-Jean, S.R., Pérez-Prieto, S. and Tafalla, C. (2010) An active DNA vaccine against infectious pancreatic necrosis virus (IPNV) with a different mode of action than fish rhabdovirus DNA vaccines. *Vaccine* 28, 3291 - 3300.
[12] Dadar, M., Memari, H. R., Vakharia, V. N., Peyghan, R., Shapouri, M. S. *et al*. (2015) Protective and immuno-genic effects of *Escherichia coli*-expressed infectious pancreatic necrosis virus (IPNV) VP2 - VP3 fusion pro-tein in rainbow trout. *Fish and Shellfish Immunology* 47, 390 - 396.
[13] Damsgard, B., Mortensen, A. and Sommer, A.-I. (1998) Effects of infectious pancreatic necrosis virus (IPNV) on appetite and growth in Atlantic salmon, *Salmo salar* L. *Aquaculture* 163, 185 - 193.
[14] de Kinkelin, P., Dorson, M. and Renault, T. (1992) Interferon and viral interference in viroses in salmonid fish. In: Kimura, T. (ed.) *Proceedings of the OJI Symposium on Salmonid Fish*. Hokkaido University Press, Sapporo, Japan, pp. 241 - 249.
[15] Debbink, K., Lindesmith, L. C., Donaldson, E. F., Swanstrom, J. and Baric, R. S. (2014) Chimeric GII. 4 norovirus virus-like-particle-based vaccines induce broadly blocking immune responses. *Journal of Virology* 88, 7256 - 7266.
[16] Delmas, B., Kibenge, F., Leong, J., Mundt, E., Vakharia, V. *et al*. (2005) *Birnaviridae*. In: Fauquet, C., Mayo, M., Maniloff, J., Desselberger, U. and Ball, L. (eds) *Virus Taxonomy: Classification and Nomenclature of Viruses. Eighth Report of the International Committee on Taxonomy of Viruses*. Elsevier/Academic Press, San Diego, California and London, pp. 561 - 569.
[17] Dhar, A.K., Bowers, R.M., Rowe, C.G. and Allnutt, F.C. (2010) Expression of a foreign epitope on infectious pancreatic necrosis virus VP2 capsid protein subviral particle (SVP) and immunogenicity in rainbow trout. *Antiviral Research* 85, 525 - 531.
[18] Dixon, P.F., Ngoh, G.H., Stone, D.M., Chang, S.F., Way, K. *et al*. (2008) Proposal for a fourth aquabirnavirus serogroup. *Archives of Virology* 153, 1937 - 1941.
[19] Ellis, A. E., Cavaco, A., Petrie, A., Lockhart, K., Snow, M. *et al*. (2010) Histology, immunocytochemistry and qRT - PCR analysis of Atlantic salmon, *Salmo salar* L., post-smolts following infection with infectious pancre-atic necrosis virus (IPNV). *Journal of Fish Diseases* 33, 803 - 818.
[20] Feldman, A.R., Lee, J., Delmas, B. and Paetzel, M. (2006) Crystal structure of a novel viral protease with a serine/lysine catalytic dyad mechanism. *Journal of Molecular Biology* 358, 1378 - 1389.
[21] Gadan, K., Sandtrø, A., Marjara, I.S., Santi, N., Munang'andu, H.M. *et al*. (2013)

Stress-induced reversion to virulence of infectious pancreatic necro-sis virus in naive fry of Atlantic salmon (*Salmo salar* L.). *PloS ONE* 8(2): e54656.

[22] Gomez-Casado, E., Estepa, A. and Coll, J. M. (2011) A comparative review on European-farmed finfish RNA viruses and their vaccines. *Vaccine* 29, 2657 - 2671.

[23] Graham, S.C., Sarin, L.P., Bahar, M.W., Myers, R.A., Stuart, D.I. *et al.* (2011) The N-terminus of the RNA polymerase from infectious pancreatic necrosis virus is the determinant of genome attachment. *PLoS Pathogens* 7(6): e1002085.

[24] Houston, R., Gheyas, A., Hamilton, A., Guy, D.R., Tinch, A.E. *et al.* (2008) Detection and confirmation of a major QTL affecting resistance to infectious pancre-atic necrosis (IPN) in Atlantic salmon (*Salmo salar*). *Developmental Biology* (*Basel*) 132, 199 - 204.

[25] ICTV (2014) *Virus Taxonomy: 2014 Release*. Available as *Virus Taxonomy: 2015 Release* at: http: //www. ictvonline. org /virusTaxonomy. asp? src = NCBI&ictv_id = 19810087 (accessed 21 October 2016).

[26] Imatoh, M., Hirayama, T. and Oshima, S. (2005) Frequent occurrence of apoptosis is not associated with patho-genic infectious pancreatic necrosis virus (IPNV) dur-ing persistent infection. *Fish and Shellfish Immunology* 18, 163 - 177.

[27] Jensen, B.B. and Kristoffersen, A.B. (2015) Risk factors for outbreaks of infectious pancreatic necrosis (IPN) and associated mortality in Norwegian salmonid farming. *Diseases of Aquatic Organisms* 114, 177 - 187.

[28] Johansen, L. and Sommer, A. (1995) Multiplication of infectious pancreatic necrosis virus (IPNV) in head kidney and blood leukocytes isolated from Atlantic salmon, *Salmo salar* (L.) *Journal of Fish Diseases* 18, 147 - 156.

[29] Johansen, L.-H. and Sommer, A.-I. (2001) Infectious pan-creatic necrosis virus infection in Atlantic salmon *Salmo salar* post-smolts affects the outcome of sec-ondary infections with infectious salmon anemia virus or *Vibrio salmonicida*. *Diseases of Aquatic Organisms* 47, 109 - 117.

[30] Johansen, L.-H., Eggset, G. and Sommer, A.-I. (2009) Experimental IPN virus infection of Atlantic salmon parr; recurrence of IPN and effects on secondary bac-terial infections in post-smolts. *Aquaculture* 290, 9 - 14.

[31] Knott, R.M. and Munro, A.L. (1986) The persistence of infectious pancreatic necrosis virus in Atlantic salmon. *Veterinary Immunology and Immunopathology* 12, 359 - 364.

[32] Kushnir, N., Streatfield, S.J. and Yusibov, V. (2012) Virus-like particles as a highly efficient vaccine platform: diversity of targets and production systems and advances in clinical development. *Vaccine* 31, 58 - 83.

[33] LaPatra, S., Lauda, K., Woolley, M. and Armstrong, R. (1993) Detection of a naturally occurring coinfection of IHNV and IPNV. *American Fisheries Society - Fish Health News* 21(1), 9 - 10.

[34] Lee, J., Feldman, A.R., Delmas, B. and Paetzel, M. (2007) Crystal structure of the VP4 protease from infectious pancreatic necrosis virus reveals the acyl-enzyme complex for an intermolecular self-cleavage reaction. *The Journal of Biological Chemistry* 282, 24928 - 24937.

[35] Leong, J., Brown, D., Dobos, P., Kibenge, F., Ludert, J. *et al.* (2000) Family *Birnaviridae*.In: VanRegenmortel, M., Bishop, D., Calisher, C., Carsten, E., Estes, M. *et al.* (eds) *Virus Taxonomy. Seventh Report of the International Committee for the Taxonomy of Viruses*. Academic Press, New York, pp. 481 - 490.

[36] Liu, M., Zhao, L.-L., Ge, J.-W., Qiao, X.-Y., Li, Y.-J. *et al.* (2012) Immunogenicity of *Lactobacillus*-expressing VP2 and VP3 of the infectious pancreatic necrosis virus (IPNV) in rainbow trout. *Fish and Shellfish Immunology* 32(1),

196 - 203.

[37] Magyar, G. and Dobos, P. (1994) Evidence for the detec-tion of the infectious pancreatic necrosis virus poly-protein and the 17-kDa polypeptide in infected cells and of the NS protease in purified virus. *Virology* 204, 580 - 589.

[38] Martinez-Alonso, S., Vakharia, V.N., Saint-Jean, S.R., Perez-Prieto, S. and Tafalla, C. (2012) Immune responses elicited in rainbow trout through the admin-istration of infectious pancreatic necrosis virus-like particles. *Developmental and Comparative Immunology* 36, 378 - 384.

[39] McAllister, P. (2007) 2.2.5 Infectious pancreatic necrosis. In: *AFS - FHS. Fish Health Section Blue Book: Sugges-ted Procedures for the Detection and Identification of Certain Finfish and Shellfish Pathogens*, 2007 edn. American Fisheries Society-Fish Health Section, Bethesda, Maryland. Available at: http://www.afs-fhs.org/perch/resources/14069231202.2.5ipnv2007ref2014.pdf (accessed 21 October 2016).

[40] McKnight, I. J. and Roberts, R. J. (1976) The pathology of infectious pancreatic necrosis. I. The sequential histopathology of the naturally occurring condition. *British Veterinary Journal* 132, 76 - 85.

[41] Mikalsen, A. B., Torgersen, J., Alestrom, P., Hellemann, A. L., Koppang, E. O. *et al*. (2004) Protection of Atlantic salmon *Salmo salar* against infectious pancreatic necrosis after DNA vaccination. *Diseases of Aquatic Organisms* 60, 11 - 20.

[42] Moen, T., Baranski, M., Sonesson, A.K. and Kjoglum, S. (2009) Confirmation and fine-mapping of a major QTL for resistance to infectious pancreatic necrosis in Atlantic salmon (*Salmo salar*): population-level asso-ciations between markers and trait. *BMC Genomics* 10: 368.

[43] Moen, T., Torgersen, J., Santi, N., Davidson, W.S., Baranski, M. *et al*. (2015) Epithelial cadherin deter-mines resistance to infectious pancreatic necrosis virus in Atlantic salmon. *Genetics* 200, 1313 - 1326.

[44] Munang'andu, H.M., Fredriksen, B.N., Mutoloki, S., Brudeseth, B., Kuo, T.Y. *et al*. (2012) Comparison of vaccine efficacy for different antigen delivery systems for infectious pancreatic necrosis virus vaccines in Atlantic salmon (*Salmo salar* L.) in a cohabitation challenge model. *Vaccine* 30, 4007 - 4016.

[45] Munang'andu, H.M., Santi, N., Fredriksen, B.N., Lokling, K.E. and Evensen, O. (2016) A systematic approach towards optimizing a cohabitation challenge model for infectious pancreatic necrosis virus in Atlantic salmon (*Salmo salar* L.). *PLoS ONE* 11 (2): e0148467.

[46] Munro, E.S. and Midtlyng, P.J. (2011) Infectious pancre-atic necrosis and associated aquabirnaviruses. In: Woo, P. and Bruno, D. (eds) *Fish Diseases and Disorders: Viral, Bacterial and Fungal infections*, 2nd edn. CAB International, Wallingford, UK, pp. 1 - 65.

[47] Norwegian Veterinary Institute (2015) *The Health Situation in Norwegian Aquaculture 2014*. Norwegian Veterinary Institute, Oslo. OIE (2003) Chapter 2.1.8. Infectious pancreatic necrosis. In: *Manual of Diagnostic Tests for Aquatic Animals*, 4th edn,. World Organisation for Animal Health, Paris, pp. 142 - 151. Available at: http://www.oie.int/doc/ged/D6505.PDF (accessed 17 November 2016).

[48] Orpetveit, I., Mikalsen, A. B., Sindre, H., Evensen, O., Dannevig, B. H. *et al*. (2010) Detection of infectious pancreatic necrosis virus in subclinically infected Atlantic salmon by virus isolation in cell culture or real-time reverse transcription polymerase chain reaction: influence of sample preservation and stor-age. *Journal of Veterinary Diagnostic Investigation* 22, 886 - 895.

[49] Ozaki, A., Araki, K., Okamoto, H., Okauchi, M., Mushiake, K. *et al*. (2012) Progress of DNA marker-assisted breed-ing in maricultured fish. *Bulletin of Fish*

Research Agencies 35, 31 - 37. Available at: https://www.fra.affrc.go.jp/bulletin/bull/bull35/35 - 5.pdf (accessed 21 October 2016).

[50] Petit, S., Lejal, N., Huet, J. C. and Delmas, B. (2000) Active residues and viral substrate cleavage sites of the protease of the birnavirus infectious pancreatic necrosis virus. *Journal of Virology* 74, 2057 - 2066.

[51] Roberts, R. J. and Pearson, M. (2005) Infectious pancre-atic necrosis in Atlantic salmon, *Salmo salar* L. *Journal of Fish Diseases* 28, 383 - 390.

[52] Ruane, N., Geoghegan, F. and Ó Cinneide, M. (2007) Infectious pancreatic necrosis virus and its impact on the Irish salmon aquacultue and wild fish sectors. Marine Environment and Health Series No. 30, Marine Institute, Oranmore Republic of Ireland.

[53] Sadasiv, E. C. (1995) Immunological and pathological responses of salmonids to infectious pancreatic necro-sis virus (IPNV). *Annual Review of Fish Diseases* 5, 209 - 223.

[54] Santi, N., Sandtro, A., Sindre, H., Song, H., Hong, J.R. *et al.* (2005) Infectious pancreatic necrosis virus induces apoptosis *in vitro* and *in vivo* independent of VP5 expression. *Virology* 342, 13 - 25.

[55] Shivappa, R. B., McAllister, P. E., Edwards, G. H., Santi, N., Evensen, O. *et al.* (2005) Development of a subunit vaccine for infectious pancreatic necrosis virus using a baculovirus insect/larvae system. *Developmental Biology (Basel)* 121, 165 - 174.

[56] Smail, D., McFarlane, L., Bruno, D. and McVicar, A. (1995) The pathology of IPN-Sp and sub-type (Sh) in farmed Atlantic salmon, *Salmo salar* L., post-smolts in the Shetland Isles, Scotland. *Journal of Fish Diseases* 18, 631 - 638.

[57] USFWS and AFS - FHS (2007) Section 2. USFWS [US Fish and Wildlife Service]/AFS-FHH Standard proce-dures for aquatic animal health inspections. In: *AFS-FHS. Fish Health Section Blue Book: Suggested Procedures for the Detection and Identification of Certain Finfish and Shellfish Pathogens*, 2007 edn. American Fisheries Society-Fish Health Section, Bethesda, Maryland.

[58] Villanueva, R.A., Galaz, J.L., Valdes, J.A., Jashes, M.M. and Sandino, A.M. (2004) Genome assembly and par-ticle maturation of the birnavirus infectious pancreatic necrosis virus. *Journal of Virology* 78, 13829 - 13838.

[59] Wood, E.M., Snieszko, S.F. and Yasutake, W.T. (1955) Infectious pancreatic necrosis in brook trout. *American Medical Association Archives of Pathology* 60, 26 - 28.

[60] Wolf, K., Snieszko, S.F., Dunbar, C.E. and Pyle, E. (1960) Virus nature of infectious pancreatic necrosis in trout. *Proceedings of Society for Experimental Biology and Medicine* 104, 105 - 108.

2

传染性造血器官坏死病病毒

Jo-Ann C. Leong* 和 Gael Kurath

2.1 引言

传染性造血器官坏死病病毒(infectious hematopoietic necrosis virus, IHNV),是一种弹状病毒(*Rhabdovirus*),可在太平洋鲑(*Oncorhynchus* spp.)、大西洋鲑、虹鳟中引起严重疾病。由其引起的传染性造血器官坏死病(infectious hematopoietic necrosis, IHN)最早在北美洲西北太平洋地区养殖的红鲑(*O. nerka*)中被发现,1969 年首次培养分离出 IHNV(Bootland and Leong, 1999)。IHNV 是弹状病毒科(Rhabdoviridae)诺拉弹状病毒属(*Novirhabdovirus*)的模式种和参考病毒。其病毒基因组是一个线状、单链、反义 RNA(长度约为 11 140 个核苷酸),具有从基因组 3′端开始的 6 个基因,依次为 N(核蛋白)、P(磷蛋白)、M(基质蛋白)、G(糖蛋白)、NV(非毒粒蛋白)和 L(RNA 聚合酶)。*Novirhabdovirus* 的名称来源于该属中特有的非病毒粒基因(Kurath and Leong, 1985; Leong and Kurath, 2011)。

IHNV 会引起宿主造血器官坏死,因此被 Amend 等(1969)命名为传染性造血器官坏死病病毒。该病毒为水传播病原,也可以通过病毒污染的精液和卵巢液进行水平传播和垂直传播(Bootland and Leong, 1999)。恢复期的虹鳟鱼苗通常能清除病毒,但有些鱼可以携带病毒长达 46 d(Drolet et al., 1995)。病毒在一些感染后幸存鱼的肾脏中可持续存在一年(Drolet et al., 1995; Kim et al., 1999)。感染后存活的虹鳟在无病毒的水体中养殖 2 年后,产卵时的精液和卵巢液中均有感染性病毒存在(Amend, 1975),这些鱼都是潜在的传染源。然而,在海水中采集并在无病毒水环境中养殖至成熟的成年红鲑在产卵时没有检测到病毒,而允许自然迁徙鱼群的患病率为 90%～100%(Amos et al., 1989)。这表明在鱼群河流迁徙期间水平传播的重要性,并且 IHNV 在不同宿主中的持留性也有所不同。最近,Müller 等(2015)证实了 IHNV 存在于幸存红鲑的脑中,而非肾脏。尽管在幸存的红鲑中没有出现疾病和死亡的

* 通信作者邮箱:joannleo@hawaii.edu。

情况，但在暴露于病毒9个月后，4%的鱼的脑中含有IHNV病毒RNA。这支持了一个假设，即一小部分受感染的鱼会成为病原携带者。如果来自鱼脑的病毒具有传染性，这个发现将对病毒控制策略产生严重影响。其他潜在的传染源包括吸附在沉积物上的病毒以及在无脊椎动物或非鲑科鱼类宿主中的病毒（尽管很少被检测到）。

IHNV是世界动物卫生组织（world organization of animal health，OIE）应具报病毒，已确诊或出现疑似病例的国家包括奥地利、比利时、玻利维亚、加拿大、中国、克罗地亚、捷克共和国、法国、德国、伊朗、意大利、日本、韩国、荷兰、波兰、俄罗斯、斯洛文尼亚、西班牙、瑞士和美国（OIE，2015；last updated 23 July 2015；and Cefas，2011，last updated 31 January）。该病毒最初在北美洲西部流行，在那里它拥有最大的宿主种类多样性和最长的疾病影响历史，可感染野生和养殖鱼类，并且最具遗传多样性。IHNV后来也蔓延到亚洲和欧洲，主要在虹鳟养殖场中出现并流行。IHNV的全球系统发育分析确定了五个主要的遗传群体（基因群）：北美和俄罗斯的U基因群；北美的M基因群和L基因群；亚洲的J基因群；欧洲的E基因群（Kurath，2012a）。

IHNV造成的经济损失可以是鱼类死亡的直接影响，也可以是与限制受IHNV感染鱼类的流通或要求销毁受感染鱼的监管条例有关的间接影响。疾病的暴发已经严重破坏北美洲西部地区的商业养殖（如虹鳟和大西洋鲑）和太平洋鲑鳟的保护/纾缓计划。由于该疾病已蔓延至欧洲和亚洲，Fofana和Baulcomb（2012）最近指出了IHNV对鲑鳟水产养殖业潜在的经济影响。虽然在英国IHNV尚未被分离出，但据估计，11年（1998—2008）中IHN的暴发理论上造成的直接损失和间接损失将为1 680万英镑（约2 550万美元）。直接损失将来自扑杀、死亡和死鱼的处置。间接损失将包括消费者反应造成的收入损失、对外出口的减少以及实施额外监管措施的费用增加。OIE建立了IHN疫情数据库，该数据库显示2012年在美国西部、德国、意大利、波兰、中国、日本、韩国和加拿大不列颠哥伦比亚省都检测到了IHNV。

2.2 临床体征与诊断

2.2.1 临床体征

尽管IHNV可以感染所有鱼龄段的鲑科鱼类，但感染IHNV会在鲑科鱼类幼鱼中引起严重疾病。通常，在开始发病时，濒死的鱼会变得嗜睡，并伴随周期性的偶尔回旋或过度亢进游动；鱼苗可能出现体色发黑、腹胀、眼球突出、鱼鳃苍白和黏液样不透明的粪管等症状[图2.1(A)]。鳍基部和肛门处会出现瘀点状出血，偶尔也发生在鳃、口、眼、皮肤和肌肉等部位[图2.1(B)(C)]。有些鱼苗的头部后面可能即刻出现皮下出血，大龄鱼的外部临床症状较少。患病的二龄红

图 2.1 (A) 幼龄虹鳟浸泡在 IHNV 中 7 d 后。颜色较浅的鱼没有感染的迹象，颜色较深的鱼有传染性造血器官坏死病的典型症状：体色变暗，有明显的眼球突出症。水箱右侧的死鱼出现出血瘀点。(B) 患有 IHN 的幼龄虹鳟，眼球突出，眼眶出血。(C) 患有 IHN 的幼龄红鲑，眼眶周围、鳃和鳍上出现点状出血

鲑会出现鳃和眼睛出血、棒状和融合的鳃瓣及皮肤损伤等症状，而 2 年左右鱼龄的红鲑(内陆红鲑，*O. nerka*)会出现不稳定的游动行为以及在鳍基部附近出血。此外，有些死于 IHN 的鱼并无明显症状(Bootland and Leong，1999)。

受感染鱼苗的肝、脾和肾因贫血而变得苍白；可能有腹水；胃内充满乳白色液体，但没有食物；肠内充盈水状黄色液体；内脏肠系膜、脂肪组织、膀胱、腹膜、脑膜和心包可能有瘀斑状出血。大鱼龄的鱼可能出现空胃、肠内充满黄色黏液和肾脏附近肌肉组织损伤等症状(Bootland and Leong，1999)。

2.2.2 诊断

在有该疾病史的地方，可根据鱼体临床症状进行初步诊断。在组织学上，消化道粒细胞坏死呈现为病理性特征(Wolf，1988)。初步诊断必须通过特异性鉴定来加以确认。最被广泛接受的诊断方法是从细胞中分离出病毒(推定诊断)，然后采用血清中和试验或基于 PCR 方法对病原体进行鉴定。OIE 在线的《水生动物诊断检测指南》(manual of diagnostic tests for aquatic animals，2015)、加拿大渔业及海洋部的《鱼类健康保护条例：合规手册》(简称《合规手册》)(fish health protection regulations：manual of compliance，department of fisheries and oceans，1984；2011 年修订）以及美国渔业协会鱼类健康部［American fisheries Society（AFS）-fish health section（FHS）］的《FHS 蓝皮书：某些鱼类和贝类病原检测和鉴定的建议程序》（简称《FHS 蓝皮书》）中的 IHNV 章节（2014）都详细描述了该疾病的免疫学和分子生物学检测方法。

许多硬骨鱼细胞系易受 IHNV 感染，但《水生动物诊断检测指南》、《合规手册》和《FHS 蓝皮书》中指定使用的是鲤鱼上皮瘤细胞系（epithelioma papulosum cyprini，EPC）和胖头鲂（fathead minnow，FHM）肌肉细胞系。《FHS 蓝皮书》还推荐了大鳞大麻哈鱼胚胎（Chinook salmon embryo，CHSE）细胞系 CHSE－214。EPC 细胞对 IHNV 最敏感（Lorenzen et al.，1999）。Bootland 和 Leong（1999）广泛讨论了其他对 IHNV 细胞致病性敏感的鱼类细胞系。IHNV 的最适生长温度约为 15 ℃（Mulcahy et al.，1984）；在 23～25 ℃的较高温度范围内病毒将停止复制。IHNV 不耐热，在 32 ℃下数小时内便会失活（Pietsch et al.，1977）。

细胞培养法检测 IHNV 需要 14 d。通常使用终点滴定法和噬菌斑测定法，其中后者更灵敏，也是量化 IHNV 的标准。用聚乙二醇预处理细胞单层可提高病毒噬菌斑测定的速度和灵敏度，并可形成较大的噬菌斑。典型的致细胞病变效应（cytopathic effect，CPE）包括葡萄状的圆形细胞簇、核染色质边聚。IHNV 的典型噬菌斑由在开口内缘收缩或堆积的细胞片层组成，中心可能含有粒状碎片。细胞应观检 14 d，如果未发生致细胞病变效应，上清液可被转移给新鲜细胞进行"盲传"。无致细胞病变效应表明样本病毒阴性（Bootland and Leong，1999）。

分离 IHNV 的首选组织是肾和脾，黏液和胸鳍剪屑也被用作非致死性样本。在亲鱼的检测中，卵巢液是首选样本，这是因为在精液内病毒的检出率较低。对产卵后的雌鱼进行采样时，可以通过保存其卵巢液或培养卵巢液细胞来提高病毒检测灵敏度。精液样本需要先进行离心，随后将离心得到的沉淀在水中孵育并对水进行病毒检测（Bootland and Leong，1999，2011）。

血清学检测需要多克隆或单克隆抗体，现已有市售的 IHNV 免疫检测试剂盒（Bio-X Diagnostics，Rochefort，France；http：//www.biox.com）。IHNV 的鉴定方法包括血清中和、间接荧光抗体法、直接碱性磷酸酶免疫细胞化学法（alkaline phosphatase immunocyto-chemistry，APIC）和酶联免疫吸附试验（ELISA），以及免疫印迹（Western blot）、斑点印迹、葡萄球菌协同凝集试验和电镜镜检等（Bootland and Leong，1999）。分子生物学方法包括逆转录 PCR（reverse transcriptase-dependent polymerase chain reaction，RT－PCR）、实时荧光定量 PCR（qRT－PCR；Bootland and Leong，2011；Purcell et al.，2013）、多重 RT－PCR（Liu et al.，2008）、环介导等温扩增法（Gunimaladevi et al.，2005）以及分子锁式探针技术（Millard et al.，2006）。除了最后一种方法外，其他

方法的初始步骤需从 IHNV 病毒 RNA(mRNA、基因组和反基因组)反转录获得 cDNA,然后使用 PCR 扩增。Bootland 和 Leong(2011)已对这些方法及其相对灵敏度(如有)进行阐述,它们的相对灵敏度情况见表 2.1。对于研究应用,改良的 qRT - PCR 法已被开发用于阳性和阴性 IHNV RNA 的特异性检测(Purcell et al., 2006a)以及混合感染中不同 IHNV 病毒株基因型的特异性检测(Wargo et al., 2010)。

表 2.1　不同诊断方法的 IHNV 检测灵敏性

方　法	报道的灵敏性	计算的物理粒子数[a]	参考文献
鱼类细胞系的感染性	1 感染单位	1 $TCID_{50}$①单位=500 1 PFU= 0.69 $TCID_{50}$ 单位=345	Durrin, 1980
葡萄球菌协同凝集试验	$1\times10^4\sim6\times10^4$ PFU	$3.45\times10^6\sim20.7\times10^6$	Bootland and Leong, 1992
免疫印迹			
过氧化物酶标记抗体	10 ng 病毒[b]	7.67×10^7	Hsu and Leong, 1985
碘[125]标记抗体	2.5 ng 病毒	1.53×10^7	Hsu and Leong, 1985
碱性磷酸酶标记二抗			
抗 IHNV N 鼠多抗	1 PFU	345	Kim et al., 2001
抗 IHNV N 鼠单抗	3×10^3 PFU	1.035×10^6	Kim et al., 2001
酶联免疫吸附试验			
抗 IHNV N 鼠多抗	10 PFU	3 450	Kim et al., 2001
抗 IHNV N 鼠单抗	10^3 PFU	3.45×10^6	Kim et al., 2001
抗 IHNV 鸡抗和兔抗	70 PFU	2.42×10^4	Medina et al., 1992
免疫印迹			
过氧化物酶标记	0.85～4.0 ng 病毒	$0.66\times10^7\sim3.07\times10^7$	McAllister and Schill, 1986
生物素标记抗体	10^2 PFU, 5.5 ng 病毒	3.45×10^4, 4.22×10^7	Schultz et al., 1989
RT - PCR,基因 N,单循环	4×10^2 PFU	1.38×10^5	Arakawa et al., 1990
RT - PCR,基因 G,单循环	2.2×10^3 G 基因 RNA 拷贝数[c]	5.9×10^2 gRNAs, 25 粒子	Emmenegger et al., 2000; Purcell et al., 2013
抗 IHNV N 单抗免疫磁珠捕获,随后进行 RT - PCR	10^6 PFU	3.45×10^6	Kim et al., 2001

① $TCID_{50}$为半数组织培养物感染量。

续 表

方　法	报道的灵敏性	计算的物理粒子数[a]	参考文献
RNA 斑点杂交(生物素或碱性磷酸酶标记)	20 pg 病毒 RNA[d] 4 pg 病毒 RNA	3.37×10^6病毒 gRNAs 4.4×10^4病毒 gRNAs	Gonzalez et al., 1997 Gonzalez et al., 1997
RT - qPCR,基因 N	7 个 N 基因 RNA 拷贝数[e]	1.1 病毒 gRNA,0.05 粒子	Purcell et al., 2013

[a]每一种方法的灵敏度用于确定用电子显微镜定量的 IHNV 病毒粒子的等效数量(Durrin, 1980)。

[b] IHNV 病毒粒子的质量来源于病毒粒子蛋白质与 RNA 的 21∶1 比例,估计病毒基因组的相对分子质量为 3.57×10^6 Da (Kurath and Leong,未发表的观察结果)。因此,病毒粒子的相对分子质量为 7.497×10^7 Da蛋白质加上 3.57×10^6 Da RNA。故一个病毒粒子的计算质量为 $1.304\ 18\times10^{-16}$ g,病毒基因组的质量为5.928×10^{-18} g。

[c] 基于 Purcell 等(2013)的数据,IHNV G 基因常规 PCR 检测方法根据病毒 RNA 拷贝数进行了校准(M. Purcell and G. Kurath,未发表),包括病毒基因组 gRNA 和 mRNA。在这个病毒 RNA 拷贝数中,gRNA∶G mRNA 的比例为 1∶2.7,gRNA∶PFU 的比例为 8 000∶1 (Purcell et al., 2006)。这些比值被用于计算病毒 gRNA 拷贝数、PFU 和物理粒子等量数值。

[d] Gonzalez 等(1997)使用部分纯化的病毒提取 RNA,因此实际病毒 RNA 数量可能会更低。

[e] 对通用的 IHNV N 基因 RT - qPCR 检测限如脚注 c 所述进行校准,但根据报道的 IHNV G mRNA∶N mRNA 的摩尔比约为 1∶2 进行了修正。因此,我们采用 gRNA∶N mRNA 的比例 1∶5.4 来计算 gRNA 的等效拷贝数。

感染了 IHNV 的鲑鱼可能会产生持续数月的强抗体反应(Lorenzen and LaPatra, 1999),而监测抗体反应是非致死性的,这可能会有助于监测有过 IHNV 暴露史的鱼群情况(Bootland and Leong, 2011)。

2.3 病理学

2.3.1 组织病理学

IHNV 的组织病理学表现包括造血组织、后肾、脾、肝、胰腺和消化道的退行性坏死。在前肾,最初的变化是由明显的巨噬细胞和退化的淋巴细胞组成的小而轻微的染色病灶区。随着病情的发展,整个肾脏的退行性变化更加明显:巨噬细胞数量增多、胞质空泡化、核染色质边集,可能存在固缩和坏死性淋巴样细胞,坏死程度可能严重到肾组织主要由坏死碎片组成。脾、胰腺、肝、肾上腺皮质和肠道细胞的局灶区表现出核多态性和染色质边集,最终坏死。各器官大面积坏死,伴有固缩、核碎裂和核溶解。IHN 的一个特异性症状是消化道固有层、致密层和颗粒层的粒细胞变性和坏死,并且肠黏膜的脱落可能会引起粪样物形成。二龄鲑和一龄幼鱼往往表现出不太严重的组织病理学。肾、脾、胰腺和肝可能出现坏死,但肠黏膜仅中度脱落,无粪样物。鱼类患有正常红细胞再生障碍性贫血,受感染鱼的血液中出现白细胞减少,伴随白细胞和血小板变性、血细胞比容和渗透压降低,以及生化特征略有改变。血涂片或肾压片中的细胞碎片(坏死生物体)是 IHN 的特异性症状(Wolf, 1988; Bootland and Leong, 1999)。

2.3.2 疾病进程

在虹鳟中，IHNV通过鳃、皮肤、鳍基部和口腔进入食道/心胃区，然后扩散到内脏。幼鱼肾脏和脾脏的造血组织受到的影响最严重，是最先出现广泛坏死的组织。通常，在浸泡暴露后的一天内，感染扩散到肾脏之前（2～4 d），可在虹鳟幼鱼的鳃、皮肤和肠中检测到低滴度的IHNV，随后广泛扩散（Bootland and leong，1999，2011）。

在感染期间，病毒感染率和滴度在5～14 d内达到峰值（Drolet et al.，1994；Penaranda et al.，2009；Purcell et al.，2009）。一些虹鳟在第28天时仍被感染，但54 d后就检测不到感染性病毒（Drolet et al.，1994）。Drolet等（1994）提出感染主要通过两种途径进行：一种是从鳃进入循环系统；另一种是先从口腔进入胃肠道（gastrointestinal，GI），再进入循环系统。两种途径都会引起全身性病毒血症。这些研究者还提出，最初的肾脏感染不是通过胃肠道，而是通过口腔内高度血管化的组织穿梭血液而传播到肾脏和造血组织。利用免疫胶体金标记的二抗检测抗N单克隆抗体与感染细胞的结合，Helmick等（1995a，b）确定了食道/贲门胃区（esophagus/cardiac stomach region，ECSR），特别是分泌黏液的贲门腺（mucus-secreting cardiac gland，MSCG）是IHNV的早期感染靶区。有证据表明感染后的1 h内IHNV在虹鳟和银大麻哈鱼（*O. kisutch*）的ECSR黏膜上皮细胞中附着和内化。银大麻哈鱼的MSCG表现出较温和的反应，这支持了先前所发现的病毒在银大麻哈鱼体内复制效率较低这一现象。在大鳞大麻哈鱼和红鲑鱼苗中，可能存在肝样肝沉积（Wood and Yasutake，1956；Yasutake，1970）。在疾病的最后阶段，坏死不仅见于肾脏的造血组织，也见于肾小球和肾小管。

在对虹鳟的另一项研究中，研究人员利用活体生物发光成像技术追踪表达荧光素酶的重组IHNV毒株在鱼体内的感染过程（Harmache et al.，2006）。除确定了鱼鳍基部是病毒入侵部位外，鳍组织和内脏中还存在持续的病毒复制。一些感染后幸存的鱼依然保持着局部病毒复制，主要是在鱼鳍中。最近，在注射感染并于24 ℃下养殖的透明斑马鱼（*Danio rerio*）幼体中监测了一株高温适应性IHNV毒株的感染进程（Ludwig et al.，2011）。尽管心脏依然跳动但是血液停止流动，肉眼可见的感染症状减缓。随后，鱼体失去触碰反应能力，在3～4 d内死亡。通过全幼体原位杂交技术，感染后6 h在主要血管中检测到第一批感染细胞，其中静脉内皮细胞是感染的主要对象。这表明感染是从受损的血管扩散到下层组织。将幼体转移至28 ℃（无病毒复制）后，在临床症状出现前一天不到就达到了导致不可逆损伤的临界阈值。

2.4 病理生理学

IHN的病理特征表现为宿主体内碱储备严重损耗和血电解质失衡，导致血液渗透压降低(Amend, 1973)。这些变化归结于病毒引起的肾组织坏死导致的肾功能丧失。在感染的 4 d 内,宿主的血细胞比容、血红蛋白、红细胞数和血浆碳酸氢盐均显著降低(Amend and Smith, 1974)。在感染后的 9 d 里,血浆氯化物、钙、磷、总蛋白和血细胞类型没有发生变化。乳酸脱氢酶 B($LDHB_4^{2'}$)同工酶水平的升高与病情的早期发展相关。在感染传染性胰腺坏死病毒(IPNV)或三种病原细菌的鱼类中均未观察到乳酸脱氢酶(LDH)升高现象。

Amend 和 Smith(1975)发现,濒死的虹鳟中血浆碳酸氢盐、氯化物、钙、磷、胆红素和渗透压降低。当鱼出现病征时,血糖和前肾抗坏血酸没有变化。受感染鱼的红细胞数、血红蛋白和血细胞比容下降,但平均红细胞体积、平均红细胞血红蛋白和平均红细胞血红蛋白浓度保持正常;未成熟红细胞百分率升高,但白细胞百分率不变;中性粒细胞减少,淋巴细胞增多,但单核细胞没有变化;血浆 pH 升高,血清蛋白 α2 和 α3 组分改变;碱储量减少,酸碱平衡和体液平衡改变。鱼死亡可能是由肾衰竭引起的严重电解质和体液失衡所致。

2.5 防控策略

疾病防控策略主要通过生物安全策略和养殖设施卫生的良好实践来防止鱼类接触病毒(Winton, 1991)。1981 年成功实施的阿拉斯加红鲑养殖政策为病毒流行地区的 IHNV 控制提供了指导方针,该方针基于三个标准:无病毒供水、严格消毒及卵孵化和幼鱼繁育的区化管理(Meyers et al., 2003)。因此,通常用碘伏(一种含有碘和增溶剂混合物的消毒剂)对鱼卵进行消毒,然后在无病毒的水中分批孵化不同鱼卵批。鱼苗也在无病毒的水中进行繁育。在整个育成期间,尽可能长时间地使用安全水源(没有易感宿主鱼)。最近,在一家大型硬头鳟孵化场实施的延迟暴露育成策略获得成功,说明了供水在最大限度减少自然放养鱼和孵化鱼之间病毒传播方面具有重要性(Breyta et al., 2016)。在非疾病流行地区,通过扑杀、消毒和检疫控制疾病暴发(McDaniel et al., 1994)。不列颠哥伦比亚省的鲑鱼养殖场通常采取其他预防措施,包括维护同龄鱼和同一品种鱼的养殖区,减少不同围栏之间的鱼类转移,在发生感染时扑杀/加速捕捞较小鱼类,以及放养期间的休耕(Saksida, 2006)。从未将运输已宰杀的鲜鱼或冷冻鱼与 IHNV 引入新的地理区域相关联,虹鳟的加工处理风险被认为可以忽略(LaPatra et al., 2001a)。

在细胞培养中,异丙肌苷可以抑制 IHNV 的复制(Siwicki et al., 2002),而氯喹可以通过阻碍病毒结合和入侵细胞来抑制 IHNV 体内复制(Hasobe and Saneyoshi, 1985; De Las Heras et al., 2008)。同样,用抗病毒药物金刚烷胺或三丁胺预处理培养细胞也能减少 IHNV 对细胞的结合和侵染。检测发现,在细胞培养中与 IHNV 基因组 RNA 5′端互补的反义磷酸二胺吗啉代寡核苷酸

(phosphorodiamidate morpholino oligomers, PMOs)对IHNV的抑制作用呈序列和剂量依赖性。该化合物与一种增强PMO进入培养细胞和活鱼组织的膜穿透肽交联(Alonso et al., 2005)。此外,一些水生环境中的细菌也被报道具有抗IHNV活性(Myouga et al., 1993)。荧光假单胞菌1生物变型(*Pseudomonas fluorescens* biovar 1)产生的一种肽(46NW-64A)完全抑制了INHV的复制(噬菌斑100%减少)。在饲料中添加类胡萝卜素、虾青素之后,虹鳟鱼苗对IHNV攻毒产生了显著的抗性(Amar et al., 2012)。在鱼单层细胞上,牛α2-CN酪蛋白水解物和总酪蛋白水解物是针对IHNV的有效抗病毒剂(Rodríguez Saint-Jean et al., 2012)。当对暴露于IHNV后当天和第3天的三月龄褐鳟(*Salmo trutta*)饲喂酪蛋白时,饲喂组鱼显示出近50%的保护作用,而对照组鱼死亡率达93%(Rodríguez Saint-Jean et al., 2013)。

自从McIntyre和Amend(1978)报道了红鲑对IHNV的抗性有30%的遗传率以来,人们一直在努力选育开发鲑鳟鱼类的IHNV抗性品种。这种抗性可能是种属特异性的,并局限于特定的病毒基因型。然而,对IHNV有抗性并不意味着对感染有抵抗力。就疾病而言,红鲑对M群病毒具有抗性,但对U群病毒敏感,而虹鳟则恰恰相反(Garver et al., 2006)。与更具抗性的品种[如银大麻哈鱼或美洲红点鲑(*Salvelinus fontinalis*)]杂交后的虹鳟杂交种对IHNV的抗性更强(LaPatra et al., 1993),而银大麻哈鱼和大鳞大麻哈鱼的杂交品种则易感IHNV而死亡(Hedrick et al., 1987)。

识别抗性基因一直是抗病虹鳟和大西洋鲑选育计划的重点。大多数关于IHNV抗性的遗传研究采用的是家族内虹鳟回交(Khoo et al., 2004),或与不易感的克氏鳟(*O. clarkia*)(Palti et al., 1999, 2001; Barroso et al., 2008)或硬头鳟回交(Rodriguez et al., 2004)。基于分子标记[微卫星和扩增片段长度多态性(AFLPs)]和数量性状位点(QTLs)的遗传连锁图谱已经被绘制出来,但此类研究尚未发现高度相关的位点(Overturf et al., 2010)或在很大的连锁不平衡区块内的显著关联(Rodríguez et al., 2004; Barroso et al., 2008)。基因组测序已经鉴定出虹鳟中与IHNV抗性相关的19个单核苷酸多态性(single-nucleotide polymorphism, SNP)标记,可用于后续的标记辅助选育(marker assisted selection, MAS)计划(Campbell et al., 2014)。据Yang等(2014)报道,抗性基因可能与MHC(主要组织相容性复合体)单倍型和T细胞免疫有关。有趣的是,Verrier等(2012)发现,来自抗病毒性出血性败血症病毒(viral haemorrhagic septicaemia virus, VHSV)克隆系的鳟鱼细胞系在体外条件下对弹状病毒感染也具有抗性。

2.5.1 疫苗

2005年,加拿大食品检验局(Canadian food inspection agency, CFIA)批准

了一种名为 APEX-IHN (Novartis Animal Health Canada Inc.)的 DNA 疫苗用于养殖大西洋鲑(CFIA, 2005)。这是世界上首个被批准用于鱼类的 DNA 疫苗,也是仅有的四种可用于动物的商品化 DNA 疫苗之一。该疫苗是基于将组成型病毒启动子控制的编码 IHNV G 基因的质粒 DNA 通过肌内注射进虹鳟鱼苗体内来诱导保护作用(Anderson et al., 1996)。此后,在美国进行的大量实验研究都证实了类似 DNA 疫苗的有效性,并对疫苗剂量、保护期、跨物种保护、给药途径和疫苗安全性等应用方面进行了考察(Kurath, 2008)。经证实,DNA 疫苗必须编码 IHNV 的 G 基因,而不是其他 IHNV 基因,才能在虹鳟鱼苗和红鲑中诱导保护和中和抗体。该疫苗对 1～2 g 虹鳟鱼苗的有效剂量极低(0.1～1 μg),而较大的鱼需要较高的剂量。接种疫苗的虹鳟鱼苗可有效抵抗来自不同地域 IHNV 的侵害,这表明该疫苗在世界范围内的可用性。DNA 疫苗可诱导快速非特异性先天免疫和 Mx 蛋白(α/β 干扰素诱导的指示蛋白)表达(LaPatra et al., 2001b; Lorenzen et al., 2002; Kim et al., 2000; Purcell et al., 2006b),随后诱导长期的特异性保护,3 个月的相对免疫保护率(RPS)大于 90%,2 年的 RPS 大于 60% (Kurath et al., 2006)。IHNV-G DNA 疫苗对虹鳟(Kurath, 2008)、大西洋鲑(Traxler et al., 1999)、大鳞大麻哈鱼、红鲑和红大麻哈鱼(*O. nerka*)都有保护作用(Garver et al., 2005a)。含有来自 M 或 U 基因群病毒 G 基因的 DNA 疫苗可以保护鱼类免受同基因群和跨基因群 IHNV 的侵染(Penaranda et al., 2011)。

DNA 疫苗是通过肌内注射给药的,是一种劳动密集型操作,难以在小鱼中实施。当用基因枪通过皮肤粒子轰击进行疫苗给药后,在虹鳟鱼苗中检测到同样的高保护水平免疫效力。然而,通过腹腔注射仅能获得部分保护效力,其余疫苗接种途径均不能诱导免疫保护(Corbeil et al., 2000)。近年来,使用海藻酸盐微球包埋 DNA 疫苗开发了一种针对 IHNV 的口服 DNA 疫苗;在虹鳟中使用此技术后,获得了显著的保护作用(Ballesteros et al., 2015)。

IHNV 的 DNA 疫苗是安全的,且鱼对其具有良好的耐受性。接种疫苗 2 年后,鱼组织中未观察到发生疫苗特异性病理变化(Garver et al., 2005b; Kurath et al., 2006)。DNA 疫苗的质粒骨架中含有人类病原体巨细胞病毒(cytomegalovirus, CMV)的即刻早期启动子。由于监管机构认为这种疫苗可能"不安全",Alonso 等(2003)测试了含有与三种不同虹鳟启动子连接的 IHNV G 基因的 DNA 疫苗,其中虹鳟干扰素调节因子 1A 启动子的疫苗提供的免疫保护与 pCMV-G 疫苗相同 (Alonso and Leong, 2012)。

已开展了大量针对 IHN 的常规疫苗研发工作,如亚单位疫苗和减毒活疫苗。有综述介绍了许多展现出不同程度保护效果的实验室 IHNV 疫苗,但尚未商业化(Winton, 1997; Bootland and Leong, 1999)。其他疫苗,如昆虫细胞内利用杆状病毒制备的 IHNV G 蛋白疫苗和在新月柄杆菌(*Caulobacter* crescentus)表面合成的重组亚单位疫苗，在实验研究中均显示出一定的保护效果(Cain et al., 1999; Simon et al., 2001)。最近对灭活的 IHNV 疫苗进行重新评估时发现灭活剂至关重要，β-丙内酯灭活的全病毒疫苗可诱导 7 d 和 56 d 的免疫保护(Anderson et al., 2008)。因此，研制出有效的灭活疫苗仍然具有可能性。

2.6 总结与研究展望

IHNV 是野生和养殖鲑科鱼类面临的全球性问题，受其影响的鲑科鱼类养殖场数量在不断增加。该病毒已经适应了新的宿主和生物环境，分别在北美、日本和欧洲进化出的鳟鱼适应性 M、J 和 E 基因群证明了这一点。因此，亟须有效的防控策略来阻止病毒向原生鱼种群和无 IHNV 地理区域传播扩散。或许可以通过开发经认证为无特定病原体(specific pathogen free, SPF)的鲑鱼或鳟鱼卵种源来控制 IHNV 的传播。简而言之，可从无 IHNV 的亲鱼及其在 SPF 水中培育的后代中获得 SPF 鱼卵。这些不含病毒的鱼生产的鱼卵和苗种可以作为无 IHNV 种源进行出售。当然，开发更快的诊断方法和定期监测可以评估野生和养殖鱼类的病毒携带情况。现在已有多种灵敏的病毒检测方法，并且可以对 IHNV 的潜在载体和传染源进行全面调查。基因分型正用于分子流行病学研究和鱼类健康管理(Breyta et al., 2016)，并已公开提供北美或欧洲病毒分离株动物流行病学和遗传信息的集中式数据库(Johnstrup et al., 2010; Emmenegger et al., 2011; Kurath, 2012b)。虽然首例鱼用 DNA 疫苗已在加拿大批准使用，但该产品尚未在其他国家获得许可。疫苗的有效性已得到充分证实，但通过注射给药往往是不切实际的，寻找新型给药方法仍将是疫苗研发的优先领域(Munang'andu and Evensen, 2015)。

参考文献

[1] AFS-FHS (2014) *FHS Blue Book: Suggested Proced-ures for the Detection and Identification of Certain Finfish and Shellfish Pathogens*, 2014 edn. American Fisheries Society-Fish Health Section, Bethesda, Maryland.

[2] Alonso, M. and Leong, J. C. (2012) Licensed DNA vac-cines against infectious hematopoietic necrosis virus (IHNV). *Recent Patents on DNA and Genes Sequences* 7, 62-65.

[3] Alonso, M., Johnson, M., Simon, B. and Leong, J.C. (2003) A fish specific expression vector containing the interferon regulatory factor 1A (IRF1A) promoter for genetic immunization of fish. *Vaccine* 21, 1591-1600.

[4] Alonso, M., Stein, D.A., Thomann, E., Moulton, H.M., Leong, J.C. *et al*. (2005) Inhibition of infectious hae-matopoietic necrosis virus in cell cultures with peptide-conjugated morpholino oligomers. *Journal of Fish Diseases* 28, 399-410.

[5] Amar, E.C., Kiron, V., Akutsu, T., Satoh, S. and Watanabe, T. (2012) Resistance of rainbow trout, *Oncorhynchus mykiss*, to infectious hematopoietic necrosis virus (IHNV) experimental infection follow-ing ingestion of natural and synthetic carotenoids. *Aquaculture* 330 – 333, 148 – 155.
[6] Amend, D.F. (1973) Pathophysiology of infectious hemat-opoietic necrosis virus disease in rainbow trout. PhD thesis, University of Washington, Seattle, Washington.
[7] Amend, D.F. (1975) Detection and transmission of infec-tious hematopoietic necrosis virus in rainbow trout. *Journal of Wildlife Diseases* 11, 471 – 478.
[8] Amend, D. F. and Smith, L. (1974) Pathophysiology of infectious hematopoietic necrosis virus disease in rain-bow trout (*Salmo gairdneri*): early changes in blood and aspects of the immune response after injection of IHN virus. *Journal of the Fisheries Research Board of Canada* 31, 1371 – 1378.
[9] Amend, D. F. and Smith, L. (1975) Pathophysiology of infectious hematopoietic necrosis virus disease in rainbow trout: hematological and blood chemical changes in moribund fish. *Infection and Immunity* 11, 171 – 179.
[10] Amend, D.F., Yasutake, W.T. and Mead, R.W. (1969) A hematopoietic virus disease of rainbow trout and sock-eye salmon. *Transactions of the American Fisheries Society* 98, 796 – 804.
[11] Amos, K. H., Hopper, K. A. and Levander, L. (1989) Absence of infectious hematopoietic necrosis virus in adult sockeye salmon. *Journal of Aquatic Animal Health* 1, 281 – 283.
[12] Anderson, E.D., Mourich, D.V., Fahrenkrug, S. and Leong, J.C. (1996) Genetic immunization of rainbow trout (*Oncorhynchus mykiss*) against infectious hematopoietic necrosis virus. *Molecular Marine Biology and Biotechnology* 5, 114 – 122.
[13] Anderson, E.D., Clouthier, S., Shewmaker, W., Weighall, A. and LaPatra, S. (2008) Inactivated infectious hae-matopoietic necrosis virus (IHNV) vaccines. *Journal of Fish Diseases* 31, 729 – 745.
[14] Arakawa, C.K., Deering, R.E., Higman, K.H., Oshima, K.H., O'Hara, P.J. and Winton, J. (1990) Polymerase chain reaction (PCR) amplification of a nucleoprotein gene sequence of infectious hematopoietic necrosis virus. *Diseases of Aquatic Organisms* 8, 165 – 170.
[15] Ballesteros, N.A., Alonso, M., Rodriguez Saint-Jean, S. and Perez-Prieto, S.I. (2015) An oral DNA vaccine against infectious haematopoietic necrosis virus (IHNV) encapsulated in alginate microspheres induces dose-dependent immune responses and significant protec-tion in rainbow trout (*Oncorhynchus mykiss*). *Fish and Shellfish Immunology* 45, 877 – 888.
[16] Barroso, R.M., Wheeler, P.A., LaPatra, S.E., Drew, R.E. and Thorgaard, G.H. (2008) QTL for IHNV resistance and growth identified in a rainbow (*Oncorhynchus mykiss*) × Yellowstone cutthroat (*Oncorhynchus clarki bouvieri*) trout cross. *Aquaculture* 277, 156 – 163.
[17] Bootland, L.M. and Leong, J. (1992) Staphylococcal coagglutination — a rapid method of identifying infec-tious hematopoietic necrosis virus. *Applied and Environmental Microbiology* 58, 6 – 13.
[18] Bootland, L.M. and Leong, J. (1999) Infectious hemat-opoietic necrosis virus. In: Woo, P.T.K. and Bruno, D.W. (eds) *Fish Diseases and Disorders. Volume 3: Viral Bacterial and Fungal Infections*, 2nd edn. CAB International, Wallingford, UK, pp. 57 – 121.
[19] Bootland, L.M. and Leong, J. (2011) Infectious haemat-opoietic necrosis virus. In: Woo, P.T.K., Bruno, D.W. (eds) *Fish Diseases and Disorders. Volume 3: Viral, Bacterial, and Fungal Infections*, 2nd edn. CAB International. Wallingford, UK, pp. 66 – 109.

[20] Breyta, R.B., Samson, C., Blair, M., Black, A. and Kurath, G. (2016) Successful mitigation of viral disease based on a delayed exposure rearing strategy at a large-scale steelhead trout conservation hatchery. *Aquaculture* 450, 213 - 224.
[21] Cain, K.D., LaPatra, S.E., Shewmaker, B., Jones, J., Byrne, K.M. and Ristow, S.S. (1999) Immunogenicity of a recombinant infectious hematopoietic necrosis virus glycoprotein produced in insect cells. *Diseases of Aquatic Organisms* 36, 62 - 72.
[22] Campbell, N.R., LaPatra, S.E., Overturf, K., Towner, R. and Narum, S.R. (2014) Association mapping of dis-ease resistance traits in rainbow trout using restric-tion site associated DNA sequencing. *G3: Genes, Genomes, Genetics* 4, 2473 - 2481.
[23] Cefas (2011) Non-OIE data for Infectious Haematopoietic Necrosis. Last updated 31 January 2014. Centre for Environment, Fisheries and Aquaculture Science, Lowestoft/Weymouth, UK. Available at: http://www.cefas.defra.gov.uk/idaad/disease.aspx?t=n&id=55 (accessed 24 October 2016).
[24] CFIA (2005) Environmental assessment for licensing infectious haematopoietic necrosis virus vaccine, DNA vaccine in Canada. Canadian Food Inspection Agency, Ottawa.
[25] Corbeil, S., Kurath, G. and LaPatra, S.E. (2000) Fish DNA vaccine against infectious hematopoietic necro-sis virus: efficacy of various routes of immunization. *Fish and Shellfish Immunology* 10, 711 - 723.
[26] De Las Heras, A.I., Rodriguez Saint-Jean, S. and Perez-Prieto, S.I. (2008) Salmonid fish viruses and cell inter-actions at early steps of the infective cycle. *Journal of Fish Diseases* 31, 535 - 546.
[27] Department of Fisheries and Oceans (1984) *Fish Health Protection Regulations: Manual of Compliance*, rev. 2011. Fisheries and Marine Service Miscellaneous Special Publication 31, Fisheries Research Directorate Aquaculture and Resource Development Branch, Ottawa.
[28] Drolet, B.S., Rohovec, J.S. and Leong, J.C. (1994) The route of entry and progression of infectious hemat-opoietic necrosis virus in *Oncorhynchus mykiss* (Walbaum): a sequential immunohistochemical study. *Journal of Fish Diseases* 17, 337 - 347.
[29] Drolet, B.S., Chiou, P.P., Heidel, J. and Leong, J.C. (1995) Detection of truncated virus particles in a persistent RNA virus infection *in vivo*. *Journal of Virology* 69, 2140 - 2147.
[30] Durrin, L. (1980) An electron micrographic study of IHNV and IPNV. Development of methods for determining particle-infectivity ratios and for rapid diagnosis. Master's thesis, Oregon State University, Corvallis, Oregon.
[31] Emmenegger, E.J., Meyer, T.R., Burton, T. and Kurath, G. (2000) Genetic diversity and epidemiology of infec-tious hematopoietic necrosis virus in Alaska. *Diseases of Aquatic Organisms* 40, 163 - 176.
[32] Emmenegger, E.J., Kentop, E., Thompson, T.M., Pittam, S., Ryan, A. *et al.* (2011) Development of an aquatic pathogen database (AquaPathogen X) and its utilization in tracking emerging fish virus pathogens in North America: *Journal of Fish Diseases* 34, 579 - 587.
[33] Fofana, A. and Baulcomb, C. (2012) Counting the costs of farmed salmonids diseases. *Journal of Applied Aquaculture* 24, 118 - 136.
[34] Garver, K.A., LaPatra, S.E. and Kurath, G. (2005a) Efficacy of an infectious hematopoietic necrosis (IHN) virus DNA vaccine in chinook *Oncorhynchus tshaw-ytscha* and sockeye *O. nerka* salmon. *Diseases of Aquatic Organisms* 64, 13 - 22.
[35] Garver, K.A., Conway, C.M., Elliott, D.G. and Kurath, G. (2005b) Analysis of DNA-vaccinated fish reveals viral antigen in muscle, kidney and thymus, and transient histopathologic changes. *Marine Biotechnology* 7, 540 - 553.
[36] Garver, K.A., Batts, W.N. and Kurath, G. (2006) Virulence comparisons of infectious

hematopoietic necrosis virus U and M genogroups in sockeye salmon and rainbow trout. *Journal of Aquatic Animal Health* 18, 232 - 243.
[37] Gonzalez, M. P., Sanchez, W., Ganga, M. A., Lopez-Lastra, M., Jashes, M. and Sandino, A.M. (1997) Detection of the infectious hematopoietic necrosis virus directly from infected fish tissues by dot blot hybridization with a non-radioactive probe. *Journal of Virological Methods* 65, 273 - 279.
[38] Gunimaladevi, I., Kono, T., LaPatra, S.E. and Sakai, M. (2005) A loop mediated isothermal amplification (LAMP) method for detection of infectious hematopoietic necro-sis virus (IHNV) in rainbow trout (*Oncorhynchus mykiss*). *Archives of Virology* 150, 899 - 909.
[39] Harmache, A., LeBerre, M., Droineau, S., Giovannini, M. and Brémont, M. (2006) Bioluminescence imaging of live infected salmonids reveals that the fin bases are the major portal of entry for *Novirhabdovirus*. *Journal of Virology* 80, 3655 - 3659.
[40] Hasobe, M. and Saneyoshi, M. (1985) On the approach to the viral chemotherapy against infectious hematopoi-etic necrosis virus (IHNV). *Bulletin of the Japanese Society of Scientific Fisheries* 49, 157 - 163.
[41] Hedrick, R.P., LaPatra, S.E., Fryer, J.L., McDowell, T. and Wingfield, W.H. (1987) Susceptibility of coho (*Oncorhynchus kisutch*) and chinook (*Oncorhynchus tshawytscha*) salmon hybrids to experimental infec-tions with infectious hemato-poietic necrosis virus (IHNV). *Bulletin of the European Association of Fish Pathologists* 7, 97 - 100.
[42] Helmick, C.M., Bailey, F.J., LaPatra, S. and Ristow, S. (1995a) Histological comparison of infectious hemat-opoietic necrosis virus challenged juvenile rainbow trout *Oncorhynchus mykiss* and coho salmon *O. kisutch* gill, esophagus/cardiac stomach region, small intestine and pyloric caeca. *Diseases of Aquatic Organisms* 23, 175 - 187.
[43] Helmick, C.M., Bailey, F.J., LaPatra, S. and Ristow, S. (1995b) The esophagus/cardiac stomach region: site of attachment and internalization of infectious hemat-opoietic necrosis virus in challenged juvenile rainbow trout *Oncorhynchus mykiss* and coho salmon *O. kisutch*. *Diseases of Aquatic Organisms* 23, 189 - 199.
[44] Hsu, Y.L. and Leong, J. (1985) A comparison of detection methods for infectious hematopoietic necrosis virus. *Journal of Fish Diseases* 8, 1 - 12.
[45] Johnstrup, S.P., Schuetze, H., Kurath, G., Gray, T., Bang Jensen, B. and Olesen, N. J. (2010) An isolate and sequence database of infectious haematopoietic necro-sis virus (IHNV). *Journal of Fish Diseases* 33, 469 - 471.
[46] Khoo, S.K., Ozaki, A., Nakamura, F., Arakawa, T. and Ishimoto, S. (2004) Identification of a novel chromo-somal region associated with IHNV resistance in rain-bow trout. *Fish Pathology* 39, 95 - 102.
[47] Kim, C.H., Dummer, D.M., Chiou, P.P. and Leong, J.C. (1999) Truncated particles produced in fish surviving infectious hematopoietic necrosis virus infection: mediators of persistence? *Journal of Virology* 73, 843 - 849.
[48] Kim, C.H., Johnson, M.C., Drennan, J.D., Simon, B.E., Thomann, E. and Leong, J. C. (2000) DNA vaccines encoding viral glycoproteins induce nonspecific immunity and Mx protein synthesis in fish. *Journal of Virology* 74, 7048 - 7054.
[49] Kim, S.J., Lee, K-Y., Oh, M.J. and Choi, T.J. (2001) Comparison of IHNV detection limits by IMS-RT-PCR, Western blot and ELISA. *Journal of Fisheries Science and Technology* 4, 32 - 38.
[50] Kurath, G. (2008) Biotechnology and DNA vaccines for aquatic animals. *Revue Scientifique et Technique (Paris)* 27, 175 - 196.
[51] Kurath, G. (2012a) Fish novirhabdoviruses. In: Dietzgen, R.G. and Kuzmin, I.V. (eds) *Rhabdoviruses: Molecular Taxonomy, Evolution, Genomics, Ecology, Host - vector*

Interactions, Cytopathology and Control. Caister Academic Press, Wymondham, UK, pp. 89 - 116.

[52] Kurath, G. (2012b) *An Online Database for IHN Virus in Pacific Salmonid Fish: MEAP-IHNV*. Fact Sheet 2012-3027, US Department of the Interior and US Geological Survey, Seattle, Washington. Available at: http://pubs.usgs.gov/fs/2012/3027/pdf/fs20123027. pdf (accessed 21 October 2016).

[53] Kurath, G. and Leong, J.C. (1985) Characterization of infectious hematopoietic necrosis virus mRNA spe-cies reveals a nonvirion rhabdovirus protein. *Journal of Virology* 53, 462 - 468.

[54] Kurath, G., Garver, K.A., Corbeil, S., Elliott, D.G., Anderson, E.D. and LaPatra, S. E. (2006) Protective immunity and lack of histopathological damage two years after DNA vaccination against infectious hematopoietic necrosis virus in trout. *Vaccine* 24, 345 - 354.

[55] LaPatra, S.E. (2014) 2.2.4 Infectious hematopoietic necrosis. In: *AFS - FHS. Fish Health Section Blue Book: Suggested Procedures for the Detection and Identification of Certain Finfish and Shellfish Pathogens*, 2014 edn. American Fisheries Society-Fish Health Section, Bethesda, Maryland. Available at: http://www.afs-fhs.org/perch/resources/14069231202.2.5ipnv2007ref2014.pdf (accessed 21 October 2010).

[56] LaPatra, S.E., Parsons, J.E., Jones, G.R. and McRoberts, W.O. (1993) Early life stage survival and susceptibility of brook trout, coho salmon, rainbow trout, and their reciprocal hybrids to infectious hematopoietic necrosis virus. *Journal of Aquatic Animal Health* 5, 270 - 274.

[57] LaPatra, S.E., Batts, W.N., Overturf, K., Jones, G.R., Shewmaker, W.D. and Winton, J.R. (2001a) Negligible risk associated with the movement of processed rain-bow trout, *Oncorhynchus mykiss* (Walbaum), from an infectious hematopoietic necrosis virus (IHNV) endemic area. *Journal of Fish Diseases* 24, 399 - 408.

[58] LaPatra, S.E., Corbeil, S., Jones, G.R., Shewmaker, W.D., Lorenzen, N. *et al*. (2001b) Protection of rain-bow trout against infectious hematopoietic necrosis virus four days after specific or semi-specific DNA vaccination. *Vaccine* 19, 4011 - 4019.

[59] Leong, J.C. and Kurath, G. (2011) Novirhabdoviruses. In: Tidona, C. and Darai, G. (eds) *The Springer Index of Viruses*. Springer, Heidelberg, Germany.

[60] Liu, Z., Teng, Y., Liu, H., Jiang, Y., Xie, X. *et al*. (2008) Simultaneous detection of three fish rhabdoviruses using multiplex real-time quantitative RT - PCR assay. *Journal of Virological Methods* 149, 103 - 109.

[61] Lorenzen, N. and LaPatra, S.E. (1999) Immunity to rhab-doviruses in rainbow trout: the antibody response. *Fish and Shellfish Immunology* 9, 345 - 360.

[62] Lorenzen, E., Carstensen, B. and Olesen, N.J. (1999) Inter-laboratory comparison of cell line for susceptibil-ity to three viruses: VHSV, IHNV and IPNV. *Diseases of Aquatic Organisms* 37, 81 - 88.

[63] Lorenzen, N., Lorenzen, E., Einer-Jensen, K. and LaPatra, S.E. (2002) DNA vaccines as a tool for ana-lysing the protective immune response against rhab-doviruses in rainbow trout. *Fish and Shellfish Immunology* 12, 439 - 453.

[64] Ludwig, M., Palha, N., Torhy, C., Briolat, V., Colucci-Guyon, E. *et al*. (2011) Whole-body analysis of a viral infection: vascular endothelium is a primary target of infectious hematopoietic necrosis virus in zebrafish larvae. *PLoS Pathogens* 7 (2): e1001269.

[65] McAllister, P.E. and Schill, W.B. (1986) Immunoblot assay: a rapid and sensitive method for identification of salmonid fish viruses. *Journal of Wildlife Diseases* 22, 468 - 474.

[66] McDaniel, T.R., Pratt, K.M., Meyers, T.R., Ellison, T.D., Follett, J.E. and Burke,

J.A. (1994) *Alaska Sockeye Salmon Culture Manual*. Special Publication 6, Alaska Department of Fish and Game, Juneau, Alaska.

[67] McIntyre, J. D. and Amend, D. F. (1978) Heritability of tol-erance for infectious hematopoietic necrosis in sock-eye salmon (*Oncorhynchus nerka*). *Transactions of the American Fisheries Society* 107, 305 - 308.

[68] Medina, D.J., Chang, P.W., Bradley, T.M., Yeh, M-T. and Sadaasiv, E.C. (1992) Diagnosis of infectious hemat-opoietic necrosis virus in Atlantic salmon (*Salmo salar*) by enzyme-linked immunosorbent assay. *Diseases of Aquatic Organisms* 13,147 - 150.

[69] Meyers, T. R., Korn, D., Burton, T. M., Glass, K., Follett, J. E. *et al.* (2003) Infectious hematopoietic necrosis virus (IHNV) in Alaskan sockeye salmon culture from 1973 to 2000: annual virus prevalences and titers in broodstocks compared with juvenile losses. *Journal of Aquatic Animal Health* 15, 21 - 30.

[70] Millard, P.J., Bickerstaff, L.E., LaPatra, S.E. and Kim, C.H. (2006) Detection of infectious haematopoietic necrosis virus and infectious salmon anaemia virus by molecular padlock amplification. *Journal of Fish Diseases* 29, 201 - 213.

[71] Mulcahy, D., Pascho, R.J. and Jenes, C.K. (1984) Comparison of *in vitro* growth characteristics of ten isolates of infectious hematopoietic necrosis virus. *Journal of General Virology* 65, 2199 - 2207.

[72] Müller, A., Sutherland, B.J.G., Koop, B.F., Johnson, S.C. and Garver, K.A. (2015) Infectious hematopoietic necrosis virus (IHNV) persistence in sockeye salmon: influence on brain transcriptome and subsequent response to the viral mimic poly(I : C). *BMC Genomics* 16: 634.

[73] Munang'andu, H.M. and Evensen, O. (2015) A review of intra and extracellular antigen delivery systems for virus vaccines of finfish. *Journal of Immunology Research* 2014: Article ID 960859.

[74] Myouga, H., Yoshimizu, M., Ezura, Y. and Kimura, T. (1993) Anti-infectious hematopoietic necrosis virus (IHNV) substances produced by bacteria from aquatic environment. *Gyobyo Kenkyu* [*Fish Pathology*] 28, 9 - 13.

[75] OIE (2015) Chapter 2. 3. 4. Infectious haematopoietic necrosis. In: *Manual of DiagnosticTests forAquaticAnimals* (*2015*). World Organisation for Animal Health, Paris. Updated 2016 version available at: http://www.oie.int/index.php?id=2439&L=0&htmfile=chapitre_ihn.htm (accessed 23 October 2016).

[76] Overturf, K., LaPatra, S.E., Towner, R., Campbell, N.R. and Narum, S.R. (2010) Relationships between growth and disease resistance in rainbow trout, *Oncorhynchus mykiss* (Walbaum). *Journal of Fish Diseases* 33, 321 - 329.

[77] Palti, Y., Parsons, J.E. and Thogaard, G.H. (1999) Identification of candidate DNA markers associated with IHN virus resistance in backcrosses of rainbow (*Oncorhynchus mykiss*) and cutthroat trout (*O. clarki*). *Aquaculture* 173, 81 - 94.

[78] Palti, Y., Nichols, K.M., Waller, K.I., Parsons, J.E. and Thorgaard, G.H. (2001) Association between DNA polymorphisms tightly linked to MHC class II genes and IHN virus resistance in backcrosses of rainbow and cutthroat trout. *Aquaculture* 194, 283 - 289.

[79] Penaranda, M. M. D., Purcell, M. K. and Kurath, G. (2009) Differential virulence mechanisms of infectious hematopoietic necrosis virus in rainbow trout (*Oncorhynchus mykiss*) include host entry and viral replica-tion kinetics. *Journal of General Virology* 90, 2172 - 2182.

[80] Penaranda, M.M.D., LaPatra, S.E. and Kurath, G. (2011) Genogroup specificity of DNA vaccines against infec-tious hematopoietic necrosis virus (IHNV) in rainbow trout (*Oncorhynchus mykiss*). *Fish and Shellfish Immunology* 31, 43 - 51.

[81] Pietsch, J. P., Amend, D. F. and Miller, C. M. (1977) Survival of infectious

hematopoietic necrosis virus held under various environmental conditions. *Journal of the Fisheries Research Board of Canada* 34, 1360 – 1364.

[82] Purcell, M.K., Alexandra Hart, S., Kurath, G. and Winton, J.R. (2006a) Strand specific, real-time RT – PCR assays for quantification of genomic and positive-sense RNAs of the fish rhabdovirus infectious hemat-opoietic necrosis virus. *Journal of Virological Methods* 132, 18 – 24.

[83] Purcell, M.K., Nichols, K.M., Winton, J.R., Kurath, G., Thorgaard, G.H. *et al.* (2006b) Comprehensive gene expression profiling following DNA vacci-nation of rainbow trout against infectious haemat-opoietic necrosis virus. *Molecular Immunology* 43, 2089 – 2106.

[84] Purcell, M.K., Garver, K.A., Conway, C., Elliott, D.G. and Kurath, G. (2009) Infectious hematopoietic necrosis virus genogroup-specific virulence mechanisms in sockeye salmon (*Oncorhynchus nerka*) from Redfish Lake Idaho. *Journal of Fish Diseases* 32, 619 – 631.

[85] Purcell, M.K., Thompson, R.L., Garver, K.A., Hawley, L.M., Batts, W.N. *et al.* (2013) Universal reverse-transcriptase real-time PCR for infectious hematopoietic necrosis virus (IHNV). *Diseases of Aquatic Organisms* 106, 103 – 115.

[86] Rodriguez, M.F., LaPatra, S., Williams, S., Famula, T. and May, B. (2004) Genetic markers associated with resistance to infectious hematopoietic necrosis in rainbow and steelhead trout (*Oncorhynchus mykiss*) backcrosses. *Aquaculture* 241, 93 – 115.

[87] Rodríguez Saint-Jean, S., Pérez Prieto, S.-L., Lopez-Exposito, I., Ramos, M., De Las Heras, A.I. and Recio, I. (2012) Antiviral activity of dairy proteins and hydrolysates on salmonid fish viruses. *International Dairy Journal* 23, 24 – 29.

[88] Rodríguez Saint-Jean, S., De las Heras, A., Carrillo, W., Recio, I., Ortiz-Delgado, J. B. *et al.* (2013) Antiviral activity of casein and alpha-s2 casein hydrolysates against the infectious hematopoietic necrosis virus, a rhabdovirus from salmonid fish. *Journal of Fish Diseases* 36, 467 – 481.

[89] Saksida, S.M. (2006) Infectious haematopoietic necrosis epidemic (2001 to 2003) in farmed Atlantic salmon *Salmo salar* in British Columbia. *Diseases of Aquatic Organisms* 72, 213 – 223.

[90] Schultz, C.L., McAllister, P.E., Schill, W.B., Lidgerding, B.C. and Hetrick, F.M. (1989) Detection of infectious hematopoietic necrosis virus in cell culture fluid using immunoblot assay and biotinylated monoclonal anti-body. *Diseases of Aquatic Organisms* 7, 31 – 31.

[91] Simon, B., Nomellini, J., Chiou, P.P., Bingle, W., Thornton, J. *et al.* (2001) Recombinant vaccines against infec-tious hematopoietic necrosis virus: production by *Caulobacter crescentus* S-layer protein secretion system and evaluation in laboratory trials. *Diseases of Aquatic Organisms* 44, 17 – 27.

[92] Siwicki, A.K., Pozet, F., Morand, M., Kazuń, B. and Trapkowska, S. (2002) *In vitro* effect of methisoprinol on salmonid rhabdoviruses replication. *Bulletin of the Veterinary Institute in Pulawy* 46, 53 – 58.

[93] Traxler, G.S., Anderson, E.A., LaPatra, S.E., Richard, J., Shewmaker, B. and Kurath, G. (1999) Naked DNA vaccination of Atlantic salmon *Salmo salar* against IHNV. *Diseases of Aquatic Organisms* 38, 183 – 190.

[94] Verrier, E.R., Langevin, C., Tohry, C., Houel, A., Ducrocq, V. *et al.* (2012) Genetic resistance to *Rhabdovirus* infec-tion in teleost fish is paralleled to the derived cell resistance status. *PLoS ONE* 7(4): e33935.

[95] Wargo, A.R., Garver, K.A. and Kurath, G. (2010) Virulence correlates with fitness *in vivo* for two M group geno-types of infectious hematopoietic necrosis virus (IHNV). *Virology* 404, 51 – 58.

[96] Winton, J.R. (1991) Recent advances in detection and control of infectious hematopoietic necrosis virus in aquaculture. *Annual Review of Fish Diseases* 1, 83 - 93.

[97] Winton, J. R. (1997) Immunization with viral antigens: infec-tious haematopoietic necrosis. In: Gudding, R., Lillehaug, A., Midtlyng, P.J. and Brown, F. (eds) *Fish Vaccinology*. Developments in Biological Standardization 90, Karger, Basel, Switzerland, pp. 211 - 220.

[98] Wolf, K. (1988) Infectious hematopoietic necrosis. In: Wolf, K. (ed.) *Fish Viruses and Fish Viral Diseases*. Cornell University Press, Ithaca, New York, pp. 83 - 114.

[99] Wood, E.M. and Yasutake, W.T. (1956) Histopathologic changes of a virus-like disease of sockeye salmon. *Transactions of the American Microscopical Society* 75, 85 - 90.

[100] Yang, J., Liu, Z., Shi, H.-N., Zhang, J.-F., Want, J.-F. *et al.* (2014) Association between MHC II beta chain gene polymorphisms and resistance to infectious haematopoietic necrosis virus in rainbow trout (*Oncorhynchus mykiss*, Walbaum, 1792). *Aquaculture Research* 47, 570 - 578.

[101] Yasutake, W. T. (1970) Comparative histopathology of epizootic salmonid virus diseases. In: Snieszko, S.F. (ed.) *A Symposium on Diseases of Fish and Shellfish*. American Fisheries Society Special Publication 5, Bethesda, Maryland, pp. 341 - 350.

3

病毒性出血性败血症病毒

John S. Lumsden*

3.1 引言

病毒性出血性败血症病毒(viral haemorrhagic septicaemia virus, VHSV)是病毒性出血性败血症(viral haemorrhagic septicaemia, VHS)的病原,是一种RNA病毒,属于单负链病毒目(Mononegavirales)的弹状病毒科(Rhabdoviridae)。这类病毒含有包裹着子弹状或圆锥形核衣壳的单股负链RNA基因组。在经济上,最重要的致病性弹状病毒为VHSV、传染性造血器官坏死病病毒(infectious haematopoietic necrosis virus, IHNV;两者都属于粒外弹状病毒属*Novirhabdovirus*)和鲤春病毒血症病毒(spring viraemia of carp virus, SVCV;新命名为*Carp sprivivirus*, *Sprivivirus*属)。详情请参阅国际病毒分类委员会相关文献(ICTV, 2015)。VHSV与IHNV共有6个相同的基因,从基因组3′端到5′端依次为N(核衣壳蛋白)、P(磷蛋白)、M(基质蛋白)、G(糖蛋白)、NV(非病毒颗粒)及L(RNA聚合酶)(Einer-Jensen et al., 2004)。而SVCV不含编码NV的基因,且存在其他微小的基因组差异(Hoffmann et al., 2002)。Smail和Snow (2011)以及LaPatra等(2016)对VHSV进行了全面综述。

血清型常被用来分离病毒亚群(Olesen et al., 1993),然而近年来则通过G和N蛋白编码基因来进行基因分型。基因型在地理上分布广泛,它们的分化已被证明对理解动物流行病的传播及发病机制非常有用(Nishizawa et al., 2002, Snow et al., 2004, Brudeseth et al., 2008)。如果我们不明确基因型和被感染物种,那么对VHSV的致病机理、宿主范围、毒力等的概括可能会产生误导。

尽管四种基因型Ⅰ～Ⅳ都在北半球被发现,但是本章讨论主要集中在基因型Ⅰ和基因型Ⅳ。基因型Ⅰ～Ⅲ存在于欧洲,其中基因型Ⅰ又分化为淡水虹鳟(*Oncorhynchus mykiss*)分离株Ⅰa和来自各种海水野生鱼类的分离株Ⅰb。Ⅰa基因型与虹鳟中VHS的原始描述相关,其造成的死亡率接近100%。基因型Ⅳ在太平洋(Ⅳa)和北美包括西大西洋(Ⅳb; Gagne et al., 2007)和五大湖(Ⅳb;

---

* 作者邮箱:jsl@uoguelph.ca。

Lumsden et al., 2007)等地被发现。随着宿主范围的扩大,在欧洲已经出现海洋病毒分离株向淡水中迁徙的趋势,而且很可能也已在北美出现(Einer-Jensen et al., 2004; Snow et al., 2004; Gagne et al., 2007; Thompson et al., 2011)。基因型对 VHSV 的管控及鱼类的迁徙都会产生影响,因为Ⅰa 基因型在欧洲以外地区的诊断比Ⅳb 基因型在北美地域的扩散更有意义。任何 VHSV 的检测发现都必须向世界动物卫生组织报告(OIE, 2015)。

VHSV 最显著的特点之一是宿主范围广;五大湖中有 28 种鱼类感染了Ⅳb 基因型(APHIS, 2009),约 80 种水生物种受到所有四种基因型的影响(OIE, 2015)。鱼类细胞中普遍存在的磷脂酰丝氨酸和纤连蛋白是 VHSV 的受体(Bearzotti et al., 1999; LaPatra et al., 2016),虽然纤连蛋白也是 IHNV 的受体,但这可能是 VHSH 宿主范围广泛的原因(Bearzotti et al., 1999)。VHSV 可通过尿液和生殖液(Eaton et al., 1991)中排出的病毒或至少通过鳃上皮细胞摄取进行水平传播(Brudeseth et al., 2008)。尽管 VHSV 存在于生殖液中,VHSV 的 RNA 在Ⅳb 型感染鱼的性腺内存在(Al-Hussinee and Lumsden, 2011a),但垂直传播尚未被证实。

虽然 VHSV 在细胞培养中的生长温度较高(Vo et al., 2015),但病毒性出血性败血症在水温 4~15 ℃就会发生(McAllister, 1990),这反映了宿主因素在疾病中的重要性。用Ⅳb 基因型毒株实验感染蓝鳃太阳鱼(*Lepomis macrochirus*),10 ℃时其死亡率为 90%,但 18 ℃以上死亡率为 0%(Goodwin and Merry, 2011)。基因型Ⅰ和基因型Ⅳ的生长温度范围及最佳温度是相似的(Goodwin and Merry, 2011; Smail and Snow, 2011),水温 20 ℃以上时仅由 VHSV 导致的死亡不太可能发生(Sano et al., 2009)。因此,应严格评估水温高于 20 ℃时进行的监测结果。在深水中,温跃层很少到达水底,因此在入夏后的底栖鱼类中仍可能检测到病毒。在宿主体外,这种病毒的传染性仍可持续一段时间。它在 4 ℃的过滤淡水中可存活一年(Hawley and Garver, 2008),而在较低温度下存活时间更长(Parry and Dixon, 1997)。该病毒在未经处理的 15 ℃原海水中的平均存活期为 4 d(Hawley and Garver, 2008)。

3.2 诊断

VHS 感染鱼类的临床症状和大体病变众多,但没有一种是特异病征性的,鉴于该疾病流行的地理区域广泛,应始终给予重视。如果温度、地理位置及鱼种类与发病率一致,并且患病鱼有出血、眼球突出和腹水性腹胀,则应考虑进行鉴别诊断。受感染的鱼还可能出现贫血、体色变暗发黑、嗜睡和异常游动行为等,并可能会迅速死亡(OIE, 2105)。然而,受感染的鱼也可能在没有大体病变的情况下死亡,这种情况在安大略湖的黑口新虾虎鱼(*Neogobius melanostomus*)中很常见(Groocock et al., 2007)，这也强调了组织病理学和辅助诊断的重要性。同样，临床表现正常的鱼可能呈 VHSV 阳性，因此需要通过 PCR 或病毒分离来确认检测。

由于病毒性出血性败血症是 OIE 应具报传染疾病，该组织需要特异性诊断技术进行确诊(OIE, 2015)。通常使用 BF-2、EPC 或其他细胞系进行病毒分离，但是耗费人力且昂贵。许多实验室使用标准RT-PCR 或定量 RT-PCR(qRT-PCR; Matejusova et al., 2008, Hope et al., 2010)，这些方法可以检测 VHSV 的所有四种基因型(Garver et al., 2011)，比病毒分离更灵敏和快速，更容易标准化并可高通量检测(Hope et al., 2010)。ELISA 和免疫荧光法[间接荧光抗体(IFA)法(IFAT)；参见 OIE, 2015]以及逆转录环介导等温扩增技术(Soliman et al., 2006)也被用于病毒诊断检测。

在使用 PCR 检测时，应保存组织以进行确证检测，包括病毒分离(OIE, 2015)。病毒分离也需要最少数量的组织和鱼群以进行阳性鉴定(AFS-FHS, 2014)。在推定阳性的情况下，定量 RT-PCR 是假定诊断或监测组织库存的理想选择。VHS 的确诊需要病毒分离和血清中和，然后进行 IFAT、ELISA 或 RT-PCR 三种之一的确证试验(OIE, 2015)。

3.3 病理学

先前已对 VHSV 的病理学进行了综述(de Kinkelin et al., 1979; Wolf, 1988; Evensen et al., 1994; Brudeseth et al., 2002, 2005; OIE, 2015; LaPatra et al., 2016)。许多急性全身性疾病的大体病变通常是典型的，包括出血、腹部肿大和眼球突出，这些是 VHSV 感染鱼典型的肉眼可见的病变(OIE, 2015)。感染了 Ⅰa 型 VHSV 的虹鳟背部肌肉组织出现瘀点出血很常见(OIE, 2015)。本章所讨论的病变都是概括性的。

除了黑口新虾虎鱼(Groocock et al., 2007)体表苍白或北美狗鱼(*Esox masquinongy*)鱼鳔出血(Elsayed et al., 2006)之外，在没有明显大体病变下依然会发生 VHS 死亡。即使在死鱼中检测到 VHSV，许多死亡事件的起因也是多因素的。野生太平洋鲱(*Clupea pallasi*)的死亡与包括 VHSV 在内的多种病原体有关(Marty et al., 1998)。在纽约州科尼瑟斯湖(Conesus Lake)中死亡的玻璃梭鲈(*Sander vitreus*)表现出与 VHSV 相符的病变，并从鱼中可分离出病毒，但感染的鱼也出现由柱状黄杆菌(*Flavobacterium columnare*)引起的典型细菌性鳃柱状病病变(Al-Hussinee et al., 2011b)(见第 16 章)。

组织学病变各不相同，但主要是坏死性的并且影响了几乎所有组织。病变的类型取决于病毒基因型和鱼种类。例如，虹鳟体内的 Ⅰa 型 VHSV 会在肾和肝中产生坏死性病变，但脾、心、脑和其他组织也可能受到影响(de Kinkelin et al., 1979; Wolf, 1988; Evensen et al., 1994)。感染 Ⅰb 基因型 VHSV 的大菱鲆(*Scophthalmus maximus*)(Brudaseth et al., 2005)和一些感染Ⅳb 型 VHSV 的五大湖地区物种的心脏损伤最为严重,但肝和造血组织也受到影响(Lumsden et al., 2007; Al-Hussinee et al., 2011b)。淡水石首鱼(*Aplodinotus grunniens*)的心脏和血管损伤非常严重,很容易就可看出是外来因素造成的(图 3.1 和图 3.2)。这种感染会在许多组织中产生严重的血管炎,但也可能很轻微(图 3.3)。淡水石首鱼在肝、肾(图 3.4)、脑和脾中形成一种泛红的、通常为纤维素样的血管炎,这种病变和内皮细胞趋向性也在Ⅰa 型感染的虹鳟和Ⅰb 型感染的大菱鲆中被证实(Brudaseth et al., 2002, 2005)。然而,被感染的北美狗鱼可形成不确定的病变。通过腹腔注射胖头鲅(*Pimephales promelas*)、虹鳟或玻璃梭鲈进行疾病实验室再现时,结果并未产生明显的血管炎,心脏病变也很

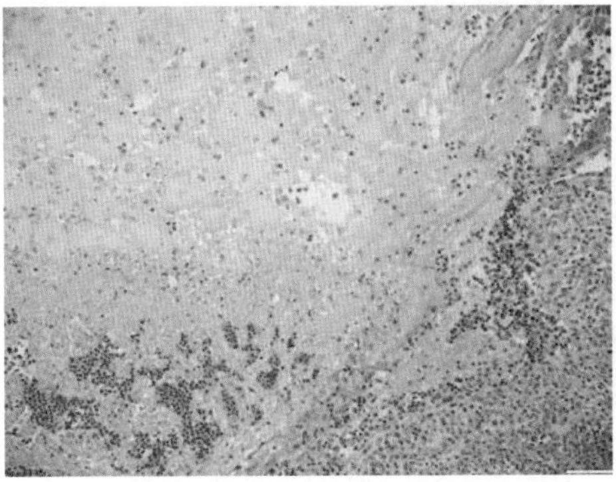

图 3.1　感染病毒性出血性败血症病毒(VHSV)的淡水石首鱼(*Aplodinotus grunniens*)心室。可见大量亚急性坏死和炎症,左下方可见残余的心肌小梁;苏木精-伊红染色;比例尺: 50 μm

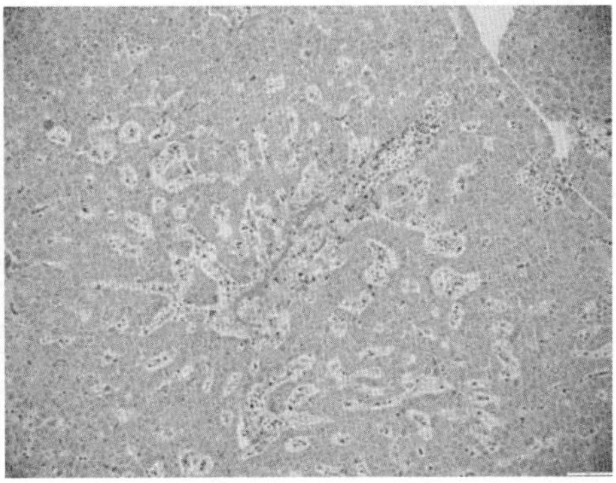

图 3.2　感染病毒性出血性败血症病毒(VHSV)的淡水石首鱼(*Aplodinotus grunniens*)肝脏中的纤维蛋白样血管炎。坏死集中于小静脉,而不是肝细胞;苏木精-伊红染色;比例尺: 50 μm

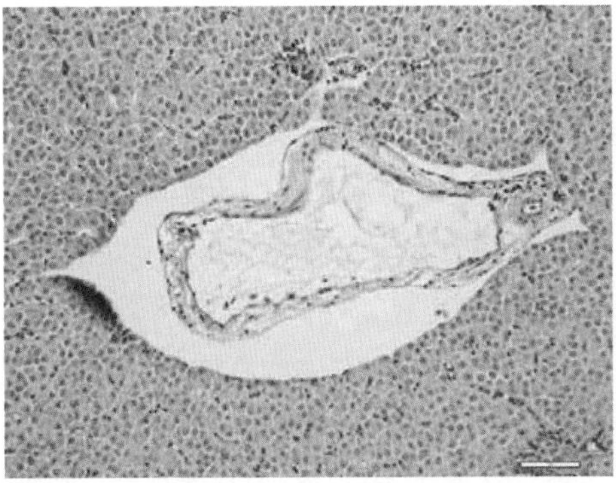

图 3.3　感染病毒性出血性败血症病毒(VHSV)的淡水石首鱼(*Aplodinotus grunniens*)肝脏轻度血管炎。苏木精-伊红染色;比例尺: 50 μm

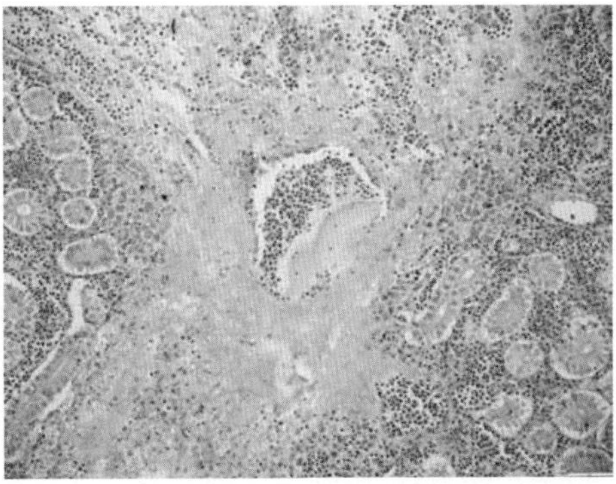

图 3.4　感染病毒性出血性败血症病毒(VHSV)的淡水石首鱼(*Aplodinotus grunniens*)肾脏严重血管炎。苏木精-伊红染色;比例尺: 50 μm

小(Al-Hussinee et al., 2010, 2011b; Grice et al., 2016)。经甲醛浸泡、健康检查并适应 18 ℃高水质井水的玻璃梭鲈,注射 VHSV 后不久即出现病毒性出血性败血症和鳃柱状病(Grice et al., 2016)。

组织病理学可以通过病毒的免疫定位来补充(Evensen et al., 1994; Brudeseth et al., 2005; Al-Hussinee et al., 2010, 2011b; Grice et al., 2016)。通过免疫组织化学(immunohistochemistry, IHC)方法将病毒与大多数病变进行共定位,巩固了病毒与发病机制的关联(Al-Hussinee et al., 2010, 2011b),但这种方法的灵敏度有限(Evensen et al., 1994)。此外,RNA 探针可用于定位 VHSV (Al-Hussinee et al., 2011a)。患有严重坏死性心肌病变的淡水石首鱼的心脏中出人意料地没有检测到病毒,至少通过 IHC 检测得到的结果是这样的(Al-Hussinee et al., 2011b),这与大菱鲆正好相反,在受感染大菱鲆的心脏中很容易检测到病毒(Brudeseth et al., 2005)。

3.4 病理生理学、发病机制与毒力因子

Smail 和 Snow (2011)对 VHSV 的病理生理学和发病机制方面进行了综述。简而言之,Ⅰa 基因型病毒对虹鳟是强毒性的(Brudeseth et al., 2002, 2008),而海洋分离株Ⅰb 对大菱鲆是强毒性的(Brudeseth et al., 2005),但对虹鳟无毒(Brudeseth et al., 2008)。许多Ⅰb 型分离株通过浸泡攻毒时对虹鳟是无毒的,但当注射攻毒时其毒力则发生变化(Skall et al., 2004)。Ⅳa 基因型病毒对太平洋鲱(*Clupea pallasi*) (Kocan et al., 1997)具有强毒性,而虹鳟对Ⅳa (Meyers and Winton, 1995)和Ⅳb(Al-Hussinee et al., 2010; Kim and Faisal, 2010)具有抗性,但Ⅳa 在虹鳟的鳃和脾巨噬细胞中表现出比Ⅳb 更强的复制能力和致细胞病变效应(CPE)(Pham et al., 2013)。

已有文献对 VHSV 通过鳃的水平传播进行了详细介绍(Brudeseth et al., 2008; Pham et al., 2013; Al-Hussinee et al., 2016)。在感染早期,可在虹鳟鳃上皮细胞中检测到Ⅰa 型病毒复制(Brudeseth et al., 2002),而在感染后期,在大菱鲆的肾脏内皮细胞中首先检测到Ⅰb 型病毒(Brudeseth et al., 2005),并仅在少数鱼的鳃部有病毒。一些研究表明 VHSV 分离株在鳃或皮肤上皮细胞中的复制能力与毒力相关。与虹鳟强毒性Ⅰa 分离株相比,虹鳟中Ⅰb 基因型的无毒性与其无法侵染和跨初级鳃上皮细胞且无法在头肾巨噬细胞中复制有关(Brudeseth et al., 2008)。在虹鳟中一株具有低毒力的Ⅳa 分离株在切除的鳍组织中可进行一定程度的复制,而在切除的鳃组织中没有进行复制(Yamamoto et al., 1992)。病毒在离体鳍组织中的复制与虹鳟对水源传染的抵抗力相关(Quillet et al., 2001)。

巨噬细胞在 VHSV 的毒力和致病机制中发挥重要作用。通过自然和实验

感染Ⅰa (Evensen et al., 1994; Brudeseth et al., 2002)、Ⅰb (Brudeseth et al., 2005)和Ⅳb (Al-Hussinee et al., 2010, 2011b)基因型病毒后,可在巨噬细胞和黑素瘤巨噬细胞中检测到病毒。病毒在体外巨噬细胞的原代培养物中也可进行复制(Brudeseth et al., 2008)。巨噬细胞会吞噬被VHSV裂解的细胞,因此,免疫化学检测到巨噬细胞内的病毒可能并不意味着病毒复制。类似地,虹鳟的原代巨噬细胞培养物是异质的,只有一部分细胞可以支持病毒复制,即使使用Ⅰa强毒分离株也是如此(Tafalla et al., 1998; Brudeseth et al., 2008),因此,显著的致细胞病变效应可能发生(Brudeseth et al., 2008),也可能不发生(Tafalla et al., 1998)。在虹鳟脾巨噬细胞RTS11细胞系中,VHSV复制失败,并且Ⅰa、Ⅳa和Ⅳb基因型分离株均未发生致细胞病变效应(Tafalla et al., 2008; Pham et al., 2013)。

这些证据大都来自实验感染或体外研究,因此,排除了可能导致鱼类死亡的环境因素。Ⅲ和Ⅳa基因型通常对鲑科鱼类毒力不高,尽管如此,来自野生鱼类的分离株还是导致了网箱养殖大西洋鲑(*Salmo salar*)的死亡(Dale et al., 2009; Garver et al., 2013)。VHSV分离株适应新宿主的机制是该病毒流行病学中的一个关键问题。一株海洋分离株在虹鳟体内连续传代五代,在不干预体外培养的情况下产生了更强的毒力,但G基因序列中没有出现差异可以解释这种毒力的增加(Snow and Cunningham, 2000)。在随后的研究中,对Ⅰb基因型分离株进行测序,这些分离株的G蛋白氨基酸序列具有大于99.4%的相似性,但对虹鳟的毒力明显不同。研究发现,仅有四个被预测的氨基酸被置换,在N、G、NV和L蛋白中各有一个,这表明基因组的细小突变可导致毒力的巨大差异性(Campbell et al., 2009)。

选择性凋亡或过度刺激组织(如心脏)的自噬,有助于解释在没有大量病毒存在时出现大面积细胞死亡的情形。病毒增强其复制能力的一种常见策略是通过产生抗细胞凋亡因子来预防或延缓细胞凋亡(Skaletskaya et al., 2001),但也有些病毒在复制后期刺激细胞凋亡,从而促进病毒扩散(Hay and Kannourakis, 2002)。虹鳟感染VHSV后,在肾间质中发生凋亡性细胞死亡,但在心脏中没有这类情况发生(Eleouet et al., 2001)。使用NV基因缺陷和NV基因敲除的重组VHSV证明,NV基因产物在病毒复制早期具有抗细胞凋亡作用,野生型VHSV感染通过诱导胱天蛋白酶(caspases)来触发细胞凋亡(Ammayappan and Vakharia, 2011)。相反地,IHNV的基质蛋白可直接诱导细胞凋亡(Chiou et al., 2000),但对于VHSV则缺乏这类数据。

自噬是一种保守的细胞过程,涉及自噬体的形成、降解及细胞成分的再循环。该过程也会影响细胞内生物包括鱼类病毒的免疫和存活(Deretic, 2011; Schiotz et al., 2010; García-Valtanen et al., 2014)。从SVCV和VHSV中纯化的G蛋白会刺激自噬体的形成,VHSV诱导的自噬抑制了斑马鱼(*Danio rerio*)胚胎成纤维细胞中这两种病毒的复制(García-Valtanen et al., 2014)。自噬主要是一种促生存机制,但如果过度的话,就会导致细胞死亡(Ouyang et al., 2012)。

环境因素引起的免疫抑制通常被认为是 VHS 的一个诱发因素,然而,VHSV 本身就能够引起免疫抑制。任何机制导致的细胞死亡,包括造血组织中的病毒裂解/坏死,都会引起免疫抑制。NV 蛋白也被认为可以引起免疫抑制,并导致脾和肾中许多获得性和先天性免疫基因的下调(Chinchilla et al., 2015)。VHSV 的毒力可能部分归因于 NV 蛋白下调干扰素(interferon, IFN)响应和肿瘤坏死因子 α (tumour necrosis factor alpha, TNFα)介导的核因子(neclear factor, NF)- kβ 激活(Kim and Kim, 2012, 2013)。

3.5 防控策略

有几种杀灭冷淡水中 VHSV 的消毒方法(Bovo et al., 2005),包括使用紫外线照射、酸性消毒剂、碱性消毒剂及碘伏(含碘与增溶剂的复合消毒剂),后者广泛用于鱼卵的消毒。疾病严重程度的显著降低及影响范围的缩小得益于对风险因素的评估、基本卫生设施和最佳策略的实施。尽管如此,这些措施必须持续不断地实践(Gregory, 2008)。与环境完全隔离是阻断 VHSV 传播的唯一最佳方法。使用地表水进行网箱养殖或陆地作业中的鱼将始终处于危险之中。在疾病流行区,基于风险管理的方法是预测疾病暴发的最明智和最具成本效益的方法(Thrush and Peeler, 2013)。OIE 给出了一些适当控制策略的细节(2015)。

Dorson 等(1995)已经进行了对Ⅰa 基因型具有高抗性欧洲虹鳟的选育工作,并且该抗性表现出高遗传力(Yanez et al., 2014)。当在疫区无法对受污染的水进行隔离作业时,鉴定现有的抗 VHSV 鱼种将是一种低成本的管理选择。现有的大眼鱼种对Ⅳb 基因型实验性感染的易感性不同(Grice et al., 2016)。一个数量性状位点与虹鳟的存活和对 VHSV 的先天免疫有关(Verrier et al., 2013)。

尽管进行了大量研究,目前还没有商业化的 VHSV 疫苗(Smail and Snow, 2011)。部分是由于疫苗接种小鱼带来的经济层面的挑战,同时因为在许多地方 VHSV 需要在确诊后才能根除(OIE, 2015)。研究人员已经开发了许多针对弹状病毒的实验性 DNA 疫苗,其中就包括 VHSV 疫苗(Boudinot et al., 1998; Lorenzen et al., 2001; McLauclan et al., 2003; Sommerset et al., 2003; Chaves-Pozo et al., 2010a; Martínez-López et al., 2013; Sepúlveda and Lorenzen, 2016)。唯一一种商业化的 DNA 疫苗在加拿大不列颠哥伦比亚省被用于预防大西洋鲑的 IHNV 感染,然而,在经济价值上,受 VHSV 影响的鱼类几乎不能与大西洋鲑相提并论。DNA 实验疫苗是基于 VHSV 糖蛋白 G 序列开发的，免疫针对的是负责细胞附着的蛋白质(Bearzotti et al., 1999)，只需注射接种一次，DNA疫苗就能获得持久的高免疫保护率。DNA疫苗相对于常规产品的优势在于，它们不仅能激发抗体反应，还能激发细胞免疫和先天免疫，不过VHSV的DNA疫苗并不总能提供无菌免疫(Chaves-Pozo et al., 2010a; Sepúlveda and Lorenzen, 2016)。此外，中和抗体反应可能在其出现前或之后就会消失，但其仍提供保护(Lorenzen et al., 1998; McLauclan et al., 2003)。由于先天免疫效应因子的强烈刺激(Byon et al., 2005; Cuesta and Tafalla, 2009)，VHSV的DNA疫苗也可对异源病毒提供短期的非特异性保护(Lorenzen et al., 2002; Sommerset et al., 2003)。

Purcell等(2012)及Pereiro等(2016)对弹状病毒的免疫性进行了综述，本章重点讨论VHSV的先天免疫。在感染VHSV和IHNV的虹鳟中，当存在中和抗体时，该抗体是具有保护性的并依赖于补体(Lorenzen and LaPatra, 1999; Lorenzen et al., 1999)。腹腔注射DNA疫苗后可诱导IgT B细胞的保护作用(Martínez-López et al., 2013)。虹鳟脾淋巴细胞表达IgM、IgD和IgT，全身感染VHSV后出现IgM和IgT的克隆扩增(Castro et al., 2013)。尽管对细胞免疫的研究已取得了重大进展(Laing and Hansen, 2011; Castro et al., 2011)，但除了细胞介导的细胞毒性(cell-mediated cytotoxicity, CMC)之外，很少有功能性证据能够体现T细胞介导的病毒抗性。VHSV感染和DNA疫苗接种可诱导T细胞的克隆扩增，并可选择基于互补决定区3(complementarity determining region 3, CDR3)的克隆型T细胞的多样性(Boudinot et al., 2001, 2004)。最近在VHSV浸浴感染的虹鳟皮肤中发现了T细胞的共受体分化群3(cluster of differentiation 3, CD3)上调和T细胞活化(Leal et al., 2016)。VHSV感染虹鳟的白细胞对与VHSV感染的主要组织相容性复合体(major histocompatibility complex, MHC) Ⅰ匹配细胞，而非异种细胞表现出CMC (Utke et al., 2007)。CD8α (CD8的α链)和自然杀伤(NK)细胞增强因子均被激发，但T细胞的瞬态响应出乎意料地先于NK细胞。用携带VHSV G和N蛋白编码基因的质粒DNA疫苗接种，通过外周血白细胞对VHSV感染的MHC Ⅰ匹配细胞产生CMC，但当使用仅含有G基因的DNA疫苗免疫时，也可激活杀死异种感染细胞的NK细胞。

干扰素的诱导和干扰素刺激基因(interferon-stimulated genes, ISGs)的产生构成了鱼类抗病毒的一个重要机制(Zou and Secombes, 2011)，但IFN是非常复杂的。IFN分为Ⅰ型和Ⅱ型(IFNγ)两种，Ⅰ型IFN又分为两种，并且许多都由多个基因组成。它们对免疫和弹状病毒的影响各不相同(Zou and Secombes, 2011; Purcell et al., 2012)。重组的Ⅰ型IFN抑制VHSV在虹鳟卵巢中的复制，VHSV感染强烈会刺激卵巢发生IFN(Chaves-Pozo et al., 2010b)和几种趋化因子基因(Chaves-Pozo et al., 2010c)反应。这种强烈的反应可能解释了 VHSV 缺乏垂直传播的原因。最近，在虹鳟中发现了细胞内 IFNs，它们在虹鳟性腺 2 型细胞系（rainbow trout gonad, RTG－2）中的过表达激活了 Mx 蛋白（α/β 干扰素诱导指示物），并产生了对 VHSV 的抗性（Chang et al., 2013）。

许多 ISGs 影响 VHSV 的响应（Verrier et al., 2011）。鱼类中研究最为清楚的 ISG 是 Mx。虹鳟具有三种 Mx 亚型，它们在响应 poly(I：C)（聚肌苷酸-聚胞苷酸，一种双链 RNA 的合成类似物）或 VHSV 后而在组织中差异上调（Tafalla et al., 2007）。虹鳟接种 VHSV DNA 疫苗 3 周后，在其注射部位检测到 Mx 蛋白（Acosta et al., 2005），同时头肾和脾中 Toll 样受体 9、TNFα 和白介素-6（IL－6）的基因转录上调（Ortega-Villaizan et al., 2009）。接种 VHSV DNA 后 Mx 表达的上调与早期保护呈正相关（Boudinot et al., 1998; McLauclan et al., 2003）。在 VHSV DNA 免疫后，VHSV 诱导的基因 *vig*－1 在虹鳟淋巴组织中高表达（Boudinot et al., 1999）。包括 *vig*－1、*vig*－2、*ISG*15/*vig*－3 和趋化因子在内的 Vigs 在暴露于 VHSV 的白细胞中上调并且在感染 VHSV 的虹鳟中表达（O'Farrell et al., 2002）。蛋白激酶 R 参与预防其他 ISG mRNA 产生 poly(I：C)依赖性转录的增高，但是当虹鳟细胞被 VHSV 感染时不能控制病毒转录（Tafalla et al., 2008）。鲑细胞系中视黄酸诱导过表达和线粒体抗病毒信号蛋白能够使其预防 VHSV 的感染（Biacchesi et al., 2009）。对 VHSV 感染或模拟感染的虹鳟白细胞进行抑制消减杂交，发现 VHSV 感染激活了 finTRIM 蛋白编码基因的表达（van der Aa et al., 2009），这些蛋白类似于三结构域（tripartite motif, TRIM）蛋白家族，其中一些负责哺乳动物抗病毒的先天免疫。这些基因和其他基因家族，例如核苷酸结合寡聚化结构域（nucleotide-binding oligomerization domain, NOD）样受体的基因可能与 VHSV 免疫相关，这还有待进一步鉴定。

大菱鲆在感染 VHSV 后，其肾巨噬细胞会产生一氧化氮气体，而外源 NO 可抑制 VHSV 的复制（Tafalla et al., 1999）。大菱鲆巨噬细胞感染 VHSV 也减少了由 TNFα 或巨噬细胞活化因子诱导的 NO 产生（Tafalla et al., 2001）。与 VHSV 抗性相关的其他可溶性因子包括抗菌肽。在 VHSV 易感家系和抗性家系大菱鲆中，抗菌蛋白 hepcidin-2 和 nk-lysin 的基因存在差异性表达（Díaz-Rosales et al., 2012）。VHSV 感染促进了金头鲷（*Sparus aurata*）中白细胞的募集及其自然细胞毒细胞样活性，同时产生了活性氧中间体和髓过氧化物酶（Esteban et al., 2008）。

VHSV 刺激虹鳟中一种鱼类急性期反应物——血清淀粉样蛋白 A（serum amyloid A, SAA）的肝脏表达（Rebl et al., 2009; Castro et al., 2014），其中，IL－6 可能是 SAA 的刺激物（Castro et al., 2014）。半乳糖凝集素 9（一种与免疫功能相关的趋化蛋白）的表达在响应 VHSV 时被上调（O'Farrell et al., 2002），但鱼类凝集素与病毒相互作用的证据有限。虹鳟 ladderlectin（而非内凝集素)，是另一种具有防御作用的凝集素，它可与固定化的纯化 VHSV 相结合(Reid et al., 2011)，但下游效应未知，因为 ladderlectin 不能固定补体。

3.6 总结与研究展望

尽管许多国家存在监管壁垒，但 DNA 疫苗等技术的前景相对光明。在加拿大销售的预防 IHNV 的 DNA 疫苗(Apex - IHN, Elanco)就是一个很好的例子，疫苗的好处证明了技术的高成本是合理的。大西洋鲑是一种高价值的水产产品，由第一章可知，IHNV 是该鱼种的一种重大疾病威胁(LaPatra et al., 2016)。用于预防 VHSV 的 DNA 疫苗能够刺激先天性和获得性免疫，并可形成足以涵盖至养殖后期的长效保护，然而，它的应用仍然面临挑战(Holvold et al., 2014)。DNA 疫苗针对 VHSV 的更广泛应用还有待探索，特别是用于如北美狗鱼等高价值的增殖放流鱼种上。此外，还需要改进接种方式、使用分子佐剂增强免疫力、显著降低成本和放宽监管限制等措施。遗憾的是，其中一些因素在未来可能仍然是重要障碍。需要继续在硬骨鱼免疫知识方面取得进展，特别是对先天性和获得性免疫系统之间的相互作用以及细胞介导免疫(cell-mediated immunity, CMI)的认知需要加强。

目前，管理 VHSV 的传播和影响仍将依靠传统方法，包括生物安全、检测和消除或限制鱼类转移等。考虑到所需要的成本，鼓励监管机构对 VHSV 的基因型进行一些区分，这将需要在 OIE 层面上实现。例如，Ⅳb 型 VHSV 在五大湖地区流行，虽然偶尔会发生死亡事件，但其影响有限，根据年度或其他检测得出的感染可能性而提出的基于风险的管理是一种行之有效的方法，目前已被加拿大和其他当局采纳。因此，希望在国际上销售产品的生产者可在 VHSV 流行区内获取一个无 VHSV 渔品资质。

参考文献

[1] Acosta, F., Petrie, A., Lockhart, K., Lorenzen, N. and Ellis, A.E. (2005) Kinetics of Mx expression in rainbow trout (Onchorhynchus mykiss W) and Atlantic salmon (*Salmo salar* L) parr in response to VHSVDNA vaccination. *Fish and Shellfish Immunology* 18, 81 - 89.

[2] AFS - FHS (2014) *FHS Blue Book: Suggested Procedures for the Detection and Identification of Certain Finfish and Shellfish Pathogens*, 2014 edn. American Fisheries Society-Fish Health Section, Bethesda, Maryland.

[3] Al-Hussinee, L. and Lumsden, J.S. (2011a) Detection of VHSV IVb within the gonads of Great Lakes fish using in situ hybridization. *Diseases of Aquatic Organisms* 95, 81 - 86.

[4] Al-Hussinee, L., Huber, P., Russell, S., Lepage, V., Reid, A. *et al*. (2010) Viral haemorrhagic septicaemia virus IVb experimental infection of rainbow trout,

Oncorhynchus mykiss (Walbaum), and fathead minnow, *Pimphales promelas* (Rafinesque). *Journal of Fish Diseases* 33, 347 - 360.

[5] Al-Hussinee, L., Lord, S., Stevenson, R.M., Casey, R.N., Groocock, G.H. *et al.* (2011b) Immunohistochemistry and pathology of multiple Great Lakes fish from mortality events associated with viral hemorrhagic septicemia virus type IVb. *Diseases of Aquatic Organisms* 93, 117 - 127.

[6] Al-Hussinee L., Tubbs L., Russell S., Pham J., Tafalla C. *et al.* (2016) Temporary protection of rainbow trout gill epithelial cells from infection with viral hemorrhagic septicemia virus IVb. *Journal of Fish Diseases* 38, 1099 - 1112.

[7] Ammayappan, A. and Vakharia, V.N. (2011) Nonvirion protein of *Novirhabdovirus* suppresses apoptosis at the early stage of virus infection. *Journal of Virology* 85, 8393 -8402.

[8] APHIS (2009) *Viral Hemorrhagic Septicemia Virus IVb: U.S. Surveillance Report 2009*. US Department of Agriculture Animal and Plant Health Inspection Service, Veterinary Services, Riverdale, Maryland. Available at: https://www.aphis.usda.gov/animal_health/animal_dis_spec/aquaculture/downloads/vhs_surv_rpt.pdf(access-ed 25 October 2016).

[9] Bearzotti, M., Delmas, B., Lamoureux, A., Loustau, A.M., Chilmonczyk, S. *et al.* (1999) Fish rhabdovirus cell entry is mediated by fibronectin. *Journal of Virology* 73, 7703 - 7709.

[10] Biacchesi S., LeBerre M., Lamoureux A., Louise Y., Lauret E. *et al.* (2009) MAVS plays a major role in induction of the fish innate immune response against RNA and DNA viruses. *Journal of Virology* 83, 7815 - 7827.

[11] Boudinot, P., Blanco M., de Kinkelin P. and Benmansour A. (1998) Combined DNA immunization with the glycoprotein gene of the viral hemorrhagic septicemia virus and infectious hematopoietic necrosis virus induces a double-specific protective immunity and a nonspecific response in rainbow trout. *Virology* 249, 297 - 306.

[12] Boudinot, P., Massin, P., Blanco, M., Riffault, S. and Benmansour A. (1999) *vig*-1, a new fish gene induced by the rhabdovirus glycoprotein, has a virus-induced homologue in humans and shares conserved motifs with the MoaA family. *Journal of Virology* 73, 1846 - 1852.

[13] Boudinot, P., Boubekeur, S. and Benmansour, A. (2001) Rhabdovirus infection induces public and private T cell responses in teleost fish. *Journal of Immunology* 167, 6202 - 6209.

[14] Boudinot, P., Bernard, D., Boubekeur, S., Thoulouze, M.I., Bremont, M. *et al.* (2004) The glycoprotein of a fish rhabdovirus profiles the virus-specific T-cell repertoire in rainbow trout. *Journal of General Virology* 85, 3099 - 3108.

[15] Bovo, G., Hill, B., Husby, A., Håstein, T., Michel, C. *et al.* (2005) *Work Package 3 Report: Pathogen Survival Outside the Host, and Susceptibility to Disinfection.* Report QLK2 - CT - 2002 - 01546: Fish Egg Trade. VESO (Veterinary Science Opportunities), Oslo. Available at: http://www.eurl-fish.eu/-/media/Sites/EURL-FISH/english/activities/scientific%20reports/fisheggtrade-wp_3.ashx?la=da (accessed 25 October 2016).

[16] Brudeseth, B.E., Castric, J. and Evensen, Ø. (2002) Studies on pathogenesis following single and double infection with viral hemorrhagic septicemia virus and infectious hematopoietic necrosis virus in rainbowtrout (*Oncorhynchus mykiss*). *Veterinary Pathology* 39, 180 - 189.

[17] Brudeseth, B., Raynard, R., King, J. and Evensen, Ø. (2005) Sequential pathology after experimental infection with marine viral hemorrhagic septicemia virus isolates of low and high virulence in turbot (*Scophthalmus maximus* L.). *Veterinary Pathology*

42, 9 - 18.
[18] Brudeseth, B.E., Skall, H.F. and Evensen, O. (2008) Differences in virulence of marine and freshwater isolates of viral hemorrhagic septicemia virus in vivo correlate with in vitro ability to infect gill epithelial cellsand macrophages of rainbow trout (*Oncorhynchus mykiss*). *Journal of Virology* 82, 10359 - 10365.
[19] Byon, J.Y., Ohira, T., Hirono, I. and Aoki, T. (2005) Use of cDNA microarray to study immunity against viral hemorrhagic septicemia (VHS) in Japanese flounder (*Paralichthys olivaceus*) following DNA vaccination. *Fish and Shellfish Immunology* 18, 135 - 147.
[20] Campbell, S., Collet, B., Einer-Jensen, K., Secombes, C.J. and Snow, M. (2009) Identifying potential virulence determinants in viral haemorrhagic septicaemia virus (VHSV) for rainbow trout. *Diseases of Aquatic Organisms* 86, 205 - 212.
[21] Castro, R., Bernard, D., Lefranc, M.P., Six, A., Benmansour, A. *et al.* (2011) T cell diversity and TcR repertoires in teleost fish. *Fish and Shellfish Immunology* 31, 644 - 654.
[22] Castro, R., Jouneau, L., Pham, H., Bouchez, O., Guidicelli, V. *et al.* (2013) Teleost fish mount complex clonal IgM and IgT responses in spleen upon systemic viral infection. *PLoS Pathogens* 9(1): e1003098.
[23] Castro, R., Abos, B., Pignatelli, J., von Gersdorff, J.L., Gonzalez Granja, A. *et al.* (2014) Early immune responses in rainbow trout liver upon viral hemorrhagic septicemia virus (VHSV) infection. PLoS ONE 9(10): e111084.
[24] Chang, M.-X., Zou, J., Nie, P., Huang, B., Yu, Z. *et al.* (2013) Intracellular interferons in fish: a unique means to combat viral infection. *PLoS Pathogens* 9 (11): e1003736.
[25] Chaves-Pozo, E., Cuesta, A. and Tafalla, C. (2010a) Antiviral DNA vaccination in rainbow trout (*Oncorhynchus mykiss*) affects the immune response in the ovary and partially blocks its capacity to support viral replication in vitro. *Fish and Shellfish Immunology* 29, 579 - 586.
[26] Chaves-Pozo, E., Zou, J., Secombes, C.J., Cuesta, A. and Tafalla, C. (2010b) The rainbow trout (*Oncorhynchus mykiss*) interferon response in the ovary. *Molecular Immunology* 47, 1757 - 1764.
[27] Chaves-Pozo, E., Montero, J., Cuesta, A. and Tafalla, C. (2010c) Viral hemorrhagic septicemia and infectious pancreatic necrosis viruses replicate differently in rainbow trout gonad and induce different chemokine transcription profiles. *Developmental and Comparative Immunology* 34, 648 - 658.
[28] Chinchilla, B., Encinas, P., Estepa, A., Coll, J.M. and Gomez-Casad, E. (2015) Transcriptome analysis of rainbow trout in response to non-virion (NV) protein of viral hemorrhagic septicemia virus (VHSV). *Applied Microbiology and Biotechnology* 99, 1827 - 1843.
[29] Chiou, P., Kim, P., Carol, H., Ormonde, P. and Leong, J.C. (2000) Infectious hematopoietic necrosis virus matrix protein inhibits host-directed gene expression and induces morphological changes of apoptosis in cell cultures. *Journal of Virology* 74, 7619 - 7627.
[30] Cuesta, A. and Tafalla, C. (2009) Transcription of immune genes upon challenge with viral hemorrhagic septicemiavirus (VHSV) in DNA vaccinated rainbow trout (*Oncorhynchus mykiss*). *Vaccine* 27, 280 - 289.
[31] Dale, O.B., Orpetveit, I., Lyngstad, T.M., Kahns, S., Skall, H.F. *et al.* (2009) Outbreak of viral haemorrhagic septicaemia (VHS) in seawater-farmed rainbow trout in Norway caused by VHS virus genotype III. *Diseases of Aquatic Organisms* 85, 93 - 103.
[32] de Kinkelin, P., Chilmonczyk, S., Dorson, M., le Berre, M. and Baudouy A.M. (1979)

Some pathogenic facets of rhabdoviral infection of salmonid fish. In: Bachmann P. A. (ed.) *Munich Symposia of Microbiologia: Mechanisms of Viral Pathogenesis and Virulence*. WHO Collaborating Centre for Collection and Education of Data on Comparative Virology, Munich, Germany, pp. 357 - 375.

[33] Deretic, V. (2011) Autophagy in immunity and cell-autonomous defense against intracellular microbes. *Immunological Reviews* 240, 92 - 104.

[34] Díaz-Rosales, P., Romero, A., Balseiro, P., Dios, S. Novoa, B. *et al.* (2012) Microarray-based identification of differentially expressed genes in families of turbot (*Scophthalmus maximus*) after infection with viral haemorrhagic septicemia virus (VHSV). *Marine Biotechnology* 14, 515 - 529.

[35] Dorson, M., Quillet, E., Hollebecq, M.G., Torhy, C. and Chevassus, B. (1995) Selection of rainbow trout resistant to viral haemorrhagic septicaemia virus and transmission of resistance by gynogenesis. *Veterinary Research* 26, 361 - 368.

[36] Eaton, W., Hulett, J., Brunson, R. and True, K. (1991) The first isolation in North America of infectious, hematopoietic necrosis virus (IHNV) and viral hemorrhagic septicemia virus (VHSV) in coho salmon from the same watershed. *Journal of Aquatic Animal Health* 3, 114 - 117.

[37] Einer-Jensen, K., Ahrens, P., Forsberg, R. and Lorenzen, N. (2004) Evolution of the fish rhabdovirus viral haemorrhagic septicaemia virus. *Journal of General Virology* 85, 1167 - 1179.

[38] Eleouet, J.F., Druesne, N., Chilmonczyk, S. and Delmas, B. (2001) Comparative study of *in-situ* cell death induced by the viruses of viral hemorrhagic septicemia (VHS) and infectious pancreatic necrosis (IPN) in rainbow trout. *Journal of Comparative Pathology* 124, 300 - 307.

[39] Elsayed, E., Faisal, M., Thomas, M., Whelan, G., Batts, W. *et al.* (2006) Isolation of viral haemorrhagic septicaemia virus from muskellunge, *Esox masquinongy* (Mitchill), in Lake St Clair, Michigan, USA reveals a new sublineage of the North American genotype. *Journal of Fish Diseases* 29, 611 - 619.

[40] Esteban, M.A., Meseguer, K., Tafalla, C. and Cuesta, A. (2008) NK-like and oxidative burst activities are the main early cellular innate immune responses activated after virus inoculation in reservoir fish. *Fish and Shellfish Immunology* 25, 433 - 438.

[41] Evensen, Ø., Meier, W., Wahli, T., Olesen, N., Vestergård Jørgensen, P. *et al.* (1994) Comparison of immunohistochemistry and virus cultivation for detection of viral haemorrhagic septicaemia virus in experimentally infected rainbow trout *Oncorhynchus mykiss*. *Diseases of Aquatic Organisms* 20, 101 - 109.

[42] Gagne, N., Mackinnon, A.M., Boston, L., Souter, B., Cook-Versloot, M. *et al.* (2007) Isolation of viral haemorrhagic septicaemia virus from mummichog, stickleback, striped bass and brown trout in eastern Canada. *Journal of Fish Diseases* 30, 213 - 223.

[43] García-Valtanen, P., Ortega-Villaizán, M.D., Martínez-López, A., Medina-Gali, R., Pérez, L. *et al.* (2014) Autophagy-inducing peptides from mammalian VSV and fish VHSV rhabdoviral G glycoproteins (G) as models for the development of new therapeutic molecules. *Autophagy* 10, 1666 - 1680.

[44] Garver, K.A., Hawley, L.M., McClure, C.A., Schroeder, T., Aldous, S. *et al.* (2011) Development and validation of a reverse transcription quantitative PCR for universal detection of viral hemorrhagic septicemia virus. *Diseases of Aquatic Organisms* 95, 97 - 112.

[45] Garver, K.A., Traxler, G.S., Hawley, L.M., Richard, J., Ross, J.P. *et al.* (2013) Molecular epidemiology of viral haemorrhagic septicaemia virus (VHSV) in British Columbia, Canada, reveals transmission from wild to farmed fish. *Diseases of Aquatic Organisms* 104, 93 - 104.

[46] Goodwin, A.E. and Merry, G.E. (2011) Mortality and carrier status of bluegills exposed to viral hemorrhagic septicemia virus genotype IVb at different temperatures. *Journal of Aquatic Animal Health* 23, 85 - 91.
[47] Gregory, A. (2008) A qualitative assessment of the risk of introduction of viral haemorrhagic septicaemia virus into the rainbow trout industry Scotland. Fisheries Research Services Internal Report No. 12/08. Fisheries Research Services, Marine Laboratory, Aberdeen, UK. Available at: http://www.gov.scot/uploads/documents/int1208.pdf (accessed 25 October 2016).
[48] Grice, J., Reid, A., Peterson, A., Blackburn, K., Tubbs, L. *et al*. (2016) Walleye *Sander vitreus* Mitchill are relatively resistant to experimental infection with viral hemorrhagic septicemia virus IVb and extant strains vary in susceptibility. *Journal of Fish Diseases* 38, 859 - 872.
[49] Groocock, G.H., Getchell, R.G., Wooster, G.A., Britt, K.L., Batts, W.N. *et al*. (2007) Detection of viral hemorrhagic septicemia in round gobies in New York State (USA) waters of Lake Ontario and the St. Lawrence River. *Diseases of Aquatic Organisms* 76, 187 - 192.
[50] Hawley, L.M. and Garver K.A. (2008) Stability of viral hemorrhagic septicemia virus (VHSV) in freshwater and seawater at various temperatures. *Diseases of Aquatic Organisms* 82, 171 - 178.
[51] Hay, S. and Kannourakis, G. (2002) A time to kill: viral manipulation of the cell death program. *Journal of General Virology* 83, 1547 - 1564.
[52] Hoffmann, B., Schutze, H. and Mettenleiter, T.C. (2002) Determination of the complete genomic sequence and analysis of the gene products of the virus of spring viremia of carp, a fish rhabdovirus. *Virus Research* 84, 89 - 100.
[53] Holvold, L.B., Myhr, A. and Dalmo, R.A. (2014) Strategies and hurdles using DNA vaccines to fish. *Veterinary Research* 45, 21.
[54] Hope, K.M., Casey, R.N., Groocock, G.H., Getchell, R.G., Bowser, P.R. et al. (2010) Comparison of quantitative RT - PCR with cell culture to detect viral hemorrhagic septicemia virus (VHSV) IVb infections in the Great Lakes. *Journal of Aquatic Animal Health* 22, 50 - 61.
[55] ICTV (2015) Virus Taxomomy: 2015 Release, EC47, London, UK, July 2015. International Committee on Taxonomy of Viruses. Available at: http://www.ictvonline.org/virusTaxonomy.asp (accessed 25 October 2016).
[56] Kim, R. and Faisal, M. (2010) Comparative susceptibility of representative Great Lakes fish species to the North American viral hemorrhagic septicemia virus sublineage IVb. *Diseases of Aquatic Organisms* 91, 23 - 34.
[57] Kim, M.S. and Kim, K.H. (2012) Effects of NV gene knock-out recombinant viral hemorrhagic septicemia virus (VHSV) on Mx gene expression in *Epithelioma papulosum cyprini* (EPC) cells and olive flounder (*Paralichthys olivaceus*). *Fish and Shellfish Immunology* 32, 459 - 463.
[58] Kim, M.S. and Kim, K.H. (2013) The role of viral hemorrhagic septicemia virus (VHSV) NV gene in TNF-α and VHSV infection-mediated NF-kB activation. *Fish and Shellfish Immunology* 34, 1315 - 1319.
[59] Kocan, R., Bradley, M., Elder, N., Meyers, T., Batts, W. *et al*. (1997) North American strain of viral hemorrhagic septicemia virus is highly pathogenic for laboratory-reared Pacific herring. *Journal of Aquatic Animal Health* 9, 279 - 290.
[60] LaPatra, S., Al-Hussinee, L., Misk, E. and Lumsden J.S. (2016) Rhabdoviruses of fish. In: Kibenge, F. and Godoy, M. (eds) *Aquaculture Virology*. Elsevier/Academic Press, Amsterdam, pp. 266 - 298.
[61] Laing, K.J. and Hansen, J.D. (2011) Fish T cells: recent advances through genomics.

Developmental and Comparative Immunology 35，1282－1295.

[62] Leal，E.，Granja，A.G.，Zarza，C. and Tafalla，C.（2016）Distribution of T cells in rainbow trout（*Oncorhynchus mykiss*）skin and responsiveness to viral infection. PLoS ONE 11(1)：e0147477.

[63] Lorenzen，N. and LaPatra，S.E.（1999）Immunity to rhabdoviruses in rainbow trout：the antibody response. *Fish and Shellfish Immunology* 9，345－360.

[64] Lorenzen，N.，Lorenzen E.，Einer-Jensen，K.，Heppell，J.，Wu，T. *et al*.（1998）Protective immunity to VHS in rainbow trout（*Oncorhynchus mykiss* Walbaum）following DNA vaccination. *Fish and Shellfish Immunology* 8，261－270.

[65] Lorenzen，N.，Olesen，N.J. and Koch，C.（1999）Immunity to VHS virus in rainbow trout. *Aquaculture* 172，41－60.

[66] Lorenzen，N.，Lorenzen，E. and Einer-Jensen，K.（2001）Immunity to viral haemorrhagic septicemia（VHS）following DNA vaccination of rainbow trout at an early lifestage. *Fish and Shellfish Immunology* 11，585－591.

[67] Lorenzen，N.，Lorenzen，E.，Einer-Jensen，K. and LaPatra，S.E.（2002）DNA vaccines as a tool for analyzing the protective immune response against rhabdoviruses in rainbow trout. *Fish and Shellfish Immunology* 12，439－453.

[68] Lumsden，J.S.，Morrison，B.，Yason，C.，Russell，S.，Young，K. *et al*.（2007）Mortality event in freshwater drum *Aplodinotus grunniens* from Lake Ontario，Canada，associated with viral haemorrhagic septicemia virus，type IV. *Diseases of Aquatic Organisms* 76，99－111.

[69] Martínez-López，A.，García-Valtanen，P.，Ortega-Villaizán，M.M.，Chico，V.，Medina-Gali，R.M. *et al*.（2013）Increasing versatility of the DNA vaccines through modification of the subcellular location of plasmid-encoded antigen expression in the in vivo transfected cells. *PLoS ONE* 8(10)：e77426.

[70] Marty，G.D.，Freiberg，E.F.，Meyers，T.R.，Wilcock，J.，Farver，T.B. *et al*.（1998）Viral hemorrhagic septicemia virus，*Ichthyophonus hoferi*，and other causes of morbidity in Pacific herring *Clupea pallasi* spawning in Prince William Sound，Alaska，USA. *Diseases of Aquatic Organisms* 32，15－40.

[71] Matejusova，I.，McKay，P.，McBeath，A.J.，Collet，B. and Snow，M.（2008）Development of a sensitive and controlled real-time RT－PCR assay for viral haemorrhagic septicaemia virus（VHSV）in marine salmonid aquaculture. *Diseases of Aquatic Organisms* 80，137－144.

[72] McAllister，P.E.（1990）*Viral Hemorrhagic Septicemia of Fishes*. *Fish Disease Leaflet* 83，US Department of the Interior Fish and Wildlife Service，Washington，DC. Available at：http://digitalcommons.unl.edu/cgi/viewcontent.cgi?article＝1141&context＝usfwspubs（accessed 25 October 2016）.

[73] McLauclan，P.E.，Collet，B.，Ingerslev，E.，Secombes，C.J.，Lorenzen，N. *et al*.（2003）DNA vaccination against viral hemorrhagic septicemia（VHS）in rainbow trout：size，dose，route of injection and duration of protection-early protection correlates with Mx expression. *Fish and Shellfish Immunology* 15，39－50.

[74] Meyers，T.R. and Winton，J.R.（1995）Viral hemorrhagic septicemia virus in North America. *Annual Review of Fish Diseases* 5，3－24.

[75] Nishizawa，T.，Iida，H.，Takano，R.，Isshiki，T.，Nakajima，K. *et al*.（2002）Genetic relatedness among Japanese，American and European isolates of viral hemorrhagic septicemia virus（VHSV）based on partial G and P genes. Diseases of Aquatic Organisms 48，143－148.

[76] O'Farrell，C.，Vaghefi，N.，Cantonnet，M.，Buteau，B.，Boudinot，P. *et al*.（2002）Survey of transcript expression in rainbow trout leukocytes reveals a major contribution of interferon-responsive genes in the early response to rhabdovirus infection. *Journal of*

Virology 76, 8040 - 8049.
[77] OIE (2015) Viral haemorrhagic septicaemia, In: *Manual of Diagnostic Tests for Aquatic Animals*, 7th edn. World Organisation for Animal Health, Paris, pp. 374 - 396. Updated 2016 version (Chapter 2.3.10.) available at: http://www.oie.int/index.php?id=2439&L=0&htmfile=chapitre_vhs.htm (accessed 25 October 2016).
[78] Olesen, N., Lorenzen, N. and Jørgensen, P. (1993) Serological differences among isolates of viral haemorrhagic septicaemia virus detected by neutralizing monoclonal and polyclonal antibodies. *Diseases of Aquatic Organisms* 16, 163 - 163.
[79] Ortega-Villaizan, M., Chico, V., Falco, A., Perez, L., Coll, J.M. *et al.* (2009) The rainbow trout TLR9 gene, its role in the immune responses elicited by a plasma encoding the glycoprotein G of the viral haemorrhagic septicaemia rhabdovirus (VHSV). *Molecular Immunology* 46, 1710 - 1717.
[80] Ouyang, L., Shi, Z., Zhao, S., Wang, F.T., Zhou, T.T. *et al.* (2012) Programmed cell death pathways in cancer: a review of apoptosis, autophagy and programmed necrosis. *Cell Proliferation* 45, 487 - 498.
[81] Parry, L. and Dixon, P.F. (1997) Stability of viral haemorrhagic septicemia virus (VHSV) isolates in seawater. *Bulletin of the European Association of Fish Pathologists* 17, 31 - 36.
[82] Pereiro, P., Figueras, A. and Novoa, B. (2016) Turbot (*Scophthalmus maximus*) vs. VHSV (viral hemorrhagic septicemia virus): a review. *Frontiers in Physiology* 7: 192.
[83] Pham, J., Lumsden, J.S., Tafalla, C., Dixon, B. and Bols, N. (2013) Differential effects of viral hemorrhagic septicemia virus (VHSV) genotypes IVa and IVb on gill epithelial and spleen macrophage cell lines from rainbow trout (*Onchorhynchus mykiss*). *Fish and Shellfish Immunology* 34, 632 - 640.
[84] Purcell, M.K., Laing, K.J. and Winton, J.R. (2012) Immunity to fish rhabdoviruses. *Viruses* 4,140 - 166.
[85] Quillet, E., Dorson, M., Aubard, G. and Torhy, C. (2001) In vitro viral hemorrhagic septicemia virus replication in excised fins of rainbow trout: correlation with resistance to waterborne challenge and genetic variation. *Diseases of Aquatic Organisms* 45, 171 - 182.
[86] Rebl, A., Goldammer, T., Fischer, U., Köllner, B. and Sayfert, H.M. (2009) Characterization of two key molecules of teleost innate immunity from rainbow trout (*Oncorhynchus mykiss*): MyD88 and SAA. *Veterinary Immunology and Immunopathology* 131, 122 - 126.
[87] Reid, A., Young, K. and Lumsden, J.S. (2011) Rainbow trout ladderlectin not intelectin binds VHSV IVb. *Diseases of Aquatic Organisms* 95, 137 - 143.
[88] Sano, M., Ito, T., Matsuyama, T., Nakayasu, C. and Kurita, J. (2009) Effect of water temperature shifting on mortality of Japanese flounder *Paralichthys olivaceus* experimentally infected with viral hemorrhagic septicemia virus. *Aquaculture* 286, 254 - 258.
[89] Schiotz, B.L., Roos, N., Rishovd, A.L. and Gjoen, T. (2010) Formation of autophagosomes and redistribution of LC3 upon *in vitro* infection with infectious salmon anemia virus. *Virus Research* 151, 104 - 107.
[90] Sepúlveda, D. and Lorenzen, N. (2016) Can VHS bypass the protective immunity induced by DNA vaccination in rainbow trout? *PLoS ONE* 11(4): e0153306.
[91] Skaletskaya, A., Bartle, L.M., Chittenden, T., McCormick, L., Mocarski, E.S. *et al.* (2001) A cytomegalovirus-encoded inhibitor of apoptosis that suppresses caspase-8 activation. *Proceedings of the National Academy of Sciences of the United States of America* 98, 7829 - 7834.
[92] Skall, H.F., Slierendrecht, W.J., King, A. and Olesen, N.J. (2004) Experimental

infection of rainbow trout *Oncorhynchus mykiss* with viral haemorrhagic septicemia virus isolates from European marine and farmed fishes. *Diseases of Aquatic Organisms* 58, 99－110.

[93] Smail, D.A. and Snow, M. (2011) Viral hemorrhagic septicemia. In: Woo, P.T.K., Bruno, D.W. (eds) *Fish Diseases and Disorders. Volume 3: Viral Bacterial and Fungal Infections*, 2nd edn. CAB International, Wallingford, UK. pp. 123－146.

[94] Soliman, H. and El-Matbouli, M. (2006) Reverse transcription loop-mediated isothermal amplification (RT-LAMP) for rapid detection of viral hemorrhagic septicaemia virus (VHS). *Veterinary Microbiology* 114, 205－213.

[95] Sommerset, I., Lorenzen, E., Lorenzen, N., Bleie, H. and Nerland, A.H. (2003) A DNA vaccine directed against a rainbow trout rhabdovirus induces early protection against a nodavirus challenge in turbot. *Vaccine* 21, 4661－4667.

[96] Snow, M. and Cunningham, C.O. (2000) Virulence and nucleotide sequence analysis of marine viral haemorrhagic septicaemia virus following in vivo passage in rainbow trout *Onchorhynchus mykiss*. Diseases of Aquatic Organisms 42, 17－26.

[97] Snow, M., Bain, N., Black, J., Taupin, V., Cunningham, C.O. *et al.* (2004) Genetic population structure of marine viral haemorrhagic septicaemia virus (VHSV). *Diseases of Aquatic Organisms* 61, 11－21.

[98] Tafalla, C., Figueras, A. and Novoa, B. (1998) In vitro interaction of viral haemorrhagic septicaemia virus and leukocytes from trout (*Oncorhynchus mykiss*) and turbot (*Scophthalmus maximus*). *Veterinary Immunology and Immunopathology* 62, 359－366.

[99] Tafalla, C., Figueras, A. and Novoa, B. (1999) Role of nitric oxide on the replication of viral haemorrhagic septicemia virus (VHSV), a fish rhabdovirus. *Veterinary Immunology and Immunopathology* 72, 249－256.

[100] Tafalla, C., Figueras, A. and Novoa, B. (2001) Viral hemorrhagic septicemia virus alters turbot *Scophthalmus maximus* macrophage nitric oxide production. *Diseases of Aquatic Organisms* 47, 101－107.

[101] Tafalla, C., Chico, V., Pérez, L., Coll, J.M. and Estepa, A. (2007) In vitro and in vivo differential expression of rainbow trout (*Oncorhynchus mykiss*) Mx isoforms in response to viral haemorrhagic septicemia virus (VHSV) G gene, poly I∶C and VHSV. *Fish and Shellfish Immunology* 23, 210－221.

[102] Tafalla, C, Sanchez, E., Lorenzen, N., DeWitte-Orr, S.J. and Bols, N.C. (2008) Effects of viral hemorrhagic septicemia virus (VHSV) on the rainbow trout (*Oncorhynchus mykiss*) monocyte cell line RTS－11. *Molecular Immunology* 45, 1439－1448.

[103] Thompson, T.M., Batts, W.N., Faisal, M., Bowser, P., Casey, J.W. *et al.* (2011) Emergence of viral hemorrhagic septicemia virus in the North American Great Lakes region is associated with low viral genetic diversity. *Diseases of Aquatic Organisms* 96, 29－43.

[104] Thrush, M.A. and Peeler, E.J. (2013) A model to approximate lake temperature from gridded daily air temperature records and its application in risk assessment for the establishment of fish diseases in the UK. *Transboundary and Emerging Diseases* 60, 460－471.

[105] Utke, K., Bergmann, S., Lorenzen, N., Kollner, B., Ototake, M. *et al.* (2007) Cell-mediated cytotoxicity in rainbow trout, *Oncorhynchus mykiss*, infected with viral haemorrhagic septicemia virus. *Fish and Shellfish Immunology* 22, 182－196.

[106] Utke, K., Kock, H., Schuetze, H., Bergmann, S.M., Lorenzen, N. *et al.* (2008) Cell-mediated immune response in rainbow trout after DNA immunization against the viral hemorrhagic septicemia virus. *Developmental and Comparative Immunology* 32,

239 - 252.
[107] van der Aa, L.M., Levraud, J., Yahmi, M., Lauret, E., Briolat, V. *et al*. (2009) A large subset of TRIM genes highly diversified by duplication and positive selection in teleost fish. *BMC Biology* 7: 7.
[108] Verrier, E.R., Langevin, C., Benmansour, A. and Boudinot, P. (2011) Early antiviral response and virus-induced genes in fish. *Developmental and Comparative Immunology* 35, 1204 - 1214.
[109] Verrier, E.R., Dorson, M., Mauger, S., Torhy, C., Ciobotaru, C. *et al*. (2013) Resistance to a rhabdovirus (VHSV) in rainbow trout: identification of a major QTL related to innate mechanisms. *PLoS ONE* 8(2): e55302.
[110] Vo, N.T.K., Bender, A.W., Lee, L.E.J., Lumsden, J.S., Lorenzen, N. *et al*. (2015) Development of a walleye cell line and use to study the effects of temperature on infection by viral hemorrhagic septicemia virus group IVb. *Journal of Fish Diseases* 38, 121 - 136.
[111] Wolf, K. (1988) *Fish Viruses and Fish Viral Diseases*. Comstock Publishing Associates, Cornell University Press, Ithaca, New York.
[112] Yamamoto, T., Batts, W.N. and Winton, J.R. (1992) In vitro infection of salmonid epidermal tissues by infectious hematopoietic necrosis virus and viral hemorrhagic septicemia virus. *Journal of Aquatic Animal Health* 4, 231 - 239.
[113] Yanez, J.M., Houston, R.D. and Newman, S. (2014) Genetics and genomics of disease resistance in salmonid species. *Frontiers in Genetics* 5, 415.
[114] Zou, J. and Secombes, C.J. (2011) Teleost fish interferons and their role in immunity. *Developmental and Comparative Immunology* 35, 1376 - 1387.

4

流行性造血器官坏死病与欧鲇病毒

Paul Hick，Ellen Ariel 和 Richard Whittington*

4.1 引言

流行性造血器官坏死病(epizootic haematopoietic necrosis，EHN)仅限于澳大利亚，是由蛙病毒属流行性造血器官坏死病病毒(epizootic haematopoietic necrosis virus，EHNV)引起的。这种系统性疾病可造成自然感染的野生河鲈(*Perca fluviatilis*)大量死亡，对养殖的虹鳟(*Oncorhynchus mykiss*)也会产生影响，并且还会对澳大利亚本土鱼类种群造成威胁(Whittington et al.，2010；Becker et al.，2013)。欧鲇病毒(European catfish virus，ECV)是一种与 EHNV 密切相关的蛙病毒，最初是在德国从欧鲇(*Silurus glanis*)中分离出来的(Ahne et al.，1989)。ECV 导致欧鲇和黑鮰(*Ameiurus melas*)的大量死亡，其他具有重要经济意义的鱼类也可能易感(Jensen et al.，2009，2011)。这两种病毒都能引起类似的疾病，呈现出非特异性的临床体征和以造血组织中最明显的广泛系统性坏死为特征的病理病变(Whittington et al.，2010)。这两种疾病在野生鱼类和水产养殖鱼类中都有发生。鉴于其致病力、缺乏宿主特异性、地理隔离以及存在可靠的诊断方法等，EHNV 是被世界动物卫生组织(OIE)列出的首个有鳍鱼类病毒(OIE，2015)。此外，世界动物卫生组织还列出了能够感染两栖动物的其他蛙病毒，因为它们具有经济和生态影响(OIE，2016a)，并且还具有类间传播的可能性(Brenes et al.，2014)。实验证据表明，除 EHNV 和 ECV 外，单一鱼种还对多种蛙病毒敏感(Jensen et al.，2009)。本章对与鱼类健康有关的新发病原 EHNV 和 ECV 进行了描述，目前对它们的动物流行病学的了解尚不完善(Gray et al.，2009)。

4.1.1 分类

EHNV 和 ECV 均属于虹彩病毒科(Iridoviridae)的蛙病毒属(*Ranavirus*)病毒(Jancovich et al.，2012)。与虹彩病毒科的其他所有病毒一样，它们都是大型双链 DNA(dsDNA)病毒，具有复杂的核质复制周期(Williams et al.，2005)。虹彩病毒

* 通信作者邮箱：richard.whittington@sydney.edu.au。

科包括许多能引起水产养殖鱼类疾病的病毒种类(Whittington et al., 2010)。

Jancovich 等(2015)介绍了五个蛙病毒属谱系,为蛙病毒属分类提供了一个框架(图 4.1)。EHNV、ECV 及两栖类病原虎纹钝口螈病毒(*Ambystoma tigrinum*

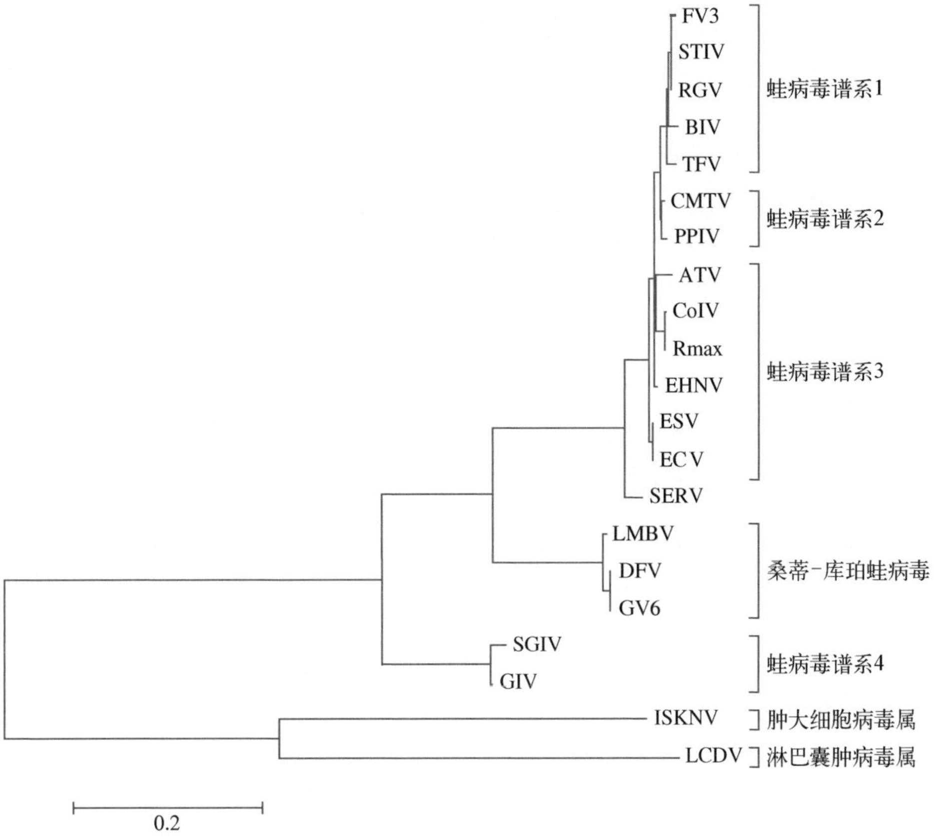

图 4.1 蛙病毒谱系 3 中的流行性造血器官坏死病病毒(EHNV)和欧鲇病毒(ECV)(Jancovich et al., 2015)。采用分子进化遗传分析软件 6.0 版(Mega6)中的最大似然法,以主要衣壳蛋白基因的 1 127 个碱基位置计算分子系统发育(Tamura et al., 2013)。按比例绘制该发育树图谱,其中分枝长度以每个位点的替换数量来度量。按照"谱系:病毒名称,Genbank 参考"的格式进行说明。谱系 1:FV3(蛙病毒 3 型),AY548484;STIV(甲鱼虹彩病毒),EU627010;RGV(美国青蛙虹彩病毒),JQ654586;BIV(饰纹汀蛙虹彩病毒),AY187046;TFV(虎纹蛙病毒),AF389451。谱系 2:CMTV(产婆蟾病毒),JQ231222;PPIV(梭鲈虹彩病毒),FJ358610。谱系 3:ATV(虎纹钝口螈病毒),AY150217;CoIV(鳕虹彩病毒),GU391284;Rmax(最大蛙病毒),GU391285;EHNV(流行性造血器官坏死病病毒),AY187045;ESV(欧洲六须鲇病毒),FJ358609;ECV(欧鲇病毒),FJ358608。SERV(短鳍鳗蛙病毒),FJ358612。*Santee-Cooper* 蛙病毒的分离株:LMBV(大口黑鲈蛙病毒),FR682503;DFV(医生鱼病毒),FR677324;GV6(虹鳉病毒 6),FR677325。谱系 4:SGIV(新加坡石斑鱼虹彩病毒),AY521625;GIV(石斑鱼虹彩病毒),AY666015。种类:肿大细胞病毒属,ISKNV(传染性脾肾坏死病毒),NC003494;淋巴囊肿病毒属,LCDV(淋巴囊肿病毒 1),NC001824。改编自文献:Jancovich et al. (2015)

Virus，ATV)组成谱系 3 中的一个分类群。每个谱系都含有鱼类病原体，在这些密切相关的病毒中，有的两栖类病原体也会引起鱼类疾病（Waltzek et al.，2014）。

EHNV 和 ECV 的基因组大约为 127 kb①(Jancovich et al.，2010；Mavian et al.，2012)，这些 dsDNA 呈环状排列并且末端冗余(Jancovich et al.，2012)。基因组中鸟嘌呤/胞嘧啶(G/C)的含量为 54%～55%，约有 25%的胞嘧啶发生甲基化(Chinchar et al.，2011)。EHNV 的基因组与 ECV 的欧洲六须鲇病毒(European sheatfish virus，ESV)分离株具有 88%的核苷酸序列同源性，并且基因组结构非常相似。这些不同但又密切相关的病毒种具有大约 100 个相似基因；然而，许多预测蛋白的功能尚不明晰（Grayfer et al.，2012）。分子进化分析表明，所有蛙病毒可能是由一个鱼类祖先病毒在相对较近的宿主间变换后进化而来的（Jancovich et al.，2010）。在此基础上，可能会出现或发现新的病毒，今后 EHNV 和 ECV 的诊断标准可能需要考虑全基因组序列数据。

4.1.2 病毒结构与复制

EHNV 和 ECV 的衣壳直径为 150～160 nm，呈所有虹彩病毒共有的二十面体对称(Ahne et al.，1989；Eaton et al.，1991)。主要衣壳蛋白(major capsid protein，MCP)占复合病毒粒子的 40%，该病毒粒子由 36 种具有结构性和酶活性的蛋白质组成(Chinchar et al.，2011)。一个外部的脂质和糖蛋白囊膜对病毒的传染性不是必需的，但似乎可以增强其传染性（Ariel et al.，1995；Jancovich et al.，2012)。已有文献综述了模式种蛙病毒 3 型（Frog virus 3，FV3)独特的细胞核和细胞质复制策略(Chinchar et al.，2011)。早期、延迟早期和晚期的基因可协调表达（Teng et al.，2008)。病毒基因组的复制最初发生在细胞核中，随后在形态不同的病毒发生基质的细胞质中形成基因组共聚体拷贝和结构蛋白，用以组装病毒颗粒(Eaton et al.，1991)。EHNV 通过细胞膜出芽释放出来，并在细胞膜上获得外包膜，从而产生直径达 200 nm 的病毒颗粒。或者，通过细胞裂解释放的裸露病毒颗粒可聚集在细胞质的次晶阵列中(Eaton et al.，1991)。

4.1.3 传播方式

需要进一步研究 EHNV 和 ECV 的传播来解释疾病反复暴发的原因。这两种病毒尚未有垂直传播的报道(Whittington et al.，2010)。EHNV 和 ECV 的水平传播是通过饮用水或摄食受感染鱼的组织而发生的（Langdon，1989；

① 1 kb=1 000 bp。

Jensen et al., 2009; Gobbo et al., 2010, Jensen et al., 2011; Becker et al., 2013; Leimbach et al., 2014)。EHN在虹鳟中的复发是由于野生河鲈的再感染(Whittington et al., 1996; Marsh et al., 2002),并且ESV可能持续存在于看似健康的鱼中(Ahne et al., 1991)。

EHNV已经从养殖的虹鳟种群中蔓延开来,这些鱼种的疾病呈低发病率活跃状态,但实际上并不明显(Whittington et al., 1999)。亚临床感染的鱼可能是该病毒的一个来源。EHNV是从自然疾病暴发后幸存下来的看似健康的河鲈中分离出来的,有人推测这种河鲈可能是病毒传染源,尽管其感染的持续时间尚不清楚(Langdon and Humphrey, 1987)。从实验攻毒后幸存的看似健康的鱼中重新分离出的感染性病毒支持了这一推测(Langdon, 1989; Jensen et al., 2009, 2011; Gobbo et al., 2010; Becker et al., 2013; Leimbach et al., 2014)。这些研究提供的证据表明,除了表现出疾病的宿主以外的那些宿主可能也是EHNV和ECV的来源。尽管蛙病毒宿主范围非常广泛并具有类间传播的能力,但目前尚无证据表明存在两栖类宿主病毒贮主(Chinchar and Waltzek, 2014)。

EHNV也可能通过病媒传播或在宿主外持久性存在,因为在某些条件下EHNV和类似蛙病毒在环境中非常稳定。EHNV在15 ℃时可保持感染效价97 d,而在4 ℃时可保持300 d。该病毒抗干燥,可在15 ℃下干燥113 d后仍具有传染性(Langdon, 1989),并在冷冻鱼中保持传染性至少2年(Whittington et al., 1996)。一种亲缘蛙病毒——饰纹汀蛙虹彩病毒(*Bohle iridovirus*, BIV),其传染性在44 ℃时降低,在52 ℃时进一步降低,但要完全丧失传染性则需要在58 ℃下处理30 min(La Fauce et al., 2012)。

有证据表明,EHNV已经通过人类活动传播,包括受感染虹鳟苗种在养殖场之间的转运(Langdon et al., 1988; Whittington et al., 1994, 1999)。随着新南威尔士州和澳大利亚首都地区马兰比吉河(Murrumbidgee River)及南澳大利亚墨累河(Murray River)等内河流域中EHNV的逐渐蔓延,在河鲈中已经发生了流行性造血器官坏死病疫情(Whittington et al., 2010)。这可能是在休闲垂钓中通过非法转移河鲈、使用冷冻河鲈诱饵或使用污染的船只和垂钓设备等病媒引起的。鸟类也可能通过在摄食感染EHNV的死鱼后几个小时内反哺食物而传播EHNV (Whittington et al., 1996)。

4.1.4 地理分布

EHNV在澳大利亚被发现,而ECV仅限于欧洲大陆(Whittington et al., 2010)。在澳大利亚境内,EHNV随时间呈不连续分布,且遍及澳大利亚东南部的淡水湖泊、河流和水库。其地理范围并未包含所有已知的易感宿主种范围(Whittington et al., 2010)。只有少数虹鳟养殖场受到感染,这些养殖场仅限于肖尔黑文河(Shoalhaven River)和马兰比吉河两个流域(Langdon et al., 1988, Whittington et al., 1999)。目前尚未对澳大利亚东南部的野生鳟鱼种群进行感染情况调查(Whittington et al., 1999)。同样,由于并非所有养殖场都进行了检测,所以 EHNV 在养殖场层面的流行情况也不清楚。根据对塔斯马尼亚和西澳大利亚鲑科鱼类的定期监测,尚未发现这种病毒。

首次报道的由 ECV(六须鲇分离株)引起的疾病暴发发生在来自东南欧的欧洲六须鲇鱼苗中,这些鱼苗在德国暖水域的一个水产养殖场饲养(Ahne et al., 1989)。随后,该疾病陆续在法国、意大利、匈牙利和波兰的鲇中发现。意大利水产养殖的鲇(Bovo et al., 1993)、匈牙利的云斑鮰(*A. nebulosus*)(Juhasz et al., 2013)和波兰的六须鲇(Borzym et al., 2015)中都暴发了疫情。在法国,天然湖泊中各种大小的鲇在夏季都受到影响。2007 年在法国发现的病原与 15 年前导致野生鲇患病的病原非常相似(Bigarre et al., 2008)。

当易感宿主存在于新的适宜环境中时,EHNV 或 ECV 就会传播到新区域(Peeler et al., 2009)。国际贸易以及观赏鱼、爬行动物、两栖动物和钓饵鱼的流通是造成某些蛙病毒传播的原因,并对原始宿主种的种群造成了严重影响(Jancovich et al., 2005; Whittington and Chong, 2007)。因其具有广泛的宿主范围和适应环境的能力强,EHNV 被 OIE(2016c)列入名录,旨在降低其对水产养殖和野生鱼类减少的进一步影响。监测和疾病控制区在疾病管控中发挥着重要作用。有关 EHNV 的国际传播尚未有文献报道,Vesely 等(2011)进行的一项调查指出,在欧洲进口的观赏鱼中未检测到蛙病毒。

4.1.5 病原对鱼类生产的影响

疾病在自然环境下的野生鱼类和水产养殖鱼类中反复暴发(Ahne et al., 1991; Whittington et al., 2010)。然而,EHNV 不是野生河鲈大规模死亡的主要原因,所有鱼龄段都可能感染 EHNV,但在疫病区,死亡仅限于种苗和幼鱼(Langdon et al., 1986; Langdon and Humphrey, 1987; Whittington et al., 1996),这种鱼类种群的减少对休闲渔业产生了影响(Whittington et al., 2010)。与野生河鲈高达 95%的死亡率相比,该疾病对养殖虹鳟的影响不到养殖总量的 5%(Whittington et al., 2010)。

由 ECV 引起的疾病也很少发生,但在个别水产养殖设施中可以反复发生。六须鲇的自然感染易造成种苗 100%的死亡和较大龄鱼 10%~30%的死亡(Ahne et al., 1991; Borzym et al., 2015)。野生鲇疾病的暴发导致了法国湖泊中六须鲇的大量死亡,而其他鱼类则没有受到影响(Bigarre et al., 2008)。但在少数养殖鲇病例中报道了包括亲鱼在内的疾病感染(Bovo et al., 1993; Juhasz et al., 2013)。由于遵守了欧盟(EU)法规和世界动物卫生组织的监测和报告指南(OIE, 2010),在欧洲尚未发现EHNV,偶尔暴发的ECV也都有完好的记录。

EHNV或ECV自然感染引起的疾病目前仅在有限的鱼种中有相关信息,而更多的鱼类在实验感染中表现出易感(Langdon, 1989; Jensen et al., 2009, 2011; Gobbo et al., 2010; Becker et al., 2013; Leimbach et al., 2014)。因此,这些病原有可能对野生鱼类、生态系统和新兴水产养殖业产生严重影响。值得注意的是,由于宿主、环境因素和病原三者间相互作用的复杂性,此类实验性试验无法评估自然疾病暴发的可能性或影响。

4.2 诊断

可使用苏木精和伊红染色标准组织学技术及病毒确认来诊断由EHNV或ECV引起的疾病。可利用针对EHNV开发的多克隆抗体,通过固定组织切片的免疫组织化学或免疫荧光染色来研究病毒与细胞病理学的关系(Reddacliff and Whittington, 1996; Bigarre et al., 2008; Jensen et al., 2009)。这些抗体可从EHNV的OIE参考实验室获得,可与所有蛙病毒属的免疫显性主要衣壳蛋白产生交叉反应,因此需要进行分子鉴定以确定病毒种类(Hyatt et al., 2000, OIE, 2016b)。电子显微镜可以检测组织中存在的虹彩病毒,但无法区分EHNV、ECV及其他蛙病毒属(Ahne et al., 1998)。病毒的检测可以应用于临床病例的组织。这些检测包括细胞培养中的病毒分离,使用免疫学技术[如酶联免疫吸附试验(ELISA)]进行抗原检测,最常见的是使用聚合酶链反应(PCR)检测特定核酸序列(Whittington et al., 2010; OIE, 2016b)。

为了防止病原体扩散,OIE通过EHNV而不是临床疾病或病理变化的发生来识别感染。具有最高验证水平的灵敏实验室检测方案可详见用于检测亚临床感染的OIE《水生动物诊断检测指南》(OIE, 2016b)。理想情况下,这些测试应在国际质量控制标准认可的实验室进行(ISO, 2005)。适于检测的样品有内脏,如来自个体鱼的肾脏、肝脏和脾脏等(Whittington and Steiner, 1993),精液和卵巢液不适合作样品。用于分子测试的组织,可以通过部分自动化的珠磨程序来快速破坏组织(这与病毒分离和PCR相容)以及使用磁珠来进行核酸纯化(Rimmer et al., 2012)

在进行无感染监测时,需要一个统计上有效的体系来计算样本量和筛选方法(OIE, 2010)。濒死的鱼用于疾病的诊断,同时建议采用偏倚或有针对性的采样进行监测。在虹鳟养殖场通过冷冻死亡的幼鱼进行后期检测以实现监测(Whittington et al., 1999)。

病毒分离是鉴定感染性病毒的唯一方法。在15~22 ℃,EHNV可从几种鱼类细胞系中分离，包括 FHM 细胞系、RTG 细胞系、BF－2 和 CHSE－214 (Langdon et al., 1986; Crane et al., 2005; OIE, 2016b)。类似地，ECV 和 ESV 可以通过使用 BF－2 或鲤鱼上皮瘤细胞、FHM 或斑点叉尾鮰卵巢细胞 (CCO)分离(Ahne et al., 1989; Pozet et al., 1992; Ariel et al., 2009)。对可以传代并引起致细胞病变效应的病毒的确认，需要免疫荧光染色(Bigarre et al., 2008)或分子鉴定(通常是 MCP 基因)。

PCR 检测最常靶向 MCP 基因(Marsh et al., 2002; Pallister et al., 2007)。在 OIE《水生动物诊断检测指南》中描述的常规 PCR 检测提供了一个可以通过限制性内切酶消化区分 EHNV、ECV 和其他蛙病毒的扩增子(Marsh et al., 2002; OIE, 2016b)。Jaramillo 等(2012)建立的实时定量 PCR(qPCR)方法，操作方便，并且有数据证实其具有诊断灵敏性和特异性。与 EHNV 反应的多克隆抗体可用于组织匀浆样品上的抗原捕获 ELISA，以作为一种低成本的方法来鉴定蛙病毒抗原；用于 EHNV 的抗体试剂和对照品可由 OIE 参考实验室提供(Whittington and Steiner, 1993)。

MCP 基因在蛙病毒属中是保守的，EHNV、ECV 和 BIV 的 MCP 序列同源性大于 97.8% (Hossain et al., 2008; Jancovich et al., 2012)。分离毒株在种水平的鉴定，需要 DNA 测序、限制性内切酶消化(Marsh et al., 2002)或物种特异性 PCR 分析(Pallister et al., 2007)。核苷酸序列分析提高了诊断水平，因为分离株在流行病学事件中聚于一簇，从而为疾病控制和预防提供指导(Jancovich et al., 2010)。除 MCP 外，还可以使用 DNA 聚合酶和神经丝三联体 H1 样蛋白基因的核苷酸序列(Holopainen et al., 2009)。今后，全基因组测序可能变得更易进行，会取代候选基因分析用于深入的蛙病毒流行病学研究(Epstein and Storfer, 2015)。

4.2.1 临床体征

由 EHNV 和 ECV 引起的疾病临床症状是非特异性的，包括食欲不振、活力下降、共济失调和多灶性出血，特别是在造血组织中(Whittington et al., 2010)。野生河鲈中的 EHN 暴发通常仅表现为大规模死亡(Langdon and Humphrey, 1987)，而实验感染鱼的体色发黑，濒死时游动失衡(Langdon, 1989)。因为患病率低，虹鳟中的自然疾病症状可能并不明显，而在实验环境中，受感染的虹鳟体色发暗、共济失调和食欲减退(Reddacliff and Whittington, 1996)。

同样，野生鲇中的蛙病毒暴发也仅表现为明显的死亡率(Bigarre et al., 2008)。养殖设施中的患病鲇表现出水肿、鳍基部和内脏出现瘀点、腹水及鳃苍白，而食欲不振在其他症状出现(Pozet et al., 1992)。在实验感染中，垂死的鲇呈现垂直姿势，头部露出水面，皮肤和鳍基部有瘀点(Gobbo et al., 2010)。患病六须鲇有时出现继发性细菌感染的并发症[嗜水气单胞菌(*Aeromonas hydrophila*)],且死亡和垂死的鱼出现皮肤溃疡(Ahne et al., 1989, 1991)。在实验感染中未发现明显的疾病临床症状,但在尸检时发现下颌区和鳍部存在弥漫性皮下出血(Leimbach et al., 2014)。

4.3 病理学

病变归因于病毒的血管内皮和造血向性,并且在所有易感宿主中都是相似的。EHNV 和 ECV 所引起的病理病变非常相似,出现广泛的系统性坏死,特别是在造血组织中(Whittington et al., 2010)。大体上,还包括鳍基部出现瘀点出血、血清性腹腔液过多、肾和脾肿胀(Langdon and Humphrey, 1987; Reddacliff and Whittington, 1996)。通常,坏死灶集中在小血管周围。河鲈肝坏死部位明显可见白色或奶油色病灶。显微镜检中,在肝坏死灶边缘,肝细胞内可见嗜碱性胞浆内包涵体(图 4.2)。这些包涵体很稀少并且难以在肾和脾中观察到。免疫组织化学染色发现 EHNV 抗原在坏死区及血管内皮和血管内单个循环白细胞中广泛分布(图 4.3)。

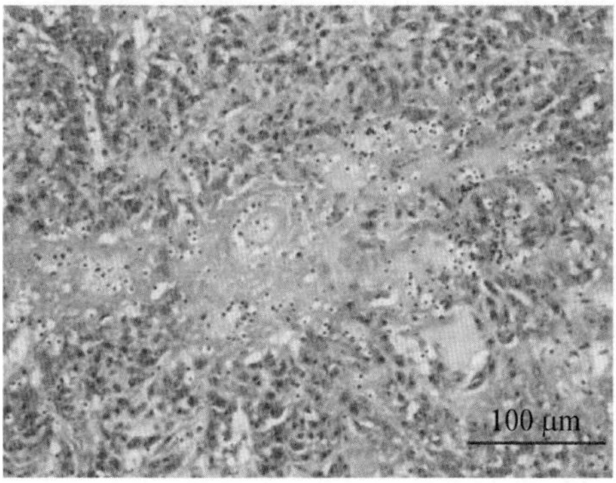

图 4.2 河鲈肝的晚期局灶性坏死。肝细胞广泛分离,其中许多处于变性的早期阶段并且含有两亲性胞质内包涵体。具有这种类型病变的肝脏颜色暗淡,并具有多个浅白色斑点。苏木精-伊红染色

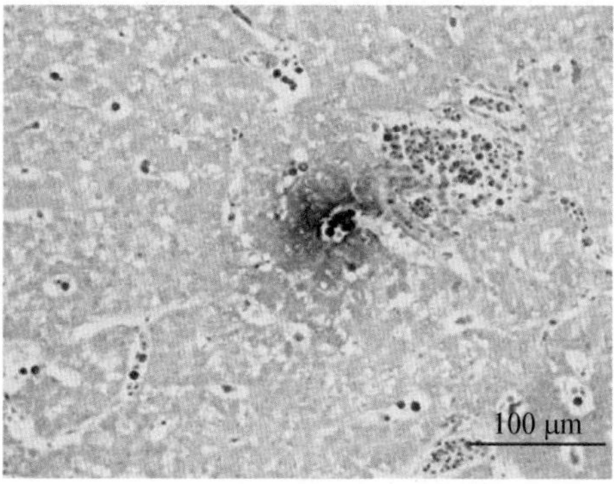

图 4.3 河鲈肝的早期局灶性坏死。在邻近大血管的肝细胞变性和坏死灶中可见 EHNV(流行性造血器官坏死病病毒)抗原的阳性染色。血管和血窦内的单个白细胞及内皮细胞也有染色。均为免疫组织化学染色

实验感染 ECV 分离株的六须鲇和鲇的显微病变特征表现为脾脏和肾脏的造血组织坏死、肾间质组织和肾小管中有退化和坏死细胞(Ogawa et al., 1990; Pozet et al., 1992)。这两种鱼的血管内皮细胞也明显普遍受损,导致内脏器官弥漫性充血和出血。对于 EHNV,可以使用间接荧光抗体或免疫组织化学染色证实病毒抗原与病变相关(图 4.4 和图 4.5)。

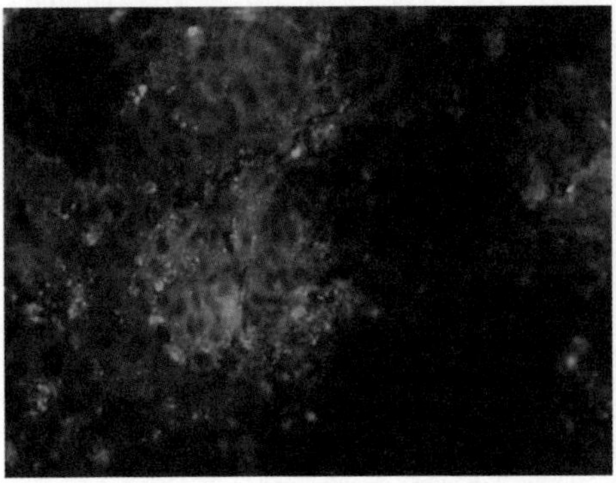

图 4.4 感染了 ECV(欧鲇病毒)的黑鲴肾脏。细胞内有与病毒抗原相关的明亮苹果绿荧光(间接荧光抗体染色)。图片由意大利 Istituto Zooprofilattico Sperimentale delle Venezie 的 Gobbo F.和 Bovo G.提供

图 4.5 感染了 ECV(欧鲇病毒)的黑鲴肠道。在黏膜下层中存在大量含有 ECV 抗原的细胞(免疫组织化学染色)。图片由意大利 Istituto Zooprofilattico Sperimentale delle Venezie 的 Gobbo F.和 Bovo G.提供

血管内常有散在的个别坏死细胞和退化的血管内皮细胞。其他病变包括弥漫性脾坏死、胃肠道上皮坏死、心房小梁坏死、鳃上皮增生和坏死、胰腺局灶性坏死、鱼鳔水肿和坏死以及溃疡性皮炎(Langdon and Humphrey，1987；Reddacliff and Whittington，1996)。在存活鱼中可能出现血管周围单核细胞浸润(Becker et al.，2013)。

仅在实验感染中发现对 EHNV 或 ECV 易感的鱼类包括几种澳大利亚本土淡水鱼类以及白斑狗鱼(*Esox lucius*)和梭鲈(*Sander lucioperca*)。实验感染鱼的病理与自然感染鱼相似(Jensen et al.，2009，2011；Becker et al.，2013)。

4.3.1 病理生理学

目前，尚未对 EHNV 和 ECV 感染的发病机制进行详细研究。疾病的自然暴发发生在幼鱼大量生长的气候温暖时期(Langdon et al.，1988；Whittington et al.，1994；Bigarre et al.，2008，Gobbo et al.，2010)。实验研究证实，病毒的传播和致病性受水温影响(Whittington and Reddacliff，1995；Jensen et al.，2009，2011；Gobbo et al.，2010；Leimbach et al.，2014)。河鲈在 12～21 ℃时发生死亡，12 ℃以下不发生疾病，水温升高不会导致这些鱼患病。水温在 10 ℃和 15 ℃时，虹鳟对 EHNV 浸泡攻毒感染不敏感，20 ℃时出现中度程度死亡(14%～24%)(Langdon，1989；Whittington and Reddacliff，1995；Ariel and Jensen，2009)。潜伏期与水温呈负相关：腹腔注射攻毒后，河鲈在 12～18 ℃时潜伏期为 10～28 d，而在

19～21 ℃时为 10～11 d；虹鳟在 8～10 ℃时潜伏期为 14～32 d，而在 19～21 ℃时为 3～10 d（Whittington and Reddacliff，1995）。研究发现，EHNV 浸泡感染狗鱼，12 ℃水温下感染 7 d 后病毒载量可达峰值，22 ℃时则仅 3 d 后就达到病毒载量峰值，这证实了 EHNV 的温度依赖性复制（Holopainen et al.，2011）。

河鲈的宿主易感性因遗传和养殖条件的不同可能存在差异。澳大利亚河鲈种群对浸泡和腹腔注射攻毒都高度易感，20 ℃时两种感染途径的死亡率均为 100%，10 ℃时腹腔注射引起的死亡率为 50%（Langdon，1989；Whittington and Reddacliff，1995）。在 EHNV 浸泡感染欧洲河鲈的试验中，水温在 10 ℃、15 ℃或 20 ℃时均未诱发死亡，而在 15 ℃下的腹腔注射感染则诱发了 47%～80%的死亡率，在 20 ℃时死亡率则为 28%～42%（Ariel and Jensen，2009）。在另一项 EHNV 浸泡攻毒试验中，河鲈幼鱼在 11～13 ℃下的死亡率为 16%，20～22 ℃时为 24%（Borzym and Maj-Paluch，2015）。实验感染表明，EHNV 与不同的 ECV 和 ESV 分离株在同一宿主鱼种中引起的疾病结果不同（Jensen et al.，2009；Gobbo et al.，2010；Leimbach et al.，2014）。从六须鲇中分离的 ESV 对其具有高致病性，在 15 ℃时的致死率为 100%，而 ECV 对六须鲇的致死率仅为 8%（Leimbach et al.，2014）。相反地，分离自欧鲇的 ECV 对欧鲇具有致病性，在 25 ℃时的死亡率为 82%，15 ℃时的死亡率为 30%，但该分离株在这些温度下仅诱导了六须鲇 5%或 1%的死亡率（Gobbo et al.，2010）。这些发现表明，尽管这两种 ECV 分离株在遗传上相似，但它们在不同宿主种中的表现不同。引发感染所需的病毒剂量取决于宿主种类。例如，澳大利亚养殖的河鲈对 EHNV 高度易感，浸泡暴露感染的剂量低至 0.08 $TCID_{50}$/mL，而虹鳟则耐浸泡感染（$10^{2.2}$ $TCID_{50}$/mL），必须通过腹腔注射感染才能引发疾病（Whittington and Reddacliff，1995）。这些研究结果与自然界中这些物种的不同临床表现相符。

自然感染 EHNV 和 ECV 的鱼类胃肠道病变突出（Ogawa et al.，1990；Pozet et al.，1992）（图 4.5）符合经口感染途径（Reddacliff and Whittington，1996）。循环白细胞和血管内皮坏死反映出病毒血症可能导致病毒扩散。这些病理过程导致小到大范围的组织坏死，并表现出大体病理变化（Langdon and Humphrey，1987；Langdon et al.，1988；Whittington et al.，1994；Reddacliff and Whittington，1996）。

目前，除了河鲈或虹鳟感染 EHNV，以及欧鲇或六须鲇感染 ECV 和 ESV 外，没有证据表明其他鱼种会自然感染这些病毒。然而，许多鱼类易受实验感染。蛙病毒属遍布全球，一些主要感染两栖动物的病毒在自然条件下或实验条件下能够进行跨类（包括对鱼类）传播（Moody and Owens，1994；Ariel and Owens，1997；Duffus et al.，2015）。例如，从同域分布、自由生长、临床感染的三刺鱼(*Gasterosteus aculeatus*)和红腿蛙(*Rana aurora*)蝌蚪中分离出了看似相同的蛙病毒(Mao et al., 1999)。然而,很少有两栖动物相关蛙病毒引起的鱼类疾病报道(Waltzek et al., 2014),EHNV 和 ECV 在非鱼类宿主范围中的数据也很有限。欧洲林蛙(*R. temporaria*)对鱼类蛙病毒和梭鲈虹彩病毒(PPIV)易感,但对 EHNV 和 ECV 则不易感(Bayley et al., 2013)。因此,实验证据及野外观察为贮主和偶发宿主交换的蛙病毒感染模型提供了支持证据和反对证据。

4.4 防控策略

目前对于由 EHNV 和 ECV 引起的疾病尚无有效的治疗方法,因此,根据通用且最佳的实践建议制定的生物安全措施是疾病控制的重点(OIE, 2009)。这些措施包括进口检定的无感染鱼类、消毒水体和控制疫区病媒。对入境鱼类的检疫期应不少于疾病的潜伏期,而最易感染鱼类的潜伏期可达 32 d (Whittington and Reddacliff, 1995)。可以进行积极的监测以确定无疫病感染区域,从而促进活鱼贸易(OIE, 2010)。针对包括 EHNV 在内的病原体,欧盟规定了水产养殖设施的病原目标性监测方法(Commission of the European Communities, 2001)。可以通过实验室检测对野生鱼类的异常死亡进行被动监测,以确定 EHN 或 ECV 疫情。

4.4.1 接种疫苗

目前还没有针对 EHNV 的商品化疫苗,而且对替代性预防措施的有效性研究也很有限。疫苗开发是可行的,并已商用于肿大细胞病毒属(*Megalocytivirus*)的虹彩病毒(Kurita and Nakajima, 2012)。目前已有检测 EHNV 特异性抗体反应的方法,但需要进一步研究免疫反应(Whittington and Reddacliff, 1995; Whittington et al., 1999)。

4.4.2 控制或清除贮主

应该防止接触看似健康的带毒鱼,包括疾病暴发后的幸存者和潜在的带毒鱼种。越来越多的证据表明不同分类类别中的许多蛙病毒可共存于相同贮主,因此水产养殖生物安全需要严格排除养殖场中的非养殖物种。

4.4.3 传播的环境调控

该疾病的发病温度范围广限制了温度控制的使用。在较冷的水域,通过行为自我选择可以降低死亡率,如河鲈成鱼,在热分层环境中可以避免这种疾病,而幼鱼由于在较温暖的浅水区域中摄食而死于该疾病(Whittington and

Reddacliff, 1995)。

对供水进行持续消毒的重要性已在EHNV感染模型中得到验证支持(Langdon, 1989; Ariel and Jensen, 2009),在该模型中,即使没有直接接触,河鲈和虹鳟之间也发生了传播。通过对水产养殖设施进行充分的清洁和消毒,可以防止疾病暴发带来的残留污染。OIE (2009)提供的消毒通用指南可以使用针对EHNV和相关蛙病毒的消毒效果数据归结为次氯酸钠200 mg/L;70%乙醇;150 mg/L氯已定(0.75%Nolvasan®)处理1 min;200 mg/L单过硫酸氢钾复合盐(1% Virkon®);或60 ℃加热处理15 min(Langdon, 1989; Bryan et al., 2009, La Fauce et al., 2012)。

4.4.4 动物养殖管理

优化水产养殖管理和维护野生鱼类的健康水道可以降低疾病在疫区的影响。降低养殖密度和改善水质降低了养殖虹鳟的死亡率(Whittington et al., 1994, 1999)。诊断和控制细菌与外部寄生虫的感染可能会改善由EHNV和ECV引起的疾病后果(Ahne et al., 1991; Whittington et al., 2010)。

4.5 总结与研究展望

EHNV和ECV可在一些对水产养殖和野生渔业很重要的淡水鱼类中引起罕见但严重的疾病。利用常规组织病理学方法和病原体特异性实验室检测很容易诊断这种疾病。由于缺乏有效的治疗和预防措施,病毒监测和有效的生物防疫是控制疾病的主要手段。因此,这些疾病在许多司法管辖区都应呈报,OIE将EHNV列入名单以监管疫病控制区的鱼类贸易。国际参考实验室、推荐和验证的实验室测试及试剂、对照品和推荐的实验操作规程的国际可用性等促进了这一工作(OIE, 2016b)。检测可以高通量形式应用,适用于在种群水平上进行免于感染的确证。

实验研究表明,EHNV和ECV可能感染许多鱼类,包括那些对食品生产和物种保护有重要意义的鱼类。此外,许多蛙病毒的广宿主性表明可能存在尚未识别的EHNV和ECV的贮主和致病宿主。因此,这些病原体对水产养殖和物种保护构成了威胁。今后需要进行进一步研究,以开发更好的疾病控制策略,包括有效的疫苗。需要应用分子技术来追踪疾病暴发的来源、传播途径以及区域和全球传播的机制。鱼类、两栖类和爬行类的病毒贮主存在的可能性有待进一步研究。目前迫切需要展开公众教育,使他们认识到通过观赏物种和钓饵鱼的无限制贸易对低等脊椎动物造成的疾病威胁(Jancovich et al., 2005; Whittington and Chong, 2007)。

参考文献

[1] Ahne, W., Schlotfeldt, H.J. and Thomsen, I. (1989) Fish viruses: isolation of an icosahedral cytoplasmic deoxyribovirus from sheatfish (*Silurus glanis*). *Zentralblatt für Veterinärmedizin, Reihe B/Journal of Veterinary Medicine, Series B* 36, 333–336.

[2] Ahne, W., Schlotfeldt, H.J. and Ogawa, M. (1991) Iridovirus infection of adult sheatfish (*Silurus glanis*). *Bulletin of the European Association of Fish Pathologists* 11, 97–98.

[3] Ahne, W., Bearzotti, M., Bremont, M. and Essbauer, S. (1998) Comparison of European systemic piscine and amphibian iridoviruses with *Epizootic haematopoietic necrosis virus and Frog virus 3*. *Journal of Veterinary Medicine, Series B: Infectious Diseases and Veterinary Public Health* [now *Zoonoses and Public Health*] 45, 373–383.

[4] Ariel, E. and Jensen, B.B. (2009) Challenge studies of European stocks of redfin perch, *Perca fluviatilis* L., and rainbow trout, *Oncorhynchus mykiss* (Walbaum), with epizootic haematopoietic necrosis virus. *Journal of Fish Diseases* 32, 1017–1025.

[5] Ariel, E. and Owens, L. (1997) Epizootic mortalities in tilapia *Oreochromis mossambicus*. *Diseases of Aquatic Organisms* 29, 1–6.

[6] Ariel, E., Owens, L. and Moody, N.J.G. (1995) A barramundi bioassay for iridovirus refractory to cell culture. In: Shariff, R.P., Subasinghe, R.P. and Arthur, J.R. (eds) *Diseases in Asian Aquaculture II. Fish Health Section*. Asian Fisheries Society, Manila, pp. 355–367.

[7] Ariel, E., Nicolajsen, N., Christophersen, M.B., Holopainen, R., Tapiovaara, H. and Jensen, B.B. (2009) Propagation and isolation of ranaviruses in cell culture. *Aquaculture* 294, 159–164.

[8] Ariel, E., Holopainen, R., Olesen, N.J. and Tapiovaara, H. (2010) Comparative study of ranavirus isolates from cod (*Gadus morhua*) and turbot (*Psetta maxima*) with reference to other ranaviruses. *Archives of Virology* 155, 1261–1271.

[9] Bayley, A.E., Hill, B.J. and Feist, S.W. (2013) Susceptibility of the European common frog *Rana temporaria* to a panel of ranavirus isolates from fish and amphibian hosts. *Diseases of Aquatic Organisms* 103, 171–183.

[10] Becker, J.A., Tweedie, A., Gilligan, D., Asmus, M. and Whittington, R.J. (2013) Experimental infection of Australian freshwater fish with *Epizootic haematopoietic necrosis virus* (EHNV). *Journal of Aquatic Animal Health* 25, 66–76.

[11] Bigarre, L., Cabon, J., Baud, M., Pozet, F. and Castric, J. (2008) Ranaviruses associated with high mortalities in catfish in France. *Bulletin of the European Association of Fish Pathologists* 28, 163–168.

[12] Borzym, E. and Maj-Paluch, J. (2015) Experimental infection with epizootic haematopoietic necrosis virus (EHNV) of rainbow trout (*Oncorhynchus mykiss* Walbaum) and European perch (*Perca fluviatilis* L.). *Bulletin of the Veterinary Institute in Pulawy* 59, 473–477.

[13] Borzym, E., Karpinska, T.A. and Reichert, M. (2015) Outbreak of ranavirus infection in sheatfish, *Silurus glanis* (L.), in Poland. *Polish Journal of Veterinary Sciences* 18, 607–611.

[14] Bovo, G., Comuzzi, M., De Mas, S., Ceschia, G., Giorgetti, G. *et al.* (1993) Isolation of an irido-like viral agent from breeding cat fish (*Ictalurus melas*). *Bollettino Societa Italiana di Patalogia Ittica* 11, 3–10.

[15] Brenes, R., Miller, D.L., Waltzek, T.B., Wilkes, R.P., Tucker, J.L. *et al.* (2014) Susceptibility of fish and turtles to three ranaviruses isolated from different ectothermic vertebrate classes. *Journal of Aquatic Animal Health* 26, 118 - 126.

[16] Bryan, L.K., Baldwin, C.A., Gray, M.J. and Miller, D.L. (2009) Efficacy of select disinfectants at inactivating ranavirus. *Diseases of Aquatic Organisms* 84, 89 - 94.

[17] Chinchar, V.G. and Waltzek, T.B. (2014) Ranaviruses: Not Just for frogs. *PLoS Pathogens* 10(1): e1003850.

[18] Chinchar, V.G., Yu, K.H. and Jancovich, J.K. (2011) The molecular biology of Frog Virus 3 and other iridoviruses infecting cold-blooded vertebrates. *Viruses* 3, 1959 - 1985.

[19] Commission of the European Communities (2001) Commission Decision of 22 February 2001. Laying down the sampling plans and diagnostic methods for the detection and confirmation of certain fish diseases and repealing Decision 92/532/EEC. 2001/183/EC. *Official Journal of the European Communities* L67, 44, 65 - 76.

[20] Crane, M.S.J., Young, J. and Williams, L. (2005) Epizootic haematopoietic necrosis virus (EHNV): growth in fish cell lines at different temperatures. *Bulletin of the European Association of Fish Pathologists* 25, 228 - 231.

[21] Duffus, A.L., Waltzek, T.B., Stöhr, A.C., Allender, M.C., Gotesman, M. *et al.* (2015) Distribution and host range of ranaviruses In: Gray, M.J. and Chincahr, V.G. (eds) *Ranaviruses: Lethal Pathogens of Ectothermic Vertebrates*. Springer, Cham, Switzerland, pp. 9 - 57.

[22] Eaton, B.T., Hyatt, A.D. and Hengstberger, S. (1991) Epizootic haematopoietic necrosis virus-purification and classification. *Journal of Fish Diseases* 14, 157 - 169.

[23] Epstein, B. and Storfer, A. (2015) Comparative genomics of an emerging amphibian virus. *G3: Genes, Genomes, Genetics* 6, 15 - 27.

[24] Gobbo, F., Cappellozza, E., Pastore, M.R. and Bovo, G. (2010) Susceptibility of black bullhead *Ameiurus melas* to a panel of ranavirus isolates. *Diseases of Aquatic Organisms* 90, 167 - 174.

[25] Gray, M.J., Miller, D.L. and Hoverman, J.T. (2009) Ecology and pathology of amphibian ranaviruses. *Diseases of Aquatic Organisms* 87, 243 - 266.

[26] Grayfer, L., Andino, F.D., Chen, G.C., Chinchar, G.V. and Robert, J. (2012) Immune evasion strategies of ranaviruses and innate immune responses to these emerging pathogens. *Viruses* 4, 1075 - 1092.

[27] He, J.G., Deng, M., Weng, S.P., Li, Z., Zhou, S.Y. et al. (2001) Complete genome analysis of the mandarin fish infectious spleen and kidney necrosis iridovirus. *Virology* 291, 126 - 139.

[28] He, J.G., Lu, L., Deng, M., He, H.H., Weng, S.P. *et al.* (2002) Sequence analysis of the complete genome of an iridovirus isolated from the tiger frog. *Virology* 292, 185 - 197.

[29] Holopainen, R., Ohlemeyer, S., Schuetze, H., Bergmann, S.M. and Tapiovaara, H. (2009) Ranavirus phylogeny and differentiation based on major capsid protein, DNA polymerase and neurofilament triplet H1-like protein genes. *Diseases of Aquatic Organisms* 85, 81 - 91.

[30] Holopainen, R., Honkanen, J., Jensen, B.B., Ariel, E. and Tapiovaara, H. (2011) Quantitation of ranaviruses in cell culture and tissue samples. *Journal of Virological Methods* 171, 225 - 233.

[31] Hossain, M., Song, J.Y., Kitamura, S.I., Jung, S.J. and Oh, M.J. (2008) Phylogenetic analysis of lymphocystis disease virus from tropical ornamental fish species based on a major capsid protein gene. *Journal of Fish Diseases* 31, 473 - 479.

[32] Huang, Y., Huang, X., Liu, H., Gong, J., Ouyang, Z. *et al.* (2009) Complete sequence determination of a novel reptile iridovirus isolated from soft-shelled turtle and

evolutionary analysis of Iridoviridae. *BMC Genomics* 10：224.

[33] Hyatt，A.D.，Gould，A.R.，Zupanovic，Z.，Cunningham，A.A.，Hengstberger，S. *et al.*（2000）Comparative studies of piscine and amphibian iridoviruses. *Archives of Virology* 145，301－331.

[34] ISO（2005）ISO/IEC 17025：2005. General requirements for the competence of testing and calibration laboratories. International Organization for Standardization，Geneva，Switzerland. Available at：http：//www.iso.org/iso/catalogue_detail?csnumber=39883（accessed 26 October 2016）.

[35] Jancovich，J.K.，Mao，J.，Chinchar，V.G.，Wyatt，C.，Case，S.T. *et al.*（2003）Genomic sequence of a ranavirus（family *Iridoviridae*）associated with salamander mortalities in North America. *Virology* 316，90－103.

[36] Jancovich，J.K.，Davidson，E.W.，Parameswaran，N.，Mao，J.，Chinchar，V.G. *et al.*（2005）Evidence for emergence of an amphibian iridoviral disease because of human-enhanced spread. *Molecular Ecology* 14，213－224.

[37] Jancovich，J.K.，Bremont，M.，Touchman，J.W. and Jacobs，B.L.（2010）Evidence for multiple recent host species shifts among the ranaviruses（Family *Iridoviridae*）. *Journal of Virology* 84，2636－2647.

[38] Jancovich，J.K.，Chinchar，V.G.，Hyatt，A.，Miyazaki，T.，Williams，T. and Zhang，Q.Y.（2012）Family *Iridoviridae*. In：King，A.M.Q.，Adams，M.J.，Carstens，E.B. and Lefkowitz，E.J.（eds）*Virus Taxonomy: Ninth Report of the International Committee on Taxonomy of Viruses*. Elsevier/Academic Press，San Diego，California，pp. 193－210.

[39] Jancovich，J.K.，Steckler，N.K. and Waltzek，T.B.（2015）Ranavirus taxonomy and phylogeny. In：Gray，M.J. and Chincahr，V.G（eds）*Ranaviruses: Lethal Pathogens of Ectothermic Vertebrates*. Springer，Cham，Switzerland，pp. 59－70.

[40] Jaramillo，D.，Tweedie，A.，Becker，J.A.，Hyatt，A.，Crameri，S. and Whittington，R.J.（2012）A validated quantitative polymerase chain reaction assay for the detection of ranaviruses（Family *Iridoviridae*）in fish tissue and cell cultures，using EHNV as a model. *Aquaculture* 356，186－192.

[41] Jensen，B.B.，Ersboll，A.K. and Ariel，E.（2009）Susceptibility of pike *Esox lucius* to a panel of ranavirus isolates. *Diseases of Aquatic Organisms* 83，169－179.

[42] Jensen，B.B.，Holopainen，R.，Tapiovaara，H. and Ariel，E.（2011）Susceptibility of pike-perch *Sander lucioperca* to a panel of ranavirus isolates. *Aquaculture* 313，24－30.

[43] Juhasz，T.，Woynarovichne，L.M.，Csaba，G.，Farkas，L.S. and Dan，A.（2013）Isolation of ranavirus causing mass mortality in brown bullheads（*Ameiurus nebulosus*）in Hungary. *Magyar Allatorvosok Lapja* 135，763－768.

[44] Kurita，J. and Nakajima，K.（2012）Megalocytiviruses. *Viruses* 4，521－538.

[45] La Fauce，K.，Ariel，E.，Munns，S.，Rush，C. and Owens，L.（2012）Influence of temperature and exposure time on the infectivity of Bohle iridovirus，a ranavirus. *Aquaculture* 354，64－67.

[46] Langdon，J.S.（1989）Experimental transmission and pathogenicity of epizootic haematopoietic necrosis virus（EHNV）in redfin perch，*Perca fluviatilis* L.，and 11 other teleosts. *Journal of Fish Diseases* 12，295－310.

[47] Langdon，J.S. and Humphrey，J.D.（1987）Epizootic haematopoietic necrosis，a new viral disease in redfin perch *Perca fluviatilis* L. in Australia. *Journal of Fish Diseases* 10，289－298.

[48] Langdon，J.S.，Humphrey，J.D.，Williams，L.M.，Hyatt，A.D. and Westbury，H.A.（1986）First virus isolation from Australian fish：an iridovirus-like pathogen from redfin perch，*Perca fluviatilis* L. *Journal of Fish Diseases* 9，263－268.

[49] Langdon，J.S.，Humphrey，J.D. and Williams，L.M.（1988）Outbreaks of an EHNV-

like iridovirus in cultured rainbow trout, *Salmo gairdneri* Richardson, in Australia. *Journal of Fish Diseases* 11, 93 - 96.

[50] Lei, X.Y., Ou, T., Zhu, R.L. and Zhang, Q.Y. (2012) Sequencing and analysis of the complete genome of Rana grylio virus (RGV). *Archives of Virology* 157, 1559 - 1564.

[51] Leimbach, S., Schutze, H. and Bergmann, S.M. (2014) Susceptibility of European sheatfish *Silurus glanis* to a panel of ranaviruses. *Journal of Applied Ichthyology* 30, 93 - 101.

[52] Mao, J. H., Green, D. E., Fellers, G. and Chinchar, V. G. (1999) Molecular characterization of iridoviruses isolated from sympatric amphibians and fish. *Virus Research* 63, 45 - 52.

[53] Marsh, I.B., Whittington, R.J., O'Rourke, B., Hyatt, A.D. and Chisholm, O. (2002) Rapid differentiation of Australian, European and American ranaviruses based on variation in major capsid protein gene sequence. *Molecular and Cellular Probes* 16, 137 - 151.

[54] Mavian, C., López-Bueno, A., Fernández Somalo, M.P., Alcamí, A. and Alejo, A. (2012) Complete genome sequence of the European sheatfish virus. *Journal of Virology* 86, 6365 - 6366.

[55] Moody, N.J.G. and Owens, L. (1994) Experimental demonstration of the pathogenicity of a frog virus, Bohle iridovirus, for a fish species, barramundi *Lates calcarifer*. *Diseases of Aquatic Organisms* 18, 95 - 102.

[56] Ogawa, M., Ahne, W., Fischer-Scherl, T., Hoffmann, R.W. and Schlotfeldt, H.J., (1990) Pathomorphological alterations in sheatfish fry *Silurus glanis* experimentally infected with an iridovirus-like agent. *Diseases of Aquatic Organisms* 9, 187 - 191.

[57] OIE (2009) Chapter 1.1.3. Methods for disinfection of aquaculture establishments. In: *Manual of Diagnostic Tests for Aquatic Animals*, 6th edn. World Organisation for Animal Health, Paris, pp. 31 - 42. Available at: http://web.oie.int/eng/normes/fmanual/1.1.3_DISINFECTION.pdf (accessed 26 October 2016).

[58] OIE (2010) Chapter 1.4. Aquatic animal health surveillance. In: *Aquatic Animal Health Code*, 2010. World Organisation for Animal Health, Paris. Available at: http://web.oie.int/eng/normes/fcode/en_chapitre_1.1.4.pdf (accessed 15 December 2015).

[59] OIE (2015) Chapter 1.2. Criteria for listing aquatic animal diseases. In: *Aquatic Animal Health Code*, 2015. World Organisation for Animal Health, Paris. Available at: http://www.oie.int/fileadmin/Home/eng/Health_standards/aahc/2010/chapitre_criteria_diseases.pdf (accessed 26 October 2016).

[60] OIE (2016a) Chapter 2.1.2 Infection with ranavirus. In: *Manual of Diagnostic Tests for Aquatic Animals*. World Organisation for Animal Health, Paris. Available at: http://www.oie.int/fileadmin/Home/eng/Health_standards/aahm/current/chapitre_ranavirus.pdf (accessed 26 October 2016).

[61] OIE (2016b) Chapter 2.3.1 Epizootic haematopoietic necrosis. In *Manual of Diagnostic Tests for Aquatic Animals* (World Organisation for Animal Health) Available at: http://www.oie.int/fileadmin/Home/eng/Health_standards/aahm/current/chapitre_ehn.pdf (accessed 26 October 2016).

[62] OIE (2016c) Chapter 10.1. Epizootic haematopoietic necrosis. In: World Organisation for Animal Health, *Aquatic Animal Health Code*, 2016. Available at: http://www.oie.int/fileadmin/Home/eng/Health_standards/aahc/current/chapitre_vhs.pdf (accessed 10 December 2015).

[63] Pallister, J., Gould, A., Harrison, D., Hyatt, A., Jancovich, J. and Heine, H. (2007) Development of real-time PCR assays for the detection and differentiation of Australian and European ranaviruses. *Journal of Fish Diseases* 30, 427 - 438.

[64] Peeler, E. J., Afonso, A., Berthe, F. C. J., Brun, E. *et al.* (2009) Epizootic haematopoietic necrosis virus-an assessment of the likelihood of introduction and establishment in England and Wales. *Preventive Veterinary Medicine* 91, 241 - 253.

[65] Pozet, F., Morand, M., Moussa, A., Torhy, C. and De, K.P. (1992) Isolation and preliminary characterization of a pathogenic icosahedral deoxyribovirus from the catfish *Ictalurus melas*. *Diseases of Aquatic Organisms* 14, 35 - 42.

[66] Reddacliff, L.A. and Whittington, R.J. (1996) Pathology of epizootic haematopoietic necrosis virus (EHNV) infection in rainbow trout (*Oncorhynchus mykiss* Walbaum) and redfin perch (*Perca fluviatilis* L). *Journal of Comparative Pathology* 115, 103 - 115.

[67] Rimmer, A.E., Becker, J.A., Tweedie, A. and Whittington, R.J. (2012) Validation of high throughput methods for tissue disruption and nucleic acid extraction for ranaviruses (family *Iridoviridae*). *Aquaculture* 338, 23 - 28.

[68] Song, W.J., Oin, Q.W., Qiu, J., Huang, C.H., Wang, F. and Hew, C.L. (2004) Functional genomics analysis of Singapore grouper iridovirus: complete sequence determination and proteomic analysis. *Journal of Virology* 78, 12576 - 12590.

[69] Tamura, K., Stecher, G., Peterson, D., Filipski, A. and Kumar, S. (2013) MEGA6: Molecular Evolutionary Genetics Analysis Version 6. 0. *Molecular Biology and Evolution* 30, 2725 - 2729.

[70] Tan, W. G., Barkman, T. J., Chinchar, V. G. and Essani, K. (2004) Comparative genomic analyses of Frog virus 3, type species of the genus Ranavirus (family *Iridoviridae*). *Virology* 323, 70 - 84.

[71] Teng, Y., Hou, Z.W., Gong, J., Liu, H., Xie, X.Y. *et al.* (2008) Whole-genome transcriptional profiles of a novel marine fish iridovirus, Singapore grouper iridovirus (SGIV) in virus-infected grouper spleen cell cultures and in orange-spotted grouper, *Epinephulus coioides*. *Virology* 377, 39 - 48.

[72] Tidona, C. A. and Darai, G. (1997) The complete DNA sequence of lymphocystis disease virus. *Virology* 230, 207 - 216.

[73] Tsai, C.T., Ting, J.W., Wu, M.H., Wu, M.F., Guo, I.C. and Chang, C.Y. (2005) Complete genome sequence of the grouper iridovirus and comparison of genomic organization with those of other iridoviruses. *Journal of Virology* 79, 2010 - 2023.

[74] Vesely, T., Cinkova, K., Reschova, S., Gobbo, F., Ariel, E. *et al.* (2011) Investigation of ornamental fish entering the EU for the presence of ranaviruses. *Journal of Fish Diseases* 34, 159 - 166.

[75] Waltzek, T.B., Miller, D.L., Gray, M.J., Drecktrah, B., Briggler, J.T. *et al.* (2014) New disease records for hatchery-reared sturgeon. I. Expansion of Frog virus 3 host range into *Scaphirhynchus albus*. *Diseases of Aquatic Organisms* 111, 219 - 227.

[76] Whittington, R. J. and Chong, R. (2007) Global trade in ornamental fish from an Australian perspective: the case for revised import risk analysis and management strategies. *Preventive Veterinary Medicine* 81, 92 - 116.

[77] Whittington, R.J. and Reddacliff, G.L. (1995) Influence of environmental temperature on experimental infection of redfin perch (*Percus fluviatilis*) and rainbow trout (*Oncorhynchus mykiss*) with *Epizootic hematopoietic necrosis virus*, an Australian iridovirus. *Australian Veterinary Journal* 72, 421 - 424.

[78] Whittington, R.J. and Steiner, K.A. (1993) *Epizootic haematopoietic necrosis virus* (EHNV): improved ELISA for detection in fish tissues and cell cultures and an efficient method for release of antigen from tissues. *Journal of Virological Methods* 43, 205 - 220.

[79] Whittington, R. J., Philbey, A., Reddacliff, G. L. and Macgown, A. R. (1994) Epidemiology of epizootic haematopoietic necrosis virus (EHNV) infection in farmed rainbow trout, *Oncorhynchus mykiss* (Walbaum): findings based on virus isolation,

antigen capture ELISA and serology. *Journal of Fish Diseases* 17, 205 – 218.

[80] Whittington, R.J., Kearns, C., Hyatt, A.D., Hengstberger, S. and Rutzou, T. (1996) Spread of Epizootic haematopoietic necrosis virus (EHNV) in redfin perch (*Perca fluviatilis*) in southern Australia. *Australian Veterinary Journal* 73, 112 – 114.

[81] Whittington, R. J., Reddacliff, L. A., Marsh, I., Kearns, C., Zupanovic, Z. and Callinan, R.B. (1999) Further observations on the epidemiology and spread of *Epizootic haematopoietic necrosis virus* (EHNV) in farmed rainbow trout *Oncorhynchus mykiss* in southeastern Australia and a recommended sampling strategy for surveillance. *Diseases of Aquatic Organisms* 35, 125 – 130.

[82] Whittington, R. J., Becker, J. A. and Dennis, M. M. (2010) Iridovirus infections in finfish-critical review with emphasis on ranaviruses. *Journal of Fish Diseases* 33, 95 – 122.

[83] Williams, T., Barbosa-Solomieu, V. and Chinchar, V.G. (2005) A decade of advances in iridovirus research. In: Maramorosch, K. and Shatkin, A.J. (eds) *Advances in Virus Research*. Elsevier/Academic Press, San Diego, California, pp. 173 – 248.

5

致瘤病毒：马苏大麻哈鱼病毒和鲤疱疹病毒

Mamoru Yoshimizu*，Hisae Kasai，Yoshihiro Sakada，
Natsumi Sano 和 Motohiko Sano

5.1 引言

因其独特的外观和明显的病理特征，早在几个世纪前鱼类肿瘤就已经被证实，有关鱼类肿瘤的科学文献更是种类繁多（Walker，1969；Anders and Yoshimizu，1994）。1965 年，华盛顿特区的史密森学会建立了世界上最大的低等动物肿瘤登记系统，病理范围涉及 300 多种鱼类的良性表皮乳头状瘤到转移性黑色素瘤和肝细胞瘤等。

根据季节和地理位置的不同，某些皮肤瘤可能在东北大西洋沿海地区的野生欧洲鳗鲡（*Anguilla anguilla*）、黄盖鲽（*Limanda limanda*）、欧洲胡瓜鱼（*Osmerus eperlanus*）（Anders，1989）和白斑狗鱼（*Esox lucius*）等鱼类中流行，而在养殖的鲤科鱼类中，同为皮肤瘤的鲤痘疮病（carp-pox）病变发生的重要性已降低（Anders and Yoshimizu，1994）。哺乳动物、鸟类、爬行动物和鱼类中都记载有疑似病毒性病因的肿瘤病变（Anders and Yoshimizu，1994）。鱼类乳头状瘤的病毒病原学最早由 Keysselitz（1908）提出。尽管病毒颗粒并不总是明显可见的，但基于流行性鱼类肿瘤频繁发生的现象，人们提出了传染性病毒病原学（Winqvist et al.，1968；Walker，1969；Mulcahy and O'Leary，1970；Anders，1989；McAllister and Herman，1989；Lee and Whitfield，1992；Anders and Yoshimizu，1994）。在这些病例中，证据主要是基于排除其他可能的病因提出的。在使用电镜和病毒学方法研究的所有病例中，约有 50%的病例在肿瘤组织中发现了病毒或病毒样颗粒。其中，良性肿瘤主要由疱疹病毒感染引起，例如表皮增生症、乳头状瘤和纤维瘤，而很少与逆转录病毒、乳多泡病毒或腺病毒有关。此外，逆转录病毒被认为可以引起恶性肿瘤，如肉瘤和淋巴肉瘤等。通常，每种肿瘤只有一种病毒类型，但在罕见病例中，同一样本可能出现与不同病毒关联的不同皮肤瘤类型。

---

* 通信作者邮箱：yosimizu@fish.hokudai.ac.jp。

这些病毒或病毒样颗粒对诱发肿瘤的意义大多是推测性的，不在本文讨论的范畴。只有来自马苏大麻哈鱼(*Onchorhynchus masou*)(见5.2节)和日本浅黄品系锦鲤[鲤鱼(*Cyprinus carpio*)的驯化变种](见5.3节)的疱疹病毒被明确证实具有致瘤性。虽然在某些病例中，通过接种无细胞滤液或肿瘤活细胞可实验性诱导肿瘤形成，但尚未成功从肿瘤的细胞培养物中分离出病毒(Peters and Waterman，1979)。然而这两种疱疹病毒的致病性和致瘤性都已通过从细胞培养物中成功分离出致病病毒被明确证实，并且符合River法则，但我们对其中可能存在的致瘤基因仍一无所知。

5.2 马苏大麻哈鱼病毒

5.2.1 引言

马苏大麻哈鱼病毒病(*Oncorhynchus masou* virus disease，OMVD)是一种在日本鲑科鱼类中发生的致瘤性皮肤溃疡病，并伴有肝炎。该疾病由鲑疱疹病毒2型(*Salmonid herpesvirus* 2，SalHV-2)引起，最早在马苏大麻哈鱼中发现并报道(Kimura et al.，1981a，b)。该病毒通常被称为马苏大麻哈鱼病毒(OMV)，其他别称包括横跨秋田县和青森县的十和田湖红大麻哈鱼病毒(NeVTA；Sano，1976)、山女鲑肿瘤病毒(YTV；Sano et al.，1983)、银大麻哈鱼病毒(OKV；Horiuchi et al.，1989)，银大麻哈鱼肿瘤病毒(COTV；Yoshimizu et al.，1995)、银大麻哈鱼疱疹病毒(CHV；Kumagai et al.，1994)、虹鳟肾疱疹病毒(RKV；Suzuki，1993)和虹鳟疱疹病毒(RHV；Yoshimizu et al.，1995)。SalHV-2为异疱疹病毒科(Alloherpesviridae)、鲑疱疹病毒属(*Salmonivirus*)病毒。

5.2.2 病原

1. 生理生化特性

在15℃左右时，感染OMV的细胞在5～7 d内会出现明显的致细胞病变效应(cytopathic effect，CPE)，其特征表现为细胞呈圆形，随后形成合胞体，并最终导致RTG-2和其他鲑科鱼类细胞系裂解[图5.1(A)]。非鲑科鱼类细胞则难以被感染(Yoshimizu et al.，1988b)。培养病毒的最大感染滴度约为10^6 $TCID_{50}$/mL，随着细胞系的不同有所差异。OMV对高温、醚和酸(pH=3)不稳定，不能凝集鲑鱼血细胞或者人O型细胞。3.0×10^3(μW·s)/cm^2(mJ)紫外线(UV)照射可使其灭活，50 μg/mL嘧啶类似物5-碘-2′-脱氧尿苷(5-iodo-2′-deoxyuridine，IUdR)或抗疱疹病毒试剂如磷乙酸(phosphonoacetate，PA)、阿昔洛韦[ACV；9-(2-羟乙氧甲基)鸟嘌呤]、(*E*)-5-(2-溴乙烯基)-2′-

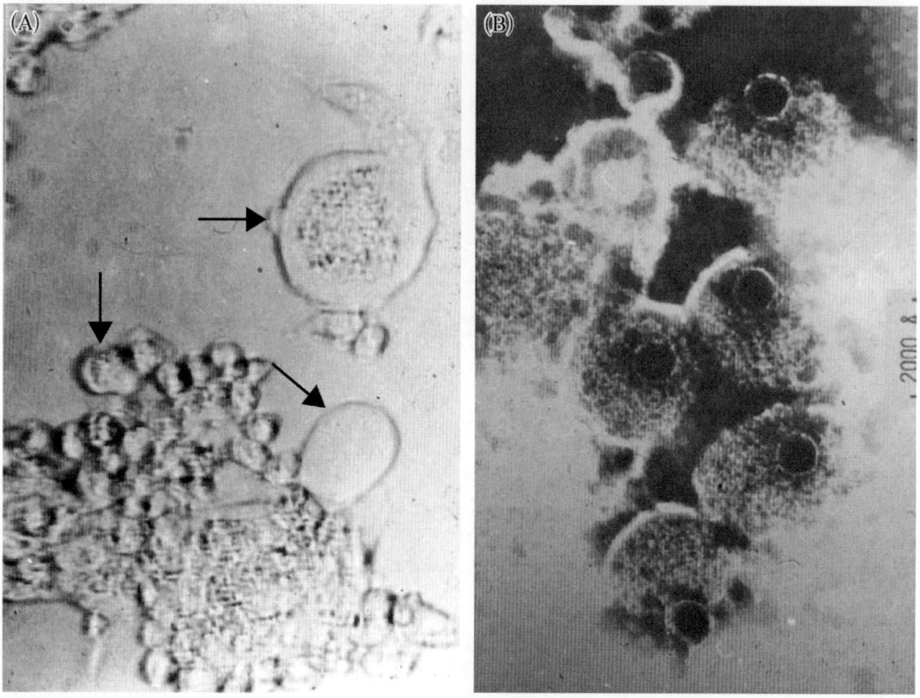

图 5.1 (A) 马苏大麻哈鱼病毒(OMV)感染的致细胞病变效应，显示在 15 ℃孵育 9 d 的 RTG－2 细胞中，OMV 会引发圆形细胞，继而形成合胞体(箭头)；(B) 负染的包膜病毒粒子电镜照片，由 T. Sano 博士提供(Yoshimizu and Kasai，2011)

脱氧尿苷[(*E*)－5－(2－bromovinyl)－2′－deoxyuridine，BVdU]和 1－β－D－阿糖基-呋喃胞嘧啶(1－β－D－arabino－furanosylcytosine，Ara－C)等可抑制该病毒复制(Kimura et al.，1981a，1983b，c；Suzuki et al.，1992)。

感染细胞的电镜镜检可见直径为 115 nm 的核内六边形衣壳，细胞质小泡表面及内部明显可见大量出芽和直径为 200 nm×240 nm 的包膜病毒粒子[图 5.1(B)]。经计算，负染的病毒粒子的衣壳粒数量为 162。OMV 的最适生长温度为 15 ℃，在 18 ℃进行复制，20 ℃及以上停止生长，OMV 的这种嗜冷性不同于斑点叉尾鮰疱疹病毒和其他两栖类疱疹病毒的温度敏感性(Kimura et al.，1981a)。

2. 血清学关系

OMV 可被同源抗血清中和，但其他鲑科鱼类病毒的抗血清对其无效，如传染性胰腺坏死病毒(IPNV)、传染性造血器官坏死病病毒(IHNV)、大麻哈鱼呼肠孤病毒(CSV)和鲑疱疹病毒等。从日本北部孵化场成熟马苏大麻哈鱼的卵巢液或肿瘤组织中分离出的 177 株 OMV 毒株(采集于 1978 年至 1986 年)，都可被 OMV 参考毒株 OO－7812 的兔抗血清中和，其 ND_{50}(半数中和剂量)为 1∶80～1∶40(Yoshimizu et al.，1988b)。通过多克隆兔抗血清的血清交叉中和试验对比了 12 株疱疹病毒：红大麻哈鱼(内陆红鲑鱼 *O. nerka*)分离株 NeVTA；三株 OMV 和山女鲑分离株 YTV；CSTV(银大麻哈鱼肿瘤病毒)，分离自银大麻哈鱼的 COTV 和两株 OKV；虹鳟分离株 RKV 和 RHV；鲑疱疹病毒。日本的疱疹病毒株与鲑疱疹病毒 2 型参考毒株 OMV OO－7812 关系密切，它们可被与其对应的抗血清中和(表 5.1)。然而，这些病毒株与鲑疱疹病毒 1 型有明显区别，马苏大麻哈鱼病毒(OMV)被认定为鲑疱疹病毒 2 型。

表 5.1 基于血清交叉中和试验 1/r(亲缘度)值①的鲑科鱼类疱疹病毒株的血清学关系

鱼种	病毒②	抗血清②								
		OMV			YTV	NeVTA	COTV	OKV(M)	RKV	HS
		Ⅰ	Ⅱ	Ⅲ						
马苏大麻哈鱼	OMV Ⅰ	1.00	1.30	0.92	1.42	1.12	1.53	0.85	1.22	>3.16
	OMV Ⅱ		1.00	0.80	1.00	1.22	1.47	0.85	0.67	>3.47
	OMV Ⅲ			1.00	1.21	0.94	1.41	1.00	0.83	>3.47
	YTV				1.00	0.84	1.33	0.70	1.19	>5.69
红大麻哈鱼	NeVTA					1.00	1.39	0.83	0.89	>4.89
银大麻哈鱼	COTV						1.00	0.96	0.51	>3.16
	OKV(M)							1.00	0.83	>3.47
虹鳟	RKV								1.00	>3.47
	HS									1.00

① 值为 1 表示血清学相同，值越大或越小表示差异越大(1/*r*：Archetti and Horsfall，1950)。
② COTV 为银大麻哈鱼肿瘤病毒；HS 为鲑疱疹病毒；NeVTA 为秋田县和青森县十和田胡红大麻哈鱼病毒；OKV(M)为银大麻哈鱼病毒；OMV 为马苏大麻哈鱼病毒；RKV 为虹鳟肾病毒；YTV 为山女鲑肿瘤病毒。

3. 病毒蛋白与基因组

OMV 的一般特性与鲑疱疹病毒 1 型类似，但在病毒粒子大小和最适生长温度上有所不同。此外，OMV 与其他已知的鱼类疱疹病毒在病毒诱导多肽模式上有所不同，在感染 OMV 的细胞中出现 34 种病毒特异性多肽，这些多肽的相对分子质量在 19 000～227 000 Da。与之相比，鲑疱疹病毒诱导 25 种多肽，相对分子质量在 19 500～250 000 Da(Kimura and Yoshimizu，1989)。CCV(斑点叉尾鮰病毒)则诱导出 32 种与 OMV 诱导的完全不同的多肽(Dixon and Farber，1980)。34 种 OMV 特异性多肽中有 2 种存在电泳迁移差异，导致 12 株 OMV 被分为 6 组(Kimura and Yoshimizu，1989)。

OMV 的 DNA 限制性内切酶图谱与鲑疱疹病毒不同。采用限制性内切酶对分离自北海道和青森县野生马苏大麻哈鱼卵巢液和肿瘤组织的7株代表性OMV毒株进行了分析，根据DNA酶切图谱将其分为4组。当用*Bam*HⅠ、*Hin*dⅢ和*Sma*Ⅰ酶切时，高代次毒株与低代次毒株的酶切图谱有所不同，但用*Eco*RI酶切时，两者未见差异（Hayashi et al.，1987）。将^{32}P标记的OMV标准株（OO－7812株）DNA作为探针，可以与其他OMV株DNA的大部分片段实现杂交（Gou et al.，1991）。从DNA同源性结果来看，OMV株和YTV株被认为是同一种病毒，而与NeVTA相似但不同（Eaton et al.，1991）。

马苏大麻哈鱼病毒OO－7812株的基因组已测序（Yoshimizu et al.，2012），并利用DNA聚合酶和末端酶基因的部分氨基酸序列预测了该病毒株与鱼类和两栖类疱疹病毒之间的系统发育关系。结果显示，OO－7812毒株与YTV和NeVTA的同源性为100%，而OMV明显区别于SalHV－1和SalHV－3（图5.2）。

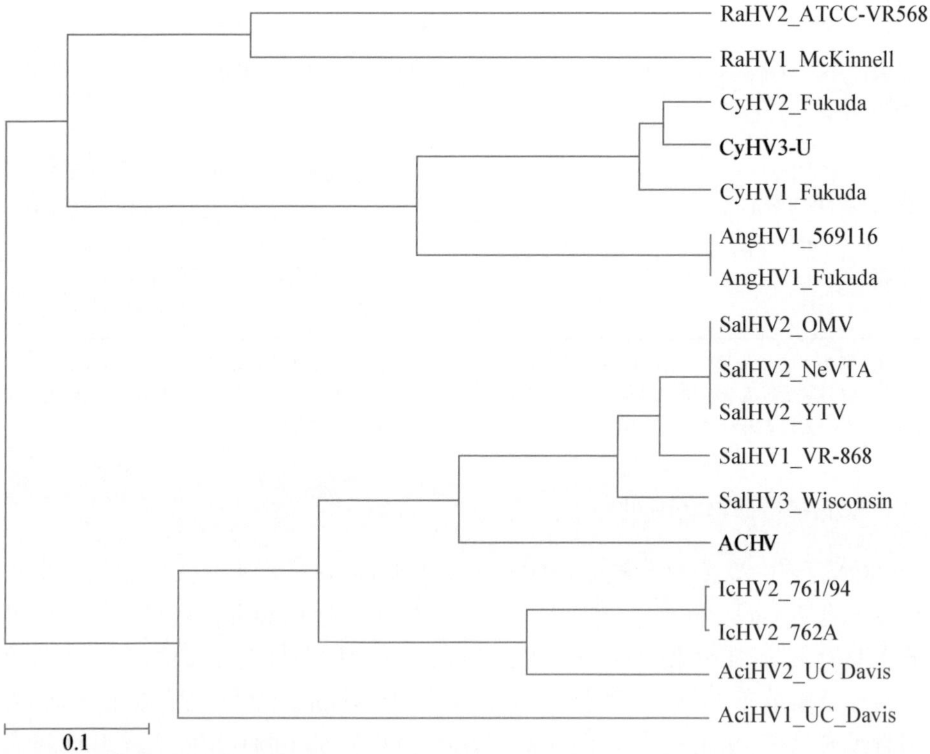

图5.2　根据串联部分推导的DNA聚合酶和端酶（包含缺口的247个氨基酸字符）基因的氨基酸序列绘制的系统发育图，描绘了鱼类和两栖类疱疹病毒之间的关系。分枝长度基于推测替换的数量，如比例尺所示。RaHV为蛙疱疹病毒（青蛙）；CyHV为鲤疱疹病毒（鲤鱼和金鱼）；AngHV为鳗鲡疱疹病毒（鳗鲡）；SalHV为鲑疱疹病毒；ACHV为大西洋鳕疱疹病毒；IcHV为鮰疱疹病毒（斑点叉尾鮰）；AciHV为鲟疱疹病毒（美洲鲟）

马苏大麻哈鱼肿瘤组织和正常组织的 DNA 聚合酶活性也已被检测研究。在 OMV 感染的肿瘤组织中检测到较高的 DNA 聚合酶 α 活性，而在正常组织中则检测不到，这说明 OMV 的 DNA 在肿瘤组织中复制良好。DNA 聚合酶活性在肿瘤组织和正常组织中是相同的，这是在疱疹病毒引起的肿瘤组织中首次检测到疱疹病毒 DNA 聚合酶。

4. 生存能力和免疫性

在 15 ℃和 10 ℃的水中，OMV 的感染滴度分别在 3 d 和 7 d 内显著降低。然而，当水温低于 5 ℃时，病毒感染性可保持 7～14 d，因为在低温下，水中产生抗病毒物质的细菌活性下降(Yoshimizu et al., 2014)。

5.2.3 地理分布

20 世纪 60 年代初，人们从北海道的日本海沿岸河流中采集马苏大麻哈鱼鱼卵，并将其运往本州。由于鱼类流通不受限制，病毒传播到了本州的岐阜、山梨和新潟等地，在这些地区首次发现了患有基底细胞癌的马苏大麻哈鱼癌性疾病(Kimura, 1976)。1972 年和 1974 年，红大麻哈鱼鱼苗出现大量死亡现象，并从濒死的鱼体中分离出病毒。该病毒被归类为疱疹病毒科的一种，在横跨秋田县和青森县的十和田湖地区被称为红大麻哈鱼病毒(NeVTA; Sano, 1976)。1978 年，在北海道日本海的乙部鲑鱼孵化场养殖的马苏大麻哈鱼卵巢液中分离出一种类似的疱疹病毒，并根据其宿主命名为马苏大麻哈鱼病毒(Kimura et al., 1981a, b)。OMV 对马苏大麻哈鱼、大麻哈鱼(*O. keta*)、银大麻哈鱼和虹鳟等鲑科鱼类都表现出致病性和致瘤性。

从 1978 年至 2015 年，在北海道和本州北部的青森县和岩手县对马苏大麻哈鱼、大麻哈鱼、驼背大麻哈鱼(*O. gorbuscha*)、红鲑鱼、红大麻哈鱼和虹鳟(46 788 条雌鱼)这 6 种鲑科鱼类的成熟鱼进行了采集并调查了它们的病毒感染情况(Yoshimizu et al., 1993; Kasai et al., 2004)。除了一个孵化场未能收集齐 60 个样本外，在 13 个孵化场的马苏大麻哈鱼中都分离出了疱疹病毒，并且所有分离株都可被兔抗 OMV 血清中和(Yoshimizu et al., 1993)。

1981 年，在本州新潟县内陆渔业试验站小出分部养殖的山女鲑(又称马苏大麻哈鱼)的口部基底细胞癌组织中分离出一种类似的疱疹病毒，这种病毒被命名为山女鲑肿瘤病毒(YTV, Sano et al., 1983)。自 1988 年以来，在本州东北地区宫城县的池塘和网箱养殖银大麻哈鱼中也分离出疱疹病毒(Kimura and Yoshimizu, 1989)。此外，1992 年以来，已从北海道的养殖虹鳟中分离出一种疱疹病毒，Suzuki(1993)初步将其命名为虹鳟肾疱疹病毒(rainbow trout kidney herpesvirus, RKV)。2000—2001 年，在日本中部(本州)长野县、静冈县和岐阜县的 18 个渔场，1.2～1.5 kg 大小的虹鳟发生了流行疫情，从病鱼中分离出的所有病毒均被鉴定为 OMV。

5.2.4 对经济的重要影响

从日本北部野生和养殖的马苏大麻哈鱼卵巢液和肿瘤中分离出的大量疱疹病毒株被鉴定为 OMV（Yoshimizu et al.，1993）。OMV 可以造成红大麻哈鱼 80％～100％的死亡率，而马苏大麻哈鱼感染 OMV 的死亡率没有相关报告，但是肿瘤的存在降低了它们的商业价值。自 1988 年以来，OMVD 一直是日本东北地区网箱养殖银大麻哈鱼面临的主要问题，这种疾病给银大麻哈鱼养殖造成了严重的经济损失。1992 年在北海道池塘养殖的虹鳟中发现了 OMV，2000 年起在日本中部出现。目前 OMVD 已在红大麻哈鱼、马苏大麻哈鱼和银大麻哈鱼中受到监测和控制(见 5.2.7 节)。直到 2005 年，OMVD 疫情仍然是虹鳟养殖场面临的一个主要问题，这从 2000 年至 2001 年长野县等 18 个养殖场的 OMVD 疫情中高比例的低体重鱼的死亡就可以看出[图 5.3(D)；Furihata et al.，2003]。

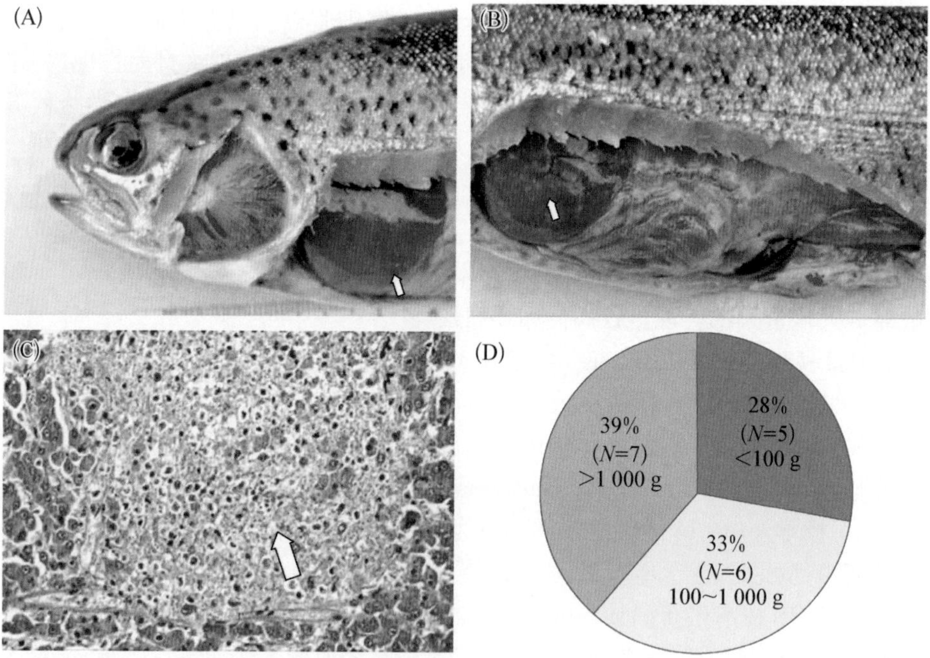

图 5.3　虹鳟感染 OMV：(A) (B) 箭头指示整条鱼的肝坏死症状，在(B)中指示肝上的白点病变；(C) 肝脏的多灶性严重坏死(用箭头指示)；(D) 死亡虹鳟的体重图表。照片和图表由 M. Furihata 博士提供(Furihata et al.，Fish Pathology，2003)

5.2.5 诊断

在 0～5 ℃下，病毒的传染性可在两周内保持不变，但在－20 ℃时，病毒在

17 d内便丧失了99.9%的传染性。病毒的分离应从冰运至实验室的鱼体中进行(Yoshimizu et al., 2005)。对于OMV的过滤,推荐使用孔径为0.4 μm的核孔滤膜(聚碳酸酯),因为醋酸纤维膜能过滤掉超过99%的病毒颗粒,故不推荐使用。若进行病毒学调查,可以参照Yoshimizu等(1985)的方法收集卵巢液,用等体积的抗生素(Amos, 1985)进行稀释,在5 ℃下孵育过夜。对于肿瘤病例,切除肿瘤组织样品,用碘伏消毒(50 mg/L,15 min),并用汉克平衡盐溶液(Hank's balanced salt solution, Hank's BSS)清洗,然后保存于抗生素溶液中运送至实验室,制备的肿瘤组织用于原代培养或者与RTG-2细胞共培养。原代细胞传代培养一次后,应检测培养基中是否有病毒(Yoshimizu, 2003b)。

采用兔抗OMV血清或单克隆抗体进行病毒的免疫荧光检测(Hayashi et al., 1993),并用DNA探针检测病毒基因组(Gou et al., 1991)。采用F10引物GTACCGAAACTCCGAGTC和R05引物AACTTGAACTACTCCGGGG,对马苏大麻哈鱼、银大麻哈鱼和虹鳟的肝、肾、脑和神经组织中的OMV分离株进行PCR扩增,可以得到439个碱基大小的DNA片段。OMV与鲑疱疹病毒扩增的DNA片段大小不同,因此可以通过它们的琼脂糖凝胶电泳图谱来区分(Aso et al., 2001)。

1. 推定诊断

通过评估一些关键特征,如鱼的生命周期、感染阶段、鱼的种类和数量,以及水温、临床体征(见5.2.6节)和养殖设施病史等,可推定诊断OMV感染。为了分离OMV,常采用标准的细胞培养技术对组织和生殖液进行培养。处理后的样本须接种到RTG-2或CHSE-214。OMV会导致细胞变圆或形成巨大合胞体等致细胞病变效应。使用甲基纤维素涂层的空斑测定法(Kamei et al., 1987),也可用于OMV的分离和计数。

2. 确诊

通过使用多克隆兔抗血清或单克隆抗体可对OMVD进行确诊。使用多克隆或单克隆抗血清的ELISA抗原检测(Yoshimizu, 2003a)以及荧光抗体技术(FAT)已被开发用于OMVD的确诊(Hayashi et al., 1993)。FAT可特异性地与OMV的所有分离株发生反应,并且确诊所需时间更短。也可采用PCR确诊在细胞培养中生长的OMV(Aso et al., 2001)。

3. 亚临床感染检测程序

在携带病毒的鱼中检测OMV比较困难,但鱼产卵时,病毒会复制并出现在卵巢液中。通过中和试验或ELISA(Yoshimizu, 2003a)检测抗体可用于动物流行病学研究。

5.2.6 病理

1. 致病性和宿主易感性

在 10 ℃下，用含有 100 $TCID_{50}$/mL OMV 的水浸泡 1 h，研究了鲑科鱼类鱼苗对 OMV 的易感性(Kimura et al., 1983a)。红大麻哈鱼(1 月龄)最易感(死亡率为 100%)；马苏大麻哈鱼和大麻哈鱼也很敏感(死亡率分别为 87%和 83%)；银大麻哈鱼和虹鳟易感性最低(死亡率分别为 39%和 29%)。刚孵化的大麻哈鱼累计死亡率为 35%，但 1～5 月龄鱼苗的累计死亡率超过 80%，3 月龄鱼苗的累计死亡率为 98%，而 6 月龄和 7 月龄时，易感性则分别降低至 7%和 2%。8 月龄鱼苗浸泡病毒后腹腔注射 200 $TCID_{50}$/尾病毒未发生死亡。相比之下，1 月龄马苏大麻哈鱼最敏感，累计死亡率可达 87%，而 3 月龄和 5 月龄马苏大麻哈鱼鱼苗的累计死亡率分别为 65%和 24%。

2. 临床症状与传播

自 1970 年以来，每年 6 月至 9 月，日本的红大麻哈鱼鱼苗都会出现很高的死亡率(约 80%)。病鱼的临床体征包括体色变暗、行动迟缓和食欲不振等。1972 年和 1974 年，在 10 ℃下孵育的 RTG－2 细胞中从濒死的患病鱼中分离出形成合胞体的病毒，该病毒在横跨秋田县和青森县的十和田湖地区被命名为红大麻哈鱼病毒(NeVTA)(Sano, 1976)。

1978 年，在北海道的乙部鲑鱼孵化场中，从一条看起来健康的成熟马苏大麻哈鱼的卵巢液中分离出一种疱疹病毒，该病毒被命名为马苏大麻哈鱼病毒(Kimura et al., 1981a, 1981b)。OMV 的一般特性与那些鲑疱疹病毒和 NeVTA 相似，但病毒粒子的大小和最适生长温度不同，与鲑疱疹病毒在病毒诱导的多肽组成图谱、血清学特性和 PCR 产物上也不相同(Kimura and Yoshimizu, 1989; Aso et al., 2001)。OMV 具有致病性，更重要的是，它对马苏大麻哈鱼和其他几种鲑科鱼类具有致瘤性(Kimura et al., 1981a, b)。1 月龄红大麻哈鱼对 OMV 最易感，马苏大麻哈鱼和大麻哈鱼对 OMV 也很易感，而银大麻哈鱼和虹鳟则不易感(Tanaka et al., 1984)。鱼体带瘤的发病率接近 60%以上，12%～100%的存活大麻哈鱼、银大麻哈鱼和马苏大麻哈鱼都有上皮瘤。虹鳟感染 4 个月后开始出现上皮瘤，并持续至少一年(Yoshimizu et al., 1987)。

自 1988 年以来，已在日本宫城县的池塘和网箱养殖银大麻哈鱼的肝、肾和发育瘤中分离出疱疹病毒(Kimura and Yoshimizu, 1989)。病症包括肝脏白斑、皮肤溃疡和嘴部或体表周围的瘤样病变等。从银大麻哈鱼中分离出的疱疹病毒被分别暂时命名为银大麻哈鱼肿瘤病毒(coho salmon tumor virus,

CSTV；Igari et al.，1991)、银大麻哈鱼病毒(*O. kisutch* virus，OKV；Horiuchi et al.，1989)、银大麻哈鱼肿瘤病毒(coho salmon tumor virus，COTV；Yoshimizu et al.，1995)，以及银大麻哈鱼疱疹病毒(coho salmon herpesvirus，CHV；Kumagai et al.，1994)。这些病毒都可以被兔抗 OMV 或 NeVTA 血清中和(Yoshimizu et al.，1995)，并且 CSTV、OKV 和 COTV 的致瘤性已被实验证实。此外，CSTV 的限制性内切酶图谱与 NeVTA 和 YTV 的相似；CHV 对银大麻哈鱼具有高致病性。

自 1992 年以来，在北海道养殖的 1 年龄虹鳟发生过死亡率在 13%～78% 的大规模死亡。除部分出现溃疡性皮肤病变外，病鱼几乎没有任何外部临床症状。剖检可见肠出血及肝脏白点，未发现细菌、真菌或寄生虫，但从肾脏、肝脏和皮肤溃疡中分离出疱疹病毒，Suzuki(1993)将这种疱疹病毒暂时命名为虹鳟肾疱疹病毒(rainbow trout kidney herpesvirus，RKV)。RKV 对市售规格的虹鳟和马苏大麻哈鱼具有高致病性(Sung et al.，1996a，b)。2000 年 2 月至 2001 年 1 月，日本长野县等 18 个养鱼场的 1.2～1.5 kg 虹鳟发生了动物流行病，从具有合胞体病变效应的 RTG－2 细胞中分离出病鱼中的一种病毒，该病毒被证实对脏器具有高感染滴度(大约为 10^8 $TCID_{50}/g$)，并在肝脏中可见多发性坏死病灶[图 5.3(A)(B)(C)]。通过血清学检测和 PCR 鉴定，该病毒为 OMV。基于这些检测结果，该疫病被诊断为 OMVD。在这些病例中，80%以上的疫情与引入的活鱼有关(Furihata et al.，2003，2004)。

OMV 的水平传播是通过与 5 月龄无特定病原的大麻哈鱼鱼苗共栖实现的，其死亡率与 3～7 月龄鱼苗的浸泡感染结果相近(见上文)。感染后鱼的临床体征包括食欲不振、眼球突出[图 5.4(A)]和体表(尤其是在下颌下方)瘀斑[图 5.4(C)]。在体内，肝脏呈现白点病变[图 5.4(B)(D)]，在晚期，整个肝脏会变成珍珠白。有些病鱼的脾脏可能会肿胀，肠道内缺乏食物(Kimura et al.，1981a，1983a)。

3. 组织病理学

感染 OMV 的 1 月龄和 3 月龄马苏大麻哈鱼、1 月龄银大麻哈鱼和 2 月龄大麻哈鱼的肾脏是病毒的主要靶器官。早期濒死的鱼苗会出现上皮细胞和肾坏死，之后，濒死的 1 月龄马苏大麻哈鱼会出现肝、脾和胰腺部分坏死，而 3 月龄马苏大麻哈鱼则出现肾脏造血组织坏死。这说明在疾病后期，主要靶器官已从肾脏转移至肝脏，并且可见明显的组织病理学变化。随着潜伏期的延长，肝脏坏死灶变得更加严重，会出现肝细胞染色质边缘化，也可以观察到脾、胰腺、心肌和脑中的细胞变性(Yoshimizu et al.，1988a)。

虹鳟、银大麻哈鱼和大麻哈鱼的组织病理学变化与马苏大麻哈鱼相似(Kumagai et al.，1994；Furihata et al.，2004)。在虹鳟的内脏器官中可以检测

图 5.4　暴露于 OMV 的 3 月龄大麻哈鱼。(A) 眼球突出；(B) 肝脏白点病变；(C) 体表瘀点；(D) 肝脏切片中多个严重坏死病灶(苏木精-伊红染色)。照片和图由 Kimura 博士提供(Kimura et al., Fish Pathology, 1981a)

到高感染滴度，并且肝脏有明显多发性坏死灶。最终的病变是脾、肾造血组织、肝脏、肠、心、鳃丝、表皮和外侧肌肉中 OMV 感染细胞的坏死。特别是，患 OMVD 的虹鳟的肠上皮和下层组织出现严重坏死和出血(Furihata et al., 2004)。

4. 肿瘤的诱导

通过 OMV 的水传染，可实验诱导肿瘤的形成。在感染后约 4 个月，12%～100%的存活马苏大麻哈鱼、大麻哈鱼、银大麻哈鱼及虹鳟会发生口腔上皮性肿瘤，并持续至少一年[图 5.5(A)(B)；Kimura et al., 1981b]。在组织病理学上，肿瘤由增生的、分化良好的上皮细胞组成，并由细小的结缔组织间质支撑。原代培养中移植的肿瘤细胞经一次传代后，可从培养基中回收 OMV(Yoshimizu et al., 1987)，口腔周围是肿瘤最常发生的部位。由于在相同条件下饲养的对照鱼没有出现肿瘤，因此 OMV 被认为是导致肿瘤发展的原因，这种瘤一旦形成可能在感染后持续一年的时间。

肿瘤细胞似乎起源于上皮细胞，在乳头状瘤细胞阵列中会出现几层由细小结缔组织间质支撑的上皮细胞，大量的有丝分裂图像表明其具有高度增殖性。出现在尾鳍、鳃盖、体表、眼角膜和肾脏的肿瘤具有与口腔肿瘤相似的特征

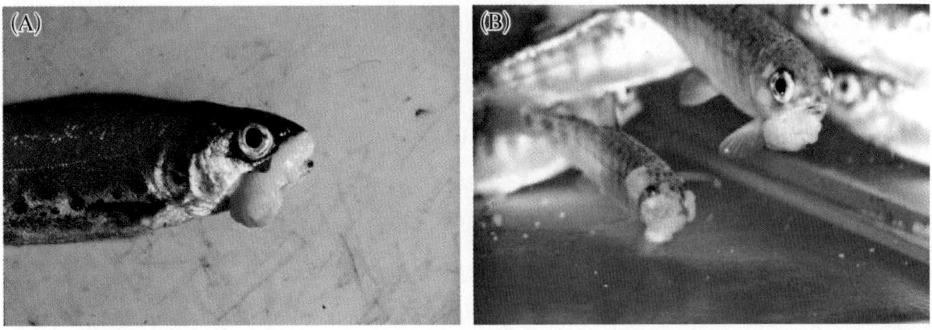

图 5.5 OMV 实验诱导的肿瘤。(A) 马苏大麻哈鱼；(B) 银大麻哈鱼。照片由 M. Yoshimizu 提供

(Kimura et al., 1981b, c)。肾脏及其他部位(如眼、鳃盖下和尾鳍)的肿瘤，也都有类似的组织病理学病变。

电镜结果显示，肿瘤细胞具有细胞核大小不等、细胞内连接松散等典型特征。不过，在肿瘤的细胞核或细胞质中均未发现 OMV 颗粒(Kimura et al., 1981b, c; Yoshimizu et al., 1987)。病毒在感染 9 个月后从一尾鱼的糜烂肿瘤组织中被分离出来，在感染 10 个月后从另一尾鱼的肿瘤细胞原代培养物中被分离出来。原代培养物显示连续生长 4 d，然后出现 CPE 样变化。此时，可从培养基中分离出 OMV。在个别鱼或者患肿瘤鱼的混合血清中可以检测到 OMV 的中和抗体。

5.2.7 防控策略

1. 流行病学

20 世纪 80 年代，OMV 在日本北部的马苏大麻哈鱼中分布广泛。1988 年，在日本中部东北地区海洋环境下网箱养殖的银大麻哈鱼中诊断发现了 OMVD。从淡水转运入海后立刻筛选濒死病鱼，并利用 FAT 检测它们的血清，成功地控制了银大麻哈鱼中的病毒。如果鱼的检测呈阳性，养殖者将在第二年发眼卵迁移前对孵卵池和设施进行消毒(Kumagai et al., 1994)。虹鳟 OMVD 于 1991 年在北海道被发现，然而自 2000 年起，OMVD 已经成为整个日本池塘养殖业的主要病害问题。自然感染和实验感染的数据表明，1～5 月龄鱼最易感。从 2000 年到 2005 年，在幼鱼、1 年龄和成熟虹鳟中均有疫情流行报道(Furihata et al., 2003)，死亡率超过 80%，大多数 OMVD 发生在 15 ℃或更低温度的淡水中。

2. 消毒剂和抗病毒化学疗法的杀毒效果

6 种消毒剂对 OMV 的杀灭效果被检测(Hatori et al., 2003)。在 15 ℃下处理 20 min，碘伏可以减少 100%的 OMV 噬菌斑。次氯酸钠、苯扎氯铵、煤酚皂、甲醛和高锰酸钾五种溶液，达到相同效果的最小浓度分别为 50 ppm、100 ppm、100 ppm、3 500 ppm 和 16 ppm。

阿昔洛韦(ACV)对 OMV、鲑疱疹病毒和 CCV 具有很好的疗效。2.5 μg/mL 的 ACV 可以抑制 100 $TCID_{50}$/mL OMV 诱导的 RTG－2 致细胞病变效应。与 9－β－D－阿腺苷(Ara－A)、碘脱氧尿苷(IUdR)和磷乙酸(PA)等化合物相比，ACV 更有效。2.5 μg/mL 的 ACV 能显著抑制 RTG－2 细胞的生长，但未见形态变化。100 $TCID_{50}$/mL 的 OMV 接种 RTG－2 细胞后，2.5 μg/mL 的 ACV 可以完全抑制细胞中 OMV 的复制。感染后 4 d 内加入 ACV 可减少 OMV 的复制。为了达到良好的抑制效果，ACV 必须持续地存在(Kimura et al.，1983a)。

研究者们利用 OMV 与大麻哈鱼鱼苗对 ACV 的治疗效果进行了评价。通过口服或浸泡 ACV 处理实验感染鱼。其中，每天通过浸泡 ACV 溶液(25 μg/mL，30 min/d，15 次)，可以降低感染鱼的死亡率，而口服给药[25 (μg/尾)/d，60 次]对大麻哈鱼的存活率没有影响。相比之下，口服 IUdR 的大麻哈鱼组的存活率比 ACV 给药组更高。这项研究表明，口服给药的鱼体内不能维持有效的 ACV 浓度。每日将被感染鱼浸泡在 ACV 溶液(25 μg/mL，30 min/d，60 次)中可显著抑制 OMV 诱导的肿瘤发展(Kimura et al.，1983b)。

3. 接种疫苗

接种 OMV 福尔马林灭活疫苗后，成熟虹鳟可以在孵化中产生具有中和活性的抗 OMV 抗体 IgM，并且降低卵巢液中 OMV 的分出率(Yoshimizu and Kasai，2011)。在鱼苗期，接种 OMV 福尔马林灭活疫苗也可以非常有效地预防 OMV 感染(Furihata，2008)。遗憾的是，日本目前还没有针对 OMVD 的商品化疫苗。

4. 控制策略

为了控制和预防鱼类疾病，孵化场实施全面卫生管理(Yoshimizu，2003b)。应特别注意避免将设备用具从一个养殖池转移到另一个养殖池，所有设备用具在使用后应定期消毒，并仔细研究孵化装置的消毒方法，以免对鱼类产生化学毒性。病原体可能通过养殖生产人员进行传播，因此，需要对他们的双手和靴子进行适当的消毒，以防止病毒传播。虽然在使用过程中很难对孵化和饲养设施进行消毒，但应对养殖鱼道和养殖池进行定期的氯消毒(Yoshimizu，2009)。无病原的水供应往往是水产养殖成功的关键，孵化场通常使用的来自河流或湖泊的水可能会含有鱼类病原体。那种没有经过过滤或消毒处理以消除和杀灭鱼类病原体的开放水源不应使用。根据对紫外线照射的敏感性，鱼类病毒可分为两组，其中敏感病毒包括 OMV、IHNV、淋巴囊肿病毒(Lymphocystis disease virus，LCDV)和牙鲆弹状病毒(*Hirame rhabdovirus*，HIRRV)，可在 10^4 (μW・s)/cm^2 紫外线下灭活

(Yoshimizu et al., 1986)。

在产卵期采集鱼卵过程中，进行有效的监测和管理非常重要。由于一些病毒可通过受污染的鱼卵或精液从亲本垂直传播给后代，因此对受精卵和发眼卵表面的消毒可阻断疱疹病毒和弹状病毒的侵染循环(Yoshimizu et al., 1989; Yoshimizu, 2009)。对成熟鱼也需要健康检查，以确保鱼体不含法定传染病原。常规检查和专门的诊断技术都是必需的，这样才能确保亲鱼无特定病原。对于鲑科鱼类，采用 Yoshimizu 等(1985)的方法采集卵巢液，并在细胞培养后进行例行检查。受精卵可以使用 25 mg/L 的碘伏消毒 20 min，或 50 mg/L 的碘伏消毒 15 min。发眼卵的卵膜内区域被认为是无病原体的(Yoshimizu et al., 1989, 2002)。自 1993 年以来，北海道所有孵卵场都采用碘伏消毒法对鱼卵进行消毒，这对消除 OMV 疫情很有帮助(Yoshimizu et al., 1993; Kasai et al., 2004; Yoshimizu, 2009)。

游动异常或有疾病征兆的鱼苗应立即移出并带到实验室进行分析。此外，应该采用细胞培养分离、FAT、免疫过氧化物酶染色技术(immunoperoxidase technique, IPT)、ELISA 抗原检测和 PCR 检测等技术定期进行健康监测。

5. 育种

在日本长野县水产试验站，将四倍体虹鳟雌鱼与性逆转二倍体褐鳟(*Salmo trutta*)雄鱼杂交，获得三倍体鲑(Kohara and Denda, 2008)，并命名为“信州鲑”(Shinsyu Salmon)。这些三倍体鲑生长快，在产卵季节存活率高，对 OMV 和 IHNV 感染均具有抗性。

5.3 鲤疱疹病毒 1 型 (*Cyprinid herpesvirus* 1, CyHV-1)

5.3.1 引言

在鲤鱼的皮肤和鳍上有时可见白色或粉红色的乳头瘤样病变，自中世纪以来该病变在欧洲被称为“鲤痘疮病”，这种病在世界各地都有分布，这些肿瘤的发病率因鲤鱼品系或品种的不同可能会有所差异。Calle 等(1999)在一份报告中指出，没有鳞片的鲤鱼肿瘤患病率更高，不过该报告是关于这方面的唯一证据。这种肿瘤组织由分化良好的细胞和表皮钉突组成，与真皮结缔组织和毛细血管的乳突相互交错(Sano et al., 1985a)。

5.3.2 病原

鲤痘疮病的病原最初被称为鲤疱疹病毒(*Cyprinid herpesvirus*, CHV)。这是一种鲤科鱼类疱疹病毒，最早从日本养殖的浅黄锦鲤乳头状瘤组织样本的胖头鲹(FHM)细胞系培养物中分离得到(Sano et al., 1985b)，目前被归类为疱疹病毒目(Herpesvirales)、异疱疹病毒科(Alloherpesviridae)、鲤疱疹病毒属(*Cyprinivirus*)的鲤疱疹病毒1型(*Cyprinid herpesvirus* 1，CyHV-1)病毒。包膜的病毒粒子直径为153～234 nm，衣壳直径为93～126 nm (Sano et al.，1985a)。15～25 ℃下，病毒可在FHM细胞中生长，但在30 ℃时不生长；15 ℃和25 ℃时的最大感染滴度为10^4～10^5 $TCID_{50}$/mL (Sano et al.，1985a，1993a)。来自鲤鱼或鲤科鱼类的其他细胞系如锦鲤鳍细胞系(KF-1)、鲤鱼脑细胞系(*C. carpio* brain，CCB)和鲤鱼上皮瘤细胞系(epithelioma papulosum cyprini，EPC)都对该病毒敏感(Sano et al.，1985a；Adkison et al.，2005)。CyHV-1引起的CPE特征是由细胞质空泡化及考德里氏A型核内包涵体形成的。病毒的全基因组序列为291 kbp的线性双链DNA，GenBank编号为JQ815363(Davison et al.，2013)。来自日本的CyHV-1分离株之间的限制性内切酶裂解图谱略有差异，这表明该病毒的基因组DNA序列存在变异(Sano et al.，1991a)。

5.3.3 诊断

鲤鱼或锦鲤的鱼鳍、皮肤或下颌骨出现的白色或粉红色乳头状瘤病变可能是CyHV-1引起的病毒性乳头瘤。使用FHM或KF-1等细胞系从细胞培养的乳头状瘤组织中分离病毒是困难的，因为感染性病毒的出现取决于乳头瘤的发展阶段。利用特异性兔抗血清进行间接免疫荧光抗体试验(IFAT)可以检测组织中的病毒抗原(Sano et al.，1991b)，使用DNA探针进行原位杂交的检测技术也已发展起来(Sano et al.，1993b)。该病毒的全基因组序列已知，包括DNA聚合酶的一些基因已被注释(Davison et al.，2013)，因此可与包括CyHV-3和CyHV-2等导致金鲫(*Carassius auratus*)和异育银鲫(*C. auratus gibelio*)死亡的其他相关鲤疱疹病毒的基因组序列进行比对，来设计特异性引物PCR对该病毒进行检测。

5.3.4 病理学

1. 致病性和致瘤性

Sano等(1985a，1990)首次报道了CyHV-1对鲤鱼的致病性和致瘤性，但是该病毒在草鱼(*Ctenopharyngodon idella*)、鲫鱼(*Carassius auratus*)或长颌须鮈(*Gnathopogon elongatus*)中均未造成死亡。实验感染中，CyHV-1对2周龄和4周龄鲤鱼鱼苗分别表现出高致死率和低致死率，但对8周龄鱼苗无致死损伤(Sano et al.，1991b)。此外，在15 ℃、20 ℃和25 ℃时感染病毒的鲤鱼鱼苗的累计死亡率分别为60%、16%和0%(Sano et al.，1993a)。在急性发病期间，在鱼的鳃、肝、肾和肠中均可检出病毒，这表明发生了全身感染，但在25 ℃时感染鱼的病毒检出率和浓度均低于 15 ℃和 20 ℃下的感染鱼。此外，在 15 ℃、20 ℃和 25 ℃下，分别有 35%、72.5%和 27.5%的存活鱼在皮肤、鳍或者下颌上诱发乳头状瘤(图 5.6)(Sano et al., 1993a)。实验诱发的乳头状瘤的组织病理学病变[图 5.7(A)]与自然感染中的相似(Sano et al., 1991b)。

图 5.6 培养病毒感染后存活鲤鱼的下颌上由鲤疱疹病毒 1 型(CyHV-1)引起的乳头状瘤。照片由 M. Sano 提供

水温在 25 ℃及以上时，使用 CyHV-1 多克隆抗体或 FHM 细胞接种 8 周后检测不到感染鱼中的病毒抗原和感染滴度。然而，利用鱼病毒基因组片段的原位杂交，可在存活鱼的脑部、脊髓、肝脏和皮下组织中检测到病毒 DNA (Sano et al., 1992, 1993b)，这表明 CyHV-1 在高水温下可潜伏感染鲤鱼。当水温降低时，感染鱼出现乳头状瘤，并且在乳头状瘤和正常表皮及皮下组织、脊髓神经和肝脏中均检测到病毒基因组。

Davison 等(2013)报道了在该病毒基因组中证实存在 JUNB 基因族，该基因族编码一种参与肿瘤发生的转录因子，但还需要进一步的研究来确定诱导乳头状瘤的基因。

2. 肿瘤的消退与复发

乳头状瘤会在春季随着水温升高而自发消退。在实验条件下，当水温从 14 ℃升高到 20 ℃、25 ℃和 30 ℃时，CyHV-1 诱导的乳头状瘤也会发生消退。在较高水温条件下(20～30 ℃)，消退速度很快，导致所有鱼的皮肤表面在 9 d 内便会脱落。

在消退之后，乳头状瘤通常会自然复发(Sano et al., 1991b)。由于病毒是潜伏感染，83%的存活鱼会在脱皮 7.5 个月后复发。随后，病毒可能在第二年重新激活并诱发乳头状瘤发展(Sano et al., 1993a)。

在乳头状瘤消退的第一阶段会出现白细胞炎症和水肿[图 5.7(B)]，随后的阶段包括大量的炎症、水肿，以及出现细胞坏死导致的海绵状组织，最后发生表皮脱落。注射抗鲤鱼外周血淋巴细胞(peripheral blood lymphocyte, PBL)的兔血清抑制正常鲤 PBLs 的体外细胞毒性，可在 20 ℃下延缓 CyHV－1 实验性诱导的乳头状瘤消退(Morita and Sano, 1990)。此外，在 20 ℃注射抗 PBL 血清后，3 个月前乳头状瘤自然脱落的 8 条鱼中有 3 条在 10 d 内乳头状瘤复发(N. Sano，未发表的数据)。这表明鲤 PBLs 的细胞毒活性对乳头状瘤的发生和消退具有重要作用(Morita and Sano, 1990)。

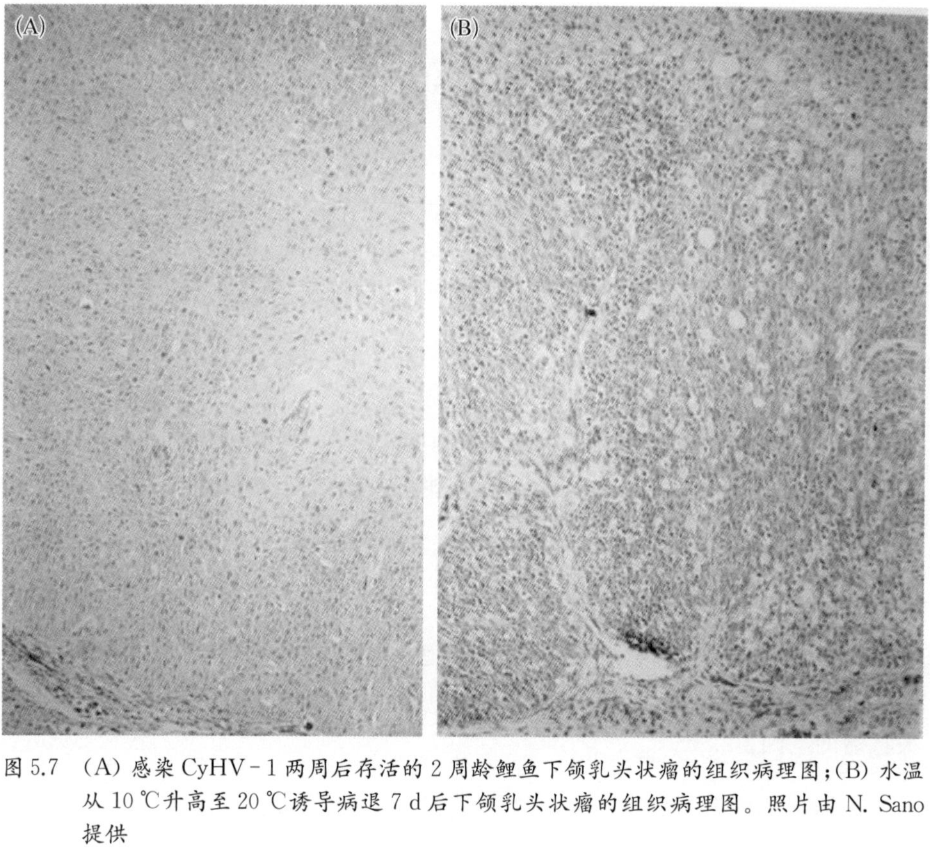

图 5.7　(A) 感染 CyHV－1 两周后存活的 2 周龄鲤鱼下颌乳头状瘤的组织病理图；(B) 水温从 10 ℃升高至 20 ℃诱导病退 7 d 后下颌乳头状瘤的组织病理图。照片由 N. Sano 提供

5.3.5 防控策略

在亚成鱼和成鱼的乳头状瘤中，病毒的传播似乎是通过带有乳头状瘤的病鱼水平传播的，因此在养殖场层面上，隔离带有乳头状瘤的病鱼是控制乳头状瘤发生的一种可行措施。此外，在锦鲤养殖场运输鱼之前，通常会进行升温处理以

诱导乳头状瘤脱落(见上文)。然而，由于 CyHV-1 没有造成养殖场鱼苗的大量死亡，因此在制订和实施以孵化为基础的控制策略方面投入的精力很少。消毒剂和生理治疗(如加热和紫外线)对感染特性和病毒存活的影响尚不清楚。

5.4 总结与研究展望

马苏大麻哈鱼病毒病是一种日本鲑科鱼类中的致瘤性和皮肤溃疡性疾病，并伴有肝炎的发生，其病原为鲑疱疹病毒 2 型(*Salmonid herpesvirus* 2，SalHV-2)，最初被认为是从马苏大麻哈鱼中分离出的一种致瘤性病毒。其他别称(缩写)包括 OMV、NeVTA、YTV、OKV、COTV、CHV、RKV 和 RHV。SalHV-2 属于异疱疹病毒科(Alloherpesviridae)，鲑疱疹病毒属(*Salmonivirus*)。其主要易感鱼类为马苏大麻哈鱼、银大麻哈鱼和虹鳟，已确认该病毒可对银大麻哈鱼和虹鳟养殖造成经济损失。使用碘伏对所有设施进行消毒和受精后立即对鱼卵进行消毒，并在发眼卵早期再次消毒，同时在使用井水或消毒河水的清洁设施中养殖，可以有效地控制 OMVD。从马苏大麻哈鱼、银大麻哈鱼和虹鳟中分离出的 OMV 毒株在致病性和致瘤性上存在差异，需要进一步的研究来确认这些毒株的异同、致病性和致瘤性，并阐明感染和肿瘤形成的机制，以及它们的基因序列。

鲤疱疹病毒 1 型(CyHV-1)的致瘤性已被证实，并在实验研究中证实了其病毒动力学，包括全身感染、潜伏感染，以及鱼乳头状瘤的发生和消退。鱼的免疫系统也可以在病毒引起的肿瘤发展和消退过程中发挥重要作用。与使用培养病毒的实验感染不同，在日本养殖场中 CyHV-1 并没有引起鱼苗的大规模死亡。在温暖的季节，鲤鱼成鱼即使没有明显的肿瘤出现，也会发生潜伏或持续感染，这些鲤鱼可能是感染的主要来源。对 CyHV-1 还需要进一步的流行病学研究，特别是其生命周期与乳头状瘤发生的相互关系。由于病毒基因组序列预测了致瘤基因的存在，因此应该对这些致瘤基因编码的蛋白进行进一步的鉴定和功能研究。乳头状瘤的发生和消退似乎依赖于宿主免疫系统的状态，因此，更准确地确定该系统在宿主-病原相互关系中的作用以及如何利用它有利于宿主将是很有价值的。根据经验，养殖场的锦鲤品系对病毒的易感性可能存在差异，因此这也将有助于更好地理解疾病发展过程，并促进开发不易感鱼品系的育种计划。

参考文献

[1] Adkison, M.A., Gilad, O. and Hedrick, R.P. (2005) An enzyme linked immunosorbent assay (ELISA) for detection of antibodies to the koi herpesvirus (KHV) in the serum of

koi Cyprinus carpio. *Fish Pathology* 40, 53 - 62.

[2] Amos, K.H. (1985) *Procedures for the Detection and Identification of Certain Fish Pathogens*, 3rd edn. Fish Health Section, American Fisheries Society, Corvallis, Oregon.

[3] Anders, K. (1989) A herpesvirus associated with an epizootic epidermal papillomatosis in European smelt (*Osmerus eperlanus*). In: Ahne, W. and Kurstak, E. (eds) *Viruses of Lower Vertebrates*. Springer, Berlin, pp. 184 - 197.

[4] Anders, K. and Yoshimizu, M. (1994) Role of viruses in the induction of skin tumours and tumour-like proliferations of fish. *Diseases of Aquatic Organisms* 19, 215 - 232.

[5] Archetti, I. and Horsfall, F. L. (1950) Persistent antigenic variation of influenza A viruses after incomplete neutralization in vivo with heterologous immune serum. *Journal of Experimental Medicine* 92, 441 - 462.

[6] Aso, Y., Wani, J., Antonio, S-K. D. and Yoshimizu, M. (2001) Detection and identification of *Oncorhynchus masou* virus (OMV) disease by polymerase chain reaction (PCR). *Bulletin of Fisheries Sciences, Hokkaido University* 52, 111 - 116.

[7] Calle, P.P., McNamara, T. and Kress, Y. (1999) Herpesvirus-associated papillomas in koi carp (*Cyprinus carpio*). *Journal of Zoo and Wildlife Medicine* 30, 165 - 169.

[8] Davison, A.J., Kurobe, T., Gatherer, D., Cunningham, C., Korf, I. *et al.* (2013) Comparative genomics of carp herpesviruses. *Journal of Virology* 87, 2908 - 2922.

[9] Dixon, R. and Farber, F. (1980) Channel catfish virus: physiochemical properties of viral genome and identification of viral polypeptides. *Virology* 103, 267 - 278.

[10] Eaton, W.D., Wingfield, W.H. and Hedrick, R.P. (1991) Comparison of the DNA homologies of five salmonid herpesvirus. *Fish Pathology* 26, 183 - 187.

[11] Furihata, M. (2008) Study on the Salmonid herpesvirus infection of rainbow trout. *Bulletin of Nagano Prefectural Fisheries Experimental Station* 10, 1 - 41.

[12] Furihata, M., Hosoe, A., Takei, K., Kohara, M., Nakamura, J. *et al.* (2003) Outbreak of salmonid herpesviral disease in cultured rainbow trout. *Fish Pathology* 38, 23 - 25.

[13] Furihata, M., Suzuki, K., Hosoe, A. and Miyazaki, T. (2004) Histopathological study on *Oncorhynchus masou* virus disease (OMVD) of cultured rainbow trout in natural outbreaks, artificial infection. *Fish Pathology* 40, 161 - 167.

[14] Gou, D.F., Kubota, H., Onuma, M. and Kodama, H. (1991) Detection of salmonid herpesvirus (*Oncorhynchus masou virus*) in fish by Southern-blot technique. *Journal of Veterinary Medical Science* 53, 43 - 48.

[15] Hatori, S., Motonishi, A., Nishizawa, T. and Yoshimizu, M. (2003) Virucidal effect of disinfectants against *Oncorhynchus masou* virus (OMV). *Fish Pathology* 38, 185 - 187.

[16] Hayashi, Y., Kodama, H., Mikami, T. and Izawa, H. (1987) Analysis of three salmonid herpesvirus DNAs by restriction endonuclease cleavage patterns. *Japanese Journal of Veterinary Science* 49, 251 - 260.

[17] Hayashi, Y., Izawa, H., Mikami, T. and Kodama, H. (1993) A monoclonal antibody cross-reactive with three salmonid herpesviruses. *Journal of Fish Diseases* 16, 479 - 486.

[18] Hedrick, R.P., McDowell, T., Eaton, W.D., Kimura, T. and Sano, T. (1987) Serological relationships of five herpesviruses isolated from salmonid fishes. *Journal of Applied Ichthyology* 3, 87 - 92.

[19] Horiuchi, M., Miyazawa, M., Nakata, M., Iida, K. and Nishimura, S. (1989) A case of herpesvirus infection of fresh water-reared coho salmon *Oncorhynchus kisutch* in Japan. *Suisan-Zoushoku* 36, 297 - 305.

[20] Igari, T., Fukuda, H. and Sano, T. (1991) The restriction endonuclease cleavage

patterns of the salmonid herpesvirus strain's DNAs. *Fish Pathology* 26, 45 - 46.

[21] Kamei, Y., Yoshimizu, M. and Kimura, T. (1987) Plaque assay of *Oncorhynchus masou* virus (OMV). *Fish Pathology* 22, 147 - 152.

[22] Kasai, H., Nomura, T. and Yoshimizu, M. (2004) Surveillance and control of salmonid viruses of wild salmonid fish returning to the northern part of Japan, from 1976 to 2002. In: *Proceedings of the Japan - Korea Joint Seminar on Fisheries Sciences*. Hokkaido University, Hakodate, Japan, pp. 142 - 147.

[23] Keysselitz, G. (1908) Über ein Epithelioma der Barden. *Archiv für Protistenkunde* 11, 326 - 333.

[24] Kimura, I. (1976) Tumor of lower vertebrates. In: Sugiyama, T. and Yamamoto, Y. (eds) *Cancer*. Iwanami-Shoten, Tokyo, pp. 270 - 283.

[25] Kimura, T. and Yoshimizu, M. (1989) Salmon herpesvirus: OMV, *Oncorhynchus masou* virus. In: Ahne, W. and Kurstak, E. (eds) *Viruses of Lower Vertebrates*. Springer, Berlin, pp. 171 - 183.

[26] Kimura, T., Yoshimizu, M., Tanaka, M. and Sannohe, H. (1981a) Studies on a new virus (OMV) from *Oncorhynchus masou* virus (OMV) I. Characteristics and pathogenicity. *Fish Pathology* 15, 143 - 147.

[27] Kimura, T., Yoshimizu, M. and Tanaka, M. (1981b) Studies on a new virus (OMV) from *Oncorhynchus masou* II. Oncogenic nature. *Fish Pathology* 15, 149 - 153.

[28] Kimura, T., Yoshimizu, M. and Tanaka, M. (1981c) Fish viruses: tumor induction in *Oncorhynchus keta* by the herpesvirus. In: Dawe, C.J., Harshbarger, J.C., Kondo, S., Sugimura, T. and Takayama, S. (eds) *Phyletic Approaches to Cancer*. Japan Scientific Societies Press, Tokyo, pp. 59 - 68.

[29] Kimura, T., Yoshimizu, M. and Tanaka, M. (1983a) Susceptibility of different fry stages of representative salmonid species to *Oncorhynchus masou* virus (OMV). *Fish Pathology* 17, 251 - 258.

[30] Kimura, T., Suzuki, S. and Yoshimizu, M. (1983b) *In vitro* antiviral effect of 9-(2-hydroxyethoxymethyl) guanine on the fish herpesvirus *Oncorhynchus masou* virus (OMV). *Antiviral Research* 3, 93 - 101.

[31] Kimura, T., Suzuki, S. and Yoshimizu, M. (1983c) *In vivo* antiviral effect of 9-(2-hydroxyethoxymethyl) guanine on experimental infection of chum salmon (*Oncorhynchus keta*) fry with *Oncorhynchus masou* virus (OMV). *Antiviral Research* 3, 103 - 108.

[32] Kohara, M. and Denda, I. (2008) Production of allotriploid Shinsyu Salmon, by chromosome manipulation. *Fish Genetics and Breeding Science* 37, 61 - 65.

[33] Kumagai, A., Takahashi, K. and Fukuda, H. (1994) Epizootics caused by salmonid herpesvirus type 2 infection in maricultured coho salmon. *Fish Pathology* 29, 127 - 134.

[34] Lee, S. and Whitfield, P.J. (1992) Virus-associated spawning papillomatosis in smelt, *Osmerus eperlanus* L., in the River Thames. *Journal of Fish Biology* 40, 503 - 510.

[35] McAllister, P.E. and Herman, R.L. (1989) Epizootic mortality in hatchery-reared lake trout *Salvelinus namaycush* caused by a putative virus possibly of the herpesvirus group. *Diseases of Aquatic Organisms* 6, 113 - 119.

[36] Morita, N. and Sano, T. (1990) Regression effect of carp, *Cyprinus carpio* L., peripheral blood lymphocytes on CHV-induced carp papilloma. *Journal of Fish Diseases* 13, 505 - 511.

[37] Mulcahy, M.F. and O'Leary, A. (1970) Cell-free transmission of lymphosarcomas in the northern pike (*Esox lucius*). *Cellular and Molecular Life Sciences* 26, 891.

[38] Peters, N. and Waterman, B. (1979) Three types of skin papillomas of flatfishes and their causes. *Marine Ecology - Progress Series* 1, 269 - 276.

[39] Sano, N., Hondo, R., Fukuda, H. and Sano, T. (1991a) *Herpesvirus cyprini*:

restriction endonuclease cleavage profiles of the viral DNA. *Fish Pathology* 26, 207 - 208.

[40] Sano, N., Sano, M., Sano, T. and Hondo, R. (1992) *Herpesvirus cyprini*: detection of the viral genome by *in situ* hybridization. *Journal of Fish Diseases* 15, 153 - 162.

[41] Sano, N., Moriwake, M. and Sano, T. (1993a) *Herpesvirus cyprini*: thermal effects on pathogenicity and oncogenicity. *Fish Pathology* 28, 171 - 175.

[42] Sano, N., Moriwake, M., Hondo, R. and Sano, T. (1993b) *Herpesvirus cyprini*: a search for viral genome in infected fish by *in situ* hybridization. *Journal of Fish Diseases* 16, 495 - 499.

[43] Sano, T. (1976) Viral diseases of cultured fishes in Japan. *Fish Pathology* 10, 221 - 226.

[44] Sano, T., Fukuda, H., Okamoto, N. and Kaneko, F. (1983) Yamame tumor virus: lethality and oncogenicity. *Bulletin of the Japanese Society of Scientific Fisheries* 49, 1159 - 1163.

[45] Sano, T., Fukuda, H. and Furukawa, M. (1985a) *Herpesvirus cyprini*: biological and oncogenic properties. *Fish Pathology* 20, 381 - 388.

[46] Sano, T., Fukuda, H., Furukawa, M., Hosoya, H. and Moriya, Y. (1985b) A herpesvirus isolated from carp papilloma in Japan. In: Ellis, A. E. (ed.) *Fish and Shellfish Pathology*. Academic Press, London, pp. 307 - 311.

[47] Sano, T., Morita, N., Shima, N. and Akimoto, M. (1990) A preliminary report of pathogenicity and oncogenicity of cyprinid herpesvirus. *Bulletin of the European Association of Fish Pathologists* 10, 11 - 13.

[48] Sano, T., Morita, N., Shima, N. and Akimoto, M. (1991b) *Herpesvirus cyprini*: lethality and oncogenicity. *Journal of Fish Diseases* 14, 533 - 543.

[49] Sung, J.T., Yoshimizu, M., Nomura, T. and Ezura, Y. (1996a) *Oncorhynchus masou* virus: serological relationships among salmonid herpesvirus isolated from kokanee salmon, masu salmon, coho salmon and rainbow trout. *Scientific Report of the Hokkaido Salmon Hatchery* 50, 139 - 144.

[50] Sung, J-T., Yoshimizu, M., Nomura, T. and Ezura, Y. (1996b) *Oncorhynchus masou* virus: pathogenicity of salmonid herpesvirus-2 strains against masu salmon. *Scientific Report of the Hokkaido Salmon Hatchery* 50, 145 - 148.

[51] Suzuki, K. (1993) A new viral disease on rainbow trout. *Shikenkenkyuwa-Ima* 165, 1 - 2.

[52] Suzuki, S., Yoshimizu, M. and Saneyoshi, M. (1992) Detection of viral DNA polymerase activity in salmon tumor tissue induced by herpesvirus, *Oncorhynchus masou* virus. *Acta Virologica* 36, 326 - 328.

[53] Tanaka, M., Yoshimizu, M. and Kimura, T. (1984) *Oncorhynchus masou* virus: pathological changes in masu salmon (*Oncorhynchus masou*), chum salmon (*O. keta*) and coho salmon (*O. kisutch*) fry infected with OMV by immersion method. *Nippon Suisan Gakkaishi* 50, 431 - 437.

[54] Walker, R. (1969) Virus associated with epidermal hyperplasia in fish. *National Cancer Institute Monograph* 31, 195 - 207.

[55] Winqvist, G., Ljungberg, O. and Hellstroem, B. (1968) Skin tumours of northern pike (*Esox lucius* L.) II. Viral particles in epidermal proliferations. *Bulletin - Office International des Épizooties* 69, 1023 - 1031.

[56] Yoshimizu, M. (2003a) Control strategy for viral diseases of salmonids and flounder. In: Lee, C.S. and Bryen, P.J.O (eds) *Biosecurity in Aquaculture Production Systems: Exclusion of Pathogens and Other Undesirables*. World Aquaculture Society, Baton Rouge, Louisiana, pp. 35 - 41.

[57] Yoshimizu, M. (2003b) Chapter 2.1.3. *Oncorhynchus masou* virus disease. In: *Manual*

of Diagnostic Tests for Aquatic Animals, 4th edn. World Organisation for Animal Health, Paris, pp. 100 – 107. Available at: http://www.oie.int/doc/ged/D6505.PDF (accessed 27 October 2016).

[58] Yoshimizu, M. (2009) Control strategy for viral diseases of salmonid fish, flounders and shrimp at hatchery and seeds production facility in Japan. *Fish Pathology* 44, 9 – 13.

[59] Yoshimizu, M. and Kasai, H. (2011) Oncogenic viruses and *Oncorhynchus masou* virus. In: Woo, P. T. K. and Bruno, D. D. (eds) *Fish Diseases and Disorders. Volume 3, Viral, Bacterial and Fungal Infections*, 2nd edn. CAB International, Wallingford, UK, pp. 276 – 301.

[60] Yoshimizu, M., Kimura, T. and Winton, J. R. (1985) An improved technique for collecting reproductive fluid samples from salmonid fishes. *Progressive Fish-Culturist* 47, 199 – 200.

[61] Yoshimizu, M., Takizawa, H. and Kimura, T. (1986) U.V. susceptibility of some fish pathogenic viruses. *Fish Pathology* 21, 47 – 52.

[62] Yoshimizu, M., Kimura, T. and Tanaka, M. (1987) *Oncorhynchus masou* virus (OMV): incidence of tumor development among experimentally infected representative salmonid species. *Fish Pathology* 22, 7 – 10.

[63] Yoshimizu, M., Tanaka, M. and Kimura, T. (1988a) Histopathological study of tumors induced by *Oncorhynchus masou* virus (OMV) infection. *Fish Pathology* 23, 133 – 138.

[64] Yoshimizu, M., Kamei, M., Dirakubusarakom, S. and Kimura, T. (1988b) Fish cell lines: susceptibility to salmonid viruses. In: Kuroda, Y., Kurstak, K. and Maramorosch, K. (eds) *Invertebrate and Fish Tissue Culture*. Japan Scientific Societies Press, Tokyo/Springer, Berlin, pp. 207 – 210.

[65] Yoshimizu, M., Sami, M. and Kimura, T. (1989) Survivability of infectious hematopoietic necrosis virus (IHNV) in fertilized eggs of masu (*Oncorhynchus masou*) and chum salmon (*O. keta*). *Journal of Aquatic Animal Health* 1, 13 – 20.

[66] Yoshimizu, M., Nomura, T., Awakura, T., Ezura, Y. and Kimura, T. (1993) Surveillance and control of infectious hematopoietic necrosis virus (IHNV) and *Oncorhynchus masou* virus (OMV) of wild salmonid fish returning to the northern part of Japan 1976 – 1991. *Fisheries Research* 17, 163 – 173.

[67] Yoshimizu, M., Fukuda, H., Sano, T. and Kimura, T. (1995) Salmonid herpesvirus 2. Epidemiology and serological relationship. *Veterinary Research* 26, 486 – 492.

[68] Yoshimizu, M., Furihata, M. and Motonishi, A. (2002) Control of *Oncorhynchus masou* virus disease (OMVD), salmonid disease notify [? notification] to OIE. In: *Report for the Disease Section, Japan Fisheries Resource Conservation*. Tokyo, pp. 101 – 108.

[69] Yoshimizu, M., Yoshinaka, T., Hatori, S. and Kasai, H. (2005) Survivability of fish pathogenic viruses in environmental water, and inactivation of fish viruses. *Bulletin of Fisheries Research Agency*, Suppl. No. 2, 47 – 54.

[70] Yoshimizu, M., Kasai, H. and Sakoda, Y. (2012) Salmonid herpesvirus disease (*Oncorhynchus masou* virus disease; OMVD). In: *Proceedings of the 46th Annual Summer Symposium of Japanese Society for Virology Hokkaido Branch*, July 21 – 22, Ootaki, Hokkaido, Japan, p. 6.

[71] Yoshimizu, M., Kasai, H. and Watanabe, K. (2014) Approaches to probiotics for aquaculture – Prevention of fish viral diseases using antiviral intestinal bacteria. *Journal of Intestinal Microbiology* 28, 7 – 14.

6

传染性鲑贫血症

Knut Falk* 和 Maria Aamelfot

6.1 引言

传染性鲑贫血症(infectious salmon anaemia, ISA)是发生在养殖大西洋鲑中的一种严重传染性病毒病,1984 年首次在挪威被报道(Thorud and Djupvik, 1988)。ISA 的暴发对大西洋鲑养殖业造成了严重的经济损失,并引发了大范围生物安全措施的实施。目前,在大多数大西洋鲑养殖地区,包括加拿大和美国东海岸、苏格兰、挪威、法罗群岛和智利等,都已报告了 ISA 的暴发(Rimstad et al., 2011)。在智利和法罗群岛,这种疾病造成了鲑鱼养殖业的重大经济损失,给整个行业的前景带来了不确定性(Mardones et al., 2009; Christiansen et al., 2011),在一定程度上如同 1989 年及随后的挪威(Håstein et al., 1999; Rimstad et al., 2011)。20 世纪 90 年代初期,ISA 被世界动物卫生组织(OIE)列为应具报疾病(Håstein et al., 1999; OIE, 2015a)。

在最初发现 ISA 之后,该疾病在挪威水产养殖业迅速蔓延,导致 1990 年出现了 90 多起疫情的暴发高峰(Håstein et al., 1999)。尽管 ISA 的致病病毒直到 1995 年才被鉴定出来(Dannevig et al., 1995),但人们已经采取了生物安全措施来防控该疾病。这些措施包括疾病的及早发现、宰杀患病种群、制定严格的运输安全条例、对屠宰场的鱼内脏和废弃物进行消毒、养殖场按鱼龄分类隔离以及改进健康管理和认证。这些措施取得显著成效,在 1994 年只报道了两起新的疫情,这表明无须使用药物和疫苗,甚至无须了解病因、疾病的流行病学与发病机理,就可以控制 ISA。目前 ISA 在挪威依然会发生,但患病率很低,而其他之前受影响的国家似乎也控制住了该疾病,没有或只有少数零星的疫情发生。有关 ISA 的流行病学信息可以在 OIE 的世界动物卫生信息系统(world animal health information system, WAHIS)数据库(WAHID)(http://www.oie.int/wahis_2/public/wahid.php/Wahidhome/Home)中找到。

ISA 通常不会在疫情开始时造成高的死亡率,如果采取适当的控制措施,

* 通信作者邮箱:knut.falk@vetinst.no。

疫情可能会在较低死亡率的情况下得到控制。即便施行了各种措施和限制，还是可能带来严重的经济损失。然而，如果不采取充分和适当的措施，该疾病的暴发可能会给有关鲑鱼养殖场及其邻近的养殖场造成严重的疾病问题(Lyngstad et al., 2008)。事实上，在挪威有很多实例表明，单一的 ISA 疫情已经发展成小规模的地方性流行病，在这种情况下，唯一的解决办法就是整个地区休耕。

ISA 的病原体为传染性鲑贫血症病毒(ISAV)，是正黏病毒科(Orthomyxoviridae)鲑传贫病毒属(*Isavirus*)的唯一成员(Palese and Shaw, 2007)。嵌入 ISAV 包膜的两种糖蛋白，即血凝素酯酶(heamagglutinin esterase, HE)糖蛋白和融合(fusion, F)糖蛋白，对病毒的摄取和细胞嗜性具有重要作用(Falk et al., 2004; Aspehaug et al., 2005; Aamelfot et al., 2012)。此外，病毒粒子由另外两种主要结构蛋白[核蛋白(nucleoprotein, NP)(Aspehaug et al., 2004; Falk et al., 2004)和基质(matrix, M)蛋白(Biering et al., 2002; Falk et al., 2004)]组成。HE 蛋白与 ISAV 的细胞受体 4 - *O* - 乙酰化唾液酸(4 - *O* - acetylated sialic acids)结合(Hellebo et al., 2004)，该受体在宿主的内皮细胞和红细胞(red blood cells, RBCs)上表达(Aamelfot et al., 2012)。

分节段的 ISAV 基因组高度保守，变异性最高的是编码 HE 和 F 蛋白的两个基因片段。基于 HE 基因对 ISAV 的分离株进行系统发育分析，揭示 ISAV 主要有两个进化枝，一个是欧洲分枝，另一个是北美分枝。此外，基于一个高度多态性区域(highly polymorphic region, HPR)的氨基酸模式对 ISAV 进行了鉴定和分型，该区域位于 HE 跨膜结构域上游，包含 11～35 个氨基酸残基(Rimstad et al., 2011)。强毒性 HPR 变种的来源可能是由推定的非毒性种全长祖先序列(HPR0)经过差异缺失形成的(Mjaaland et al., 2002)。HPR0 变种最初是在苏格兰野生鲑中被发现的(Cunningham et al., 2002)。尽管从 ISA 疫情中分离出来的所有 ISAV 分离株在 HPR 都有缺失，通常被表示为 HPR 缺失，但 HPR0 亚型却与 ISA 的临床或病理体征无关(Cunningham et al., 2002; Mjaaland et al., 2002; Cook-Versloot et al., 2004; McBeath et al., 2009; Christiansen et al., 2011)。流行病学研究表明，HPR0 变种常出现在海水养殖的大西洋鲑中。HPR0 株在自然环境下似乎更具有季节性和短暂性，并表现出细胞和组织趋向性，在鳃上皮细胞中普遍存在(Christiansen et al., 2011; Lyngstad et al., 2012; Aamelfot et al., 2016)，也可能在皮肤上(Aamelfot et al., 2016)。与 HPR 缺失型 ISAV 不同，HPR0 型的一个特性是不能在当前可用的培养细胞中复制(Christiansen et al., 2011)。在 HPR0 型 ISAV 贮主中出现致病性 HPR 缺失型 ISAV 变种的风险较低，但也不可忽略[Christiansen et al., 2011; EFSA Panel on Animal Health and Welfare(AHAW), 2012;

Lyngstad et al., 2012]。

其他基因片段也可能对 ISA 疫情的发展起作用。在 F 蛋白中已鉴定出一个可能的毒力标志，在推定的蛋白切割激活位点附近的单个氨基酸取代或序列插入是毒力的先决条件（Kibenge et al., 2007; Markussen et al., 2008; Fourrier et al., 2015）。事实上，Fourrier 等（2015）最近的研究表明，HPR 缺失和 F 蛋白中某些氨基酸取代的共同作用下，可能会影响 F 蛋白的蛋白水解激活和蛋白活性。

ISA 疫情仅在养殖大西洋鲑中被发现，而且大多数病例发生在鱼生命周期的海水阶段。然而，感染实验表明，不同发育阶段的大西洋鲑在淡水和海水中都容易受到感染。实验证实，该病毒可在其他一些鱼类中复制但无临床疾病表现，这包括褐鳟（*S. trutta* L.）、虹鳟、北极红点鲑（*Salvelinus alpinus* L.）、大麻哈鱼、银大麻哈鱼、大西洋鲱（*Clupea harengus* L.）和大西洋鳕（*Gadus morhua* L.）（Rimstad et al., 2011）。在健康的野生大西洋鲑和褐鳟中也检测到 ISAV（Raynard et al., 2001; Plarre et al., 2005）。这些野生鱼类或其他鱼类可能成为病毒的携带者或贮主。

ISAV 的传染性预计在宿主之外可保持很长时间。在一次实验中使用不同温度下的淡水和海水，Tapia 等（2013）发现病毒在 20 ℃的海水中可以存活 5 d，而在 10 ℃的淡水中可以存活 70 d，但在自然条件下，病毒会附着在一些有机物上并被其保护，人们并不了解此时的病毒存活情况。共栖实验证实 ISAV 可以通过水媒传播，这可能是 ISA 在邻近渔场内部和之间传播的重要途径（Thorud and Djupvik, 1988; Lyngstad et al., 2008）。病毒可以通过各种途径进入水中，包括皮肤、黏液、粪便、尿液和血液，以及来自死鱼的废弃物（Totland et al., 1996）。病毒侵入鱼体内的主要途径被认为是鳃（Rimstad et al., 2011），尽管最近的研究表明病毒的早期复制在鳃和皮肤中均有发生（Aamelfot et al., 2015）。

在养殖场内部和养殖场间暴发疾病期间，病毒的水平传播都已经被完整记录在案，但在大多数情况下，调查人员并不知道这种感染最初是如何传入的。某些病例中，感染或疫情发生在被感染的二龄幼鲑转入海水养殖之后，而各种养殖设备（包括活鱼舱船）的转运输送也是重要的传播途径（Vågsholm et al., 1994; Jarp and Karlsen et al., 1997; Murray et al., 2002）。

已有人提出了从频繁出现的 HPR0 型转化成 HPR 缺失的强毒型病毒的假设（Cunningham et al., 2002; Mjaaland et al., 2002; Lyngstad et al., 2012），不过这种现象的显著性和频率还没有被记载。海洋环境中的病毒贮主也是一种可能的来源，因为一些鱼类会发生亚临床感染。

最后，一些报道提出了垂直传播的可能性（Melville and Griffiths, 1999; Nylund et al., 2007; Vike et al., 2009; Marshall et al., 2014），但是并未得到

证实(Rimstad et al., 2011)。挪威鲑鱼养殖业的经验资料表明,垂直传播问题并不值得重点关注。

6.2 诊断

大西洋鲑养殖场中暴发的 ISA 在病程发展、临床症状和组织学病变存在相当大的差异,这反映了病毒、宿主和环境之间复杂的相互作用。ISA 的一个特点是,临床疾病往往以非系统性方式传播,在养殖场中从一个网箱慢慢地传播到另一个网箱,这可能反映了从感染到发展为严重的贫血和临床疾病需要较长的时间,而自然暴发的潜伏期从几周到几个月不等(Vågsholm et al., 1994; Jarp and Karlsen, 1997)。

患病鱼通常嗜睡,经常表现出不正常的游动行为,日死亡率一般为 0.05%~0.1%。如果不采取措施限制疾病的发展,疾病可能会蔓延,养殖场中的累计死亡率可在几个月内达到 80%以上。在低死亡率的疫情中,临床症状和宏观病理变化可能仅限于贫血和包括出血在内的循环障碍。这种低死亡率的慢性疾病阶段很容易被忽视。有时,几周内便有可能随之发生急性发作、高死亡率疫情,特别是在不采取任何措施的情况下。接下来的病理更加严重,以腹水和出血为主。该疾病全年均会出现,但在春季、夏初和秋末更容易发生。

ISA 最显著的临床症状包括白鳃(鳃瘀血的情况除外)、眼球突出、腹胀、眼球前房出血,有时还有皮肤出血,尤其是腹部皮肤(图 6.1),以及存鳞袋(scale pockets)水肿。也常出现血细胞比容值低于 10%的严重贫血(Thorud and Djupvik, 1988; Evensen et al., 1991; Thorud, 1991; Rimstad et al., 2011)。

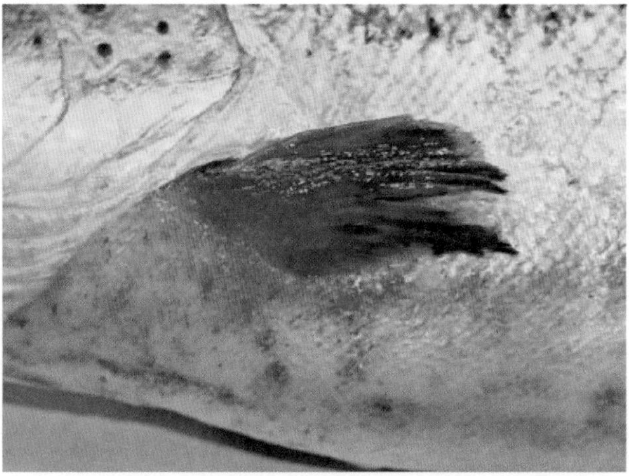

图 6.1 患有传染性鲑贫血症(ISA)的大西洋鲑表现出典型的皮肤出血症状

ISA 的诊断基于临床体征、肉眼可见的损伤和组织学病变的综合评价，并辅以内皮感染的免疫组织化学检查。免疫组织化学检查尤为重要，因为这种方法可能建立起病毒和疾病之间的直接联系。之后，通过实时定量反转录 PCR 和病毒分离可以证实 IHC 的阳性结果。通常，也会对 HE 和 F 基因进行测序，以用于流行病学评估和确定病毒类型。OIE 的《水生动物诊断检测指南》(OIE, 2015b)中对这些诊断方法进行了说明。

鉴别诊断其他贫血和出血性疾病，如红细胞包涵体综合征、冬季溃疡和黏性放线菌(*Moritella viscosa*)感染引起的败血症。在大西洋鲑病例中，血细胞比容值低于 10%并不是 ISA 的独特病理结果，然而，应始终对没有任何明显原因的低血细胞比容病例进行 ISAV 检测(OIE, 2015b)。

6.3 病理学

根据感染剂量的大小、病毒株、水温、鱼龄和免疫状况，感染 HPR 缺失型 ISAV 的鱼可能没有明显病理变化也可能有严重病变。ISA 最显著的症状是贫血(通常血细胞比容值低于 10%)和血液循环障碍。外部症状包括鳃苍白、眼睛和皮肤局部出血(图 6.1)、眼球突出和鳞片水肿。然而，该疾病可能以不同的表现形式出现，病程发展可能为急性或以慢性病形式缓慢发展(Rimstad et al., 2011)。

腹水、脾肿大、水肿和浆膜瘀点出血为常见症状，更多变的症状是肝、肾、肠或鳃的严重出血病变，但出现时非常明显。ISA 引发的“经典”肝脏病变表现特征为出血性坏死引起的深色肝(Evensen et al., 1991)。

缓慢发展的慢性 ISA 表现出的临床症状和病理可能更加微妙。肝脏会呈现灰白或微黄，并且贫血可能不像急性疾病那样严重。与急性型相比，腹水较少，但皮肤和鱼鳔出血、存鳞袋和鱼鳔水肿比急性病鱼更明显(Evensen et al., 1991; Rimstad et al., 2011)。

肝脏组织学病变包括肝细胞带状变性和坏死(图 6.2)(Evensen et al., 1991; Speilberg et al., 1995; Simko et al., 2000)。病变导致肝脏广泛充血，并伴有血窦扩张，后期出现充血腔。Speilberg 等(1995)研究了实验性感染鲑的肝脏病理，发现肝细胞变性和多灶坏死之前的窦内皮细胞变性和丢失，但在受感染的肝细胞内并未发现病毒。这些观察结果都得到了 IHC 检查的证实，在病理损伤的肝细胞中并未检测到病毒(Aamelfot et al., 2012)。

出血在肝、肾、肠和鳃等部位的临床表现各不相同。脾常表现为肿大、暗沉(Rimstad et al., 2011; Aamelfot, 2012)；肾脏病理表现特征为中度肿胀，并伴有间质性出血和部分肾小管坏死(Byrne et al., 1998; Simko et al., 2000)(图 6.3)。肠道则因肠壁而非肠腔内(新鲜样本)出血呈暗红色，由于血液已经在鳃

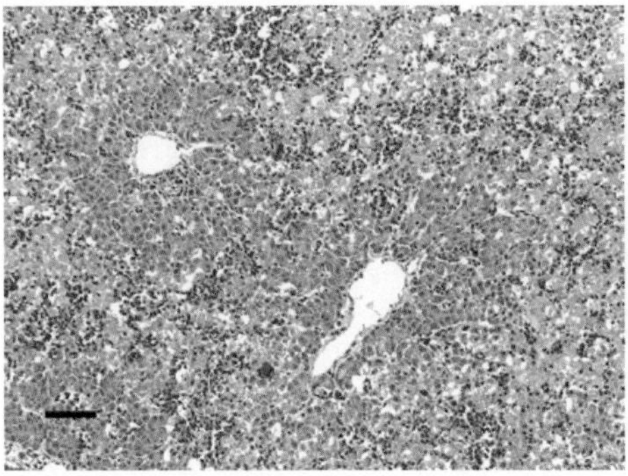

图 6.2 患传染性鲑贫血症的大西洋鲑组织切片显示肝脏带状出血坏死(苏木精-伊红染色)。标尺: 50 μm。图片由 Agnar Kvellestad 博士提供

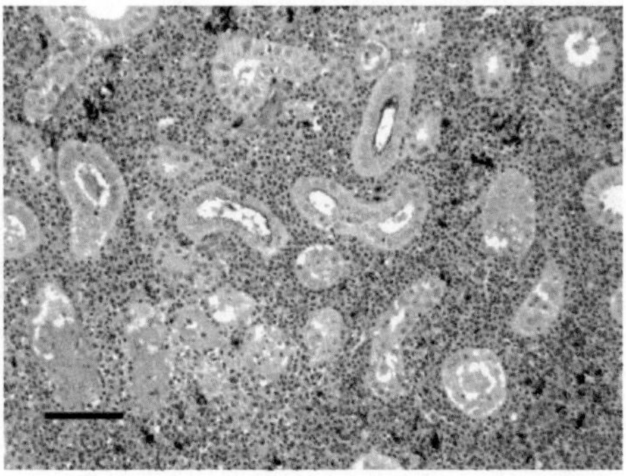

图 6.3 患传染性鲑贫血症的大西洋鲑组织切片显示肾出血和肾小管坏死(苏木精-伊红染色)。标尺: 50 μm

部尤其是鳃丝中央静脉窦中积累,因此鳃表现为苍白贫血鳃的一个例外。尸检中可见肝和肠等器官的出血性病灶,但在鳃和肾中并不明显。在一些特定的 ISA 疫情中,一种出血性器官病变表现可能占主导地位,而在其他疫情中,即便在同一鱼体内也可以发现所有病症。以肝或肾病变表现为主的疫情最为常见。然而,器官出血性病变在疫情暴发的最初阶段可能不存在或非常罕见,只留下贫血和更轻微的血液循环障碍作为病因学的初步线索。

6.4 发病机制

ISA 是养殖大西洋鲑的一种全身性致命疾病,其晚期特征为贫血、出血和血液循环障碍。尽管可能的因素开始浮现,但该疾病病理背后的确切机制目前仍不清楚(Aamelfot et al., 2014)。即便对于假定相同的分离株,其疾病表现和严重程度也常存在显著差异。因此,在评价 ISA 的发病机制时,必须考虑宿主、致病因子和环境(病因学三要素)之间的相互作用。

Thorud(1991)对养殖场疫情中的贫血鱼进行了检查,发现红细胞脆性增加、未成熟红细胞数量增加和红细胞血影或破碎细胞比例增加,所有这些都表明发生了溶血性贫血。然而,并没有观察到黄疸或血红蛋白过度分解的其他迹象,这可能是因为贫血发生在疾病的晚期。观察到白细胞减少,包括淋巴细胞和血小板减少,这是由疾病的一般应激所致。其他血浆参数显示谷草转氨酶(又称天冬氨酸转氨酶)、谷丙转氨酶(又称丙氨酸转氨酶)和乳酸脱氢酶含量增多,表明存在器官损伤。最终,血浆渗透压升高表明渗透压调节受损。

病毒毒力是由多种因素决定的,取决于受体结合、细胞摄取、复制速度和新病毒粒子的脱落、宿主免疫反应的调节以及向新宿主传播的能力等因素

(McBeath et al., 2007; Purcell et al., 2009; Medina and Garcia-Sastre, 2011; Peñaranda et al., 2011; Wargo and Kurath, 2012; Cauldwell et al., 2014; McBeath et al., 2015)。因此,除了宿主和环境因素外,评估发病机制还可能涉及多个特性。与相关的流感病毒相比,ISAV 具有同样的功能特性,包括病毒受体结合、融合活性、受体破坏活性、病毒聚合酶促进复制效率,以及调节宿主免疫应答的能力,这些都是疾病发展的重要因素(Palese and Shaw, 2007; Rimstad et al., 2011)。

ISAV 的主要靶细胞是所有器官血管内壁的内皮细胞,包括血窦、心内膜(图 6.4)、前肾的内皮清除细胞,可能还有辅助血管系统的内皮细胞(Rummer et al., 2014)和红细胞。Evensen 等(1991)早前便已提出内皮细胞可能是靶细胞。电镜观察证实,血管内皮受损继而产生肝细胞变性,是 ISAV 感染鲑鱼的先兆。此外,Evensen 等(1991)还发现脾脏的吞噬作用增强,Aamelfot 等(2012)也证实了这一点,他们除了发现红细胞对 ISAV 感染的内皮细胞的原位吸附外,还观察到广泛的血细胞吞噬现象。Aamelfot 等(2012)还证实了没有致细胞病变效应或明显血管周围白细胞浸润的完整内皮,即未检测到炎症反应和细胞凋亡,这表明 ISAV 感染不会在感染细胞中引起炎症反应或导致细胞病变。这与家禽中报道的系统性流感不一致,因为在流感中细胞凋亡和炎症很常见(Kobayashi et al., 1996; Schultz-Cherry et al., 1998)。据报道,在其他嗜内皮细胞病毒中,如丝状病毒(埃博拉病毒和马尔堡病毒)也有类似发现,其内皮细胞的感染也不会破坏细胞结构(Geisbert et al., 2003)。然而,受感染的内皮细胞的渗透性仍可能受到损伤。

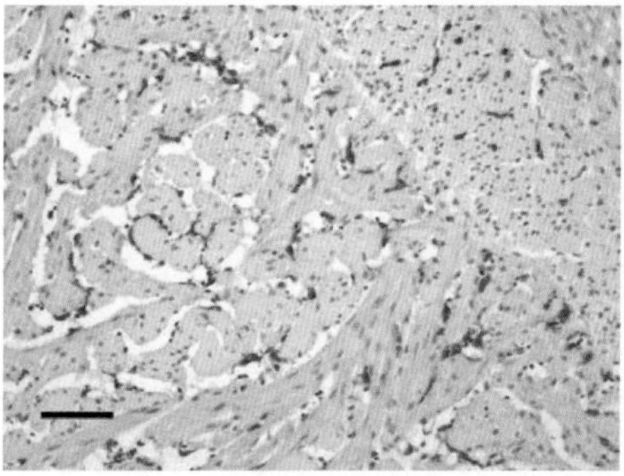

图 6.4 患传染性鲑贫血症的大西洋鲑心脏免疫组织化学切片。红色为感染 ISAV 的内皮细胞。标尺: 50 μm

Aamelfot 等(2012)也考察了 ISA 确诊疫情中严重贫血的濒死大西洋鲑的肝脏和肾脏的病理病变,其实质坏死灶中似乎并未出现 ISAV 感染。一种可能的解释是,这种病变不是由感染直接引起的,而是间接的或继发性的,是由宿主反应破坏红细胞并在严重贫血的鱼体中造成缺氧引起的。这种缺氧状态可能会引发恶性循环,导致失液性休克和死亡,类似于埃博拉病毒的内皮感染(Schnittler et al., 1993)。

虽然典型 ISA 的特征为循环系统的内皮感染,但在感染非致病性的 HPR0 型 ISAV 时,细胞向性是不同的。虽然在这种感染过程中病毒存在于鳃和皮肤中,但只有上皮细胞受到感染(Aamelfot et al., 2016)。从 HPR0 型 ISAV 向

HPR 缺失型病毒的转变修饰了融合蛋白的活性(Fourrier et al., 2015),使融合蛋白除了具有 HE 蛋白的受体结合功能外,还决定了 ISAV 感染期间病毒的细胞嗜性和毒力。此外,病毒融合活性还影响病毒的细胞摄取和复制效率。

通过强毒株和低毒株的病毒浸泡实验研究了细胞对 ISAV 的摄取(Aamelfot et al., 2015; McBeath et al., 2015),发现在这两种情况下,上皮细胞均呈现出短暂的初期感染,而低毒力病毒的早期上皮细胞感染更明显,持续时间更长,并且还能更快地扩散到内脏器官,更迅速地引起宿主的全身性免疫反应。这些免疫应答可能为鱼类提供一些保护,这可能是低毒分离株毒力和致死率较低的部分原因。

正黏病毒科病毒可以结合红细胞或凝集红细胞。ISAV 可以凝集大西洋鲑、虹鳟、马、驴和兔子的红细胞,但不凝集褐鳟的红细胞。除了大西洋鲑(Falk et al., 1997),可在所有被检测物种的红细胞上观察到预期的由病毒酯酶(即 HE 蛋白上的病毒受体破坏酶)引起的体外血细胞凝集反应洗脱。

ISAV 可与循环红细胞广泛结合,而红细胞会吸附在受感染的内皮细胞上,这可能被认为是体内血细胞吸附(Aamelfot et al., 2012),这些观察发现可能对疾病发病机制具有重要意义,并且可以解释一些现象。第一,病毒附着在红细胞上可能破坏细胞膜的完整性,使红细胞的渗透脆性增加,就像 Thorud(1991)观察到的那样。实验数据证实,红细胞上出现大量病毒与红细胞渗透脆性增加相一致(K. Falk, 2016,未发表的数据)。第二,病毒包裹的红细胞可能会被宿主的清除系统认为是外来物,这就解释了红细胞吞噬增多的现象。实际上,附着流感病毒的犬红细胞只能存活其预期寿命的一小部分(Stewart et al., 1995)。第三,ISA 病理学的早期报告中观察到的血栓(Evensen et al., 1991)和红细胞会附着到 ISAV 感染的细胞上(Aamelfot et al., 2012),也可以通过原位血凝和血细胞吸附来解释。第四,正如 Baum 等(2002)指出的那样,红细胞表面的病毒受体可能作为诱饵受体,并与病毒相互作用。这是否会像 HIV 的相关报道那样(Baum et al., 2002)促进感染的传播,或像细小病毒的相关报道那样(Traving and Schauer, 1998)推迟感染,还需要进一步观察。此外,ISAV 在体外条件下不能从大西洋鲑红细胞中释放出来(Falk et al., 1997),尽管这一现象的意义尚不清楚,但我们已经鉴定出了两株 ISAV 分离株,它们都是在体外条件下从大西洋鲑红细胞中洗脱出来的。其中一个分离株在感染实验测试中表现出低毒力,与强毒株相比,其致死率更低、病理更少(McBeath et al., 2015)。

简单来说,这些观察表明 ISAV 和红细胞之间的相互作用在其发病机制中非常重要,尤其是解释了贫血,但也可能解释了其他临床症状。对这种相互作用的进一步研究可能有助于理解其他血凝病毒的发病机制。

针对 ISAV 的先天性和获得性细胞与体液免疫反应都已经在大西洋鲑中被实验证实。这包括特异性抗体反应(Falk and Dannevig, 1995; Lauscher et al.,

2011)和干扰素相关反应(McBeath et al., 2007; LeBlanc et al., 2010; McBeath et al., 2015)。此外,ISAV 非结构蛋白具有干扰素拮抗活性(McBeath et al., 2006; Garcia-Rosado et al., 2008),不过这对发病机制和毒力方面的意义尚不清楚。

6.5 防控策略

ISA 的疫情往往发展缓慢,并且呈现时间和空间上分隔的局部性扩散,这表明该疾病通过水平传播(Lyngstad et al., 2008, 2011)。然而,养殖鲑科鱼类种群中的传染源尚不清楚,未知的病原库、鲑疮痂鱼虱(*Lepeophtheirus salmonis*)等病媒的存在(Krøyer, 1837)、亚临床感染或垂直传播都有可能。此外,还有可能是从普遍存在的 ISAV HPR0 无毒株转变成致病性 HPR 缺失株。

通过实施常规性生物安全措施,降低水平转移和感染压力,可大大降低 ISA 的发病率和影响。这些措施包括早期发现、隔离和宰杀病鱼、一般性运输限制、对来自屠宰场的内脏和废弃物进行消毒、在养殖场按鱼龄隔离,以及加强健康管理和认证(Håstein et al., 1999)。在挪威,这些措施成功地改善了水产养殖业的卫生状况。随着养殖管理的显著改善,实验室鉴定的改进以及随后对发生 ISA 疫情的渔场实施的限制,都显著而迅速地减少了 ISA 的暴发(Håstein et al., 1999)。

因此,如果及早发现、认真对待并采取正确措施,就可以控制 ISA 疫情。有许多例子表明,如果在疫情初期得不到正确处理,它们将发展为毁灭性的流行病,代价非常高。由于 ISA 的暴发,法罗群岛和智利大西洋鲑养殖业的生产和经济都曾遭遇了重大挫折(Rimstad et al., 2011)。

挪威、苏格兰、加拿大东海岸、法罗群岛和智利都经历了 ISA 大流行,这些国家为控制和防治这一疾病而采取的措施略有不同。在挪威,ISA 在大多数地区被认为是地方病,每年暴发 2～20 次,一旦发现,将采取严格的控制措施,包括宰杀和分区,以限制传播。近年来,疫苗接种变得越来越普遍。继 1998 年 ISA 在苏格兰流行之后,便实施了根除计划,目前苏格兰仍然是无 ISA 疫病国家。加拿大、法罗群岛和智利在 ISA 疫情暴发之后,组合实施了生物安全措施和疫苗接种,目前没有疫情或者偶尔暴发。

由于 ISA 病情发展缓慢,生物安全措施对疫情控制十分重要,有效疫苗的使用将作为其他控制策略的补充。目前可用的疫苗是将细胞培养的全病毒灭活后添加矿物油佐剂制备而成的灭活疫苗,这些疫苗已普遍在加拿大东海岸、法罗群岛和智利得到应用,并在某种程度上也用于挪威。欧盟规定,在已宣布无 ISA 的欧洲地区,水产养殖中不得接种 ISA 疫苗,例如苏格兰。

科学文献中关于 ISA 疫苗的报道很少。Jones 等(1999)报道了在实验性试验中使用一种添加佐剂的细胞培养灭活 ISA 病毒，获得了 84%～95%的免疫保护率(relative percentage survival, RPS)。最近，Lauscher 等(2011)也测试了病毒细胞培养制品，针对实验感染得到了 86%的 RPS。免疫保护率取决于疫苗抗原注射剂量，也与抗 ISAV 的抗体相关。一些会议的摘要和互联网出版物也支持了上述的这些研究结果。

加拿大于 1999 年首次尝试接种 ISAV 疫苗，并与控制计划和常规生物安全措施相结合。然而，在免疫接种鱼群中发生了 ISA 疫情，因此该疫苗的有效性受到质疑，目前没有关于 ISAV 疫苗养殖现场表现的报告。另一个例子是在法罗群岛，该地区在 2000 年至 2005 年多次暴发 ISA 疫情，之后通过使用 ISA 疫苗接种计划成功地控制了疫情。几乎所有法罗群岛的渔场都进行了休耕，并实施了基于常规卫生准则的严格控制制度。大多数鱼接种了细胞培养制备的油佐剂 ISA 灭活疫苗(Christiansen et al., 2011)。尽管广泛的监测计划下经常发现无毒性的 HPR0 型 ISAV，但自 2005 年以来已经没有 ISA 暴发记录。也有人推测，接种疫苗可能阻止了 ISAV HPR0 向致病性的 HPR 缺失型病毒的转变(Christiansen et al., 2011)。

虽然商业疫苗的开发主要集中在灭活的 ISAV 细胞培养制剂上，但利用分子技术进行疫苗的研究也已有了报道。Mikalsen 等(2005)将编码 HE 蛋白的质粒每隔 3 周进行 3 次肌内(intramuscularly, IM)注射，可产生 RPS 为 40%～60%的适度保护效果。鱼接种疫苗后，通过腹腔(intraperitoneal, IP)注射进行了攻毒，这可能是免疫保护效果差的原因。Wolf 等(2013)通过肌内(IM)注射一种表达 ISAV 血凝素酯酶的鲑甲病毒复制子，随后采取共栖方式实施攻毒，RPS 达到了 65%～69%，而作为对照组的 ISAV 细胞培养抗原灭活制剂的 RPS 为 80%。

更有趣的是来自智利研究小组的两份关于口服疫苗的报告，两组疫苗都是基于壳聚糖的胶囊化疫苗。Rivas-Aravena 等(2015)使用了细胞培养的病毒，还在壳聚糖颗粒中添加了一种编码甲病毒复制酶的 DNA 新型佐剂，获得 77%的 RPS。Caruffo 等(2016)将重组表达的 ISAV 的 HE 蛋白和 F 蛋白用于制备壳聚糖胶囊，结果 RPS 达到了 64%。在这两份报告中，疫苗都是靶向肠黏膜表面。不过，在这两项研究中，鱼都是通过注射来攻毒的，这可能降低了疫苗的效力。口服疫苗的优点是：(1) 避免了与油佐剂疫苗有关的严重副作用；(2) 可以进行增强免疫接种。

目前还没有使用抗病毒药物、化疗药物或其他药物治疗 ISA 的报道。

6.6 总结与研究展望

传染性鲑贫血症是一种病程缓慢、全身性和致命性的疾病，可对养殖大西洋鲑造成有害影响。它被 OIE 列为应具报疾病，在大多数大西洋鲑养殖区都已实施了控制计划。诊断通常基于病理检查，包括免疫组织化学染色和逆转录聚合酶链反应（RT - PCR）。

该疾病的主要特征为贫血，整个循环系统的内皮细胞都受到了感染。其发病机制尚不清楚，但观察到的病毒与红细胞结合可能是一个重要的致病因素。

如果及早发现，并立即采取适当和严格的生物安全措施，疫情相对容易控制。尽管实验性疫苗在实验室条件下效果良好，但其有效力尚未在田间得到证实。基于病毒培养的疫苗生产具有局限性，因为细胞培养中的病毒产量通常较低，这限制了疫苗中的抗原量。应开发基于 ISAV 毒力因子的疫苗，并对其在实际应用中的效力进行测试和记录。

ISA 是如何引入并持留在鲑鱼种群中，还需要进一步研究。可能的机制包括海洋中的病原贮主库、垂直或水平传播、低毒力病毒引起的未被发现的亚临床感染，以及从经常发生和无毒 HPR0 型发展成强毒性 ISA 病毒。

目前控制 ISA 所面临的主要挑战是与非致病性 HPR0 株向 HPR 缺失的强毒性 ISAV 转变的风险和影响因素相关问题。这些问题不仅对研究 ISA 的流行病学很重要，对大西洋鲑的国际贸易也很重要，因为这两种病毒类型都是必须呈报的（OIE，2015a）。因此，对 ISAV HPR0 的检测可以并已经用于限制鲑鱼的出口。为了解决无毒性 HPR0 非毒株向 HPR 缺失的 ISAV 毒株转化的关键问题，我们需要明确转变的原因和方式。由于这种转变可能是一个涉及多个突变的循序渐进的过程，因此我们也有必要深入了解 ISAV 毒力致病机制。

参考文献

[1] Aamelfot, M. (2012) Tropism of Infectious salmon anaemia virus and distribution of the 4-O-acetylated sialic acid receptor. PhD thesis, Norwegian School of Veterinary Science, Oslo, Norway.

[2] Aamelfot, M., Dale, O. B., Weli, S. C., Koppang, E. O. and Falk, K. (2012) Expression of the infectious salmon anemia virus receptor on Atlantic salmon endothelial cells correlates with the cell tropism of the virus. *Journal of Virology* 86, 10571 - 10578.

[3] Aamelfot, M., Dale, O.B. and Falk, K. (2014) Infectious salmon anaemia-pathogenesis and tropism. *Journal of Fish Diseases* 37, 291 - 307.

[4] Aamelfot, M., McBeath, A., Christiansen, D.H., Matejusova, I. and Falk, K. (2015) Infectious salmon anaemia virus (ISAV) mucosal infection in Atlantic salmon. *Veterinary Research* 46, 120.

[5] Aamelfot, M., Christiansen, D. H., Dale, O. B., McBeath, A., Benestad, S. L. and Falk, K. (2016) Localised infection of Atlantic salmon epithelial cells by HPR0 infectious salmon anaemia virus. *PLoS ONE* 11(3): e0151723.

[6] Aspehaug, V., Falk, K., Krossoy, B., Thevarajan, J., Sanders, L. et al. (2004) Infectious salmon anemia virus (ISAV) genomic segment 3 encodes the viral nucleoprotein (NP), an RNA-binding protein with two monopartite nuclear localization

signals (NLS). *Virus Research* 106, 51 - 60.
[7] Aspehaug, V., Mikalsen, A. B., Snow, M., Biering, E. and Villoing, S. (2005) Characterization of the infectious salmon anemia virus fusion protein. *Journal of Virology* 79, 12544 - 12553.
[8] Baum, J., Ward, R.H. and Conway, D.J. (2002) Natural selection on the erythrocyte surface. *Molecular Biology and Evolution* 19, 223 - 229.
[9] Biering, E., Falk, K., Hoel, E., Thevarajan, J., Joerink, M. *et al*. (2002) Segment 8 encodes a structural protein of infectious salmon anaemia virus (ISAV); the collinear transcript from Segment 7 probably encodes a non-structural or minor structural protein. *Diseases of Aquatic Organisms* 49, 117 - 122.
[10] Byrne, P.J., MacPhee, D.D., Ostland, V.E., Johnson, G. and Ferguson, H.W. (1998) Haemorrhagic kidney syndrome of Atlantic salmon, *Salmo salar* L. *Journal of Fish Diseases* 21, 81 - 91.
[11] Caruffo, M., Maturana, C., Kambalapally, S., Larenas, J. and Tobar, J.A. (2016) Protective oral vaccination against infectious salmon anaemia virus in *Salmo salar*. Fish and Shellfish Immunology 54, 54 - 59.
[12] Cauldwell, A. V., Long, J. S., Moncorge, O. and Barclay, W. S. (2014) Viral determinants of influenza A host range. *Journal of General Virology* 95, 1193 - 1210.
[13] Christiansen, D.H., Østergaard, P.S., Snow, M., Dale, O.B. and Falk, K. (2011) A low-pathogenic variant of infectious salmon anemia virus (ISAV-HPR0) is highly prevalent and causes a non-clinical transient infection in farmed Atlantic salmon (*Salmo salar* L.) in the Faroe Islands. *Journal of General Virology* 92, 909 - 918.
[14] Cook-Versloot, M., Griffiths, S., Cusack, R., McGeachy, S. and Ritchie, R. (2004) Identification and characterization of infectious salmon anaemia virus (ISAV) haemagglutinin gene highly polymorphic region (HPR) type 0 in North America. *Bulletin of the European Association of Fish Pathologists* 24, 203 - 208.
[15] Cunningham, C.O., Gregory, A., Black, J., Simpson, I. and Raynard, R.S. (2002) A novel variant of the infectious salmon anaemia virus (ISAV) haemagglutinin gene suggests mechanisms for virus diversity. *Bulletin of the European Association of Fish Pathologists* 22, 366 - 374.
[16] Dannevig, B. H., Falk, K. and Namork, E. (1995) Isolation of the causal virus of infectious salmon anaemia (ISA) in a long-term cell line from Atlantic salmon head kidney. *Journal of General Virology* 76, 1353 - 1359.
[17] EFSA Panel on Animal Health and Welfare (AHAW) (2012) Scientific opinion on infectious salmon anaemia (ISA). *EFSA Journal* 10(11): 2971.
[18] Evensen, O., Thorud, K. E. and Olsen, Y. A. (1991) A morphological study of the gross and light microscopic lesions of infectious anaemia in Atlantic salmon (*Salmo salar*). *Research in Veterinary Science* 51, 215 - 222.
[19] Falk, K. and Dannevig, B.H. (1995) Demonstration of a protective immune response in infectious salmon anaemia (ISA)-infected Atlantic salmon *Salmo salar*. *Diseases of Aquatic Organisms* 21, 1 - 5.
[20] Falk, K., Namork, E., Rimstad, E., Mjaaland, S. and Dannevig, B. H. (1997) Characterization of infectious salmon anemia virus, an orthomyxo-like virus isolated from Atlantic salmon (*Salmo salar* L.). *Journal of Virology* 71, 9016 - 9023.
[21] Falk, K., Aspehaug, V., Vlasak, R. and Endresen, C. (2004) Identification and characterization of viral structural proteins of infectious salmon anemia virus. *Journal of Virology* 78, 3063 - 3071.
[22] Fourrier, M., Lester, K., Markussen, T., Falk, K., Secombes, C.J. et al. (2015) Dual mutation events in the haemagglutinin-esterase and fusion protein from an infectious salmon anaemia virus HPR0 genotype promote viral fusion and activation by an

ubiquitous host protease. *PLoS ONE* 10(10)：e0142020.
[23] Garcia-Rosado，E.，Markussen，T.，Kileng，O.，Baekkevold，E. S.，Robertsen，B. *et al*. (2008) Molecular and functional characterization of two infectious salmon anaemia virus (ISAV) proteins with type I interferon antagonizing activity. *Virus Research* 133，228 - 238.
[24] Geisbert，T.W.，Young，H.A.，Jahrling，P.B.，Davis，K.J.，Larsen，T. *et al*. (2003) Pathogenesis of Ebola hemorrhagic fever in primate models-evidence that hemorrhage is not a direct effect of virus-induced cytolysis of endothelial cells. *American Journal of Pathology* 163，2371 - 2382.
[25] Håstein，T.，Hill，B.J. and Winton，J.R. (1999) Successful aquatic animal disease emergency programmes. *Revue Scientifique et Technique* 18，214 - 227.
[26] Hellebo，A.，Vilas，U.，Falk，K. and Vlasak，R. (2004) Infectious salmon anemia virus specifically binds to and hydrolyzes 4-O-acetylated sialic acids. *Journal of Virology* 78，3055 - 3062.
[27] Jarp，J. and Karlsen，E. (1997) Infectious salmon anaemia (ISA) risk factors in sea-cultured Atlantic salmon *Salmo salar*. *Diseases of Aquatic Organisms* 28，79 - 86.
[28] Jones，S.R.M.，Mackinnon，A.M. and Salonius，K. (1999) Vaccination of freshwater-reared Atlantic salmon reduces mortality associated with infectious salmon anaemia virus. *Bulletin of the European Association of Fish Pathologists* 19，98 - 101.
[29] Kibenge，F.S.，Kibenge，M.J.，Wang，Y.，Qian，B.，Hariharan，S. and McGeachy，S. (2007) Mapping of putative virulence motifs on infectious salmon anemia virus surface glycoprotein genes. *Journal of General Virology* 88，3100 - 3111.
[30] Kobayashi，Y.，Horimoto，T.，Kawaoka，Y.，Alexander，D.J. and Itakura，C. (1996) Pathological studies of chickens experimentally infected with two highly pathogenic avian influenza viruses. *Avian Pathology* 25，285 - 304.
[31] Lauscher，A.，Krossoy，B.，Frost，P.，Grove，S.，Konig，M. *et al*. (2011) Immune responses in Atlantic salmon (*Salmo salar*) following protective vaccination against Infectious salmon anemia (ISA) and subsequent ISA virus infection. *Vaccine* 29，6392 - 6401.
[32] LeBlanc，F.，Laflamme，M. and Gagne，N. (2010) Genetic markers of the immune response of Atlantic salmon (*Salmo salar*) to infectious salmon anemia virus (ISAV). *Fish and Shellfish Immunology* 29，217 - 232.
[33] Lyngstad，T.M.，Jansen，P.A.，Sindre，H.，Jonassen，C.M.，Hjortaas，M.J. *et al*. (2008) Epidemiological investigation of infectious salmon anaemia (ISA) outbreaks in Norway 2003 - 2005. *Preventive Veterinary Medicine* 84，213 - 227.
[34] Lyngstad，T.M.，Hjortaas，M.J.，Kristoffersen，A.B.，Markussen，T.，Karlsen，E.T. *et al*. (2011) Use of molecular epidemiology to trace transmission pathways for infectious salmon anaemia virus (ISAV) in Norwegian salmon farming. *Epidemics* 3，1 - 11.
[35] Lyngstad，T.M.，Kristoffersen，A.B.，Hjortaas，M.J.，Devold，M.，Aspehaug，V. *et al*. (2012) Low virulent infectious salmon anaemia virus (ISAV-HPR0) is prevalent and geographically structured in Norwegian salmon farming. *Diseases of Aquatic Organisms* 101，197 - 206.
[36] Mardones，F.O.，Perez，A.M. and Carpenter，T.E. (2009) Epidemiologic investigation of the re-emergence of infectious salmon anemia virus in Chile. *Diseases of Aquatic Organisms* 84，105 - 114.
[37] Markussen，T.，Jonassen，C.M.，Numanovic，S.，Braaen，S.，Hjortaas，M. *et al*. (2008) Evolutionary mechanisms involved in the virulence of infectious salmon anaemia virus (ISAV)，a piscine orthomyxovirus. *Virology* 374，515 - 527.
[38] Marshall，S.H.，Ramírez，R.，Labra，A.，Carmona，M. and Muñoz，C. (2014) Bona fide evidence for natural vertical transmission of infectious salmon anemia virus (ISAV)

in freshwater brood stocks of farmed Atlantic salmon (*Salmo salar*) in southern Chile. *Journal of Virology* 88, 6012 - 6018. 10.1128/JVI.03670 - 13.

[39] McBeath, A.J.[A.], Collet, B., Paley, R., Duraffour, S., Aspehaug, V. *et al*. (2006) Identification of an interferon antagonist protein encoded by segment 7 of infectious salmon anaemia virus. *Virus Research* 115, 176 - 184.

[40] McBeath, A.J.[A.], Snow, M., Secombes, C.J., Ellis, A.E. and Collet, B. (2007) Expression kinetics of interferon and interferon-induced genes in Atlantic salmon (*Salmo salar*) following infection with infectious pancreatic necrosis virus and infectious salmon anaemia virus. *Fish and Shellfish Immunology* 22, 230 - 241.

[41] McBeath, A.J.A., Bain, N. and Snow, M. (2009) Surveillance for infectious salmon anaemia virus HPR0 in marine Atlantic salmon farms across Scotland. *Diseases of Aquatic Organisms* 87, 161 - 169.

[42] McBeath, A.[J.A.], Aamelfot, M., Christiansen, D.H., Matejusova, I., Markussen, T. *et al*. (2015) Immersion challenge with low and highly virulent infectious salmon anaemia virus reveals different pathogenesis in Atlantic salmon, *Salmo salar* L. *Journal of Fish Diseases* 38, 3 - 15.

[43] Medina, R. A. and Garcia-Sastre, A. (2011) Influenza A viruses: new research developments. *Nature Reviews: Microbiology* 9, 590 - 603.

[44] Melville, K.J. and Griffiths, S.G. (1999) Absence of vertical transmission of infectious salmon anemia virus (ISAV) from individually infected Atlantic salmon *Salmo salar*. *Diseases of Aquatic Organisms* 38, 231 - 234.

[45] Mikalsen, A.B., Sindre, H., Torgersen, J. and Rimstad, E. (2005) Protective effects of a DNA vaccine expressing the infectious salmon anemia virus hemagglutininesterase in Atlantic salmon. *Vaccine* 23, 4895 - 4905.

[46] Mjaaland, S., Hungnes, O., Teig, A., Dannevig, B.H., Thorud, K. and Rimstad, E. (2002) Polymorphism in the infectious salmon anemia virus hemagglutinin gene: importance and possible implications for evolution and ecology of infectious salmon anemia disease. *Virology* 304, 379 - 391.

[47] Murray, A.G., Smith, R.J. and Stagg, R.M. (2002) Shipping and the spread of infectious salmon anemia in Scottish aquaculture. *Emerging Infectious Diseases* 8, 1 - 5.

[48] Nylund, A., Plarre, H., Karlsen, M., Fridell, F., Ottem, K.F. *et al*. (2007) Transmission of infectious salmon anaemia virus (ISAV) in farmed populations of Atlantic salmon (*Salmo salar*). *Archives of Virology* 152, 151 - 179.

[49] OIE (2015a) Chapter 10.4. Infection with infectious salmon anaemia virus. In: *Aquatic Animal Health Code*. World Organisation for Animal Health, Paris. Updated 2016 version available at: http://www.oie.int/index.php?id=171&L=0&htmfile=chapitre_isav.htm (accessed 28 October 2016).

[50] OIE (2015b) Chapter 2.3.5. Infection with infectious salmon anaemia virus. In: *Manual of Diagnostic Tests for Aquatic Animals*. World Organisation for Animal Health, Paris. Updated 2016 version available at: http://www.oie.int/fileadmin/Home/eng/Health_standards/aahm/current/chapitre_isav.pdf (accessed 28 October 2016).

[51] Palese, P. and Shaw, M.L. (2007) Orthomyxoviridae: the viruses and their replication. In: Knipe, D.M. and Howley, P.M. (eds) *Fields Virology*, 5th edn. Lippincott Williams & Wilkins, Philadelphia, Pennsylvania, pp. 1647 - 1689.

[52] Peñaranda, M.M., Wargo, A.R. and Kurath, G. (2011) *In vivo* fitness correlates with host-specific virulence of Infectious hematopoietic necrosis virus (IHNV) in sockeye salmon and rainbow trout. *Virology* 417, 312 - 319.

[53] Plarre, H., Devold, M., Snow, M. and Nylund, A. (2005) Prevalence of infectious salmon anaemia virus (ISAV) in wild salmonids in western Norway. *Diseases of Aquatic Organisms* 66, 71 - 79.

[54] Purcell, M.K., Garver, K.A., Conway, C., Elliott, D.G. and Kurath, G. (2009) Infectious haematopoietic necrosis virus genogroup-specific virulence mechanisms in sockeye salmon, *Oncorhynchus nerka* (Walbaum), from Redfish Lake, Idaho. *Journal of Fish Diseases* 32, 619 - 631.
[55] Raynard, R.S., Murray, A.G. and Gregory, A. (2001) Infectious salmon anaemia virus in wild fish from Scotland. *Diseases of Aquatic Organisms* 46, 93 - 100.
[56] Rimstad, E., Dale, O.B., Dannevig, B.H. and Falk, K. (2011) Infectious salmon anaemia. In: Woo, P.T.K. and Bruno, D.W. (eds) *Fish Diseases and Disorders, Volume 3: Viral, Bacterial and Fungal Infections*, 2nd edn. CAB International, Wallingford, UK, pp. 143 - 165.
[57] Rivas-Aravena, A., Fuentes, Y., Cartagena, J., Brito, T., Poggio, V. *et al.* (2015) Development of a nanoparticlebased oral vaccine for Atlantic salmon against ISAV using an alphavirus replicon as adjuvant. *Fish and Shellfish Immunology* 45, 157 - 166.
[58] Rummer, J.L., Wang, S., Steffensen, J.F. and Randall, D.J. (2014) Function and control of the fish secondary vascular system, a contrast to mammalian lymphatic systems. *Journal of Experimental Biology* 217, 751 - 757.
[59] Schnittler, H.J., Mahner, F., Drenckhahn, D., Klenk, H.D. and Feldmann, H. (1993) Replication of Marburg virus in human endothelial cells. A possible mechanism for the development of viral hemorrhagic disease. *Journal of Clinical Investigation* 91, 1301 - 1309.
[60] Schultz-Cherry, S., Krug, R.M. and Hinshaw, V.S. (1998) Induction of apoptosis by influenza virus. *Seminars in Virology* 8, 491 - 495.
[61] Simko, E., Brown, L.L., MacKinnon, A.M., Byrne, P.J., Ostland, V.E. and Ferguson, H.W. (2000) Experimental infection of Atlantic salmon, Salmo salar L., with infectious salmon anaemia virus: a histopathological study. *Journal of Fish Diseases* 23, 27 - 32.
[62] Speilberg, L., Evensen, O. and Dannevig, B.H. (1995) A sequential study of the light and electron microscopic liver lesions of infectious anaemia in Atlantic salmon (*Salmo salar* L.). *Veterinary Pathology* 32, 466 - 478.
[63] Stewart, W.B., Petenyi, C.W. and Rose, H.M. (1955) The survival time of canine erythrocytes modified by influenza virus. *Blood* 10, 228 - 234.
[64] Tapia, E., Monti, G., Rozas, M., Sandoval, A., Gaete, A. *et al.* (2013) Assessment of the in vitro survival of the infectious salmon anaemia virus (ISAV) under different water types and temperature. *Bulletin of the European Association of Fish Pathologists* 33, 3 - 12.
[65] Thorud, K.E. (1991) Infectious salmon anaemia-transmission trials. Haematological, clinical and morphological investigations. PhD thesis, Norwegian College of Veterinary Medicine, Oslo, Norway.
[66] Thorud, K. and Djupvik, H.O. (1988) Infectious anaemia in Atlantic salmon (*Salmo salar* L.). *Bulletin of the European Association of Fish Pathologists* 8, 109 - 111.
[67] Totland, G.K., Hjeltnes, B.K. and Flood, P.R. (1996) Transmission of infectious salmon anaemia (ISA) through natural secretions and excretions from infected smelts of Atlantic salmon *Salmo salar* during their presymptomatic phase. *Diseases of Aquatic Organisms* 26, 25 - 31.
[68] Traving, C. and Schauer, R. (1998) Structure, function and metabolism of sialic acids. *Cellular and Molecular Life Sciences* 54, 1330 - 1349.
[69] Vågsholm, I., Djupvik, H.O., Willumsen, F.V., Tveit, A.M. and Tangen, K. (1994) Infectious salmon anaemia (ISA) epidemiology in Norway. *Preventive Veterinary Medicine* 19, 277 - 290.
[70] Vike, S., Nylund, S. and Nylund, A. (2009) ISA virus in Chile: evidence of vertical

transmission. *Archives of Virology* 154，1－8.

[71] Wargo，A.R. and Kurath，G.（2012）Viral fitness：definitions，measurement，and current insights. *Current Opinion in Virology* 2，538－545.

[72] Wolf，A.，Hodneland，K.，Frost，P.，Braaen，S. and Rimstad，E.（2013）A hemagglutinin-esteraseexpressing salmonid alphavirus replicon protects Atlantic salmon（*Salmo salar*）against infectious salmon anemia（ISA）. *Vaccine* 31，661－669.

7

鲤春病毒血症

Peter Dixon 和 David Stone*

7.1 引言

鲤春病毒血症(spring viremia of carp, SVC)是一种常见的鲤、锦鲤(*Cyprinus carpio*)和其他鱼类的致死性出血性疾病。该疾病由鲤春病毒血症病毒(spring viremia of carp virus, SVCV)引起,该病毒是一种弹状病毒(图 7.1)的模式种,属春季血症病毒属(*Sprivivirus*),学名为 *Carp sprivivirus*(Stone et al., 2013; Adams et al., 2014)。SVC 被世界动物卫生组织(the world organisation for animal health, OIE)列为应具报疾病,但该病害病毒与其他非具报病毒存在血清学关联。

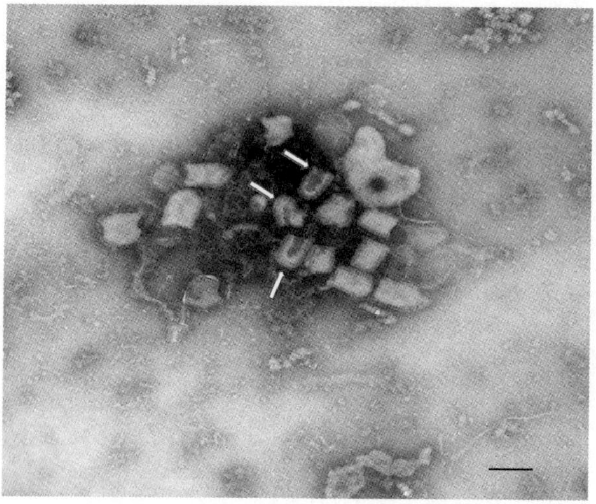

图 7.1 用透射电子显微镜观察到的负染 SVCV,图中可见大量子弹状颗粒,病毒核衣壳被表面有刺突的包膜包围。比例尺: 100 nm。图片由 J. V. Warg 博士提供

* 通信作者邮箱: david.stone@cefas.co.uk。

7.2 宿主

尽管一些早期的疾病报告可能实际上是细菌性鲤鱼红皮炎，但鲤春病毒血症已在许多宿主种类中被描述。除了鲤和锦鲤(*C. carpio*)外，鲤春病毒血症的宿主还包括以下鲤科鱼类(未列引用的可查阅 Dixon，2008)：鳙(*Aristichthys nobilis*)、鲫(*Carassius carassius*)、草鱼(*Ctenopharyngodon idella*)、鲢(*Hypophthalmichthys molitrix*)、金鱼(*Carassius auratus*)、圆腹雅罗鱼(*Leuciscus idus*)、丁鲥(*Tinca tinca*)、拟鲤(*Rutilus rutilus*)、欧鳊(*Abramis brama*)(Basic et al.，2009)、翡翠夏纳鱼(*Notropis atherinoides*)(Cipriano et al.，2011)。鲤春病毒血症病毒的非鲤科鱼宿主有六须鲇(又称欧鲇，*Silurus glanis*)、梭子鱼(*Esox lucius*)、西伯利亚鲟(*Acipenser baerii*)(Vicenova et al.，2011)、大口黑鲈(*Micropterus salmoides*)、蓝鳃太阳鱼(*Lepomis macrochirus*)(Cipriano et al.，2011；Phelps et al.，2012)。

SVCV 核苷酸已在鳟鱼中被发现(Shchelkunov et al.，2005)，并且已从虹鳟(*Oncorhynchus mykiss*)中分离出 SVCV 病毒株(Stone et al.，2003；Jeremic et al.，2006；Haghighi Khiabanian Asl et al.，2008a；I. Shchelkunov，个人通信)。这些病毒分离株对鲤鱼都有致病性，但是这些病毒株无论是通过浸泡感染(Jeremic et al.，2006)还是腹腔注射(P.F. Dixon，J. Munro and D.M. Stone，未发表的数据)感染都对虹鳟无致病性。与上述结果相反，近期的一项研究表明，一株 SVCV 北美分离株腹腔注射后对虹鳟表现出了中等程度的致病性(Emmenegger et al.，2016)。

此外，有研究通过免疫组织化学染色确认了从黑鮰(*Ictalurus melas*，首选名称为 *Ameiurus melas*)(Selli et al.，2002)和尼罗罗非鱼(*Sarotherodon niloticus*，首选名称为 *Oreochromis niloticus*)中分离到的某个病毒株为 SVCV(Soliman et al.，2008)。但是由于 SVCV 的抗体可能和其他病毒发生交叉反应，因此在没有进一步的可靠证据前，上述物种不应被认为是 SVCV 的宿主。有研究通过逆转录聚合酶链反应(RT - PCR)在三种患病的印度鲤科鱼类[南亚野鲮(*Labeo rohita*)，印度鲮(*Cirrhinus mrigala*)，喀拉鲃(*Catla catla*)]组织中鉴定出了 SVCV 的核苷酸序列，但是并未从病鱼中分离出病毒(Haghighi Khiabanian Asl et al.，2008b)，然而，上传到 Genbank 上的核酸序列与已知 SVCV 的核酸序列只有不到 80%的匹配度。

有研究者在夏威夷细角滨对虾(*Litopenaeus stylirostris*)和南美白对虾(*Litopenaeus vannamei*)中分离出了一种对虾弹状病毒，并在这两种对虾中进行了复制(Dixon，2008)。经鉴定，该病毒的 G 蛋白(糖蛋白)基因的核酸序列与 SVCV 相同，这两种病毒都应被视为 SVCV。这说明 SVCV 曾经在夏威夷出现过(尽管夏威夷尚未有 SVC 的报道)，目前还不清楚该病毒是否仍然存在于夏威夷。

还有一些其他鱼类在实验室条件下也对 SVCV 敏感，但是由于支持数据不足，暂不予讨论。目前，斑马鱼(*Danio rerio*)正越来越多地用作研究 SVCV 和宿主间相互作用的动物模型并发挥重要作用。

7.3 地理分布

长期以来，SVC 只在欧洲的鲤鱼生产国有暴发记录，但自 2000 年以来，发现 SVCV 分离株的地理范围有所扩大。然而，分离的数量并没有反映出疾病的发生情况。例如，在中国养殖的鲤鱼、锦鲤及加拿大的野生鲤鱼中都发现了 SVCV 分离株，但这两个国家都没有观察到疾病的暴发(Garver et al., 2007; Zhang et al., 2009)。然而，SVCV 与美国的鲤鱼和锦鲤的死亡率有关(Warg et al., 2007; Phelps et al., 2012)。在巴西，使用酶联免疫吸附试验(ELISA)在金鱼组织中检测到了 SVCV 抗原，但并未尝试进行病毒分离(Alexandrino et al., 1998)，因此巴西是否可以纳入 SVCV 的地理分布范围需要进一步的确认。

7.4 传播方式

鲤春病毒血症的流行温度范围很广，但是鲤鱼的田间死亡通常发生在 5～18 ℃(Fijan, 1988; Ahne et al., 2002)。该疾病大多发生在寒冬过后的春季，在气温高于 10 ℃时病情会迅速发展。在实验室条件下，温度的升高并不一定引起疾病，但是越冬的鱼感染后却比秋季感染的鱼死亡率更高或病程更快，而越冬鱼的健康状况不佳可能是其中的一个风险因素(Baudouy et al., 1980a,b,c)。

在水产养殖中，虽然所有鱼龄的鲤鱼都易受到影响，但 9～12 月龄和 21～24 月龄的鲤鱼通常更容易感染 SVCV (Fijan, 1988)。在实验室条件下，草鱼和鲤鱼对 SVCV 的抗性随着鱼龄的增长而增强(Shchelkunov and Shchelkunova, 1989)，但成年野生鲤鱼仍会被感染(Marcotegui et al., 1992; Dikkeboom et al., 2004; Phelps et al., 2012)，并且野生鲤鱼比养殖鲤鱼更易感(Hill, 1977)。

SVCV 通过水平方式传播(Fijan, 1988)，已从卵巢液中分离出了该病毒，但是垂直传播方式尚未得到证实(Békési and Csontos, 1985)。SVCV 似乎首先经由腮进入鱼宿主(Dixon, 2008)，然后扩散到肾、肝、心、脾和消化道。在疾病暴发期间，在受感染鱼的肝和肾中可以检测到高滴度的病毒，而在鳃、脾和脑中只发现较低滴度的病毒(Fijan et al., 1971)。

SVCV 可通过死鱼尸体和受感染鱼的排泄物释放到环境中(Dixon, 2008)。病毒在宿主外的存活时间与温度呈反比，例如在 10 ℃的河水中，病毒可存活

35 d；而在 4 ℃的池塘淤泥中可存活 42 d(Ahne，1982a，b)。

幸存鱼体脱落散播的 SVCV 可能是病毒传播的主要途径，但是对于病毒在受感染鱼中的持留时间、病毒脱落的持续时间和脱落数量，我们仍知之甚少。病毒可能会在应激事件后脱落，特别是在严冬后的春季，从健康状况不佳的鱼身上大量脱落(Fijan，1988)。

SVCV 通过鹭科鸟类(herons)(Peters and Neukirch，1986)、叶状鱼虱(*Argulus foliaceus*)和尺蠖鱼蛭(*Piscicola geometra*)(Ahne，1978，1985a；Pfeil-Putzien，1978)的机械性传播已被实验证实，但尚不清楚这种传播在自然界中是否存在。Pfeil-Putzien 和 Baath(1978)从自然患病鱼上取得的鲤虱中分离出 SVCV，这表明鲤虱确实会从受感染宿主中摄取 SVCV。捕食了感染 SVCV 的梭子鱼苗的梭子鱼也会被染病(Ahne，1985b)。SVCV 还可能通过诱饵鱼的迁移而转移(Goodwin et al.，2004；Misk et al.，2016)，此外，无脊椎动物也可能是 SVCV 的传染源，但是这两种传播方式都没有明确的证据支持。

7.5 对鱼类生产的影响

鲤春病毒血症会造成鲤鱼产量的重大损失，特别是在欧洲国家和苏联(Ahne et al.，2002)。不过，由于动物流行病的暴发比较零散且每年损失都有波动，病害造成的具体损失数额难以统计(Fijan，1999)。据 1980 年的一份历史数据估计，SVC 在欧洲每年造成 4 000 t 1 年龄鲤鱼(占该鱼龄组的 10%～15%)的损失(Sano et al.，2011)。在一些国家，养殖鲤鱼用于游钓渔业，这是一项利润丰厚的休闲产业。1997 年英国整理的估算数据显示，养鱼场的病害损失约为 2 万～23 万英镑，渔业损失超过 3 万英镑，零售店的损失约为 2 万～3 万英镑(Taylor et al.，2013)。此外，对宰杀受感染鱼的控制措施也将造成进一步的生产损失(Taylor et al.，2013)，任何此类控制疾病扩散的措施都会产生不利的财务负担(Dixon，2008)。

7.6 疾病临床体征

SVC 没有特异性的临床病征，而且并不是所有感染的鱼都会表现出所有症状。在感染初期，通常发病迅速且死亡率逐渐增加。受感染的鱼可能游动缓慢而不稳定，失去平衡、侧身泳动。两个最明显和最一致的特征是腹胀和出血，出血可能发生在皮肤、鳍基部、眼睛和鳃等部位上，并导致出血处变得苍白(图 7.2)。感染鱼的皮肤可能变暗，常见眼球突出。此外，患病鱼的肛门可能肿胀、发炎并有拖尾黏液样管型物。病鱼的腹部通常充满清亮液体，有时带有血迹；脾常肿大，且大部分内脏器官存在水肿，脏器相互粘连并黏附于腹膜。此外，

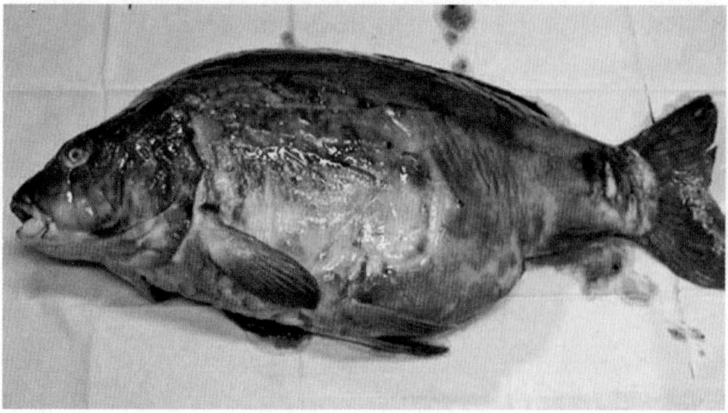

图 7.2　SVCV 感染的鲤鱼出现腹胀、大面积出血和眼球突出等病征

肌肉组织和鱼鳔中会存在瘀点性出血(Dixon，2008)，但是这些症状在 SVCV 亚洲株引起的 SVC 中并不常见(Goodwin，2003；Dikkeboom et al.，2004)。

Gaafar 等(2011)对鲤鱼进行了详细的组织病理学研究，并将他们的发现与其他研究者的研究进行了比较，此外，Misk 等(2016)还报道了三个实验宿主的鲤春病毒血症组织病理学结果。组织病理学研究通常显示肝、胰腺、肠、脾、肾、心、肌肉和鱼鳔出现水肿、炎症、出血和坏死的症状，但并非所有的鱼都会出现上述每种症状。

在疾病发展早期，病鱼脾脏特别是网状内皮组织出现增生，髓内可充满红细胞，而在疾病后期则出现营养不良和细胞退化变性。由于含铁巨噬细胞的数量增多[图 7.3(D)]，病鱼出现色素沉着增加，且肾脏发生退行性和增生性变化。在严重疾病中，病鱼会出现明显水肿，并伴有肾小球内细胞和液体的解离及部分小管堵塞。同时，病鱼可见尿路上皮局灶性剥脱、肾变病和肾小管周围水肿。此外，造血组织也会发生水肿，细胞表现为营养不良和变性出血[图 7.3(B)]。在较轻的病例中，病鱼表现出肾充血增加和造血组织增生，并伴有黑色素巨噬细胞增多。

病鱼的胰腺胰岛发生伴有白细胞浸润的腺周水肿，同时，病鱼还可能存在肝细胞增生和异常肝细胞，也可发生增生性变化。在肝实质水肿中，常可见局灶性或弥漫性红细胞渗出症(红细胞外渗)和肝细胞脂肪变性，肝细胞增大并呈现颗粒状或空泡化营养不良。病鱼肠道可能发生退行性和增生性变化[图 7.3(C)]，可见壁内血管周围炎症、黏液变性、坏死和上皮脱落。在感染后期，肠道绒毛可能会萎缩。此外，病鱼心脏的心外膜和心肌可能会发生局灶性细胞浸润，并进一步恶化为局灶性变性和坏死[图 7.3(A)]，随着心肌条纹的消失，肌束变得松散。

病鱼脑部血管可能随着神经元细胞周水肿而增大，少数神经胶质细胞中可

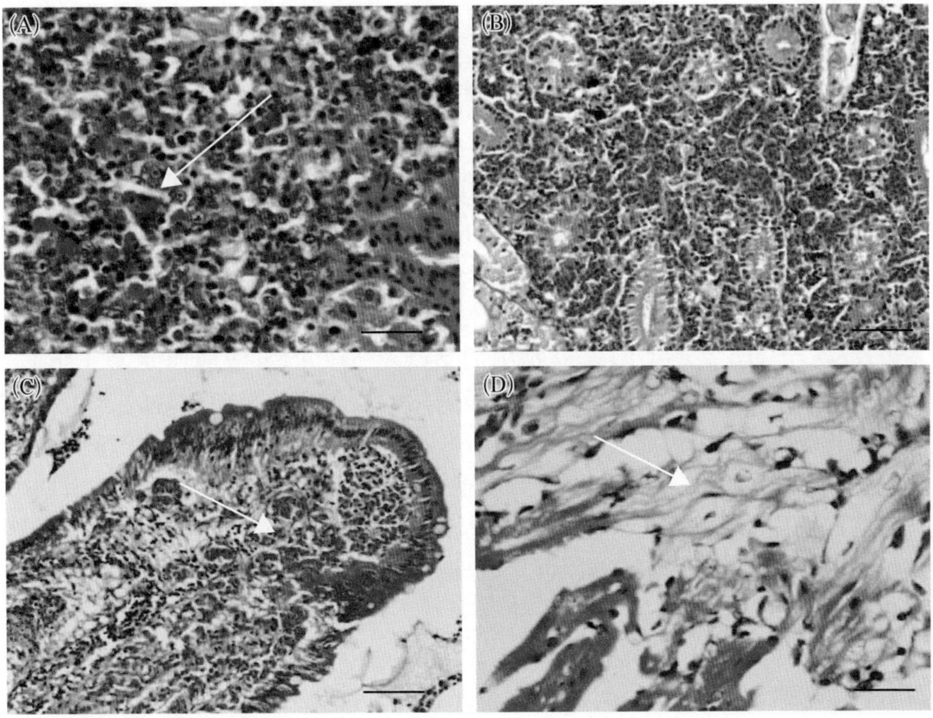

图 7.3 苏木精和伊红染色的感染鲤春病毒血症的鲤鱼组织切片显示：(A) 心肌细胞坏死区(箭头)，伴有胞质染色丧失和核变性；(B) 肾间质内弥漫性出血；(C) 肠绒毛固有层出血(箭头)；(D) 肾间质组织中黑色素巨噬细胞增多(箭头)。(A)～(C)比例尺：50 μm，(D)比例尺：25 μm。图片由 S. W. Feist 博士提供

见嗜酸性包涵体。病鱼腹膜可能发生炎症，腹膜上可见充满碎屑、巨噬细胞和淋巴细胞的扩张淋巴管。病鱼的内部黏液可能增厚，并伴有组织退化病灶，在这些病灶中上皮剥落和脱屑。此外，病鱼鳃可能表现出退行性和增生性变化、坏死和浸润，鳃部还可能发生弥漫性的鳃瓣融合、细胞板层基部的上皮衬里增生及弥漫性鳃坏死，黑色素巨噬细胞的数量可能会增加。

7.7 诊断

对 SVC 的确诊需要从细胞培养中分离出病毒，然后再使用血清学和(或)分子生物学技术进行鉴定。世界动物卫生组织(OIE)在其《水生动物诊断检测指南》中列入了诊断和鉴定 SVC 的综合方法(OIE，2015a)，并推荐使用鲤鱼上皮细胞系(epithelioma papulosum cyprini，EPC)(Fijan et al.，1983)和胖头鲅肌肉细胞系(ATCC CCL－42)进行病毒的分离。当临床鱼类中存在高水平病毒时，可以使用血清学或分子方法直接检测其组织，以便在等待病毒分离结果的同时提供快速推定诊断。从幸存的鱼或亚致死感染的鱼身上分离

SVCV并不容易，这使得追踪接触过病毒的鱼的分布变得困难。通过对表面健康的鲤鱼进行病毒筛查后，中国和加拿大两国均发现了SVCV(Liu et al., 2004; Garver et al., 2007)。虽然没有证据表明病毒传播需要超过7 d，但为了得出群体病毒感染阴性的结论，需要连续两个7 d周期细胞培养后的病毒筛查。此外，也可以使用分子方法检查细胞培养物以增强阴性结果的可靠性。

必须鉴定从细胞培养物中分离出的病毒，但有些抗血清可能会与其他病毒发生交叉反应[如梭子鱼鱼苗弹状病毒(pike fry rhabdovirus, PFRV)]，此外，抗血清也可能和SVCV的某些分离株发生反应(Dixon, 2008)，而不与其他分离株发生反应。因此，鉴定细胞培养物中的SVCV最好的方法应当是基于RT-PCR的方法。世界动物卫生组织(OIE)采用的检测方法靶向广泛的SVCV分离株和其他血清学上可能会被误诊为SVCV的近缘病毒分离株的G基因的部分核苷酸序列(Stone et al., 2003)。根据Stone等(2003)的报道，SVCV有四个基因群：基因群Ⅰ包含SVCV分离株、基因群Ⅱ包含单一的草鱼分离株、基因群Ⅲ包含PFRV参考分离株、基因群Ⅳ(也被称为丁鲹弹状病毒群)包含一些未分类的分离株和以前鉴定为PFRV的分离株。然而，国际病毒分类委员会(the international committee on taxonomy of viruses, ICTV)仅认可其中SVCV和PFRV两个种，而草鱼弹状病毒和丁鲹弹状病毒被认为属于PFRV种(Adams et al., 2014)。

根据Stone等(2003)的研究，基因群Ⅰ的分离株之间具有显著的核苷酸序列差异，并且可以按地理来源分为四个亚群，其中，Ⅰa亚群分离株来自亚洲、美国和加拿大(Garver et al., 2007; Phelps et al., 2012)；Ⅰb亚群来自摩尔多瓦和乌克兰；Ⅰc亚群来自乌克兰和俄罗斯；Ⅰd亚群来自欧洲，并包括一个来自摩尔多瓦和一个来自乌克兰的分离株。此外，还可能存在由奥地利分离株组成的第五个亚群Ⅰe(D.M. Stone，未发表的数据)。基于G基因的分离株系统发育比较可能成为追溯SVCV分离株来源的有价值的工具(Stone et al., 2003)。

此外，常规RT-PCR、实时定量聚合酶链反应(PCR)和环介导等温扩增分析也可用于对SVCV感染的检测和确认(Koutná et al., 2003; Liu et al., 2008a,b; Yue et al., 2008; Zhang et al., 2009; Shimahara et al., 2016)。除了Shimahara等(2016)开发的常规PCR检测方法外，其他检测方法都没有针对SVCV所有基因群的代表分离株进行验证，并且使用实时定量PCR检测时(Liu et al., 2008a; Yue et al., 2008)，也出现了无法检测来自至少一个SVCV亚群分离株的情况(D.M. Stone，未发表的数据)。基于聚合酶基因的一组通用引物可用于鉴定来自春季血症病毒属(*Sprivivirus*)和鲈弹状病毒属(*Perhabdovirus*)的病毒，并可用于筛选病毒培养物(Ruane et al., 2014)。

7.8 病理生理学

受感染的鲤鱼针对 SVCV 的体液免疫依赖于温度。与 25 ℃相比，在 13～14 ℃时感染鲤鱼的体液免疫应答反应更慢，中和抗体滴度更低(Fijan et al., 1977)。在 10～12 ℃时，鲤鱼中 SVCV 引起的死亡率可达 90%，并且检测不到中和抗体；而在 20～22 ℃时，SVCV 不会导致鲤鱼的死亡，并在感染 30 d 后可检测到相应的中和抗体(Ahne, 1980)。干扰素在 SVCV 感染后 24 h 内即可产生(Baudouy, 1978)。在一些独立的实验中，鲤鱼腹腔注射 SVCV 可导致：(1) 几种抗病毒分子的上调(Feng et al., 2011)；(2) 血细胞、鳃、肠和脾中自然杀伤细胞增强因子 β 基因的表达显著增加(Huang et al., 2009)；(3) 肾、外周血、脾和肠中胸腺素 α 转录物显著上调，以及肠、外周血、肝和脾中胸腺素 β 转录物上调，这两种肽均能刺激免疫反应(Xiao et al., 2015)。

SVC 的病理生理学尚未得到很好的研究。目前对其病理生理学的了解来自对自然感染和一个实验感染的研究。Řehulka(1996)研究了来自捷克共和国的某个国家渔场越冬池捕集的鲤鱼中 SVC 病情的发展情况，发现它们在冬眠结束时易受 SVCV 感染。这些鲤鱼在 4 月初被转移到一个实验池塘，饲喂颗粒料，疾病发展过程的研究持续到 6 月初。在患病鲤鱼中(通过病毒分离和血清中和证实)，淋巴细胞减少，单核细胞增多，嗜中性粒细胞增多，中幼粒细胞和晚幼粒细胞的增多尤为明显。这些变化都与鱼的健康状况恶化有关，并伴随着病鱼细胞的形态改变，特别是细胞核和细胞质的空泡化。此外，病鱼还存在贫血的迹象。总体而言，该疾病导致近 60%的鲤鱼死亡。在疾病进程中，Osadchaya 和 Rudenko(1981)发现肝脏中糖原合成显著减少，表现为肝细胞中几乎完全没有糖原颗粒，并且脾髓中含铁色素非常少。Fijan(1999)的研究认为，SVCV 在毛细血管上皮以及造血和排泄肾组织的增殖中影响渗透调节，这对病鱼来说可能是致命的。

Jeney 等(1990)的研究表明，实验条件下使用 SVCV 感染欧鲇幼鱼后，其血细胞比容值和血红蛋白浓度显著下降，表明感染鱼的造血功能受到了影响。感染后，鱼体内的血清谷氨酸草酰乙酸转氨酶和谷氨酸丙酮酸转氨酶水平显著升高，表明发生了组织坏死。在注射感染 33 h 后，青鱼(*Mylopharyngodon piceus*)的肠、肌肉和肝中的线粒体抗病毒信号蛋白(先天免疫反应的一部分) mRNA 均有增加，但在脾中则减少(Zhou et al., 2015)。

7.9 预防和控制

目前，SVC 的预防和控制都是通过立法和良好的管理来实现的。此外，还曾尝试通过疫苗接种和使用鲤抗病品种来预防 SVC。例如，Kirpichnikov 等(1993)报道，与其他鲤鱼种或品系相比，乌克兰-罗普沙鲤(Ukrainian-Ropsha carp)杂交系对 SVC 的抗性增强。

20 世纪 70 年代和 80 年代，捷克斯洛伐克进行了 SVCV 商业疫苗的田间试验(Fijan，1984，1988；Dixon，1997；Sano et al.，2011)。然而，这个疫苗如今已经不再提供，Dixon(1997)讨论了该疫苗田间试验存在的问题。有研究尝试使用低毒力 SVCV 或通过细胞培养减毒的 SVCV 进行口服疫苗接种。实验室结果显示出良好结果，但在池塘养殖鱼中进行的大规模田间试验未获成功(Fijan，1988)。此外，Kölbl(1980，1990)考察了利用不同减毒病毒的口服疫苗接种，田间试验的存活率让大家备受鼓舞(接种鱼的死亡率为 15%，而对照鱼的死亡率为 49%)。随后，这种疫苗通过未指明的方法被进一步改造并取得了田间试验的成功，试验结果表明接种鱼未发生 SVC 并且养殖产量有所增加。例如，某地因 SVC 的流行导致养殖产量只有 15 t，在接种疫苗后的第一年产量提高到 50 t，次年提高到 70 t。不过，该疫苗没有继续商业化。

随后，开发者们又尝试了一系列疫苗制造方法，但是这些方法都没有成功商业化应用。将重组 SVCV 糖蛋白注入鲤鱼体内可产生中和抗体，但不能保护鲤鱼抵抗 SVC (Dixon，1997)。将一种表面表达 SVCV 糖蛋白和鲤疱疹病毒- 3 (CyHV - 3)ORF81 蛋白的植物乳杆菌(*Lactobacillus plantarum*)基因工程口服疫苗加入颗粒饲料(Cui et al.，2015)，对实验室条件下的 SVCV 攻毒，接种鲤鱼的死亡率为 29%，而对照鱼的死亡率高达 78%和 89%。同时，该疫苗也能保护锦鲤免受 CyHV - 3 的侵害(见第 10 章)。此外，目前有研究进行了基于 SVCV 的 G 基因的 DNA 疫苗实验室试验。还有研究对含有部分或完整 G 基因片段的 SVCV DNA 质粒的不同组合进行疫苗效果比较(Kanellos et al.，2006)，实验结果表明两组接种了不同疫苗的鲤鱼的免疫保护率(RPS)分别为 33%和 48%，而对照组的死亡率超过 60%。Emmenegger 和 Kurath(2008)尝试 DNA 疫苗单次接种免疫锦鲤，对 SVCV 攻毒的 RPS 为 50%～70%。

在 SVCV 攻毒前，给鲤鱼注射干扰素诱导剂双链 RNA 可将死亡率从 100%降低到 22%～40% (Masycheva et al.，1995；Alikin et al.，1996)。在 SVCV 感染前，以含有 poly(I∶C)(聚肌苷酸-聚胞苷酸，一种合成的干扰素诱导剂)和细菌脂多糖的脂质体作为免疫刺激剂，采用浸泡或腹腔注射接种脂质体也可降低斑马鱼的死亡率(RPS 分别为 33.3%和 42.3%) (Ruyra et al.，2014)。此外，有研究证实异丙肌苷(一种抗病毒剂)可在体外抑制 SVCV 的复制，但在鱼体内的效果未被测试(Siwicki et al.，2003)。

其他预防或治疗 SVC 的思路还包括：① 鉴定实验性 SVCV 感染中存活斑马鱼依赖的多通路基因以开发可能用于预防疾病的药物(Encinas et al.，2013)；② 研究自噬或 RNA 抑制作为抑制 SVC 复制的方法(Garcia-Valtanen et al.，

2014a；Gotesman et al.，2015；Liu et al.，2015)；③ 研究使用表达病毒性出血性败血症病毒糖蛋白的质粒作为 SVCV G 基因 DNA 疫苗分子佐剂(MartinezLopez et al.，2014)；④ 研究斑马鱼 β-防御素的免疫调节特性及其作为 DNA 疫苗分子佐剂的作用(Garcia-Valtanen et al.，2014b)。

在没有商业疫苗或治疗药物的情况下，生物安全策略仍然是预防 SVC 最有效的方法。这些策略涵盖从国家或国际层面的立法或标准建立到养殖场层面的良好管理实践。后者包括对新鱼种群的现场检疫、设施消毒、使用蘸足消毒、减少胁迫和其他疾病等。

一些研究确定了消毒剂和其他化学或物理手段灭活病毒的效果(Kiryu et al.，2007；Dixon，2008；Dixon et al.，2012)，但是必须谨慎对待这些研究的结果。例如，99.9%的 SVCV 在 pH=3.0 的盐酸中 2 h 后被灭活(Ahne，1976)，但在 pH=4.0 的甲酸中(模拟处理死鱼的程序)，病毒至少能够存活 28 d(Dixon et al.，2012)。

通过立法控制疾病在一个国家的传入和传播是预防和控制 SVC 最重要的途径。这些立法往往通过国际标准或协议得到加强。世界动物卫生组织(OIE)发布的《水生动物卫生法典》(*Aquatic Animal Health Code*，OIE，2015b)概述了从国家到生产现场各层级可实施的生物安全措施。Håstein 等(2008)概述了生物安全策略的实施概况，其中将 SVC 作为一个具体的疾病案例研究。Oidtmann 等(2011)综述了与鱼类和鱼类产品流通有关的国际标准，着重研究了一些防范鱼类疾病的国家措施。英国实施了一项防范 SVC 传播和扩散的鱼类健康行动计划(Taylor et al.，2013)。通过立法管控，克罗地亚可能已经摆脱了 SVC 的困扰，并减少了匈牙利的 SVC 发病率(Molnár and Csaba，2005；Oraic and Zrncic，2005)。然而，病毒在个别国家的流行病学将决定建立一个无 SVC 疫病国家的可行性。例如，在美国，SVC 主要发生在野生鱼类中(Phelps et al.，2012)，在那里很难实现无疫病的目标。此外，由于 SVC 在塞尔维亚的野生鲤鱼和养殖鲤鱼间频繁发生交叉感染，因此 Jeremic 等(2004)也对彻底消除塞尔维亚的 SVC 持悲观态度。然而，即使 SVC 在一个国家内广泛传播，只要严格控制鱼类的流通，仍有可能建立无疫病区。立法防止通过受污染的鱼类或鱼类产品引入 SVC 是一项重要措施。所有的流通，无论是涉及活鱼还是受污染的加工产品，都应受到严格控制。未经批准的受污染产品的贸易流通可能已经将 SVC 引入或重新引入美国和英国(Håstein et al.，2008；Taylor et al.，2013)。因此，对进口鱼类和鱼类产品的持续监测至关重要。

7.10 总结与研究展望

SVC 造成了鲤鱼生产的重大损失，但由于疾病的散发性和损失的年度波动，准确的病害损失数额统计非常复杂。该疾病没有特异性临床表现，因此诊断SVC需要分离出病毒株并随后运用血清学或分子学方法进行鉴定。世界动物卫生组织（OIE）为SVC的诊断和鉴定提供了一套全面的方法，即先使用在20 ℃下培养的EPC或FHM细胞系分离病毒，然后采用基于RT－PCR的方法对分离到的病毒进行鉴定和序列分析，从而确诊SVCV。此外，根据G基因部分序列的分析结果可将鉴定出的SVCV进一步分型至四个不同基因群。

目前，立法监控和良好的管理措施相结合是有效预防和控制这一疾病的最佳办法。接种疫苗和使用鲤抗病品种防控SVC是可能的，但目前尚无商业化疫苗可用。此外，未经批准的鱼类产品流通可能是将SVC引入或重新引入美国和英国的原因，因此持续监测进口的鱼类和鱼类产品非常必要。

中国尚未有SVC暴发的报告，但已多次检测发现该病毒。此外，在加拿大也发现了外表健康的SVCV带毒鲤鱼，这与美国鲤鱼和锦鲤感染SVCV亚洲株的高死亡率形成了鲜明的对比。这表明不同SVCV亚洲株的致病性存在显著差异，需要进一步的研究。

对SVCV的传播方式和允许温度的了解一般都很清楚，但SVCV的存活和持留情况却十分多变，需要进一步的研究。宿主的年龄、遗传特性、水温和环境都可能决定疾病的发展速度。

对留档样本（如用福尔马林浸泡保存的器官）使用原位杂交和实时qPCR技术检测病毒，将有助于今后调查研究带病毒鱼体内SVCV的持留性和疾病再度出现的可能性。这些工具对于筛查潜在的病毒宿主也非常有价值。

参考文献

[1] Adams, M.J., Lefkowitz, E.J., King, A.M.Q. and Carstens, E.B. (2014) Ratification vote on taxonomic proposals to the International Committee on Taxonomy of Viruses (2014). *Archives of Virology* 159, 2831－2841.

[2] Ahne, W. (1976) Untersuchungen über die Stabilität des karpfenpathogenen Virusstammes 10/3. *Fisch und Umwelt* 2, 121－127.

[3] Ahne, W. (1978) Untersuchungsergebnisse über die akute Form der Bauchwassersucht (Frühlingsvirämie) der Karpfen. *Fischwirt* 28, 46－47.

[4] Ahne, W. (1980) Rhabdovirus carpio-Infektion beim Karpfen (Cyprinus carpio): Untersuchungen über Reaktionen des Wirtsorganismus. *Fortschritte in der Veterinärmedizin* 30, 180－183.

[5] Ahne, W. (1982a) Untersuchungen zur Tenazität der Fischviren. *Fortschritte in der Veterinärmedizin* 35, 305－309.

[6] Ahne, W. (1982b) Vergleichende Untersuchungen über die Stabilität von vier fischpathogenen Viren (VHSV, PFR, SVCV, IPNV). *Zentralblatt für Veterinarmedizin, Reihe B* 29, 457－476.

[7] Ahne, W. (1985a) *Argulus foliaceus* L. and *Piscicola geometra* L. as mechanical vectors of spring viraemia of carp virus (SVCV). *Journal of Fish Diseases* 8, 241－242.

[8] Ahne, W. (1985b) Viral infection cycles in pike (*Esox lucius* L.). *Journal of Applied*

Ichthyology 1, 90 - 91.
[9] Ahne, W., Bjorklund, H.V., Essbauer, S., Fijan, N., Kurath, G. *et al.* (2002) Spring viremia of carp (SVC). *Diseases of Aquatic Organisms* 52, 261 - 272.
[10] Alexandrino, A.C., Ranzani-Paiva, M.J.T. and Romano, L.A. (1998) Identificación de viremia primaveral de la carpa (VPC) Carrassius auratus en San Pablo, Brasil. *Revista Ceres* 45, 125 - 137.
[11] Alikin, Y.S., Shchelkunov, I.S., Shchelkunova, T.I., Kupinskaya, O.A., Masycheva, V.I. *et al.* (1996) Prophylactic treatment of viral diseases in fish using native RNA linked to soluble and corpuscular carriers. *Journal of Fish Biology* 49, 195 - 205.
[12] Basic, A., Schachner, O., Bilic, I. and Hess, M. (2009) Phylogenetic analysis of spring viraemia of carp virus isolates from Austria indicates the existence of at least two subgroups within genogroup Id. *Diseases of Aquatic Organisms* 85, 31 - 40.
[13] Baudouy, A.-M. (1978) Relation hôte-virus au cours de la virémie printanière de la carpe. Comptes Rendus Hebdomadaires des Séances de l'Académie des Sciences. *Serie D: Sciences Naturelles* 286, 1225 - 1228.
[14] Baudouy, A.-M., Danton, M. and Merle, G. (1980a) Experimental infection of susceptible carp fingerlings with spring viremia of carp virus, under wintering environmental conditions. In: Ahne, W. (ed.) *Fish Diseases. Third COPRAQ-Session.* Springer, Berlin, pp. 23 - 27.
[15] Baudouy, A.-M., Danton, M. and Merle, G. (1980b) Virémie printanière de la carpe. Résultats de contaminations expérimentales effectuées au printemps. *Annales de Recherches Vétérinaires* 11, 245 - 249.
[16] Baudouy, A.-M., Danton, M. and Merle, G. (1980c) Virémie printanière de la carpe: étude expérimentale de l'infection évoluant à différentes températures. *Annales de l'Institut Pasteur/Virologie* 131E, 479 - 488.
[17] Békési, L. and Csontos, L. (1985) Isolation of spring viraemia of carp virus from asymptomatic broodstock carp, *Cyprinus carpio* L. *Journal of Fish Diseases* 8, 471 - 472.
[18] Cipriano, R. C., Bowser, P. R., Dove, A., Goodwin, A. and Puzach, C. (2011) Prominent emerging diseases within the United States. In: *Bridging America and Russia with Shared Perspectives on Aquatic Animal Health. Proceedings of the Third Bilateral Conference between Russia and the United States, 12 - 20 July, 2009 Shepherdstown*, W V. Khaled bin Sultan Living Oceans Foundation, Landover, Maryland, pp. 6 - 17.
[19] Cui, L.C., Guan, X.T., Liu, Z.M., Tian, C.Y. and Xu, Y.G. (2015) Recombinant lactobacillus expressing G protein of spring viremia of carp virus (SVCV) combined with ORF81 protein of koi herpesvirus (KHV): a promising way to induce protective immunity against SVCV and KHV infection in cyprinid fish via oral vaccination. *Vaccine* 33, 3092 - 3099.
[20] Dikkeboom, A.L., Radi, C., Toohey-Kurth, K., Marcquenski, S.V., Engel, M. *et al.* (2004) First report of spring viremia of carp virus (SVCV) in wild common carp in North America. *Journal of Aquatic Animal Health* 16, 169 - 178.
[21] Dixon, P.F. (1997) Immunization with viral antigens: viral diseases of carp and catfish. In: Gudding, R., Lillehaug, A., Midtlyng, P. J. and Brown, F. (eds) *Fish Vaccinology*. Karger, Basel, pp. 221 - 232.
[22] Dixon, P.F. (2008) Virus diseases of cyprinids. In: Eiras, J.C., Segner, H., Wahli, T. and Kapoor, B.G. (eds) *Fish Diseases*. Science Publishers, Enfield, New Hampshire, pp. 87 - 184.
[23] Dixon, P.F., Smail, D.A., Algoet, M., Hastings, T.S., Bayley, A. *et al.* (2012) Studies on the effect of temperature and pH on the inactivation of fish viral and bacterial

pathogens. Journal of Fish Diseases 35, 51 - 64.

[24] Emmenegger, E. J. and Kurath, G. (2008) DNA vaccine protects ornamental koi (*Cyprinus carpio koi*) against North American spring viremia of carp virus. *Vaccine* 26, 6415 - 6421.

[25] Emmenegger, E.J., Sanders, G.E., Conway, C.M., Binkowski, F.P., Winton, J.R. *et al.* (2016) Experimental infection of six North American fish species with the North Carolina strain of spring viremia of carp virus. *Aquaculture* 450, 273 - 282.

[26] Encinas, P., Garcia-Valtanen, P., Chinchilla, B., Gomez-Casado, E., Estepa, A. *et al.* (2013) Identification of multipath genes differentially expressed in pathway targeted microarrays in zebrafish infected and surviving spring viremia carp virus (SVCV) suggest preventive drug candidates. *PLoS ONE* 8(9): e73553.

[27] Feng, H., Liu, H., Kong, R.Q., Wang, L., Wang, Y.P. *et al.* (2011) Expression profiles of carp IRF-3 /-7 correlate with the up-regulation of RIG-I /MAVS /TRAF3 / TBK1, four pivotal molecules in RIG-I signaling pathway. *Fish and Shellfish Immunology* 30, 1159 - 1169.

[28] Fijan, N. (1984) Vaccination of fish in European pond culture: prospects and constraints. In: Oláh, J. (ed.) *Fish, Pathogens and Environment in European Polyculture*. Akadémiai Kiadó, Budapest, pp. 233 - 241.

[29] Fijan, N. (1988) Vaccination against spring viraemia of carp. In: Ellis, A.E. (ed.) *Fish Vaccination*. Academic Press, London, pp. 204 - 215.

[30] Fijan, N. (1999) Spring viraemia of carp and other viral diseases and agents of warm-water fish. In: Woo, P. T. K. and Bruno, D. W. (eds) *Fish Diseases and Disorders, Volume 3: Viral, Bacterial and Fungal Infections*, 1st edn. CAB International, Wallingford, UK, pp. 177 - 244.

[31] Fijan, N., Petrinec, Z., Sulimanovic, D. and Zwillenberg, L.O. (1971) Isolation of the viral causative agent from the acute form of infectious dropsy of carp. *Veterinarsky Arhiv* 41, 125 - 138.

[32] Fijan, N., Petrinec, Z., Stancl, Z., Kezic, N. and Teskeredzic, E. (1977) Vaccination of carp against spring viraemia: comparison of intraperitoneal and peroral application of live virus to fish kept in ponds. *Bulletin de l'Office International des Epizooties* 87, 441 - 442.

[33] Fijan, N., Sulimanovic, D., Bearzotti, M., Muzinic, D., Zwillenberg, L. O. *et al.* (1983) Some properties of the *Epithelioma papulosum cyprini* (EPC) cell line from carp *Cyprinus carpio*. *Annales de l'Institut Pasteur /Virologie* 134, 207 - 220.

[34] Gaafar, A. Y., Vesely, T., Nakai, T., El-Manakhly, E. M., Soliman, M. K. *et al.* (2011) Histopathological and ultrastructural study of experimental spring viraemia of carp (SVC) infection of common carp with comparison between different immunohistodignostic techniques efficacy. *Life Science Journal* 8, 523 - 533.

[35] Garcia-Valtanen, P., Ortega-Villaizan, M.D., Martinez-Lopez, A., Medina-Gali, R., Perez, L. *et al.* (2014a) Autophagy-inducing peptides from mammalian VSV and fish VHSV rhabdoviral G glycoproteins (G) as models for the development of new therapeutic molecules. *Autophagy* 10, 1666 - 1680.

[36] Garcia-Valtanen, P., Martinez-Lopez, A., Ortega-Villaizan, M., Perez, L., Coll, J.M. *et al.* (2014b) In addition to its antiviral and immunomodulatory properties, the zebrafish beta-defensin 2 (zfBD2) is a potent viral DNA vaccine molecular adjuvant. *Antiviral Research* 101, 136 - 147.

[37] Garver, K.A., Dwilow, A.G., Richard, J., Booth, T.F., Beniac, D.R. *et al.* (2007) First detection and confirmation of spring viraemia of carp virus in common carp, *Cyprinus carpio* L., from Hamilton Harbour, Lake Ontario, Canada. *Journal of Fish Diseases* 30, 665 - 671.

[38] Goodwin, A.E. (2003) Differential diagnosis: SVCV vs KHV in koi. *Fish Health Newsletter*, *American Fisheries Society*, *Fish Health Section* 31, 9 - 13.
[39] Goodwin, A.E., Peterson, J.E., Meyers, T.R. and Money, D.J. (2004) Transmission of exotic fish viruses: the relative risks of wild and cultured bait. *Fisheries* 29, 19 - 23.
[40] Gotesman, M., Soliman, H., Besch, R. and El-Matbouli, M. (2015) Inhibition of spring viraemia of carp virus replication in an *Epithelioma papulosum cyprini* cell line by RNAi. *Journal of Fish Diseases* 38, 197 - 207.
[41] Haghighi Khiabanian Asl, A., Bandehpour, M., Sharifnia, Z. and Kazemi, B. (2008a) The first report of spring viraemia of carp in some rainbow trout propagation and breeding by pathology and molecular techniques in Iran. *Asian Journal of Animal and Veterinary Advances* 3, 263 - 268.
[42] Haghighi Khiabanian Asl, A., Azizzadeh, M., Bandehpour, M., Sharifnia, Z. and Kazemi, B. (2008b) The first report of SVC from Indian carp species by PCR and histopathologic methods in Iran. *Pakistan Journal of Biological Sciences* 11, 2675 - 2678.
[43] Håstein, T., Binde, M., Hine, M., Johnsen, S., Lillehaug, A. *et al*. (2008) National biosecurity approaches, plans and programmes in response to diseases in farmed aquatic animals: evolution, effectiveness and the way forward. *Revue Scientifique et Technique* 27, 125 - 145.
[44] Hill, B.J. (1977) Studies on SVC virulence and immunization. *Bulletin de l'Office International des Epizooties* 87, 455 - 456.
[45] Huang, R., Gao, L.-Y., Wang, Y.-P., Hu, W. and Guo, Q. L. (2009) Structure, organization and expression of common carp (*Cyprinus carpio* L.) NKEF-B gene. *Fish and Shellfish Immunology* 26, 220 - 229.
[46] Jeney, G., Jeney, Z., Oláh, J. and Fijan, N. (1990) Effect of rhabdovirus infections on selected blood parameters of wels (Silurus glanis L.). *Aquacultura Hungarica* 6, 153 - 160.
[47] Jeremic, S., Dobrila, J.-D. and Radosavljevic, V. (2004) Dissemination of spring viraemia of carp (SVC) in Serbia during the period 1992 - 2002. *Acta Veterinaria* (Beograd) 54, 289 - 299.
[48] Jeremic, S., Ivetic, V. and Radosavljevic, V. (2006) Rhabdovirus carpio as a causative agent of disease in rainbow trout (*Oncorhynchus mykiss*-Walbaum). *Acta Veterinaria* (Beograd) 56, 553 - 558.
[49] Kanellos, T., Sylvester, I.D., D'Mello, F., Howard, C.R., Mackie, A. *et al*. (2006) DNA vaccination can protect *Cyprinus carpio* against spring viraemia of carp virus. *Vaccine* 24, 4927 - 4933.
[50] Kirpichnikov, V.S., Ilyasov, J.I., Shart, L.A., Vikhman, A.A., Ganchenko, M.V. et al. (1993) Selection of Krasnodar common carp (*Cyprinus carpio* L.) for resistance to dropsy: principal results and prospects. Aquaculture 111, 7 - 20.
[51] Kiryu, I., Sakai, T., Kurita, J. and Iida, T. (2007) Virucidal effect of disinfectants on spring viremia of carp virus. *Fish Pathology* 42, 111 - 113.
[52] Kölbl, O. (1980) Diagnostic de la virémie printanière de la carpe et essais d'immunisation contre cette maladie. *Bulletin de l'Office International des Epizooties* 92, 1055 - 1068.
[53] Kölbl, O. (1990) Entwicklung eines Impfstoffes gegen die Fruhjahrsviramie der Karpfen (spring viraemia of carp, SVC); Feldversuchserfahrungen. *Tierärztliche Umschau* 45, 624 - 649.
[54] Koutná, M., Vesely, T., Psikal, I. and Hulová, J. (2003) Identification of spring viraemia of carp virus (SVCV) by combined RT - PCR and nested PCR. *Diseases of Aquatic Organisms* 55, 229 - 235.

[55] Liu, H., Gao, L., Shi, X., Gu, T., Jiang, Y. et al. (2004) Isolation of spring viraemia of carp virus (SVCV) from cultured koi (*Cyprinus carpio koi*) and common carp (*C. carpio carpio*) in PR China. *Bulletin of the European Association of Fish Pathologists* 24, 194 - 202.

[56] Liu, L.Y., Zhu, B.B., Wu, S.S., Lin, L., Liu, G.X. *et al.* (2015) Spring viraemia of carp virus induces autophagy for necessary viral replication. *Cellular Microbiology* 17, 595 - 605.

[57] Liu, Z., Teng, Y., Liu, H., Jiang, Y., Xie, X. *et al.* (2008a) Simultaneous detection of three fish rhabdoviruses using multiplex real-time quantitative RT - PCR assay. *Journal of Virological Methods* 149, 103 - 109.

[58] Liu, Z., Teng, Y., Xie, X., Li, H., Lv, J. *et al.* (2008b) Development and evaluation of a one-step loopmediated isothermal amplification for detection of spring viraemia of carp virus. *Journal of Applied Microbiology* 105, 1220 - 1226.

[59] Marcotegui, M.A., Estepa, A., Frías, D. and Coll, J.M. (1992) First report of a rhabdovirus affecting carp in Spain. *Bulletin of the European Association of Fish Pathologists* 12, 50 - 52.

[60] Martinez-Lopez, A., Garcia-Valtanen, P., Ortega-Villaizan, M., Chico, V., Gomez-Casado, E. *et al.* (2014) VHSV G glycoprotein major determinants implicated in triggering the host type I IFN antiviral response as DNA vaccine molecular adjuvants. *Vaccine* 32, 6012 - 6019.

[61] Masycheva, V.I., Alikin, Y.S., Klimenko, V.P., Fadina, V.A., Shchelkunov, I.S. *et al.* (1995) Comparative antiviral effects of dsRNA on lower and higher vertebrates. *Veterinary Research* 26, 536 - 538.

[62] Misk, E., Garver, K.A., Nagy, E., Isaacs, S., Tubbs, L. *et al.* (2016) Pathogenesis of spring viremia of carp virus in emerald shiner *Notropis atherinoides* Rafinesque, fathead minnow *Pimephales promelas* Rafinesque and white sucker *Catostomus commersonii* (Lacepede). *Journal of Fish Diseases* 39, 729 - 739.

[63] Molnár, K. and Csaba, G. (2005) Sanitary management in Hungarian aquaculture. *Veterinary Research Communications* 29, 143 - 146.

[64] Oidtmann, B.C., Thrush, M.A., Denham, K.L. and Peeler, E.J. (2011) International and national biosecurity strategies in aquatic animal health. *Aquaculture* 320, 22 - 33.

[65] OIE (2015a) Spring viraemia of carp. In: *Manual of Diagnostic Tests for Aquatic Animals*. World Organisation for Animal Health, Paris. Updated 2016 version (Chapter 2.3.9.) available at: http://www.oie.int/index.php?id=2439&L=0&htmfile=chapitre_svc.htm (accessed 31 October 2016).

[66] OIE (2015b) Spring viraemia of carp. In: *Aquatic Animal Health Code*. World Organisation for Animal Health, Paris. Updated 2016 version (Chapter 10.9.) available at: http://www.oie.int/index.php?id=171&L=0&htmfile=chapitre_svc.htm (accessed 31 October 2016).

[67] Oraic, D. and Zrncic, S. (2005) An overview of health control in Croatian aquaculture. *Veterinary Research Communications* 29, 139 - 142.

[68] Osadchaya, E.F. and Rudenko, A.P. (1981) Patogennost virusov, vydelennyh pri krasnuhe (vesennej viremii) karpov i kliniko-morfologicheskaja harakteristika estestvennogo techenija bolezni i v eksperimente [Pathogenicity of viruses isolated from carp with infectious dropsy (red-spot disease, spring viraemia) and clinical and morphological characteristics of the course of the natural and experimental disease]. *Rybnoe Khozyaistvo* (Kiev) 32, 66 - 71.

[69] Peters, F. and Neukirch, M. (1986) Transmission of some fish pathogenic viruses by the heron, *Ardea cinerea*. *Journal of Fish Diseases* 9, 539 - 544.

[70] Pfeil-Putzien, C. (1978) Experimentelle Übertragung der Frühjahrsvirämie (spring

viraemia) der Karpfen durch Karpfenläuse (Argulus foliaceus). *Zentralblatt für Veterinarmedizin, Reihe B* 25, 319 - 323.

[71] Pfeil-Putzien, C. and Baath, C. (1978) Nachweis einer Rhabdovirus-carpio-Infektion bei Karpfen im Herbst. *Berliner und Münchener Tierärztliche Wochenschrift* 91, 445 - 447.

[72] Phelps, N.B.D., Armien, A.G., Mor, S.K., Goyal, S.M., Warg, J.V. *et al.* (2012) Spring viremia of carp virus in Minnehaha Creek, Minnesota. *Journal of Aquatic Animal Health* 24, 232 - 237.

[73] Řehulka, J. (1996) Blood parameters in common carp with spontaneous spring viremia (SVC). *Aquaculture International* 4, 175 - 182.

[74] Rexhepi, A., Bërxholi, K., Scheinert, P., Hamidi, A. and Sherifi, K. (2011) Study of viral diseases in some freshwater fish in the Republic of Kosovo. *Veterinarsky Arhiv* 81, 405 - 413.

[75] Ruane, N.M., Rodger, H.D., McCarthy, L.J., Swords, D., Dodge, M. *et al.* (2014) Genetic diversity and associated pathology of rhabdovirus infections in farmed and wild perch *Perca fluviatilis* in Ireland. *Diseases of Aquatic Organisms* 112, 121 - 130.

[76] Ruyra, A., Cano-Sarabia, M., Garcia-Valtanen, P., Yero, D., Gibert, I. *et al.* (2014) Targeting and stimulation of the zebrafish (*Danio rerio*) innate immune system with LPS/dsRNA-loaded nanoliposomes. *Vaccine* 32, 3955 - 3962.

[77] Sanders, G.E., Batts, W.N. and Winton, J.R. (2003) Susceptibility of zebrafish (*Danio rerio*) to a model pathogen, spring viremia of carp virus. *Comparative Biochemistry and Physiology - Part C: Toxicology and Pharmacology* 53, 514 - 521.

[78] Sano, M., Nakai, K. and Fijan, N. (2011) Viral diseases and agents of warmwater fish. In: Woo, P.T.K. and Bruno, D.W. (eds) *Fish Diseases and Disorders, Volume 3: Viral, Bacterial and Fungal Infections*, 2nd edn. CAB International, Wallingford, UK, pp. 166 - 244.

[79] Selli, L., Manfrin, A., Mutinelli, F., Giacometti, P., Cappellozza, E. *et al.* (2002) Isolamento, da pesce gatto (*Ictalurus melas*), di un agente virale sierologicamente correlato al virus della viremia primaverile della carpa (SVCV). Bollettino Societa Italiana di Patologia Ittica 14, 3 - 13.

[80] Shchelkunov, I.S. and Shchelkunova, T.I. (1989) Rhabdovirus carpio in herbivorous fishes: isolation, pathology and comparative susceptibility of fishes. In: Ahne, W. and Kurstak, E. (eds) *Viruses of Lower Vertebrates*. Springer, Berlin, pp. 333 - 348.

[81] Shchelkunov, I.S., Oreshkova, S.F., Popova, A.G., Nikolenko, G.N., Shchelkunova, T.I. *et al.* (2005) Development of PCR-based techniques for routine detection and grouping of spring viraemia of carp virus. In: Cipriano, R.C., Shchelkunov, I.S. and Faisal, M. (eds) *Health and Diseases of Aquatic Organisms: Bilateral Perspectives*. Michigan State University Press, East Lansing, Michigan, pp. 260 - 284.

[82] Shimahara, Y., Kurita, J., Nishioka, T., Kiryu, I., Yuasa, K. *et al.* (2016) Development of an improved RT - PCR for specific detection of spring viraemia of carp virus. *Journal of Fish Diseases* 39, 269 - 275.

[83] Siwicki, A.K., Pozet, F., Morand, M., Kazun, B., Trapkowska, S. *et al.* (2003) Influence of methisoprinol on the replication of rhabdoviruses isolated from carp (*Cyprinus carpio*) and catfish (*Ictalurus melas*): in vitro study. *Polish Journal of Veterinary Sciences* 6, 47 - 50.

[84] Soliman, M.K., Aboeisa, M.M., Mohamed, S.G. and Saleh, W.D. (2008) First record of isolation and identification of spring viraemia of carp virus from Oreochromis niloticus in Egypt. In: *Eighth International Symposium on Tilapia in Aquaculture. Proceedings of the ISTA 8, October 12 - 14, 2008, Cairo, Egypt. Final Papers Submitted to ISTA 8*, pp. 1287 - 1306. Available at: http://ag.arizona.edu/azaqua/ista/ISTA8/FinalPapers/12/4%20Viral%20Infection/Magdy%20Khalil%20paper%20final.doc

(accessed 31 October 2016).

[85] Stone, D.M., Ahne, W., Denham, K.D., Dixon, P.F., Liu, C.T.Y. *et al.* (2003) Nucleotide sequence analysis of the glycoprotein gene of putative spring viraemia of carp virus and pike fry rhabdovirus isolates reveals four genogroups. *Diseases of Aquatic Organisms* 53, 203 - 210.

[86] Stone, D.M., Kerr, R.C., Hughes, M., Radford, A.D. and Darby, A.C. (2013) Characterisation of the genomes of four putative vesiculoviruses: tench rhabdovirus, grass carp rhabdovirus, perch rhabdovirus and eel rhabdovirus European X. *Archives of Virology* 158, 2371 - 2377.

[87] Taylor, N.G.H., Peeler, E.J., Denham, K.L., Crane, C.N., Thrush, M.A. *et al.* (2013) Spring viraemia of carp (SVC) in the UK: the road to freedom. Preventive Veterinary Medicine 111, 156 - 164.

[88] Vicenova, M., Reschova, S., Pokorova, D., Hulova, J. and Vesely, T. (2011) First detection of pike fry-like rhabdovirus in barbel and spring viraemia of carp virus in sturgeon and pike in aquaculture in the Czech Republic. *Diseases of Aquatic Organisms* 95, 87 - 95.

[89] Warg, J.V., Dikkeboom, A.L., Goodwin, A.E., Snekvik, K. and Whitney, J. (2007) Comparison of multiple genes of spring viremia of carp viruses isolated in the United States. *Virus Genes* 35, 87 - 95.

[90] Xiao, Z.G., Shen, J., Feng, H., Liu, H., Wang, Y.P. *et al.* (2015) Characterization of two thymosins as immunerelated genes in common carp (*Cyprinus carpio* L.). *Developmental and Comparative Immunology* 50, 29 - 37.

[91] Yue, Z., Teng, Y., Liang, C., Xie, X., Xu, B. *et al.* (2008) Development of a sensitive and quantitative assay for spring viremia of carp virus based on real-time RT - PCR. *Journal of Virological Methods* 152, 43 - 48.

[92] Zhang, N.Z., Zhang, L.F., Jiang, Y.N., Zhang, T. and Xia, C. (2009) Molecular analysis of spring viraemia of carp virus in China: a fatal aquatic viral disease that might spread in East Asian. *PLoS ONE* 4(7): e6337.

[93] Zhou, W., Zhou, J.J., Lv, Y., Qu, Y.X., Chi, M.D. *et al.* (2015) Identification and characterization of MAVS from black carp *Mylopharyngodon piceus*. *Fish and Shellfish Immunology* 43, 460 - 468.

8

斑点叉尾鮰病毒病

Larry A. Hanson* 和 Lester H. Khoo

8.1 引言

斑点叉尾鮰病毒病(channel catfish viral disease, CCVD)是一种急性病毒血症，主要发生在水产养殖的幼龄(0～4 个月)斑点叉尾鮰(*Ictalurus punctatus*)中。CCVD 暴发几乎只发生在夏季水温超过 25 ℃时，不到 2 周的时间内死亡率可能超过 90%。较大的鱼可能会经历更慢性的暴发，并通常伴随继发性柱状黄杆菌(*Flavobacterium columnare*)或气单胞菌(*Aeromonas*)感染，这些感染可能会掩盖潜在的 CCVD 疫情(Plumb, 1978)。在苗种生产设施中经常出现养殖池互相之间的传播。Fijan 等(1970)首先报道了这种疾病，其最显著的临床症状是眼球突出、腹胀、游动迷失方向和死亡率迅速上升。

Wolf 和 Darlington(1971)发现并表征了斑点叉尾鮰病毒病的致病因子为 1 型鮰疱疹病毒(Ictalurid herpesvirus 1，IcHV1，通常被称为斑点叉尾鮰病毒，channel catfish virus，CCV)。该病毒是鮰疱疹病毒属(*Ictalurivirus*)中的模式种，并且是异疱疹病毒科(Alloherpesviridae)中最具特点的成员之一，异疱疹病毒科包括大多数鱼类和两栖动物的疱疹病毒。IcHV1 的病毒粒子结构和复制方式与疱疹病毒目的其他病毒成员相似(Booy et al., 1996)。然而，在分子水平上(基因组序列)，IcHV1 与鱼疱疹病毒科其他属成员有着明显的不同之处，并且其基因组序列与疱疹病毒科的成员几乎没有同源性(Davison, 1992; Waltzek et al., 2009; Doszpoly et al., 2011a)。

鮰疱疹病毒属中与 IcHV1 密切相关的病毒为 2 型鮰疱疹病毒(Ictalurid herpesvirus 2, IcHV2)，它会引起一种类似于 CCVD 的疾病，该疾病曾摧毁了意大利的黑鮰养殖业(Alborali et al., 1996; Doszpoly et al., 2008, 2011b; Roncarati et al., 2014)。斑点叉尾鮰鱼苗和幼鱼对 IcHV 2 极为敏感，在实验感染中，鱼苗和幼鱼在 24～25 ℃时会患上类似 CCVD 的疾病，这比常见的 IcHV1 感染温度更低(Hedrick et al., 2003)。

* 通信作者邮箱：hanson@cvm.msstate.edu。

IcHV1具有宿主特异性，仅感染斑点叉尾鮰、蓝鲇（*Ictalurus furcatus*）及它们的杂交种（Plumb，1989）。早期关于CCVD会限制鲇鱼产业发展的担忧是多余的，养殖生产者已通过降低养殖密度和在关键的第一个夏季生产期间尽量减少胁迫影响来管控该病害的暴发。因此，整个行业的CCVD死亡率的影响已经变得可控。不过，偶发的CCVD疫情确实发生时，对个体养殖生产造成的损失可能是毁灭性的。这种疫情往往发生在多个池塘，这可能导致灾难性的生产损失。此外，CCVD疫情后幸存鱼的生长会减速（McGlamery and Gratzek，1974）。

8.1.1 流行和传播

大多数商业化生产养殖斑点叉尾鮰的地方都存在IcHV1。该病毒具有潜伏期，垂直传播并能保持多代不被发现（无病征）。使用聚合酶链反应（PCR），Thompson等（2005）评估了密西西比州商业孵化场3～5日龄鱼苗中IcHV1的流行情况，他们发现调查的五个孵化场的繁育种群都有这种病毒，而这些孵化场为美国的叉尾鮰养殖业提供了20%的苗种。上述孵化场鱼苗中潜隐IcHV1病毒的患病率为11.7%～26.7%，但研究者通过细胞培养未能检测到任何病毒的复制，研究期间在生产设施内也没有发生斑点叉尾鮰病毒病。跟踪考察进入第一个生产季节的鱼种群后发现，种群中IcHV1潜伏携带率从13%增加到35%以上。然而，疾病仍然没有发生，并且细胞培养中仍未检测到病毒的复制。这些结果表明，病毒垂直传播多发生在亲本和子代之间，亚临床水平传播多发生在苗种群体内。

8.1.2 促进CCVD暴发的因素

影响CCVD暴发的因素包括水温、胁迫、鱼群密度和鱼龄（Plumb，1978，1989）。关于感染研究的实验证明了温度的影响：将水温从19 ℃升高到28 ℃，极大促进了CCVD的暴发，而将温度从28 ℃降低到19 ℃则基本上遏制了CCVD造成的损失（Plumb，1973a）。

间接证据表明，鱼对CCVD的易感性与鱼龄呈负相关，商业养殖系统中几乎所有的CCVD疫情都发生在幼龄种群中。鱼龄对CCVD易感性的明显影响可能与鱼生长过程中亚致死性的病毒暴露引起的免疫力提高或随着鱼龄而变化的行为模式有关。在最易感时期，实验证据表明，刚出生的鱼苗比1月龄鱼对CCVD更具抗性，这可能与母体抗体的存在有关（Hanson et al.，2004）。此外，不同斑点叉尾鮰品系的易感性也存在差异（Plumb et al.，1975）。

胁迫对CCVD暴发的影响尚不清楚。环境证据表明，水质差和操作胁迫与自然环境中的CCVD暴发相关，但较差水质和口服皮质醇并没有影响实验条件下诱发的CCVD(Davis et al.，2002，2003)。应激激素的作用可能更多地与病毒复发的数量相关,而与鱼对浸泡感染的易感性无关。在暴露于低温环境下的斑点叉尾鮰成鱼中,使用合成皮质类固醇地塞米松可从分离的白细胞中回收更多的病毒(Bowser et al., 1985)。Arnizaut 和 Hanson(2011)发现,地塞米松提高了带毒的斑点叉尾鮰幼鱼的病毒 DNA 载量和病毒基因表达水平。虽然没有在上述幼鱼中检测到传染性病毒,但循环 IcHV1 特异性抗体增加表明鱼体内出现了一些病毒复发。此外,许多 CCVD 的暴发都没有明显地诱发应激事件。

人们对斑点叉尾鮰大龄幼鱼和成鱼的 IcHV1 感染的重要性尚不清楚。Hedrick 等(1987)的浸泡感染实验研究证明,刚成年的斑点叉尾鮰对 CCVD 易感。此外,鱼体中抗病毒抗体的季节性增加现象表明至少在夏季时成鱼体内存在一些抗原表达(Bowser and Munson, 1986)。当冬季水温为 8 ℃时,从患病的斑点叉尾鮰成鱼和鱼群中的未患病鱼中都能分离出 IcHV1 (Bowser et al., 1985)。在冬季,从鱼体中基本检测不到病毒复制,相关的疾病对经济的影响也是微乎其微。

8.2 诊断

CCVD 的推定诊断通常基于幼龄斑点叉尾鮰死亡率的快速增长,并表现出一系列典型的临床症状(腹胀、眼球突出和游动障碍)、大体病理学、组织病理学及在细胞培养中产生 IcHV1 的典型致细胞病变效应(CPE)。

IcHV1 易于在鮰细胞系中培养,但在大多数非鮰细胞系中不产生 CPE。简单地说,首先要在无血清细胞培养基中,将整个鱼苗、幼鱼内脏或来自较大幼鱼的后肾匀浆处理,然后使用 0.22 μm 过滤器去除匀浆中的微生物,并将该滤液在细胞培养基中以最终浓度(组织:培养基=1:100)稀释至28~30 ℃下的斑点叉尾鮰卵巢(CCO)或棕鲇(BB)细胞培养液中。两种细胞系均可从美国典型培养物保藏中心(ATCC, Manassas, Virginia)获得。急性感染 CCV 的鱼的培养物可以产生高浓度的感染性病毒,并在 24~48 h 快速发生 CPE。CPE 阳性的典型表征是由于细胞融合而形成合胞体,这些合胞体随后从培养瓶中脱离,留下黏附在附着点的放射状细胞质(图 8.1)。无 CPE 的培养物应连续观察 7 d,然后盲传后再连续观察7 d,如果此时仍未出现 CPE 则可判定 IcHV1 阴性。

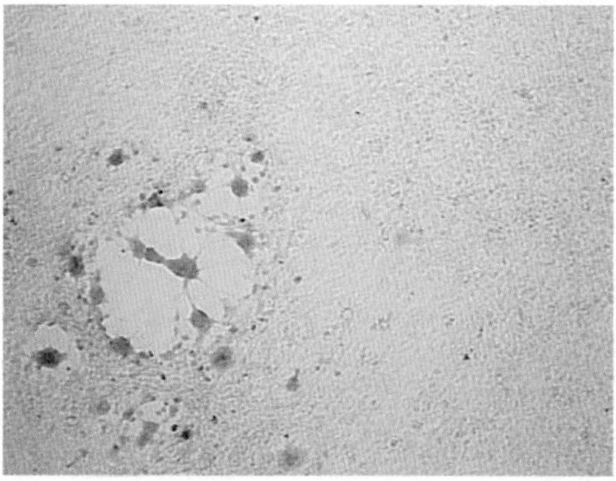

图 8.1 由 IcHV1 在单层斑点叉尾鮰卵巢细胞上引起的空斑。注意收缩的合胞体留下细胞质突起(100 倍放大)

斑点叉尾鮰呼肠孤病毒(Amend

et al.，1984)和 IcHV2 均能感染北美鲇种，并且还会在 28～30 ℃时引起 CCO 和 BB 细胞系的致细胞病变效应。斑点叉尾鮰呼肠孤病毒引起有限的合胞体，但 CPE 传播较慢，该病毒与斑点叉尾鮰养殖业的严重病害损失无关。相比之下，IcHV2 更值得关注，因为它会导致北美鲇产生实验感染中的 CCVD 样疾病，还会产生类似于 IcHV1 的 CPE(Hedrick et al.，2003)。在 IcHV2 疾病最初暴发的描述中，CPE 可见于蓝鳃太阳鱼细胞系(BF - 2)和源自胖头鲹的鲤上皮瘤(epithelioma papulosum cyprini，EPC)细胞系(Alborali et al.，1996)。这表明 IcHV2 具有更广泛的细胞宿主范围，因此区分 IcHV2 和 IcHV1 可能具有诊断价值。

细胞培养分离病毒后，需要使用血清学或分子分析等手段进行确认。使用中和单克隆抗体(Arkush et al.，1992)可以将 IcHV1 与 IcHV2 区分开来(Hedrick et al.，2003)。此外，多克隆抗体也可用于鉴定，但 IcHV1 在兔中诱导产生的抗体效果很差。此外，研究者还开发了其他的中和单克隆抗体和单特异性多克隆抗体(Wu et al.，2011；Liu et al.，2012)，但是这些抗体是否能区分 IcHV1 和 IcHV2 尚未可知，并且这些诊断抗体都无市售产品。

确诊 IcHV1 最常见的分子方法是 PCR 检测。研究者已经开发了几种灵敏的传统 PCR 测定法(Boyle and Blackwell，1991；Baek and Boyle，1996；Gray et al.，1999；Thompson et al.，2005)，但它们区分 IcHV1 和 IcHV2 的能力尚未报道。有研究者专门针对 IcHV2 开发了一种基于水解探针(TaqMan)的定量 PCR(Goodwin and Marecaux，2010)。还有研究测试了两种不同的基于 TaqMan 的定量 PCR(qPCR)，可用来区分 IcHV1 的两种基因群，而不检测 IcHV2(表 8.1)(Hanson，2016)。与所有 PCR 检测方法一样，这些方法对先前 PCR 产物的基因组污染非常敏感，因此检测应在不同的场所进行并使用单独的设备。此外，必须设置阴性对照培养物以确保在样品制备期间没有发生污染。

表 8.1　IcHV1 和 IcHV2 qPCR 检测的引物、探针和扩增参数[a]

方法、引物、探针	序　列	循环参数
IcHV1 TaqMan		
上游引物	CTCCGAGCGATGACACCAC	在 64 ℃时 60 s
下游引物	TGTGTTCAGAGGAGCGTCG	在 72 ℃时 15 s
IcHV1a 探针[b]	FAM - CCCATCCCTTCCCTCCTCCCTG - BHQ - 1[c]	在 95 ℃时 15 s
IcHV1b 探针[b]	FAM - CCCATCCTTCCCCTCCTCCCTG - BHQ - 1	
IcHV2 TaqMan		
上游引物	ATACATCGGTCTCACTCAAGAGG	在 59 ℃时 45 s
下游引物	TAATGGGTATTGGTACAAATCTTCATC	在 72 ℃时 45 s
探针	FAM - CGC+CTG+AGA+ACC+GAGCA - BHQ - 1[d]	在 95 ℃时 30 s

[a] IcHV2 检测方法来自 Goodwin 和 Marecaux(2010)。
[b] IcHV1a 探针检测 A 型基因组，IcHV1b 探针检测 B 型基因组。两种方法使用相同的引物和反应条件。
[c] FAM 表示 6 - FAM(6 -羧基荧光素)荧光染料，BHQ - 1 表示黑洞猝灭剂。
[d] “+”表示使用锁定的碱基来升高熔点。

此外，还需要一些方法来检测潜伏感染以识别鱼类的带毒群体。来自带毒鱼样本的直接细胞培养没有诊断价值，因为当病毒潜伏时不会产生感染性病毒。虽然使用地塞米松注射诱导 IcHV1 复发，随后提取白细胞与 CCO 细胞的共培养已成功证实斑点叉尾鮰成鱼中存在 IcHV1（Bowser et al.，1985），但这种方法仅在冬季冷水养殖的鱼中获得成功。其他研究者使用相同的方案未能成功地在已知带毒鱼中重新激活可培养的潜伏 IcHV1，因此这不是一种可靠的诊断方法（Arnizaut and Hanson，2011）。检测亲鱼中的 IcHV1 抗体已成功用于带毒种群的识别（Plumb，1973b；Amend and McDowell，1984；Bowser and Munson，1986；Crawford et al.，1999）。此外，酶联免疫吸附试验（ELISA）比血清中和更灵敏（Crawford et al.，1999）。血清学检测是一种间接检测方法，它们的可靠性不仅取决于病原体的存在，还取决于鱼的免疫状态。带毒鱼中的 IcHV1 特异性抗体浓度在整个种群内的鱼之间存在很大差异（Arnizaut and Hanson，2011）。更重要的是，鱼体内的 IcHV1 特异性抗体的平均浓度在一年中变化很大，最高抗体水平主要出现在秋季（Bowser and Munson，1986）。此外，由于垂直传播而携带潜伏 IcHV1 但未经历病毒抗原表达的斑点叉尾鮰没有可检测的抗体。因此，应当首先直接检测带毒鱼中的潜伏病毒基因组。

上述常规 PCR 和 qPCR 方法都非常灵敏，可以检测携带者组织中潜伏的 IcHV1。然而，尽管使用 PCR 检测带毒种群的检测方法非常敏感，但感染组织中潜伏 IcHV1 的基因组浓度可能接近最低检测限。进行 PCR 检测的组织采样取决于鱼的生长阶段和健康状况。卵黄囊苗应该在去除卵黄后整体处理（Thompson et al.，2005）。为了评估临床患病鱼而直接用于 CCVD 诊断的，应当优选其后肾组织，而对于大龄鱼的潜伏期评估，尾鳍活组织的检测是可靠的（Arnizaut，2002）。可以使用市售的 DNA 提取试剂盒制备样品，如 Gentra Puregene 组织试剂盒（Qiagen）。当对来自鱼体组织的 1 μg DNA 进行 qPCR 检测时，CCVD 临床感染的鱼或受感染细胞培养物的循环数阈值一般低于 27，而对于潜伏感染的鱼，其循环数阈值应当为 33～38。在对亲鱼的研究中，研究者发现 PCR 阴性亲本仍然可以产生 IcHV1 阳性后代（Y. Habte and L. Hanson，未发表）。因此，尽管 PCR 检测可用于检测带毒种群，但其灵敏度尚不足以从种群中剔除所有阳性个体。

8.3 病理学

8.3.1 临床体征和大体病变

对 CCVD 的一些原始描述和综述（Fijan et al.，1970；Plumb，1978，1986；Hanson et al.，2011；Plumb and Hanson，2011）提供了关于 CCVD 临床症状及大体病理学的详细描述。CCVD 的临床症状包括游动异常（沿纵轴回旋），以及患病鱼在水体底部快速呼吸并抽搐，然后停止呼吸。濒死鱼会在水中垂直悬挂，头部露出水面。这种鱼在水中的垂直姿势曾被认为是该疾病的诊断特征。大体上，患病鱼可能会出现鳍基部和下腹侧外出血、眼球突出、腹胀和鳃苍白或出血等症状。也有感染鱼的临床报告中记载通常有双侧眼球突出症和腹部肿胀，但往往缺乏外部出血性病变[图 8.2(A)]。

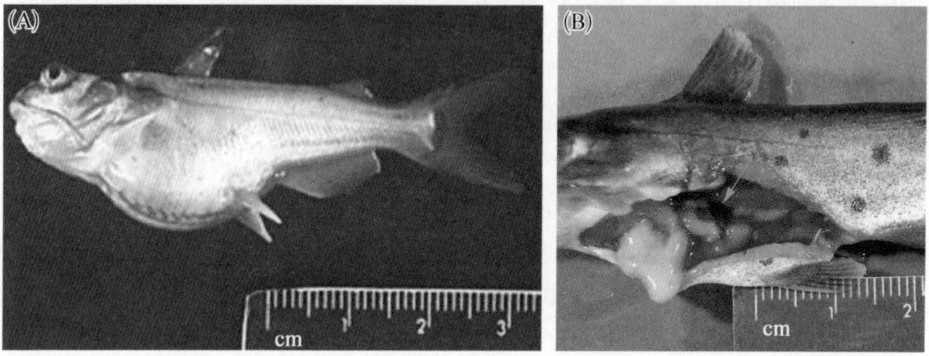

图 8.2　患斑点叉尾鮰病毒病(CCVD)的鲇幼鱼的照片。(A) 外部大体病变，注意腹胀明显。(B) 内部大体病变；箭头所示为脾脏充血，与轻度肿胀的后肾相邻，胃肠道轻度充血，腹水从腹部切口中流出，其中的脏器闪闪发光

遗憾的是，根据临床资料，通常没有准确的方法来确定疾病的长期发展进度，缺乏出血性外部病变可能反映了疾病连续发展的某个时间点或阶段的特征。历史上描述的大体内部病变包括肌肉组织、肝和肾区域的出血。此外，CCVD 还会导致受感染鱼出现肝灰白、后肾苍白肿大、脾充血肿大以及胃肠消化道内无食物但充满黄色黏液样液体等病变。Plumb(1986)描述了受感染的鱼出现了整体性内脏或体腔充血，并且存在稻草色/黄色腹水导致的腹胀。在临床提交的材料中几乎总是可以看到腹水、眼球突出和脾脏充血的记录。肌肉组织、肝和后肾出血及肾肿大等病变可能由于现象不太明显，并且因被感染鱼体型小而可能未得到重视[图 8.2(B)]。

8.3.2　组织病理学

早期对组织病理学病变的描述主要来自注射病毒的实验性感染鱼(Wolf et al, 1972; Plumb et al, 1974)。肾脏(前部和后部)通常是受影响最严重的组织，其次是胃肠道、肝和骨骼肌。Plumb 等(1974)发现病鱼除了近端小管上皮细胞出血和坏死外，后肾中淋巴样细胞数量也在增多。此外，肾脏的造血组织也存在广泛坏死，胰腺腺泡细胞周围的上皮细胞坏死明显，伴随有限或极少的胰腺腺泡细胞坏死。脾脏明显的出血或充血使白髓变得模糊。在胃肠道的黏膜下层和绒毛中也可见出血。

Major 等(1975)研究了实验感染和自然感染的鱼类,发现自然感染的幼鱼病变更严重。除了 Wolf 等(1972)和 Plumb 等(1974)报道过的病变,他们还发现鱼脑和胰腺中的病变。此外,Major 等(1975)描述了伴有周围神经纤维水肿的神经元空泡化及胰腺腺泡细胞坏死。水肿病变不仅限于大脑,还存在于心脏、脊髓、鳃、肾脏和胃肠道。在自然感染的幼鱼中还存在肝细胞坏死,而水肿和多灶性坏死更常见于大鱼。他们还在实验感染鱼的肝细胞中检测到嗜酸性胞质内包涵体,这在自然感染鱼中很少被检测到。

CCV 自然感染鱼的组织病理学损伤与 Major 等(1975)所描述的一致。另外,这些鱼的前肾内还存在造血元件的多处局部广泛坏死(图 8.3),但与实验感染的鱼相比,这些样本中很难发现淋巴成分的增加(Plumb et al., 1974)。和实验感染的鱼一样,自然感染的鱼后肾间质部分也存在中度至重度多灶性坏死(图 8.4),坏死仅偶尔会延伸至肾小管上皮。这些坏死病灶中有时还会存在罕见的多核合胞体[图 8.4(A)]。嗜酸性核内包涵体有时存在于后肾肾血管内被推测为淋巴细胞的细胞中(细胞核明显膨大,仅有一薄层嗜碱性细胞,这些细胞似乎缺乏细胞质),而在脾脏等其他器官中则很少存在[图 8.4(B)]。此外,核内包涵体很少见。

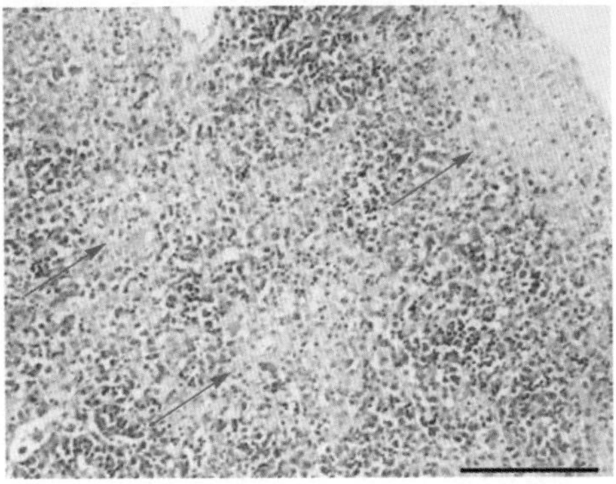

图 8.3 患斑点叉尾鮰病毒病(CCVD)的斑点叉尾鮰幼鱼的前肾组织病理学。箭头所示为嗜酸性坏死病灶,伴随致密核和核裂碎片[苏木精-伊红(HE)染色;标尺约为 100 μm]

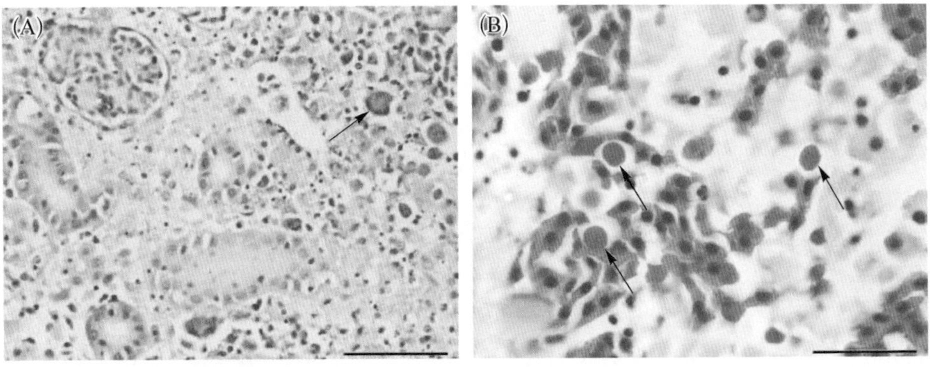

图 8.4 患斑点叉尾鮰病毒病(CCVD)的斑点叉尾鮰幼鱼的前肾组织病理学显示:(A) 如箭头所示,间质坏死伴多核合胞体[苏木精-伊红(HE)染色,标尺约为 50 μm];(B) 如箭头所示,具有嗜酸性包涵体的细胞(可能是淋巴细胞)和只有薄轮缘的染色质(HE 染色,标尺约为 20 μm)

病鱼的肝细胞坏死是多灶性和轻度的(图 8.5)，胰腺受累很小。严重的脾充血或出血掩盖了正常的脾脏结构及可能存在的坏死情况(图 8.6)。胃腺部存在坏死[图 8.7(A)]，并可以延伸至近端肠。在肠黏膜中，坏死呈多灶性并广泛分布于黏膜下层区域。有时在绒毛的基底部可见单核炎性细胞的小聚集体[图 8.7(B)]。部分胃肠道中也可能存在局部广泛充血。包括脑、肌肉和心在内的其他组织中均未见明显的微观病变。临床病例中的病变与文献中已经报道的基本一致(Plumb et al., 1974; Wolf et al., 1972; Major et al., 1975)，差异可能是由于疾病发展处于不同的时期、环境条件不同抑或病毒毒株不同。

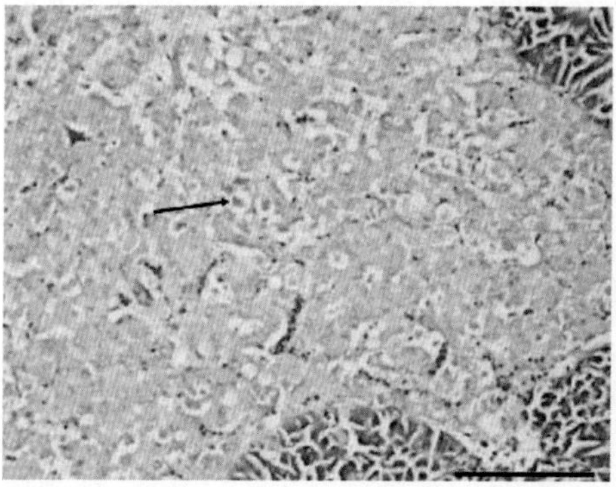

图 8.5　患斑点叉尾鮰病毒病(CCVD)的斑点叉尾鮰幼鱼肝脏坏死灶。箭头所指为其中一个稍微偏嗜酸性的萎缩肝细胞[苏木精-伊红(HE)染色；标尺约为 100 μm]

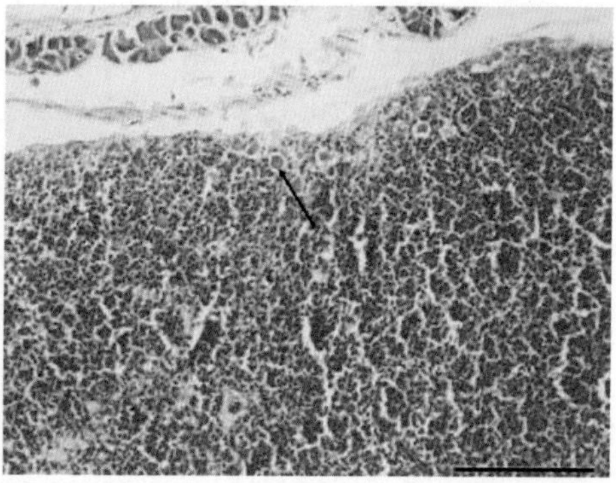

图 8.6　患斑点叉尾鮰病毒病(CCVD)的斑点叉尾鮰幼鱼严重充血脾脏的组织病理学。箭头指向一个带有核内包涵体的细胞，如图 8.4(A)所示。白髓被红细胞遮挡[苏木精-伊红(HE)染色；标尺约为 100 μm]

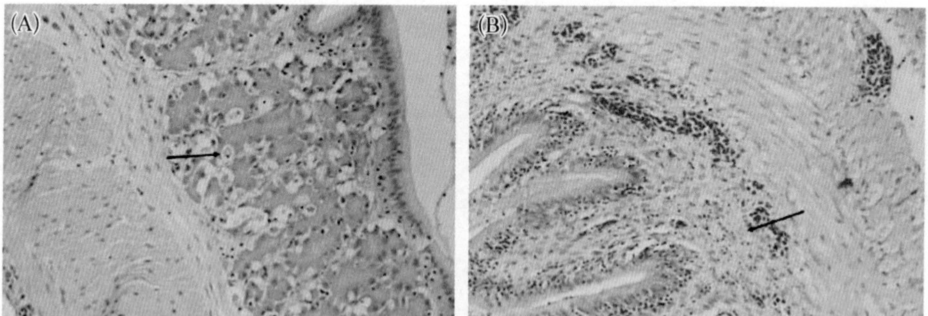

图 8.7　患斑点叉尾鮰病毒病(CCVD)的斑点叉尾鮰幼鱼胃肠道病理病变。(A) 胃切片，箭头指向胃腺部内众多坏死细胞中的一个[苏木精-伊红(HE)染色；标尺约为100 μm]。(B) 近端肠部分的切片，箭头指向黏膜下层的炎性坏死灶。黏膜中坏死区域位于更多空泡的区域内，绒毛基底部有小的单核炎性细胞聚集。图中几条血管也出现了充血(HE 染色，标尺约为 100 μm)

8.4 病理生理学

肾脏似乎是受CCVD影响的主要器官之一，肾脏损害导致体液失调，这可以解释病鱼出现的一些大体病变(例如腹水引起的腹胀和眼球突出)。然而，腹水的颜色更能表明这可能是一种从血管系统渗漏的改良渗出液。由于吸收和蛋白质合成的减少可降低血浆的胶体渗透压，胃肠道和肝脏中的病变也可能导致腹水和水肿。鳃苍白很可能是多个器官(尤其是脾脏)出血和充血的结果。另外，病鱼的前肾内及后肾间质中发生坏死，这将影响造血功能并导致贫血。虽然尚未确定IcHV1的细胞嗜性，但大多数异疱疹病毒科的病毒是嗜上皮的(Hanson et al., 2011)。胃肠道黏膜上皮和胰腺周围上皮的坏死表明IcHV1具有一些上皮嗜性。病毒定量和示踪实验证实，肾组织中最早产生病毒并有着最高的病毒滴度，其次是肝、肠和皮肤(Plumb, 1971; Nusbaum and Grizzle, 1987; Kancharla and Hanson, 1996)。细胞培养检测表明，IcHV1在成纤维细胞衍生细胞系及B淋巴细胞中复制良好，并能在其他白细胞中引起CPE(Bowser and Plumb, 1980; Chinchar et al., 1993)。在28℃下暴露于亚致死剂量和致死剂量的病毒环境下3～4 d后，肾脏中的病毒滴度达到峰值，这也是死亡率的高峰时期(Kancharla and Hanson, 1996)。

IcHV1的具体致病机制尚未明确。IcHV1基因组编码超过76个基因(Davison, 1992)，其中大多数基因对病毒的复制是非必需的。在毒力中起作用的非必需基因包括胸苷激酶(thymidine kinase, TK)基因和基因50。病毒编码的TK基因可能促进病毒在非复制细胞中的复制(Zhang and Hanson, 1995)。斑点叉尾鮰病毒的TK突变株的复制水平与野生型病毒相似，并在实验感染鱼中也具有与野生型病毒类似的早期动力学特征，但该病毒突变株会更快地被清除(Zhang and Hanson, 1995; Kancharla and Hanson, 1996)。基因50的产物是一种功能未知的分泌型黏蛋白，该基因的突变会导致病毒的毒力减弱(Vanderheijden et al., 1996, 1999, 2001)。

鱼群密度过高引起的温度和剂量效应是影响CCVD暴发的最重要因素。温度的影响似乎是调节异疱疹病毒科病毒基因表达和促进或限制相关病毒疾病的一个共同因素。温度也已被证实可以调节由1型鲤疱疹病毒(Cyprinid herpesvirus 1)、3型鲤疱疹病毒(Cyprinid herpesvirus 3)、3型鲑疱疹病毒(Salmonid herpesvirus 3)(导致湖鳟流行性上皮病)、1型鲈疱疹病毒(Percid herpesvirus 1)(引起大眼鰤鲈弥漫性表皮增生)和1型狗鱼疱疹病毒(Esocid herpesvirus 1)(引起北方狗鱼蓝斑病)引起的疾病暴发(Hanson et al., 2011; Plumb and Hanson, 2011)。温度的影响在3型鲤疱疹病毒引起的疾病中最具特征性：锦鲤疱疹病毒病只限于18～28℃发生，超出此温度范围似乎会导致病毒产生潜伏感染，这一点已在细胞培养和宿主中得到证实（Gilad et al.，2003；Dishon et al.，2007；Ilouze et al.，2012）。

8.5 防控策略

由于积极和有意识的管理，在斑点叉尾鮰苗种养殖生产中实现了相对较低的 CCVD 发病率。鱼苗生产养殖池的放养量通常低于 250 000 尾/公顷。当水温高于 25 ℃时，苗种生产者会避免鱼苗密度过高或收集鱼苗。此外，生产者还需要仔细监测养殖池并充气以保持溶氧水平高于 3 mg/L。同时，池中氯化物含量通常维持在 100 mg/L 以上，以避免产生毒性亚硝酸盐，并只投喂优质的新鲜饲料。当生产者为鱼苗接种用以预防其他疾病的疫苗时，须仔细监测水温和溶氧水平。通常只在凉爽的早晨时段投放鱼苗，任何批次出现异常死亡率升高的鱼苗都应被丢弃，并对装载过鱼苗的水槽进行消毒。许多生产者避免在养殖季节后期收集鱼卵和孵化鱼苗，以防止鱼苗投放时池塘的水温升高。

疫苗接种、抗病品种选育和建立无 IcHV1 亲鱼系是有望预防和控制 CCVD 的方法，但目前尚未真正实施。病毒减毒疫苗候选株包括在蟾胡鲇（*Clarias batrachus*）细胞系中培养的 IcHV1 自然突变株（Noga and Hartmann，1981）、TK 缺失的重组 IcHV1（Zhang and Hanson，1995）和基因 50 缺失的重组 IcHV1（Vanderheijden et al.，2001）。其中两种重组减毒病毒都显示出表达外源基因并诱导免疫的潜力，表明它们可以用作疫苗载体（Zhang and Hanson，1996；Vanderheijden et al.，2001）。同时，利用插入 IcHV1 的细菌人工染色体开发技术让研究人员可使用高效的基于细菌的重组和筛选方法来生产重组 IcHV1 疫苗（Kunec et al.，2008，2009）。接种 CCVD 疫苗的一个限制因素是需要在具有免疫能力并且可以产生保护性免疫力的鱼龄期实施接种（Petrie-Hanson and Ainsworth，2001），但仍要在最易感 CCVD 的幼早鱼龄期尽可能接种。此外，目前的养殖生产管理方法不允许在 8 日龄浸泡接种疫苗，因为这正是鱼苗投放鱼塘的时候。其他增强斑点叉尾鮰抗病力的免疫增强措施有应用干扰素诱导剂，如 poly(I：C)（聚肌苷酸-聚胞苷酸）（Plant et al.，2005），或通过将鱼苗预先暴露于鮰呼肠孤病毒（Chinchar et al.，1998）的方法，这已被细胞培养实验证实有效。尚未有研究涉及这些免疫增强剂的实际应用方法及其在斑点叉尾鮰群体中的效用。

由于一些斑点叉尾鮰品系对 CCVD 具有较强抗性，因此，可以通过选择性育种控制疾病的暴发（Plumb et al.，1975）。早期研究表明，斑点叉尾鮰和蓝鲇的杂交种可能对 CCVD 更具抗性（Plumb and Chappell，1978），但后来的研究并不支持这一发现；有人提出，选育中使用的斑点叉尾鮰品系会影响杂交种的易感性（Silverstein et al.，2008）。使用严格的筛选和剔除方法可以获得无 IcHV1

亲本以生产无病毒的苗种。但是,这样得到的苗种必须在比现有水产养殖设施更具生物安全性的环境中饲养。IcHV1 很可能是地方流行性的,因为生产者不会对亲鱼进行例行的病毒筛查。

如果暴发斑点叉尾鮰病毒病,有一些可用的实际措施能够减少损失。如果条件允许,可通过降低水温来降低鱼的死亡率。另外,疾病暴发后的管理仅限于采取防控措施来防止病毒传播。例如,来自受感染区域的鱼、水或设施不应在另一个池塘中使用;应限制病鱼与其他野生动物,特别是捕食者和食腐动物的接触;应迅速从养殖池中清除死鱼,并将其尸体填埋或焚烧以防止食腐动物重新传播疾病。如果鱼在转移到池塘之前在孵化场中发生疾病,养殖者通常对受感染水槽中的所有鱼进行安乐死并对设施进行消毒,此时投资相对较少,养殖者倾向于采取这种果断行动,而不是冒着疾病在孵化场内扩散的风险。CCVD 暴发后幸存下来的鱼常被养殖到上市而没有明显的负面影响,但是在亲鱼选育时应当避免使用这些鱼,因为它们可能携带病毒并促进垂直传播。在疫情暴发结束后,应在重新放养前清空池塘,并通过在潮湿的地方使用干燥的熟石灰来消毒(Camus, 2004)。

8.6 总结与研究展望

斑点叉尾鮰病毒病在斑点叉尾鮰种苗养殖中遭成巨大损失的时间已经超过45 年,该病毒在整个斑点叉尾鮰产业中以亚临床状态存在。鉴于 CCVD 暴发与水温升高和高养殖密度相关,该行业已经优化了水产养殖方法,以最大限度地减少疾病在关键的首个夏季生长期的影响。现在叉尾鮰生产大都使用高密度养殖系统,再加上不断上升的气温和多变的天气模式,高密度养殖可能会使 CCVD 成为更大的威胁。为了最大限度地减少 CCVD 的影响,确定病毒激活的触发因素从而制定有效的控制措施至关重要。研究免疫系统调控和病毒重组系统的分子基础(Kunec et al., 2008)可以阐明疱疹病毒典型而错综复杂的宿主/病原关系。此外,建立无病毒鱼种资源和使用有效的生物安全措施有助于更有效地预防 CCVD。为了建立无病毒种群,应继续研究防止病毒垂直传播的方法并将这些方法纳入生产杂交鮎的体外受精方案中。确定蓝鮎对 CCVD 的抗性机制,并利用这些信息优化杂交种遗传基因,可以进一步减轻 CCVD 对叉尾鮰产业的有害影响。

参考文献

[1] Alborali, L., Bovo, G., Lavazza, A., Cappellaro, H. and Guadagnini, P.F. (1996) Isolation of an herpesvirus in breeding catfish (*Ictalurus melas*). *Bulletin of the*

European Association of Fish Pathologists 16, 134 - 137.

[2] Amend, D.F. and Mcdowell, T. (1984) Comparison of various procedures to detect neutralizing antibody to the channel catfish virus in California brood channel catfish. *Progressive Fish Culturist* 46, 6 - 12.

[3] Amend, D.F., McDowell, T. and Hedrick, R.P. (1984) Characteristics of a previously unidentified virus from channel catfish (*Ictalurus punctatus*). *Canadian Journal of Fisheries and Aquatic Sciences* 41, 807 - 811.

[4] Arkush, K.D., Mcneill, C. and Hedrick, R.P. (1992) Production and characterization of monoclonal antibodies against channel catfish virus. *Journal of Aquatic Animal Health* 4, 81 - 89.

[5] Arnizaut, A. (2002) *Comparison of Channel Catfish Virus and Thymidine Kinase Negative Recombinant Channel Catfish Virus Latency and Recrudescence*. PhD, Mississippi State University, Starkville, Mississippi.

[6] Arnizaut, A.B. and Hanson, L.A. (2011) Antibody response of channel catfish after channel catfish virus infection and following dexamethasone treatment. *Diseases of Aquatic Organisms* 95, 189 - 201.

[7] Baek, Y.-S. and Boyle, J.A. (1996) Detection of channel catfish virus in adult channel catfish by use of nested polymerase chain reaction. *Journal of Aquatic Animal Health* 8, 97 - 103.

[8] Booy, F.P., Trus, B.L., Davison, A.J. and Steven, A.C. (1996) The capsid architecture of channel catfish virus, an evolutionarily distant herpesvirus, is largely conserved in the absence of discernible sequence homology with herpes simplex virus. *Virology* 215, 134 - 141.

[9] Bowser, P.R. and Munson, A.D. (1986) Seasonal variation in channel catfish virus antibody titers in adult channel catfish. *The Progressive Fish Culturist* 48, 198 - 199.

[10] Bowser, P.R. and Plumb, J.A. (1980) Channel catfish virus: comparative replication and sensitivity of cell lines from channel catfish ovary and the brown bullhead. *Journal of Wildlife Diseases* 16, 451 - 454.

[11] Bowser, P.R., Munson, A.D., Jarboe, H.H., Francis-Floyd, R. and Waterstrat, P.R. (1985) Isolation of channel catfish virus from channel catfish, *Ictalurus punctatus* (Rafinesque), broodstock. *Journal of Fish Diseases* 8, 557 - 561.

[12] Boyle, J. and Blackwell, J. (1991) Use of polymerase chain reaction to detect latent channel catfish virus. American *Journal of Veterinary Research* 52, 1965 - 1968.

[13] Camus, A.C. (2004) *Channel Catfish Virus Disease. SRAC Publication No. 4702*, Southern Regional Aquaculture Center, Stoneville, Mississippi.

[14] Chinchar, V.G., Rycyzyn, M., Clem, L.W. and Miller, N.W. (1993) Productive infection of continuous lines of channel catfish leukocytes by channel catfish virus. *Virology* 193, 989 - 992.

[15] Chinchar, V.G., Logue, O., Antao, A. and Chinchar, G.D. (1998) Channel catfish reovirus (CRV) inhibits replication of channel catfish herpesvirus (CCV) by two distinct mechanisms: viral interference and induction of an antiviral factor. *Diseases of Aquatic Organisms* 33, 77 - 85.

[16] Crawford, S.A., Gardner, I.A. and Hedrick, R.P. (1999) An enzyme linked immunosorbent assay (ELISA) for detection of antibodies to channel catfish virus (CCV) in channel catfish. *Journal of Aquatic Animal Health* 11, 148 - 153.

[17] Davis, K.B., Griffin, B.R. and Gray, W.L. (2002) Effect of handling stress on susceptibility of channel catfish *Ictalurus punctatus* to *Ichthyophthirius multifiliis* and channel catfish virus infection. *Aquaculture* 214, 55 - 66.

[18] Davis, K.B., Griffin, B.R. and Gray, W.L. (2003) Effect of dietary cortisol on resistance of channel catfish to infection by *Ichthyopthirius multifiliis* and channel

catfish virus disease. *Aquaculture* 218, 121–130.
[19] Davison, A.J. (1992) Channel catfish virus: a new type of herpesvirus. *Virology* 186, 9–14.
[20] Dishon, A., Davidovich, M., Ilouze, M. and Kotler, M. (2007) Persistence of cyprinid herpesvirus 3 in infected cultured carp cells. *Journal of Virology* 81, 4828–4836.
[21] Doszpoly, A., Kovacs, E.R., Bovo, G., Lapatra, S.E., Harrach, B. *et al.* (2008) Molecular confirmation of a new herpesvirus from catfish (*Ameiurus melas*) by testing the performance of a novel PCR method, designed to target the DNA polymerase gene of alloherpesviruses. *Archives of Virology* 153, 2123–2127.
[22] Doszpoly, A., Somogyi, V., Lapatra, S.E. and Benkö, M. (2011a) Partial genome characterization of Acipenserid herpesvirus 2: taxonomical proposal for the demarcation of three subfamilies in Alloherpesviridae. *Archives of Virology* 156, 2291–2296.
[23] Doszpoly, A., Benko, M., Bovo, G., Lapatra, S.E. and Harrach, B. (2011b) Comparative analysis of a conserved gene block from the genome of the members of the genus *Ictalurivirus*. *Intervirology* 54, 282–289.
[24] Fijan, N.N., Welborn, T.L.J. and Naftel, J.P. (1970) An acute viral disease of channel catfish. Technical Paper 43, US Fish and Wildlife Service, Washington, DC.
[25] Gilad, O., Yun, S., Adkison, M.A., Way, K., Willits, N.H. *et al.* (2003) Molecular comparison of isolates of an emerging fish pathogen, koi herpesvirus, and the effect of water temperature on mortality of experimentally infected koi. *Journal of General Virology* 84, 2661–2667.
[26] Goodwin, A.E. and Marecaux, E. (2010) Validation of a qPCR assay for the detection of *Ictalurid herpesvirus-2* (IcHV-2) in fish tissues and cell culture supernatants. *Journal of Fish Diseases* 33, 341–346.
[27] Gray, W.L., Williams, R.J., Jordan, R.L. and Griffin, B.R. (1999) Detection of channel catfish virus DNA in latently infected catfish. *Journal of General Virology* 80, 1817–1822.
[28] Hanson, L. (2016) Ictalurid herpesvirus 1. In: Liu, D. (ed.) *Molecular Detection of Animal Viral Pathogens*. CRC Press/Taylor and Francis Group, Boca Raton, Florida, pp. 797–806.
[29] Hanson, L.A., Rudis, M.R. and Petrie-Hanson, L. (2004) Susceptibility of channel catfish fry to channel catfish virus (CCV) challenge increases with age. *Diseases of Aquatic Organisms* 62, 27–34.
[30] Hanson, L., Dishon, A. and Kotler, M. (2011) Herpesviruses that infect fish. *Viruses* 3, 2160–2191.
[31] Hedrick, R.P., Groff, J.M. and Mcdowell, T. (1987) Response of adult channel catfish to waterborne exposures of channel catfish virus. *The Progressive Fish Culturist* [now *North American Journal of Aquaculture*] 49, 181–187.
[32] Hedrick, R.P., Mcdowell, T.S., Gilad, O., Adkison, M. and Bovo, G. (2003) Systemic herpes-like virus in catfish *Ictalurus melas* (Italy) differs from Ictalurid herpesvirus 1 (North America). *Diseases of Aquatic Organisms* 55, 85–92.
[33] Ilouze, M., Dishon, A. and Kotler, M. (2012) Downregulation of the cyprinid herpesvirus-3 annotated genes in cultured cells maintained at restrictive high temperature. *Virus Research* 169, 289–295.
[34] Kancharla, S.R. and Hanson, L.A. (1996) Production and shedding of channel catfish virus (CCV) and thymidine kinase negative CCV in immersion exposed channel catfish fingerlings. *Diseases of Aquatic Organisms* 27, 25–34.
[35] Kunec, D., Hanson, L.A., van Haren, S., Nieuwenhuizen, I.F. and Burgess, S.C. (2008) An over-lapping bacterial artificial chromosome system that generates vectorless progeny for channel catfish herpesvirus. *Journal of Virology* 82, 3872–3881.

[36] Kunec, D., van Haren, S., Burgess, S.C. and Hanson, L.A. (2009) A Gateway® recombination herpesvirus cloning system with negative selection that produces vectorless progeny. *Journal of Virological Methods* 155, 82-86.
[37] Liu, Y., Yuan, J., Wang, W., Chen, X., Tang, R. *et al.* (2012) Identification of envelope protein ORF10 of channel catfish herpesvirus. *Canadian Journal of Microbiology* 58, 271-277.
[38] Major, R.D., McCraren, J.P. and Smith, C.E. (1975) Histopathological changes in channel catfish (*Ictalurus punctatus*) experimentally and naturally infected with channel catfish virus disease. *Journal of the Fisheries Board of Canada* 32, 563-567.
[39] McGlamery, M.H. Jr and Gratzek, J. (1974) Stunting syndrome associated with young channel catfish that survived exposure to channel catfish virus. *The Progressive Fish-Culturist* [now *North American Journal of Aquaculture*] 36, 38-41.
[40] Noga, E.J. and Hartmann, J.X. (1981) Establishment of walking catfish (*Clarias batrachus*) cell lines and development of a channel catfish (*Ictalurus punctatus*) virus vaccine. *Canadian Journal of Fisheries and Aquatic Sciences* 38, 925-929.
[41] Nusbaum, K.E. and Grizzle, J.M. (1987) Uptake of channel catfish virus from water by channel catfish and bluegills. *American Journal of Veterinary Research* 48, 375-377.
[42] Petrie-Hanson, L. and Ainsworth, A.J. (2001) Ontogeny of channel catfish lymphoid organs. *Veterinary Immunology and Immunopathology* 81, 113-127.
[43] Plant, K.P., Harbottle, H. and Thune, R.L. (2005) Poly I : C induces an antiviral state against *Ictalurid herpesvirus 1* and Mx1 transcription in the channel catfish (*Ictalurus punctatus*). *Developmental and Comparative Immunology* 29, 627-635.
[44] Plumb, J.A. (1971) Tissue distribution of channel catfish virus. *Journal of Wildlife Diseases* 7, 213-216.
[45] Plumb, J.A. (1973a) Effects of temperature on mortality of fingerling channel catfish (*Ictalurus punctatus*) experimentally infected with channel catfish virus. *Journal of the Fisheries Research Board of Canada* 30, 568-570.
[46] Plumb, J.A. (1973b) Neutralization of channel catfish virus by serum of channel catfish. *Journal of Wildlife Diseases* 9, 324-330.
[47] Plumb, J.A. (1978) Epizootiology of channel catfish virus disease. *Marine Fisheries Review* 3, 26-29.
[48] Plumb, J.A. (1986) *Channel Catfish Virus Disease*. Fish Disease Leaflet 73, US Department of the Interior, Washington, DC.
[49] Plumb, J. (1989) Channel catfish herpesvirus. In: Ahne, W. and Kurstak, E. (eds) *Viruses of Lower Vertebrates*. Springer, Berlin, pp. 198-216.
[50] Plumb, J.A. and Chappell, J. (1978) Susceptibility of blue catfish to channel catfish virus. *Proceedings of the Annual Conference of the Southeastern Association of Fish and Wildlife Agencies 32*, 680-685.
[51] Plumb, J.A. and Hanson, L.A. (2011) Health Maintenance and Principal Microbial *Diseases of Cultured Fishes*. Wiley, Ames, Iowa.
[52] Plumb, J.A., Gaines, J.L., Mora, E.C. and Bradley, G.G. (1974) Histopathology and electron microscopy of channel catfish virus in infected channel catfish, *Ictalurus punctatus* (Rafinesque). *Journal of Fish Biology* 6, 661-664.
[53] Plumb, J.A., Green, O.L., Smitherman, R.O. and Pardue, G.B. (1975) Channel catfish virus experiments with different strains of channel catfish. *Transactions of the American Fisheries Society* 104, 140-143.
[54] Roncarati, A., Mordenti, O., Stocchi, L. and Melotti, P. (2014) Comparison of growth performance of 'Common Catfish *Ameiurus melas*, Rafinesque 1820', reared in pond and in recirculating aquaculture system. Journal of Aquaculture Research and Development 5: 218.

[55] Silverstein, P.S., Bosworth, B.G. and Gaunt, P.S. (2008) Differential susceptibility of blue catfish, *Ictalurus furcatus* (Valenciennes), channel catfish, *I. punctatus* (Rafinesque), and blue×channel catfish hybrids to channel catfish virus. *Journal of Fish Diseases* 31, 77 - 79.
[56] Thompson, D.J., Khoo, L. H., Wise, D.J. and Hanson, L.A. (2005) Evaluation of channel catfish virus latency on fingerling production farms in Mississippi. *Journal of Aquatic Animal Health* 17, 211 - 215.
[57] Vanderheijden, N., Alard, P., Lecomte, C. and Martial, J.A. (1996) The attenuated V60 strain of channel catfish virus possesses a deletion in ORF50 coding for a potentially secreted glycoprotein. *Virology* 218, 422 - 426.
[58] Vanderheijden, N., Hanson, L.A., Thiry, E. and Martial, J.A. (1999) Channel catfish virus gene 50 encodes a secreted, mucin-like glycoprotein. *Virology* 257, 220 - 227.
[59] Vanderheijden, N., Martial, J. A. and Hanson, L. A. (2001) *Channel Catfish Virus Vaccine*. US Patent 6, 322, 793. Available at: https://www.google.com.au/patents/US6322793 (accessed 1 November 2016).
[60] Waltzek, T.B., Kelley, G.O., Alfaro, M.E., Kurobe, T., Davison, A.J. *et al*. (2009) Phylogenetic relationships in the family *Alloherpesviridae*. *Diseases of Aquatic Organisms* 84, 179 - 194.
[61] Wolf, K. and Darlington, R. W. (1971) Channel catfish virus: a new herpesvirus of Ictalurid fish. *Journal of Virology* 8, 525 - 533.
[62] Wolf, K., Herman, R.L. and Carlson, C.P. (1972) Fish viruses: histopathologic changes associated with experimental channel catfish virus disease. *Journal of the Fisheries Research Board of Canada* 29, 149 - 150.
[63] Wu, B., Zhu, B., Luo, Y., Wang, M., Lu, Y. *et al*. (2011) Generation and characterization of monoclonal antibodies against channel catfish virus. *Hybridoma* 30, 555 - 558.
[64] Zhang, H.G. and Hanson, L.A. (1995) Deletion of thymidine kinase gene attenuates channel catfish herpesvirus while maintaining infectivity. *Virology* 209, 658 - 663.
[65] Zhang, H.G. and Hanson, L.A. (1996) Recombinant channel catfish virus (*Ictalurid herpesvirus 1*) can express foreign genes and induce antibody production against the gene product. *Journal of Fish Diseases* 19, 121 - 128.

9

大口黑鲈虹彩病毒病

Rodman G. Getchell* 和 Geoffrey H. Groocock

9.1 引言

第一种蛙病毒是从美国东部的豹蛙(*Rana pipiens*)中分离出来的(Granoff et al., 1965)。30 年后,Plumb 等(1996)报道了南卡罗来纳州 Santee-Cooper 水库首次暴发的大口黑鲈虹彩病毒病,他们将死鱼事件中采集的两尾受感染的大口黑鲈(*Micropterus salmoides*)成鱼的过滤匀浆接种胖头鲹[fathead minnow (*Pimephales promelas*), FHM]细胞,并从中分离出相关病毒。随后通过实验注射感染,将二十面体病毒颗粒(直径约为 174 nm 的包膜病毒颗粒)传播给未感染的大口黑鲈。病毒分离株暂时归类为虹彩病毒科(Iridoviridae),并拟命名为大口黑鲈虹彩病毒(largemouth bass virus, LMBV)(Plumb et al., 1996)。

以往,鱼类中虹彩病毒感染仅限于淋巴囊肿病毒属(*Lymphocystivirus*)的模式种淋巴囊肿病毒 1 型(*Lymphocystis disease virus* 1, LCDV - 1)(Weissenberg, 1965; Wolf, 1988)。Mao 等(1997)的研究表明,六种鱼类虹彩病毒与蛙病毒(*Ranavirus*)的关系比 LCDV 更为密切;蛙病毒最初被认为只感染两栖动物(Hedrick et al., 1992)。这些新的鱼类虹彩病毒在澳大利亚、日本和欧洲的鱼类中引起了高发病率和死亡率的全身性疾病(Langdon et al., 1986, 1988; Ahne et al., 1989; Inouye et al., 1992; Hedrick and McDowell, 1995; Go et al., 2006)。

1997 年 8 月至 1998 年 11 月期间,美国东南部鱼类疾病合作实验室(奥本大学,亚拉巴马州)调查了大口黑鲈虹彩病毒病的暴发流行并考察了美国 8 个州的 78 个地点。调查人员从 4 个不同水系的 6 个水库中采集的大口黑鲈中分离出病毒(Plumb et al., 1999)。感染 LMBV 的鱼的大体病理体征为鱼鳔肿大(图 9.1)和红斑性气腺(Plumb et al., 1996, 1999; Hanson et al., 2001a, b)。基因序列分析表明,这些调查中分离出的病毒与早前在南卡罗来纳州 Santee-Cooper 水库发现的 LMBV 相同。

* 通信作者邮箱: rgg4@cornell.edu。

LMBV在五种鱼细胞系中生长的最适复制温度为30 ℃(Piaskoski et al., 1999)。其中在BF-2和FHM细胞系中病毒感染效果最好,表现出早发性致细胞病变效应、病毒复制快和LMBV滴度高的特点。在最初的研究中,人工感染LMBV的大口黑鲈成鱼未见临床症状或死亡现象(Plumb et al., 1996),但幼鱼易感且死亡率高达60%(Plumb and Zilberg, 1999b)。类似地,LMBV感染的条纹鲈(*Morone saxatilis*)幼鱼的死亡率为63%。

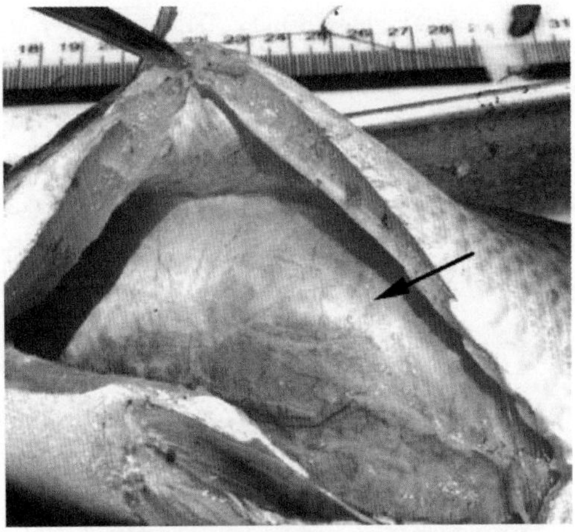

图9.1 感染大口黑鲈虹彩病毒(LMBV)的大口黑鲈呈现出过度膨胀的鱼鳔(箭头所示)。照片原件由奥本大学的John Grizzle提供

Mao等(1999)的研究证实,LMBV属于蛙病毒属,并且与来自裂唇鱼(*Labroides dimidiatus*)的病毒DFV(doctor fish virus)和孔雀鱼(*Poecilia reticulata*)的病毒GV6(guppy virus 6)相类似(Hedrick and McDowell, 1995)。Grizzle等(2002)的回顾性研究,从遗传学层面对1991年分离自佛罗里达州湖堰临床正常的大口黑鲈虹彩病毒与1995年的Santee-Cooper分离株进行了比较。湖堰分离株与该湖中零星的鱼类死亡有关。限制性片段长度多态性(restriction fragment length polymorphism, RFLP)分析和主要衣壳蛋白(major capsid protein, MCP)部分基因的DNA序列测定表明,这两个病毒株是相同的。同时,作者还指出,尚不清楚LMBV是一种在美国东南部广泛传播的新引入病原,还是一种因增加检测而被发现的原住病毒。LMBV在整个美国东部持续被发现(Grizzle and Brunner, 2003)。许多被感染的鱼在临床上是健康的,因此LMBV感染并不总是致死的(Plumb et al., 1999; Hanson et al., 2001b)。

9.1.1 LMBV的相关描述

虹彩病毒科(*Iridoviridae*)家族包含五个属的双链DNA病毒:感染无脊椎动物的虹彩病毒属(*Iridovirus*)、绿虹彩病毒属(*Chloriridovirus*),感染变温动物的蛙病毒属(*Ranavirus*)、肿大细胞病毒属(*Megalocytivirus*)和淋巴囊肿病毒属(*Lymphocystivirus*)。蛙病毒属和肿大细胞病毒属均为海水和淡水鱼类的重要病原体(MacLachlan and Dubovi, 2011)。蛙病毒具有遗传多样性,主要感染硬骨鱼类、两栖动物和爬行动物(Chinchar, 2002)。LMBV只是在20世纪80年代中期和90年代,鱼类死亡病例中新发现的众多蛙病毒属中的一种

(Williams et al., 2005)。国际病毒分类委员会(ICTV)指定以其最初分离出毒株的地点命名，大口黑鲈虹彩病毒目前公认的名称为桑蒂·库珀蛙病毒(*Santee-Cooper Ranavirus*)。

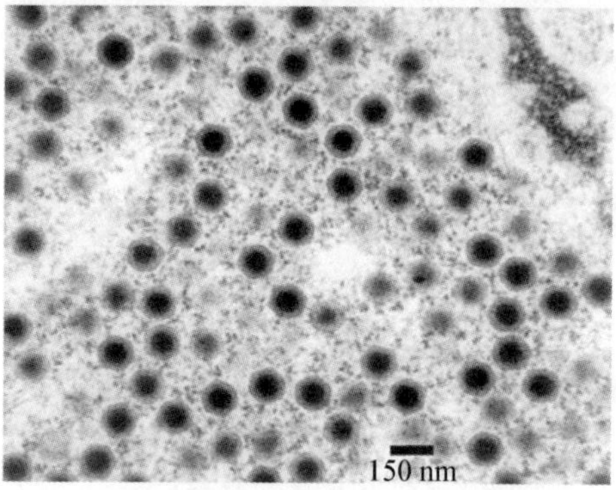

图 9.2　感染的细胞培养中 LMBV 的透射电子显微镜图像。照片由 Andrew Goodwin 提供，原版来自奥本大学 John Plumb 的收藏

通过电子显微镜观察 LMBV 感染的 FHM 细胞(图 9.2)，可见细胞质内二十面体病毒颗粒，其大小为面间平均距离为 132 nm，角间平均距离为 145 nm (Plumb et al., 1996)，符合虹彩病毒直径为 120～200 nm 的大小范围特征(MacLachlan and Dubovi, 2011)。感染性的病毒颗粒有无包膜，取决于它们是通过裂解从细胞中释放，还是从质膜中萌发而出。蛙病毒感染伴有明显的早发性致细胞病变效应，感染蛙病毒的培养细胞则发生细胞凋亡(Chinchar et al., 2005)。

9.1.2　传播方式

当易感鱼浸入带病毒的水中(Plumb and Zilberg, 1999b)或饲喂感染 LMBV 的活饵时(Woodland et al., 2002a)，LMBV 的传播就会发生。用感染 LMBV 的孔雀鱼管饲(强制喂食)24 尾大口黑鲈幼鱼后，其中有 5 尾病毒检测呈阳性，不过没有一尾鱼表现出疾病临床症状。LMBV 的垂直传播尚未得到证实，也没有证据表明在孵化场的鱼类中存在这种情况(Woodland et al., 2002b)。该病毒有时存在于受感染实验鱼的皮肤黏液中(Woodland et al., 2002a)，这可能发生直接传播。诸如小口黑鲈(*Micropterus dolomieu*)、暗色狗鱼(*Esox niger*)、蓝鳃太阳鱼(*Lepomis macrochirus*)和小冠太阳鱼(*Lepomis microlophus*)等暖水性鱼类，都可感染 LMBV 但没有临床症状(Grizzle et al., 2003)。因为病毒可在水中存活，例如在渔船的养鱼舱中或运输活鱼卡车的水箱中，所以人类可能在不知不觉中参与了传播 LMBV(Grizzle and Brunner, 2003)。Johnson 和 Brunner (2014)提出，微生物和浮游动物群落可以迅速灭活池塘水中的蛙病毒，从而最大限度地减少环境传播。宿主行为、密度和接触率在病毒传播动力学的形成中也起着关键作用。

9.1.3　地理分布

自从首次报告发现该病毒(Plumb et al., 1996, 1999)以来，整个美国东部都有了 LMBV 病例记录。1998 年在密西西比州的萨迪斯水库(Hanson et al.,

2001b)，1999 年在图尼卡运河、弗格森湖（密西西比州）、得克萨斯州和路易斯安那州的水域中都相继发现了与 LMBV 相关的死亡病例。这些新的疫情表明 LMBV 正在蔓延，或者环境条件已变得更适于 LMBV 引起的疾病。在两年时间里，对密西西比州 17 个不同渔场的超过 1 400 尾大口黑鲈进行了 LMBV 感染情况评估，发现其中有 7 个渔场的 LMBV 患病率在第二年有所上升，而同期其他渔场的患病率则有所下降（Hanson et al.，2001a）。该研究最明显的结果是 LMBV 在密西西比州普遍存在，该州 17 个大口黑鲈种群中有 12 个种群可检出病毒（Hanson et al.，2001a）。

LMBV 的外来源头可由美国北部鱼类死亡和病毒分离的模式推理出，该病毒于 1991 年首次在佛罗里达州的威尔湖被发现（Grizzle et al.，2002），随后在北部和西部各州也分离出 LMBV。2000 年，一项得克萨斯州水域的调查显示，对 2 876 尾大口黑鲈成鱼进行检测，其中 50 个水库中有 15 个和 13 大流域中有 8 个检测到了 LMBV（Southard et al.，2009）。在受感染水库中检出 LMBV 的患病率为 1.7%～13.3%。在对小口黑鲈亲鱼死亡率的长期研究中也发现了该病毒（Southard et al.，2009）。2002 年，在佛蒙特州的尚普兰湖发现了感染 LMBV 的大口黑鲈。尽管五大湖流域没有大量鱼类死亡的报道，但分别于 2003 年在位于加拿大安大略省和美国密歇根州之间的圣克莱尔湖以及 2000 年在加拿大安大略湖昆特湾中的大口黑鲈中发现了这种病毒（美国国家环境保护局和加拿大环境部，2009）。对纽约 37 个水体的调查（2004 和 2005）证实，当使用定量 PCR 检测时，283 尾大口黑鲈中的 23 尾和 8 尾小口黑鲈中的 5 尾存在 LMBV（Groocock et al.，2008）。鱼龄、性别和季节与 LMBV 患病率无显著相关性。2006 年夏天，在阿肯色州一个面积为 2.2×10^5 m^2 的私人蓄水池中，长达 3 个月的因 LMBV 感染导致的持续死亡疫情表明，在其所收集的大口黑鲈中约有 7%为 LMBV 阳性（Neal et al.，2009）。

迄今为止，已在 30 个州的至少 17 种鱼类中检测到 LMBV。已尝试在目前已知范围以西的州中检测 LMBV，但仅 2010 年在亚利桑那州发现了感染的鱼（Silverwood and McMahon，2012）。在外来入侵性乌鳢（*Channa argus*）中检测到 LMBV 的首次报告涉及来自切萨皮克湾内潮汐支流的收集样本，该支流与马里兰州和弗吉尼亚州接壤（Iwanowicz et al.，2013）。淡水小鱼是乌鳢食谱中的一部分，因此捕食或接触受 LMBV 感染的猎物可能是其感染源。在萨斯奎汉纳河和波托马克河流域的小口黑鲈大规模死鱼中，LMBV 与细菌病原体都被检测出。LMBV 在这些鱼类死亡中的作用尚不明晰（Blazer et al.，2010）。

从中国广东养殖的大口黑鲈死鱼中分离出蛙病毒，这些死鱼伴有皮肤和肌肉溃疡、脾脏和肾脏肿胀（Deng et al.，2011）。研究人员可通过肌内注射蛙病毒分离株到健康的鲈鱼和鳜鱼（*Siniperca chuatsi*）体内实现传播感染。对分离株 MCP 基因的扩增、测序和系统发育分析显示，该病毒与 DFV 相同，并与 LMBV

密切相关(Deng et al., 2011)。此外,通过病毒分离、电子显微镜观察、PCR、MCP 测序(99.9%同一性)和传播研究,George 等(2015)证实 LMBV 是印度东南沿海养鱼场中锦鲤(*Cyprinus carpio*)死亡的病因。

科学家报告了其他虹彩病毒引起的疾病暴发,这可能表明 LMBV 的流行率有所增加。然而,目前尚不清楚这是由于虹彩病毒感染病例的实际增加,还是由于在更多脊椎动物中检测到虹彩病毒感染的采集能力的结果(Williams et al., 2005)。首次在美国以外的大口黑鲈中发现虹彩病毒的报告来自中国台湾地区,那里的死亡率很少超过 30%～50%。通过 MCP 序列比较,初步显示它们与中国台湾地区的石斑鱼虹彩病毒(TGIV)密切相关,但与 LMBV 不同(Chao et al., 2002)。随后的研究将该病毒归类为肿大细胞病毒属(*Megalocytivirus*)(Chao et al., 2004; Huang et al., 2011)。

9.1.4 LMBV 对鱼类种群的影响

1999 年,美国东南部 10 个州的 15 个州孵化场接受了 LMBV 流行情况的调查。在佛罗里达州、路易斯安那州、田纳西州和得克萨斯州的 5 个孵化场的大口黑鲈亲鱼中发现了该病毒。这些孵化场的管理措施存在差异,因此不可能辨别影响传播的因素。大口黑鲈幼鱼在得克萨斯州的某个地方被感染(Woodland et al., 2002b),但尚未确证受感染亲鱼的 LMBV 垂直传播。在孵化场的幼鱼中发现感染 LMBV 确实表明,鱼类放养有助于病毒传播。调查期间,在任一孵化场都没有发现 LMBV 病的迹象。

另一项病毒影响研究是在密西西比州萨迪斯水库进行重复采样,当时地表水温为 29～32 ℃,大约 3 000 尾大口黑鲈成鱼相继死亡。在随后一年中的五次(第 1、2、4、7 和 13 个月)抽样显示,171 尾鲈鱼中分别有 53%、57%、42%、57%和 32%呈现 LMBV 阳性。尽管有证据表明 LMBV 持续存在,但是研究者不认为萨迪斯水库种群中 3 000 尾大口黑鲈成鱼的减少会产生重大影响(Hanson et al., 2001b)。该湖每年的大口黑鲈捕捞损失超过 30 000 尾。研究者建议需要进一步的评估以确定 LMBV 对大口黑鲈种群的影响。虽然经病毒培养发现 36 尾存在鱼鳔病变的鲈鱼中有 97%为 LMBV 阳性,但没有发现病毒与性别、大小、操作损伤或皮肤损伤的相关性。从大口黑鲈幼鱼中分离出 LMBV,证实该病毒的感染不仅限于成鱼。在疫情暴发后的 4～7 个月,Hanson 等(2001b)未在同水域鱼类中发现感染 LMBV,其中包括蓝鳃太阳鱼、白刺盖太阳鱼(*Pomoxis annularis*)、金眼狼鲈(*Morone chrysops*)和美洲真鰶(*Dorosoma cepedianum*)。

Grizzle 和 Brunner(2003)的研究中强调了 LMBV 与 24 尾大口黑鲈的相继死亡有关,但发现该病毒感染有时没有引发临床表现。由于没有关于鲈鱼死亡后其种群数量减少的文献记录,因此需要进一步的研究来确定 LMBV 感染是否使得垂钓运动中奖品规格的鲈鱼数量减少。LMBV 的不同地理分离株可能在致病性和毒力方面存在重大差异(Goldberg et al., 2003)。在实验中存活下来的鱼,其 LMBV 感染死亡率和病毒滴度因分离株来源不同而有所不同。注射三种不同分离株的鲈鱼的中位生存时间分别为 11.0 d、7.5 d 和 4.5 d。毒力最强的毒株在鱼中的复制也达到最高水平(Goldberg et al., 2003)。研究者认为毒株变异可以解释鲈鱼群体对 LMBV 反应的临床差异。

9.2 诊断

由蛙病毒引起的大体病变并非该病毒所特有的。在被感染细胞的细胞质内可见 120～300 nm 的无包膜二十面体颗粒,可以确定可能的虹彩病毒感染。美国渔业协会(American fisheries society, AFS)鱼类健康组(fish health section, FHS)蓝皮书《检测和鉴定某些有鳍鱼类和贝类病原体的建议程序》概述了分离 LMBV 的程序(AFS-FHS, 2014)。用于 LMBV 检测的器官包括鱼鳔、躯干肾和脾脏。Beck 等(2006)的研究表明,在浸泡感染后的最初几天,鳃和鱼鳔中可能含有 LMBV,并且鱼鳔和躯干肾将具有更高的病毒滴度。因为单个器官可能并不总是具有可检测水平的 LMBV,所以建议采用合并的器官样本(Beck et al., 2006)。目前,LMBV 是从细胞培养中分离出来的,然后采用 PCR 进行分子鉴定确认。几种鱼类细胞系可促进 LMBV 的复制(Piaskoski et al., 1999; McClenahan et al., 2005b)。FHM 细胞系和 BF-2 最适合从细胞培养液和匀浆器官样本中分离 LMBV。若鱼在冷冻时或在 5 个月内检测时是新鲜的,则保存在−10 ℃的死鱼或垂死鱼的冷冻组织是稳定的(Plumb and Zilberg, 1999a)。BF-2 细胞中的大口黑鲈虹彩病毒 CPE 感染后首先出现细胞核固缩、细胞沿单层变圆,然后细胞裂解和脱离。具有 CPE 的 LMBV 毒株接种 FHM 细胞后有感染灶,其会变成没有细胞的斑块(图 9.3)。使用 BF-2 或 FHM 细胞可在 24 h

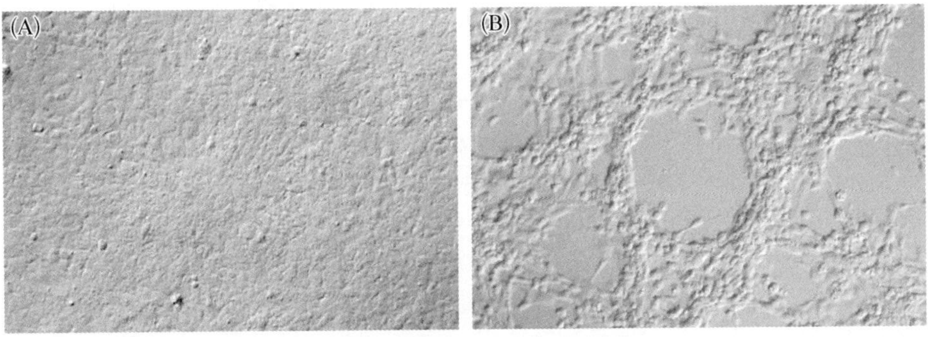

图 9.3 (A) 作为对照的胖头鲹肌肉细胞和(B) 大口黑鲈虹彩病毒(LMBV)诱导 FHM 细胞形成空斑。照片原件由奥本大学的 John Grizzle 提供

或更短时间内达到高病毒滴度。近期有研究开发并鉴定了一种源自大口黑鲈卵巢的新细胞系(Getchell et al., 2014)，不过其 LMBV 滴度比 BF-2 细胞中的低 10 倍。

细胞质包涵体不染色则不可见(AFS-FHS, 2014)。Hanson 等(2001b)使用细胞培养法评估了鱼鳃、脾脏和鱼鳔的病毒感染情况。鳃组织很少被感染，而从鱼鳔中分离出病毒的频率最高。与单独培养鱼鳔样本相比，同时对鱼鳔和脾脏进行检测，病毒阳性检出率增加了 10%。经口接触感染 LMBV 的大口黑鲈，其皮肤黏液、鱼鳔、头肾、躯干肾、脾、性腺和肠中具有不同的病毒滴度(Woodland et al., 2002a)。鱼鳔产生的滴度最高可达 $10^{5.5}$～$10^{9.5}$ $TCID_{50}$/g，但未观察到其中的病变。

Mao 等(1997, 1999)利用 PCR 和 RT-PCR 均成功扩增了病毒特异性核酸，并证明所检测到的那些鱼类虹彩病毒属于蛙病毒属。病毒感染细胞的蛋白质电泳分析和病毒 DNA 的 RFLP 分析阐明了这些分离株的相关性。后来的研究使用 PCR 来确认 LMBV 感染(Grizzle et al., 2002; Grizzle and Brunner, 2003)。当没有足够的 LMBV 用于细胞培养时，可以使用 PCR 检测，因为它具有更高的灵敏度(Grizzle et al., 2003; McClenahan et al., 2005b)。通过将假定阳性细胞培养物中的稀释培养液添加到 PCR 混合物中，而不是使用提取的 DNA 用于 PCR 模板，可以简化 PCR 方法(McClenahan et al., 2005a)。Grizzle 等开发的特定 LMBV 引物对 288F-535R(2003)从 Santee-Cooper 水库的 LMBV 样本中扩增得到一个 248 bp 的 DNA 片段，但没有从其他四种蛙病毒(青蛙病毒 3，FV3；DFV；GV6；EHNV)或真鲷虹彩病毒中成功扩增出 DNA 片段。这对引物可从 30 株病毒分离株中扩增出正确大小的片段，这些分离株根据病毒分离和其他方法被推定为 LMBV。后来，Ohlemeyer 等(2011)克隆并测序了 DFV、GV6 和 LMBV 的完整 MCP 基因。DFV 和 GV6 序列相同，而 LMBV 序列同源性为 99.21%。这些研究者随后开发了一种不会扩增相关石斑鱼虹彩病毒的 Santee-Cooper 特异性 PCR。他们得出结论，这三种病毒可能不属于蛙病毒属(Hyatt et al., 2000; Whittington et al., 2010)。

Goldberg 等(2003)和 Getchell 等(2007)开发了用于检测大口黑鲈虹彩病毒 MCP 基因片段的实时定量 PCR (qPCR)技术。qPCR 检测为监测鱼类种群中的病原体流行提供了显著优势，包括高通量能力和减少污染问题。qPCR 法与噬斑法比较，证实了检测 LMBV 稀释液时 qPCR 的线性范围扩大，并显示当检测感染细胞培养液时，实时 qPCR 法比噬斑法的灵敏度大约提高 100 倍。另一种用于检测许多鱼类蛙病毒的 qPCR 检测方法没有针对 LMBV 进行测试，但用于 DFV 和 GV6 的检测效果很好。

由于广泛的交叉反应性，血清学检测不足以鉴定密切相关的蛙病毒(Hedrick et al., 1992; Marsh et al., 2002)。尽管黑带石斑鱼(*Epinephelus malabaricus*)的组织用福尔马林固定,但靶向 MCP 基因序列的原位杂交被成功地用于鉴定一种虹彩病毒(Huang et al., 2004)。PCR 和 RFLP 技术用于蛙病毒属的种间鉴别(Marsh et al., 2002)。最近,扩增片段长度多态性(AFLP)和 qPCR 被用于 LMBV 地理分离株的检测和鉴别(Goldberg et al., 2003)。Miller 等(2015)在应使用哪种诊断手段来实现特定的研究结果方面提出了很好的建议。理想情况下,应提交多份样本并进行多次测试。

9.3 病理

肿大的鱼鳔和红斑气腺是 LMBV 感染的大体病理特征(Plumb et al., 1996)。Zilberg 等(2000)描述了腹腔(IP)注射 LMBV 的几尾大口黑鲈幼鱼的病变。其临床体征和外部病变包括注射部位发炎、腹胀、螺旋式泳动、侧卧和嗜睡。内部病变包括局灶性苍白的肝脏、鲜红的脾脏和发红的盲肠。显微病变包括急性纤维素性腹膜炎和存在于鱼鳔腹侧的渗出物。在另一项研究中(Hanson et al., 2001b),LMBV 感染似乎仅涉及鱼鳔并导致鱼鳔内腔积聚黄色蜡状物(图 9.4)。该物质由纤维蛋白凝块中的红细胞、细胞碎片和嗜酸性粒细胞组成。Grant 等(2003)报道了大口黑鲈内部大体病理,如渗出性多浆膜炎、肺囊炎和各种内脏器官(尤其是肝脏)的颜色变化。在垂死的感染锦鲤幼鱼中观察到的临床症状包括皮肤变暗、鳞片脱落、垂悬,不协调游动、倒置、侧外旋,间歇浮出、侧卧水底和死亡(George et al., 2015)。

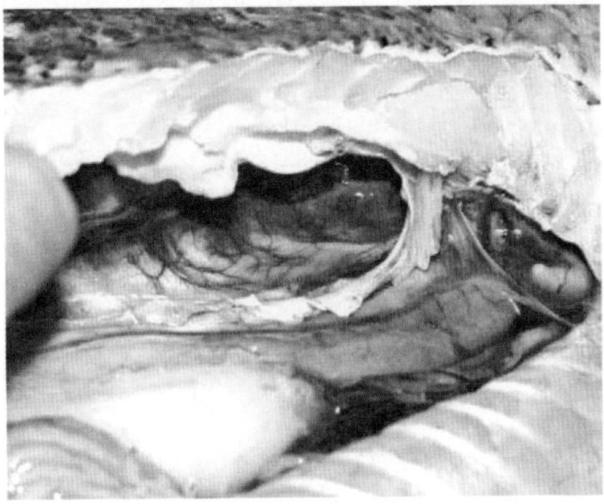

图 9.4 感染大口黑鲈虹彩病毒(LMBV)的大口黑鲈鱼鳔中的黄色渗出液。照片原件由美国鱼类及野生动植物管理局的 Andrew Goodwin 提供

9.4 病理生理学

针对大口黑鲈的钓鱼锦标赛在美国 50 个州中的 48 个州举行(Schramm et al., 1991)。休闲渔业和环境压力因素的具体影响及其对大口黑鲈的生理学影响可能与 LMBV 暴发有关(Schramm et al., 2006)。人为和环境应激因素与病毒感染的相互作用可能导致鱼类死亡。这些生理变化可能会增加 LMBV 的致病性。已经证明与 LMBV 相关的鱼类死亡通常发生在炎热的夏季(Plumb et al.，1996；Hanson et al.，2001b）。在夏季，影响生理的环境压力因素包括水温升高、溶解氧含量低和垂钓压力增加（Grant et al.，2003）。这些因素可能会阻碍免疫反应并使病毒复制增加。确定引发 LMBV 导致鱼类死亡的具体因素对控制疾病非常重要。超出正常范围的温度对鱼类有免疫抑制作用（Bly and Clem，1992）。Pickering 和 Pottinger（1989）还证明了低慢性应激可以降低疾病抵抗力。

9.4.1 对内分泌系统和渗透调节的影响

钓获-放流的垂钓法与影响体内平衡的生理变化相关（Gustaveson et al.，1991；Suski et al.，2003；Cooke et al.，2004），并可能导致免疫抑制（Anderson，1990）。Gustaveson 等（1991）确定了大口黑鲈在上岸之前垂钓遛鱼的时长与血液皮质醇和血浆乳酸的增加相关。垂钓时，在冷水（11～13 ℃）中上钩溜鱼 1～5 min，大口黑鲈不会产生应激激素，在 16～20 ℃时激素产生较为平缓，而在28～30 ℃上钩持续 5 min 时应激激素量急剧上升。Suski 等（2003）的研究显示，比赛捕获的鲈鱼血浆皮质醇和葡萄糖浓度及血浆渗透压显著高于对照组。他们认为，在称重期间的活鱼槽限制、搬运操作和空气接触等因素可能在代谢紊乱中起重要作用。

9.4.2 对生长的影响

在钓鱼比赛之后称重时取样的大口黑鲈的代谢状态发生了很大变化，包括肌肉能量储存的大幅减少和乳酸浓度的大幅增加（Suski et al.，2003）。这些变化可能使大口黑鲈更容易受到 LMBV 感染。得克萨斯州公园和野生动物部门的工作人员测量了得克萨斯州的两个水库中鲈鱼的生长参数，在那里发生了由 LMBV 导致的死亡（Bister et al.，2006）。在两个水库中都没有发现三龄鲈鱼的平均体长、相对质量或垂钓成功率有所下降的情况，而在亚拉巴马州的五个水库中，Maceina 和 Grizzle（2006）注意到，与 LMBV 感染相关的大口黑鲈生长缓慢且相对质量较低。与此同时，锦标赛和电钓渔获记录中的鲈鱼体长也发生了令人难忘的下降。Beck 等（2006）认为，LMBV 引起的超级浮力可能导致鲈鱼试图保持潜水状态时精疲力竭。

9.4.3 LMBV 的发病机制和生物能量消耗

Grant 等（2003）的研究表明，30 ℃下腹腔注射感染 LMBV 的大口黑鲈的死亡率高于注射感染并保持在 25 ℃的大口黑鲈。他们使用 qPCR 在濒死鱼和死鱼中检测到比试验中存活鱼更高的病毒载量，并认为较高的温度可能更适合病毒复制，对宿主的压力更大。病毒载量与内部病变之间存在显著关联。然而，在相同来源大口黑鲈的类似攻毒试验中虽具有高 LMBV 滴度，但却没有发现相同的关联(Goldberg et al., 2003)。与低密度下饲养的接种 LMBV 的鲈鱼相比,高密度圈养并感染 LMBV 的大口黑鲈幼鱼具有更高的死亡率,病毒载量更高,身体健康状况更差(Inendino et al., 2005)。由此得出结论:最大限度地提高鱼类生活的自然环境质量,同时尽量减少群居行为压力应该是最有效的提高鱼群健康水平和生产力的策略。

最近,通过研究 LMBV 引起的细胞死亡所涉及的细胞信号传导事件,确定了对 LMBV 分子致病机制的深入认知(Huang et al., 2014)。LMBV 感染鲤鱼上皮瘤细胞产生了涉及 caspase-8 和 caspase-9 的内源性和外源性通路介导的细胞凋亡。肌醇脂-3-激酶[又称磷脂酰肌醇 3-激酶(phosphatidylinositol 3-kinase, PI3K)]和胞外信号调节激酶(extracellularsignal-regulated kinase, ERK)信号通路参与了 LMBV 复制及感染诱导的细胞凋亡。确定这些机制将有助于开发所有虹彩病毒感染的潜在治疗靶点(Reshi et al., 2016)。

9.5 防控策略

Oh 等(2015)的研究表明,尽管 DNA 疫苗的保护性较弱,但在 poly(I:C)(聚肌苷酸-聚胞苷酸)给药后接种真鲷虹彩病毒活疫苗对鱼类有保护作用。目前尚没有可用于预防 LMBV 的商业疫苗。

即捕即放的垂钓练习中,将未感染和感染的大口黑鲈置于同一个活鱼舱或储槽中加速了 LMBV 的传播(Grant et al., 2005)。对 LMBV 的快速摄取(1 h),加上鱼槽内同伴的病毒脱落,可能意味着即使很短的暴露时间也可能感染活鱼舱内的其他鱼类(Beck et al., 2006)。将这些储槽分区或减少其使用可降低传播率,降低活鱼舱中的温度将是有益的,因为较高的温度将促进病毒复制(Grant et al., 2003)。Schramm 等(2006)建议,不应保留锦标赛捕获的鲈鱼或者将其与受感染鱼群隔离开来。荧光假单胞菌(*Pseudomonas fluorescens*)的生物被膜可保护 LMBV 免受次氯酸盐和碘伏的杀伤,但对含酒精的消毒剂不起作用(Nath et al., 2010)。在两个月前含 LMBV 阳性鱼的池塘中,其生物被膜或水中未检测到病毒。

渔业管理者、垂钓者和垂钓运动业界对 LMBV 影响的担忧,促使鲈鱼垂钓者运动员协会安排了一个多机构、多部门的合作,以促进控制 LMBV 传播的研究和指导行动(Terre et al., 2008)。专题研讨会的建议提醒垂钓者和划船者采取以下步骤,以有助于防止病毒传播扩散。

(1) 在两次钓鱼航次之间,要彻底清洗船只、拖车和其他垂钓用具。

(2) 不要将鱼或鱼的身体部分从一个水域移到另一个水域,也不要向任何水体中释放活饵。

(3) 如果您打算释放它们,请尽可能温和地处理它们并尽快释放。

(4) 如果您打算放生，请不要在活鱼池里长时间拖拉鱼。

(5) 在7月中旬到8月中旬期间，特别是在异常炎热的天气条件下，尽量减少以大口黑鲈为目标。

(6) 向自然资源机构办事处报告死亡或垂死的大口黑鲈成鱼情况。

(7) 志愿者帮助机构收集鲈鱼用于LMBV监测。

(8) 教育其他垂钓者了解LMBV。

9.6 总结与研究展望

9.6.1 发现我们知识上的空白与欠缺

缺乏完整的LMBV基因组序列限制了将对LMBV防控措施产生影响的各项进展。了解LMBV的表型和基因型变异如何影响毒力，对于理解这种疾病的生物学、进化和控制具有重要意义。不过，Goldberg等(2003)得出的研究结论是除了病原体固有毒力之外的其他因素，如环境和宿主相关因素等，一定会对LMBV现场感染的临床表现产生显著影响。

9.6.2 对未来研究的展望

回顾过去关于FV3的研究成果和目前其他蛙病毒的研究工作，以及其他属虹彩病毒(如肿大细胞病毒)的研究表明，研究人员将确定在毒力中起重要作用的病毒基因，并最终促进有效抗蛙病毒疫苗的创制。了解这些病毒如何逃避低等脊椎动物的抗病毒免疫反应，将有望取得丰硕成果(Jancovich et al., 2015)。

参考文献

[1] AFS-FHS (2014) Section 2. USFWS standard procedures for aquatic animal health inspections. In: *FHS Blue Book: Suggested Procedures for the Detection and Identification of Certain Finfish and Shellfish Pathogens*, 2014 edn. American Fisheries Society-Fish Health Section, Bethesda, Maryland. Updated 2016 edition available at: http://afs-fhs.org/bluebook/inspection-index.php (accessed 1 November 2016).

[2] Ahne, W., Schlotfeldt, H.J. and Thomsen, I. (1989) Fish viruses: isolation of an icosahedral cytoplasmic deoxyribovirus from sheatfish (*Silurus glanis*). *Journal of Veterinary Medicine Series B* [now *Zoonoses and Public Health*] 36, 333-336.

[3] Anderson, D.P. (1990) Immunological indicators: effects of environmental stress on immune protection and disease outbreaks. In: Adams, S.M. (ed.) *Biological Indicators of Stress in Fish*. American Fisheries Society Symposium 8. American Fisheries Society, Bethesda, Maryland, pp. 38-50.

[4] Beck, B.H., Bakal, R.S., Brunner, C.J. and Grizzle, J.M. (2006) Virus distribution and signs of disease after immersion exposure to largemouth bass virus. *Journal of Aquatic Animal Health* 18, 176-183.

[5] Bister, T.J., Myers, R.A. Driscoll, M.T. and Terre, D.R. (2006) Largemouth bass population trends in two Texas reservoirs with LMBV-attributed die-offs. *Proceedings of the Annual Conference of Southeastern Association of Fish and Wildlife Agencies* 60, 101 - 105.

[6] Blazer, V.S., Iwanowicz, L.R., Starliper, C.E., Iwanowicz, D.D., Barbash, P. *et al*. (2010) Mortality of centrarchid fishes in the Potomac drainage: survey results and overview of potential contributing factors. *Journal of Aquatic Animal Health* 22, 190 - 218.

[7] Bly, J.E. and Clem, L.W. (1992) Temperature and teleost immune functions. *Fish and Shellfish Immunology* 2, 159 - 171.

[8] Chao, C.B., Yang, S.C., Tsai, H.Y., Chen, C.Y., Lin, C.S. *et al*. (2002) A nested PCR for the detection of grouper iridovirus in Taiwan (TGIV) in cultured hybrid grouper, giant sea perch and largemouth bass. *Journal of Aquatic Animal Health* 14, 104 - 113.

[9] Chao, C.B., Chen, C.Y., Lai, Y.Y., Lin, C.S. and Huang, H.T. (2004) Histological, ultrastructural, and *in situ* hybridization study on enlarged cells in grouper *Epinephelus* hybrids infected by grouper iridovirus in Taiwan (TGIV). *Diseases of Aquatic Organisms* 58, 127 - 142.

[10] Chinchar, V. G. (2002) Ranaviruses (family *Iridoviridae*): emerging cold-blooded killers. *Archives of Virology* 147, 447 - 470.

[11] Chinchar, V.G., Essbauer, S., He, J. G., Hyatt, A, Miyazaki, T. *et al*. (2005) Iridoviridae. In: Fauquet, C.M., Mayo, M.A., Maniloff, J., Desselberger, U. and Ball, L.A. (eds) *Virus Taxonomy: Eighth Report of the International Committee on the Taxonomy of Viruses*. Elsevier, San Diego, California and London, pp. 163 - 175.

[12] Cooke, S.J., Bunt, C.M., Ostrand, K.G., Philipp, D.P. and Wahl, D.H. (2004) Angling-induced cardiac disturbance of free-swimming largemouth bass (*Micropterus salmoides*) monitored with heart rate telemetry. *Journal of Applied Ichthyology* 20, 28 - 36.

[13] Deng, G., Li, S., Xie, J., Bai, J., Chen, K. et al. (2011) Characterization of a ranavirus isolated from cultured largemouth bass (*Micropterus salmoides*) in China. Aquaculture 312, 198 - 204.

[14] George, M.R., John, K.R., Mansoor, M.M., Saravanakumar, R., Sundar, P. *et al*. (2015) Isolation and characterization of a ranavirus from koi, *Cyprinus carpio* L., experiencing mass mortalities in India. *Journal of Fish Diseases* 38, 389 - 403.

[15] Getchell, R.G., Groocock, G.H., Schumacher, V.L., Grimmett, S.G., Wooster, G.A. *et al*. (2007) Quantitative polymerase chain reaction assay for largemouth bass virus. *Journal of Aquatic Animal Health* 19, 226 - 233.

[16] Getchell, R.G., Groocock, G.H., Cornwell, E.R., Schumacher, V.L., Glasner, L.I. *et al*. (2014) Development and characterization of a largemouth bass cell line. *Journal of Aquatic Animal Health* 26, 194 - 201.

[17] Go, J., Lancaster, M., Deece, K., Dhungyel, O. and Whittington, R. (2006) The molecular epidemiology of iridovirus in Murray cod (*Maccullochella peelii peelii*) and dwarf gourami (*Colisa lalia*) from distant biogeographical regions suggests a link between trade in ornamental fish and emerging iridoviral diseases. *Molecular and Cellular Probes* 20, 212 - 222.

[18] Goldberg, T.L., Coleman, D.A., Inendino, K.R., Grant, E.C. and Philipp, D.P. (2003) Strain variation in an emerging iridovirus of warm water fishes. *Journal of Virology* 77, 8812 - 8818.

[19] Granoff, A., Came, P.E., Keen, A. and Rafferty, J. (1965) The isolation and properties of viruses from Rana pipiens: their possible relationship to the renal

adenocarcinoma of the leopard frog. *Annals of the New York Academy of Sciences* 126, 237 - 255.

[20] Grant, E.C., Philipp, D.P., Inendino, K.R. and Goldberg, T.L. (2003) Effects of temperature on the susceptibility of largemouth bass to largemouth bass virus. *Journal of Aquatic Animal Health* 15, 215 - 220.

[21] Grant, E.C., Philipp, D.P., Inendino, K.R. and Goldberg, T.L. (2005) Effects of practices related to catch-and-release angling on mortality and viral transmission in juvenile largemouth bass infected with largemouth bass virus. *Journal of Aquatic Animal Health* 17, 315 - 322.

[22] Grizzle, J.M. and Brunner, C.J. (2003) Review of largemouth bass virus. *Fisheries* 28, 10 - 14.

[23] Grizzle, J.M., Altinok, I., Fraser, W.A. and Francis-Floyd, R. (2002) First isolation of largemouth bass virus. *Diseases of Aquatic Organisms* 50, 233 - 235.

[24] Grizzle, J.M., Altinok, I. and Noyes, A.D. (2003) PCR method for detection of largemouth bass virus. *Diseases of Aquatic Organisms* 54, 29 - 33.

[25] Groocock, G.H., Grimmett, S.G., Getchell, R.G., Wooster, G.A. and Bowser, P.R. (2008) A survey to determine the presence and distribution of largemouth bass virus in wild freshwater bass in New York State. *Journal of Aquatic Animal Health* 20, 158 - 164.

[26] Gustaveson, A.W., Wydoski, R.S. and Wedemeyer, G.A. (1991) Physiological response of largemouth bass to angling stress. *Transactions of the American Fisheries Society* 120, 629 - 636.

[27] Hanson, L.A., Hubbard, W.D. and Petrie-Hanson, L. (2001a) Distribution of largemouth bass virus may be expanding in Mississippi. *Fish Health Newsletter, Fish Health Section/American Fisheries Society* (Bethesda, Maryland) 29(4), 10 - 12. Available at: http://www.afsfhs.org/communications/newsletter/V29 - 4_2001.PDF (accessed 29 November 2015).

[28] Hanson, L.A., Petrie-Hanson, L., Meals, K.O., Chinchar, V.G. and Rudis, M. (2001b) Persistence of largemouth bass virus infection in a northern Mississippi reservoir after a die-off. *Journal of Aquatic Animal Health* 13, 27 - 34.

[29] Hedrick, R.P. and McDowell, T.S. (1995) Properties of iridoviruses from ornamental fish. *American Journal of Veterinary Research* 26, 423 - 437.

[30] Hedrick, R.P., McDowell, T.S., Ahne, W., Torhy, C. and de Kinkelin, P. (1992) Properties of three iridovirus-like agents associated with systemic infections of fish. *Diseases of Aquatic Organisms* 13, 203 - 209.

[31] Holopainen, R., Honkanen, J., Bang Jensen, B., Ariel, E. and Tapiovaara, H. (2011) Quantitation of ranaviruses in cell culture and tissue samples. *Journal of Virological Methods* 171, 225 - 233.

[32] Huang, C., Zhang, X., Gin, K.Y.H. and Qin, Q.W. (2004) *In situ* hybridization of a marine fish virus, Singapore grouper iridovirus with a nucleic acid probe of major capsid protein. *Journal of Virological Methods* 117, 123 - 128.

[33] Huang, S.-M., Tu, C., Tseng, C.-H., Huang, C.-C., Chou, C.-C. *et al*. (2011) Genetic analysis of fish iridoviruses isolated in Taiwan during 2001 - 2009. *Archives of Virology* 156, 1505 - 1515.

[34] Huang, X., Wang, W., Huang, Y., Xu, L. and Qin, Q. (2014) Involvement of the PI3K and ERK signaling pathways in largemouth bass virus-induced apoptosis and viral replication. *Fish and Shellfish Immunology* 41, 371 - 379.

[35] Hyatt, A.D., Gould, A.R., Zupanovic, Z., Cunningham, A.A., Hengstberger, S. *et al*. (2000) Comparative studies of piscine and amphibian iridoviruses. *Archives of Virology* 145, 301 - 331.

[36] Inendino, K.R., Grant, E.C., Philipp, D.P. and Goldberg, T.L. (2005) Effects of factors related to water quality and population density on the sensitivity of juvenile largemouth bass to mortality induced by viral infection. *Journal of Aquatic Animal Health* 17, 304 - 314.

[37] Inouye, K., Yamano, K., Maeno, Y., Nakajima, K., Matsuoka, M. *et al*. (1992) Iridovirus infection of cultured red sea bream, *Pagrus major*. *Fish Patholog*y 27, 19 - 27 (in Japanese with English abstract).

[38] Iwanowicz, L., Densmore, C., Hahn, C., McAllister, P. and Odenkirk, J. (2013) Identification of largemouth bass virus in the introduced northern snakehead inhabiting the Chesapeake Bay watershed. Journal Aquatic Animal Health 25, 191 - 196.

[39] Jancovich, J. K., Qin, Q., Zhang, C.-Y. and Chinchar, V. G. (2015) Ranavirus replication: molecular, cellular, and immunological events. In: Gray, M. J. and Chinchar, V. G. (eds) *Ranaviruses: Lethal Pathogens of Ectothermic Vertebrates*. Springer, New York, pp. 105 - 139.

[40] Johnson, A. F. and Brunner, J. L. (2014) Persistence of an amphibian ranavirus in aquatic communities. *Diseases of Aquatic Organisms* 111, 129 - 138.

[41] Langdon, J.S., Humphrey, J.D., Williams, L.M., Hyatt, A.D. and Westbury, H.A. (1986) First virus isolation from Australian fish: an iridovirus-like pathogen from redfin perch, Perca fluviatilis L. *Journal of Fish Diseases* 9, 263 - 268.

[42] Langdon, J.S., Humphrey, J.D. and Williams, L.M. (1988) Outbreaks of an EHNV-like iridovirus in cultured rainbow trout, *Salmo gairdneri* Richardson, in Australia. *Journal of Fish Diseases* 11, 93 - 96.

[43] Maceina, M.J. and Grizzle, J.M. (2006) The relation of largemouth bass virus to largemouth bass population metrics in five Alabama reservoirs. *Transactions of the American Fisheries Society* 135, 545 - 555.

[44] MacLachlan, N. J. and Dubovi, E. J. (2011) Asfarviridae and Iridoviridae. In: MacLachlan, N. J. and Dubovi, E. J. (eds) *Fenner's Veterinary Virology*, 4th edn. Elsevier/Academic Press, London, pp. 272 - 277.

[45] Mao, J. Hedrick, R.P. and Chinchar, V.G. (1997) Molecular characterization, sequence analysis and taxonomic position of newly isolated fish iridoviruses. *Virology* 229, 212 - 220.

[46] Mao, J., Wang, J., Chinchar, G. D. and Chinchar, V. G. (1999) Molecular characterization of a ranavirus isolated from largemouth bass *Micropterus salmoides*. *Diseases of Aquatic Organisms* 37, 107 - 114.

[47] Marsh, I.B., Whittington, R.J., O'Rouke, B., Hyatt, A.D. and Chisholm, O. (2002) Rapid identification of Australian, European, and American ranaviruses based on variation in major capsid protein gene sequences. *Molecular and Cellular Probes* 16, 137 - 151.

[48] McClenahan, S.D., Grizzle, J.M. and Schneider, J.E. (2005a) Evaluation of unpurified cell culture supernatant as template for the polymerase chain reaction (PCR) with largemouth bass virus. *Journal of Aquatic Animal Health* 17, 191 - 196.

[49] McClenahan, S.D., Beck, B.H. and Grizzle, J.M. (2005b) Evaluation of cell culture methods for detection of largemouth bass virus. *Journal of Aquatic Animal Health* 17, 365 - 372.

[50] Miller, D.L., Pessier, A.P., Hick, P. and Whittington, R.J. (2015) Comparative pathology of ranaviruses and diagnostic techniques. In: Gray, M.J. and Chinchar, V.G. (eds) *Ranaviruses: Lethal Pathogens of Ectothermic Vertebrates*. Springer, New York, pp. 171 - 208.

[51] Nath, S., Aron, G.M., Southard, G.M. and McLean, R.J.C. (2010) Potential for largemouth bass virus to associate with and gain protection from bacterial biofilms.

Journal of Aquatic Animal Health 22，95－101.
[52] Neal，J.W.，Eggleton，M.A. and Goodwin，A.E. (2009) The effects of largemouth bass virus on a quality largemouth bass population in Arkansas. *Journal of Wildlife Diseases* 45，766－771.
[53] Oh，S.-Y.，Kim，W.-S.，Oh，M.-J. and Nishizawa，T. (2015) Quantitative change of red seabream iridovirus (RSIV) in rock bream *Oplegnathus fasciatus*，following Poly (I：C) administration. *Aquaculture International* 23，93－98.
[54] Ohlemeyer，S.，Holopainen，R.，Tapiovaara，H.，Bergmann，S.M. and Schütze，H. (2011) Major capsid protein gene sequence analysis of the Santee-Cooper ranaviruses DFV，GV6，and LMBV. *Diseases of Aquatic Organisms* 96，195－207.
[55] Piaskoski，T.O.，Plumb，J.A. and Roberts，S.R. (1999) Characterization of the largemouth bass virus in cell culture. *Journal of Aquatic Animal Health* 11，45－51.
[56] Pickering，A.D. and Pottinger，T.G. (1989) Stress response and disease resistance in salmonid fish：effects of chronic elevation of plasma cortisol. *Fish Physiology and Biochemistry* 7，253－258.
[57] Plumb，J.A. and Zilberg，D. (1999a) Survival of largemouth bass iridovirus in frozen fish. *Journal of Aquatic Animal Health* 11，94－96.
[58] Plumb，J.A. and Zilberg，D. (1999b) The lethal dose of largemouth bass virus in juvenile largemouth bass and the comparative susceptibility of striped bass. *Journal of Aquatic Animal Health* 11，246－252.
[59] Plumb，J.A.，Grizzle，J.M.，Young，H.E. and Noyes，A.D. (1996) An iridovirus isolated from wild largemouth bass. *Journal of Aquatic Animal Health* 8，265－270.
[60] Plumb，J.A.，Noyes，A.D.，Graziano，S.，Wang，J.，Mao，J. and Chinchar，V.G. (1999) Isolation and identification of viruses from adult largemouth bass during a 1997－1998 survey in the Southeastern United States. *Journal of Aquatic Animal Health* 11，391－399.
[61] Reshi，L.，Wu，J.-L.，Wang，H.-V. and Hong，J.-R. (2016) Aquatic viruses induce host cell death pathways and its application. *Virus Research* 211，133－144.
[62] Schramm，H.L.，Armstrong，M.L.，Funicelli，N.A.，Green，D.M.，Lee，D.P. *et al.* (1991) The status of competitive sport fishing in North America. *Fisheries* 16，4－12.
[63] Schramm，H.L.，Walters，A.R.，Grizzle，J.M.，Beck，B.H.，Hanson，L.A. *et al.* (2006) Effects of live-well conditions on mortality and largemouth bass virus prevalence in largemouth bass caught during summer tournaments. *North American Journal of Fisheries Management* 26，812－825.
[64] Silverwood，K. and McMahon，T. (2012) *2012－Arizona Risk Analysis: Largemouth Bass Virus (LMBV)*. Arizona Game and Fish Department，Phoenix，Arizona. Available at：https：//portal.azgfd.stagingaz.gov/PortalImages/files/fishing/InvasiveSpecies/RA/Largemouth%20Bass%20Virus%202012RAforAZ.pdf(accessed 2 November 2016).
[65] Southard，G.M.，Fries，L.T. and Terre，D.R. (2009) Largemouth bass virus in Texas：distribution and management issues. *Journal of Aquatic Animal Health* 21，36－42.
[66] Suski，C.D.，Killen，S.S.，Morrissey，M.B.，Lund，S.G. and Tufts，B.L. (2003) Physiological changes in the largemouth bass caused by live-release angling tournaments in southeastern Ontario. *North American Journal of Fisheries Management* 23，760－769.
[67] Terre，D.R.，Schramm，H.L. and Grizzle，J.M. (2008) Dealing with largemouth bass virus：benefits of multisector collaboration. *Proceedings of the Annual Conference of Southeastern Association of Fish and Wildlife Agencies* 62，115－119.
[68] US EPA and Environment Canada (2009) Nearshore Areas of the Great Lakes，2009. *State of the Lakes Ecosystem Conference 2008: Background Paper*. [Prepared for the SOLEC Conference at Niagara Falls，Ontario in 2008] by the Governments of Canada

and the United States of America. Prepared by Environment Canada and the US Environmental Protection Agency. Available at: https://binational.net//wp-content/uploads/2014/05/SOGL_2009_nearshore_en.pdf (accessed 1 November 2016).
[69] US Fish and Wildlife Service (2015) National Wild Fish Health Survey, US Fish and Wildlife Service, Washington, DC. Available at: http://ecos.fws.gov/wildfishsurvey/database/nwfhs/(accessed 2 November 2016).
[70] Weissenberg, R. (1965) Fifty years of research on the lymphocystis virus disease of fishes (1914 - 1964). *Annals of the New York Academy of Sciences* 126, 362 - 374.
[71] Whittington, R.J., Becker, J.A. and Dennis, M.M. (2010) Iridovirus infections in finfish: critical review with emphasis on ranaviruses. *Journal of Fish Diseases* 33, 95 - 122.
[72] Williams, T., Barbarosa-Solomieu, V. and Chinchar, V.G. (2005) A decade of advances in iridovirus research. *Advanced Virus Research* 65, 173 - 248.
[73] Wolf, K. (1988) *Fish Viruses and Fish Viral Diseases*. Cornell University Press, Ithaca, New York.
[74] Woodland, J.E., Brunner, C.J., Noyes, A.D. and Grizzle, J.M. (2002a) Experimental oral transmission of largemouth bass virus. *Journal of Fish Diseases* 25, 669 - 672.
[75] Woodland, J.E., Noyes, A.D. and Grizzle, J.M. (2002b) A survey to detect largemouth bass virus among fish from hatcheries in the southeastern USA. *Transactions of the American Fisheries Society* 131, 308 - 311.
[76] Zilberg, D., Grizzle, J.M. and Plumb, J.A. (2000) Preliminary description of lesions in juvenile largemouth bass injected with largemouth bass virus. *Diseases of Aquatic Organisms* 39, 143 - 146.

10

锦鲤疱疹病毒病

Keith Way* 和 Peter Dixon

10.1 引言

锦鲤疱疹病毒病(koi herpesvirus disease, KHVD)是一种由疱疹病毒感染(Hedrick et al., 2000)可在鲤鱼(*Cyprinus carpio*)和鲤鱼变种[例如锦鲤和鬼鲤(koi × common carp)]中引起具有高度传染性的致死性急性病毒血症(Haenen et al., 2004)。其病原体被归类为鲤疱疹病毒3型(*Cyprinid herpesvirus* 3, CyHV-3),属异疱疹病毒科(Alloherpesviridae)中的成员,是10种感染鱼类的异疱疹病毒之一(Boutier et al., 2015a)。

CyHV-3为水平传播,可以直接或间接发生。Uchii等(2014)的研究认为,当水温升高时,感染恢复的鱼体内的CyHV-3会被周期性重新激活,并在与幼稚鱼密切接触时(如产卵时)传播给它们。感染和未感染鲤鱼之间通过皮肤接触以及鲤鱼间同类相食和食尸行为都会引起该病毒的直接传播。包括粪便、水生沉积物、浮游生物和水生无脊椎动物在内的一些媒介可以促进CyHV-3的间接传播(Boutier et al., 2015a)。受污染的水是主要的非生物传播媒介,因为它可能含有随尿液排出以及从粪便、鳃和皮肤黏液中脱落排出的强毒病毒。受污染的水是一种高效的传播途径(Boutier et al., 2015b),不过在没有宿主的情况下,CyHV-3在水中会迅速失活(Boutier et al., 2015a)。目前还没有垂直传播的公开证据。

继1997年德国、1998年以色列和美国首次报告KHVD以来,该疾病的地理范围变得越来越广泛。锦鲤的全球贸易是导致CyHV-3传播的主要原因,目前至少有30个国家的鱼类进口出现或已报告了这种疾病(OIE, 2012; Boutier et al., 2015a)。在亚洲,2002年在印度尼西亚首次暴发了KHVD,并导致养殖锦鲤大规模死亡。2003年,日本曾报道了网箱养殖鲤鱼的大量死亡(Haenen et al., 2004)。此后,在中国大陆和台湾地区、韩国、新加坡、马来西亚以及泰国都发现了CyHV-3 (Boutier et al., 2015a)。在非洲,KHVD仅在南

* 通信作者邮箱:kman71@live.co.uk。

非有过相关报道(Haenen et al., 2004)。

在北美,1998 年和 1999 年的 KHVD 首次暴发与锦鲤经销商有关(Haenen et al., 2004),但随后,CyHV-3 在美国和加拿大造成了野生鲤鱼的大规模死亡(Boutier et al., 2015a)。在欧洲,该疾病已在 18 个国家被报道记录,其中在德国、波兰和英国报告了鲤鱼的大面积死亡(OIE, 2012; Boutier et al., 2015a)。

CyHV-3 能够引发高致病性和传染性的疾病,在鲤鱼和锦鲤中造成大量死亡和重大经济损失。自 2007 年以来,锦鲤疱疹病毒病已被世界动物卫生组织(OIE)列为应具报疾病,并且它的流行具有季节性,主要发病水温为 17～28 ℃。感染后 22 d 死亡,8～12 d 死亡率最高(Ilouze et al., 2006)。仅在鲤鱼和其变种(如锦鲤)中报告了天然发生的 CyHV-3 感染(OIE, 2012)。但是,与鲤鱼杂交的鲫属(*Carassius* spp.)品种可能对 KHVD 易感。实验性感染报告中,鲫鱼×锦鲤杂交种的死亡率在 91%～100%;金鱼(*Carassius auratus*)×锦鲤杂交种的死亡率在 35%～42%(Bergmann et al., 2010a);金鱼×鲤鱼杂交种的死亡率只有 5%(Hedrick et al., 2006)。

金鱼易受 CyHV-3 感染(OIE, 2012),但不会发病。越来越多的证据表明,其他鲤科鱼类和非鲤科鱼类[如草鱼(*Ctenopharyngodon idella*)、鲇鱼、鲇形目]是将 CyHV-3 传染给鲤鱼的潜在病毒携带者(Boutier et al., 2015a)。

1998 年至 2000 年,KHVD 扩散到 90%的以色列鲤鱼养殖场,每年的病害损失约 300 万美元。在印度尼西亚,2002 年 4 月至 11 月期间,该病从东爪哇蔓延到其他四个主要岛屿,到 2003 年损失超过 1 500 万美元(Haenen et al., 2004)。毫无疑问,在监管机构意识到这种疾病前并且尚无检测方法的情况下,CyHV-3 已在全球范围内传播。1996 年在英国(Haenen et al., 2004)和 1998 年在韩国(Lee et al., 2012)发生了不明原因的鲤鱼大量死亡,在此期间收集存档的鲤鱼组织学标本中都检测到 CyHV-3 的 DNA。在以色列,食鱼鸟类被怀疑在养鱼场间传播 CyHV-3(Ilouze et al., 2011)。其他传播途径包括在锦鲤展的同一鱼缸中混放鱼、以低于市场价格出售受感染的鱼、将受感染的鱼放流到公共水域(Boutier et al., 2015a)。

10.2 诊断

对临床受影响鱼类的 KHVD 诊断使用了一系列测试,但很少可以完全确诊。《水生动物诊断检测指南》(OIE, 2012)中关于 KHVD 的章节建议,KHVD 的诊断应该依靠综合多种测试,包括临床检查和病毒检测。KHVD 的最终诊断必须依赖于病毒 DNA 的直接检测或使用免疫学和分子技术来分离和鉴定 CyHV-3。

10.2.1 行为变化

KHVD 最明显的行为症状是嗜睡。病鱼表现为离群，躺在水槽或池塘底部或头朝下悬垂。它们也可能在靠近进水口和其他充气区域或池塘边的水面喘息（OIE，2012）。有些鱼可能会失去平衡和迷失方向，但与此同时，可能会变得过度活跃（Hedrick et al.，2000）。

10.2.2 外部大体病理学

KHVD 没有明显的病理性大体病灶。最一致的大体病理学表现为皮肤呈现苍白、不规则斑块，伴有黏液分泌过多或黏液分泌不足，以及由此导致的呈砂纸样质地的皮肤斑块（Haenen et al.，2004）[图 10.1(A)]。皮肤也可能表现出充血、出血和溃疡（Boutier et al.，2015a）[图 10.1(B)]。随着病情的发展，鱼可能会出现局灶性或广泛的皮肤上皮组织损失。在病情较轻的鲤鱼中，常见的临床表现包括厌食和眼球内陷（眼睛陷入眼窝内）（Boutier et al.，2015a）[图 10.1(C)]。鳃中的大体病理学是 KHVD 临床中最一致的特征。病理变化从苍白的坏死斑块到伴有严重坏死和炎症的广泛性褪色（褪色变白）（OIE，2012）（图10.2）。

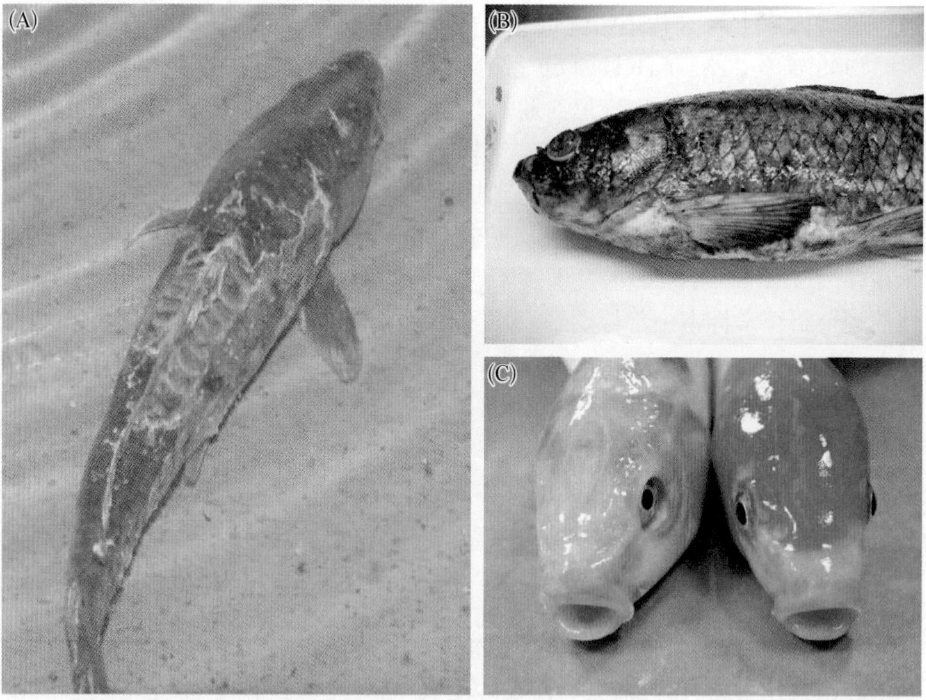

图 10.1　患有锦鲤疱疹病毒病（KHVD）的鱼的外部大体病理学：(A) 皮肤上的不规则斑块主要与黏液的分泌不足有关，其中皮肤斑块具有砂纸样纹理；(B) 鲤鱼皮肤上的充血和出血；(C) 锦鲤的眼球内陷

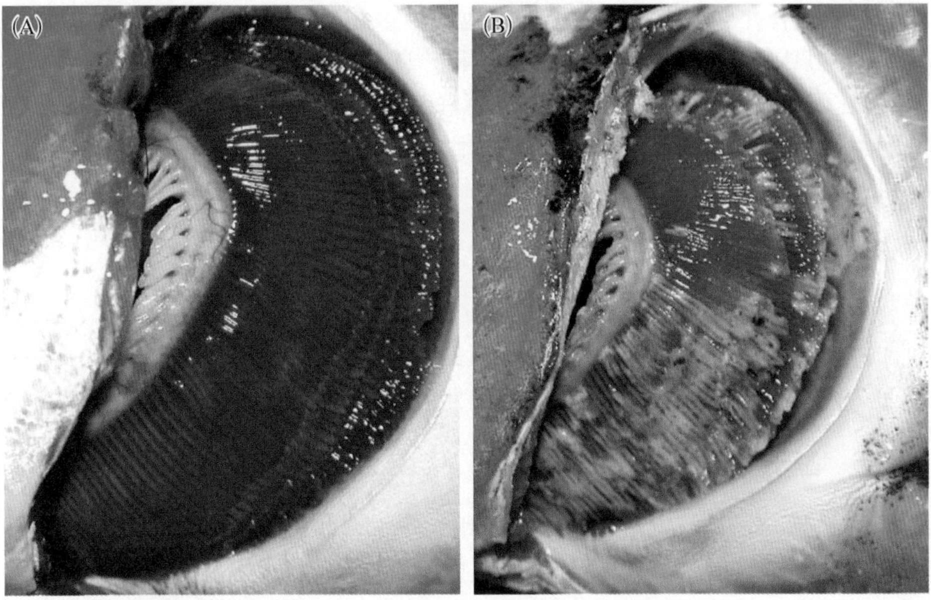

图 10.2 去除鳃盖的鲤鱼鳃的外部大体病理学：(A) 正常的健康鳃；(B) 患有锦鲤疱疹病毒病(KHVD)的鱼的鱼鳃出现炎症和坏死

10.2.3 内部大体病理学

可能存在腹腔积液和腹腔粘连，内脏器官可能肿大，发黑或斑驳，但这些病变并非 KHVD 的病理特征(Boutier et al.，2015a)。在患病鱼中，特别是在鲤鱼中，肉眼可见的病变可能因体外寄生虫感染变得复杂。八个寄生虫属[如三代虫(*Gyrodactylus* sp.)]通常与受 KHVD 影响的鲤鱼有关，还报道了伴随鳃部的单殖吸虫继发感染和一系列细菌[如黄杆菌(*Flavobacterium* sp.)]感染(OIE，2012)。

10.2.4 采样

鱼龄 1 年以下的幼鲤通常临床上对 KHVD 易感，应选择为采样对象。鲤鱼或锦鲤、鬼鲤等变种最容易感染该病，其次是任何鲤鱼和鲫属鱼类的杂交种(OIE，2012)。具有临床症状的濒死或新鲜死亡的鱼适合于目前可用的大多数免疫学和分子测试。已经腐烂的鲤鱼组织样本可能仅适用于使用基于聚合酶链反应(PCR)的方法进行检测。当样品来自可疑群体中看似健康的鲤鱼时，使用基于 PCR 的检测方法进行检测最可靠(OIE，2012)。

应在选择好鲤鱼后立即收集器官样本(OIE，2012)。可将整条鱼包裹在冰中，组织则保存在病毒运送液或 80%～100%的乙醇中送去检测。保存在酒精中或冷冻的样品仅适用于基于 PCR 的检测。

应避免合并样本或限制每个样本池中最多放两尾鱼。2014 年在哥本哈根举行的 KHV 疾病监测与确诊的采样和诊断程序会议的报告(Olesen et al., 2014)建议,在急性病例中,可以将五尾鱼的组织合并为一个样本。推荐的采样组织包括鳃、肾脏和脾脏(OIE, 2012),因为它们含有最大的 DNA 浓度(Gilad et al., 2004)。

10.2.5 直接免疫诊断方法

可以采用间接荧光抗体法(indirect fluorescent antibody test, IFAT)从肾脏印迹中直接检测 CyHV-3(OIE, 2012)。免疫过氧化物酶染色也被用于检测 CyHV-3 抗原,但这种方法易产生假阳性染色(Pikarsky et al., 2004)。

10.2.6 基于 PCR 的检测

基于 PCR 的检测方法通常是检测组织中 CyHV-3 的最灵敏和可靠的方法。OIE(2012)推荐了一种结合一组靶向胸苷激酶(TK)基因引物的诊断方法(Bercovier et al., 2005)和另一种改进的 Gray-SpH 引物组测定方案(Yuasa et al., 2005)

相对于常规 PCR,许多诊断医师倾向于实时定量 PCR(qPCR)测定。检测 CyHV-3 最常用的 qPCR 是 Gilad Taqman 实时定量 PCR 法(Gilad et al., 2004),可检测并可定量评估极低拷贝数的靶核酸序列(Boutier et al., 2015a)。实时定量 PCR 避免了传统 PCR 中固有的许多污染风险(OIE, 2012),它还靶向较短的 DNA 序列,更有可能检测到分解组织中降解的病毒 DNA。

环介导等温扩增(loop-mediated isothermal amplification, LAMP)是一种快速的单步 PCR 检测方法,由于不需要热循环仪,被广泛应用于田间的现场诊断。已开发出检测 CyHV-3 的 TK 基因 LAMP 法,其灵敏度与常规 PCR 相当或更高(Yoshino et al., 2009)。还开发了将 DNA 杂交技术和抗原-抗体反应与 LAMP 相结合的测定方法,并提高了灵敏度和特异性(Soliman and El-Matbouli, 2010)。LAMP 具有对临床感染鱼类进行无损检测的潜力,但 OIE 并没有详细说明 LAMP 检测 KHVD 的方法,因为尚未提交任何评估文件,也未进行登记说明。

最灵敏的 PCR 方法是上面提到的 Gilad Taqman 实时定量 PCR 法(OIE, 2012)。在基于 PCR 的测定的比较中,Bergmann 等的研究(2010b)显示,包括第二轮嵌套引物的常规 PCR 检测与实时检测一样灵敏。然而,巢式 PCR 更容易受到先前检测 PCR 产物的交叉污染,从而产生假阳性。

10.2.7 组织病理学

该疾病的组织病理学是非特异性和可变的。受 KHVD 影响的鱼中,在其鳃、皮肤、肾、心、脾、肝、肠和脑中都发现了变化(OIE, 2012; Boutier et al., 2015a)。在鱼的皮肤、鳃和肾中最容易观察到感染疱疹病毒的证据,这部分将在10.3节中讲述。

10.2.8 电子显微镜检测法

除非有严重感染,否则使用透射电子显微镜(TEM)检查临床感染鲤鱼组织中的病毒颗粒是不可靠的(OIE, 2012)。

10.2.9 细胞培养中的病毒分离

建议用鲤鱼脑(common carp brain, CCB)和锦鲤鳍条1型(koi fin, KF-1)细胞系来分离CyHV-3,但前者更易感(OIE,2012; J. Savage, UK,个人通信)。Boutier等(2015a)列出了对CyHV-3易感的其他细胞系。在出现临床症状之前,鱼鳃中的病毒水平高于肾组织(Gilad et al., 2004; Yuasa et al., 2012),从鳃组织中分离出病毒最为可靠。然而,细胞培养中的病毒分离方法不如基于PCR检测CyHV-3 DNA的方法可靠或灵敏(OIE, 2012)。

10.2.10 包括非致死性检测的其他诊断方法

原位杂交(*in situ* hybridization, ISH)和IFAT已被用于检测和鉴定鱼白细胞中的CyHV-3(Boutier et al., 2015a)。虽然这些方法尚未与其他技术进行充分比较,但它们是非致死性的,可能有助于诊断(OIE, 2012)。血液、鳃拭子、鳃活检和黏液刮片等非致死样本也是诊断的合适材料(Olesen et al., 2014)。其他非致死性免疫诊断方法包括一种用于检测鱼粪便中CyHV-3抗原的ELISA(Dishon et al., 2005)。此外,还有一种可在田间15 min内检测CyHV-3糖蛋白(FASTest Koi HV试剂盒)的侧向流检测装置,并且最适用于鳃拭子样本的检测(Vrancken et al., 2013)。

10.3 病理学

感染CyHV-3的鱼的鳃和皮肤表现出明显的临床症状,这在组织病理学中得到反映。鳃组织的增生和肥大很常见(Boutier et al., 2015a)[图10.3(A)]。Pikarsky等(2004)研究观察了实验感染2 d后鱼鳃中的病理变化,包括鳃瓣缺损和一些鳃丝上的混合炎性浸润。感染6 d时,增生、严重炎症和中央静脉窦充血变得更加明显。鳃耙上皮下发炎和血管充血,并伴有高度的降低和表面上皮脱落;有时在鳃片毛细血管中可见出血。增大的鳃上皮细胞有核肿胀,伴有染色质边聚和浅色弥散性嗜酸性包涵体(呈“印章环”外观),也称为核内包涵体[图10.3(B)]。当使用TEM观察时,这些细胞通常含有CyHV-3。在心脏、

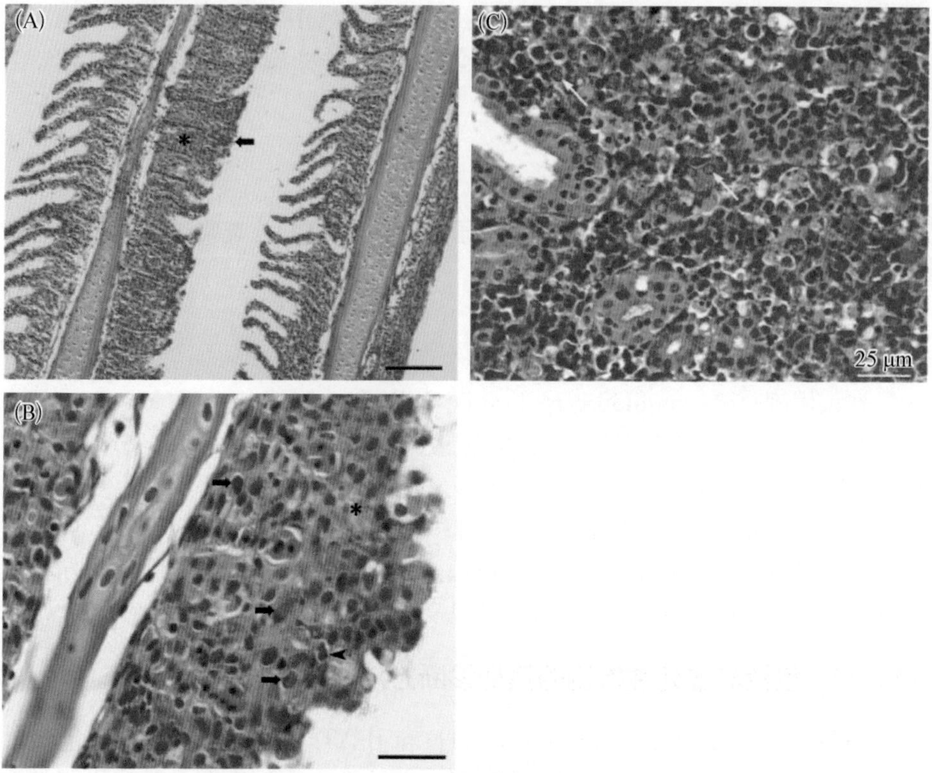

图 10.3　锦鲤疱疹病毒病(KHVD)的鲤鱼组织切片(苏木精-伊红染色)中的显微病变，(A) 鳃上皮细胞肥大和增生(*)与鳃次级鳃片(箭头所示)融合；(B) 鳃坏死和炎症(*)，细胞凋亡(箭簇所示)和核内包涵体(箭头所示)；(C) 肾间质造血细胞坏死和含有内含物(箭头所示)的细胞核。比例尺：(A) 100 μm；(B) 20 μm；(C) 25 μm

肾脏[图 10.3(C)]、脾脏、肝脏、肠、胃、脑和鳍表皮中也观察到包涵体。次级鳃片常与增生性鳃上皮融合，并且经常可见坏死的情况，特别是在顶端处(Boutier et al., 2015a; Miwa et al., 2015)。

感染第 1 天时可观察到鳍表皮下的细胞浸润，感染第 2 天时表皮完整性开始被破坏。感染第 3 天时，也会开始失去表皮，并且头部表皮经常脱落。受感染鱼皮肤杯状细胞数量减少了 50%，并且它们看起来很单薄，这表明黏液被释放且没有被补充(Adamek et al., 2013)。在感染后期，经常可见患病鱼皮肤表皮的侵蚀，这通常是观察到的最严重的病理变化(Adamek et al., 2013; Miwa et al., 2015)。

患病鱼的肾脏表现出明显的病理变化。从感染第 2 天开始出现肾小管周围炎性浸润；感染第 6 天时，出现重度间质炎性浸润并伴有血管充血；感染达到或超过 8 d 时，肾小管上皮细胞出现增生或变性，并存在上皮内淋巴细胞。可见坏死细胞，严重时造血细胞坏死。可观察到慢性肾小球炎、肾小球周围纤维化、间质性肾炎伴有造血细胞的减少等病变(Boutier et al., 2015a; Miwa et al., 2015)。

肝脏表现出的轻度炎性浸润主要在肝实质中发生,部分病鱼伴有局灶性坏死。胆管柱状上皮细胞可能发生增生(Pikarsky et al., 2004; Cheng et al., 2011)。胃腺上皮内膜增生阻塞胃腔,肠绒毛也有增生。肠上皮细胞脱落到肠腔内(El Din, 2011)。脾实质中有坏死灶,脾、肾中存在胰腺腺泡细胞坏死(Hedrick et al., 2000; Cheng et al., 2011)。

在脑部,存在局部病灶性脑膜炎和脑膜外炎症,以及毛细血管和小静脉的充血(Boutier et al., 2015a)。许多心肌细胞表现出核变性,肌原纤维扩张或凝固并伴随横纹消失。在疾病的后期,巨噬细胞和淋巴细胞渗入心肌并观察到心肌坏死(Miyazaki et al., 2008; Cheng et al., 2011)。

10.4 病理生理学

研究已经证实 CyhV-3 侵染鲤鱼的主要入口是皮肤,而不是最初提出的鳃和肠道(Costes et al., 2009)。然而,在感染早期就可观察到鳃损伤(Hedrick et al., 2000; Pikarsky et al., 2004),并早在感染的第一天就在鳃和肠道中检测到病毒 DNA(Gilad et al., 2004)。Costes 等(2009)使用表达萤光素酶(luciferase, LUC)的重组 CyHV-3 诱导了鲤鱼的 KHVD,与亲本野生毒株诱导的疾病没有区别。此外,对使用这种表达 LUC 的重组病毒自然感染的鲤鱼进行成像显示,CyHV-3 通过皮肤而非鳃进入鱼体。重组 CyHV-3 的感染结果显示 CyHV-3 通过皮肤进入鱼而不是鳃。早期病毒在皮肤上皮中复制,主要是在鱼鳍中(图 10.4)。感染 12 h 后(12 hours postinfection, hpi)在皮肤中可检测到 CyHV-3 的 RNA 表达(Adamek et al., 2013),并在感染 2 d 后在鳍上皮感染细胞中检测到病毒 DNA(Miwa et al., 2015)。用减毒的 CyHV-3 浸泡感染鲤鱼也证实了病毒从皮肤向其他器官的扩散(Boutier et al., 2015a)。使用类似的技术,Fournier 等(2012)的研究表明,咽部牙周黏膜是经口感染后的主要侵染入口(图 10.4)。更重要的是,病毒向其他器官的传播和临床疾病的进展与浸泡感染情况相当。

感染后,CyHV-3 迅速扩散到其他器官,这些器官是病毒继发感染的部位。CyHV-3 对白细胞的嗜性可能是病毒通过血液迅速传播的原因(Boutier et al., 2015a)。感染 1 d 后,可在血液、鳃、肝、脾、肾、肠和脑组织中检测到病毒 DNA(Gilad et al., 2004; Pikarsky et al., 2004)。在急性感染的新死鲤鱼中,CyHV-3 在鳃、肠和肾组织中的 DNA 拷贝数为每 10^6 个宿主细胞 10^9 ~ 10^{11}(Gilad et al., 2004)。这些渗透调节器官在疾病过程中发生了显著的病理变化,Gilad 等(2004)认为,渗透调节功能的丧失导致患病鱼的死亡。

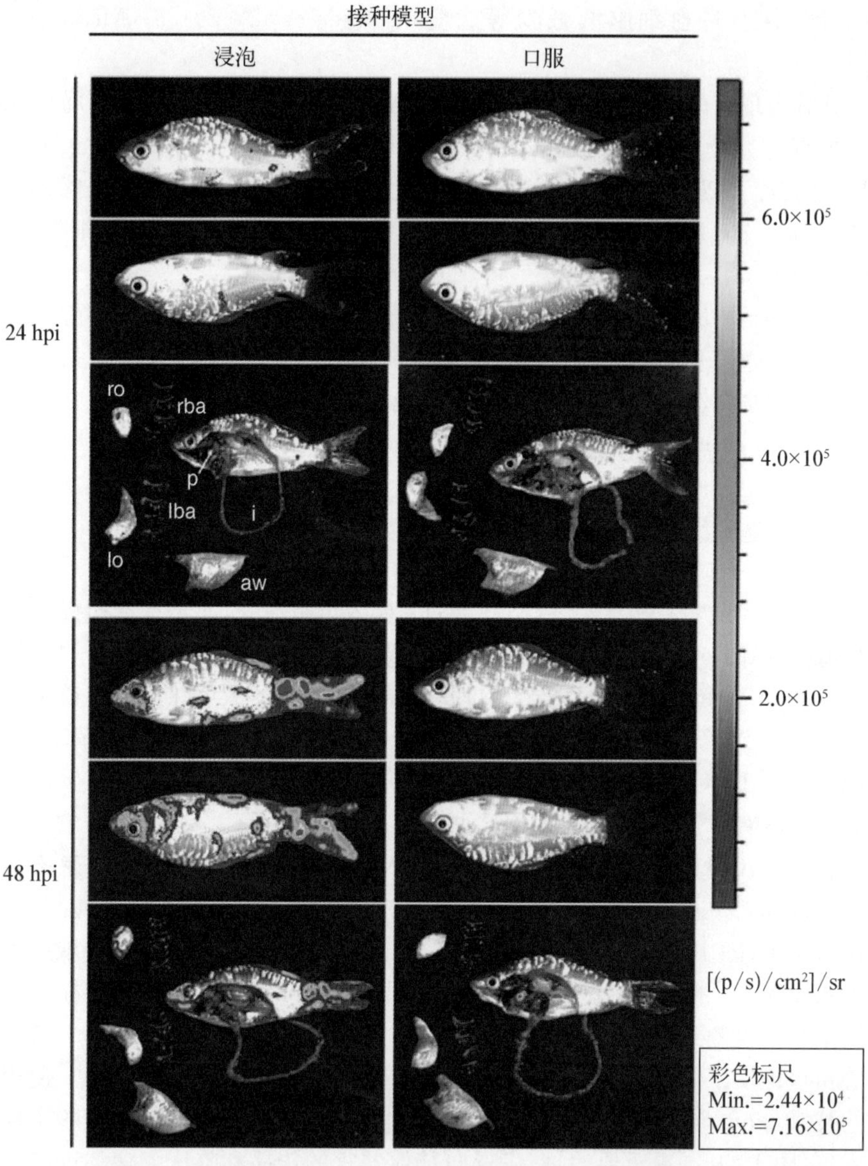

图 10.4 CyHV-3侵染入口的活体生物发光成像分析。以萤光素酶为报告基因的重组CyHV-3菌株感染两组鱼(平均体重为10 g)，感染的方式为将鱼浸泡在含有病毒的水中(浸泡，左侧)或者用含有病毒颗粒的食物颗粒投喂(口服，右侧)。在感染后的指定时间(hpi)，通过生物发光体内成像系统(*in vivo* imaging system，IVIS)分析每组的六尾鱼，分析了每尾鱼的左右两侧。在安乐死和解剖后对其内部信号进行分析。对解剖的鱼和分离的器官进行了体外生物发光分析。每个时间点和接种模式显示一条代表性鱼。使用软件 Living Image 3.2在从紫色(最低强度)到红色(最强烈)的相同伪色标度下，对实验过程中收集的图像进行归一化。aw为腹壁；i为肠；lba为左鳃弓；lo为左鳃盖；p为咽；rba为右鳃弓；ro为右鳃盖。Min.和Max.分别为光子通量的最大阈值和最小阈值；[(p/s)/cm²]/sr=[(光子/秒)/平方厘米]/球面度。经Fournier等(2012)许可转载，原始出版商是BioMed Central

Negenborn 等(2015)的研究发现,在实验感染期间电解质水平(主要是尿液中的钠离子)和血清电解质水平随之下降。电解质水平的变化与肾和鳃中的严重病理改变相对应,进一步证明严重的渗透压调节功能障碍可能导致死亡。在实验性 CyHV-3 感染中,Miwa 等(2015)报道了感染 5～8 d 时在皮肤表皮造成广泛损伤。死亡前血清渗透压也很低,揭示皮肤损伤引起的低渗性休克(hypo-osmotic shock)可能是导致死亡的原因之一。

10.4.1 先天免疫反应

研究人员认为,鱼类对 CyHV-3 感染具有强烈而快速的先天免疫反应,这可以通过早期上调补体相关蛋白和 C 反应蛋白来证明(Pionnier et al., 2014)。干扰素(interferon, IFN)起重要的调节作用,对皮肤和肠道反应的研究揭示了 IFN 型Ⅰ类通路的激活(Adamek et al., 2013; Syakuri et al., 2013)。然而,与鲤春病毒血症病毒不同,CyHV-3 可以抑制 IFN 型Ⅰ类通路并抑制受刺激巨噬细胞的活性和淋巴细胞的增殖反应(Boutier et al., 2015a)。此外,CyHV-3 不会诱导细胞凋亡,并且细胞凋亡内源性途径的刺激被延迟(Miest et al., 2015)。在 CyHV-3 感染期间,编码紧密连接蛋白、黏蛋白和 β 防御素抗菌肽的基因表达下调。皮肤黏膜屏障这些重要组分的破坏加速了皮肤的崩解(Adamek et al., 2013)。

10.4.2 适应性免疫反应和免疫逃逸

鲤鱼能够产生一种强烈的、温度依赖性的针对 CyHV-3 的保护性抗体。与 31 ℃下的快速反应相比,对病毒的抗体反应在 12～14 ℃下较慢。在病毒感染的允许温度下,感染 7～14 d 可以检测到抗体,感染 20～40 d 时抗体水平达到峰值,并且至少 65 周时仍能检测到抗体(Perelberg et al., 2008; St-Hilaire et al., 2009)。

在浸泡感染的鲤鱼中,皮肤黏膜中可检测到一种减毒的 CyHV-3 重组病毒,并诱导针对野生型 CyHV-3 的强保护性黏膜免疫应答(Boutier et al., 2015b)。这可能与刺激分泌 IgT 的 B 细胞有关,IgT 是一种参与硬骨鱼类黏膜免疫的免疫球蛋白同种型(Boutier et al., 2015a)。

计算机模拟、体外和体内的研究表明,CyHV-3 可能表达参与免疫逃避的蛋白(Boutier et al., 2015a),这可能能够解释与 KHVD 相关的急性和显著的临床症状。

10.4.3 潜伏感染

在暴发 KHVD 的野生鲤鱼和养殖鲤鱼的幸存者中持续存在 CyHV-3 感染情况(Baumer et al., 2013; Uchii et al., 2014),尽管该病毒的潜伏期尚未被证实(Boutier et al., 2015a)。为了解释 CyHV－3 如何在恢复期的鲤鱼群体中持续存在，有研究者提出了季节性再激活。

在血清中检测到病毒阳性的鱼的脑中检测到病毒复制相关基因的 RNA 表达，而其他鱼中仅表达潜伏相关基因(Uchii et al., 2014)。这表明再激活是鲤鱼群体中的一种暂态现象。温度胁迫(Eide et al., 2011)和网捕胁迫(Bergmann and Kempter, 2011)可能触发 CyHV－3 的再激活。在没有临床表现和可检测到的感染性病毒颗粒的情况下，在之前接触过 CyHV－3 的锦鲤及既往无 KHVD 史的野生鲤鱼(Eide et al., 2011；Xu et al., 2013)的白细胞(white blood cells, WBCs)中检测到病毒 DNA。涉及病毒持久存在的主要 WBC 类型是 IgM^{+} B 细胞，其中发现的病毒 DNA 拷贝数比其余 WBC 部分多 20 倍(Reed et al, 2014)。在长期感染鱼的各种组织中都已发现了该病毒，特别是在脑部。与其他疱疹病毒一样，神经系统可能意味着另一个潜伏位点(Boutier et al., 2015a)。

10.5 预防和控制

10.5.1 鲤鱼抗病品系

不同鲤鱼品系对 CyHV－3 的易感性不同。在杂交育种实验中，一些野生鲤鱼品系，如 Sassan 和 Amur 品系，当与家养品系杂交时，杂交种会产生对 KHVD 的抗性(Shapira et al., 2005；Dixon et al., 2009；Piackova et al., 2013)。以色列家养品系 Dor－70 也表现出很强的感染抗性(Zak et al., 2007)。相比之下，在日本，本地鲤鱼受 KHVD 的影响比驯养的欧亚鲤鱼和锦鲤更严重(Ito et al., 2014)。分子水平上的抗病分析已经确定了其与参与免疫应答基因的多态性有关，包括主要组织相容性复合体(major histocompatibility complex, MHC)Ⅱ类 B 基因(Rakus et al., 2012)和鲤鱼 IL－10(白细胞介素 10)基因(Kongchum et al., 2011)。这些研究的结果尚未定论，但确实表明存在可以加强选择育种的抗性遗传标记。

10.5.2 疫苗接种

早期对鲤鱼进行疫苗接种的尝试包括将实验鱼与感染 CyHV－3 的鱼在 22～23 ℃下同养 3～5 d，然后将它们转移到 30 ℃下的池塘中养殖 30 d(Ronen et al., 2003)。用 CyHV－3 攻毒后，这些鱼对病毒的抵抗力更强。这种方法降低了以色列境内 CyHV－3 流行地区鲤鱼的死亡率(H. Bercovier 和 Dixon 的个人交流，2008)，但缺点包括成本高和疾病复发、处理过的鱼是感染源、死亡率仍然很高。

将通过细胞培养连续传代减毒的 CyHV－3 腹腔注射鲤鱼后，当它们与感染 CyHV－3 的鱼共处时获得了完全抗感染保护(Ronen et al., 2003)。减毒病毒浸泡 10 min 可诱导保护作用，但其保护效果随时间而下降，在水中 4 h 后无效(Perelberg et al., 2005)。用紫外线(UV)光照射减毒病毒以诱导可能阻止回复毒力的突变。紫外线处理过的病毒能够保护鲤鱼免受 CyHV-3 感染，但没有数据证实减毒株会保持无毒。开发减毒疫苗的研究仍在继续(Perelberg et al., 2008; O'Connor et al., 2014; Weber et al., 2014)。美国曾对这种疫苗的有效性和安全性进行了检验(Weber et al., 2014)，结果发现这种疫苗是安全有效的，特别对体重超过 87 g 的鱼。疫苗接种 13 个月后，具有免疫力的锦鲤(未注明体重)的攻毒死亡率为 36%，而未接种的对照组死亡率为 100%(O'Connor et al., 2014)。

一种由 KoVax 公司生产制备的减毒疫苗 Cavoy 在以色列上市，并于 2012 年作为浸泡疫苗在美国获批使用，用于体重大于 100 g 的鲤鱼的免疫接种。然而，该疫苗在一年后停止销售(Boutier et al., 2015b)。后者的作者制备了一种缺失 ORF56 和 ORF57 的重组 CyHV-3，与亲本病毒相比，其在宿主体内复制水平较低且传播效率较低。鲤鱼通过浸泡接种重组病毒，20 d 后无一例死亡。在用亲本病毒感染的鲤鱼进行共处感染攻毒后，接种组存活率为 80%，而模拟接种组的存活率为 0%。接种鱼与前哨鱼同栖时，重组病毒存在较低水平的传播，但仅靠水不能传播。

Yasumoto 等(2006)用福尔马林灭活 CyHV-3 并将其包埋在脂质体中。将 CyHV-3-脂质体喷涂到干燥的颗粒饲料上，投喂鲤鱼 3 d 后喂以正常颗粒饲料。然后用 CyHV-3 感染实验鱼的鳃，导致两组接种鱼的死亡率均为 23%，两组对照鱼的死亡率分别为 66%和 80%以上。

研究小组正在尝试开发没有残留毒力甚至无毒力回复可能的 DNA 疫苗或重组疫苗。基于 CyHV-3 糖蛋白基因(Nuryati et al., 2010)和 ORF81 和 ORF25 基因(Boutier et al., 2015a)的原型 DNA 疫苗对注射接种后的鲤鱼提供了不同程度的保护，但均没有进行田间试验。基于 TK、核糖核酸还原酶或脱氧尿苷三磷酸酶基因缺失的重组病毒疫苗(Fuchs et al., 2011)在浸泡免疫试验中给出了不一致的结果。最近，有研究组将表达 CyHV-3 的 ORF81 蛋白和 SVCV 糖蛋白的基因工程植物乳杆菌掺入颗粒饲料中以制备口服疫苗(Cui et al., 2015)。在实验条件下，CyHV-3 攻毒后，具有免疫力的鲤鱼的死亡率为 47%，而对照组的死亡率为 85%和 93%。该疫苗还能保护锦鲤抗 SVCV(见第 7 章)。

10.5.3 管理和生物安全策略

在缺乏经过临床验证的商业疫苗或治疗药物的情况下，病害管理措施和生物安全策略是防治 CyHV-3 传播的主要途径。这些策略涵盖从国家或国际立法和标准到养殖场层面的良好管理规范。后者包括新鱼种群的现场检疫、设施消毒和使用足浴消毒、减少应激和其他疾病等。养殖场新引进的鱼在与易感鱼混合之前应至少隔离 4 周至 2 个月，但所采取的程序及其影响将取决于水产养殖任务的规模。尽管将水温提高至 26～28 ℃以上和减少养殖密度可能会降低死亡率，但传染和死亡的速度可能很快，并且限制了管理措施的使用(Gilad et al.，2003；Ronen et al.，2003；Sunarto et al.，2005)。

仅有一项关于通过化学和物理方法灭活 CyHV－3 的研究(Kasai et al.，2005)，但结果尚无定论。分别在 15 ℃下处理 30 s 和处理 20 min 后实现 100% 的噬斑减少的最小化学浓度：碘伏为 200 mg/L(两个时间段)，次氯酸钠为 400 mg/L 和 200 mg/L，苯扎氯铵为 60 mg/L(两个时间段)和乙醇为 40%和 30%。在含有 1×10^5 PFU/mL(暴露时间未说明)和 1.6×10^4 PFU/mL 的病毒溶液中产生 100%的噬斑减少的紫外线剂量为 4×10^3 (μW • s)/cm^2。CyHV－3 在高于 50 ℃的温度下 1 min 灭活。

Yoshida 等(2013)制备了一种实验室规模的废水处理系统，该系统将抑制 CyHV－3 的细菌物质吸附到包埋在柱内的多孔载体上。将接种 CyHV－3 的出水样品通过处理柱之前和之后分别注射到鲤鱼中，其死亡率分别为 80%以上和 0%。研究者正致力于设计一个可用于处理水产养殖废水的操作系统。

大多数国家已颁布立法，以控制包括鱼类疾病在内的各种疾病的引入和传播。这是目前预防和控制 KHVD 最重要的方法。这种立法往往得到国际协议的加强。世界动物卫生组织制定了"国际水生动物卫生法典"(OIE，2015)，其中概述了可以全国性和生产现场本地实施的生物安全策略。Håstein 等(2008)概述了实施生物安全策略的方法，Oidtmann 等(2011)综述了有关鱼和鱼产品流通的国际标准。

10.6 总结与研究展望

KHVD 在世界范围内都有发生，是世界动物卫生组织规定的应具报疾病。CyHV－3 病毒可引起高度传染性的急性病毒血症，对鲤鱼和锦鲤养殖业造成严重的经济损失。

尽管没有病理性肉眼可见的病变，但外部临床体征的迅速发作和严重程度及受感染鲤鱼的数量提供了 KHVD 的迹象。可能的死亡原因是渗透调节功能障碍或渗透性休克。使用基于 PCR 的方法可以快速鉴定病毒，并且已经开发出田间非致死性检测方法。该病毒可诱导鲤鱼产生强烈的先天免疫和保护性抗体反应，但是商业疫苗的可用性有限。

KHVD 的存活鱼会持续感染 CyHV－3，并且有证据表明其存在季节性和应激性诱导的再激活。需要更多的研究来确认和鉴定潜伏期的位点及与潜伏期相关的病毒基因转录产物，这些病毒转录产物可能是病毒监测的有用靶标。此外，还需要进行研究以评估温度如何影响并可能调节裂解性感染和潜伏性感染之间的转换。

一些鲤鱼品系对 KHVD 有抗性。对 CyHV－3 抗性/敏感性的遗传基础的继续研究可能会加强对 KHVD 抗性的选择育种计划。

良好的管理和生物安全策略是防止 CyHV－3 传播的主要途径。在过去的 15 年中，全球冷水水产养殖行业中的这些措施得到了改善，很多都是源于 KHVD 的威胁。

参考文献

[1] Adamek, M., Syakuri, H., Harris, S., Rakus, K.L., Brogden, G. *et al*. (2013) Cyprinid herpesvirus 3 infection disrupts the skin barrier of common carp (*Cyprinus carpio* L.). *Veterinary Microbiology* 162, 456－470.

[2] Baumer, A., Fabian, M., Wilkens, M.R., Steinhagen, D. and Runge, M. (2013) Epidemiology of cyprinid herpesvirus-3 infection in latently infected carp from aquaculture. *Diseases of Aquatic Organisms* 105, 101－108.

[3] Bercovier, H., Fishman, Y., Nahary, R., Sinai, S., Zlotkin, A. *et al*. (2005) Cloning of the koi herpesvirus (KHV) gene encoding thymidine kinase and its use for a highly sensitive PCR based diagnosis. *BMC Microbiology* 5: 13.

[4] Bergmann, S.M. and Kempter, J. (2011) Detection of koi herpesvirus (KHV) after reactivation in persistently infected common carp (*Cyprinus carpio* L.) using non-lethal sampling methods. *Bulletin of the European Association of Fish Pathologists* 31, 92－100.

[5] Bergmann, S.M., Sadowski, J., Kielpinski, M., Bartlomiejczyk, M., Fichtner, D. *et al*. (2010a) Susceptibility of koi × crucian carp and koi × goldfish hybrids to koi herpesvirus (KHV) and the development of KHV disease (KHVD). *Journal of Fish Diseases* 33, 267－272.

[6] Bergmann, S.M., Riechardt, M., Fichtner, D., Lee, P. and Kempter, J. (2010b) Investigation on the diagnostic sensitivity of molecular tools used for detection of koi herpesvirus. *Journal of Virological Methods* 163, 229－233.

[7] Boutier, M., Ronsmans, M., Rakus, K., Jazowiecka-Rakus, J., Vancsok, C. *et al*. (2015a) Cyprinid herpesvirus 3, an archetype of fish alloherpesviruses. In: Kielian, M., Maramorosch, K. and Mettenleiter, T.C. (eds) *Advances in Virus Research*, *Volume 93*. Elsevier/Academic Press, Waltham, Massachussetts/San Diego, California/London/Kidlington, Oxford, UK, pp. 161－256.

[8] Boutier, M., Ronsmans, M., Ouyang, P., Fournier, G., Reschner, A. *et al*. (2015b) Rational development of an attenuated recombinant cyprinid herpesvirus 3 vaccine using prokaryotic mutagenesis and *in vivo* bioluminescent imaging. *PLoS Pathogens* 11 (2): e1004690.

[9] Cheng, L., Chen, C.Y., Tsai, M.A., Wang, P.C., Hsu, J.P. *et al*. (2011) Koi herpesvirus epizootic in cultured carp and koi, *Cyprinus carpio* L., in Taiwan. *Journal of Fish Diseases* 34, 547－554.

[10] Costes, B., Stalin Raj, V., Michel, B., Fournier, G., Thirion, M. *et al*. (2009) The major portal of entry of koi herpesvirus in *Cyprinus carpio* is the skin. *Journal of Virology* 83, 2819－2830.

[11] Cui, L.C., Guan, X.T., Liu, Z.M., Tian, C.Y. and Xu, Y.G. (2015) Recombinant

lactobacillus expressing G protein of spring viremia of carp virus (SVCV) combined with ORF81 protein of koi herpesvirus (KHV): a promising way to induce protective immunity against SVCV and KHV infection in cyprinid fish via oral vaccination. *Vaccine* 33, 3092 - 3099.

[12] Dishon, A., Perelberg, A., Bishara-Shieban, J., Ilouze, M., Davidovich, M. *et al.* (2005) Detection of carp interstitial nephritis and gill necrosis virus in fish droppings. *Applied and Environmental Microbiology* 71, 7285 - 7291.

[13] Dixon, P.F., Joiner, C.L., Way, K., Reese, R.A., Jeney, G. *et al.* (2009) Comparison of the resistance of selected families of common carp, *Cyprinus carpio* (L.), to koi herpesvirus: preliminary study. *Journal of Fish Diseases* 32, 1035 - 1039S.

[14] Eide, K.E., Miller-Morgan, T., Heidel, J.R., Kent, M.L., Bildfell, R.J. *et al.* (2011) Investigation of koi herpesvirus latency in koi. *Journal of Virology* 85, 4954 - 4962.

[15] El Din, M.M.M. (2011) Histopathological studies in experimentally infected koi carp (*Cyprinus carpio koi*) with koi herpesvirus in Japan. *World Journal of Fish and Marine Science* 3, 252 - 259.

[16] Fournier, G., Boutier, M., Raj, V.S., Mast, J., Parmentier, E. *et al.* (2012) Feeding Cyprinus carpio with infectious materials mediates cyprinid herpesvirus 3 entry through infection of pharyngeal periodontal mucosa. *Veterinary Research* 43: 6.

[17] Fuchs, W., Fichtner, D., Bergmann, S.M. and Mettenleiter, T.C. (2011) Generation and characterization of koi herpesvirus recombinants lacking viral enzymes of nucleotide metabolism. *Archives of Virology* 156, 1059 - 1063.

[18] Gilad, O., Yun, S., Adkison, M.A., Way, K., Willits, N.H. *et al.* (2003) Molecular comparison of isolates of an emerging fish pathogen, koi herpesvirus, and the effect of water temperature on mortality of experimentally infected koi. *Journal of General Virology* 84, 2661 - 2668.

[19] Gilad, O., Yun, S., Zagmutt-Vergara, F.J., Leutenegger, C.M., Bercovier, H. *et al.* (2004) Concentrations of a koi herpesvirus (KHV) in tissues of experimentally infected Cyprinus carpio koi as assessed by real-time TaqMan PCR. *Diseases of Aquatic Organisms* 60, 179 - 187.

[20] Haenen, O.L.M., Way, K., Bergmann, S.M. and Ariel, E. (2004) The emergence of koi herpesvirus and its significance to European aquaculture. *Bulletin of the European Association of Fish Pathologists* 24, 293 - 307.

[21] Håstein, T., Binde, M., Hine, M., Johnsen, S., Lillehaug, A. *et al.* (2008) National biosecurity approaches, plans and programmes in response to diseases in farmed aquatic animals: evolution, effectiveness and the way forward. *Revue Scientifique et Technique* 27, 125 - 145.

[22] Hedrick, R.P., Gilad, O., Yun, S., Spangenberg, J.V., Marty, G.D. *et al.* (2000) A herpesvirus associated with mass mortality of juvenile and adult koi, a strain of common carp. *Journal of Aquatic Animal Health* 12, 44 - 57.

[23] Hedrick, R.P., Waltzek, T.B. and McDowell, T.S. (2006) Susceptibility of koi carp, common carp, goldfish, and goldfish × common carp hybrids to cyprinid herpesvirus-2 and herpesvirus-3. *Journal of Aquatic Animal Health* 18, 26 - 34.

[24] Ilouze, M., Dishon, A. and Kotler, M. (2006) Characterization of a novel virus causing a lethal disease in carp and koi. *Microbiology and Molecular Biology Reviews* 70, 147 - 156.

[25] Ilouze, M., Davidovich, M., Diamant, A., Kotler, M. and Dishon, A. (2011) The outbreak of carp disease caused by CyHV - 3 as a model for new emerging viral diseases in aquaculture: a review. *Ecological Research* 28, 885 - 892.

[26] Ito, T., Kurita, J. and Yuasa, K. (2014) Differences in the susceptibility of Japanese indigenous and domesticated Eurasian common carp (*Cyprinus carpio*), identified by

mitochondrial DNA typing, to cyprinid herpesvirus 3 (CyHV - 3). *Veterinary Microbiology* 171, 31 - 40.
[27] Kasai, H., Muto, Y. and Yoshimizu, M. (2005) Virucidal effects of ultraviolet, heat treatment and disinfectants against koi herpesvirus (KHV). *Fish Pathology* 40, 137 - 138.
[28] Kongchum, P., Sandel, E., Lutzky, S., Hallerman, E.M., Hulata, G. *et al.* (2011) Association between IL - 10a single nucleotide polymorphisms and resistance to cyprinid herpesvirus-3 infection in common carp (*Cyprinus carpio*). *Aquaculture* 315, 417 - 421.
[29] Lee, N.S., Jung, S.H., Park, J.W. and Do, J.W. (2012) In situ hybridization detection of koi herpesvirus in paraffin-embedded tissues of common carp *Cyprinus carpio* collected in 1998 in Korea. *Fish Pathology* 47, 100 - 103.
[30] Miest, J. J., Adamek, M., Pionnier, N., Harris, S., Matras, M. *et al.* (2015) Differential effects of alloherpesvirus CyHV - 3 and rhabdovirus SVCV on apoptosis in fish cells. *Veterinary Microbiology* 176, 19 - 31.
[31] Miwa, S., Kiryu, I., Yuasa, K., Ito, T. and Kaneko, T. (2015) Pathogenesis of acute and chronic diseases caused by cyprinid herpesvirus-3. *Journal of Fish Diseases* 38, 695 - 712.
[32] Miyazaki, T., Kuzuya, Y., Yasumoto, S., Yasuda, M. and Kobayashi, T. (2008) Histopathological and ultrastructural features of koi herpesvirus (KHV)-infected carp *Cyprinus carpio*, and the morphology and morphogenesis of KHV. *Diseases of Aquatic Organisms* 80, 1 - 11.
[33] Negenborn, J., van der Marel, M.C., Ganter, M. and Steinhagen, D. (2015) Cyprinid herpesvirus-3 (CyHV - 3) disturbs osmotic balance in carp (*Cyprinus carpio* L.) — a potential cause of mortality. *Veterinary Microbiology* 177, 280 - 288.
[34] Nuryati, S., Alimuddin, Sukenda, Soejoedono, R. D., Santika, A. *et al.* (2010) Construction of a DNA vaccine using glycoprotein gene and its expression towards increasing survival rate of KHV-infected common carp (*Cyprinus carpio*). *Jurnal Natur Indonesia* 13, 47 - 52.
[35] O'Connor, M.R., Farver, T.B., Malm, K.V., Yun, S.C., Marty, G.D. *et al.* (2014) Protective immunity of a modified-live cyprinid herpesvirus 3 vaccine in koi (*Cyprinus carpio koi*) 13 months after vaccination. *American Journal of Veterinary Research* 75, 905 - 911.
[36] Oidtmann, B.C., Thrush, M.A., Denham, K.L. and Peeler, E.J. (2011) International and national biosecurity strategies in aquatic animal health. *Aquaculture* 320, 22 - 33.
[37] OIE (2012) Chapter 2.3.6. Koi herpesvirus disease. In: *Manual of Diagnostic Tests for Aquatic Animals*. World Organisation for Animal Health, Paris. Updated 2016 version available as Chapter 2.3.7. at: http://www.oie.int/index.php?id=2439&L=0&htmfile=chapitre_koi_herpesvirus.htm (accessed 2 November 2016).
[38] OIE (2015) *International Aquatic Animal Health Code*. World Organisation for Animal Health, Paris. Updated 2016 version available at: http://www.oie.int/en/international-standard-setting/aquatic-code/accessonline/(accessed 2 November 2016).
[39] Olesen, N. J., Mikkelsen, S. S., Vendramin, N., Bergmann, S., Way, K. and Engelsma, M. (2014) *Report from Meeting on Sampling and Diagnostic Procedures for the Surveillance and Confirmation of KHV Disease, Copenhagen, February 25 - 26th 2014*. Available at: http://www.eurl-fish.eu/-/media/Sites/EURL-FISH/english/diagnostic%20 manuals/khv_disease/MEETING-REPORT-11 - 03 - 14-Final.ashx?la=da (accessed 2 November 2016).
[40] Perelberg, A., Ronen, A., Hutoran, M., Smith, Y. and Kotler, M. (2005) Protection of cultured *Cyprinus carpio* against a lethal viral disease by an attenuated virus vaccine. *Vaccine* 23, 3396 - 3403.

[41] Perelberg, A., Ilouze, M., Kotler, M. and Steinitz, M. (2008) Antibody response and resistance of *Cyprinus carpio* immunized with cyprinid herpes virus 3 (CyHV－3). *Vaccine* 26, 3750－3756.
[42] Piackova, V., Flajshans, M., Pokorova, D., Reschova, S., Gela, D. *et al*. (2013) Sensitivity of common carp, *Cyprinus carpio* L., strains and crossbreeds reared in the Czech Republic to infection by cyprinid herpesvirus 3 (CyHV－3; KHV). *Journal of Fish Diseases* 36, 75－80.
[43] Pikarsky, E., Ronen, A., Abramowitz, J., Levavi-Sivan, B., Hutoran, M. *et al*. (2004) Pathogenesis of acute viral disease induced in fish by carp interstitial nephritis and gill necrosis virus. *Journal of Virology* 78, 9544－9551.
[44] Pionnier, N., Adamek, M., Miest, J.J., Harris, S.J., Matras, M. *et al*. (2014) C-reactive protein and complement as acute phase reactants in common carp *Cyprinus carpio* during CyHV－3 infection. *Diseases of Aquatic Organisms* 109, 187－199.
[45] Rakus, K.L., Irnazarow, I., Adamek, M., Palmeira, L., Kawana, Y. *et al*. (2012) Gene expression analysis of common carp (*Cyprinus carpio* L.) lines during cyprinid herpesvirus 3 infection yields insights into differential immune responses. *Developmental and Comparative Immunology* 37, 65－76.
[46] Reed, A. N., Izume, S., Dolan, B. P., LaPatra, S., Kent, M. *et al*. (2014) Identification of B cells as a major site for cyprinid herpesvirus 3 latency. *Journal of Virology* 88, 9297－9309.
[47] Ronen, A., Perelberg, A., Abramowitz, J., Hutoran, M., Tinman, S. *et al*. (2003) Efficient vaccine against the virus causing a lethal disease in cultured *Cyprinus carpio*. *Vaccine* 21, 4677－4684.
[48] Shapira, Y., Magen, Y., Zak, T., Kotler, M., Hulata, G. *et al*. (2005) Differential resistance to koi herpes virus (KHV)/carp interstitial nephritis and gill necrosis virus (CNGV) among common carp (*Cyprinus carpio* L.) strains and crossbreds. *Aquaculture* 245, 1－11.
[49] Soliman, H. and El-Matbouli, M. (2010) Loop mediated isothermal amplification combined with nucleic acid lateral flow strip for diagnosis of cyprinid herpes virus-3. *Molecular and Cellular Probes* 24, 38－43.
[50] St-Hilaire, S., Beevers, N., Joiner, C., Hedrick, R.P. and Way, K. (2009) Antibody response of two populations of common carp, *Cyprinus carpio* L., exposed to koi herpesvirus. *Journal of Fish Diseases* 32, 311－320.
[51] Sunarto, A., Rukyani, A. and Itami, T. (2005) Indonesian experience on the outbreak of koi herpesvirus in koi and carp (*Cyprinus carpio*). *Bulletin of Fisheries Research Agency* [Japan], Supplement No. 2, 15－21.
[52] Syakuri, H., Adamek, M., Brogden, G., Rakus, K.L., Matras, M. *et al*. (2013) Intestinal barrier of carp (*Cyprinus carpio* L.) during a cyprinid herpesvirus 3-infection: molecular identification and regulation of the mRNA expression of claudin encoding genes. *Fish and Shellfish Immunology* 34, 305－314.
[53] Uchii, K., Minamoto, T., Honjo, M.N. and Kawabata, Z. (2014) Seasonal reactivation enables cyprinid herpesvirus 3 to persist in a wild host population. *Fems Microbiology Ecology* 87, 536－542.
[54] Vrancken, R., Boutier, M., Ronsmans, M., Reschner, A., Leclipteux, T. *et al*. (2013) Laboratory validation of a lateral flow device for the detection of CyHV－3 antigens in gill swabs. *Journal of Virological Methods* 193, 679－682.
[55] Weber, E.P.S., Malm, K.V., Yun, S.C., Campbell, L.A., Kass, P.H. *et al*. (2014) Efficacy and safety of a modified-live cyprinid herpesvirus 3 vaccine in koi (Cyprinus carpio koi) for prevention of koi herpesvirus disease. *American Journal of Veterinary Research* 75, 899－904.

[56] Xu, J.R., Bently, J., Beck, L., Reed, A., Miller-Morgan, T. *et al.* (2013) Analysis of koi herpesvirus latency in wild common carp and ornamental koi in Oregon, USA. *Journal of Virological Methods* 187, 372 - 379.
[57] Yasumoto, S., Kuzuya, Y., Yasuda, M., Yoshimura, T. and Miyazaki, T. (2006) Oral immunization of common carp with a liposome vaccine fusing koi herpesvirus antigen. *Fish Pathology* 41, 141 - 145.
[58] Yoshida, N., Sasaki, R.K., Kasai, H. and Yoshimizu, M. (2013) Inactivation of koi-herpesvirus in water using bacteria isolated from carp intestines and carp habitats. *Journal of Fish Diseases* 36, 997 - 1005.
[59] Yoshino, M., Watari, H., Kojima, T., Ikedo, M. and Kurita, J. (2009) Rapid, sensitive and simple detection method for koi herpesvirus using loop-mediated isothermal amplification. *Microbiology and Immunology* 53, 375 - 383.
[60] Yuasa, K., Sano, M., Kurita, J., Ito, T. and Iida, T. (2005) Improvement of a PCR method with the Sph 1 - 5 primer set for the detection of koi herpesvirus (KHV). *Fish Pathology* 40, 37 - 39.
[61] Yuasa, K., Sano, M. and Oseko, N. (2012) Effective procedures for culture isolation of koi herpesvirus (KHV). *Fish Pathology* 47, 97 - 99.
[62] Zak, T., Perelberg, A., Magen, I., Milstein, A. and Joseph, D. (2007) Heterosis in the growth rate of Hungarian - Israeli common carp crossbreeds and evaluation of their sensitivity to koi herpes virus (KHV) disease. *Israeli Journal of Aquaculture* 59, 63 - 72.

11

病毒性脑病和视网膜病

Anna Toffan*

11.1 引言

病毒性脑病和视网膜病(viral encephalopathy and retinopathy, VER),又称为病毒性神经坏死病(viral nervous necrosis, VNN),是由野田病毒科(Nodaviridae)β-野田病毒属(*Betanodavirus*) RNA 病毒引起的一种严重神经病理学疾病。该病毒在 20 世纪 80 年代末被发现,现已传播至世界各地,成为地方性流行病,是一些国家海水养殖业的主要制约因素。最近,由于宿主范围扩大和缺乏适当有效的预防措施,该疾病已被列为有鳍鱼类中最重要的病毒性病害之一 (Rigos and Katharios, 2009; Walker and Winton, 2010; Shetty et al., 2012)。

11.2 传染病原

VER 的病原体是一种小的球形(直径为 25～30 nm)无包膜病毒颗粒,具有由两个单链正义 RNA 分子组成的双分段基因组。野田病毒的命名来源于日本的野田(Nodamura)村,原型病毒最先从当地的三带喙库蚊(*Culex tritaeniorhynchus*)中分离出来。1992 年,从黄带拟鲹(*Pseudocaranx dentex*)幼苗中分离出一种不同类型的野田病毒,该病毒被称为条纹鲹神经坏死病毒(Mori et al., 1992)。随后的分子研究将这些病毒分为两个不同的属: α-野田病毒(*Alphanodavirus*)和 β-野田病毒(*Betanodavirus*),分别感染昆虫和鱼类(King et al., 2011)。最近,在印度的罗氏沼虾(*Macrobranchium rosenbergii*)中检测到第三种野田病毒属(*Gammanodavirus*)的病毒,但尚未被国际病毒分类委员会(ICTV) 接受(NaveenKumar et al., 2013)。

β-野田病毒基因组由两个可读框(ORF)构成: RNA1 区段(3.1 kb)负责编码 RNA 依赖性 RNA 聚合酶(RNA-dependent RNA-polymerase, RdRp),

* 作者邮箱: atoffan@izsvenezie.it。

RNA2 区段(1.4 kb)负责编码病毒衣壳蛋白(King et al., 2011)。RNA1 区段的转录明显发生在病毒周期的起始阶段,而衣壳的表达和产生及感染性病毒颗粒的增加发生在后期(Lopez-Jimena et al., 2011)。另一个 0.4 kb 的亚基因组转录体被称为 RNA3,在活跃的病毒复制过程中从 RNA1 分子上被切割出来,编码 B1 和 B2 蛋白,以拮抗宿主细胞 RNA 干扰机制(Iwamoto et al., 2005; Fenner et al., 2006a; Chen et al., 2009)。在感染后较早的时间点,细胞培养中 RNA3 的合成量比 RNA1 丰富得多(Sommerset and Nerland, 2004)。在受感染的细胞培养中以及最近在大西洋庸鲽(*Hippoglossus hippoglossus*)中,都只能够在感染早期阶段检测出 B2 非结构蛋白,而 β-野田病毒衣壳蛋白也存在于慢性感染的鱼中(Mézeth et al., 2009)。

根据 Nishizawa 等(1997)基于 RNA2 片段内 T4 可变区的系统发生分析,β-野田病毒有四种:条纹鲹神经坏死病毒(striped jack nervous necrosis virus, SJNNV)、红鳍东方鲀神经坏死病毒(tiger puffer nervous necrosis virus, TPNNV)、条斑星鲽神经坏死病毒(barfin flounder nervous necrosis virus, BFNNV)和赤点石斑鱼神经坏死病毒(redspotted grouper nervous necrosis virus,RGNNV)。β-野田病毒的基因型间和基因型内的基因重配都曾被检测到(Panzarin et al., 2012; He and Teng, 2015),其中两种重配病毒株是 RGNNV/SJNNV 和 SJNNV/RGNNV(Toffolo et al., 2007; Olveira et al., 2009; Panzarin et al., 2012)。这些重配病毒株可能是由两种不同病毒共存于同一宿主(很可能是野生鱼类)中而产生的,也可能是在 20 世纪 80 年代的一次单交换重组事件中造成的(Sakamoto et al., 2008; Lopez-Jimena et al., 2010; He and Teng, 2015)。其他野田病毒,如大西洋鳕神经坏死病毒(Atlantic cod nervous necrosis virus, ACNNV)、大西洋庸鲽神经坏死病毒(Atlantic halibut nervous necrosis virus, AHNV)、大菱鲆野田病毒(turbot nodavirus, TNV)以及其他许多病毒,研究者们都已有过研究并进行了叙述,但它们仍需要得到 ICTV 的认可(King et al., 2011)。

当使用多克隆抗体分型时,β-野田病毒可分为三种不同的血清型:血清型 A(SJNNV 基因型)、血清型 B(TPNNV 基因型)和血清型 C(RGNNV 和 BFNNV 基因型)(Mori et al., 2003)。有研究认为 RGNNV 与 SJNNV 之间存在抗原差异(Skliris et al., 2001; Chi et al., 2003; Costa et al., 2007),最近这一假设通过反向遗传病毒得到证实(Panzarin et al., 2016)。衣壳蛋白的 C 末端突出结构域似乎与不同的免疫反应性有关,并可能含有宿主特异性决定簇(Iwamoto et al., 2004; Ito et al., 2008; Bandín and Dopazo, 2011; Souto et al., 2015a)。上述重组病毒仍与 RNA2 供体毒株属于相同的抗原组。BFNNV 基因型最近被划归为血清型 B(Panzarin et al., 2016),这与 Mori 等

(2003)先前的报道相悖。需要进一步的研究来确定β-野田病毒的免疫反应性和分子决定簇。

遗传基因不同的β-野田病毒对环境温度的表现不同：BFNNV和TPNNV被认为是“冷水病毒性神经坏死病毒”，因为它们最嗜冷，最适培养温度为15～20℃。然而，SJNNV的最佳复制温度为25℃，RGNNV则具有极好的耐温性(15～35℃)，最适培养温度在25～30℃(Iwamoto et al., 2000; Hata et al., 2010)。重组病毒行为类似于RNA1供体毒株，这证明通过编码聚合酶，RNA1基因调节了鱼类β-野田病毒的温度依赖性(Panzarin et al., 2014)。

11.2.1 地理分布、宿主范围和传播途径

自1985年首次报道该病害以来，VNN在20世纪90年代几乎同时出现在亚洲、澳大利亚和南欧。到2000年年初，这种疾病已蔓延到北美和北欧，现在VNN几乎遍布世界各地，差不多影响了所有海洋鱼类养殖品种(Shetty et al., 2012)。

不同β-野田病毒的地理分布反映了它们的温度依赖性。RGNNV是最常见的VNN病毒，已在世界范围内引发临床疾病(Ucko et al., 2004; Sakamoto et al., 2008; Chérif et al., 2009; Gomez et al., 2009; Panzarin et al., 2012; Ransangan and Manin, 2012; Shetty et al., 2012; Binesh and Jithendran, 2013)。SJNNV为第二常见病毒株(Maeno et al., 2004; García-Rosado et al., 2007; Sakamoto et al., 2008)，其次是RGNNV/SJNNV病毒重配株，该病毒在伊比利亚半岛和地中海地区很常见(Olveira et al., 2009; Panzarin et al., 2012; Souto et al., 2015a)。值得注意的是，仅在地中海地区发现了由重配株引起的VNN暴发(He and Teng, 2015)。BFNNV仅发现于北大西洋、北海和日本海的寒冷水域中(Nguyen et al., 1994; Grotmol et al., 2000; Nylund et al., 2008)，而TPNNV仅在日本被发现(Nishizawa et al., 1995; Furusawa et al., 2007)。

已在24目79科的160种鱼类中检测到VNN(表11.1)。表11.1中列出的易感鱼种中，最常报道的科为鲹科(Carangidae)、鲈科(Percichthyidae)、鮨科(Serranidae)、石首鱼科(Sciaenidae)、鲽科(Pleuronectidae)、鲻科(Mugilidae)、平鲉科(Sebastidae)和鳕科(Gadidae)。事实上，受影响最普遍和最严重的鱼种是海鲈(*Dicentrarchus labrax* 和 *Lates calcarifer*)、石斑鱼(*Ephinephelus* spp.)、鲆鲽鳎类(*Solea* spp., *Scophtalmus maximus*, *Paralichthys olivaceus*)、条纹鲹(*Pseudocaranx dentex*, *Trachinotus* spp.)和石首鱼(*Umbrina cirrosa*, *Argyrososmus regius*, *Scienops ocellatus*, *Atractoscion nobilis*)。

表 11.1 病毒性神经坏死病(VNN)易感鱼类[a]

目	科	种	参考文献
鲟形目 Acipenseriformes	鲟科 Acipenseridae	俄罗斯鲟[b] *Acipenser gueldenstaedti*	Maltese and Bovo, 2007
鳗鲡目 Anguilliformes	鳗鲡科 Anguillidae	欧洲鳗鲡[b] *Anguilla Anguilla*	
	海鳗科 Muraenesocidae	海鳗[c] *Muraenesox cinereus*	Baeck et al., 2007
	鯙科 Muraenidae	五彩鳗[c] *Rhinomuraena quaesita*	Gomez et al., 2006
银汉鱼目 Atheriniformes	黑带银汉鱼科 Melanotaeniidae	薄唇虹银汉鱼[b,d] *Melanotaenia praecox*	Furusawa et al., 2007
		伊岛银汉鱼[b,d] *Iriatherina werneri*	
	沼银汉鱼科 Telmatherinidae	七彩霓虹[b,d] *Marosatherina ladigesi*	
蟾鱼目 Batrachoidiformes	蟾鱼科 Batrachoididae	腋孔蟾鱼[c] *Halobatrachus didactylus*	Moreno et al., 2014
颌针鱼目 Beloniformes	青鳉科 Adrianichthyidae	青鳉[b,d] *Oryzias latipes*	Furusawa et al., 2007
	颌针鱼科 Belonidae	颌针鱼[c] *Belone belone*	Ciulli et al., 2006a
金眼鲷目 Beryciformes	松球鱼科 Monocentridae	日本松球鱼[c] *Monocentris japonica*	Gomez et al., 2006
	棘鲷科 Trachichthyidae	红棘胸鲷 *Hoplostethus mediterraneus*	Giacopello et al., 2013
脂鲤目 Characiformes	锯脂鲤科 Serrasalmidae	红腹锯鲑脂鲤[b] *Pygocentrus nattereri*	Gomez et al., 2006
鲱形目 Clupeiformes	鲱科 Clupeidae	沙丁鱼[c] *Sardina pilchardus*	Ciulli et al., 2006a
	鳀科 Engraulidae	日本鳀[c] *Engraulis japonicus*	Gomez et al., 2006
鲤形目 Cypriniformes	鲤科 Cyprinidae	金鱼[b] *Carassius auratus*	Binesh, 2013
		斑马鱼[b] *Danio rerio*	Lu et al., 2008

续 表

目	科	种	参考文献
鳉形目 Cyprinodontiformes	花鳉科 Poeciliidae	孔雀花鳉 *Poecilia reticulate*	Maltese and Bovo, 2007
鳕形目 Gadiformes	鳕科 Gadidae	大西洋鳕 *Gadus morhua*	Munday et al., 2002
		太平洋鳕 *Gadus macrocephalus*	
		黑线鳕[c] *Melanogrammus aeglefinus*	Maltese and Bovo, 2007
		欧洲无须鳕[c] *Merluccius merluccius*	Ciulli et al., 2006a
		牙鳕[c] *Merlangius merlangus*	
		细长臀鳕[c] *Trisopterus minutus*	
	长尾鳕科 Macrouridae	大西洋膜首鳕[c] *Hymenocephalus italicus*	Giacopello et al., 2013
		多棘腔吻鳕[c] *Caelorinchus multispinulosus*	Baeck et al., 2007
鼠鱚目 Gonorynchiformes	虱目鱼科 Chanidae	虱目鱼 *Chanos chanos*	OIE, 2013
虎鲨目 Heterodontiformes	虎鲨科 Heterodontidae	宽纹虎鲨[c] *Heterodontus japonicus*	Gomez et al., 2004
鮟鱇目 Lophiiformes	鮟鱇科 Lophiidae	黄鮟鱇[c] *Lophius litulon*	Baeck et al., 2007
背棘鱼目 Notacanthiformes	背棘鱼科 Notacanthidae	波氏背棘鱼[c] *Notacanthus Bonaparte*	Giacopello et al., 2013
鲈形目 Perciformes	刺尾鲷科 Acanthuridae	横带刺尾鱼 *Acanthurus triostegus*	OIE, 2013
		黄高鳍刺尾鱼[c] *Zebrasoma flavescens*	Gomez et al., 2006
	发光鲷科 Acropomatidae	赤鲑[c] *Doederleinia berycoides*	Baeck et al., 2007
	攀鲈科 Anabantidae	龟壳攀鲈[b,d] *Anabas testudineus*	Furusawa et al., 2007

续 表

目	科	种	参考文献
鲈形目 Perciformes	狼鳚科 Anarhichadidae	花狼鱼 *Anarhichas minor*	OIE，2013
	天竺鲷科 Apogonidae	单线天竺鲷 *Apogon exostigma*	OIE，2013
		细条天竺鲷[c] *Apogon lineatus*	Baeck et al.，2007
	鳚科 Blenniidae	河鳚[b] *Salaria fluviatilis*	Vendramin et al.，2012
	鲹科 Carangidae	黄带拟鲹 *Pseudocaranx dentex*	Munday et al.，2002
		高体鰤 *Seriola dumerili*	
		镰鳍鲳鲹 *Trachinotus falcatus*	
		布氏鲳鲹 *Trachinotus blochii*	Maltese and Bovo，2007
		竹荚鱼属[c] *Trachurus* spp.	Ciulli et al.，2006a
		竹荚鱼[c] *Trachurus japonicus*	Baeck et al.，2007
		蓝圆鲹[c] *Decapterus maruadsi*	
		突颌月鲹[c] *Selene vomer*	
	鼠衔鱼科 Callionymidae	月斑鲔[c] *Callionymus lunatus*	Baeck et al.，2007
	太阳鱼科 Centrarchidae	大口黑鲈[b] *Micropterus salmoides*	Bovo et al.，2011
	丽鱼科 Cichlidae	尼罗罗非鱼[b] *Oreochromis niloticus*	OIE，2013
		神仙鱼[b,d] *Pterophyllum scalare*	Furusawa et al.，2007
		淡黑镊丽鱼[b,d] *Labidochromis caeruleus*	

续 表

目	科	种	参考文献
鲈形目 Perciformes	丽鱼科 Cichlidae	纵带黑丽鱼[b,d] *Melanochromis auratus*	
		隆氏拟丽鱼[b,d] *Maylandia lombardoi*	
	塘鳢科 Eleotridae	线纹尖塘鳢 *Oxyeleotris lineolata*	Munday et al., 2002
	白鲳科 Ephippidae	圆燕鱼 *Platax orbicularis*	OIE, 2013
	深海天竺鲷科 Epigonidae	少耙后竺鲷[c] *Epigonus telescopus*	Giacopello et al., 2013
	中国东部虾虎鱼科 Gobidae	黑虾虎鱼[c] *Gobius niger*	Ciulli et al., 2006a
	舵鱼科 Kyphosidae	柴鱼[c] *Microcanthus strigatus*	Gomez et al., 2004
	花鲈科 Lateolabracidae	花鲈 *Lateolabrax japonicus*	Maltese and Bovo, 2007
	尖吻鲈科 Latidae	尖吻鲈 *Lates calcarifer*	Munday et al., 2002
	婢鳊科 Latridae	条纹婢鳊 *Latris lineata*	
	鲾科 Leiognathidae	颈带鲾[c] *Leiognathus nuchalis*	Baeck et al., 2007
	笛鲷科 Lutjanidae	红鳍笛鲷 *Lutjanus erythropterus*	Maltese and Bovo, 2007
		紫红笛鲷[d] *Lutjanus argentimaculatus*	Maeno et al., 2007
	软棘鱼科 Malacanthidae	日本方头鱼 *Branchiostegus japonicus*	OIE, 2013
	狼鲈科 Moronidae	条纹鲈×金眼狼鲈[b] *Morone saxatilis* × *M. chrysops*	Bovo et al., 2011
		舌齿鲈 *Dicentrarchus labrax*	Munday et al., 2002

续 表

目	科	种	参考文献
鲈形目 Perciformes	鲻科 Mugilidae	鲻鱼 *Mugil cephalus*	Maltese and Bovo, 2007
		金鲮 *Liza aurata*	
		羊鱼 *Mullus barbatus*	OIE, 2013
		纵带羊鱼[c] *Mullus surmuletus*	Panzarin et al., 2012
		厚唇鲻[c] *Chelon labrosus*	
		跳鲮 *Liza saliens*	Zorriehzahra et al., 2014
		薄唇鲮[c] *Liza ramada*	Ciulli et al., 2006a
	石鲷科 Oplegnathidae	条石鲷 *Oplegnathus fasciatus*	Munday et al., 2002
		斑石鲷 *Oplegnathus punctatus*	
	丝足鲈科 Osphronemidae	恒河毛足鲈[b,d] *Trichogaster chuna*	Furusawa et al., 2007
		三星攀鲈[b,d] *Trichopodus trichopterus*	
		条纹短攀鲈[b,d] *Trichopsis pumila*	
		暹罗斗鱼[b,d] *Betta splendens*	
	真鲈科 Percichthydae	九斑麦氏鲈 *Macquaria novemaculeata*	Moody et al., 2009
		澳洲麦氏鲈[d] *Macquaria australasica*	Munday et al., 2002
		虫纹麦鳕鲈[d] *Maccullochella peelii*	
	河鲈科 Percidae	白梭吻鲈[b] *Sander lucioperca*	Bovo et al., 2011

续 表

目	科	种	参考文献
鲈形目 Perciformes	叶鲈科 Polycentridae	叶形鱼[b] *Monocirrhus polyacanthus*	Gomez et al., 2006
	雀鲷科 Pomacentridae	黑双带小丑鱼 *Amphiprion sebae*	Binesh et al., 2013
		霓虹雀鲷[c] *Pomacentrus coelestis*	Gomez et al., 2004
		三斑圆雀鲷[c] *Dascyllus trimaculatus*	Gomez et al., 2008b
	军曹鱼科 Rachycentridae	军曹鱼 *Rachycentron canadum*	Maltese and Bovo, 2007
	石首鱼科 Sciaenidae	美国红鱼 *Sciaenops ocellatus*	
		波纹短须石首鱼 *Umbrina cirrosa*	
		有名锤形石首鱼 *Atractoscion nobilis*	Munday et al., 2002
		大西洋白姑鱼 *Argyrosomus regius*	Thiéry et al., 2004
		银彭纳石首鱼[c] *Pennahia argentata*	Baeck et al., 2007
	鲭科 Scombridae	太平洋黑鲔 *Thunnus orientalis*	OIE, 2013
		日本鲭[c] *Scomber japonicus*	Baeck et al., 2007
	平鲉科 Sebastidae	椭圆平鲉 *Sebastes oblongus*	Maltese and Bovo, 2007
		许氏平鲉 *Sebastes schlegeli*	Gomez et al., 2004
		厚头平鲉 *Sebastes pachycephalus*	
		无备平鲉[c] *Sebastes inermis*	
		褐菖鲉[c] *Sebastiscus marmoratus*	

续 表

目	科	种	参考文献
鲈形目 Perciformes	鮨科 Serranidae	赤点石斑鱼 *Epinephelus akaara*	Munday et al., 2002
		青石斑鱼 *Epinephelus awoara*	
		七带石斑鱼 *Epinephelus septemfasciatus*	
		老虎斑 *Epinephelus fuscoguttatus*	
		黑带石斑鱼 *Epinephelus malabaricus*	
		乌鳍石斑鱼 *Epinephelus marginatus*	
		云纹石斑鱼 *Epinephelus moara*	
		鲈滑石斑鱼 *Epinephelus tauvina*	
		龙胆石斑鱼 *Epinephelus lanceolatus*	
		老鼠斑 *Chromileptes altivelis*	
		青铜石斑鱼 *Epinephelus aeneus*	OIE, 2013
		点带石斑鱼 *Epinephelus coioides*	
		地中海石斑鱼 *Epinephelus costae*	Vendramin et al., 2013
		斑刺棘鲈 *Plectropomus maculatus*	Pirarat et al., 2009a
	鱚科 Sillaginidae	少鳞鱚[c] *Sillago japonica*	Baeck et al., 2007
	鲷科 Sparidae	金头鲷 *Sparus aurata*	Munday et al., 2002
		绯小鲷[c] *Pagellus erythrinus*	Ciulli et al., 2006a

续 表

目	科	种	参考文献
鲈形目 Perciformes	鲷科 Sparidae	腋斑小鲷[c] *Pagellus acarne*	
		赤鲷 *Pagrus pagrus*	García-Rosado et al., 2007
		三长棘赤鲷 *Pagrus auriga*	
		真鲷[c] *Pagrus major*	Baeck et al., 2007
		重牙鲷 *Diplodus sargus*	Dalla Valle et al., 2005
		黑椎鲷[c] *Spondyliosoma cantharus*	Moreno et al., 2014
		牛眼鲷[c] *Boops boops*	Ciulli et al., 2006a
	蝲科 Terapontidae	银锯眶蝲[d] *Bidyanus bidyanus*	Munday et al., 2002
	镰鱼科 Zanclidae	镰鱼[c] *Zanclus cornutus*	Gomez et al., 2004
	绵鳚科 Zoarcidae	吉氏绵鳚[c] *Zoarces gilli*	Baeck et al., 2007
鲽形目 Pleuronectiformes	舌鳎科 Cynoglossidae	半滑舌鳎 *Cynoglossus semilaevis*	Li et al., 2014
		粗体舌鳎[c] *Cynoglossus robustus*	Baeck et al., 2007
	牙鲆科 Paralichthyidae	褐牙鲆 *Paralichthys olivaceus*	Munday et al., 2002
		五眼斑鲆[c] *Pseudorhombus pentophthalmus*	Baeck et al., 2007
	鲽科 Pleuronectidae	条斑星鲽 *Verasper moseri*	Munday et al., 2002
		美洲拟鲽 *Pseudopleuronectes americanus*	
		大西洋庸鲽 *Hippoglossus hippoglossus*	
		横滨鲽[c] *Pleuronectes yokohamae*	Baeck et al., 2007

续 表

目	科	种	参考文献
鲽形目 Pleuronectiformes	菱鲆科 Scophthalmidae	大菱鲆 *Scophthalmus maximus*	Munday et al., 2002
	鳎科 Soleidae	欧鳎 *Solea solea*	Maltese and Bovo, 2007
		塞内加尔鳎 *Solea senegalensis*	Ito et al., 2008
鳐形目 Rajiformes	鳐科 Rajidae	斑瓮鳐[c] *Okamejei kenojei*	Baeck et al., 2007
鲑形目 Salmoniformes	鲑科 Salmonidae	海鳟[c] *Salmo trutta trutta*	Panzarin et al., 2012
		大西洋鲑[d] *Salmo salar*	Korsnes et al., 2005
鲉形目 Scorpaeniformes	鲬科 Platycephalidae	鲬 *Platycephalus indicus*	Munday et al., 2002
	须蓑鲉科 Scorpaenidae	龙须蓑鲉[c] *Pterois lunulata*	Gomez et al., 2004
	毒鲉科 Synanceiidae	日本鬼鲉 *Inimicus japonicus*	
	鲂科 Triglidae	小眼绿鳍鱼[c] *Chelidonichthys spinosus*	Baeck et al., 2007
		细鳞绿鳍鱼[c] *Chelidonichthys lucerna*	Ciulli et al., 2006a
鲇形目 Siluriformes	鲇科 Siluridae	花鲇[b] *Parasilurus asotus*	Maltese and Bovo, 2007
		澳洲鳗鲇[b] *Tandanus tandanus*	Munday et al., 2002
海龙目 Syngnathiformes	虾鱼科 Centriscidae	条纹虾鱼[c] *Aeoliscus strigatus*	Gomez et al., 2006
鲀形目 Tetraodontiformes	鳞鲀科 Balistidae	钩鳞鲀属[c] *Balistapus* spp.	Panzarin et al., 2010
	二齿鲀科 Diodontidae	六斑二齿鲀[c] *Diodon holocanthus*	Gomez et al., 2004
	单角鲀科 Monacanthidae	绿鳍马面鲀[c] *Thamnaconus modestus*	

续　表

目	科	种	参考文献
鲀形目 Tetraodontiformes	单角鲀科 Monacanthidae	丝背细鳞鲀 *Stephanolepis cirrhifer*	Pirarat et al., 2009b
	四齿鲀科 Tetraodontidae	红鳍东方鲀 *Takifugu rubripes*	Munday et al., 2002
		星点东方鲀[c] *Takifugu niphobles*	Baeck et al., 2007
		豹纹多纪鲀[c] *Takifugu pardalis*	Gomez et al., 2004
		月尾兔头鲀[c] *Lagocephalus lunaris*	Baeck et al., 2007
海鲂目 Zeiformes	海鲂科 Zeidae	远东海鲂[c] *Zeus faber*	Baeck et al., 2007

[a]基于 www.fishbase.org 网站的鱼类分类。

[b]淡水物种。

[c]在无症状的野生鱼类中检测到 β-野田病毒。

[d]实验感染。

最近在淡水鱼类中出现了越来越多的关于 VNN 暴发的报道(Vendramin et al., 2012; Binesh, 2013; Pascoli et al., 2016)。β-野田病毒可引发疾病并感染大量鱼类,这与水的盐度无关(Furusawa et al., 2007; Maeno et al., 2007; Bovo et al., 2011)。在没有临床症状的野生鱼类中检测到 β-野田病毒的情况也很常见(如表 11.1 中列出的大量鱼的种类所示),这可能是其在自然界普遍存在的原因。患有神经系统疾病的野生鱼类[特别是石斑鱼(*Epinephelus* spp.)]的死亡率增加,并在这些鱼中检测出病毒,这是一个值得关注的问题(Gomez et al., 2009, Vendramin et al., 2013; Haddad-Boubaker et al., 2014; Kara et al., 2014)。

该疾病通过直接接触受感染的鱼、受污染的水和养殖设施进行水平传播。通过浸泡感染及肌肉或腹腔注射可以很容易地在实验室再现该疾病(Munday et al., 2002; Maltese and Bovo, 2007)。有证据表明,该疾病可通过摄入受感染的鱼或受污染的饲料经口传播(Shetty et al., 2012)。该病毒已在数种海洋无脊椎动物中被发现(Gomez et al., 2006, 2008a; Panzarin et al., 2012; Fichi et al., 2015)。事实上,该病毒对化学和物理因素(即加热、酸性和碱性消毒剂和杀毒剂)具有很高的抵抗力,因此,它不仅会污染海水、无脊椎动物和微生物,还会污染网箱、围网、水槽(池)和其他养殖设施(Maltese and Bovo, 2007)。

在病害中幸存下来的野生和养殖鱼类是"无 VNN"养殖场中最可能的感染

源。由于胁迫或水温的变化，无症状感染鱼的疾病和病毒脱落可能被多次激活(Johansen et al., 2004; Rigos and Katharios, 2009; Lopez-Jimena et al., 2010; Souto et al., 2015b)。已有研究报道了最易感鱼类中存在垂直传播(Shetty et al., 2012)，并在其性腺和精液中发现了该病毒。

11.3 感染诊断

11.3.1 临床症状

VNN 感染的临床症状包括皮肤颜色变化、厌食、嗜睡、游泳行为异常以及由脑部和视网膜病变引起的神经症状。一般来说，幼鱼最容易受到感染。在仔鱼/稚鱼中，疾病的发作可能是超急性的，唯一明显的症状是死亡率急剧上升。在大龄鱼中，疾病的发作可能较慢，累计死亡率较低。受感染的鱼类生长速度减慢，导致其体重/大小不均(Vendramin et al., 2014)，这会间接导致严重的经济损失，而这种损失往往被低估。鱼鳔过度膨胀是另一种常见的临床症状(Maltese and Bovo, 2007; Pirarat et al., 2009a; Hellberg et al., 2010; Vendramin et al., 2013)。

患病舌齿鲈(*D. labrax*)的临床症状包括厌食、体表变黑，以及旋转、转圈移动与阶段性嗜睡交替出现等特征性异常游动行为，另外，在水层中的深度和垂直位置均出现异常。同时，病鱼还会出现失明，在受到刺激时表现出过度兴奋(Péducasse et al., 1999; Athanassopoulou et al., 2003)。鱼的下颌、头部、眼睛和鼻子出现创伤性损伤(图 11.1)是其游泳能力受损的自然后果(Shetty et al., 2012)。鱼鳔的过度膨胀也会造成其游泳行为异常，导致鱼下沉或漂浮(Lopez-Jimena et al., 2011)。在鲈鱼仔鱼和稚鱼中，可以观察到其大脑(有时可通过头骨感知)或整个头部处于充血状态，以及与高死亡率相关的由肌肉过度收缩导致的典型"镰状体位"(Bovo, 2010)。死亡率因水温和鱼龄而异，疾病在孵化场中暴发可能带来毁灭性的后果，死亡率极高(80%～100%)。尽管成鱼中也曾遭受过严重的病害损失，但大龄鱼受到的影响通常较小(Munday et al., 2002; Chérif et al., 2009)。在海鲈中，该疾病几乎完全是由 RGNNV 基因型引起的，在水温高于 23～25 ℃时该基因型病毒会引发疾病，当温度低于 18～22 ℃时死亡率会下降(Bovo et al., 1999; Breuil et al.,

图 11.1 患有病毒性神经坏死病的养殖欧洲舌齿鲈(*Dicentrachus labrax*)幼鱼出现皮肤糜烂和头部充血症状

2001；Chérif et al.，2009)。欧洲鲈也可能感染其他VNN病毒株，当感染发生时，临床表现较轻微(Vendramin et al.，2014；Souto et al.，2015a)。

在自然感染的亚洲海鲈仔鱼(*Lates calcarifer*)中，可观察到其体表变黑，鱼在水面聚集，累计死亡率可达60%～100%。仔鱼也表现出厌食、身体灰白色素沉着、失去平衡及死前的螺旋状泳动等症状。所有鱼龄段的死亡率一般都在50%以上(Maeno et al.，2004；Azad et al.，2005；Parameswaran et al.，2008)。

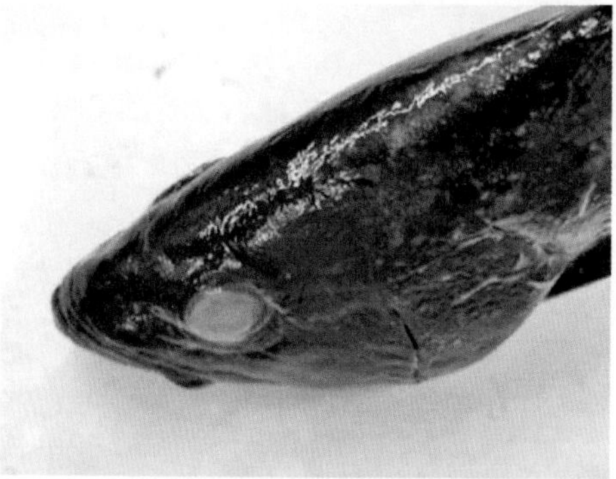

图11.2 地中海野生乌鰭石斑鱼(*Epinephelus marginatus*)的病毒性神经坏死病临床症状：头部皮肤糜烂、角膜混浊和全眼炎(包括眼内结构的所有眼外膜炎症)

石斑鱼是最易受VNN感染的鱼类之一，该病可发生在养殖和野生鱼类的任何鱼龄段，主要由RGNNV基因型引起。疾病的典型症状包括失去平衡、螺旋式泳动、嗜睡并伴随着对刺激的异常反应。鱼鳔的过度膨胀和角膜混浊是最相关的临床症状(图11.2)。脊柱畸形和突眼症也有相关报道(Sohn et al.，1998；Gomez et al.，2009；Pirarat et al.，2009a；Vendramin et al.，2013；Kara et al.，2014)。

鲆鲽类[鲽形目(Pleuronectiformes)]的临床症状不太明显。患病鱼会停留在水箱底部，身体弯曲伴随头部和尾部抬起，有时倒置在底部。他们可能会颤抖或像“落叶”一样掉落到水箱底部(Maltese and Bovo，2007)。据报道，患病大西洋庸鲽(*H. hippoglossus*)会出现皮肤脱色(Grotmol et al.，1997)。在自然暴发期间，欧鳎(*Solea solea*)表现出比平常更黑或更苍白的体色，间或出现厌食症，累计死亡率可达100%(Starkey et al.，2001)。塞内加尔鳎(*S. senegalensis*)对RGNNV/SJNNV极度易感，可能有皮肤溃疡(Olveira et al.，2008)，在22℃时死亡率为100%，但在16℃时死亡率会降低至8%。VNN也可造成鳎鱼的持续感染，将水温升高至22℃时可重新激活疾病(Souto et al.，2015a，b)。

石首鱼科鱼类，如波纹短须石首鱼(*Umbrina cirrosa*)和美国红鱼(*Sciaenops ocellatus*)对该病高度易感，幼鱼尤其易感(Bovo et al.，1999；Oh et al.，2002；Katharios and Tsigenopoulos，2010)。除了典型的临床症状外，有名锤形石首鱼(*Atractoscion nobilis*)在12～15℃时有明显的鱼鳔过度膨胀(Curtis et al.，2003)，相比之下，受感染的野生大西洋白姑鱼(*Argyrosomus regius*)并没有出现任何特殊的临床症状(Lopez-Jimena et al.，2010)。

一些冷水鱼类，例如鳕(*Gadus morhua*)和大西洋庸鲽(*H. hippoglossus*)，可在6～15℃时被感染，并表现出嗜睡、厌食、体表变黑、神经性临床症状、鱼鳔炎症和角膜混浊等体征(Grotmol et al., 1997; Patel et al., 2007; Hellberg et al., 2010)。

鲷科鱼类通常对临床疾病有抵抗力。例如,与感染β-野田病毒的欧洲鲈同居共养的金头鲷(*Sparus aurata*)从未出现死亡或临床症状(Castric et al., 2001; Aranguren et al., 2002; Ucko et al., 2004)。然而,海鲷易受实验感染(肌内注射 RGNNV 病毒),这可能会引起幼鱼死亡或者使其成为海鲈的非临床传染性宿主(Castric et al., 2001; Aranguren et al., 2002)。近来可观察到 VER 在鲷鱼仔鱼中的暴发增多,并伴有较高的死亡率(Beraldo et al., 2011; Toffan, 2015, 未发表的结果)。从海鲷中分离出的病毒株始终是 RGNNV / SJNNV,这表明这种重配株对海鲷的特殊适应性,这与在塞内加尔鳎中的发现一致(Souto et al., 2015a)。

自然感染或实验感染的淡水鱼类的临床症状与海水鱼类相似。

11.3.2 实验室诊断

VNN 病毒的分离一般是通过连续细胞培养进行的,最常用的细胞系是来自线鳢(*Ophicephalus striatus*)的 SSN-1 细胞(Frerichs et al., 1996)及其衍生的克隆细胞系 E-11(Iwamoto et al., 2000)。这两种细胞系的高度易感性归因于它们持续感染线鳢 SnRV 逆转录病毒(Lee et al., 2002; Nishizawa et al., 2008)。其他鱼类的连续细胞系(例如来自斜带石斑鱼的 GF-1 和其他细胞系,来自尖吻鲈的 SISS 和 ASBB 细胞系以及来自金头鲷的 SAF-1 细胞系)已成功开发并应用,但其中大部分尚未商业化,它们的诊断性能和应用还不得而知(Hasoon et al., 2011; Sano et al., 2011)。

适宜的孵育温度取决于 VNN 的基因型(Iwamoto et al., 2000; Ciulli et al., 2006b; Hata et al., 2010; Panzarin et al., 2014)。典型的致细胞病变效应(CPE)是 VNN 感染细胞的变暗和收缩。典型的迹象是感染细胞的细胞质中出现成簇的空泡,细胞逐渐聚集和从单层细胞上脱离,并演变为扩展的坏死灶。培养的病毒可以通过单克隆或多克隆抗体血清中和(Mori et al., 2003; V.Panzarin, Italy, 2015,个人通信)、荧光素结合抗体(Péducasse et al., 1999; Thiéry et al., 1999; Castric et al., 2001; Mori et al., 2003)、酶联免疫吸附试验(ELISA)(Fenner et al., 2006b)和分子生物学技术进行鉴定。ELISA 可以直接用于检测感染鱼的中枢神经系统(central nervous system, CNS)组织(Breuil et al., 2001; Nuñez-Ortiz et al., 2016),IFAT 同样也可以(Nguyen et al., 1997; Totland et al., 1999; Curtis et al., 2001; Johansen et al., 2003)。

现在可以使用逆转录聚合酶链反应(RT-PCR)和实时 RT-PCR(qRT-PCR)快速诊断 VNN。靶向 RNA2 区段 T4 可变区的 F2-R3 引物组已被广泛用于诊断目的(Nishizawa et al., 1994)。其他数种基于PCR的方案已被开发出来，以提高β-野田病毒诊断检测的灵敏度和特异性(Grotmol et al., 2000; Gomez et al., 2004; Dalla Valle et al., 2005; Cutrín et al., 2007)。经验证的qRT-PCR方法也可用于检测所有已知的VNN基因型(Grove et al., 2006a; Panzarin et al., 2010; Hick et al., 2011; Hodneland et al., 2011; Baud et al., 2015)。

病毒分离、组织病理学或免疫染色可用于VNN的诊断和确认，但在首次确诊该疾病时，建议组合采用至少两种不同的分子检测方法(如靶向病毒基因组不同区域的RT-PCR方案，或RT-PCR后测序)(OIE, 2013)。

可以使用ELISA或血清中和试验来检测实验免疫或感染鱼体内的抗体。ELISA是最常用的检测方法(Breuil et al., 2000; Watanabe et al., 2000; Grove et al., 2006b; Scapigliati et al., 2010)，而血清中和试验法通常用于评估鱼的体液免疫(Skliris and Richards, 1999; Tanaka et al., 2001; Kai et al., 2010)。

虽然不同鱼种中存在较大差异，但疫苗接种和自然暴露的鱼体内在一年后仍可检测到抗体(Breuil and Romestand, 1999; Breuil et al., 2000; Grove et al., 2006b; Kai et al., 2010)。然而，接种疫苗75 d后石斑鱼血清中和抗体滴度显著下降(Pakingking et al., 2010)。在温带鱼类中，夏季抗体滴度高于冬季，这表明应在夏季采集血清样本(Breuil et al., 2000)。由于缺乏可靠的研究，特异性抗体的检测尚未被视为评估鱼类病毒状态的常规筛查方法(OIE, 2013)。

11.4 病理学和病理生理学

11.4.1 肉眼可见的病变与微观病变

欧洲海鲈(*D. labrax*)感染VNN的临床表现包括鼻、下颌和头部充血、擦伤，有时甚至出现坏死。据报道，数种鱼类中呈现角膜混浊及游泳能力受损引起的身体和鳍上的皮肤糜烂等临床症状(Maltese and Bovo, 2007; Shetty et al., 2012)。剖检时，鱼鳔的过度膨胀现象在几乎所有易感种类中都很常见，CNS和脑膜充血是最相关的内部病变(图11.3)。

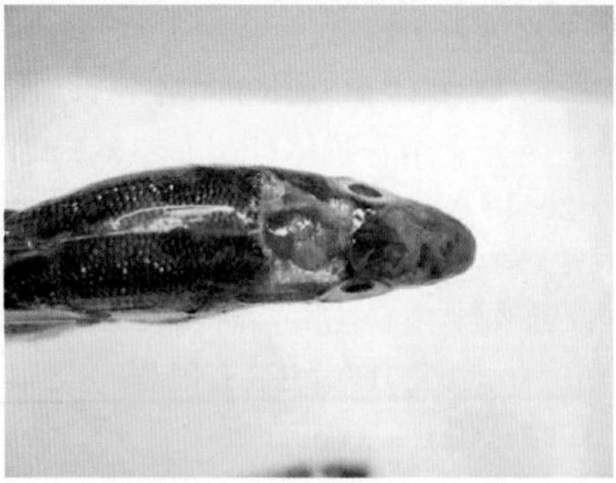

图11.3 患病毒性神经坏死病的欧洲海鲈(*Dicentrachus labrax*)呈现的脑部和脑膜充血

CNS脑炎以多个胞质内空泡化为特征，嗅球、端脑、间脑、中脑、小脑、延髓、脊髓和视网膜的灰质中均可见大量直径为5～10 μm的空白区，与周围区域明显分开(Munday et al., 2002; Grove et al., 2003; Mladineo, 2003; Maltese and Bovo, 2007; Lopez-Jimena et al., 2011)。空泡化的严重程度取决于鱼的种类、鱼龄和感染阶段,但值得注意的是,在 VNN 的病例中这是一致性的发现(几乎是病征性的)。感染鱼的所有神经组织均有核固缩、核破裂、神经元变性和炎性浸润的相关报道。嗅球、中脑视顶盖、小脑颗粒层和浦肯野细胞层及脊髓的运动神经元通常是受影响最大的区域(Totland et al., 1999; Pirarat et al., 2009a, b)。在亚洲海鲈感染的临床前期,CNS 和脊髓中存在深染和活跃分裂的神经元区域,未见空泡化。空泡在感染后期伴随着该疾病的神经症状一起出现(Azad et al., 2006)。脑实质和脑膜中的血管充血很常见,并可能演变为脑内轻微或大量出血(Mladineo, 2003; Korsnes et al., 2005; Pirarat et al., 2009b)。

免疫组织化学(immunohistochemistry, IHC)或原位杂交(*in situ* hybridization, ISH)可用于检测 CNS、视顶盖和小脑中空泡周围的病毒抗原(图 11.4),上述方法也可用于空泡不存在的情况(Grove et al., 2003; Mladineo, 2003; Pirarat et al., 2009a, b; Katharios and Tsigenopoulos, 2010; Lopez-Jimena et al., 2011)。TEM 下通常可观察到感染的脑和视网膜细胞胞质中膜结合野田病毒的晶体阵列(Grotmol et al., 1997; Tanaka et al., 2004; Ucko et al., 2004)。已经在脊髓,特别是颅部的树突状细胞中观察到了空泡病变(经免疫组织化学证实)(Grotmol et al., 1997)。脊髓病变可在颅内病变发生前或之后出现(Nguyen et al., 1996; Pirarat et al., 2009b)。

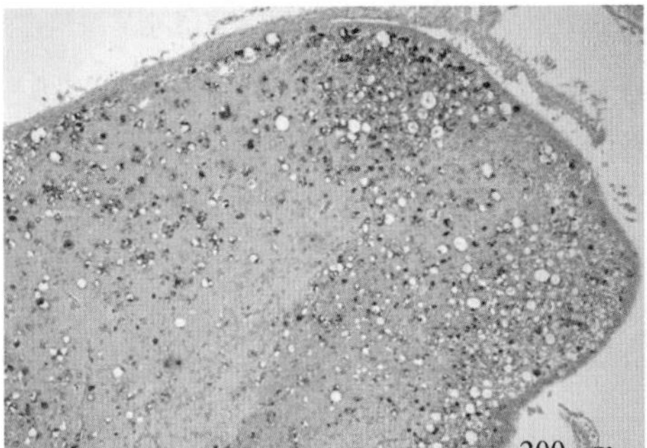

图 11.4 免疫组织化学检测欧洲海鲈(*Dicentrachus labrax*)脑中的 β-野田病毒(10 倍放大),可见大量的免疫沉淀和严重的嗅叶空泡化。照片由 Pretto Tobia 博士提供

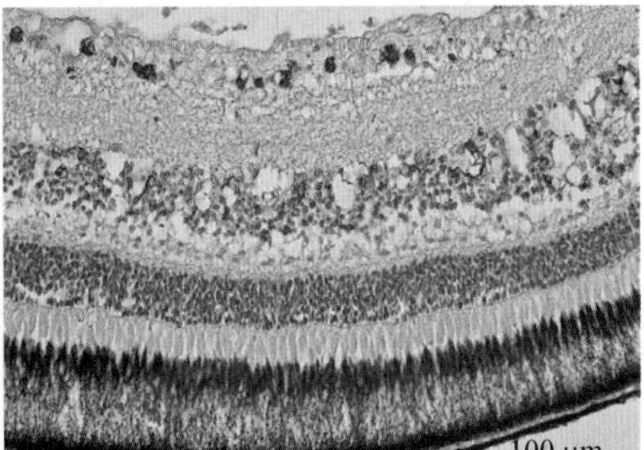

图 11.5 免疫组织化学检测欧洲海鲈(*Dicentrachus labrax*)视网膜中的 β-野田病毒(25 倍放大),内核层和神经节细胞层可见明显的免疫沉淀和空泡化。照片由 Pretto Tobia 博士提供

受感染鱼的视网膜也可能发生重大病变,这些病变具有显著的组织病理学变化,例如内外核层以及神经节细胞层中小圆形细胞的大量坏死和海绵状空泡化(图 11.5)。在某些情况下,内部网状层中也存在空泡(Pirarat et al., 2009a;

Katharios and Tsigenopoulos, 2010; Lopez-Jimena et al., 2011)。视神经可能表现出从广泛的空泡化到无明显病变的不同程度的改变(Mladineo, 2003; Lopez-Jimena et al., 2011)。

使用 IHC/ISH 或 PCR 在多种鱼类的鳃、鳍、心、前后肠、胃、脾、肝、肾和性腺等非神经组织中检测到病毒颗粒(Grotmol et al., 1997; Nguyen et al., 1997; Johansen et al., 2003; Mladineo, 2003; Grove et al., 2006a; Korsnes et al., 2009; Lopez-Jimena et al., 2011; Mazelet et al., 2011)。与神经组织不同,上述这些器官不是病毒复制的主要场所,因而认为它们不在疾病发病机制中起重要作用。消化道也可以在疾病传播中发挥作用(Nguyen et al., 1996; Totland et al., 1999)。β-野田病毒抗原在感染的丝背细鳞鲀(*Stephanolepis cirrhifer*)胃肠道细胞中强有力的存在,表明此处可能是病毒复制和传播特别活跃的部位(Pirarat et al., 2009b)。此外,卵巢等生殖组织中的免疫染色研究结果证实 VNN 存在垂直传播(Nguyen et al., 1997; Azad et al., 2005, 2006)。

尽管具有 VNN 临床症状的鱼的鱼鳔反复出现不可逆的过度膨胀,但这并不被认为是野田病毒的靶器官,因为除了轻微充血或存在嗜酸性核内包涵体外,从未观察到其他组织发生病理学病变(Pirarat et al., 2009a)。

11.4.2 致病机制

尽管最近对 VNN 的发病机制进行了一些研究,但该疾病发生的重要病毒机制仍不清楚。GF-1 细胞系通常表达的热激蛋白 GHSC70 是良好的 VNN 受体(Chang and Chi, 2015),但是,可能还有其他膜蛋白充当病毒受体。关于β-野田病毒的神经趋性无可辩驳,但其通往神经组织的途径仍是有待讨论的问题。鼻腔、肠上皮和完整皮肤等组织可能是侵染入口。Tanaka 等(2004)经鼻感染七带石斑鱼(*Epinephelus septemfasciatus*)的研究,假设病毒首先穿透鼻上皮,到达嗅神经和嗅球,最后侵入嗅叶并在那里进行复制。在自然感染的七带石斑鱼、斑鳃棘鲈(*Plectropomus maculatus*)和欧洲海鲈仔鱼中也提出了这种侵入途径,它们的病理变化首先在嗅叶中发生 (Mladineo, 2003; Banu et al., 2004; Pirarat et al., 2009a)。

皮肤上皮细胞和肠上皮细胞也可能是 VNN 的侵染入口(Totland et al., 1999;Grove et al., 2003)。皮肤和体侧线出现的 IHC 阳性反应令一些学者认为 VNN 可以通过皮肤的上皮细胞进入鱼体内(Nguyen et al., 1997; Péducasse et al., 1999; Totland et al., 1999; Azad et al., 2006; Kuo et al., 2011),但这个假设仍需进一步的研究。

一旦进入其宿主,β-野田病毒便会增殖并通过病毒血症阶段扩散到靶器官,这一点已被血液中的病毒检测所广泛证实(Lu et al., 2008; Olveira et al., 2008; Korsnes et al., 2009; Lopez-Jimena et al., 2011)。在大西洋庸鲽中报道了可能由病毒血症引发的复发性心内膜病变(Grotmol et al., 1997)。另一种假设是该病毒通过颅神经的轴突运输直接到达中枢神经系统(Grotmol et al., 1999; Húsgaro et al., 2001; Tanaka et al., 2004),然后病毒主动复制到靶神经器官中,在这里空泡化的表现通常与临床症状的显现相关(Azad et al., 2006)。

β-野田病毒感染可强烈诱导先天免疫(Lu et al., 2008; Scapigliati et al., 2010; Chen et al., 2014),继此首次短期的非特异性免疫反应之后,在不同鱼类中检测到特异性的循环VNN病毒抗体的产生。这些诱导产生的抗体可中和病毒活性,但是它们是否能够克服血脑屏障以保护中枢神经系统免受损伤并清除感染仍存在争议。事实上,β-野田病毒试图通过隐藏在神经组织中来躲避宿主的保护系统,在神经组织中病毒会产生潜伏感染(Chen et al., 2014)。因此,携带者状态的出现需要进一步的研究。

有报道称石斑鱼亲鱼接种疫苗后,母源性免疫可转移至鱼卵和鱼苗中,在欧洲海鲈中也提出了这种假设(Breuil et al., 2001; Kai et al., 2010),但是,这同样需要进一步的研究以了解其持续时间和有效性。

11.5 防控策略

在实验室阶段已获得了一些有前景的VNN疫苗,涵盖全病毒灭活、重组衣壳蛋白和病毒样颗粒(由不同的载体微生物产生)等多种产品形式,这些疫苗通过肌内注射和腹腔注射方式接种(Gomez-Casado et al., 2011; Shetty et al., 2012; Chen et al., 2014)。

在17℃的低温下养殖的石斑鱼中使用活疫苗取得了良好的效果。然而,由于β-野田病毒具有持续感染的能力,因此疫苗接种不能确保潜伏感染的清除(Nishizawa et al., 2012; Oh et al., 2013)。

如第11.4.1节所述,幼稚鱼和仔鱼对VNN病毒更易感,在鱼类的这些生长阶段进行疫苗注射接种是最不切实际的。为了克服这个问题,近来有人进行了浸泡和口服免疫接种的开发测试,但是现有数据并未证实这两种接种方式的有效性(Gomez-Casado et al., 2011; Nishizawa et al., 2012; Kai et al., 2014; Wi et al., 2015)。

由于尚无针对β-野田病毒的市售疫苗,该疾病在鱼类密集养殖区(例如地中海盆地、日本和中国台湾地区海岸及东南亚)的传播呈指数级增长,并且正如在中国台湾的石斑鱼种群中观察到的那样,非携带病毒鱼类的可获得性可能在降低(Chen et al., 2014)。

在缺乏有效疫苗的情况下,防止该疾病在养鱼场引入和传播的最佳方法是遵守严格的生物安全制度。在孵化场中,生物安全必须发挥主要作用,VNN的意外引入可能摧毁整个亲鱼种群。最重要的安全措施包括进水消毒、将孵化场划分为有专门人员和设施的若干区域、定期对养殖水槽（池）和设备进行消毒、定期休耕、对新引进的亲鱼进行隔离检疫和检测，以及对冷冻或活饵料严格控制（Munday et al.，2002；Maltese and Bovo，2007；Shetty et al.，2012）。显然，这些措施并不适用于潟湖和海上网箱等开放环境。尽管如此，在开放环境中，生物安全预防措施（例如快速和适当地处置尸体、降低养殖密度、降低饲喂率、定期休耕）和细心的管理可能有助于减轻防疫压力并最大限度减少损失。

需要强调的是，β-野田病毒对各种环境条件都具有极强的抵抗力（Arimoto et al.，1996；Munday et al.，2002；Liltved et al.，2006）。它们可以耐受极端的pH变化，在pH为3～7的情况下可存活6周，在pH为2～11时可存活24 h以上；β-野田病毒还可忍受极端温度条件，在25 ℃时可存活超过一个月，在15 ℃时能够存活超过一年（Frerichs，2000）。需要在60 ℃或更高温度下热处理至少30 min才有可能使VNN病毒失活（Arimoto et al.，1996），因此热灭活不宜作为首选的清洁方法。碘伏和次氯酸盐溶液具有最佳的杀毒效果，在20 ℃下使用浓度为50 ppm的溶液处理10 min，可完全有效地灭活病毒（Munday et al.，2002），而福尔马林的效率较低，它需要浓度、温度和处理时间的合适组合才能达到杀毒效果，因而在实践中难以应用（Arimoto et al.，1996；Frerichs，2000；Maltese and Bovo，2007；Kai and Chi，2008）。

紫外线照射和臭氧处理是可行的消毒措施（Arimoto et al.，1996；Frerichs，2000）。应特别注意的是对鱼卵进行消毒。使用臭氧（4 mg /L，0.5 min）处理实验感染的庸鲽鱼卵可中和病毒，对孵化能力几乎没有影响（Grotmol and Totland，2000）。使用臭氧还成功地对实验感染的黑线鳕（*Melanogrammus aeglefinus*）鱼卵进行了消毒（Buchan et al.，2006），但臭氧的安全剂量因鱼种而异（Grotmol et al.，2003）。然而，臭氧化对于VNN病毒的完全灭活并不总是有效的，实施灭活实验和海水中氧化剂残留测定等标准化程序的缺乏导致了不同实验结果之间存在差异（Liltved et al.，2006）。因此，需要在不同环境条件下进行更多试验，以获得可在养殖现场常规应用的安全有效的消毒方案。

由于在多种鱼类中均存在病毒的垂直传播，因此将新的亲鱼引入无疫病的养殖场时需要慎重考虑。卵巢活检或卵本身的血清学和分子生物学技术的某种组合可以为孵化场人员选择无病毒亲鱼提供标准（Mushiake et al.，1992；Breuil and Romestand，1999；Watanabe et al.，2000；Mazelet et al.，2011）。

抗病基因的遗传选育已经大规模应用于淡水鱼类（如鲑鱼）的抗病毒感染中，但针对β-野田病毒的相关研究还很少。尽管数据结构对于获得结论性结果并不是最优的，但仍观察到了VNN病毒攻毒后的大西洋鳕品系内存活遗传力的一个极高的评估值（Ødegård et al.，2010）。

11.6 总结与研究展望

目前,控制 VNN 的主要制约因素是缺乏免疫策略。迫切需要安全和廉价的商用疫苗,特别是针对最流行的 β-野田病毒基因型(RGNNV 和 SJNNV)的疫苗。

能否获得可靠的血清学检测手段对于更清楚地揭示鱼类体液反应和评估任何疫苗的免疫保护程度至关重要。它还将作为一种非侵入性诊断测试,以补充确认无病毒亲鱼种群的生物分子方案。需要制定、分享和应用标准化的采样、无特定病原种群认证和养殖场消毒的规程。

应更多关注养殖和野生鱼类种群之间病原体的相互作用和交换,以评估病毒从一种环境传播到另一种环境的风险,并更好地了解野生鱼类在 β-野田病毒生态学中的作用。

最后,需要使用遗传选育方案来增强不同鱼类对这种疾病的天然抗性。这种选育方案应被视为疫苗使用的一种协同手段,以有效遏制该疾病的传播和影响。

参考文献

[1] Aranguren, R., Tafalla, C., Novoa, B. and Figueras, A. (2002) Experimental transmission of encephalopathy and retinopathy induced by nodavirus to sea bream, *Sparus aurata* L., using different infection models. *Journal of Fish Diseases* 25, 317 - 324.

[2] Arimoto, M., Sato, J., Maruyama, K., Mimura, G. and Furusawa, I. (1996) Effect of chemical and physical treatments on the inactivation of striped jack nervous necrosis virus (SJNNV). *Aquaculture* 143, 15 - 22.

[3] Athanassopoulou, F., Billinis, C., Psychas, V. and Karipoglou, K. (2003) Viral encephalopathy and retinopathy of *Dicentrarchus labrax* (L.) farmed in fresh water in Greece. *Journal of Fish Diseases* 26, 361 - 365.

[4] Azad, I.S., Shekhar, M.S., Thirunavukkarasu, A.R., Poornima, M., Kailasam, M. *et al*. (2005) Nodavirus infection causes mortalities in hatchery produced larvae of *Lates calcarifer*: first report from India. *Diseases of Aquatic Organisms* 63, 113 - 118.

[5] Azad, I.S., Jithendran, K.P. and Shekhar, M.S. (2006) Immunolocalisation of nervous necrosis virus indicates vertical transmission in hatchery produced Asian sea bass (*Lates calcarifer* Bloch) - a case study. *Aquaculture* 255, 39 - 47.

[6] Baeck, G.W., Gornez, D.K., Oh, K.S., Kim, J.H., Choresca, C.H. *et al*. (2007) Detection of piscine nodaviruses from apparently healthy wild marine fish in Korea. *Bulletin of the European Association of Fish Pathologists* 27, 116 - 122.

[7] Bandín, I. and Dopazo, C.P. (2011) Host range, host specificity and hypothesized host shift events among viruses of lower vertebrates. *Veterinary Research* 42: 67.

[8] Banu, G., Mori, K., Arimoto, M., Chowdhury, M. and Nakai, T. (2004) Portal entry and progression of betanodaviruses causing viral nervous necrosis in sevenband grouper

Ephinephelus septemfasciatus. *Bangladesh Journal of Veterinary Medicine* 2, 83–87.
[9] Baud, M., Cabon, J., Salomoni, A., Toffan, A., Panzarin, V. and Bigarré, L. (2015) First generic one step real-time Taqman RT–PCR targeting the RNA1 of betanodaviruses. *Journal of Virological Methods* 211, 1–7.
[10] Beraldo, P., Panzarin, V., Galeotti, M. and Bovo, G. (2011) Isolation and molecular characterization of viral encephalopathy and retinopathy virus from gilthead sea bream larvae (*Sparua aurata*) showing mass mortalities. In *15th EAFP International Conference on Diseases of Fish and Shellfish, Split, Croatia. Conference Abstract Book*, p. 351. European Association of Fish Pathologists. Conference website available at: https://eafp.org/15th-eafp-spilt-2011/(accessed 3 November 2016).
[11] Binesh, C.P. (2013) Mortality due to viral nervous necrosis in zebrafish *Danio rerio* and goldfish *Carassius auratus*. *Diseases of Aquatic Organisms* 104, 257–260.
[12] Binesh, C.P. and Jithendran, K.P. (2013) Genetic characterization of betanodavirus isolates from Asian seabass *Lates calcarifer* (Bloch) in India. *Archives of Virology* 158, 1543–1546.
[13] Binesh, C.P., Renuka, K., Malaichami, N. and Greeshma, C. (2013) First report of viral nervous necrosis-induced mass mortality in hatchery-reared larvae of clownfish, *Amphiprion sebae* Bleeker. *Journal of Fish Diseases* 36, 1017–1020.
[14] Bovo, G., Nishizawa, T., Maltese, C., Borghesan, F., Mutinelli, F. *et al*. (1999) Viral encephalopathy and retinopathy of farmed marine fish species in Italy. *Virus Research* 63, 143–146.
[15] Bovo, G., Gustinelli, A., Quaglio, F., Gobbo, F., Panzarin, V. *et al*. (2011) Viral encephalopathy and retinopathy outbreak in freshwater fish farmed in Italy. *Diseases of Aquatic Organisms* 96, 45–54.
[16] Breuil, G. and Romestand, B. (1999) A rapid ELISA method for detecting specific antibody level against nodavirus in the serum of the sea bass, *Dicentrarchus labrax* (L.): application to the screening of spawners in a sea bass hatchery. *Journal of Fish Diseases* 22, 45–52.
[17] Breuil, G., Pepin, J.F., Castric, J., Fauvel, C. and Thiéry, R. (2000) Detection of serum antibodies against nodavirus in wild and farmed adult sea bass: application to the screening of broodstock in sea bass hatcheries. *Bulletin of the European Association of Fish Pathologists* 20, 95–100.
[18] Breuil, G., Mouchel, O., Fauvel, C. and Pepin, J.F. (2001) Sea bass *Dicentrarchus labrax* nervous necrosis virus isolates with distinct pathogenicity to sea bass larvae. *Diseases of Aquatic Organisms* 45, 25–31.
[19] Buchan, K.A.H., Martin-Robichaud, D.J., Benfey, T.J., MacKinnon, A.M. and Boston, L. (2006) The efficacy of ozonated seawater for surface disinfection of had-dock (*Melanogrammus aeglefinus*) eggs against piscine nodavirus. *Aquacultural Engineering* 35, 102–107.
[20] Castric, J., Thiéry, R., Jeffroy, J., de Kinkelin, P. and Raymond, J.C. (2001) Sea bream *Sparus aurata*, an asymptomatic contagious fish host for nodavirus. *Diseases of Aquatic Organisms* 47, 33–38.
[21] Chang, J.S. and Chi, S.C. (2015) GHSC70 is involved in the cellular entry of nervous necrosis virus. *Journal of Virology* 89, 61–70.
[22] Chen, L.J., Su, Y.C. and Hong, J.R. (2009) Betanodavirus non-structural protein B1: a novel anti-necrotic death factor that modulates cell death in early replication cycle in fish cells. *Virology* 385, 444–454.
[23] Chen, Y.M., Wang, T.Y. and Chen, T.Y. (2014) Immunity to betanodavirus infections of marine fish. *Developmental and Comparative Immunology* 43, 174–183.
[24] Chérif, N., Thiéry, R., Castric, J., Biacchesi, S., Brémont, M. *et al*. (2009) Viral

encephalopathy and retinopathy of *Dicentrarchus labrax* and *Sparus aurata* farmed in Tunisia. *Veterinary Research Communications* 33, 345 - 353.
[25] Chi, S. C., Shieh, J. R. and Lin, S. J. (2003) Genetic and antigenic analysis of betanodaviruses isolated from aquatic organisms in Taiwan. *Diseases of Aquatic Organisms* 55, 221 - 228.
[26] Ciulli, S., Di Marco, P., Natale, A., Galletti, E., Battilani, M. *et al.* (2006a) Detection and characterization of *Betanodavirus* in wild fish from Sicily, Italy. *Ittiopatologia* 3, 101 - 112.
[27] Ciulli, S., Gallardi, D., Scagliarini, A., Battilani, M., Hedrick, R.P. *et al.* (2006b) Temperature-dependency of *Betanodavirus* infection in SSN-1 cell line. *Diseases of Aquatic Organisms* 68, 261 - 265.
[28] Costa, J.Z., Adams, A., Bron, J.E., Thompson, K.D., Starkey, W.G. *et al.* (2007) Identification of B-cell epitopeson the betanodavirus capsidprotein. *Journal of Fish Diseases* 30, 419 - 426.
[29] Curtis, P.A., Drawbridge, M., Iwamoto, T., Nakai, T., Hedrick, R.P. *et al.* (2001) Nodavirus infection of juvenile white seabass, *Atractoscion nobilis*, cultured in southern California: first record of viral nervous necrosis (VNN) in North America. *Journal of Fish Diseases* 24, 263 - 271.
[30] Curtis, P.A., Drawbridge, M., Okihiro, M.S., Nakai, T., Hedrick, R.P. *et al.* (2003) Viralnervousnecrosis (VNN) in white seabass, *Atractoscion nobilis*, cultured in Southern California, and implications for marine fish aquaculture. *Journal of Fish Diseases* 24, 263 - 271.
[31] Cutrín, J. M., Dopazo, C. P., Thiéry, R., Leao, P., Olveira, J. G. *et al.* (2007) Emergence of pathogenic betanod-aviruses belonging to the SJNNV genogroup in farmed fish species from the Iberian Peninsula. *Journal of Fish Diseases* 30, 225 - 232.
[32] Dalla Valle, L., Toffolo, V., Lamprecht, M., Maltese, C., Bovo, G. *et al.* (2005) Development of a sensitive and quantitative diagnostic assay for fish nervous necrosis virus based on two-target real-time PCR. *Veterinary Microbiology* 110, 167 - 179.
[33] Fenner, B., Thiagarajan, R., Chua, H., and Kwang, J. (2006a) Betanodavirus B2 is an RNA interference antagonist that facilitates intracellular viral RNA accumulation. *Journal of Virology* 80, 85 - 94.
[34] Fenner, B. J., Du, Q., Goh, W., Thiagarajan, R., Chua, H. K. *et al.* (2006b) Detection of betanodavirus in juvenile barramundi, *Lates calcarifer* (Bloch), by antigen capture ELISA. *Journal of Fish Diseases* 29, 423 - 432.
[35] Fichi, G., Cardeti, G., Perrucci, S., Vanni, A., Cersini, A. *et al.* (2015) Skin lesion-associated pathogens from *Octopus vulgaris*: first detection of *Photobacterium swingsii*, *Lactococcus garvieae* and betanodavirus. *Diseases of Aquatic Organisms* 115, 147 - 156.
[36] Frerichs, G. (2000) Temperature, pH and electrolyte sensitivity, and heat, UV and disinfectant inactivation of sea bass (*Dicentrarchus labrax*) neuropathy nodavirus. *Aquaculture* 185, 13 - 24.
[37] Frerichs, G.N., Rodger, H.D. and Peric, Z. (1996) Cell culture isolation of piscine neuropathy nodavirus from juvenile sea bass, *Dicentrarchus labrax*. *Journal of General Virology* 77, 2067 - 2071.
[38] Furusawa, R., Okinaka, Y., Uematsu, K. and Nakai, T. (2007) Screening of freshwater fish species for their susceptibility to a betanodavirus. *Diseases of Aquatic Organisms* 77, 119 - 125.
[39] García-Rosado, E., Cano, I., Martín-Antonio, B., Labella, A., Manchado, M. *et al.* (2007) Co-occurrence of viral and bacterial pathogens in disease outbreaks affecting newly cultured sparid fish. *International Microbiology* 10, 193 - 199.
[40] Giacopello, C., Foti, M., Bottari, T., Fisichella, V. and Barbera, G. (2013) Detection

of viral encephalopathy and retinopathy virus (VERV) in wild marine fish species of the south Tyrrhenian Sea (centralMediterranean). *Journal of Fish Diseases* 36, 819 - 821.
[41] Gomez, D.K., Sato, J., Mushiake, K., Isshiki, T., Okinaka, Y. *et al.* (2004) PCR-based detection of betanodaviruses from cultured and wild marine fish with no clinical signs. *Journal of Fish Diseases* 27, 603 - 608.
[42] Gomez, D. K., Lim, D. J., Baeck, G. W., Youn, H. J., Shin, N. S. *et al.* (2006) Detection of betanodaviruses in apparently healthy aquarium fishes and invertebrates. *Journal of Veterinary Science* 7, 369 - 374.
[43] Gomez, D.K., Baeck, G.W., Kim, J.H., Choresca, C.H. and Park, S.C. (2008a) Molecular detection of betan-odaviruses from apparently healthy wild marine invertebrates. *Journal of Invertebrate Pathology* 97, 197 - 202.
[44] Gomez, D.K., Baeck, G.W., Kim, J.H., Choresca, C.H. and Park, S.C. (2008b) Molecular detection of betanodavirus in wild marine fish populations in Korea. *Journal of Veterinary Diagnostic Investigation* 20, 38 - 44.
[45] Gomez, D.K., Matsuoka, S., Mori, K., Okinaka, Y., Park, S.C. *et al.* (2009) Genetic analysis and pathogenicity of betanodavirus isolated from wild redspotted grouper *Epinephelus akaara* with clinical signs. *Archives of Virology* 154, 343 - 346.
[46] Gomez-Casado, E., Estepa, A. and Coll, J. M. (2011) A comparative review on European-farmed finfish RNA viruses and their vaccines. *Vaccine* 29, 2657 - 2671.
[47] Grotmol, S. and Totland, G. K. (2000) Surface disinfection of Atlantic halibut *Hippoglossus hippoglossus* eggs with ozonated sea-water inactivates nodavirus and increases survival of the larvae. *Diseases of Aquatic Organisms* 39, 89 - 96. Available at: http://www.ncbi. nlm.nih.gov/pubmed/10715814.
[48] Grotmol, S., Totland, G.K., Thorud, K. and Hjeltnes, B.K. (1997) Vacuolating encephalopathy and retinopathy associated with a nodavirus-like agent: a probable cause of mass mortality of cultured larval and juvenile Atlantic halibut *Hippoglossus hippoglossus*. *Disease of Aquatic Organisms* 29, 85 - 97.
[49] Grotmol, S., Bergh, O. and Totland, G. K. (1999) Transmission of viral encephalopathy and retinopathy (VER) to yolk-sac larvae of the Atlantic halibut *Hippoglossus hippoglossus*: occurrence of nodavirus in various organs and a possible route of infec-tion. *Diseases of Aquatic Organisms* 36, 95 - 106.
[50] Grotmol, S., Nerland, A.H., Biering, E., Totland, G.K. and Nishizawa, T. (2000) Characterization of the capsid protein gene from a nodavirus strain affecting the Atlantic halibut *Hippoglossus hippoglossus* and design of an optimal reverse-transcriptase polymerase chain reaction (RT - PCR) detection assay. *Diseases of Aquatic Organisms* 39, 79 - 88.
[51] Grotmol, S., Dahl-Paulsen, E. and Totland, G.K. (2003) Hatchability of eggs from Atlantic cod, turbot and Atlantic halibut after disinfection with ozonated sea-water. *Aquaculture* 221, 245 - 254.
[52] Grove, S., Johansen, R., Dannevig, B.H., Reitan, L.J. and Ranheim, T. (2003) Experimental infection of Atlantic halibut *Hippoglossus hippoglossus* with nodavirus: tissue distribution and immune response. *Diseases of Aquatic Organisms* 53, 211 - 221.
[53] Grove, S., Faller, R., Soleim, K.B. and Dannevig, B.H. (2006a) Absolute quantitation of RNA by a competitive real-time RT - PCR method using piscine nodavirus as a model. *Journal of Virological Methods* 132, 104 - 112.
[54] Grove, S., Johansen, R., Reitan, L.J., Press, C.M. and Dannevig, B.H. (2006b) Quantitative investigation of antigen and immune response in nervous and lymphoid tissues of Atlantic halibut (*Hippoglossus hip- poglossus*) challenged with nodavirus. *Fish Shellfish Immunology* 21, 525 - 539.
[55] Haddad-Boubaker, W., Sondès Boughdir, W., Sghaier, S., Souissi, J., Aida Megdich,

B. *et al*. (2014) Outbreak of viral nervous necrosis in endangered fish species *Epinephelus costae* and *E. marginatus* in Northern Tunisian coasts. *Fish Pathology* 49, 53 - 56.

[56] Hasoon, M.F., Daud, H.M., Abdullah, A.A., Arshad, S.S. and Bejo, H.M. (2011) Development and partial characteri-zation of new marine cell line from brain of Asian sea bass *Lates calcarifer* for virus isolation. *In Vitro Cellular and Developmental Biology - Animal* 47, 16 - 25.

[57] Hata, N., Okinaka, Y., Iwamoto, T., Kawato, Y., Mori, K. I. *et al*. (2010) Identification of RNA regions that deter-mine temperature sensitivities in betanodaviruses. *Archives of Virology* 155, 1597 - 1606.

[58] He, M. and Teng, C.B. (2015) Divergence and codon usage bias of *Betanodavirus*, a neurotropic pathogen in fish. *Molecular Phylogenetics and Evolution* 83, 137 - 142.

[59] Hellberg, H., Kvellestad, A., Dannevig, B., Bornø, G., Modahl, I. *et al*. (2010) Outbreaks of viral nervous necrosis in juvenile and adult farmed Atlantic cod, *Gadus morhua* L., in Norway. *Journal of Fish Diseases* 33, 75 - 81.

[60] Hick, P., Tweedie, A. and Whittington, R.J. (2011) Optimization of *Betanodavirus* culture and enumeration in striped snakehead fish cells. *Journal of Veterinary Diagnostic Investigation* 23, 465 - 475.

[61] Hodneland, K., García, R., Balbuena, J.A., Zarza, C. and Fouz, B. (2011) Real-time RT - PCR detection of betan-odavirus in naturally and experimentally infected fish from Spain. *Journal of Fish Diseases* 34, 189 - 202.

[62] Húsgaro, S., Grotmol, S., Hjeltnes, B.K., Rødseth, O.M. and Biering, E. (2001) Immune response to a recombinant capsid protein of *Striped jack nervous necrosis virus* (SJNNV) in turbot *Scophthalmus maximus* and Atlantic halibut *Hippoglossus hippoglossus*, and evaluation of a vaccine against SJNNV. *Diseases of Aquatic Organisms* 45, 33 - 44.

[63] Ito, Y., Okinaka, Y., Mori, K.-I., Sugaya, T., Nishioka, T. et al. (2008) Variable region of betanodavirus RNA2 is sufficient to determine host specificity. *Diseases of Aquatic Organisms* 79, 199 - 205.

[64] Iwamoto, T., Nakai, T., Mori, K., Arimoto, M. and Furusawa, I. (2000) Cloning of the fish cell line SSN-1 for piscine nodaviruses. *Diseases of Aquatic Organisms* 43, 81 - 89.

[65] Iwamoto, T., Okinaka, Y., Mise, K., Mori, K.-I., Arimoto, M. *et al*. (2004) Identification of host-specificity determinants in betanodaviruses by using reassortants between striped jack nervous necrosis virus and sevenband grouper nervous necrosis virus. *Journal of Virology* 78, 1256 - 1262.

[66] Iwamoto, T., Mise, K., Takeda, A., Okinaka, Y., Mori, K. I. *et al*. (2005) Characterization of *Striped jack nervous necrosis virus* subgenomic RNA3 and biological activities of its encoded protein B2. *Journal of General Virology* 86, 2807 - 2816.

[67] Johansen, R., Amundsen, M., Dannevig, B.H. and Sommer, A.I. (2003) Acute and persistent experimen-tal nodavirus infection in spotted wolffish *Anarhichas minor*. *Diseases of Aquatic Organisms* 57, 35 - 41.

[68] Johansen, R., Sommerset, I., Tørud, B., Korsnes, K., Hjortaas, M.J. *et al*. (2004) Characterization of nodavirus and viral encephalopathy and retinopathy in farmed turbot, *Scophthalmus maximus* (L.). *Journal of Fish Diseases* 27, 591 - 601.

[69] Kai, Y.-H. and Chi, S.C. (2008) Efficacies of inactivated vaccines against betanodavirus in grouper larvae (*Epinephelus coioides*) by bath immunization. *Vaccine* 26, 1450 - 1457.

[70] Kai, Y. H., Su, H. M., Tai, K. T. and Chi, S. C. (2010) Vaccination of grouper broodfish (*Epinephelus tukula*) reduces the risk of vertical transmission by nervous

necrosis virus. *Vaccine* 28, 996 - 1001.
[71] Kai, Y.H., Wu, Y.C. and Chi, S.C. (2014) Immune gene expressions in grouper larvae (*Epinephelus coioides*) induced by bath and oral vaccinations with inacti-vated betanodavirus. *Fish and Shellfish Immunology* 40, 563 - 569.
[72] Kara, H.M., Chaoui, L., Derbal, F., Zaidi, R., de Boisséson, C. *et al.* (2014) Betanodavirus-associated mortalities of adult wild groupers *Epinephelus margi-natus* (Lowe) and *Epinephelus costae* (Steindachner) in Algeria. *Journal of Fish Diseases* 37, 237 - 238.
[73] Katharios, P. and Tsigenopoulos, C.S. (2010) First report of nodavirus outbreak in cultured juvenile shi drum, *Umbrina cirrosa* L., in Greece. *Aquaculture Research* 42, 147 - 152.
[74] King, A., Adams, M., Carstens, E. and Lefkowitz, E. (2011) *Virus Taxonomy. Ninth Report of the International Committee on Taxonomy of Viruses*. Elsevier / Academic Press, London /Waltham, Massachussetts, San Diego, California, pp. 1061 - 1066.
[75] Korsnes, K., Devold, M., Nerland, A.H. and Nylund, A. (2005) Viral encephalopathy and retinopathy (VER) in Atlantic salmon *Salmo salar* after intraperitoneal challenge with a nodavirus from Atlantic halibut *Hippoglossus hippoglossus*. *Diseases of Aquatic Organisms* 68, 7 - 15.
[76] Korsnes, K., Karlsbakk, E., Devold, M., Nerland, A.H. and Nylund, A. (2009) Tissue tropism of nervous necrosis virus (NNV) in Atlantic cod, *Gadus morhua* L., after intraperitoneal challenge with a virus isolate from diseased Atlantic halibut, *Hippoglossus hippo-glossus* (L.). *Journal of Fish Diseases* 32, 655 - 665.
[77] Kuo, H.C., Wang, T.Y., Chen, P.P., Chen, Y.M., Chuang, H.C. *et al.* (2011) Real-time quantitative PCR assay for monitoring of nervous necrosis virus infection in grouper aquaculture. *Journal of Clinical Microbiology* 49, 1090 - 1096.Lee, K.W., Chi, S.C. and Cheng, T.M. (2002) Interference of the life cycle of fish nodavirus with fish retrovirus. *Journal of General Virology* 83, 2469 - 2474.
[78] Li, J., Shi, C.Y., Huang, J., Geng, W.G., Wang, S.Q. and Su, Z.D. (2014) Complete genome sequence of a betanodavirus isolated from half-smooth tongue sole (*Cynoglossus semilaevis*). *Genome Announcements* 2, 3 - 4.
[79] Liltved, H., Vogelsang, C., Modahl, I. and Dannevig, B.H. (2006) High resistance of fish pathogenic viruses to UV irradiation and ozonated seawater. *Aquacultural Engineering* 34, 72 - 82.
[80] Lopez-Jimena, B., Cherif, N., Garcia-Rosado, E., Infante, C., Cano, I. *et al.* (2010) A combined RT - PCR and dot-blot hybridization method reveals the coexistence of SJNNV and RGNNV betanodavirus genotypes in wild meagre (*Argyrosomus regius*). *Journal of Applied Microbiology* 109, 1361 - 1369.
[81] Lopez-Jimena, B., Alonso, M.D.C., Thompson, K.D., Adams, A., Infante, C. *et al.* (2011) Tissue distribution of red spotted grouper nervous necrosis virus (RGNNV) genome in experimentally infected juvenile European seabass (*Dicentrarchus labrax*). *Veterinary Microbiology* 154, 86 - 95.
[82] Lu, M.W., Chao, Y.M., Guo, T.C., Santi, N., Evensen, O. *et al.* (2008) The interferon response is involved in nervous necrosis virus acute and persistent infection in zebrafish infection model. *Molecular Immunology* 45, 1146 - 1152.
[83] Maeno, Y., de La Peña, L.D. and Cruz-Lacierda, E.R. (2004) Mass mortalities associated with viral nervous necrosis in hatchery-reared sea bass *Lates calcarifer* in the Philippines. *Japan Agricultural Research Quarterly* 38, 69 - 73.
[84] Maeno, Y., de La Peña, L.D. and Cruz-Lacierda, E.R. (2007) Susceptibility of fish species cultured in man-grove brackish area to piscine nodavirus. *Japan Agricultural*

Research Quarterly 41, 95 - 99.
[85] Maltese, C. and Bovo, G. (2007) Monografie. Viral encephalopathy and retinopathy / Encefalopatia e retinopatia virale. *Ittiopatologia* 4, 93 - 146.
[86] Mazelet, L., Dietrich, J. and Rolland, J. L. (2011) New RT-qPCR assay for viral nervous necrosis virus detection in sea bass, *Dicentrarchus labrax* (L.): application and limits for hatcheries sanitary control. *Fish and Shellfish Immunology* 30, 27 - 32.
[87] Mézeth, K.B., Patel, S., Henriksen, H., Szilvay, A.M. and Nerland, A.H. (2009) B2 protein from betanodavirus is expressed in recently infected but not in chronically infected fish. *Diseases of Aquatic Organisms* 83, 97 - 103.
[88] Mladineo, I. (2003) The immunohistochemical study of nod-avirus changes in larval, juvenile and adult sea bass tissue. *Journal of Applied Ichthyology* 19, 366 - 370.
[89] Moody, N.J.G., Horwood, P.F., Reynolds, A., Mahony, T.J., Anderson, I.G. *et al.* (2009) Phylogenetic analy-sis of betanodavirus isolates from Australian finfish. *Diseases of Aquatic Organisms* 87, 151 - 160.
[90] Moreno, P., Olveira, J. G., Labella, A., Cutrín, J. M., Baro, J. C. *et al.* (2014) Surveillance of viruses in wild fish populations in areas around the Gulf of Cadiz (South Atlantic Iberian Peninsula). *Applied and Environmental Microbiology* 80, 6560 - 6571.
[91] Mori, K., Nakai, T., Muroga, K., Arimoto, M., Mushiake, K. *et al.* (1992) Properties of a new virus belonging to nodaviridae found in larval striped jack (*Pseudocaranx dentex*) with nervous necrosis. *Virology* 187, 368 - 371.
[92] Mori, K., Mangyoku, T., Iwamoto, T., Arimoto, M., Tanaka, S. *et al.* (2003) Serological relationships among gen-otypic variants of betanodavirus. *Diseases of Aquatic Organisms* 57, 19 - 26.
[93] Munday, B. L., Kwang, J. and Moody, N. (2002) Betanodavirus infections of teleost fish: a review. *Journal of Fish Diseases* 25, 127 - 142.
[94] Mushiake, K., Arimoto, M., Furusawa, T., Furusawa, I., Nakai, T. *et al.* (1992) Detection of antibodies against striped jack nervous necrosis virus (SJNNV) from broodstocks of striped jack. *Nippon Suisan Gakkaishi* 58, 2351 - 2356.
[95] NaveenKumar, S., Shekar, M., Karunasagar, I. and Karunasagar, I. (2013) Genetic analysis of RNA1 and RNA2 of *Macrobrachium rosenbergii* nodavirus (MrNV) isolated from India. *Virus Research* 173, 377 - 385.
[96] Nguyen, H. D., Mekuchi, T., Imura, K., Nakai, T., Nishizawa, T. *et al.* (1994) Occurrence of viral nervous necrosis (VNN) in hatchery-reared juvenile Japanese flounder *Paralichthys olivaceus*. *Fisheries Science* 60, 551 - 554.
[97] Nguyen, H.D., Nakai, T. and Muroga, K. (1996) Progression of striped jack nervous necrosis virus (SJNNV) infection in naturally and experimentally infected striped jack *Pseudocaranx dentex* larvae. *Diseases of Aquatic Organisms* 24, 99 - 105.
[98] Nguyen, H.D., Mushiake, K., Nakai, T. and Muroga, K. (1997) Tissue distribution of striped jack nervous necrosis virus (SJNNV) in adult striped jack. *Diseases of Aquatic Organisms* 28, 87 - 91.
[99] Nishizawa, T., Toshihiro, K.M., Iwao, N. and Muroga, K. (1994) Polymerase chain reaction (PCR) amplifica-tion of RNA of striped jack nervous necrosis virus (SJNNV). *Diseases of Aquatic Organisms* 18, 103 - 107.
[100] Nishizawa, T., Mori, K., Furuhashi, M., Nakai, T., Furusawa, I. *et al.* (1995) Comparison of the coat protein genes of five fish nodaviruses, the causative agents of viral nervous necrosis in marine fish. *Journal of General Virology* 76, 1563 - 1569.
[101] Nishizawa, T., Furuhashi, M., Nagai, T., Nakai, T. and Muroga, K. (1997) Genomic classification of fish nodaviruses by molecular phylogenetic analysis of the coat protein gene. *Applied and Environmental Microbiology* 63, 1633 - 1636.
[102] Nishizawa, T., Kokawa, Y., Wakayama, T., Kinoshita, S. and Yoshimizu, M.

(2008) Enhanced propagation of fish nodaviruses in BF - 2 cells persistently infected with snakehead retrovirus (SnRV). *Diseases of Aquatic Organisms* 79, 19 - 25.

[103] Nishizawa, T., Gye, H.J., Takami, I. and Oh, M.J. (2012) Potentiality of a live vaccine with nervous necrosis virus (NNV) for sevenband grouper *Epinephelus septemfasciatus* at a low rearing temperature. *Vaccine* 30, 1056 - 1063.

[104] Nuñez-Ortiz, N., Stocchi, V., Toffan, A, Pascoli, F., Sood, N. *et al*. (2016) Quantitative immunoenzymatic detection of viral encephalopathy and retinopathy virus (betan-odavirus) in sea bass *Dicentrarchus labrax*. *Journal of Fish Diseases* 39, 821 - 831.

[105] Nylund, A., Karlsbakk, E., Nylund, S., Isaksen, T.E., Karlsen, M. *et al*. (2008) New clade of betanodavi-ruses detected in wild and farmed cod (*Gadus morhua*) in Norway. *Archives of Virology* 153, 541 - 547.

[106] Ødegård, J., Sommer, A.-I. and Præbel, A.K. (2010) Heritability of resistance to viral nervous necrosis in Atlantic cod (*Gadus morhua* L.). *Aquaculture* 300, 59 - 64.

[107] Oh, M.-J., Jung, S.-J., Kim, S.-R., Rajendran, K.V, Kim, Y.-J. *et al*. (2002) A fish nodavirus associated with mass mortality in hatchery-reared red drum, *Sciaenops ocellatus*. *Aquaculture* 211, 1 - 7.

[108] Oh, M.-J., Gye, H.J. and Nishizawa, T. (2013) Assessment of the sevenband grouper *Epinephelus septemfas-ciatus* with a live nervous necrosis virus (NNV) vaccine at natural seawater temperature. *Vaccine* 31, 2025 - 2027.

[109] OIE (2013) Chapter 2. 3. 11. Viral encephalopathy and retinopathy. In: *Manual of Diagnostic Tests for Aquatic Animals*. World Organization for Animal Health, Paris. Updated 2016 version available as Chapter 2.3.12. at: http://www.oie.int/fileadmin/Home/eng/Health_standards/aahm/current/chapitre_viral_encephalopathy_retinopathy.pdf (accessed 4 November 2016).

[110] Olveira, J. G., Soares, F., Engrola, S., Dopazo, C. P. and Bandín, I. (2008) Antemortem versus postmortem methods for detection of betanodavirus in Senegalese sole (*Solea senegalensis*). *Journal of Veterinary Diagnostic Investigation* 20, 215 - 219.

[111] Olveira, J. G., Souto, S., Dopazo, C. P., Thiéry, R., Barja, J. L. *et al*. (2009) Comparative analysis of both genomic segments of betanodaviruses isolated from epizootic outbreaks in farmed fish species provides evidence for genetic reassortment. *Journal of General Virology* 90, 2940 - 2951.

[112] Pakingking, R., Bautista, N. B., de Jesus-Ayson, E. G. and Reyes, O. (2010) Protective immunity against viral nervous necrosis (VNN) in brown-marbled grouper (*Epinephelus fuscogutattus*) following vaccination with inactivated betanodavirus. *Fish and Shellfish Immunology* 28, 525 - 533.

[113] Panzarin, V., Patarnello, P., Mori, A., Rampazzo, E., Cappellozza *et al*. (2010) Development and validation of a real-time TaqMan PCR assay for the detection of betanodavirus in clinical specimens. *Archives of Virology* 155, 1193 - 1203.

[114] Panzarin, V., Fusaro, A., Monne, I., Cappellozza, E., Patarnello, P. *et al*. (2012) Molecular epidemiology and evolutionary dynamics of betanodavirus in south-ern Europe. *Infection, Genetics and Evolution* 12, 63 - 70.

[115] Panzarin, V., Cappellozza, E., Mancin, M., Milani, A., Toffan, A. *et al*. (2014) *In vitro* study of the replication capacity of the RGNNV and the SJNNV betanodavirus genotypes and their natural reassortants in response to temperature. *Veterinary Research* 45, 56.

[116] Panzarin, V., Toffan, A., Abbadi, M., Buratin, A., Mancin, M. *et al*. (2016) Molecular basis for antigenic diversity of genus *Betanodavirus*. *PLoS One* 11 (7): e0158814.

[117] Parameswaran, V., Kumar, S.R., Ahmed, V.P.I. and Hameed, A.S.S. (2008) A fish nodavirus associated with mass mortality in hatchery-reared Asian Sea bass, *Lates calcarifer*. *Aquaculture* 275, 366 - 369.

[118] Pascoli, F., Serra, M., Toson, M., Pretto, T. and Toffan, A. (2016) Betanodavirus ability to infect juvenile European sea bass, *Dicentrarchus labrax*, at different water salinity. *Journal of Fish Diseases* 39, 1061 - 1068.

[119] Patel, S., Korsnes, K., Bergh, Ø., Vik-Mo, F., Pedersen, J. *et al*. (2007) Nodavirus in farmed Atlantic cod *Gadus morhua* in Norway. *Diseases of Aquatic Organisms* 77, 169 - 173.

[120] Péducasse, S., Castric, J., Thiéry, R., Jeffroy, J., Le Ven, A. *et al*. (1999) Comparative study of viral encephalopathy and retinopathy in juvenile sea bass *Dicentrarchus labrax* infected in different ways. *Diseases of Aquatic Organisms* 36, 11 - 20.

[121] Pirarat, N., Ponpornpisit, A., Traithong, T., Nakai, T., Katagiri, T. *et al*. (2009a) Nodavirus associated with pathological changes in adult spotted coralgroupers (*Plectropomus maculatus*) in Thailand with viral nervous necrosis. *Research in Veterinary Science* 87, 97 - 101.

[122] Pirarat, N., Katagiri, T., Maita, M., Nakai, T. and Endo, M. (2009b) Viral encephalopathy and retinopathy in hatchery-rearedjuvenilethread-sailfilefish (*Stephanolepis cirrhifer*). *Aquaculture* 288, 349 - 352.

[123] Ransangan, J. and Manin, B. O. (2012) Genome analysis of *Betanodavirus* from cultured marine fish species in Malaysia. *Veterinary Microbiology* 156, 16 - 44.

[124] Rigos, G. and Katharios, P. (2009) Pathological obstacles of newly-introduced fish species in Mediterranean mariculture: a review. *Reviews in Fish Biology and Fisheries* 20, 47 - 70.

[125] Sakamoto, T., Okinaka, Y., Mori, K.I., Sugaya, T., Nishioka, T. *et al*. (2008) Phylogenetic analysis of Betanodavirus RNA2 identified from wild marine fish in oceanic regions. *Fish Pathology* 43, 19 - 27.

[126] Sano, M., Nakai, T. and Fijan, N. (2011) Viral diseases and agents of warmwater fish. In: Woo, P. and Bruno, D.W. (eds) *Fish Disease and Disorders. Volume 3: Viral, Bacterial and Fungal Infections*. CAB International, Wallingford, UK, pp. 198 - 207.

[127] Scapigliati, G., Buonocore, F., Randelli, E., Casani, D., Meloni, S. *et al*. (2010) Cellular and molecular immune responses of the sea bass (*Dicentrarchus labrax*) experimentally infected with betanodavirus. *Fish and Shellfish Immunology* 28, 303 - 311.

[128] Shetty, M., Maiti, B., Shivakumar Santhosh, K., Venugopal, M. N. and Karunasagar, I. (2012) Betanodavirus of marine and freshwater fish: distribu-tion, genomic organization, diagnosis and control measures. *Indian Journal of Virology* 23, 114 - 123.

[129] Skliris, G.P. and Richards, R.H. (1999) Induction of nodavirus disease in seabass, *Dicentrarchus labrax*, using different infection models. *Virus Research* 63, 85 - 93.

[130] Skliris, G.P., Krondiris, J.V, Sideris, D.C., Shinn, A.P., Starkey, W.G. *et al*. (2001) Phylogenetic and antigenic characterization of new fish nodavirus isolates from Europe and Asia. *Virus Research* 75, 59 - 67.

[131] Sohn, S.-G., Park, M.-A., Oh, M.-J. and Cho, S.-K. (1998) A fish nodavirus isolated from cultured sevenband grouper, *E. septemfasciatus*.*Journal of Fish Pathology* 11, 97 - 104.

[132] Sommerset, I. and Nerland, A.H. (2004) Complete sequence of RNA1 and subgenomic RNA3 of Atlantic halibut nodavirus (AHNV). *Diseases of Aquatic Organisms* 58,

117 - 125.
[133] Souto, S., Lopez-Jimena, B., Alonso, M. C., García-Rosado, E. and Bandín, I. (2015a) Experimental susceptibility of European sea bass and Senegalese sole to different betanodavirus isolates. *Veterinary Microbiology* 177, 53 - 61.
[134] Souto, S., Olveira, J. G. and Bandín, I. (2015b) Influence of temperature on Betanodavirus infection in Senegalese sole (*Solea senegalensis*). *Veterinary Microbiology* 179, 162 - 167.
[135] Starkey, W.G., Ireland, J.H., Muir, K.F., Jenknis, M.E., Roy, W.J. *et al*. (2001) Nodavirus infection in Atlantic cod and Dover sole in the UK. *Veterinary Record* 149, 179 - 181.
[136] Tanaka, S., Mori, K., Arimoto, M., Iwamoto, T. and Nakai, T. (2001) Protective immunity of sevenband grouper, *Epinephelus septemfasciatus* Thunberg, against exper-imental viral nervous necrosis. *Journal of Fish Diseases* 24, 15 - 22. doi: 10.1046/j.1365 - 2761.2001.00259.x.
[137] Tanaka, S., Takagi, M. and Miyazaki, T. (2004) Histopathological studies on viral nervous necrosis of sevenband grouper, *Epinephelus septemfasciatus* Thunberg, at the grow-out stage. *Journal of Fish Diseases* 27, 385 - 399. doi: 10.1111/j.1365 - 2761.2004.00559.x.
[138] Thiéry, R., Raymond, J. C. and Castric, J. (1999) Natural outbreak of viral encephalopathy and retinopathy in juvenile sea bass, *Dicentrarchus labrax*: study by nested reverse transcriptase-polymerase chain reac-tion. *Virus Research* 63, 11 - 17.
[139] Thiéry, R., Cozien, J., de Boisséson, C., Kerbart-Boscher, S. and Névarez, L. (2004) Genomic classifi-cation of new betanodavirus isolates by phylogenetic analysis of the coat protein gene suggests a low host-fish species specificity. *Journal of General Virology* 85, 3079 - 3087.
[140] Toffolo, V., Negrisolo, E., Maltese, C., Bovo, G., Belvedere, P. *et al*. (2007) Phylogeny of betanodaviruses and molecular evolution of their RNA polymerase and coat proteins. *Molecular Phylogenetics and Evolution* 43, 298 - 308.
[141] Totland, G. K., Grotmol, S., Morita, Y., Nishioka, T. and Nakai, T. (1999) Pathogenicity of nodavirus strains from striped jack *Pseudocaranx dentex* and Atlantic halibut *Hippoglossus hippoglossus*, studied by waterborne challenge of yolk-sac larvae of both teleost species. *Diseases of Aquatic Organisms* 38, 169 - 175.
[142] Ucko, M., Colorni, A. and Diamant, A. (2004) Nodavirus infections in Israeli mariculture. *Journal of Fish Diseases* 27, 459 - 469.
[143] Vendramin, N., Padrós, F., Pretto, T., Cappellozza, E., Panzarin, V. *et al*. (2012) Viral encephalopathy and retinopathy outbreak in restocking facilities of the endangered freshwater species, *Salaria fluviatilis* (Asso). *Journal of Fish Diseases* 35, 867 - 871.
[144] Vendramin, N., Patarnello, P., Toffan, A., Panzarin, V., Cappellozza, E. *et al*. (2013) Viral encephalopathy and retinopathy in groupers (*Epinephelus* spp.) in southern Italy: a threat for wild endangered species? *BMC Veterinary Research* 9: 20.
[145] Vendramin, N., Toffan, A., Mancin, M., Cappellozza, E., Panzarin, V. *et al*. (2014) Comparative pathogenicity study of ten different betanodavirus strains in experimentally infected European sea bass, *Dicentrarchus labrax* (L.). *Journal of Fish Diseases* 37, 371 - 383.
[146] Walker, P.J. and Winton, J.R. (2010) Emerging viral dis-eases of fish and shrimp. *Veterinary Research* 41, 51. Watanabe, K. I., Nishizawa, T. and Yoshimizu, M. (2000) Selection of brood stock candidates of barfin flounder using an ELISA system with recombinant protein of barfin flounder nervous necrosis virus. *Diseases of Aquatic Organisms* 41, 219 - 223.

[147] Wi, G.R., Hwang, J.Y., Kwon, M.G., Kim, H.J., Kang, H.A. *et al.* (2015) Protective immunity against nervous necrosis virus in convict grouper *Epinephelus septemfasciatus* following vaccination with virus-like par-ticles produced in yeast *Saccharomyces cerevisiae*. *Veterinary Microbiology* 177, 214 - 218.
[148] Zorriehzahra, M.J., Nazari, A., Ghasemi, M., Ghiasi, M., Karsidani, S.H. *et al.* (2014) Vacuolating encephalopathy and retinopathy associated with a nodavirus-like agent: a probable cause of mass mortality of wild Golden grey mullet (*Liza aurata*) and Sharpnose grey mullet (*Liza saliens*) in Iranina waters of the Caspian sea. *Virus Diseases* 25, 430 - 436.

12

虹彩病毒病：真鲷虹彩病毒和高首鲟虹彩病毒

Kawato Yasuhiko，Kuttichantran Subramaniam，Kazuhiro Nakajima，
Thomas Waltzek 和 Richard Whittington *

12.1 真鲷虹彩病毒

12.1.1 引言

真鲷虹彩病毒(red sea bream iridovirus，RSIV)是直径为 200 nm 的二十面体双链 DNA 病毒。根据国际病毒分类委员会(ICTV)的分类，它属于虹彩病毒科(Iridoviridae)的细胞肿大病毒属(*Megalocytivirus*)，尽管其尚未被批准为该属中的一个种(Jancovich et al.，2012)。

1990 年夏天，在日本首次发现了导致养殖真鲷(*Pagrus major*)大量死亡的真鲷虹彩病毒病(red sea bream iridoviral disease，RSIVD)(Inouye et al.，1992)。疫情于 11 月结束，此时养殖场中水温低于 20 ℃。该病在第二年夏季复发，不仅发生在真鲷中，而且还在其他海水养殖品种中出现，如五条鰤(*Seriola quinqueradiata*)、高体鰤(*S. dumerili*)、海鲈(*Lateolabrax* sp.)、黄带拟鲹(*Pseudocaranx dentex*)和条石鲷(*Oplegnathus fasciatus*)(Matsuoka et al.，1996)。从那时起，RSIVD 已经蔓延到 30 多种海水养殖鱼类，并且每年夏季都会发生，给日本海水养殖业造成了巨大的经济损失(Kawakami and Nakajima，2002)。自 20 世纪 90 年代以来，在东亚和东南亚的许多养殖鱼类中都报道了类似的病毒性疾病，而致病病毒也是细胞肿大病毒属(Kurita and Nakajima，2012)。这些病毒包括来自中国台湾地区的一种石斑鱼(*Epinephelus* sp.)的石斑鱼虹彩病毒(Chou et al.，1998；Chao et al.，2004)、来自淡水鳜鱼(*Siniperca chuatsi*)的传染性脾肾坏死病毒(Infectious spleen and kidney necrosis virus，ISKNV)(He et al.，2000)、来自条石鲷的条石鲷虹彩病毒(rock bream iridovirus，RBIV)(Jung and Oh，2000)和来自大菱鲆的大菱鲆红体病虹彩病毒(turbot reddish body iridovirus，TRBIV)(Shi et al.，2004)。传染性脾肾坏死病毒是细胞肿大病毒属的模式种(Jancovich

* 通信作者邮箱：richard.whittington@sydney.edu.au。

et al.，2012）。

根据主要衣壳蛋白基因的核苷酸序列，这些病毒在遗传上被分为三类（Kurita and Nakajima，2012）。它们是 RSIV 型，主要见于海洋鱼类中（Inouye et al.，1992；Jung and Oh，2000；Chen et al.，2003；Gibson-Kueh et al.，2004；Lü et al.，2005；Dong et al.，2010；Huang et al.，2011）；ISKNV 型，在海水和淡水鱼类中均有过相关报道（He et al.，2000，2002；Weng et al.，2002；Chao et al.，2004；Wang et al.，2007；Fu et al.，2011；Huang et al.，2011）；TRBIV 型，鲆鲽鱼类中有过相关报道（Shi et al.，2004；Kim et al.，2005；Do et al.，2005a；Oh et al.，2006）。RSIV 型病毒广泛分布于东亚和东南亚，在日本出现的所有病毒均为 RSIV 型，ISKNV 型病毒来自东南亚、中国，有关 TRBIV 型的报道仅限于中国和韩国（Kurita and Nakajima，2012）。目前，由 RSIV 和 ISKNV 型病毒引起的 RSIVD 被世界动物卫生组织（OIE）列入名录，但 TRBIV 型病毒没有包括在内，因为该病毒的致病性仍有争议（OIE，2012）。

RSIVD 的暴发通常从一个网箱传播到相邻的网箱。已经通过实验证实，病毒传播是通过同栖发生的（He et al.，2002）。然而，在孵化场的鱼中还没有 RSIVD 的相关报道。因此，通过卵子或精子垂直传播的风险很低（Nakajima and Kurita，2005）。一旦疾病发生，每年夏季都会在同一地区再次暴发。这与 RSIV 的最适体外生长温度为 25 ℃的观察结果一致（Nakajima and Sorimachi，1994）。感染 RSIV 的鱼在低水温（<18 ℃）下饲养 100 d 以上可以康复，存活鱼会对 RSIVD 产生抗性（Oh et al.，2014；Jung et al.，2015）。此外，RSIVD 最常见于许多鱼类的一龄鱼中（Matsuoka et al.，1996；Kawakami and Nakajima，2002）。每年新引入网箱的鱼中暴发 RSIVD 可能是由痊愈后仍携带病毒的存活大龄鱼的水平传播引起的。Ito 等（2013）在实验感染 RSIV 的存活五条鰤的脾、肾、心、鳃、肠和尾鳍中发现了病毒基因组。在水产养殖区周围，使用 PCR 方法从许多没有临床症状的野生鱼类（6 目，18 科和 39 种）中检测到病毒基因组（Wang et al.，2007），这意味着这些鱼也可能是病毒携带者。Jin 等（2014）的研究指出，一些生活在鱼类养殖场周围的双壳类动物是潜在的病毒载体。

细胞肿大病毒属虹彩病毒不是宿主特异性的（表 12.1）。例如，它们能感染非洲蓝眼灯（*Aplocheilichthys normani*）和东南亚的丽丽鱼（拉利毛足鲈）（*Colisa lalia*）等观赏鱼类（Sudthongkong et al.，2002a；Jeong et al.，2008a；Weber et al.，2009；Kim et al.，2010；Sriwanayos et al.，2013；Nolan et al.，2015）。这些鱼类产生的疾病的组织病理学类似于 RSIVD，致病因子属于 ISKNV 类型（Sudthongkong et al.，2002b；Weber et al.，2009；Kurita and Nakajima，2012；Sriwanayos et al.，2013）。另外，从三刺鱼（*Gasterosteus aculeatus*）中报道了一种新的细胞肿大虹彩病毒，被正式称为三刺鱼虹彩病毒（threespine stickleback iridovirus，TSIV）（Waltzek et al.，2012）。

表 12.1　真鲷虹彩病毒(RSIV)和其他细胞肿大病毒的宿主范围[a]

目	科	种数	代表鱼种	病　毒	OIE[b]	参考文献
脂鲤目 Characiformes	锯脂鲤科 Serrasalmidae	1	银板鱼	未知		Nolan et al.，2015
鳉形目 Cyprinodontiformes	假鳃鳉科 Nothobranchiidae	1	茄氏旗鳉	未知		Nolan et al.，2015
	花鳉科 Poeciliidae	4	蓝眼灯	ISKNV，未知		Paperna et al.，2001；Sudthongkong et al.，2002a；Nolan et al.，2015
鲻形目 Mugiliformes	鲻科 Mugilidae	1	鲻鱼	ISKNV	+	Gibson-Kueh et al.，2004
鲈形目 Perciformes	天竺鲷科 Apogonidae	1	考氏鳍竺鲷	ISKNV		Weber et al.，2009
	斗鱼科 Belontiidae	2	珍珠毛足鲈	ISKNV		Jeong et al.，2008a
	鲹科 Carangidae	7	五条鰤	RSIV	+	Kawakami and Nakajima，2002
	太阳鱼科 Centrarchidae	1	大口黑鲈	ISKNV	+	He et al.，2002
	慈鲷科 Cichlidae	8	神仙鱼	未知		Armstrong and Ferguson，1989；Rodger et al.，1997；Nolan et al.，2015
	塘鳢科 Eleotridae	1	褐塘鳢	ISKNV		Wang et al.，2011
	白鲳科 Ephippidae	1	圆眼燕鱼	ISKNV		Sriwanayos et al.，2013
	石鲈科 Haemulidae	2	三线矶鲈	RSIV	+	Kawakami and Nakajima，2002
	齿唇鲈科 Helostomatidae	1	吻鲈	未知		Nolan et al.，2015
	舵鱼科 Kyphosidae	1	斑鮀	RSIV	+	Kawakami and Nakajima，2002
	花鲈科 Lateolabracidae	2	花鲈	RSIV	+	Kawakami and Nakajima，2002；Do et al.，2005b
	尖吻鲈科 Latidae	1	尖吻鲈	RSIV，ISKNV	+	Huang et al.，2011

续 表

目	科	种数	代表鱼种	病　毒	OIE[b]	参考文献
鲈形目 Perciformes	裸颊鲷科 Lethrinidae	2	红鳍裸颊鲷	RSIV	+	Kawakami and Nakajima，2002
	梦鲈科 Moronidae	1	条纹鲈×金眼狼鲈	RSIV	+	Kurita and Nakajima，2012
	石鲷科 Oplegnathidae	2	条石鲷	RSIV	+	Kawakami and Nakajima，2002
	丝足鲈科 Osphronemidae	3	拉利毛足鲈	ISKNV，未知		Paperna et al.，2001；Sudthongkong et al.，2002a；Kim et al.，2010
	真鲈科 Percichthyidae	2	鳜鱼	ISKNV	+	He et al.，2000；Go et al.，2006
	军曹鱼科 Rachycentridae	1	军曹鱼	RSIV	+	Kawakami and Nakajima，2002
	石首鱼科 Sciaenidae	2	大黄鱼	RSIV，ISKNV	+	Weng et al.，2002；Chen et al.，2003
	鲭科 Scombridae	3	蓝鳍金枪鱼	RSIV	+	Kawakami and Nakajima，2002
	鮨科 Serranidae	10	七带石斑鱼	RSIV，ISKNV	+	Kawakami and Nakajima，2002；Sudthongkong et al.，2002b；Chao et al.，2004；Gibson-Kueh et al.，2004；Huang et al.，2011
	鲷科 Sparidae	4	真鲷	RSIV	+	Inouye et al.，1992；Kawakami and Nakajima，2002；Kurita and Nakajima，2012
鲽形目 Pleuronectiformes	牙鲆科 Paralichthyidae	1	牙鲆	RSIV，TRBIV	+	Kawakami and Nakajima，2002；Do et al.，2005a
	鲽科 Pleuronectidae	2	圆斑星鲽	RSIV，TRBIV		Kawakami and Nakajima，2002；Won et al.，2013
	菱鲆科 Scophthalmidae	1	大菱鲆	TRBIV		Shi et al.，2004

续　表

目	科	种数	代表鱼种	病　毒	OIE[b]	参考文献
鲉形目 Scorpaeniformes	平鲉科 Sebastidae	2	许氏平鲉	RSIV	＋	Kim et al.，2002
鲀形目 Tetraodontiformes	单棘鲀科 Monacanthidae	1	冠鳞单棘鲀	RSIV		本章
	四齿鲀科 Tetraodontidae	1	红鳍东方鲀	RSIV	＋	Kawakami and Nakajima，2002

[a] ISKNV，传染性脾肾坏死病毒；RSIV，真鲷虹彩病毒；TRBIV，大菱鲆红体病虹彩病毒。

[b] ＋表示 OIE 指南中已列出（OIE，2012）。

12.1.2　诊断

被感染的鱼体色变深、嗜睡，并且由于严重贫血而经常有明显的呼吸运动。偶尔会观察到轻微的眼球突出和皮肤出血。常见的临床症状包括严重贫血、鳃部瘀点、脾肿大、心包腔内有出血性渗出物和内脏苍白（Inouye et al.，1992）。在自然感染中，这种疾病通常是慢性的，累计死亡率通常在 1、2 个月内达到 20％～60％（Nakajima et al.，1999）。

RSIVD 的典型特点为细胞的异常增大，这可以通过脾脏的吉姆萨染色印迹（Giemsa-stained impression）涂片的显微镜镜检来确认。这是推定诊断中最简单快速的方法（图 12.1），需要注意的是，在一些受感染的鱼中很少发现增大的细胞。

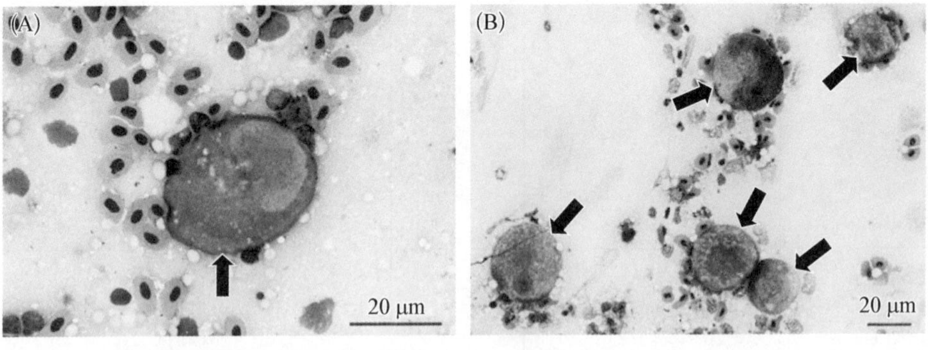

图 12.1　真鲷虹彩病毒（RSIV）感染鱼的脾脏涂片中增大的病毒感染细胞：(A) 真鲷；(B) 高体鰤。箭头所指为推断的病毒感染细胞。Giemsa 染色

使用特异性单克隆抗体（M10）的免疫荧光染色法用于诊断 RSIVD（Nakajima and Sorimachi，1995；Nakajima et al.，1995）。尽管尚未检查出

TRBIV 型病毒，但 M10 抗体与包括 RSIV 和 ISKNV 的许多细胞肿大虹彩病毒能够发生反应，但不与其他虹彩病毒反应，如蛙病毒 3 型（*Frog virus* 3，FV3）、流行性造血器官坏死病病毒、欧洲鲶病毒和新加坡石斑鱼虹彩病毒（Singapore grouper iridovirus，SGIV）（Nakajima and Sorimachi，1995；Nakajima et al.，1998）。使用 M10 抗体的间接荧光抗体法（IFAT）作为 RSIVD 的确诊方法被列入 OIE《水生动物诊断检测指南》（OIE，2012），并因其快速可靠而在日本得到广泛应用（Matsuoka et al.，1996；Kawakami and Nakajima，2002）。

研究者们已经开发了几种 PCR 引物（Kurita et al.，1998；Oshima et al.，1998；Jeong et al.，2004；Go et al.，2006；Rimmer et al.，2012），并将 PCR 检测技术应用于野生鱼类的诊断和调查（Wang et al.，2007）。用于定量检测 RSIV 的实时 PCR 具有快速、特异性和灵敏性的特点（Caipang et al.，2003；Wang et al.，2006；Gias et al.，2011；Rimmer et al.，2012；Oh and Nishizawa，2013）。

RSIV 首先使用 RTG-2（虹鳟性腺细胞）、CHSE-214（鲑鱼胚胎细胞）、FHM（胖头鲅肌肉细胞）、BF-2（蓝鳃太阳鱼细胞）和 KRE-3（海带和赤点石斑鱼胚胎细胞）从患病鱼的脾脏匀浆中分离出来（Inouye et al.，1992）。RSIV 的致细胞病变效应表现为感染细胞的变圆和增大。然而，这些细胞系的敏感性较低，并且通过连续传代病毒滴度逐渐降低（Nakajima and Sorimachi，1994）。石鲈鳍（Grunt fin，GF）细胞系对 RSIV 更敏感，它通常用于病毒的分离和增殖（Jung ct al.，1997；Nakajima and Maeno，1998；Ito et al.，2013）。尽管如此，RSIV 接种的细胞（即使是 GF 细胞），也可能不发生完全的 CPE，当病毒浓度较低时，很难识别任何 CPE（Ito et al.，2013）。其他几种鱼类细胞系对 RSIV 和 ISKNV 敏感（Imajoh et al.，2007；Dong et al.，2008；Wen et al.，2008），但不能从美国典型培养物保藏中心（ATCC）或欧洲认证细胞培养物收藏中心（ECACC）获得这些细胞系，并且也没有其他公开获得渠道。

12.1.3 病理学

最典型的组织病理学变化是脾、心、肾、肝或鳃中存在增大的细胞。在石蜡切片中用苏木精对细胞进行嗜碱性染色（图 12.2），而在印模涂片中，通常对它们用迈格吉（May-Grunwald-Giemsa）染色法进行异质染色（图 12.1）。在中度增大的细胞中常可见大的核样结构。这些异常增大的细胞被感染了。这种症状在 RSIVD（Inouye et al.，1992；Jung et al.，1997；He et al.，2000；Jung and Oh，2000）和其他细胞肿大虹彩病毒感染的疾病（Sudthongkong et al.，2002a；Shi et al.，2004；Weber et al.，2009；Sriwanayos et al.，2013）中已有过报道。最

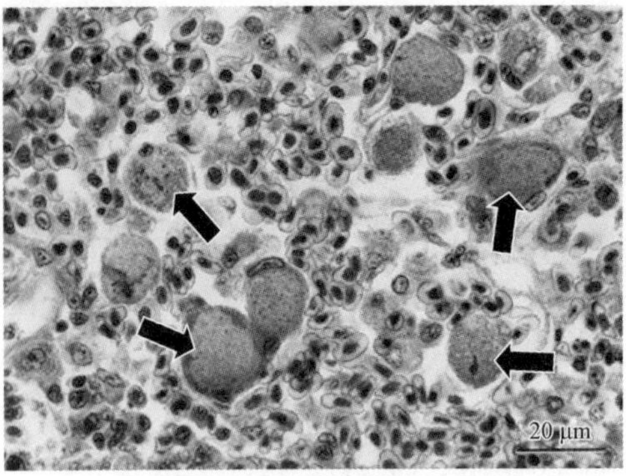

图 12.2　自然患真鲷虹彩病毒病(RSIVD)的真鲷脾脏。苏木精-伊红染色，箭头所指为推定的病毒感染细胞

严重的病变通常在脾脏中，那里通常有大量病毒感染的肿大细胞、弥漫性组织坏死和黑色素巨噬细胞中心减少。肾间质、肾小球、心内膜和中心静脉窦的淋巴组织及鳃丝上皮下也有许多异常增大的细胞(Inouye et al.，1992)。这些增大的细胞使用 IFAT(Nakajima et al.，1995)或原位杂交(*in situ* hybridization，ISH)(Chao et al.，2004)染色呈阳性。在 ISH 中，杂交信号首先出现在 RSIV 感染细胞的细胞核中，并且在感染后期细胞质也呈阳性(Chao et al.，2004)。

在电子显微镜下，肿大细胞的细胞质中可见二十面体病毒粒子(直径约为 200 nm)。每个病毒粒子由中央电子致密核(120 nm)和电子半透明区域组成。在病毒感染的细胞中，在病毒粒子形成之前观察到细胞的增大和核变性(Inouye et al.，1992)。

12.1.4　病理生理学

RSIV 对脾脏有明显的组织嗜性，脾脏是肿大细胞首先出现(Chao et al.，2004)和病毒基因组数量先增加的地方(Zhang et al.，2012；Ito et al.，2013)，但是在脑、脊髓和神经节中很少观察到肿大细胞(Jung et al.，1997；Chao et al.，2004)。在疾病的后期，当在脾、肾、心、鳃、鳍和肠中检测到大量病毒基因组时，可能发生多器官衰竭(Ito et al.，2013)。尽管如此，除了脾脏外，这些器官的实质看起来完好无损。病毒感染的肿大细胞通常分布于造血组织中，如脾脏和肾脏，并且遍布各种器官的血管。这些发现，以及在构成实质组织的细胞中未发现病毒感染细胞的事实表明，RSIV 的靶标是造血细胞。通过破坏这些细胞，RSIV 可能引起再生障碍性贫血，从而导致死亡。需要进一步研究以确认 RSIV 的靶细胞，并阐明死亡原因。

12.1.5　防控策略

目前可提供针对 RSIVD 的疫苗接种，但尚无可用的化疗药物。自 1999 年以来，真鲷虹彩病毒疫苗已在日本销售，这是第一种针对海水鱼类病毒性疾病的商用疫苗。该疫苗由接种 RSIV 的 GF 细胞的福尔马林灭活培养上清液组成。疫苗的有效性已在实验和田间试验中得到证实(Nakajima et al.，1997，1999)。目前，该疫苗在日本已用于真鲷、黄条鰤、高体鰤、条纹鲹、玛拉巴石斑鱼（*Epinephelus malabaricus*）、点带石斑鱼（*E. coioides*）、褐带石斑鱼（*E. bruneus*）和七带石斑鱼（*E. septemfasciatus*）的养殖病害防控。自疫苗批准上市以来，日本 RSIVD 的暴发率已显著下降。疫苗的成本增加了鱼类养殖生产的费用，因此，针对 RSIVD 更经济的重组亚单位疫苗和 DNA 疫苗得到了研究，但它们尚未商业化（Caipang et al., 2006；Kim et al., 2008；Shimmoto et al., 2010；Fu et al., 2012, 2014），并且这些疫苗的效果并不令人满意。最近，人们研发了另一种灭活疫苗，它是由能产生更多病毒的鱼类细胞系（来自鲑鱼）生产的（Dong et al., 2013a, b）。目前在日本销售的疫苗还有另一个问题，其对条石鲷和斑石鲷（*O. punctatus*）的免疫效力仍显不足。为了解决这个问题，有人建议使用活疫苗，因为接种病毒的鱼在低水温（<18 ℃）下养殖时会产生免疫力（Oh et al., 2014；Jung et al., 2015）。

感染的鱼禁食 10 d 以上可降低 RSIVD 死亡率（Tanaka et al., 2003）。尽管这种降低的机制仍不清楚，但在 RSIVD 发生后，养鱼场经常采取禁食措施，不过这会导致相当大的鱼体重减轻。

RSIVD 最常发生在一龄鱼中，而它们长大后似乎会获得一定的抵抗力（Matsuoka et al., 1996；Kawakami and Nakajima, 2002）。因此，夏季来临前，幼鱼的快速生长有助于避免暴发严重的 RSIVD。

12.1.6 总结

1990 年，日本某小范围养殖的真鲷突然发生 RSIVD。在随后的几年里，该疾病迅速蔓延至其他地区和其他鱼类（Matsuoka et al., 1996）。日本的所有病毒分离株均为 RSIV 型，表明该病毒很可能是通过进口鱼传入日本的。目前，RSIVD 流行于日本西南部沿海及东亚和东南亚地区，由于其缺乏宿主特异性，当水温有利于病毒传播时，就会引起许多种类水生动物患病。

在许多鱼类中已报道过细胞肿大虹彩病毒感染（表 12.1），包括来自东南亚的观赏鱼类（Sudthongkong et al., 2002a；Jeong et al., 2008a；Weber et al., 2009；Kim et al., 2010；Sriwanayos et al., 2013；Subramaniam et al., 2014；Nolan et al., 2015）。在日本发现 RSIVD 之前，在观赏鱼中发现过类似的病毒感染（特征是嗜碱性肿大细胞的形成），不过没有鉴定出这种病毒（Armstrong and Ferguson, 1989）。因此，通过观赏鱼的国际贸易传播细胞肿大虹彩病毒的可能性令人担忧（Jeong et al., 2008a）。例如，在养殖墨瑞鳕（*Maccullochella peelii*）中暴发的细胞肿大虹彩病毒感染就是通过向澳大利亚进口丽丽鱼引发的（Go et al., 2006；Go and Whittington, 2006）。此外，来自珍珠毛足鲈（*Trichogaster leeri*）的细胞肿大病毒对条石鲷具有高致病性（Jeong et al., 2008b）。

12.2 高首鲟虹彩病毒

12.2.1 引言

在 20 世纪，对鱼子酱(传统上来自黑海和里海的野生鲟)需求的增加和相关栖息地的退化(如污染和河流筑坝)，导致世界自然保护联盟(International Union for Conservation of Nature, IUCN)将一些物种，特别是鲟鱼和高首鲟列为极度濒危物种(Birstein, 1993; Pikitch et al., 2005)。全球鲟鱼养殖业的发展提高了食品和渔业种群资源的供应，旨在缓解这些野生种群减少的压力。

自 1998 年以来，在养殖高首鲟(*Acipenser transmontanus*)幼鱼的体表、鳃和上消化道出现了一种致命的病毒性疾病，即高首鲟虹彩病毒(white sturgeon iridovirus, WSIV)，对北美西北太平洋地区的商业养殖场造成了严重影响(Hedrick et al., 1990; LaPatra et al., 1994, 1999; Raverty et al., 2003; Drennan et al., 2006; Kwak et al., 2006)。WSIV 最早是在加利福尼亚州北部和中部农场养殖场的高首鲟幼鱼中发现的，在那里每年造成的病害损失高达 95%(Hedrick et al., 1990, 1992)。随后，在俄勒冈州和华盛顿州的哥伦比亚河下游、爱达荷州南部的斯内克河和北部的库特奈河的养殖高首鲟中都发现了 WSIV(LaPatra et al., 1994, 1999)。2001 年，加拿大不列颠哥伦比亚省的一个水产养殖场中的高首鲟种苗受到感染，这些鱼苗很可能产自该省弗雷泽河流域的种鱼(Raverty et al., 2003)。

实验研究表明，WSIV 可以通过受污染的水进行水平传播(Hedrick et al., 1990, 1992)。高首鲟幼鱼的死亡开始于接触病毒感染后第 10 天，并在第 15～30 天达到高峰。尽管高首鲟成鱼具有对临床疾病的抗性，但持续感染的成鱼会将 WSIV 垂直传播给其后代。LaPatra 等(1994)的研究认为，爱达荷州和俄勒冈州高首鲟孵化场中流行的 WSIV 来源于野生亲鱼。对加利福尼亚州鲟鱼养殖场 WSIV 的流行病学研究表明，垂直传播比池到池的水平传播更重要(Georgiadis et al., 2001)。湖鲟(*A. fluvescens*)可以通过实验感染，但不表现出临床疾病(Hedrick et al., 1992)。

在捕获的密苏里铲鲟(*Scaphirhynchus albus*)和扁吻铲鲟(*S. platorynchus*)幼鱼中发现一种被称为密苏里河鲟鱼虹彩病毒(Missouri River sturgeon iridovirus, MRSIV)的类似致命病毒，对密苏里河流域内孵化场的种群增殖计划产生了负面影响(Kurobe et al., 2010, 2011)。在其他养殖鲟品种中，包括北欧的俄罗斯鲟(*A. gueldenstaedtii*)(Adkison et al., 1998)、加拿大曼尼托巴省的湖鲟(Clouthier et al., 2013)和加拿大新不伦瑞克省的短吻鲟(*A. brevirostrum*)(LaPatra et al., 2014)，也已报道了具有惊人相似病理特征的其他动物流行病。

根据其与虹彩病毒的组织病理学和超微结构相似性，WSIV 暂时归类为虹彩病毒科(Hedrick et al., 1990; Jancovich et al., 2012)。不过,WSIV 的主要衣壳蛋白(MCP)与虹彩病毒缺乏明显的序列同源性(Kwak et al., 2006)。仅依靠 MCP 的初步系统发育分析支持 WSIV 和其他鲟虹彩样病毒作为拟菌病毒科(*Mimiviridae*)的一个新分支(Kwak et al., 2006; Kurobe et al., 2010; Clouthier et al., 2013, 2015; Waltzek et al.,未发表)。

12.2.2 诊断

养殖的高首鲟鱼苗和幼稚鱼在水温为 12～20 ℃时极易感染 WSIV(Hedrick et al., 1992; LaPatra et al., 1994; Watson et al., 1998b)。典型的 WSIV 流行病表现为一种慢性消耗综合征,厌食的幼鱼在水槽底部徘徊并逐渐消瘦,导致生长受损,死亡率高达 95%(Hedrick et al., 1990; Raverty et al., 2003)。WSIV 具有上皮细胞趋向性,导致包括外皮、鳃和上消化道等外部组织中出现特殊病征性微观损伤(Hedrick et al., 1990; LaPatra et al., 1994; Watson et al., 1998a)。组织学检查显示特征性肥大上皮细胞,染色为双嗜性至嗜碱性(Hedrick et al., 1990)。使用针对 WSIV 的单克隆抗体免疫组织化学检测可以在组织切片中确认病毒(OIE, 2003)。

由于 WSIV 在高首鲟脾细胞(WSS-2)和皮肤细胞(WSSK-1)中生长缓慢,病毒的分离具有挑战性(Hedrick et al., 1992)。将合并的鳃和皮肤组织匀浆接种细胞,在 20 ℃下培养,并连续检查 30 d 的 CPE,如果没有观察到 CPE,则进行盲传再观察 15 d(OIE, 2003)。尽管最早可以在接种后 1 周检测到 CPE,但在感染后 2～3 周内可能观察不到变化。受感染的细胞通常单独或以小群的形式出现,并且可以通过其变圆、增大和折光率的变化进行识别(Hedrick et al., 1992; Watson et al., 1998a)(图 12.3)。受感染的细胞最终会死亡并脱离,但单

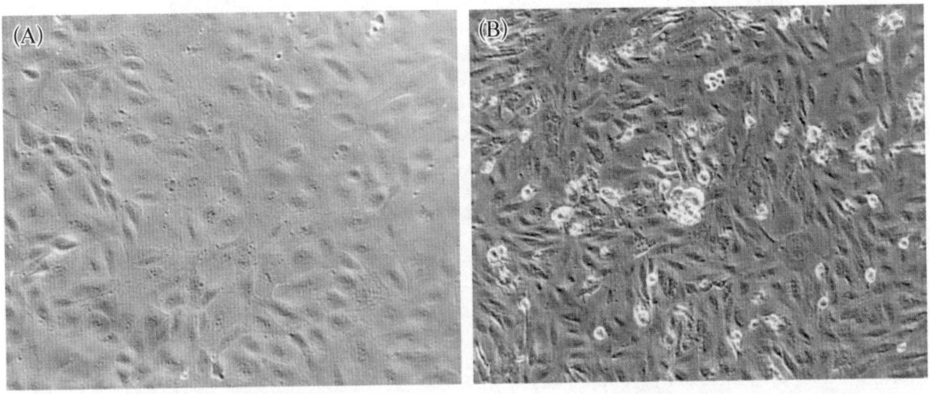

图 12.3 感染 WSIV 的高首鲟脾脏细胞(WSS-2):(A) 未感染的对照图;(B) 单层培养的呈单个或小聚集的感染细胞,通过其变圆、增大和折光率改变的特点进行识别。图片由 Ronald P. Hedrick 提供

层细胞的完全破坏仅发生在感染最严重的培养物中。可以使用血清中和试验、间接荧光抗体法或 TEM 观察来确定有可疑 CPE 的培养细胞(Hedrick et al., 1990；OIE，2003)。

针对 WSIV 主要衣壳蛋白的常规 PCR 检测方法的开发也提供了一种急需的特异性、灵敏和快速的诊断工具(Kwak et al., 2006)。该检测方法不扩增其他虹彩病毒,可检测低至 1 fg 的靶质粒 DNA。胸鳍组织碎片的取样是非致死性的,因此不需要牺牲大量珍贵的养殖或受到威胁的野生高首鲟种群(Drennan et al., 2007a)。最近,针对 MCP 保守区域的一种常规 PCR 方法和两种单独的定量 PCR 方法被设计用于扩增包括 WSIV 在内的所有北美鲟核质大 DNA 病毒(Clouthier et al., 2015)。

12.2.3 病理学

WSIV 的嗜上皮性导致皮肤和鳃的肉眼可见病变(Hedrick et al., 1990; LaPatra et al., 1994)。受感染鱼的腹侧鳞甲通常发红或出血[图 12.4(A)],鳃

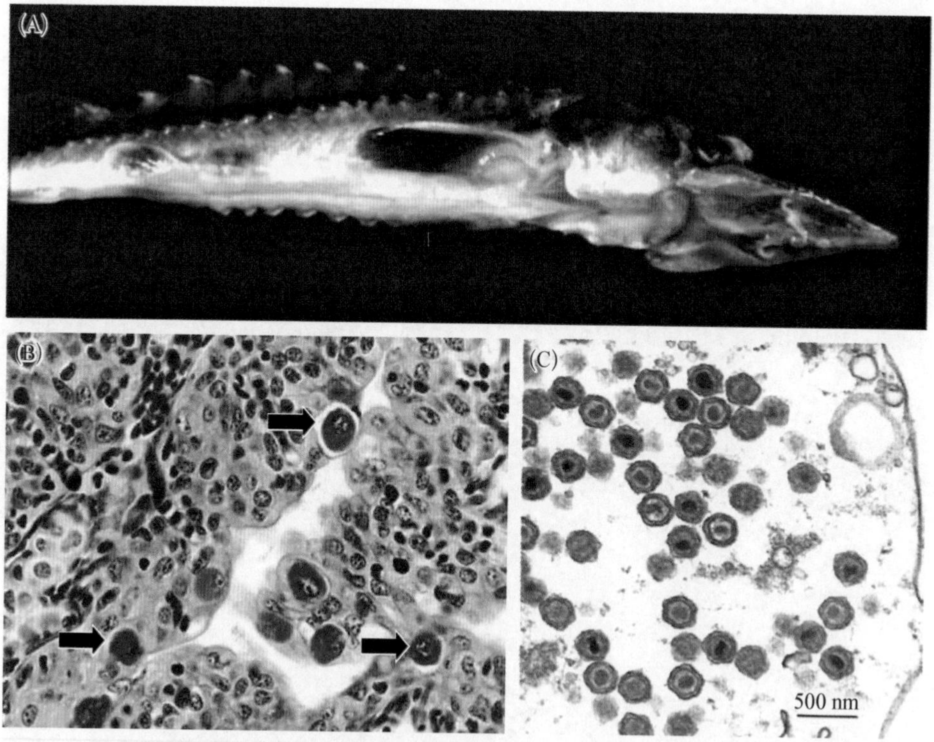

图 12.4 感染高首鲟虹彩病毒(WSIV)的高首鲟幼鱼：(A) 沿腹侧表面可见的小出血,包括口腔区和腹侧鳞甲;(B) WSIV 感染的鳃组织切片显示大量增大的两染性细胞(箭头所示),苏木精-伊红染色;(C) 透射电子显微镜照片：直接从高首鲟的鳃中感染的细胞里发现的 WSIV 复杂病毒粒子。图片由 Joseph M. Groff 和 Ronald P. Hedrick 提供

可能会显得肿胀和苍白。受感染的幼鱼通常显得消瘦和发育不良。在体内，可以观察到苍白的肝脏、空的胃肠道和减少的脂肪储存。

皮肤和鳃的组织病理学检查通常显示局灶性至弥漫性上皮增生，并伴随一些上皮细胞的增大，细胞染色双嗜性到嗜碱性[图 12.4.(B)]。这些肥大上皮细胞是 WSIV 感染的病征细胞，并且已经在体表（皮肤和鳍）、嗅觉器官（触须和鼻孔）、口咽腔（硬腭、舌、嘴唇、鳃盖瓣）、上消化道（食道）和鳃中观察到（Hedrick et al., 1990; Watson et al., 1998a）。在疾病晚期，可观察到鳃上皮的增生和鳃瓣血管通道内层支柱细胞坏死导致的小出血（Hedrick et al., 1990）。对 WSIV 阳性鱼的组织学或细胞学检查可发现次生病原体，包括细菌[黄杆菌属（*Flavobacterium* sp.）]、原生生物[车轮虫属（*Trichodina* sp.）]和卵菌[水霉属（*Saprolegnia* sp.）]（Hedrick et al., 1990; Watson et al., 1998a）。

感染组织（如皮肤或鳃）或培养物的 TEM 可以确诊 WSIV 感染。肥大上皮细胞的细胞质内，在病毒装配位点周围可以观察到独特的大而复杂的二十面体状病毒颗粒的聚集（Watson et al., 1998a）。这些病毒颗粒的对侧平均直径为 262～273 nm，相对顶点之间的平均直径为 299～302 nm（Hedrick et al., 1990; Watson et al., 1998a）[图 12.4(C)]。

12.2.4 病理生理学

已对高首鲟 WSIV 感染的病理生理学进行了研究（Watson et al., 1998a, b）。皮肤和鳃的病变被认为改变了感染鲟鱼的呼吸和渗透调节能力。实验发现，在较高水温下，受感染的高首鲟幼鱼的血细胞比容降低，血浆蛋白浓度下降（Watson et al., 1998b）。WSIV 感染高首鲟幼鱼的感觉上皮（鼻孔、触须、嘴唇、舌头）被认为会造成厌食，导致致命的消耗综合征（Watson et al., 1998a）。Watson 等(1998b)发现实验感染 WSIV 的高首鲟幼鱼在较低的水温下无法保持正常的肝重指数，这表明存在食物缺乏（Hung et al., 1997）和膳食葡萄糖的减少（Fynn-Aikens et al., 1992）。最后，推测 WSIV 造成的黏膜屏障破坏使高首鲟幼鱼易于受到次生感染（Watson et al., 1998b）。

12.2.5 防控策略

在孵化场条件下，导致 WSIV 流行病发生或严重程度的因素包括高养殖密度、生理和养殖过程应激压力、水温的波动或较低及生物安全漏洞（LaPatra et al., 1994, 1996; Watson et al., 1998b; Georgiadis et al., 2001; Drennan et al., 2005）。降低养殖密度和增加水流量可降低死亡率（LaPatra et al., 1994）。实验研究表明，相对于低养殖密度组，高养殖密度组的累计死亡率显著提高（LaPatra et al., 1996; Drennan et al., 2005）。可能增加 WSIV 流行病暴发频率的生理和养殖过程应激源包括搬运、运输和温度波动（LaPatra et al.,

1994，1996）。应激可以引发亚临床带毒鱼的疾病发作，随后将病毒传播给其他个体（Georgiadis et al.，2001）。一项实验感染研究（Watson et al.，1998b）证实，与较高水温（23 ℃）下饲养的鱼相比，在较低水温（10 ℃）下饲养的幼鲟的累计死亡率明显更高，并且具有更严重的微观病变。较低的水温能够促进 WSIV 的体外复制（Hedrick et al.，1992），并且该病毒可能抑制鲟鱼的免疫应答（Watson et al.，1998b）。虽然在较高的水温下养殖高首鲟幼鱼会降低死亡率，但在较高的温度下，细菌（*Flavobacterium* sp.）的次生感染问题更大（Watson et al.，1998b）。

在 WSIV 病暴发后存活下来的高首鲟幼鱼似乎即使有胁迫情况发生也会受到保护。存活鱼会产生个体或群体免疫力，能够在它们再次暴露于 WSIV 时保护它们（Georgiadis et al.，2001）。在存活鱼体内已经证实存在抗 WSIV 血清和黏膜抗体（Drennan et al.，2007b）。因此，有人提出了培育具有抗性亲鱼或与疾病暴发后存活鱼一起混养的生产策略（Georgiadis et al.，2000，2001）。此外，养殖者应该实行严格的生物安全措施，包括在单独的建筑物内隔离新的亲鱼种群，并在产卵前对其进行 WSIV 筛查。需要开发经济型 WSIV 疫苗和能够刺激黏膜免疫的给药系统，以减轻 WSIV 流行病在鲟鱼养殖中造成的影响。

12.2.6 总结

在过去的几十年中，我们已在如下两个方面取得进展：(1) 对 WSIV 病理学、病理生理学和流行病学的了解；(2) 检测 WSIV 的诊断工具的可用性。生理和养殖过程应激源影响流行病的发生和严重程度。通过适当的生物安全和管理措施，可以最大限度地减少 WSIV 的影响，然而，经济性的疫苗将有助于高首鲟的水产养殖。

参考文献

[1] Adkison, M.A., Cambre, M. and Hedrick, R.P. (1998) Identification of an iridovirus in Russian sturgeon (*Acipenser guldenstadi*) from northern Europe. *Bulletin of the European Association of Fish Pathologists* 18, 29 - 32.

[2] Armstrong, R.D. and Ferguson, H.W. (1989) Systemic viral disease of the chromide cichlid *Etropus maculatus*. *Diseases of Aquatic Organisms* 7, 155 - 157.

[3] Birstein, V. J. (1993) Sturgeons and paddlefishes: threatened fishes in need of conservation. *Conservation Biology* 7, 773 - 787.

[4] Caipang, C.M., Hirono, I. and Aoki, T. (2003) Development of a real-time PCR assay for the detection and quantification of red seabream iridovirus (RSIV). *Fish Pathology* 38, 1 - 7.

[5] Caipang, C.M., Takano, T., Hirono, I. and Aoki, T. (2006) Genetic vaccines protect red seabream, *Pagrus major*, upon challenge with red seabream iridovirus (RSIV). *Fish and Shellfish Immunology* 21, 130 - 138.

[6] Chao, C.B., Chen, C.Y., Lai, Y.Y., Lin, C.S. and Huang, H.T. (2004) Histological,

ultrastructural, and *in situ* hybridization study on enlarged cells in grouper *Epinephelus* hybrids infected by grouper iridovirus in Taiwan (TGIV). *Diseases of Aquatic Organisms* 58, 127 - 142.

[7] Chen, X.H., Lin, K.B. and Wang, X.W. (2003) Outbreaks of an iridovirus disease in maricultured large yellow croaker, *Larimichthys crocea* (Richardson), in China. *Journal of Fish Diseases* 26, 615 - 619.

[8] Chou, H.Y., Hsu, C.C. and Peng, T.Y. (1998) Isolation and characterization of a pathogenic *Iridovirus* from cultured grouper (*Epinephelus* sp.) in Taiwan. *Fish Pathology* 33, 201 - 206.

[9] Clouthier, S.C., VanWalleghem, E., Copeland, S., Klassen, C., Hobbs, G. *et al.* (2013) A new species of nucleo-cytoplasmic large DNA virus (NCLDV) associated with mortalities in Manitoba lake sturgeon *Acipenser fulvescens*. *Disease of Aquatic Organisms* 102, 195 - 209.

[10] Clouthier, S.C., VanWalleghem, E. and Anderson, E.D. (2015) Sturgeon nucleo-cytoplasmic large DNA virus phylogeny and PCR tests. *Diseases of Aquatic Organisms* 117, 93 - 106.

[11] Do, J.W., Cha, S.J., Kim, J.S., An, E.J., Lee, N.S. *et al.* (2005a) Phylogenetic analysis of the major capsid protein gene of iridovirus isolates from cultured flounders *Paralichthys olivaceus* in Korea. *Diseases of Aquatic Organisms* 64, 193 - 200.

[12] Do, J.W., Cha, S.J., Kim, J.S., An, E.J., Park, M.S. *et al.* (2005b) Sequence variation in the gene encoding the major capsid protein of Korean fish iridoviruses. *Archives of Virology* 150, 351 - 359.

[13] Dong, C., Weng, S., Shi, X., Xu, X., Shi, N. and He, J. (2008) Development of a mandarin fish *Siniperca chuatsi* fry cell line suitable for the study of infectious spleen and kidney necrosis virus (ISKNV). *Virus Research* 135, 273 - 281.

[14] Dong, C., Weng, S., Luo, Y., Huang, M., Ai, H. *et al.* (2010) A new marine megalocytivirus from spotted knifejaw, *Oplegnathus punctatus*, and its pathogenicity to freshwater mandarinfish, *Siniperca chuatsi*. *Virus Research*, 147, 98 - 106.

[15] Dong, C.F., Xiong, X.P., Luo, Y., Weng, S., Wang, Q. and He, J. (2013a) Efficacy of a formalin-killed cell vaccine against infectious spleen and kidney necrosis virus (ISKNV) and immunoproteomic analysis of its major immunogenic proteins. *Veterinary Microbiology* 162, 419 - 428.

[16] Dong, Y., Weng, S., He, J. and Dong, C. (2013b) Field trial tests of FKC vaccines against RSIV genotype *Megalocytivirus* in cage-cultured mandarin fish (*Siniperca chuatsi*) in an inland reservoir. *Fish and Shellfish Immunology* 35, 1598 - 1603.

[17] Drennan, J.D., Ireland, S., LaPatra, S.E., Grabowski, L., Carrothers, T.K. and Cain, K.D. (2005) High-density rearing of white sturgeon *Acipenser transmontanus* (Richardson) induces white sturgeon iridovirus disease among asymptomatic carriers. *Aquaculture Research* 36, 824 - 827.

[18] Drennan, J.D., LaPatra, S.E., Siple, J.T., Ireland, S. and Cain, K.D. (2006) Transmission of white sturgeon iri-dovirus in Kootenai River white sturgeon *Acipenser transmontanus*. *Diseases of Aquatic Organisms* 70, 37 - 45.

[19] Drennan, J.D., LaPatra, S.E., Samson, C.A., Ireland, S., Eversman, K.F. and Cain, K.D. (2007a) Evaluation of lethal and non-lethal sampling methods for the detection of white sturgeon iridovirus infection in white stur-geon, *Acipenser transmontanus* (Richardson). *Journal of Fish Diseases* 30, 367 - 379.

[20] Drennan, J.D., LaPatra, S.E., Swan, C.M., Ireland, S. and Cain, K.D. (2007b) Characterization of serum and mucosal antibody responses in white sturgeon (*Acipenser transmontanus* Richardson) following immunization with WSIV and a protein hapten antigen. *Fish and Shellfish Immunology* 23, 657 - 669.

[21] Fu, X., Li, N., Liu, L., Lin, Q., Wang, F. *et al*. (2011) Genotype and host range analysis of infectious spleen and kidney necrosis virus (ISKNV). *Virus Genes* 42, 97 - 109.

[22] Fu, X., Li, N., Lai, Y., Liu, L., Lin, Q., Shi, C., Huang, Z. and Wu, S. (2012) Protective immunity against iridovirus disease in mandarin fish, induced by recombinant major capsid protein of infectious spleen and kidney necrosis virus. *Fish & Shellfish Immunology* 33, 880 - 885.

[23] Fu, X., Li, N., Lin, Q., Guo, H., Zhang, D. *et al*. (2014) Protective immunity against infectious spleen and kidney necrosis virus induced by immunization with DNA plasmid containing mcp gene in Chinese perch *Siniperca chuatsi*. *Fish & Shellfish Immunology* 40, 259 - 266.

[24] Fynn-Aikens, K., Hung, S. S. O., Liu, W. and Hongbin, L. (1992) Growth, lipogenesis and liver composition of juvenile white sturgeon fed different levels of D-glucose. *Aquaculture* 105, 61 - 72.

[25] Georgiadis, M.P., Hedrick, R.P., Johnson, W.O., Yun, S. and Gardner, I.A. (2000) Risk factors for outbreaks of disease attributable to white sturgeon iridovirus and white stur-geon herpesvirus-2 at a commercial sturgeon farm. *American Journal of Veterinary Research* 61, 1232 - 1240.

[26] Georgiadis, M.P., Hedrick, R.P., Carpenter, T.E. and Gardner, I.A. (2001) Factors influencing transmission, onset and severity of outbreaks due to white sturgeon iridovirus in a commercial hatchery. *Aquaculture* 194, 21 - 35.

[27] Gias, E., Johnston, C., Keeling, S., Spence, R. P. and McDonald, W. L. (2011) Development of real-time PCR assays for detection of megalocytiviruses in imported ornamental fish. *Journal of Fish Diseases* 34, 609 - 618.

[28] Gibson-Kueh, S., Ngoh-Lim, G.H., Netto, P., Kurita, J., Nakajima, K. and Ng, M. L. (2004) A systematic iridoviral disease in mullet, *Mugil cephalus* L., and tiger grouper, *Epinephelus fuscoguttatus* Forsskal: a first report and study. *Journal of Fish Diseases* 27, 693 - 699.

[29] Go, J. and Whittington, R. (2006) Experimental transmis-sion and virulence of a megalocytivirus (Family *Iridoviridae*) of dwarf gourami (*Colisa lalia*) from Asia in Murray cod (*Maccullochella peelii peelii*) in Australia. *Aquaculture* 258, 140 - 149.

[30] Go, J., Lancaster, M., Deece, K., Dhungyel, O. and Whittington, R. (2006) The molecular epidemiology of iridovirus in Murray cod (*Maccullochella peelii peelii*) and dwarf gourami (*Colisa lalia*) from distant biogeographical regions suggests a link between trade in ornamental fish and emerging iridoviral diseases. *Molecular and Cellular Probes* 20, 212 - 222.

[31] He,J.G.,Wang,S.P.,Zeng,K.,Huang,Z.J.andChan,S.-M. (2000) Systemic disease caused by an iridovirus-like agent in cultured mandarin fish, *Siniperca chuatsi* (Basilewsky), in China. *Journal of Fish Diseases* 23, 219 - 222.

[32] He, J.G., Zeng, K., Weng, S.P. and Chan, S.-M. (2002) Experimental transmission, pathogenicity and physicalchemical properties of infectious spleen and kidney necrosis virus (ISKNV). *Aquaculture* 204, 11 - 24.

[33] Hedrick, R.P., Groff, J.M., McDowell, T. and Wingfield, W.H. (1990) An iridovirus infection of the integument of the white sturgeon *Acipenser transmontanus*. *Diseases of Aquatic Organisms* 8, 39 - 44.

[34] Hedrick, R.P., McDowell, T.S., Groff, J.M., Yun, S. and Wingfield, W.H. (1992) Isolation and some properties of an iridovirus-like agent from white sturgeon *Acipenser transmontanus*. *Diseases of Aquatic Organisms* 12, 75 - 81.

[35] Huang, S.M., Tu, C., Tseng, C.H., Huang, C.C., Chou, C.C. *et al*. (2011) Genetic analysis of fish iridoviruses isolated in Taiwan during 2001 - 2009. *Archives of Virology*

156, 1505 - 1515.

[36] Hung, S.O., Liu, W., Hongbin, L., Storebakken, S. and Cui, Y. (1997) Effect of starvation on some morpho-logical and biochemical parameters in white sturgeon *Acipenser transmontanus*. *Aquaculture* 150, 357 - 363.

[37] Imajoh, M., Ikawa, T. and Oshima, S. (2007) Characterization of a new fibroblast cell line from a tail fin of red sea bream, *Pagrus major*, and phylogenetic relationships of a recent RSIV isolate in Japan. *Virus Research* 126, 45 - 52.

[38] Inouye, K., Yamano, K., Maeno, Y., Nakajima, K., Matsuoka, M. *et al.* (1992) Iridovirus infection of cultured red sea bream, *Pagrus major*. *Fish Pathology* 27, 19 - 27.

[39] Ito, T., Yoshiura, Y., Kamaishi, T., Yoshida, K. and Nakajima, K. (2013) Prevalence of red sea bream iridovirus (RSIV) among organs of Japanese amber-jack (*Seriola quinqueradiata*) exposed to cultured RSIV. *Journal of General Virology* 94, 2094 - 2101.

[40] Jancovich, J.K., Chinchar, V.G., Hyatt, A., Miyazaki, T., Williams, T. and Zhang, Q.Y. (2012) Family *Iridoviridae*. In: King, A.M.Q., Adams, M.J., Carstens, E.B. and Lefkowitz, E. J. (eds) *Virus Taxonomy: Ninth Report of the International Committee on Taxonomy of Viruses*. Elsevier /Academic Press, London /Waltham, Massachussetts /San Diego, California, pp. 193 - 210. Available at: http: //www.trevorwilliams.info/ictv_iridoviridae_2012.pdf (accessed 7 November 2016).

[41] Jeong, J.B., Park, K.H., Kim, H.Y., Hong, S.H., Chung, J.-K. *et al.* (2004) Multiplex PCR for the diagnosis of red sea bream iridoviruses isolated in Korea. *Aquaculture* 235, 139 - 152.

[42] Jeong, J.B., Kim, H.Y., Jun, L.J., Lyu, J.H., Park, N.G. *et al.* (2008a) Outbreaks and risks of infectious spleen and kidney necrosis virus disease in freshwa-ter ornamental fishes. *Diseases of Aquatic Organisms* 78, 209 - 215.

[43] Jeong, J.B., Cho, H.J., Jun, L.J., Hong, S.H., Chung, J.K. and Jeong, H.D. (2008b) Transmission of iridovirus from freshwater ornamental fish (pearl gourami) to marine fish (rock bream). *Diseases of Aquatic Organisms* 82, 27 - 36.

[44] Jin, J.W., Kim, K.I., Kim, J.K., Park, N.G. and Jeong, H.D. (2014) Dynamics of megalocytivirus transmission between bivalve molluscs and rock bream *Oplegnathus fasciatus*. *Aquaculture* 428/429, 29 - 34.

[45] Jung, M.H., Jung, S.J., Vinay, T.N., Nikapitiya, C., Kim, J.O. *et al.* (2015) Effects of water temperature on mortality in *Megalocytivirus*-infected rock bream *Oplegnathus fasciatus* (Temminck et Schlegel) and development of protective immunity. *Journal of Fish Diseases* 38, 729 - 737.

[46] Jung, S., Miyazaki, T., Miyata, M., Danayadol, Y. and Tanaka, S. (1997) Pathogenicity of iridovirus from Japan and Thailand for the red sea bream *Pagrus major* in Japan, and histopathology of experimentally infected fish. *Fisheries Science* 63, 735 - 740.

[47] Jung, S.J. and Oh, M.J. (2000) Iridovirus-like infection associated with high mortalities of striped beakperch, *Oplegnathus fasciatus* (Temminck et Schlegel), in southern coastal areas of the Korean Peninsula. *Journal of Fish Diseases* 23, 223 - 226.

[48] Kawakami, H. and Nakajima, K. (2002) Cultured fish species affected by red sea bream iridoviral disease from 1996 to 2000. *Fish Pathology* 37, 45 - 47.

[49] Kim, T.J., Jang, E.J. and Lee, J.I. (2008) Vaccination of rock bream, *Oplegnathus fasciatus* (Temminck and Schlegel), using a recombinant major capsid protein of fish iridovirus. *Journal of Fish Diseases* 31, 547 - 551.

[50] Kim, W. S., Oh, M. J., Jung, S. J., Kim, Y. J. and Kitamura, S. I. (2005) Characterization of an iridovirus detected from cultured turbot *Scophthalmus maximus*

in Korea. *Diseases of Aquatic Organisms* 64, 175 - 180.

[51] Kim, W.S., Oh, M.J., Kim, J.O., Kim, D., Jeon, C.H. and Kim, J.H. (2010) Detection of megalocytivirus from imported tropical ornamental fish, paradise fish *Macropodus oper-cularis*. *Diseases of Aquatic Organisms* 90, 235 - 239.

[52] Kim, Y.J., Jung, S.J., Choi, T.J., Kim, H.R., Rajendran, K.V. and Oh, M.J. (2002) PCR amplification and sequence analysis of irido-like virus infecting fish in Korea. *Journal of Fish Diseases* 25, 121 - 124.

[53] Kurita, J. and Nakajima, K. (2012) Megalocytiviruses: a review. *Viruses* 4, 521 - 538.

[54] Kurita, J., Nakajima, K., Hirono, I. and Aoki, T. (1998) Polymerase chain reaction (PCR) amplification of DNA of red sea bream iridovirus (RSIV). *Fish Pathology* 33, 17 - 23.

[55] Kurobe, T., Kwak, K.T., MacConnell, E., McDowell, T.S., Mardones, F.O. and Hedrick, R.P. (2010) Development of PCR assays to detect iridovirus infections among captive and wild populations of Missouri River sturgeon. *Diseases of Aquatic Organisms* 93, 31 - 42.

[56] Kurobe, T., MacConnell, E., Hudson, C., Mardones, F.O. and Hedrick, R.P. (2011) Iridovirus infections among Missouri River sturgeon: initial characterization, transmission and evidence for establishment of a carrier state. *Journal of Aquatic Animal Health* 23, 9 - 18.

[57] Kwak, K.T., Gardner, I.A., Farver, T.B. and Hedrick, R.P. (2006) Rapid detection of white sturgeon iridovirus (WSIV) using a polymerase chain reaction (PCR) assay. *Aquaculture* 254, 92 - 101.

[58] LaPatra, S.E., Groff, J.M., Jones, G.R., Munn, B., Patterson, T.L. *et al*. (1994) Occurrence of white sturgeon iridovirus infections among cultured white sturgeon in the Pacific Northwest. *Aquaculture* 126, 201 - 210.

[59] LaPatra, S.E., Groff, J.M., Patterson, T.L., Shewmaker, W.K., Casten, M. *et al*. (1996) Preliminary evidence of sturgeon density and other stressors on manifestation of white sturgeon iridovirus disease. *Journal of Applied Aquaculture* 6, 51 - 58.

[60] LaPatra, S.E., Ireland, S.C., Groff, J.M., Clemens, K.M. and Siple, J.T. (1999) Adaptive disease management strategies for the endangered population of Kootenai River white sturgeon. *Fisheries* 24, 6 - 13.

[61] LaPatra, S.E., Groff, J.M., Keith, I., Hogans, W.E. and Groman, D. (2014) Case report: concurrent herpesviral and presumptive iridoviral infection associated with disease in cultured shortnose sturgeon, *Acipenser brevirostrum* (L.), from the Atlantic coast of Canada. *Journal of Fish Diseases* 37, 141 - 147.

[62] Lü, L., Zhou, S.Y., Chen, C., Weng, S.P., Chan, S.-M. and He, J.G. (2005) Complete genome sequence analysis of an iridovirus isolated from the orange-spotted grouper, *Epinephelus coioides*. *Virology* 339, 81 - 100.

[63] Matsuoka, S., Inouye, K. and Nakajima, K. (1996) Cultured fish species affected by red sea bream iridoviral disease from 1991 to 1995. *Fish Pathology* 31, 233 - 234.

[64] Nakajima, K. and Kurita, J. (2005) Red sea bream iridovi-ral disease. *Uirusu* 55, 115 - 126.

[65] Nakajima, K. and Maeno, Y. (1998) Pathogenicity of red sea bream iridovirus and other fish iridoviruses to red sea bream. *Fish Pathology* 33, 143 - 144.

[66] Nakajima, K. and Sorimachi, M. (1994) Biological and physicochemical properties of the iridovirus isolated from cultured red sea bream, *Pagrus major*. *Fish Pathology* 29, 29 - 33.

[67] Nakajima, K. and Sorimachi, M. (1995) Production of monoclonal antibodies against red sea bream iridovirus. *Fish Pathology* 30, 47 - 52.

[68] Nakajima, K., Maeno, Y., Fukudome, M., Fukuda, Y., Tanaka, S. *et al*. (1995)

Immunofluorescence test for the rapid diagnosis of red sea bream iridovirus infection using monoclonal antibody. *Fish Pathology* 30, 115 – 119.

[69] Nakajima, K., Maeno, Y., Kurita, J. and Inui, Y. (1997) Vaccination against red sea bream iridoviral disease in red sea bream. *Fish Pathology* 32, 205 – 209.

[70] Nakajima, K., Maeno, Y., Yokoyama, K., Kaji, C. and Manabe, S. (1998) Antigen analysis of red sea bream iridovirus and comparison with other fish iridoviruses. *Fish Pathology* 33, 73 – 78.

[71] Nakajima, K., Maeno, Y., Honda, A., Yokoyama, K., Tooriyama, T. and Manabe, S. (1999) Effectiveness of a vaccine against red sea bream iridoviral disease in a field trial test. *Diseases of Aquatic Organisms* 36, 73 – 75.

[72] Nolan, D., Stephens, F., Crockford, M., Jones, J.B. and Snow, M. (2015) Detection and characterization of viruses of the genus *Megalocytivirus* in ornamental fish imported into an Australian border quarantine premises: an emerging risk to national biosecurity. *Journal of Fish Diseases* 38, 187 – 195.

[73] Oh, M.J., Kitamura, S.I., Kim, W.S., Park, M.K., Jung, S.J. *et al*. (2006) Susceptibility of marine fish species to a megalocytivirus, turbot iridovirus, isolated from turbot, *Psetta maximus* (L.). *Journal of Fish Diseases* 29, 415 – 421.

[74] Oh, S.Y. and Nishizawa, T. (2013) Optimizing the quanti-tative detection of the red seabream iridovirus (RSIV) genome from splenic tissues of rock bream *Oplegnathus fasciatus*. *Fish Pathology* 48, 21 – 24.

[75] Oh, S.Y., Oh, M.J. and Nishizawa, T. (2014) Potential for a live red seabream iridovirus (RSIV) vaccine in rock bream *Oplegnathus fasciatus* at a low rearing temperature. *Vaccine*, 32, 363 – 368.

[76] OIE (2003) *Manual of Diagnostic Tests for Aquatic Animals*. World Organisation for Animal Health, Paris. Updated version 2016 available at: http://www.oie.int/international-standard-setting/aquatic-manual/access-online/(accessed 7 November 2016).

[77] OIE (2012) Red sea bream iridoviral disease. In: *Manual of Diagnostic Tests for Aquatic Animals 2012*. World Organisation for Animal Health, Paris, pp. 345 – 356. Updated version 2016 available at: http://www.oie.int/index.php? id=2439&L=0&htmfile=chapitre_rsbid.htm (accessed 7 November 2016).

[78] Oshima, S., Hata, J., Hirasawa, N., Ohtaka, T., Hirono, I. *et al*. (1998) Rapid diagnosis of red sea bream iridovirus infection using the polymerase chain reaction. *Diseases of Aquatic Organisms* 32, 87 – 90.

[79] Paperna, I., Vilenkin, M. and de Matos, A.P. (2001) Iridovirus infections in farm-reared tropical ornamental fish. *Diseases of Aquatic Organisms* 48, 17 – 25.

[80] Pikitch, E.K., Doukakis, P., Lauck, L., Chakrabarty, P. and Erickson, D.L. (2005) Status, trends and management of sturgeon and paddlefish fisheries. *Fish and Fisheries* 6, 233 – 265.

[81] Raverty, S., Hedrick, R., Henry, J. and Saksida, S. (2003) Diagnosis of sturgeon iridovirus infection in farmed white sturgeon in British Columbia. *Canadian Veterinary Journal* 44, 327 – 328.

[82] Rimmer, A.E., Becker, J.A., Tweedie, A. and Whittington, R.J. (2012) Development of a quantitative polymerase chain reaction (qPCR) assay for the detection of dwarf gourami iridovirus (DGIV) and other megalocytiviruses and comparison with the Office International des Epizooties (OIE) reference PCR protocol. *Aquaculture* 358/359, 155 – 163.

[83] Rodger, H.D., Kobs, M., Macartney, A. and Frerichs, G.N. (1997) Systemic iridovirus infection in freshwater angelfish, *Pterophyllum scalare* (Lichtenstein). *Journal of Fish Diseases* 20, 69 – 72.

[84] Shi, C.Y., Wang, Y.G., Yang, S.L., Huang, J. and Wang, Q.Y. (2004) The first report of an iridovirus-like agent infec-tion in farmed turbot, *Scophthalmus maximus*, in China. *Aquaculture* 236, 11 - 25.
[85] Shimmoto, H., Kawai, K., Ikawa, T. and Oshima, S. (2010) Protection of red sea bream *Pagrus major* against red sea bream iridovirus infection by vaccination with a recombinant viral protein. *Microbiology and Immunology* 54, 135 - 142.
[86] Sriwanayos, P., Francis-Floyd, R., Stidworthy, M.F., Petty, B.D., Kelley, K. and Waltzek, T.B. (2013) Megalocytivirus infection in orbiculate batfish *Platax orbicularis*. *Diseases of Aquatic Organisms* 105, 1 - 8.
[87] Subramaniam, K., Shariff, M., Omar, A.R., Hair-Bejo, M. and Ong, B.L. (2014) Detection and molecular characterization of infectious spleen and kidney necrosis virus from major ornamental fish breeding states in Peninsular Malaysia. *Journal of Fish Diseases* 37, 609 - 618.
[88] Sudthongkong, C., Miyata, M. and Miyazaki, T. (2002a) Iridovirus disease in two ornamental tropical freshwater fishes: African lampeye and dwarf gourami. *Diseases of Aquatic Organisms* 48, 163 - 173.
[89] Sudthongkong, C., Miyata, M. and Miyazaki, T. (2002b) Viral DNA sequences of genes encoding the ATPase and the major capsid protein of tropical iridovirus isolates which are pathogenic to fishes in Japan, South China Sea and Southeast Asian countries. *Archives of Virology* 147, 2089 - 2109.
[90] Tanaka, S., Aoki, H., Inoue, M. and Kuriyama, I. (2003) Effectiveness of fasting against red sea bream iridoviral disease in red sea bream. *Fish Pathology* 38, 67 - 69.
[91] Waltzek, T.B., Marty, G.D., Alfaro, M.E., Bennett, W.R. Garver, K.A. *et al*. (2012) Systemic iridovirus from threespine stickleback *Gasterosteus aculeatus* represents a new megalocytivirus species (family *Iridoviridae*). *Diseases of Aquatic Organisms* 98, 41 - 56.
[92] Wang, Q., Zeng, W.W., Li, K.B., Chang, O.Q., Liu, C. *et al*. (2011) Outbreaks of an iridovirus in marbled sleepy goby, *Oxyeleotris marmoratus* (Bleeker), cultured in southern China. *Journal of Fish Diseases* 34, 399 - 402.
[93] Wang, X.W., Ao, J.Q., Li, Q.G. and Chen, X.H. (2006) Quantitative detection of a marine fish iridovirus isolated from large yellow croaker, *Pseudosciaena crocea*, using a molecular beacon. *Journal of Virological Methods* 133, 76 - 81.
[94] Wang, Y.Q., Lu, L., Weng, S.P., Huang, J.N., Chan, S.M. and He, J.G. (2007) Molecular epidemiology and phylogenetic analysis of a marine fish infectious spleen and kidney necrosis virus-like (ISKNV-like) virus. *Archives of Virology* 152, 763 - 773.
[95] Watson, L.R., Groff, J.M. and Hedrick, R.P. (1998a) Replication and pathogenesis of white sturgeon iridovirus (WSIV) in experimentally infected white sturgeon *Acipenser transmontanus* juveniles and sturgeon cell lines. *Diseases of Aquatic Organisms* 32, 173 - 184.
[96] Watson, L.R., Milani, A. and Hedrick, R.P. (1998b) Effects of water temperature on experimentally-induced infections of juvenile white sturgeon (*Acipenser transmontanus*) with the white sturgeon iridovirus (WSIV). *Aquaculture* 166, 213 - 228.
[97] Weber, E.S., Waltzek, T.B., Young, D.A., Twitchell, E.L., Gates, A.E. *et al*. (2009) Systemic iridovirus infection in the Banggai cardinalfish (*Pterapogon kauderni* Koumans 1933). *Journal of Veterinary Diagnostic Investigation* 21, 306 - 320.
[98] Wen, C.M., Lee, C.W., Wang, C.S., Cheng, Y.H. and Huang, H.Y. (2008) Development of two cell lines from *Epinephelus coioides* brain tissue for charac-terization of betanodavirus and megalocytivirus infec-tivity and propagation. *Aquaculture* 278, 14 - 21.
[99] Weng, S.P., Wang, Y.Q., He, J.G., Deng, M., Lu, L. *et al*. (2002) Outbreaks of an

iridovirus in red drum, *Sciaenops ocellata* (L.), cultured in southern China. *Journal of Fish Diseases* 25, 681 - 685.

[100] Won, K.M, Cho, M.Y., Park, M.A., Jee, B.Y., Myeong, J.I. and Kim, J.W. (2013) The first report of a megalocytivirus infection in farmed starry flounder, *Platichthys stel-latus*, in Korea. *Fisheries and Aquatic Sciences* 16, 93 - 99.

[101] Zhang, M., Xiao, Z. Z., Hu, Y. H. and Sun, L. (2012) Characterization of a megalocytivirus from cultured rock bream, *Oplegnathus fasciatus* (Temminck & Schlege), in China. *Aquaculture Research* 43, 556 - 564.

13 鲑甲病毒

Marius Karlsen* 和 Renate Johansen

13.1 引言

甲病毒(*Alphavirus*)是披膜病毒科(Togaviridae)的一个 RNA 病毒属。大多数已知的甲病毒是蚊媒传播的,并在诸如鸟类、啮齿动物和包括人类在内的大型哺乳动物等陆生宿主中引发疾病(Strauss and Strauss, 1994)。感染甲病毒可导致多种症状,如皮疹、胃肠疾病、关节炎、肌肉炎和脑炎(Kuhn, 2007; Steele and Twenhafel, 2010)。鲑鱼胰腺病病毒,俗称鲑甲病毒(Salmonid Alphavirus, SAV)(Weston et al., 2002),是目前唯一已知的以鱼类为天然宿主的甲病毒(Powers et al., 2001)。SAV 与该属中其他成员的亲缘关系较远,但它仍然会导致一些类似于哺乳动物感染中的病理表现(McLoughlin and Graham, 2007; Biacchesi et al., 2016)。1995 年,从爱尔兰患胰腺病(pancreas disease, PD)的海水养殖大西洋鲑(*Salmo salar*)的细胞培养物中首次分离出 SAV(Nelson et al., 1995)。与此同时,在法国患有昏睡病(sleeping disease, SD)的淡水虹鳟中分离出一种类似的病毒(Castric et al., 1997)。尽管 PD 和 SD 是由同种病毒引起的,并且具有相同的组织病理学特征(Boucher and Baudin Laurencin, 1996; Weston et al., 2002),但这两个不同的名称仍然被沿用。这可能是历史原因造成的,因为 SD 主要与淡水中较小体型(约 0.3~2 kg)虹鳟的生产有关,而 PD 则与海水养殖业中大西洋鲑和虹鳟的感染有关。

SAV 的基因组是一条长 11.9 kb 的大的单链 RNA 分子,具有两个大的可读框(Weston et al., 2002)。它们编码复制基因组的四种非结构蛋白(NSP1~NSP4),以及结构蛋白(衣壳、E1、E2、E3 和 6K/TF)。这些结构蛋白构成了一个二十面体状的病毒颗粒,其包含一个来源于宿主的膜包围的内衣壳(Villoing et al., 2000a)。在该膜上嵌入大量糖蛋白刺突,其中颗粒表面的大部分由糖蛋白 E2 构成(Moriette et al., 2005),同时,糖蛋白 E2 也是免疫系统的主要抗原靶标。

* 通信作者邮箱:marius.karlsen@pharmq.no。

来自欧洲不同地区的 SAV 表现出相当大的遗传多样性,已发现有 SAV1～SAV6 六种遗传亚型(Fringuelli et al., 2008)。在不列颠群岛周围的大西洋鲑和虹鳟中发现了亚型 1、2、4、5 和 6(Graham et al., 2012; Karlsen et al., 2014);对 SAV3 的报道仅见于挪威(Hodneland et al., 2005);SAV2 似乎比其他亚型更具多样性,并且包含两个不同的谱系,一个在苏格兰和挪威的海洋养殖场中引起 PD,另一个则导致了欧洲大陆 SD,偶尔在苏格兰和英格兰地区暴发(Graham et al., 2012;Hjortaas et al., 2013;Karlsen et al., 2014)。

尽管尚不清楚 SAV 最初是如何被引入养殖鱼类中的,但在北海或其周围海域可能存在多种多样的 SAV 野生传染源。系统发育断代研究表明,每个亚型都必定代表着这个感染源的单独引入事件(Karlsen et al., 2014)。已有证据表明,在野生欧洲黄盖鲽(*Limanda limanda*)、美洲拟庸鲽(*Hippoglossoides platessoides*)和扁海鲽(*Platessa platessa*)中存在 SAV (Snow et al., 2010; Bruno et al., 2014; McCleary et al., 2014; Simons et al., 2016),且在不列颠群岛周围的欧洲黄盖鲽中 SAV 的流行率似乎相对较高。目前尚不清楚这些鲆鲽鱼类是否构成 SAV 的原始野生宿主,或者它们是否因鲑感染的外溢而受到感染(Karlsen, 2015; Simons et al., 2016)。在褐鳟(*Salmo trutta*)中的病原传播实验表明,这种鱼也容易感染 SAV(Boucher et al., 1995)。

SAV 感染可以发展成高滴度的病毒血症(Desvignes et al., 2002),在病毒血症期,鱼会将大量感染性病毒颗粒排入水中(Andersen et al., 2010; Andersen et al., 2012)。当宿主产生强烈的抗体反应来清除病毒血症时,感染性病毒颗粒向水中的脱落会显著减少或停止。脱落的病毒颗粒可以感染附近的鱼类(McLoughlin et al., 1996)。通过同一海流网络中养鱼场之间的水平传播,SAV 可能在养殖鱼类的种群中持续存在(Viljugrein et al., 2009)。在这种网络中的鱼和设备的运输会增强病毒传输,这也是偶尔发生的长距离病毒传输最可能的解释(Karlsen et al., 2014)。在以感染鱼为食的鲑虱(*Lepeophtheirus salmonis*)中也检测到了病毒 RNA,但其在海虱中的活跃复制尚未得到证实(Karlsen et al., 2006; Petterson et al., 2009),且在实验室条件下也不会感染海虱(Karlsen et al., 2015)。SAV2 /SD 在较小规模的虹鳟养殖业中的传播可能与活鱼或鱼卵贸易运输有关,并且也存在垂直传播的可能性(Castric et al., 2005; Borzym et al., 2014)。目前对 SD 动物流行病学的研究很少。

在欧洲,SAV 对经济造成的损失是巨大的(Aunsmo et al., 2010)。在比较挪威鲑鱼产业中与病原相关的损失时,SAV 感染造成的损失仅次于海虱感染。死亡导致的养殖量减少是损失的一部分,但鱼体生长不良和鱼片质量下降也是感染的主要后果。在挪威,据估计 2007 年 SAV 感染使大西洋鲑的生产成本每千克增加了 5 挪威克朗(从 25 挪威克朗 / 千克增加到 30 挪威克朗 / 千克)。据估计,在一个有 50 万尾二龄鲑的养殖生产基地,SAV 暴发造成的损失约为 1 440 万挪威克朗(Aunsmo et al., 2010)。因 SAV 主要感染幼鱼,因此这些损失数据不应被视为虹鳟养殖业中 SAV2 流行病的代表。由于 SD 的暴发没有定期报告,因此虹鳟由于 SAV 流行病暴发造成的病害损失难以估算。

13.2 诊断

对于 SAV 的检测,可以通过观察其对宿主的影响来间接诊断,或者通过鉴定病毒特异性分子(如宿主体内的病毒基因组、病毒蛋白或传染性病毒颗粒)直接检测。初步诊断和间接诊断可以基于大体临床症状和组织病理学(McLoughlin et al., 2002; Taksdal et al., 2007),但由于其他病原可能产生类似的临床体征,因此通过鉴定病毒分子来确诊 SAV 感染始终是较为可靠的。

临床疾病全年都有发生,但在夏季水温较高的海水养殖场中更常见,这可能反映出该病毒的最适流行温度为 10～15 ℃(Villoing et al., 2000a; Graham et al., 2008)。被感染鱼的第一个临床症状通常是食欲下降,随后聚集在养殖网箱表面、游动缓慢,有时背鳍露出水面。这种行为通常从一个网箱开始,然后蔓延至其他网箱。在疾病的临床暴发期,经常可见水中有大量淡黄色拖曳粪便(McVicar, 1987; McLoughlin et al., 2002; Taksdal et al., 2007)。大多数 SAV 感染会导致死亡率增加。

一项对 2006—2007 年挪威 23 个地区因感染 SAV3 而暴发 PD 的研究表明,在疫情暴发期间的平均死亡率为 6.9%,死亡率上升平均持续了 2.8 个月(Jansen et al., 2010)。在挪威,SAV2 造成的死亡率被认为是较低的(Jansen et al., 2015)。据报道,不列颠群岛周围暴发的死亡率变化很大(Crockford et al., 1999),但在极端情况下很难排除其他病原的影响。

SAV 引起的大体病理包括肝脏变色、幽门盲囊和内脏脂肪上出现瘀斑,以及腹腔内有腹水(图 13.1)。此外,肠道是空的或充满淡黄色粪便(McLoughlin and Graham, 2007)。鱼片损伤、体表暗沉和体色斑驳不均是 SAV 感染进行期间或前次感染引发的其他症状(Taksdal et al., 2012)。在暴发后期,可以观察到 K 因子(相对体长的低体重)非常低的鱼(McLoughlin and Graham, 2007)。

SAV 并不总是表现为急性临床疾病,感染可能在急性暴发后持续存在(Graham et al., 2010),也可能在没有急性期或几乎没有观察到临床体征的情况下发生(Graham et al., 2006b; Jansen et al., 2010)。这种亚临床感染可能传播病毒,这突显了不依赖临床症状而是通过更具体的方法检测养殖场中 SAV 的重要性(Graham et al., 2006b, 2010; Hodneland and Endresen, 2006)。

对 PD 的确诊通常基于组织病理学和病毒 RNA 的检测。由 SAV 引起的大体病理和组织病理学与鲑科鱼类的其他病毒性疾病具有相似性,如心脏和骨骼

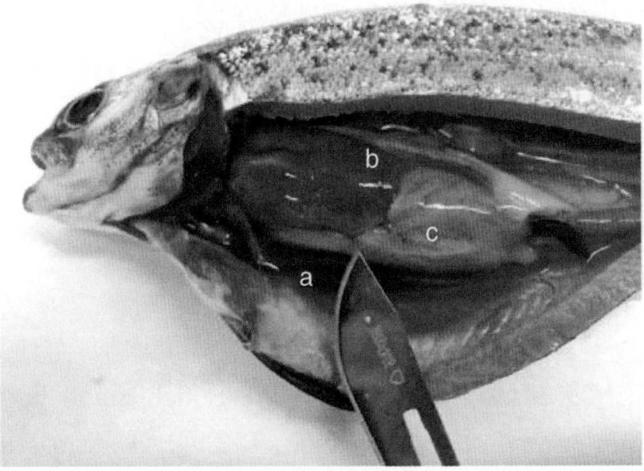

图 13.1　鲑甲病毒(SAV)实验感染后患有胰腺疾病(PD)的大西洋鲑。临床上 PD 的典型病理学包括：(a) 腹水；(b) 肝脏变色；(c) 幽门盲囊上的瘀点。图片由 Rolf Hetlelid Olsen 和 PHARMAQ AS 提供

肌炎症(heart and skeletal muscle inflammation, HSMI)、心肌病综合征(cardiomyopathy syndrome, CMS)和传染性胰腺坏死(McLoughlin and Graham, 2007)。而胰腺病变和心脏及骨骼肌病理学的结合确实能够将 PD 与这些疾病区分开来。在鱼类的感染过程中很容易检测到 SAV，RT－PCR 因其灵敏度高、具有特异性和快速检测大量样品的能力而被广泛使用(Graham et al., 2006a; Hodneland and Endresen, 2006)。挪威和英国的几个商业实验室(如 Patogen Analyse、PHARMAQ Analytiq、Fish Vet Group)都提供这项服务。在病毒血症期间，即使不是全部，也可在大多数器官中检测到病毒 RNA，但心脏中病毒 RNA 的含量通常较高(Andersen et al., 2007)，在该器官中 PCR 信号的持续时间也更长。基于 PCR 的方法可用于分离不同的 SAV 亚型(Hodneland and Endresen, 2006)，因此，这些方法确实提供了比大多数其他检测方法更详细的动物流行病学信息。

其他诊断方法包括病毒血症阶段，在抗体反应被激活之前从血清或组织匀浆中分离出细胞培养物中的病毒(Nelson et al., 1995; Castric et al., 1997; Desvignes et al., 2002)。大多数鱼类细胞系似乎对该病毒敏感(Graham et al., 2008)，但致细胞病变效应(CPE)在某些情况下可能很弱或不存在。因此建议通过特异性分子方法，如 IFAT，进一步确认 SAV 的体外复制(Todd et al., 2001; Moriette et al., 2005)。病毒感染也可以使用抗体介导的检测方法进行原位检测，如免疫组织化学(Villoing et al., 2000b)或 RNA 探针杂交(Cano et al., 2015)。这些方法可用于研究组织趋向性和发病机制，但比 PCR 的灵敏度低且费力。因此，它们不用于常规诊断。感染晚期或完全清除的感

染可以使用中和抗体法进行检测，目前已经开发了一种病毒中和检测方法，尤其是在爱尔兰和英国被用于筛查以前接触过 SAV 的鱼(Graham et al., 2003)。

13.3 病理学

13.3.1 实验室攻毒的死亡率

早期的 SAV 实验感染研究并没有导致死亡率的增加，而且类似于有时从田间报道的亚临床感染(Boucher et al., 1995; Boucher and Baudin Laurencin, 1996; McLoughlin et al., 1996; Andersen et al., 2007; Christie et al., 2007)。然而，后来的研究表明，实验感染后死亡率显著增加并伴有临床症状(Boscher et al., 2006; Moriette et al., 2006; Karlsen et al., 2012; Xu et al., 2012; Hikke et al., 2014a)。因此，SAV 的实验感染确实足以涵盖田间报道的临床体征和观察范围。临床疾病可以通过已测试的所有感染途径重现：腹腔和肌内注射、浸泡和共栖感染。如果 Jansen 等(2010)计算的 SAV3 田间平均死亡率为 6.9%具有代表性，那么就可以预计一组 30 尾鱼中只有 2 尾死亡。因此，文献中报道的许多实验感染缺乏检测这种小规模死亡率的统计能力。实验室研究中，宿主状态、攻毒滴度和病毒株差异等因素都会影响疾病的严重程度。然而，这有可能诱导甚至比极端现场病例更高的死亡率，特别是在使用高剂量毒株的注射攻毒模型中(Karlsen et al., 2012)。

13.3.2 组织病理学变化

胰腺疾病的名称来源于对该疾病的最初描述，即早先报道的胰腺坏死(Munro et al., 1984)。后来发现心脏、骨骼肌和食道的肌病也是该疾病的重要特征(Ferguson et al., 1986)。这些病变依次发生发展，胰腺坏死后是心肌病和心脏炎症(Boucher and Baudin Laurencin, 1994; McLoughlin et al., 1996; Weston et al., 2002)，骨骼肌的病变发展较晚。

实验感染的第一个组织学观察是心脏[图 13.2(B)]和胰腺[图 13.3(C)]中的单个细胞坏死(Mcloughlin and Graham, 2007)。坏死的心肌细胞通过苏木精-伊红(haematoxylin-eosin, HE)染色形成特征性透明嗜酸性外观，通常被称为 PD 细胞，这一过程被称为单细胞渐进性坏死(single cell necrosis, SCN)。在这个阶段，很少或没有临床症状。在田间的临床暴发期间，最常见的是心脏心室出现严重炎症[图 13.2(C)]和胰腺坏死或萎缩[图 13.3(B)和 13.3(C)]，还可以观察到骨骼肌炎症，尤其是鱼体侧红色肌肉中的炎症[图 13.2(D)]。在血氧合能力下降的鱼中常可见肝脏坏死，如果心脏受到严重影响，则在患 PD 的鱼中也可能观察到肝脏坏死[图 13.3(D)]。

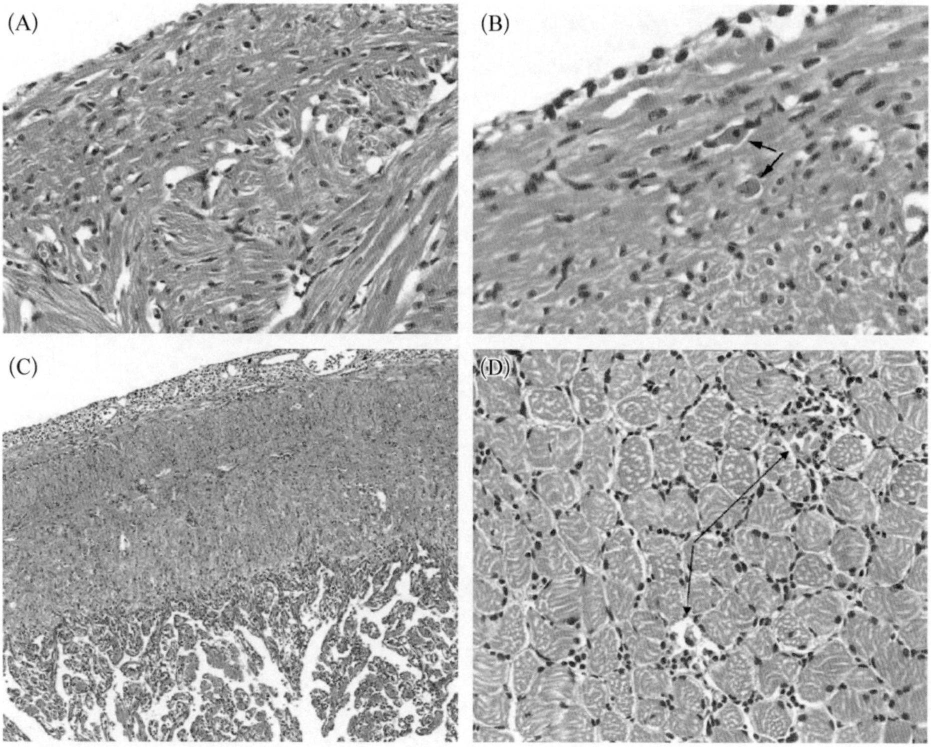

图 13.2 使用苏木精-伊红染色的鲑组织学图像：(A) 正常心脏组织；(B) 鲑甲病毒(SAV)感染早期伴有单细胞坏死(箭头所示)的心室；(C) 胰腺病(PD)临床暴发阶段患严重炎症的心室；(D) 胰腺病鲑红色骨骼肌中的局灶性炎症(箭头所示)

在疾病中存活下来的鱼的胰腺可以完全再生，而心脏和骨骼肌的损伤需要更长时间才能愈合，并且会降低鱼片质量(Taksdal et al., 2012)。

13.3.3 组织向性和体内增殖

一般来说，甲病毒不是宿主细胞特异性的。SAV 的复制装置可以在不同条件下以及在鱼、哺乳动物、昆虫和甲壳类动物的细胞中发挥作用(Graham et al., 2008; Olsen et al., 2013; Hikke et al., 2014b)。虽然宿主细胞的选择可能受复制装置和宿主蛋白之间相互作用的影响，但组织趋向性可能更多取决于结构蛋白，特别是覆盖病毒颗粒大部分表面的 E2(Kuhn, 2007; Voss et al., 2010; Karlsen et al., 2015)。基于与其他甲病毒的同源性，推测 E2 是 SAV 的受体结合蛋白(Villoing et al., 2000a)。尽管不同的 SAV 亚型中 E2 具有相当大的遗传多样性(Fringuelli et al., 2008)，但它们似乎都具有性质上相同的组织趋向性(Graham et al., 2011)。不过，它们在选择靶器官方面可能在数量上有所不同。

SAV 进入宿主的途径尚不清楚，但病毒通过水接触传播(McLoughlin

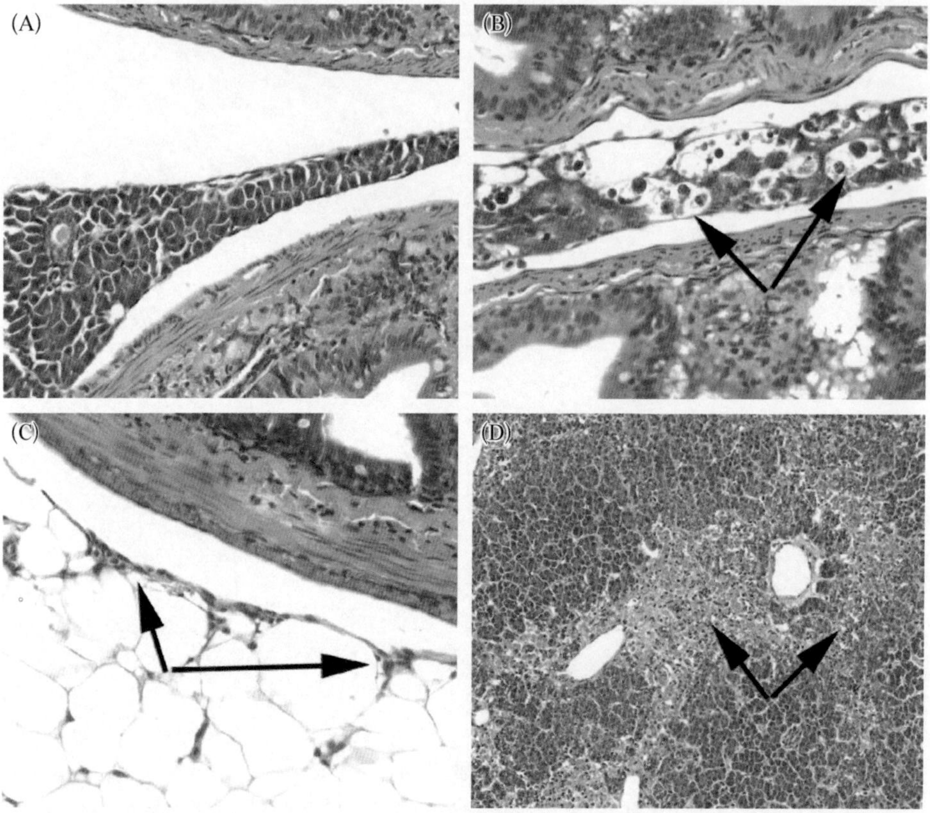

图 13.3　使用苏木精-伊红染色的鲑组织学图像：(A) 小肠之间的正常胰腺组织；(B) 胰腺病早期的胰腺细胞坏死(箭头所示)；(C) 胰腺病临床阶段中的胰腺萎缩导致胰腺组织缺失(箭头所示)；(D) 鲑甲病毒感染致心力衰竭的肝脏坏死(箭头所示)

et al., 1996)。因此，肠、鳃和皮肤中的细胞是可能的进入位点。在最初感染后，SAV在宿主中迅速扩增，并通过高滴度病毒血症全身性扩散到多个器官(Desvignes et al., 2002; Andersen et al., 2007)。在此阶段，在所有富血器官中都能检测到病毒RNA。Andersen等(2007)的研究证实，在感染的后期(腹腔注射后190 d)，仍可在伪鳃、鳃、心脏和肾脏中检测到病毒RNA。目前尚不清楚这种持续性是否会导致感染颗粒的有限脱落，但实验室实验已将主要脱落期与感染的病毒血症阶段联系起来(Andersen, 2012)。在此类携带者中诱发临床暴发的复发的尝试并未成功(Andersen et al., 2007)。已证实在胰腺组织、心脏、骨骼肌和白细胞中都有活跃的复制(Houghton, 1995; Villoing et al., 2000b; Moriette et al., 2005; Andersen, 2012; Cano et al., 2015)。最近，还有研究证明在虹鳟中，肌肉卫星细胞是病毒的靶细胞(Biacchesi et al., 2016)。一些甲病毒与神经营养性感染和神经病理学有关(Strauss and Strauss, 1994)，但这似乎不是鲑科鱼类中SAV感染的重要特征(McLoughlin and Graham，2007）。

13.3.4 毒株之间的毒力差异

SAV 毒株之间在病理学上似乎没有质的差异，但它们确实在引起组织损伤的数量上存在显著差异（Weston et al.，2002；Christie et al.，2007；Graham et al.，2011；Taksdal et al.，2015）。虽然尚不清楚毒力的差异是不是亚型特异性的，抑或是每个亚型内存在较大的变异，但 SAV2 和 SAV3 之间的初步比较表明，SAV3 通常比 SAV2 更具毒性（Taksdal et al.，2015）。

SAV 通过导致鱼体内毒力降低的突变来适应细胞培养，这已在 SAV2 和 SAV3 毒株中观察到（Moriette et al.，2006；Merour et al.，2013；Petterson et al.，2015）。毒力的变化是由 E2 中的突变引起的，单次的基因替代足以减弱 SAV2 对虹鳟的致病性（Merour et al.，2013）。这表明，尽管 SAV 在其病理上作为一种相当稳定和可预测的病毒，但它有能力迅速改变其特征。在 SAV3 的野外分离株中也观察到 E2 的替代突变。现在，E2 蛋白 206 位的丝氨酸取代脯氨酸的突变在当前 SAV3 亚型流行株中占主导优势，但有研究认为这种突变影响的是病毒的传播速度而不是其毒力（Karlsen et al.，2015）。

13.4 病理生理学

SAV 的主要靶器官（外分泌胰腺、心肌和骨骼肌）可能受到感染的严重影响。胰腺组织坏死可能会降低消化食物的能力，这可能是感染暴发期间经常观察到生长减缓和产生大量粪便拖尾的原因（McLoughlin and Graham，2007）。生长减缓的另一个原因是食欲不振，这是鱼类对严重感染的普遍反应。

心脏炎症导致血液循环能力下降，这可以解释一些尸检结果，如肝脏变色、出血和有腹水。肝脏坏死可能是由于血氧合能力降低的结果，因此可能是由心力衰竭导致的继发性表现。患有 PD 的鱼通常也会在鳃中携带外源病原并发生感染，这会减少鱼的气体交换（Nylund et al.，2011）。对于患有这种合并感染的鱼来说，血液循环衰竭尤其危险。

SAV 炎症反应可能会导致加工鱼片的变色。红色肌肉对低氧水平也很敏感，因此，血液循环衰竭可以解释为什么红色肌肉通常比白色无氧肌肉受到更大影响。

SAV 感染在细胞水平上一个有趣的方面是病毒调节转录和翻译的能力（Xu et al.，2010）。甲病毒有几种不同的策略来实现这一点，包括 NSP2 和衣壳与细胞蛋白的相互作用。已有研究发现，表达 SAV 衣壳的鱼细胞无法继续进行细胞分裂（Karlsen et al.，2010b）。这些转录组变化有很多可能是短暂性的，目前尚不清楚这是如何影响生物体的。

13.5 防控策略

在挪威，SAV3 和 SAV2 的主导传播源似乎都是养殖鲑鱼（Viljugrein et al.，2009；Karlsen et al.，2014；Hjortaas et al.，2016）。因此，疾病缓解策略的重点是这个传播源。海水养殖场之间的水接触可能是传播的主要途径之一（Viljugrein et al.，2009）。另一种重要性未知的途径是通过载体进行机械性传染，例如活鱼舱船或养殖场之间共享的其他设施（Karlsen et al.，2014）。2007年，挪威加大了减少 SAV3 传播的力度，PD 成为应具报的传染病，并在西海岸划定了一个流行区。SAV 阳性或表现出其他 PD 迹象的鱼不允许离开该区域。SAV 阳性的鱼允许保留在这个区域直至屠宰，在重新引种之前必须对养殖场地进行消毒和实施休耕。区域外受感染的鱼被立即宰杀，养殖场实施休耕，以防止疾病进一步蔓延。随着 SAV2 在 SAV3 流行区以北的引入（Hjortaas et al.，2013），一项 SAV2 监管规定被制定，以界定 SAV3 流行区以北的 SAV2 流行区。从该区域接收二龄鲑的海水养殖场必须在其转移出该区域后 2 个月和 4 个月，通过 PCR 检测证实无 SAV 感染。该法规还规定了其他的措施，如限制运输 SAV2 阳性的鱼类，以及对活鱼舱船和养殖网箱进行消毒的详细规定。

为了减少因 PD 造成的经济损失，业界采用的一种策略是提前收获。在这一策略中，养殖者可以利用基于 PCR 的农场感染状况监测所产生的疾病信息。随着疾病流行率的增加，可以预期临床疫情的临近，养殖者可以选择宰杀鱼类，以避免因死亡、生长减缓和鱼片质量差造成的相关损失。据估计，当鱼的质量大于 3.2 kg 时，这种策略是划算的。

13.5.1 疫苗

一种基于 SAV1 型病毒株 F93－125 的油包水剂型单价全病毒灭活疫苗，已由默沙东动物保健公司（MSD Animal Health）于 2003 年在爱尔兰和挪威以及 2005 年在英国上市销售。该疫苗已广泛应用于 PD 多发地域。尽管据报道它对一些生产参数有积极影响，但其预防效果还不足以消除养殖鱼类种群中的 SAV，接种疫苗的鱼仍会继续经历 PD 的暴发（Bang et al.，2012）。2015 年，两种基于 F93－125 毒株的多价疫苗上市销售（MSD Animal Health 公司产品）。

实验室实验及商业规模的田间试验，已经确定了基于全病毒灭活技术（Karlsen et al.，2012）、DNA 质粒（Hikke et al.，2014a；Simard and Horne，2014）、活减毒株（Moriette et al.，2006）和重组亚单位抗原（Xu et al.，2012）等有前景的新候选疫苗（Karlsen et al.，2012）。全病毒灭活疫苗具有在以后相对可预测的监管审批优势。可以通过诸如在制剂中添加 CpG 寡核苷酸或 poly（I：C）（聚肌苷酸-聚胞苷酸）激活细胞免疫应答的手段来提高这些疫苗的免疫功效(Strandskog et al., 2011; Thim et al., 2012, 2014)。与其他疫苗技术相比,灭活抗原还有一个优势,即它们更容易与预防其他疾病的细菌和病毒抗原一起配伍在多价疫苗中。

表达 SAV 结构蛋白的 DNA 疫苗似乎可提供显著的抗感染保护作用,前提是所有糖蛋白共同表达以确保正确折叠(Xu et al., 2012; Hikke et al., 2014a; Simard and Horne, 2014)。在加拿大,这种疫苗技术已被用于保护鲑科鱼类免受 IHNV 的侵害,但在欧洲,食用动物 DNA 疫苗的注册过程不如全病毒灭活疫苗那样明确。尽管如此,最近一种针对 SAV 的 DNA 疫苗成为在欧洲获得上市许可的推荐产品。

SAV 感染对继发感染提供了强有力的保护(Lopez-Doriga et al., 2001)。因此,活疫苗可能会提供非常好的功效,并且已经证实一种 SAV2 减毒株可保护虹鳟鱼苗免受野生型病毒的继发感染(Moriette et al., 2006)。这种技术的挑战在于确保安全减毒而不损失免疫效力。上述的 SAV2 减毒株仅通过一到两个突变而减毒(Merour et al., 2013),更安全的疫苗构建策略可能是删除部分基因组,或用其他甲病毒的同源部分替换它们。这种策略已经用于对人类具有致病性的甲病毒,并且可能强力阻止毒力回复。针对 SAV2 和 SAV3 毒株,已经建立了可用于构建此类减毒株的反向遗传系统(Moriette et al., 2006; Karlsen et al., 2010a; Guo et al., 2014)。除了监管要求,一些专利涉及 SAV 疫苗的开发,这可能会减少制药公司的自由运作空间。

13.5.2 减轻 SAV 感染的其他措施

已采取措施开始培育对 SAV 感染具有更强抗性的鲑鱼。这一策略已成功应用于改善对传染性胰腺坏死病毒(IPNV)的控制,这可能是近年来 IPNV 疫情减少的重要原因。在大西洋鲑中,对 PD 的抗性似乎也存在较高的遗传力,并且最近发现了影响该物种对 SAV 感染抗性的数量性位点(quantitative trait locus, QTL)(Gonen et al., 2015)。现在就断定将这种 QTL 引入养殖鲑鱼种群将使 PD 降低到与改良 IPN 抗性育种相同的程度还为时尚早,但主要的鱼卵生产商提供的卵是根据 PD 抗性选择的。

EWOS、Skretting 和 Biomar 等饲料公司销售了旨在病毒感染期间产生有益效果的"功能性饲料",这些饲料是用来减少炎症反应的或更容易消化(Alne et al., 2009; Martinez-Rubio et al., 2012)。尽管养殖业界使用了大量此类饲料,但很少有发表的研究表明它们对 PD 的影响,因此尚不清楚它们的效果如何。

据报道,水产养殖中常用的消毒剂如 Virkon S、Virex、Halamid、FAM30 和 Buffodine 一般都可有效灭活 SAV,但 Virkon S 和 Virex 在有机负荷增加或温度变化的条件下表现更好(Graham et al., 2007)。

13.6 总结与研究展望

SAV是影响欧洲大西洋鲑养殖的一个主要病毒，同时，它也影响虹鳟养殖业。感染SAV会引起胰腺、心脏和骨骼肌的病理变化，并可能导致鱼类生长缓慢、鱼片质量下降，甚至死亡。如下工作可能有助于疾病的减轻：改进预防方法，如更有效的疫苗或抗病力更强的鱼类品种，将可能通过抵御区域输送传播压力，减少病毒传播和改善疾病控制。与此相关的是，更好地记录这些疾病传输网络并使用PCR和测序等分子技术更密切地监测病毒传播将非常有用。这可以极大地帮助决策何时何地转移新的鱼群。

未来SAV和其他RNA病毒诊断的一个可能发展是对基因序列的关注增加，基因序列包含比PCR信号更丰富的信息量。与大多数RNA病毒一样，SAV进化时具有相当高的替换率(Karlsen et al.，2014)。A养殖场的SAV亲本株可以在B养殖场和C养殖场中产生SAV毒株，在最初的传播事件发生一年后，这些毒株的基因组序列就会经过几次基因碱基替换而区别开来(Karlsen，2015)。这种系统发育信号对于跟踪传播链和更准确地识别传播背后的机制非常有用。

参考文献

[1] Alne, H., Thomassen, M.S., Takle, H., Terjesen, B., Grammes, B.F. et al. (2009) Increased survival by dietary tetradecylthioacetic acid (TTA) during a natural outbreak of heart and skeletal muscle inflammation (HSMI) in S0 Atlantic salmon. Journal of Fish Diseases 32, 953 - 961.

[2] Andersen, L. (2012) *Alphavirus* Infection in *Atlantic Salmon*, *Salmo salar L.* - Viral Pathogenesis. University of Bergen, Bergen, Norway.

[3] Andersen, L., Bratland, A., Hodneland, K. and Nylund, A. (2007) Tissue tropism of *Salmonid Alphaviruses* (subtypes SAV1 and SAV3) in experimentally challenged Atlantic salmon (*Salmo salar L.*). Archives of Virology 152, 1871 - 1883.

[4] Andersen, L., Hodneland, K. and Nylund, A. (2010) No influence of oxygen levels on pathogenesis and virus shedding in *Salmonid Alphavirus* (SAV)-challenged Atlantic salmon (*Salmo salar L.*). Virology Journal 7: 198.

[5] Aunsmo, A., Valle, P.S., Sandberg, M., Midtlyng, P.J. and Bruheim, T. (2010) Stochastic modelling of direct costs of pancreas disease (PD) in Norwegian farmed Atlantic salmon (*Salmo salar L.*). Preventive Veterinary Medicine 93, 233 - 241.

[6] Bang, J.B., Kristoffersen, A.B., Myr, C. and Brun, E. (2012) Cohort study of effect of vaccination on pancreas disease in Norwegian *salmon* aquaculture. Diseases of Aquatic Organisms 102, 23 - 31.

[7] Biacchesi, S., Jouvion, G., Mérour, E., Boukadiri, A., Desdouits, M. etal. (2016) Rainbow trout (*Oncorhynchus mykiss*) muscle satellite cells are targets of *Salmonid Alphavirus* infection. Veterinary Research 47: 9.

[8] Borzym, E., Maj-Paluch, J., Stachnik, M., Matras, M. and Reichert, M. (2014) First

laboratory confirmation of *Salmonid Alphavirus* type 2 (SAV2) infection in Poland. Bulletin of the Veterinary Institute in Pulawy 58, 341 - 345.

[9] Boscher, S.K., McLoughlin, M., Le Ven, A., Cabon, J., Baud, M. et al. (2006) Experimental transmission of sleeping disease in one-year-old rainbow trout, *Oncorhynchus mykiss* (*Walbaum*), induced by sleeping disease virus. Journal of Fish Diseases 29, 263 - 273.

[10] Boucher, P. and Baudin Laurencin, F. (1994) Sleeping disease (SD) of *Salmonids*. Bulletin of the European Association Fish Pathologists 14, 179 - 180.

[11] Boucher, P. and Baudin Laurencin, F. (1996) Sleeping disease and pancreas disease: comparative histopathology and acquired cross protection. Journal of Fish Diseases 19, 303 - 310.

[12] Boucher, P., Raynard, R. S., Hughton, G. and Baudin Laurencin, F. (1995) Comparative experimental transmission of pancreas disease in Atlantic salmon, rainbow trout and brown trout. Diseases of Aquatic Organisms 22, 19 - 24.

[13] Bruno, D.W., Noguera, P.A., Black, J., Murray, W., Macqueen, D.J. et al. (2014) Identification of a wild reservoir of *Salmonid Alphavirus* in common dab *Limanda Limanda*, with emphasis on virus culture and sequencing. Aquaculture Environment Interactions 5, 89 - 98.

[14] Cano, I., Joiner, C., Bayley, A., Rimmer, G., Bateman, K. et al. (2015) An experimental means of transmitting pancreas disease in Atlantic salmon *Salmo salar L.* fry in freshwater. Journal of Fish Diseases 38, 271 - 281.

[15] Castric, J., Baudin Laurencin, F., Brémont, M., Jeffroy, J., Le Ven, A. et al. (1997) Isolation of the virus responsible for sleeping disease in experimentally infected rainbow trout (*Oncorhynchus mykiss*). Bulletin of the European Association Fish Pathologists 17, 27 - 30.

[16] Castric, J., Cabon, J. and Le Ven, A. (2005) Experimental study of vertical transmission of sleeping disease virus (SDV) in rainbow trout (*Oncorhynchus mykiss*). Conference poster presented at: EAFP European Association of Fish Pathologists, 12th International Conference "Diseases of Fish and Shellfish", 11 - 16 September 2005, Copenhagen, Denmark, Volumes 1 - 2. DIS Congress Service Copenhagen A/S, p. 95 [vol. not known].

[17] Christie, K.E., Graham, D.A., McLoughlin, M.F., Villoing, S., Todd, D. et al. (2007) Experimental infection of Atlantic salmon *Salmo salar* pre-smolts by i. p. injection with new Irish and Norwegian *Salmonid Alphavirus* (SAV) isolates: a comparative study. Diseases of Aquatic Organisms 75, 13 - 22.

[18] Crockford, T., Menzies, F.D., McLoughlin, M.F., Wheatley, S.B. and Goodall, E.A. (1999) Aspects of the epizootiology of pancreas disease in farmed Atlantic salmon *Salmo salar* in Ireland.Diseases of Aquatic Organisms 36, 113 - 119.

[19] Desvignes, L., Quentel, C., Lamour, F. and Le, V. A. (2002) Pathogenesis and immune response in Atlantic salmon (*Salmo salar L.*) parr experimentally infected with salmon pancreas disease virus (SPDV). Fish and Shellfish Immunology 12, 77 - 95.

[20] Ferguson, H.W., Roberts, R.J., Richards, R.H., Collins, R.O. and Rice, D.A. (1986) Severe degenerative cardiomyopathy associated with pancreas disease in Atlantic salmon, *Salmo salar L.* Journal of Fish Diseases 20, 95 - 98.

[21] Fringuelli, E., Rowley, H.M., Wilson, J.C., Hunter, R., Rodger, H. et al. (2008) Phylogenetic analyses and molecular epidemiology of European *Salmonid Alphaviruses* (SAV) based on partial E2 and nsP3 gene nucleotide sequences. Journal of Fish Diseases 31, 811 - 823.

[22] Fringuelli, E., Rowley, H.M., Wilson, J.C., Hunter, R., Rodger, H. et al. (2008) Phylogenetic analyses and molecular epidemiology of European *Salmonid Alphaviruses*

(SAV) based on partial E2 and nsP3 gene nucleotide sequences. Journal of Fish Diseases 31, 811 - 823.

[23] Gonen, S., Baranski, M., Thorland, I., Norris, A., Grove, H. et al. (2015) Mapping and validation of a major QTL affecting resistance to pancreas disease (*Salmonid Alphavirus*) in Atlantic salmon (*Salmo salar*). Heredity 115, 405 - 414.

[24] Graham, D.A., Jewhurst, V.A., Rowley, H.M., McLoughlin, M.F. and Todd, D. (2003) A rapid immunoperoxidase-based virus neutralization assay for *Salmonid Alphavirus* used for a serological survey in Northern Ireland. Journal of Fish Diseases 26, 407 - 413.

[25] Graham, D.A., Taylor, C., Rodgers, D., Weston, J., Khalili, M. et al. (2006a) Development and evaluation of a one-step real-time reverse transcription polymerase chain reaction assay for the detection of *Salmonid Alphaviruses* in serum and tissues. Diseases of Aquatic Organisms 70, 47 - 54.

[26] Graham, D.A., Jewhurst, H., McLoughlin, M.F., Sourd, P., Rowley, H.M. et al. (2006b) Sub-clinical infection of farmed Atlantic salmon *Salmo salar* with *Salmonid Alphavirus*-a prospective longitudinal study. Diseases of Aquatic Organisms 72, 193 - 199.

[27] Graham, D.A., Cherry, K., Wilson, C.J. and Rowley, H.M. (2007) Susceptibility of *Salmonid Alphavirus* to a range of chemical disinfectants. Journal of Fish Diseases 30, 269 - 277.

[28] Graham, D. A., Wilson, C., Jewhurst, H. and Rowley, H. (2008) Cultural characteristics of *Salmonid Alphaviruses*-influence of cell line and temperature. Journal of Fish Diseases 31, 859 - 868.

[29] Graham, D.A., Fringuelli, E., Wilson, C., Rowley, H.M., Brown, A. et al. (2010) Prospective longitudinal studies of *Salmonid Alphavirus* infections on two Atlantic salmon farms in Ireland; evidence for viral persistence. Journal of Fish Diseases 33, 123 - 135.

[30] Graham, D.A., Frost, P., McLaughlin, K., Rowley, H.M., Gabestad, I. et al. (2011) A comparative study of marine *Salmonid Alphavirus* subtypes 1 - 6 using an experimental cohabitation challenge model. Journal of Fish Diseases 34, 273 - 286.

[31] Graham, D.A., Fringuelli, E., Rowley, H.M., Cockerill, D., Cox, D.I. et al. (2012) Geographical distribution of *Salmonid Alphavirus* subtypes in marine farmed Atlantic salmon, *Salmo salar L.*, in Scotland and Ireland. Journal of Fish Diseases 10, 755 - 765.

[32] Guo, T.C., Johansson, D.X., Haugland, O., Liljestrom, P. and Evensen, Ø. (2014) A 6K-deletion variant of *Salmonid Alphavirus* is non-viable but can be rescued through RNA recombination.PLoS ONE 9(7): e100184.

[33] Hikke, M.C., Braaen, S., Villoing, S., Hodneland, K., Geertsema, C. et al. (2014a) *Salmonid Alphavirus* glycoprotein E2 requires low temperature and E1 for virion formation and induction of protective immunity. Vaccine 32, 6206 - 6212.

[34] Hikke, M. C., Verest, M., Vlak, J. M. and Pijlman, G. P. (2014b) *Salmonid Alphavirus* replication in mosquito cells: towards a novel vaccine production system. Microbial Biotechnology 7, 480 - 484.

[35] Hjortaas, M.J., Skjelstad, H.R., Taksdal, T., Olsen, A.B., Johansen, R. et al. (2013) The first detections of subtype 2-related *Salmonid Alphavirus* (SAV2) in Atlantic salmon, *Salmo salar L.*, in Norway. Journal of Fish Diseases 36, 71 - 74.

[36] Hjortaas, M.J., Bang, J.B., Taksdal, T., Olsen, A.B., Lillehaug, A. et al. (2016) Genetic characterization of *Salmonid Alphavirus* in Norway. Journal of Fish Diseases 39, 249 - 257.

[37] Hodneland, K. and Endresen, C. (2006) Sensitive and specific detection of *Salmonid Alphavirus* using real-time PCR (TaqMan). Journal of Virological Methods 131,

184 - 192.
[38] Hodneland, K., Bratland, A., Christie, K.E., Endresen, C. and Nylund, A. (2005) New subtype of *Salmonid Alphavirus* (SAV), Togaviridae, from Atlantic salmon *Salmo salar* and rainbow trout *Oncorhynchus mykiss* in Norway. Diseases of Aquatic Organisms 66, 113 - 120.
[39] Houghton, G. (1995) Kinetics of infection of plasma, blood leucocytes and lymphoid tissue from Atlantic salmon *Salmo salar* experimentally infected with pancreas disease. Diseases of Aquatic Organisms 22, 193 - 198.
[40] Jansen, M.D., Taksdal, T., Wasmuth, M.A., Gjerset, B., Brun, E. et al. (2010) *Salmonid Alphavirus* (SAV) and pancreas disease (PD) in Atlantic salmon, *Salmo salar L.*, in freshwater and seawater sites in Norway from 2006 to 2008. Journal of Fish Diseases 33, 391 - 402.
[41] Jansen, M.D., Jensen, B.B. and Brun, E. (2015) Clinical manifestations of pancreas disease outbreaks in Norwegian marine salmon farming-variations due to *Salmonid Alphavirus* subtype. Journal of Fish Diseases 38, 343 - 353.
[42] Karlsen, M. (2015) *Salmonid Alphavirus* Subtype 3-Characterization by Phylogenetics and Reverse Genetics. University of Bergen. Bergen, Norway.
[43] Karlsen, M., Hodneland, K., Endresen, C. and Nylund, A. (2006) Genetic stability within the Norwegian subtype of *Salmonid Alphavirus* (family Togaviridae). Archives of Virology 151, 861 - 874.
[44] Karlsen, M., Villoing, S., Ottem, K. F., Rimstad, E. and Nylund, A. (2010a) Development of infectious Cdna clones of *Salmonid Alphavirus* subtype 3. BMC Research Notes 3: 241.
[45] Karlsen, M., Yousaf, M.N., Villoing, S., Nylund, A. and Rimstad, E. (2010b) The amino terminus of the *Salmonid Alphavirus* capsid protein determines subcellular localization and inhibits cellular proliferation. Archives of Virology 155, 1281 - 1293.
[46] Karlsen, M., Tingbo, T., Solbakk, I. T., Evensen, O., Furevik, A. et al. (2012) Efficacy and safety of an inactivated vaccine against *Salmonid Alphavirus* (family *Togaviridae*). Vaccine 30, 5688 - 5694.
[47] Karlsen, M., Gjerset, B., Hansen, T. and Rambaut, A. (2014) Multiple introductions of *Salmonid Alphavirus* from a wild reservoir have caused independent and self-sustainable epizootics in aquaculture. Journal of General Virology 95, 52 - 59.
[48] Karlsen, M., Andersen, L., Blindheim, S.H., Rimstad, E. and Nylund, A. (2015) A naturally occurring substitution in the E2 protein of *Salmonid Alphavirus* subtype 3 changes viral fitness. Virus Research 196, 79 - 86.
[49] Kuhn, R.J. (2007) Togaviridae: the viruses and their replication. In: Knipe, D.M. and Howley, P.M. (eds) Field's Virology, 5th edn. Wolters Kluwer Health /Lippincott Williams & Wilkins, Philadelphia, Pennsylvania, pp. 1001 - 1022.
[50] Lopez-Doriga, M.V., Smail, D.A., Smith, R.J., Domenech, A., Castric, J. et al. (2001) Isolation of salmon pancreas disease virus (SPDV) and its ability to protect against infection by the "wild-type" agent. Fish and Shellfish Immunology 11, 505 - 522.
[51] Martinez-Rubio, L., Morais, S., Evensen, O., Wadsworth, S., Ruohonen, K. et al. (2012) Functional feeds reduce heart inflammation and pathology in Atlantic salmon (*Salmo salar L.*) following experimental challenge with Atlantic salmon reovirus (ASRV) PLoS ONE 7(11): e40266.
[52] McCleary, S., Giltrap, M., Henshilwood, K. and Ruane, N.M. (2014) Detection of *Salmonid Alphavirus* RNA in Celtic and Irish Sea flatfish. Diseases of Aquatic Organisms 109, 1 - 7.
[53] McLoughlin, M.F. and Graham, D.A. (2007) Alphavirus infections in *Salmonids*-a review. Journal of Fish Diseases 30, 511 - 531.

[54] McLoughlin, M.F., Nelson, R.T., Rowley, H.M., Cox, D.I. and Grant, A.N. (1996) Experimental pancreas disease in Atlantic salmon *Salmo salar* post-smolts induced by salmon pancreas disease virus (SPDV). Diseases of Aquatic Organisms 26, 117–124.
[55] McLoughlin, M.F., Nelson, R.T., McCormack, J.E., Rowley, H.M. and Bryson, D.B. (2002) Clinical and histopathological features of naturally occurring pancreas disease in farmed Atlantic salmon, *Salmo salar L.* Journal of Fish Diseases 25, 33–43.
[56] McVicar, A.H. (1987) Pancreas disease of farmed Atlantic Salmon, *Salmo salar*, in Scotland: epidemiology and early pathology. Aquaculture 67, 71–78.
[57] Merour, E., Lamoureux, A., Bernard, J., Biacchesi, S. and Brémont, M. (2013) A fully attenuated recombinant *Salmonid Alphavirus* becomes pathogenic through a single amino acid change in the E2 glycoprotein. Journal of Virology 87, 6027–6030.
[58] Moriette, C., LeBerre, M., Boscher, S. K., Castric, J. and Brémont, M. (2005) Characterization and mapping of monoclonal antibodies against the sleeping disease virus, an aquatic *Alphavirus*. Journal of General Virology 86, 3119–3127.
[59] Moriette, C., LeBerre, M., Lamoureux, A., Lai, T. L. and Brémont, M. (2006) Recovery of a recombinant Salmonid Alphavirus fully attenuated and protective for rainbow trout. Journal of Virology 80, 4088–4098.
[60] Munro, A. L. S., Ellis, A. E., McVicar, A. H., McLay, H. A. and Needham, E. A. (1984) An exocrine pancreas disease of farmed Atlantic salmon in Scotland. Helgoländer Meeresuntersuchungen 37, 571–586.
[61] Nelson, R.T., McLoughlin, M.F., Rowley, H.M., Platten, M.A. and McCormick, J.I. (1995) Isolation of a toga-like virus from farmed Atlantic salmon *Salmo salar* with pancreas disease. Diseases of Aquatic Organisms 22, 25–32.
[62] Nylund, S., Andersen, L., Saevareid, I., Plarre, H., Watanabe, K. et al. (2011) Diseases of farmed Atlantic salmon *Salmo salar* associated with infections by the microsporidian *Paranucleospora theridion*. Diseases of Aquatic Organisms 16, 41–57. doi: 10.3354/dao02313.
[63] Olsen, C.M., Pemula, A.K., Braaen, S., Sankaran, K. and Rimstad, E. (2013) *Salmonid Alphavirus* replicon is functional in fish, mammalian and insect cells and in vivo in shrimps (*Litopenaeus vannamei*). Vaccine 31, 5672–5679.
[64] Pettersen, J.M., Rich, K.M., Jensen, B.B. and Aunsmo, A. (2015) The economic benefits of disease triggered early harvest: a case study of pancreas disease in farmed Atlantic salmon from Norway. Preventive Veterinary Medicine 121, 314–324.
[65] Petterson, E., Sandberg, M. and Santi, N. (2009) *Salmonid Alphavirus* associated with *Lepeophtheirus salmonis* (*Copepoda: Caligidae*) from Atlantic salmon, *Salmo salar L.* Journal of Fish Diseases 32, 477–479.
[66] Petterson, E., Guo, T.C., Evensen, O., Haugland, O. and Mikalsen, A.B. (2015) In vitro adaptation of SAV3 in cell culture correlates with reduced in vivo replication capacity and virulence to Atlantic salmon (*Salmo salar L.*) parr. Journal of General Virology 96, 3023–3034.
[67] Powers, A.M., Brault, A.C., Shirako, Y., Strauss, E.G., Kang, W. et al. (2001) Evolutionary relationships and systematics of the Alphaviruses. Journal of Virology 75, 10118–10131.
[68] Simard, N. and Horne, M. (2014) *Salmonid Alphavirus* and uses thereof. Patent Publication No. WO2014041189 A1, Application No. PCT/EP2013/069241. Available at: https://www.google.com/patents/WO2014041189A1?cl=zh-TW (accessed 8 November 2016).
[69] Simons, J., Bruno, D. W., Ho, Y. M., Murray, W. and Matejusova, I. (2016) Common dab, Limanda limanda (L.), as a natural carrier of *Salmonid Alphavirus* (SAV) from waters off north-west Ireland. Journal of Fish Diseases 39, 507–510.

[70] Snow, M., Black, J., Matejusova, I., McIntosh, R., Baretto, E. et al. (2010) Detection of *Salmonid Alphavirus* RNA in wild marine fish: implications for the origins of salmon pancreas disease in aquaculture. Diseases of Aquatic Organisms 91, 177 - 188.
[71] Steele, K.E. and Twenhafel, N.A. (2010) Review paper: pathology of animal models of *Alphavirus* encephalitis. Veterinary Pathology 47, 790 - 805.
[72] Strandskog, G., Villoing, S., Iliev, D.B., Thim, H.L., Christie, K.E. et al. (2011) Formulations combining CpG containing oliogonucleotides and poly I : C enhance the magnitude of immune responses and protection against pancreas disease in Atlantic salmon. Developmental and Comparative Immunology 35, 1116 - 1127.
[73] Strauss, J. H. and Strauss, E. G. (1994) The *Alphaviruses*: gene expression, replication, and evolution. Microbiological Reviews 58, 491 - 562.
[74] Taksdal, T., Olsen, A.B., Bjerkas, I., Hjortaas, M.J., Dannevig, B.H. et al. (2007) Pancreas disease in farmed Atlantic salmon, *Salmo salar L.*, and rainbow trout, *Oncorhynchus mykiss* (*Walbaum*), in Norway. Journal of Fish Diseases 30, 545 - 558.
[75] Taksdal, T., Wiik-Nielsen, J., Birkeland, S., Dalgaard, P. and Morkore, T. (2012) Quality of raw and smoked fillets from clinically healthy Atlantic salmon, *Salmo salar L.*, following an outbreak of pancreas disease (PD). Journal of Fish Diseases 35, 897 - 906.
[76] Taksdal, T., Bang, J.B., Bockerman, I., McLoughlin, M.F., Hjortaas, M.J. et al. (2015) Mortality and weight loss of Atlantic salmon, *Salmo salar L.*, experimentally infected with *Salmonid Alphavirus* subtype 2 and subtype 3 isolates from Norway. Journal of Fish Diseases 38, 1047 - 1061.
[77] Thim, H.L., Iliev, D.B., Christie, K.E., Villoing, S., McLoughlin, M.F. et al. (2012) Immunoprotective activity of a *Salmonid Alphavirus* vaccine: comparison of the immune responses induced by inactivated whole virus antigen formulations based on CpG class B oligonucleotides and poly I : C alone or combined with an oil adjuvant. Vaccine 30, 4828 - 4834.
[78] Thim, H.L., Villoing, S., McLoughlin, M., Christie, K.E., Grove, S. et al. (2014) Vaccine adjuvants in fish vaccines make a difference: comparing three adjuvants (Montanide ISA763A Oil, CpG/Poly I : C Combo and VHSV Glycoprotein) alone or in combination formulated with an inactivated whole *Salmonid Alphavirus* antigen. Vaccines (Basel) 2, 228 - 251.
[79] Todd, D., Jewhurst, V.A., Welsh, M.D., Borghmans, B.J., Weston, J.H. et al. (2001) Production and characterisation of monoclonal antibodies to salmon pancreas disease virus. Diseases of Aquatic Organisms 46, 101 - 108.
[80] Viljugrein, H., Staalstrom, A., Molvaelr, J., Urke, H.A. and Jansen, P.A. (2009) Integration of hydrodynamics into a statistical model on the spread of pancreas disease (PD) in salmon farming. Diseases of Aquatic Organisms 88, 35 - 44.
[81] Villoing, S., Béarzotti, M., Chilmonczyk, S., Castric, J. and Brémont, M. (2000a) Rainbow trout sleeping disease virus is an atypical *Alphavirus*. Journal of Virology 74, 173 - 183.
[82] Villoing, S., Castric, J., Jeffroy, J., Le Ven, A., Thiery, R. et al. (2000b) An RT - PCR-based method for the diagnosis of the sleeping disease virus in experimentally and naturally infected Salmonids. Diseases of Aquatic Organisms 40, 19 - 27.
[83] Voss, J.E., Vaney, M.C., Duquerroy, S., Vonrhein, C., Girard-Blanc, C. et al. (2010) Glycoprotein organization of Chikungunya virus particles revealed by X-ray crystallography. Nature 468, 709 - 712.
[84] Weston, J., Villoing, S., Brémont, M., Castric, J., Pfeffer, M. et al. (2002) Comparison of two aquatic *Alphaviruses*, salmon pancreas disease virus and sleeping disease virus, by using genome sequence analysis, monoclonal reactivity, and cross-

infection. Journal of Virology 76，6155 - 6163.
[85] Xu，C.，Guo，T.C.，Mutoloki，S.，Haugland，O.，Marjara，I.S. et al.（2010）Alpha interferon and not gamma interferon inhibits *Salmonid Alphavirus* subtype 3 replication in vitro. Journal of Virology 84，8903 - 8912.
[86] Xu，C.，Mutoloki，S. and Evensen，O.（2012）Superior protection conferred by inactivated whole virus vaccine over subunit and DNA vaccines against *Salmonid Alphavirus* infection in Atlantic salmon（*Salmo salar L.*）. Vaccine 30，3918 - 3928.

14

杀鲑气单胞菌和嗜水气单胞菌

Bjarnheidur K. Gudmundsdottir* 和 Bryndis Bjornsdottir

14.1 引言

气单胞菌(*Aeromonas*)属于γ-变形菌纲(Gammaproteobacteria),气单胞菌科(Aeromonadales)(Colwell et al., 1986)。气单胞菌存在于淡水、河口和海洋环境,以及无脊椎动物、脊椎动物和土壤中(Janda and Abbott, 2010)。模式种气单胞菌为运动型嗜水气单胞菌(*A. hydrophila*),是一种动物病原菌;相比之下,作为鱼类病原,杀鲑气单胞菌(*A. salmonicida*)不具有运动性。气单胞菌可引起鱼类疖疮病(furunculosis)、非典型疖疮病、溃疡性疾病、运动型气单胞菌败血症(motile *Aeromonas* septicaemia, MAS)和烂鳍烂尾病(Cipriano and Austin, 2011)。嗜水气单胞菌和其他运动型菌种(如*A. veronii* biovar. *sobria*、*A. bestiarum*、*A. dhakensis*)可引发水产养殖疾病,是潜在的人畜共患病原(Rahman et al., 2002; Janda and Abbott, 2010; Austin and Austin, 2012a; Colston et al., 2014)。杀鲑气单胞菌和嗜水气单胞菌的种内分类需要重新评估(Martin-Carnahan and Joseph, 2005),对一些气单胞菌的基因组测序揭示了基因水平转移的潜力,这增强了细菌对不同环境和宿主的适应能力(Piotrowska and Popowska, 2015)。

杀鲑气单胞菌是目前世界范围内最重要的鱼类病原之一。它由四个嗜冷亚种[杀鲑亚种(*A. s. salmonicida*)、无色亚种(*A. s. achromogenes*)、杀日本鲑亚种(*A. s. masoucida*)和史氏亚种(*A. s. smithia*)]和嗜中温亚种[溶果胶亚种(*A. s. pectinolytica*)]组成(Pavan et al., 2000; Martin-Carnahan and Joseph, 2005)。其中,杀鲑气单胞菌杀鲑亚种是杀鲑气单胞菌的典型亚种,可导致许多冷水鱼类系统性疖疮病。该病害对国际上的鲑鳟养殖造成严重经济损失,直到油佐剂疖疮病疫苗上市才得以好转(Gudding and van Muiswinkel, 2013)。其他非典型杀鲑气单胞菌菌株在遗传和表型上都有差异(Austin et al., 1998)。

非典型杀鲑气单胞菌也会引起水产养殖中的重大问题(Gudmundsdottir,

---

* 通信作者邮箱:bjarngud@hi.is。

1998；Wiklund and Dalsgaard，1998）。杀鲑气单胞菌溶果胶亚种是唯一的非致病性亚种（Austin and Austin，2012b）。由于非典型疖疮病的疫苗预防效果不佳，该疾病目前仍是大西洋鳕（*Gadus morhua*）养殖中的主要威胁。该疾病问题同样存在于养殖北极红点鲑（*Salvelinus alpinus*）、大西洋庸鲽（*Hippoglossus hippoglossus*）、大西洋狼鱼（*Anarhichas lupus*）、花狼鱼（*Anarhichas minor*）及各种"清洁鱼"——隆头鱼（*Labridae*）和吸盘圆鳍鱼（*Cyclopterus lumpus*）中（Gulla et al.，2015b）。非典型杀鲑气单胞菌会引起金鱼（*Carassius auratus*）、鲤（*Cyprinus carpio*）、欧川鲽（*Platichthys flesus*），以及其他各种鱼类发生溃疡性疾病（Trust et al.，1980；Wiklund and Dalsgaard，1998）。

杀鲑气单胞菌可以通过受污染的水、表面感染的鱼卵、脊椎动物和非脊椎动物携带者及养殖设施和覆盖物等水平传播，垂直传播尚未得到证实（Nomura，1993；Wiklund，1995）。杀鲑气单胞菌附着在鱼的鳃和皮肤/黏液区域（Ferguson et al.，1998）。除新西兰外，非典型杀鲑气单胞菌的感染在世界范围内都有发生。澳大利亚、智利和新西兰尚未有典型杀鲑气单胞菌的感染病例报告。

本章涉及的嗜水气单胞菌能感染温血和冷血动物，包括人类，同时它会导致温水性养殖鱼类的严重死亡（Janda and Abbott，2010；Cipriano and Austin，2011；Colston et al.，2014）。自2009年以来，源自亚洲的高致病性嗜水气单胞菌已在美国东南部的斑点叉尾鮰（*Ictalurus punctatus*）中造成动物流行疫情（Hossain et al.，2014）。嗜水气单胞菌之前有两个亚种：嗜水气单胞菌嗜水亚种（*A. h. hydrophila*）（Seshadri et al.，2006）和嗜水气单胞菌蛙亚种（*A. h. ranae*）（Huys et al.，2003），嗜水气单胞菌脱色亚种（*A. h. decolorationis*）是被提名的第三个亚种（Ren et al.，2006）。此外，嗜水气单胞菌属于三个DNA杂交组（HGs）（Martin-Carnahan and Joseph，2005）。大多数关于嗜水气单胞菌感染的报道都没有定义亚种分类。嗜水气单胞菌是机会致病菌，疾病的发生通常与环境胁迫有关（如高养殖密度、水温升高）。然而，死亡率也可以在低温下达到峰值（Cipriano and Austin，2011）。由于菌株变异和不同宿主的特异性，疫苗接种防治存在一定困难（Austin and Austin，2012b）。嗜水气单胞菌通过受污染的水、带菌鱼、外部寄生虫、养殖设施和覆盖物等实现水平传播（Rusin et al.，1997；Udeh，2004；Austin and Austin，2012a）。

14.2 诊断

疾病诊断依据是临床症状、养殖设施的使用情况以及相关细菌的分离和鉴定。感染通常会导致败血症，在许多器官及皮肤溃疡中发现细菌或细菌产物，但也会在没有任何可检测到的病理学情况下发生死亡。疾病症状包括体色变暗、嗜睡、游动异常、食欲不振、鳃苍白、皮肤溃疡、鳍及尾部腐烂、红皮炎、出血、败血症、腹胀，以及鳞片突出或脱落。细菌鉴定是基于表型特征和分子生物学技术(Martin-Carnahan and Joseph, 2005; Tenover, 2007; Austin and Austin, 2012a,b)。

气单胞菌是直径为0.3～1.0 μm、长为1.0～3.5 μm的革兰氏阴性双杆菌，可呈球形及染色双极性，具有细胞色素氧化酶活性，兼性厌氧，发酵葡萄糖产气或不产气，并且对弧菌抑菌剂O/129有抗性。气单胞菌最初通常分离自头肾、皮肤损伤处或鳃。杀鲑气单胞菌是非运动性的并且在37 ℃下不生长，而嗜水气单胞菌是运动性的且可在37 ℃下生长(Martin-Carnahan and Joseph, 2005)。气单胞菌的培养通常使用胰蛋白胨大豆琼脂(tryptone soya agar, TSA)或脑心浸液琼脂(brain heart infusion agar, BHIA)，但许多非典型杀鲑气单胞菌则需要血琼脂培养(Cipriano and Bertolini, 1988; Wiklund, 1990; Gulla et al., 2015b)。非典型杀鲑气单胞菌生长条件非常苛刻，其生长很容易被其他微生物覆盖。因此，已经开发了其他多种技术来进行这些细菌的鉴定，如酶联免疫吸附试验(ELISA)、特异性抗体凝集(Adams and Thompson, 1990; Gilroy and Smith, 2003; Saleh et al., 2011)和基于16S rRNA基因序列的DNA测序方法、PCR扩增和实时定量PCR(qRT－PCR)等技术(Gustafson et al., 1992; Byers et al., 2002; Clarridge, 2004; Balcázar et al., 2007; Beaz-Hidalgo et al., 2013)。杀鲑气单胞菌和嗜水气单胞菌的种内分组是十分复杂的，需要对基因或基因组进行生物信息学比较(Colston et al., 2014)。

杀鲑气单胞菌具有表面A－层蛋白(VapA)，赋予该菌增强的疏水性和自凝集特性(Chart et al., 1984; Belland and Trust, 1985)。在培养基中添加考马斯亮蓝R－250染料有助于鉴定A－层阳性菌株(Cipriano and Bertolini, 1988)，并建议在15～20 ℃下进行孵育，因为温度高于20 ℃可能会增加A－层损失(Moki et al., 1995)。最近的研究发现，基于序列变异的VapA分型可分离杀鲑气单胞菌亚种，并识别到多种尚未明确鉴定的亚型(Moki et al., 1995)。杀鲑气单胞菌的最适生长温度为22～25 ℃，大多数菌株在37 ℃下不生长(Martin-Carnahan and Joseph, 2005)。然而，有一个运动型生物群可在37 ℃下生长，并在30～37 ℃下培养时诱导运动性(McIntosh and Austin, 1991)。此外，有些菌株可能是氧化酶阴性(Wiklund et al., 1994)。典型杀鲑气单胞菌在胰蛋白胨培养基中生长时通常会产生水溶性棕色色素，但也存在无色菌株(Wiklund et al., 1993)。如果非典型菌株产生棕色色素，则色素的产生是缓慢的。Schwenteit等(2011)报道，杀鲑气单胞菌色素的产生受群体感应系统调控(图14.1)，这可能与毒力相关。

嗜水气单胞菌具有运动性和极生鞭毛。它是一个异质性群体，其成员在血清学和基因型上存在差异。因此，除了表型特征外，16S rRNA基因序列是鉴定

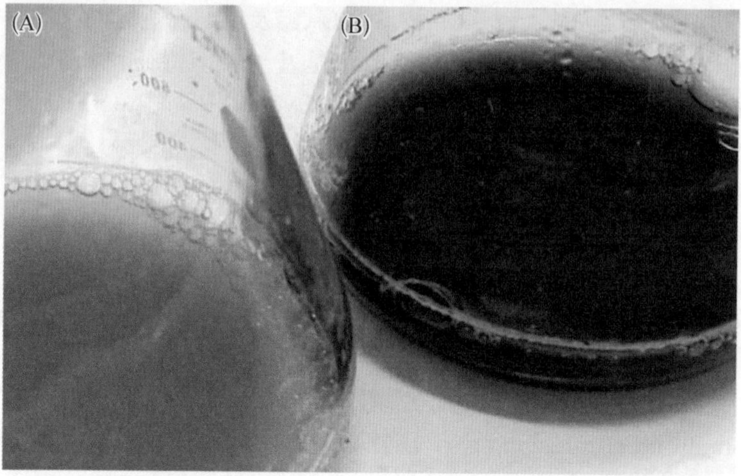

图 14.1　杀鲑气单胞菌色素的产生受群体感应系统调控。该图显示了杀鲑气单胞菌无色亚种菌株 Keldur 265－87 16 ℃条件下在脑心浸液琼脂(BHIA)中生长 190 h 的培养液：(A) Δ*asaI* 突变株(群体感应阴性)；(B) 野生型群体感应阳性菌株。照片由 Bjarnheidur K. Gudmundsdottir 提供

所必需的(MacInnes et al.，1979；Valera and Esteve，2002；Martin-Carnahan and Joseph，2005)。

14.3　病理学

疖疮病是由杀鲑气单胞菌杀鲑亚种引起的全身性疾病(McCarthy and Roberts，1980)，而非典型疖疮病则是由其他亚种引起的(Gudmundsdottir，1998)。皮肤溃疡病是由一些非典型杀鲑气单胞菌菌株引发的另一种气单胞菌疾病。相对而言，溃疡病患病鱼的死亡率通常没有全身性疖疮病严重(Austin and Austin，2012b)。

由杀鲑气单胞菌引起的全身性疾病可分为过急性、急性、亚急性或慢性疾病(McCarthy，1975)。过急性疾病常见于幼鱼，死鱼除了体色发黑和眼球突出外，没有其他症状。存活下来的鱼常出现充血和点状出血性病灶，可能有明显的局灶性出血和鳃充血。病鱼的前肾、鳃、脾和心肌中可见病原菌，心肌组织可能坏死。疾病进展迅速，宿主反应极小。急性疖疮病表现为全身性败血症，死亡率很高，鱼可能在没有明显病症的情况下就死亡了。在所有组织浆膜表面都有充血，脾常肿大发红；肾变软、易碎或液化；肝包膜下出血呈苍白色，或局灶性坏死伴体腔腹水导致肝表面呈斑驳状。皮肤病变可能会发展为沿身有出血性斑块或典型的疖肿。

急性毒性败血症的病变比亚急性和慢性形式的全身性疾病更为严重。无论病原菌定植在鱼体组织何处,都可见毒性造血坏死及心肌和肾小管变性。亚急性和慢性疖疮病在大龄鱼中更为常见,患病鱼通常可以存活并恢复健康。受感染的鱼皮肤发暗、食欲不振,嗜睡和疖疮等症状很常见。如图 14.2 所示,由典型杀鲑气单胞菌引发的溃疡深入肌肉组织,导致液化性坏死更加突出,皮肤出血比感染非典型菌株的鱼更严重。亚急性疖疮病的体体病理学与急性疖疮病相似,但慢性疖疮病的全身性内脏充血和腹膜炎更明显。细菌感染的靶器官多为心脏和脾脏。在疫情中存活下来的鱼成为可能传播疾病的带菌者(McCarthy, 1975; Hellberg et al., 1996; Gudmundsdottir, 1998; Bjornsdottir et al., 2005)。

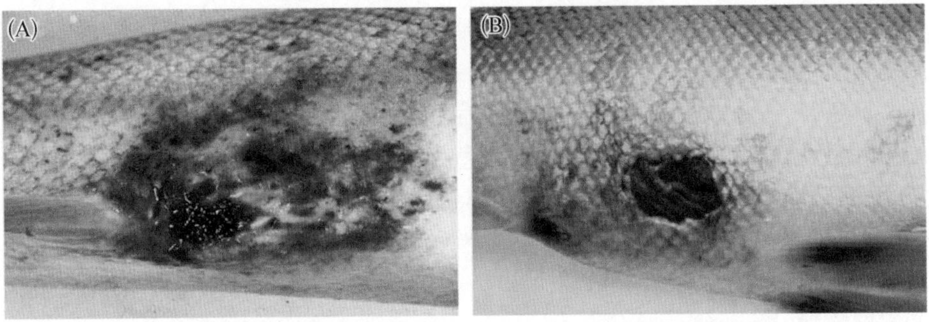

图 14.2 冰岛埃利达尔河野生大西洋鲑(平均体重 2.5 kg)的慢性疖疮病临床表现:(A) 由杀鲑气单胞菌杀鲑亚种引起;(B) 由杀鲑气单胞菌无色亚种引起。照片由 Ragnar Th. Sigurdsson 提供

大西洋鳕系统性杀鲑气单胞菌无色亚种感染后的感染病理学特征可以通过鳕独特的免疫系统来解释。鳕缺乏Ⅱ类主要组织相容性复合体(MHC Ⅱ)基因,该基因是体液免疫反应的基础(Star et al., 2011)。因此,鳕主要依赖于细胞和固有免疫应答进行免疫,其典型的免疫反应特征是形成肉芽肿(图 14.3)(Magnadottir et al., 2002, 2013)。

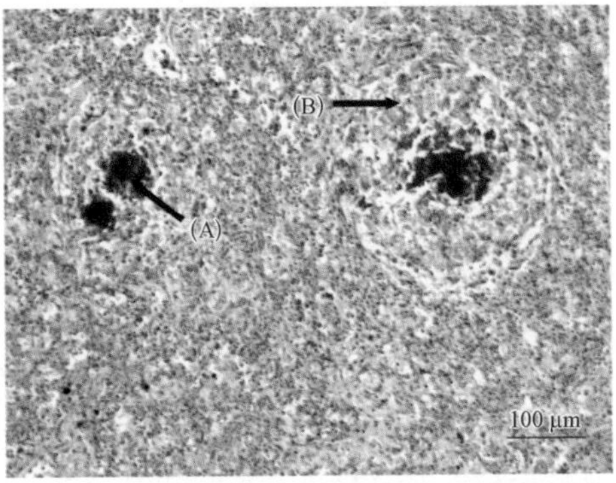

图 14.3 用抗杀鲑气单胞菌无色亚种抗体免疫染色处理的大西洋鳕非典型疖疮病中的脾肉芽肿图。肉芽肿中心由 (A) 细菌菌落及 (B) 被上皮细胞包围的细胞碎片组成。苏木精染色与碱性磷酸酶染色结合抗体。照片由 Bergljot Magnadottir 提供

非典型杀鲑气单胞菌引起的溃疡性疾病涉及浅表病理学。最初症状是皮肤上的小出血,逐渐发展为常见的两侧多灶性损伤。由其他病原菌引发

的继发感染也很常见。鱼死亡时可能没有可检测到的菌血症，但随着感染的发展，病原菌将定植于全身和内脏器官（Mawdesley-Thomas，1969；Bootsma et al.，1977）。Wiklund 和 Bylund（1993）将比目鱼的溃疡发展分为三个阶段：首先，皮肤有轻微出血；之后会发展成周围有出血性炎症组织的白色病灶；最后，溃疡表现为皮肤糜烂，肌肉暴露。其他杀鲑气单胞菌引起的溃疡病包括鲤赤皮病（Bootsma et al.，1977），鲑溃疡病（Paterson et al.，1980；Bullock et al.，1983）、金鱼溃疡病（Elliott and Shotts，1980）和真鲅（*Phoxinus phoxinus*）（Hastein et al.，1978）、日本鳗鲡（*Anguilla japonica*）（Kitao et al.，1984）、许氏平鲉（*Sebastes schlegeli*）（Han et al.，2011）中的各种溃疡病。

鱼感染嗜水气单胞菌的病理学各不相同。病变可能局限于皮肤，但也会发生全身性感染。感染可导致急性、慢性和隐性疾病（Cipriano and Austin，2011）。在急性 MAS 中，鱼可能会在疾病症状显现前死亡。临床体征可能包括眼球突出，皮肤、鳍和口腔的出血和坏死。皮肤溃疡可能发展为坏死并延伸至肌肉。肝瘀点、脾肿大和肾肿胀也是常见症状。鳃可能会出血并且鳞片竖起。组织病理学上，皮肤和肌肉组织可表现为急性至慢性皮炎和中性粒细胞浸润性肌炎，病变常涉及肾、肝、脾和心坏死（Cipriano et al.，1984；Cipriano and Austin，2011）。鲤鱼的病理症状在肝和肾中最严重，肠和心上皮细胞的变性及器官间质组织出血也很常见（Stratev et al.，2015）。斑点叉尾鮰中的感染可引起全身性和皮肤疾病（Ventura and Grizzle，1988）。在斑点叉尾鮰皮肤感染中，可能有几种由细菌毒素引起的隐匿性病变（Ventura and Grizzle，1988）。

14.4 病理生理学

14.4.1 病原基因组学

一些气单胞菌属的基因组已经完成测序，可以在美国国家生物技术信息中心（National Center for Biotechnology Information，NCBI）数据库中获得。已有 10 个杀鲑气单胞菌基因组数据可用，其中杀鲑亚种非致病性 A449 株的基因组是完整的（Reith et al.，2008），由 1 个环状染色体和 5 个质粒组成。该数据库还提供了 5 个典型菌株和 4 个非典型菌株的未完成基因组，其中包括杀鲑气单胞菌溶果胶亚种（Pavan et al.，2013）、杀鲑气单胞菌杀日本鲑亚种和杀鲑气单胞菌无色亚种（Han et al.，2013）3 个亚种。基于微阵列的杀鲑气单胞菌毒力基因比较显示了亚种之间的变异性（Nash et al.，2006）。Seshadri 等（2006）报道了嗜水气单胞菌模式菌株 ATCC 7966^{T}的基因组（单个染色体），分析表明，该菌株具有广泛的代谢能力和众多推定的毒力基因和系统，使其能够在多种生态系统中生存。Reith 等（2008）在杀鲑气单胞菌杀鲑亚种 A449 株中发现并报道了大量功能失调的基因，这些基因仅限于该亚种，其中包括几个假定的毒力基因。A449 株的基因组显示有大量的插入序列(IS)转座酶,这些转座酶大多数位于杀鲑气单胞菌的特异性基因内,表明它们可能通过水平转移或非生殖基因转移进行 DNA 交换(Studer et al., 2013)。假基因的积累降低了杀鲑气单胞菌杀鲑亚种产生一些细胞器和合成某些代谢酶的能力。这些特征是物种形成的关键,并反映了其对鲑科鱼类的宿主特异性(Reith et al., 2008)。

14.4.2 毒力因子

以下是杀鲑气单胞菌和嗜水气单胞菌毒力因子的概要,有关更多详情可参见 Tomás (2012), Beaz-Hidalgo 和 Figueras (2013),以及 Dallaire-Dufresne 等(2014)的相关文献。

1. 表面黏附素

气单胞菌的结晶表层是由自组装的单一蛋白质构成的最外层细胞包膜,在杀鲑气单胞菌中,它是由 *vapA* 编码的 A-层(Chu et al., 1991),在嗜水气单胞菌中则是由 *ahsA* 编码的 S-层(Dooley and Trust, 1988)。A-层与脂多糖(lipopolysaccharide, LPS)相连,抵抗吞噬作用和保护免受补体介导的裂解,还能促进对宿主细胞和免疫球蛋白的黏附(Munn et al., 1982; Phipps and Kay, 1988; Garduno et al., 2000)。此外,还鉴定了其他几种外膜蛋白(outer membrane proteins, OMPs)(Ebanks et al., 2005),包括补体激活孔蛋白(Yu et al., 2005)。

Tomás (2012)对杀鲑气单胞菌和嗜水气单胞菌脂多糖的结构、生物合成和基因组学进行了综述。杀鲑气单胞菌的 O-抗原具有免疫异质性,并且每个亚种的脂多糖核心是独特的(Wang et al., 2007; Jimenez et al., 2009)。Merino 等(2015)证明所有杀鲑气单胞菌菌株共享相同的 O-抗原生物合成基因簇,并且除了溶果胶亚种外,所有亚种都能产生 A-层。运动型气单胞菌,如嗜水气单胞菌,有一个单端鞭毛(Canals et al., 2007),但其他菌株也可在黏性环境中表达侧生鞭毛。两种鞭毛类型都参与黏附、侵入和生物被膜的形成(Kirov et al., 2004)。非运动型杀鲑气单胞菌中两种鞭毛基因都受到了破坏,这解释了它们为何不具有运动性(Reith et al., 2008)。尽管没有检测到菌毛,但在典型杀鲑气单胞菌中发现了一套Ⅰ型和三套Ⅳ型菌毛系统的编码基因(Boyd et al., 2008; Reith et al., 2008)。在杀鲑气单胞菌中,Ⅰ型菌毛有助于附着(Dacanay et al., 2010),Ⅳ型菌毛增强毒力(Masada et al., 2002; Boyd et al., 2008)。在嗜水气单胞菌中也发现了Ⅳ型菌毛系统(Seshadri et al., 2006)。

2. 分泌型毒力因子

气单胞菌产生并分泌毒素、脂酶和肽酶(Pemberton et al., 1997)。虽然典型和非典型杀鲑气单胞菌的大部分细胞相关因子是相同的，但它们分泌的胞外毒力因子各不相同(Gudmundsdottir et al., 2003a; Nash et al., 2006)。

嗜水气单胞菌同时产生细胞毒素和肠毒素，如气溶素、细胞毒性肠毒素(cytotonic enterotoxin, Act)、耐热细胞肠毒素(heat-stable cytotonic enteroxin, Ast)和不耐热细胞肠毒素(heat-labile cytotonic enteroxin, Alt)(Seshadri et al., 2006; Pang et al., 2015)。作为一种成孔毒素，杀鲑气单胞菌和嗜水气单胞菌中的气溶素(AerA)赋予嗜水气单胞菌溶血和肠毒性活性(Chakraborty et al., 1987; Bucker et al., 2011)。一种敲除嗜水气单胞菌腹泻临床分离株(SSU)中三种肠毒素基因(Axt、Alt 和 Ast)的缺陷菌株已被构建，这三种肠毒素每种都可通过引起小鼠体液分泌而导致胃肠炎(Sha et al., 2002)。几种非典型杀鲑气单胞菌会产生一种天青霉素金属内肽酶(Gudmundsdottir et al., 2003a)——AsaP1，它是杀鲑气单胞菌无色亚种的主要分泌毒力因子。虽然 AsaP1 是致死因子，但是该基因的敲除会使菌株的毒力减弱(Arnadottir et al., 2009)，杀鲑气单胞菌模式菌株 A449 株不产生功能性 AsaP1(Boyd et al., 2008)。

几种类型的溶血素和肽酶与毒力有关，包括丝氨酸肽酶(AspA 或 P1)，其会导致肌肉液化并促成疖疮形成(Fyfe et al., 1987)。肽酶还能激活甘油磷脂脂肪酶：胆固醇酰基转移酶(GCAT)，GCAT 可攻击膜磷脂，溶解鱼红细胞(Ellis, 1991)。杀鲑气单胞菌和嗜水气单胞菌均可产生 GCAT(Chacón et al., 2002)。此外，GCAT 与脂多糖的复合物(GCAT - LPS)对鲑科鱼类有毒性(Lee and Ellis, 1990)。然而，GCAT 缺陷型突变株仍具有毒力(Vipond et al., 1998)。

3. 铁摄取

杀鲑气单胞菌杀鲑亚种，有两套完整基因簇用于编码铁载体合成和铁摄取系统。一种编码推定的儿茶酚类铁载体(anguibactin 或 acinetobactin)，另一种编码曾在嗜水气单胞菌中发现的 amonabactin(Stintzi and Raymond, 2000; Reith et al., 2008)。最近，Balado 等(2015)研发发现一些杀鲑亚种产生 acinetobactin 和 amonabactin，但是一些致病菌株只产生 acinetobactin。他们进一步得出结论，acinetobactin 铁载体仅限于杀鲑气单胞菌，而 amonabactin 铁载体则在整个气单胞菌属中普遍存在。到目前为止，只有典型杀鲑气单胞菌产生铁载体，而一些非典型菌株具有不依赖铁载体的铁螯合系统(Hirst et al., 1991; Hirst and Ellis, 1996)。这种不依赖铁载体的机制涉及特异性结合宿主蛋白(如血红素)的外膜蛋白或受体，然后通过依赖于 TonB(与 OMPs 相互作用的转运蛋白)的内化过程。杀鲑气单胞菌和嗜水气单胞菌都具有血红素摄取基因(Ebanks et al., 2004; Najimi et al., 2008; Reith et al., 2008)。许多涉及铁摄取的基因表达，无论是否依赖铁载体，都受铁摄取调节子 Fur(ferric uptake regulator)(一种铁结合抑制蛋白)的调节(Braun et al., 1998)。当铁营养受限时，杀鲑气单胞菌会表达铁调节外膜蛋白(IROMPs)，包括铁载体和血红素受体(Ebanks et al.，2004；Menanteau-Ledouble et al.，2014)。Najimi 等(2009)发现杀鲑气单胞菌杀鲑亚种的铁调节基因具有明显的遗传多样性。

4. 分泌系统

革兰氏阴性菌已进化发展出至少六种分泌系统(T1SS～T6SS)，可将蛋白质/毒力因子转移至宿主组织(Costa et al.，2015)。杀鲑气单胞菌拥有 T2SS 和 T3SS 两套分泌系统，其中 A449 株中存在编码非功能性 T6SS 的基因(Reith et al.，2008；Vanden Bergh and Frey，2014)。在杀鲑气单胞菌无色亚种(ATCC 33659)中鉴定的分泌系统基因簇与模式菌株中的非常相似(Nash et al.，2006)。嗜水气单胞菌拥有 T2SS、T3SS 和 T6SS 三套分泌系统，但 T3SS 并非存在于所有菌株中(Seshadri et al.，2006；Pang et al.，2015)。

T2SS 是一个两步系统，可分泌毒素和肽酶(Nivaskumar and Francetic，2014)。气单胞菌菌株利用 T2SS 分泌包括气溶素和 GCAT 在内的多种因子(Brumlik et al.，1997；Schoenhofen et al.，1998)。

T3SS 一步即可分泌和转运效应蛋白或毒素(Portaliou et al.，2015)。效应蛋白破坏肌动蛋白细胞骨架，诱导细胞凋亡，并阻止信号转导和吞噬作用(Sierra et al.，2010；Tosi et al.，2013)。已经发现鉴定了几种杀鲑气单胞菌 T3SS 效应蛋白(Vanden Bergh and Frey，2014)，并证实它们对致病性至关重要(Burr et al.，2005；Dacanay et al.，2006)。编码杀鲑气单胞菌 T3SS 结构和调节蛋白的基因位于大质粒 pAsa5 上，当生长温度高于 22 ℃时，会导致 T3SS 功能丧失和无毒力(Daher et al.，2011)。

与 T3SS 一样，T6SS 直接将效应蛋白导入宿主细胞质或竞争细菌中(Cianfanelli et al.，2016)。嗜水气单胞菌拥有的 T6SS 可诱导细胞毒性并有助于先天免疫逃逸(Suarez et al.，2010a，b；Sha et al.，2013)。

5. 群体感应

气单胞菌中的群体感应系统与 LuxI/R 同源 (Jangid et al.，2007)，在杀鲑气单胞菌和嗜水气单胞菌中分别命名为 AsaI/R 和 AhyI/R。两种菌中产生的主要自诱导物为丁酰基- L -高丝氨酸内酯(CH_4 - AHL)(Swift et al.，1997；Schwenteit et al.，2011；Tan et al.，2015)。在嗜水气单胞菌中，生物被膜的形成(Lynch et al.，2002)、丝氨酸蛋白酶和金属蛋白酶活性的产生(Swift et al.，1999)都受到群体感应调控。在杀鲑气单胞菌无色亚种中，群体感应调节 AsaP1 毒性金属蛋白酶、细胞毒性和棕色色素的产生(Schwenteit et al.，2011)。杀鲑气单胞菌和嗜水气单胞菌对江鳕(*Lota lota*)的毒力受群体感应调控(Natrah et al.，2012)。

14.5 防控策略

14.5.1 良好的养殖规范和消毒措施

优质的养殖用水、人员培训，良好的卫生条件和养殖规范能够限制渔场养殖病害的发生。应定期对鱼类进行已知病原的检疫和筛查，建议在孵化场使用循环水进行消毒。臭氧化、过滤和紫外线照射等消毒措施均可有效抑制气单胞菌(Bullock and Stuckey，1977；Colberg and Lingg，1978)。当养殖水槽和其他设施干燥时，休耕制对于控制疫病蔓延也是有效的。另外，应对鱼卵进行消毒以杀灭其表面的细菌(Bergh et al.，2001)。当气单胞菌暴发时，有必要启动实施消毒程序，并在重新引入种苗前需要进行隔离防控(Torgersen and Hastein，1995)。在5～22 ℃，Mainous等(2011)发现，与多种消毒剂(2%的戊二醛；50%或70%的乙醇；1%的苄基-4-氯苯酚/苯基苯酚；1%的过氧化单硫酸钾+氯化钠；50 mg/L、100 mg/L、200 mg/L或50 000 mg/L的次氯酸钠；1∶256的*N*-烷基二甲基苄基氯化铵；50 mg/L或100 mg/L的聚维酮碘)接触1 min即可对杀鲑气单胞菌和嗜水气单胞菌产生杀伤活性，而福尔马林(250 mg/L)在这些温度下无效。氯化作用(每升软水含0.1 mg Cl_2)在30 min内可杀灭99.9%的杀鲑气单胞菌，但在硬水中Cl_2的浓度需要加倍(Wedermeyer and Nelson，1977)。

14.5.2 抗生素与细菌致病性抑制

气单胞菌感染的控制有赖于谨慎使用有效的抗生素(Cabello，2006)。对于抗生素的使用各国的管理要求各不相同，当需要抗生素治疗时，养殖者必须咨询本国的监管机构。四环素类(土霉素)、喹诺酮类(噁喹酸、氟甲喹和恩诺沙星)和氯霉素类(氟苯尼考)是水产中常用的抗生素类药物，阿莫西林和磺胺类药物也被使用(Rodgers and Furones，2009；Cipriano and Austin，2011)。

抗生素耐药性的产生推动了对新药的需求。海藻(*Gracilaria folifera*和*Sargassum longifolium*)对杀鲑气单胞菌具有抗菌活性(Thanigaivel et al.，2015)。茴芹(*Pimpinella anisum*)对杀鲑气单胞菌和嗜水气单胞菌具有拮抗作用(Parasa et al.，2012)。植物(*Dorycnium pentaphyllum*)可产生对嗜水气单胞菌的抗菌活性，但不抑制杀鲑气单胞菌(Turker and Yildirim，2015)。

群体感应信号分子的抑制或失活(称为群体猝灭)，可以减弱由群体感应调节的毒力过程，并有可能控制疾病(Defoirdt et al.，2011)。体外研究表明，群体猝灭可降低杀鲑气单胞菌和嗜水气单胞菌的致病性(Swift et al.，1999；Rasch et al.，2007；Schwenteit et al.，2011)，但这还需要进一步研究。

14.5.3 非特异性免疫刺激

1. 益生菌

益生菌不仅可以改善鱼类的生长和健康状况,还能对抗气单胞菌感染。通过竞争游离铁营养,荧光假单胞菌菌株可以限制杀鲑气单胞菌的生长(Smith and Davey, 1993),而革兰氏阳性乳酸杆菌(LAB)可降低感染杀鲑气单胞菌无色亚种的虹鳟死亡率(Nikoskelainen et al., 2001)。嗜水气单胞菌、肉食杆菌(*Carnobacterium* sp.)、河流弧菌(*Vibrio fluvialis*),以及在鲑科鱼类和大菱鲆肠道中的一种未鉴定的细菌对虹鳟的杀鲑气单胞菌感染具有拮抗作用(Irianto and Austin, 2002),用死的益生菌细胞投喂虹鳟能够抑制疖疮病的发展(Irianto and Austin, 2003)。益生性溶藻弧菌(*V. alginolyticus*)可诱导大西洋鲑对疖疮病的抗性(Austin et al., 1995)。

Vine 等(2004)研究了使用三带双锯鱼(*Amphiprion percula*)的肠道菌群来对抗嗜水气单胞菌感染。饲喂草鱼(*Ctenopharyngodon idellus*)由厦门希瓦氏菌(*Shewanella xiamenensis*)和维氏气单胞菌(*Aeromonas veronii*)组成的益生菌 28 d 后,降低了嗜水气单胞菌感染后的死亡率(Wu et al., 2015)。铜绿假单胞菌(*Pseudomonas aeruginosa*)、枯草芽孢杆菌(*Bacillus subtilis*)和植物乳杆菌(*Lactobacillus plantarum*)的细胞产物也可增强露斯塔野鲮(*Labeo rohita*)对嗜水气单胞菌感染的保护作用(Giri et al., 2015b)。

2. 益生元和其他免疫增强剂

用甘露寡糖(MOS)强化的饲料(每千克饲料含 2.5 g MOS)投喂虹鳟可改善其生长并增强对杀鲑气单胞菌的防护能力(Rodriguez-Estrada et al., 2013),富含 β-葡聚糖的饲料可显著增强杀鲑气单胞菌感染时的急性期反应(Pionnier et al., 2013)。

鲁氏耶尔森菌(*Yersinia ruckeri*)鞭毛蛋白已被用作非佐剂亚单位疫苗,可在虹鳟中诱发对杀鲑气单胞菌的非特异性免疫保护作用(Scott et al., 2013)。草药来源的免疫增强剂也受到越来越多的关注,如南非醉茄又称印度人参(*Withania somnifera*)具有多种药用特性(Mirjalili et al., 2009),当用其根粉饲喂时,野鲮能抵抗嗜水气单胞菌感染(Sharma et al., 2010)。含大蒜提取物(0.5% 或 1.0%)的饲料增加了虹鳟对杀鲑气单胞菌感染的抵抗力,但是更高的剂量没有效果且对鱼类健康有害(Breyer et al., 2015)。用番石榴(*Psidium guajava*)叶喂养野鲮能促进其生长并增强对嗜水气单胞菌的抵抗力(Giri et al., 2015a)。

富热休克蛋白(Hsp)细菌在水产养殖中具有防治病害的作用(Sung,

2014)，在冬季反复处理增强 Hsp70 响应，增强了异育银鲫(*Carassius auratus gibelio*)和斑点叉尾鮰对嗜水气单胞菌的抵抗力(Yang et al., 2015)。有研究报道了在野鲮感染嗜水气单胞菌期间七种热休克蛋白编码基因的表达分析情况(Das et al., 2015)；Hsp70 的表达被下调，但 HspA5 在肝、脾和前肾中表达上调，表明其潜在的免疫调节作用。热休克蛋白 HspA8 和 Hsp70 还可能介导齐口裂腹鱼(*Schizothorax prenanti*)针对嗜水气单胞菌的免疫应答(Li et al., 2015)。

14.5.4 疫苗接种

自 20 世纪 90 年代以来，随着杀鲑气单胞菌杀鲑亚种油乳佐剂灭活疫苗的商业化，疫苗接种已经控制了疖疮病。目前，商业化的鲑鱼疫苗大都为灭活疫苗。疖疮病疫苗一般通过注射方式接种于鲑、大菱鲆、锦鲤以及其他几种需预防典型和非典型疖疮病的鱼类；由于非典型杀鲑气单胞菌的抗原多样性和鱼类免疫反应的差异，这些疫苗的成功率各不相同。在一些地区(如挪威、法罗群岛和冰岛)，预防鲑科鱼类和海水鱼类(包括吸盘圆鳍鱼)的非典型疖疮病疫苗已经商品化(Hastein et al., 2005; Gudmundsdottir and Bjornsdottir, 2007; Coscelli et al., 2015; 见 http://www.pharmaq.no/)。大西洋鳕不能通过接种疫苗来有效地预防那些具有 T 细胞依赖性的保护性抗原的病原体，而有效非典型疖疮病疫苗的缺乏限制了鳕养殖业的发展(FAO, 2014; Magnadottir, 2014)。

杀鲑气单胞菌循环抗体与免疫保护之间存在显著相关性，这在大西洋鲑(Romstad et al., 2013)、虹鳟(Villumsen et al., 2012)、北极红点鲑(Schwenteit et al., 2014)，庸鲽(Gudmundsdottir et al., 2003b)、花狼鱼(Ingilae et al., 2000)和大菱鲆(Bjornsdottir et al., 2005; Coscelli et al., 2015)中都已被证实。通过浸泡接种福尔马林灭活疫苗(Villumsen and Raida, 2013)和腹腔(IP)注射接种油乳佐剂疫苗(Villumsen et al., 2015)，能诱导虹鳟对水源性杀鲑气单胞菌的长期免疫保护。在浸泡免疫的鳟鱼中没有检测到循环抗体升高，但在腹腔接种的鱼中抗体水平与保护力相关。

嗜水气单胞菌的异质性使有效疫苗的开发复杂化，尚无商业化的疫苗，目前应用的疫苗是只在限制区域内获得许可的自家疫苗(Yanong, 2008 /9, rev. 2014)。基于嗜水气单胞菌天然和重组蛋白的疫苗已被证实具有免疫保护作用，这些保护性抗原包括黏附素(Fang et al., 2004)，LPS(Sun et al., 2012)、OMPs(Khushiramani et al., 2012; Maiti et al., 2012; Sharma and Dixit, 2015)和溶血素共调节蛋白(Hcp) (Wang et al., 2015)。

已构建开发了嗜水气单胞菌减毒活疫苗，并且在同源毒株攻毒实验中产生了良好的保护效力(Liu and Bi, 2007; Pridgeon et al., 2013)。此外，通过活菌载体(减毒的鳗弧菌)递送嗜水气单胞菌抗原也产生了良好的免疫接种结果

(Zhao et al., 2011)。Tu 等(2010)研究了使用嗜水气单胞菌菌蜕灭活疫苗口服免疫鲤鱼,研究发现菌蜕可诱导产生优异的保护性黏膜和全身性免疫反应。

14.6 总结与研究展望

目前,全世界养殖的有鳍鱼类约有 350 种(FAO, 2014),这其中许多鱼类都易感气单胞菌。气单胞菌的栖息环境、宿主易感性和致病性质具有多样性。可持续水产养殖业的持续发展,需要采取环境友好的疾病控制措施防止疾病在野生鱼类和养殖鱼类之间传播,以及防止疾病进入新的地区。此外,运动型气单胞菌的人畜共患传播和抗生素耐药性的分布是主要关注的问题。

疾病感染的诊断有赖于准确的病原鉴定,但非典型杀鲑气单胞菌和运动型气单胞菌的分类和鉴定尚不完全确定,需要进一步厘清明确。因此,正确鉴定和分类这些细菌的快速可靠的方法对于疾病诊断、流行病学和病害预防非常重要。需要更多的全基因组测序来建立分类和诊断工具标准化的跨国方法。此外,需要有关杀鲑气单胞菌自然栖息地的信息,以揭示它是否具有除鱼类以外的重要分布来源。

必须采取多学科措施,加强对气单胞菌疾病的预防和控制。对其致病性方面的了解仍存在相当大的差距,希望今后随着人们对病原和宿主基因组序列、基因功能及其相互作用认知的快速增长将有助于弥合这些差距。

参考文献

[1] Adams, A. and Thompson, K. (1990) Development of an enzyme-linked immunosorbent assay (ELISA) for the detection of *Aeromonas salmonicida* in fish tissue. Journal of Aquatic Animal Health 2, 281 - 288.

[2] Arnadottir, H., Hvanndal, I., Andresdottir, V., Burr, S. E., Frey, J. and Gudmundsdottir, B.K. (2009) The AsaP1 peptidase of *Aeromonas salmonicida* subsp. *achromogenes* is a highly conserved deuterolysin metalloprotease (family M35) and a major virulence factor. *Journal of Bacteriology* 191, 403 - 410.

[3] Austin, B. and Austin, D. A. (2012a) 4 *Aeromonadaceae* representatives (motile aeromonads). In: Austin, B. and Austin, D.A. *Bacterial Fish Pathogens: Disease of Farmed and Wild Fish*, 5th edn. Springer, Dordrecht, The Netherlands/Heidelberg, Germany/New York/London, pp. 119 - 146.

[4] Austin, B. and Austin, D. A. (2012b) 5 *Aeromonadaceae* representative (*Aeromonas salmonicida*). In: Austin, B. and Austin, D.A. *Bacterial Fish Pathogens: Disease of Farmed and Wild Fish*, 5th edn. Springer, Dordrecht, The Netherlands/Heidelberg, Germany/New York/London, pp. 145 - 212.

[5] Austin, B., Stuckey, L.F., Robertson, P.A.W., Effendi, I. and Griffith, D.R.W. (1995) A probiotic strain of *Vibrio alginolyticus* effective in reducing diseases caused by *Aeromonas salmonicida*, *Vibrio anguillarum* and *Vibrio ordalii*. *Journal of Fish Diseases* 18, 93 - 96.

[6] Austin, B., Austin, D. A., Dalsgaard, I., Gudmundsdottir, B. K., Hoie, S. *et al.* (1998) Characterization of atypical *Aeromonas salmonicida* by different methods. *Systematic and Applied Microbiology* 21, 50 - 64.

[7] Balado, M., Souto, A., Vences, A., Careaga, V. P., Valderrama, K. *et al.* (2015) Two catechol siderophores, acinetobactin and amonabactin, are simultaneously produced by *Aeromonas salmonicida* subsp. *salmonicida* sharing part of the biosynthetic pathway. *ACS Chemical Biology* 10, 2850 - 2860.

[8] Balcázar, J.L., Vendrell, D., de Blas, I., Ruiz-Zarzuela, I., Girones, O. and Muzquiz, J.L. (2007) Quantitative detection of *Aeromonas salmonicida* in fish tissue by real-time PCR using self-quenched, fluorogenic primers. *Journal of Medical Microbiology* 56, 323 - 328.

[9] Beaz-Hidalgo, R. and Figueras, M. J. (2013) *Aeromonas* spp. whole genomes and virulence factors implicated in fish disease. *Journal of Fish Diseases* 36, 371 - 388.

[10] Beaz-Hidalgo, R., Latif-Eugenin, F. and Figueras, M.J. (2013) The improved PCR of the *fstA* (ferric siderophore receptor) gene differentiates the fish pathogen *Aeromonas salmonicida* from other *Aeromonas* species. *Veterinary Microbiology* 166, 659 - 663.

[11] Belland, R.J. and Trust, T.J. (1985) Synthesis, export, and assembly of *Aeromonas salmonicida* A-layer analyzed by transposon mutagenesis. *Journal of Bacteriology* 163, 877 - 881.

[12] Bergh, O., Nilsen, F. and Samuelsen, O.B. (2001) Diseases, prophylaxis and treatment of the Atlantic hali-but *Hippoglossus hippoglossus*: a review. *Diseases of Aquatic Organisms* 48, 57 - 74. Bjornsdottir, B., Gudmundsdottir, S., Bambir, S. H. and Gudmundsdottir, B. K. (2005) Experimental infection of turbot, *Scophthalmus maximus* (L.), by *Aeromonas salmonicida* subsp *achromogenes* and evaluation of cross protection induced by a furunculosis vaccine. *Journal of Fish Diseases* 28, 181 - 188.

[13] Bootsma, R., Fijan, N. and Blommaert, J. (1977) Isolation and identification of the causative agent of carp erythrodermatitis. *Veterinarski Arhiv* 47, 291 - 302.

[14] Boyd, J.M., Dacanay, A., Knickle, L.C., Touhami, A., Brown, L.L. *et al.* (2008) Contribution of type IV pili to the virulence of *Aeromonas salmonicida* subsp *salmonicida* in Atlantic salmon (*Salmo salar* L.). *Infection and Immunity* 76, 1445 - 1455.

[15] Braun, V., Hantke, K. and Koster, W. (1998) Bacterial iron transport: mechanisms, genetics, and regulation. *Metal Ions in Biological Systems* 35, 67 - 145.

[16] Breyer, K.E., Getchell, R.G., Cornwell, E.R. and Wooster, A.A. (2015) Efficacy of an extract from garlic, *Alliumsativum*, against infection with the furunculosis bacterium, *Aeromonas salmonicida*, in rainbow trout, *Oncorhynchus mykiss*. *Journal of the World Aquaculture Society* 46, 569 - 569.

[17] Brumlik, M.J., van der Goot, F.G., Wong, K.R. and Buckley, J. T. (1997) The disulfide bond in the *Aeromonas hydrophila* lipase /acyltransferase stabilizes the structure but is not required for secretion or activity. *Journal of Bacteriology* 179, 3116 - 3121.

[18] Bucker, R., Krug, S.M., Rosenthal, R., Gunzel, D., Fromm, A. *et al.* (2011) Aerolysin from *Aeromonas hydrophila* perturbs tight junction integrity and cell lesion repair in intestinal epithelial HT-29 /B6 cells. *Journal of Infectious Diseases* 204, 1283 - 1292.

[19] Bullock, G.L. and Stuckey, H.M. (1977) Ultraviolet treatment of water for destruction of five Gram-negative bacteria pathogenic to fishes. *Journal of the Fisheries Research Board of Canada* 34, 1244 - 1249.

[20] Bullock, G. L., Cipriano, R. C. and Snieszko, S. F. (1983) Furunculosis and other diseases caused by Aeromonas salmonicida. Paper 133, US Fish and Wildlife

Publications, US Fish and Wildlife Service, Washington, DC. Available at: http://digitalcommons.unl.edu/cgi/viewcontent.cgi? article = 1132&context = usfwspubs (accessed 9 November 2016). Revised (2001) version available at: https://articles.extension.org/sites/default/files/w/b/b7/Furunculosis.pdf (accessed 9 November 2016)

[21] Burr, S.E., Pugovkin, D., Wahli, T., Segner, H. and Frey, J. (2005) Attenuated virulence of an *Aeromonas salmonicida* subsp. *salmonicida* type III secretion mutant in a rainbow trout model. *Microbiology* 151, 2111 - 2118.

[22] Byers, H.K., Cipriano, R.C., Gudkovs, N. and Crane, M.S. (2002) PCR-based assays for the fish pathogen *Aeromonas salmonicida*. II. Further evaluation and validation of three PCR primer sets with infected fish. *Diseases of Aquatic Organisms* 49, 139 - 144.

[23] Cabello, F.C. (2006) Heavy use of prophylactic antibiotics in aquaculture: a growing problem for human and animal health and for the environment. *Environmental Microbiology* 8, 1137 - 1144.

[24] Canals, R., Vilches, S., Wilhelms, M., Shaw, J.G., Merino, S. *et al.* (2007) Non-structural flagella genes affecting both polar and lateral flagella-mediated motility in *Aeromonas hydrophila*. *Microbiology* 153, 1165 - 1175.

[25] Chacón, M.R., Castro-Escarpulli, G., Soler, L., Guarro, J. and Figueras, M.J. (2002) A DNA probe specific for *Aeromonas* colonies. *Diagnostic Microbiology and Infectious Disease* 44, 221 - 225.

[26] Chakraborty, T., Huhle, B., Hof, H., Bergbauer, H. and Goebel, W. (1987) Marker exchange mutagenesis of the aerolysin determinant in *Aeromonas hydrophila* demonstrates the role of aerolysin in *A. hydrophila*-associated systemic infections. *Infection and Immunity* 55, 2274 - 2280.

[27] Chart, H., Shaw, D. H., Ishiguro, E. E. and Trust, T. J. (1984) Structural and immunochemical homogeneity of *Aeromonas salmonicida* lipopolysaccharide. *Journal of Bacteriology* 158, 16 - 22.

[28] Chu, S., Cavaignac, S., Feutrier, J., Phipps, B.M., Kostrzynska, M. *et al.* (1991) Structure of the tetrago-nal surface virulence array protein and gene of *Aeromonas salmonicida*. *The Journal of Biological Chemistry* 266, 15258 - 15265.

[29] Cianfanelli, F.R., Monlezun, L. and Coulthurst, S.J. (2016) Aim, load, fire: the type VI secretion system, a bacterial nanoweapon. *Trends in Microbiology* 24, 51 - 62.

[30] Cipriano, R.C. and Austin, B. (2011) Furunculosis and other aeromoniosis diseases. In: Woo, P.T.K. and Bruno, D.W. (eds) *Fish Diseases and Disorders, Volume 3: Viral, Bacterial and Fungal Infections*, 2nd edn. CAB International, Wallingford, UK.

[31] Cipriano, R. C. and Bertolini, J. (1988) Selection for virulence in the fish pathogen *Aeromonas salmonicida*, using Coomassie Brilliant Blue agar. *Journal of Wildlife Diseases* 24, 672 - 678.

[32] Cipriano, R.C., Bullock, G. L. and Pyle, S. W. (1984) *Aeromonas hydrophila* and motile aeromonad septicemias of fish. Paper 134, *US Fish and Wildlife Publications*, US Fish and Wildlife Service, Washington, DC. Available at: http://digitalcommons.unl.edu/usfwspubs/134/(accessed 9 November 2016). Revised (2001) version available at: https://articles.extension.org/sites/default/files/w/1/1e/Aeromonas_hydrophila.pdf (accessed 9 November 2016).

[33] Clarridge, J.E. 3rd (2004) Impact of 16S rRNA gene sequence analysis for identification of bacteria on clinical microbiology and infectious diseases. *Clinical Microbiology Reviews* 17, 840 - 862.

[34] Colberg, P.J. and Lingg, A.J. (1978) Effect of ozoniation on microbial fish pathogens, ammonia, nitrate, nitrite and biological oxygen demand in simulated reuse hatchery water. *Journal of the Fisheries Research Board of Canada* 35, 1290 - 1296.

[35] Colston, S.M., Fullmer, M.S., Beka, L., Lamy, B., Gogarten, J.P. *et al*. (2014) Bioinformatic genome comparisons for taxonomic and phylogenetic assignments using *Aeromonas* as a test case. *Mbio* 5(6): e02136.
[36] Colwell, R.R., Macdonell, M.T. and Deley, J. (1986) Proposal to recognize the family *Aeromonadaceae* famnov. *International Journal of Systematic Bacteriology* 36, 473 - 477.
[37] Coscelli, G.A., Bermudez, R., Losada, A.P., Santos, A. and Quiroga, M.I. (2015) Vaccination against *Aeromonas salmonicida* in turbot (*Scophthalmus maximus* L.): study of the efficacy, morphological changes and antigen distribution. *Aquaculture* 445, 22 - 32.
[38] Costa, T.R., Felisberto-Rodrigues, C., Meir, A., Prevost, M.S., Redzej, A. *et al*. (2015) Secretion systems in Gram-negative bacteria: structural and mechanistic insights. *Nature Reviews Microbiology* 13, 343 - 359.
[39] Dacanay, A., Knickle, L., Solanky, K.S., Boyd, J.M., Walter, J.A. *et al*. (2006) Contribution of the type III secretion system (TTSS) to virulence of *Aeromonas salmonicida* subsp. *salmonicida*. *Microbiology* 152, 1847 - 1856.
[40] Dacanay, A., Boyd, J.M., Fast, M.D., Knickle, L.C. and Reith, M.E. (2010) *Aeromonas salmonicida* type I pilus system contributes to host colonization but not invasion. *Diseases of Aquatic Organisms* 88, 199 - 206.
[41] Daher, R.K., Filion, G., Tan, S.G., Dallaire-Dufresne, S., Paquet, V.E. *et al*. (2011) Alteration of virulence factors and rearrangement of pAsa5 plasmid caused by the growth of *Aeromonas salmonicida* in stressful conditions. *Veterinary Microbiology* 152, 353 - 360.
[42] Dallaire-Dufresne, S., Tanaka, K.H., Trudel, M.V., Lafaille, A. and Charette, S.J. (2014) Virulence, genomic features, and plasticity of *Aeromonas salmonicida* subsp. *salmonicida*, the causative agent of fish furunculosis. *Veterinary Microbiology* 169, 1 - 7.
[43] Das, S., Mohapatra, A. and Sahoo, P.K. (2015) Expression analysis of heat shock protein genes during *Aeromonas hydrophila* infection in rohu, *Labeo rohita*, with special reference to molecular characteriza-tion of Grp78. *Cell Stress Chaperones* 20, 73 - 84.
[44] Defoirdt, T., Sorgeloos, P. and Bossier, P. (2011) Alternatives to antibiotics for the control of bacterial disease in aquaculture. *Current Opinion in Microbiology* 14, 251 - 258.
[45] Dooley, J.S. and Trust, T.J. (1988) Surface protein com-position of *Aeromonas hydrophila* strains virulent for fish: identification of a surface array protein. *Journal of Bacteriology* 170, 499 - 506.
[46] Ebanks, R.O., Dacanay, A., Goguen, M., Pinto, D.M. and Ross, N.W. (2004) Differential proteomic analysis of *Aeromonas salmonicida* outer membrane proteins in response to low iron and *in vivo* growth conditions. *Proteomics* 4, 1074 - 1085.
[47] Ebanks, R.O., Goguen, M., Mckinnon, S., Pinto, D.M. and Ross, N.W. (2005) Identification of the major outer membrane proteins of *Aeromonas salmonicida*. *Diseases of Aquatic Organisms* 68, 29 - 38.
[48] Elliott, D.G. and Shotts, E.B. (1980) Aetiology of an ulcerative disease in goldfish *Carassius auratus* (L): microbiological examination of diseased fish from seven locations. *Journal of Fish Diseases* 3, 133 - 143.
[49] Ellis, A.E. (1991) An appraisal of the extracellular toxins of *Aeromonas salmonicida* ssp. *salmonicida*. *Journal of Fish Diseases* 14, 265 - 277.
[50] Fang, H.M., Ge, R.W. and Sin, Y.M. (2004) Cloning, characterisation and expression of *Aeromonas hydrophila* major adhesin. *Fish and Shellfish Immunology* 16,

645 - 658.

[51] FAO (2014) *The State of World Fisheries and Aquaculture: Opportunities and Challenges*. Food and Agriculture Organization of the United Nations, Rome.

[52] Ferguson, Y., Bricknell, I. R., Glover, L. A., Macgregor, D. M. and Prosser, J. I. (1998) Colonisation and transmission of luxmarked and wild-type *Aeromonas salmonicida* strains in Atlantic salmon (*Salmo salar* L.). *FEMS Microbiology Ecology* 27, 251 - 260.

[53] Fyfe, L., Coleman, G. and Munro, A. L. S. (1987) Identification of major common extracellular proteins secreted by *Aeromonas salmonicida* strains isolated from diseased fish. *Applied and Environmental Microbiology* 53, 722 - 726.

[54] Garduno, R.A., Moore, A.R., Olivier, G., Lizama, A.L., Garduno, E. *et al*. (2000) Host cell invasion and intracellular residence by *Aeromonas salmonicida*: role of the S-layer. *Canadian Journal of Microbiology* 46, 660 - 668.

[55] Gilroy, D. and Smith, P. (2003) Application-dependent, laboratory-based validation of an enzyme-linked immunosorbent assay for *Aeromonas salmonicida*. *Aquaculture* 217, 23 - 38.

[56] Giri, S.S., Sen, S.S., Chi, C., Kim, H.J., Yun, S. *et al*. (2015a) Effect of guava leaves on the growth performance and cytokine gene expression of *Labeo rohita* and its susceptibility to *Aeromonas hydrophila* infection. *Fish and Shellfish Immunology* 46, 217 - 224.

[57] Giri, S.S., Sen, S.S., Chi, C., Kim, H.J., Yun, S. *et al*. (2015b) Effect of cellular products of potential probiotic bacteria on the immune response of *Labeo rohita* and susceptibility to *Aeromonas hydrophila* infection. *Fish and Shellfish Immunology* 46, 716 - 722.

[58] Gudding, R. and van Muiswinkel, W.B. (2013) A history of fish vaccination: science-based disease prevention in aquaculture. *Fish and Shellfish Immunology* 35, 1683 - 1688.

[59] Gudmundsdottir, B.K. (1998) Infections by atypical strains of the bacterium *Aeromonas salmonicida*: a review. *Icelandic Agricultural Sciences* 12, 61 - 72.

[60] Gudmundsdottir, B. K. and Bjornsdottir, B. (2007) Vaccination against atypical furunculosis and winter ulcer disease of fish. *Vaccine* 25, 5512 - 5523.

[61] Gudmundsdottir, B. K., Hvanndal, I., Bjornsdottir, B. and Wagner, U. (2003a) Analysis of exotoxins produced by atypical isolates of *Aeromonas salmonicida*, by enzymatic and serological methods. *Journal of Fish Diseases* 26, 15 - 29.

[62] Gudmundsdottir, S., Lange, S., Magnadottir, B. and Gudmundsdottir, B.K. (2003b) Protection against atypical furunculosis in Atlantic halibut, *Hippoglossus hippoglossus*, comparison of fish vaccinated with commercial furunculosis vaccine and an autogenous vaccine based on the challenge strain. *Journal of Fish Diseases* 26, 331 - 338.

[63] Gulla, S., Lund, V., Kristoffersen, A.B., Sørum, H. and Colquhoun, D.J. (2015a) *vapA* (A-layer) typing differentiates *Aeromonas salmonicida* subspecies and identifies a number of previously undescribed subtypes. *Journal of Fish Diseases* 39, 329 - 342.

[64] Gulla, S., Duodu, S., Nilsen, A., Fossen, I. and Colquhoun, D.J. (2015b) *Aeromonas salmonicida* infection levels in pre-and post-stocked cleaner fish assessed by culture and an amended qPCR assay. *Journal of Fish Diseases* 39, 867 - 877.

[65] Gustafson, C. E., Thomas, C. J. and Trust, T. J. (1992) Detection of *Aeromonas salmonicida* from fish by using polymerase chain reaction amplification of the virulence surface array protein gene. *Applied and Environmental Microbiology* 58, 3816 - 3825.

[66] Han, H.J., Kim, D.Y., Kim, W.S., Kim, C.S., Jung, S.J. *et al*. (2011) Atypical *Aeromonas salmonicida* infection in the black rockfish, *Sebastes schlegeli* Hilgendorf, in Korea. *Journal of Fish Diseases* 34, 47 - 55.

[67] Han, J.E., Kim, J.H., Shin, S.P., Jun, J.W., Chai, J.Y. *et al*. (2013) Draft genome sequence of *Aeromonas salmonicida* subsp. *achromogenes* AS03, an atypical strain isolated from crucian carp (*Carassius carassius*) in the Republic of Korea. *Genome Announcements* 1(5): e00791-13.
[68] Hastein, T., Saltveit, S.J. and Roberts, R.J. (1978) Mass mortality among minnows *Phoxinus phoxinus* (L) in Lake Tveitevatn, Norway, due to an aberrant strain of *Aeromonas salmonicida*. *Journal of Fish Diseases* 1, 241-249.
[69] Hastein, T., Gudding, R. and Evensen, O. (2005) Bacterial vaccines for fish: an update of the current situation worldwide. *Developmental Biology* 121, 55-74.
[70] Hellberg, H., Moksness, E. and Hoie, S. (1996) Infection with atypical *Aeromonas salmonicida* in farmed common wolffish, *Anarhichas lupus* L. *Journal of Fish Diseases* 19, 329-332.
[71] Hirst, I.D. and Ellis, A.E. (1996) Utilization of transferrin and salmon serum as sources of iron by typical and atypical strains of *Aeromonas salmonicida*. *Microbiology* 142, 1543-1550.
[72] Hirst, I.D., Hastings, T.S. and Ellis, A.E. (1991) Siderophore production by *Aeromonas salmonicida*. *Journal of Genetic Microbiology* 137, 1185-1192.
[73] Hossain, M.J., Sun, D.W., McGarey, D.J., Wrenn, S., Alexander, L.M. *et al*. (2014) An Asian origin of virulent *Aeromonas hydrophila* responsible for disease epidemics in United States-farmed catfish. *Mbio* 5, e00848-00814.
[74] Huys, G., Pearson, M., Kampfer, P., Denys, R., Cnockaert, M. *et al*. (2003) *Aeromonas hydrophila* subsp. *ranae* subsp. nov., isolated from septicaemic farmed frogs in Thailand. *International Journal of Systematic and Evolutionary Microbiology* 53, 885-891.
[75] Ingilae, M., Arnesen, J.A., Lund, V. and Eggset, G. (2000) Vaccination of Atlantic halibut *Hippoglossus hippoglossus* L., and spotted wolffish *Anarhichas minor* L., against atypical *Aeromonas salmonicida*. *Aquaculture* 183, 31-44.
[76] Irianto, A. and Austin, B. (2002) Use of probiotics to control furunculosis in rainbow trout, *Oncorhynchus mykiss* (Walbaum). *Journal of Fish Diseases* 25, 333-342.
[77] Irianto, A. and Austin, B. (2003) Use of dead probiotic cells to control furunculosis in rainbow trout, *Oncorhynchus mykiss* (Walbaum). *Journal of Fish Diseases* 26, 59-62.
[78] Janda, J.M. and Abbott, S.L. (2010) The genus *Aeromonas*: taxonomy, pathogenicity, and infection. *Clinical Microbiology Reviews* 23, 35-73.
[79] Jangid, K., Kong, R., Patole, M.S. and Shouche, Y.S. (2007) *LuxRI* homologs are universally present in the genus *Aeromonas*. *BMC Microbiology* 7: 93.
[80] Jimenez, N., Lacasta, A., Vilches, S., Reyes, M., Vazquez, J. *et al*. (2009) Genetics and proteomics of *Aeromonas salmonicida* lipopolysaccharide core biosynthesis. *Journal of Bacteriology* 191, 2228-2236.
[81] Khushiramani, R.M., Maiti, B., Shekar, M., Girisha, S.K., Akash, N. *et al*. (2012) Recombinant *Aeromonas hydrophila* outer membrane protein 48 (Omp48) induces a protective immune response against *Aeromonas hydrophila* and *Edwardsiella tarda*. *Research in Microbiology* 163, 286-291.
[82] Kirov, S.M., Castrisios, M. and Shaw, J.G. (2004) *Aeromonas* flagella (polar and lateral) are enterocyte adhesins that contribute to biofilm formation on surfaces. *Infection and Immunity* 72, 1939-1945.
[83] Kitao, T., Yoshida, T., Aoki, T. and Fukudome, M. (1984) Atypical *Aeromonas salmonicida*, the causative agent of an ulcer disease of eel occurred in Kagoshima Prefecture. *Fish Pathology* 19, 113-117.
[84] Lee, K.K. and Ellis, A.E. (1990) Glycerophospholipid: cholesterol acyltransferase

complexed with lipopolysaccharide (LPS) is a major lethal exotoxin and cytolysin of *Aeromonas salmonicida*: LPS stabilizes and enhances toxicity of the enzyme. *Journal of Bacteriology* 172, 5382 - 5393.

[85] Li, J.X., Zhang, H.B., Zhang, X.Y., Yang, S.Y., Yan, T.M. *et al.* (2015) Molecular cloning and expression of two heat-shock protein genes (HSC70/HSP70) from Prenant's schizothoracin (*Schizothorax prenanti*). *Fish Physiology and Biochemistry* 41, 573 - 585.

[86] Liu, Y.J. and Bi, Z.X. (2007) Potential use of a transposon Tn916-generated mutant of *Aeromonas hydrophila* J - 1 defective in some exoproducts as a live attenuated vaccine. *Preventive Veterinary Medicine* 78, 79 - 84.

[87] Lynch, M.J., Swift, S., Kirke, D.F., Keevil, C.W., Dodd, C.E.R. *et al.* (2002) The regulation of biofilm development by quorum sensing in *Aeromonas hydrophila*. *Environmental Microbiology* 4, 18 - 28.

[88] MacInnes, J.I., Trust, T.J. and Crosa, J.H. (1979) Deoxyribonucleic-acid relationships among members of the genus *Aeromonas*. *Canadian Journal of Microbiology* 25, 579 - 586.

[89] Magnadottir, B. (2014) The immune response of Atlantic cod, *Gadus morhua* L. *Icelandic Agricultural Sciences* 27, 41 - 61.

[90] Magnadottir, B., Bambir, S.H., Gudmundsdottir, B.K., Pilstrom, L. and Helgason, S. (2002) Atypical *Aeromonas salmonicida* infection in naturally and experimentally infected cod, *Gadus morhua* L. *Journal of Fish Diseases* 25, 583 - 597.

[91] Magnadottir, B., Gudmundsdottir, B.K. and Groman, D. (2013) Immunohistochemical determination of humoral immune markers within bacterial induced granuloma formation in Atlantic cod (*Gadus morhua* L.). *Fish and Shellfish Immunology* 34, 1372 - 1375.

[92] Mainous, M.E., Kuhn, D.D. and Smith, S.A. (2011) Efficacy of common aquaculture compounds for dis- infection of *Aeromonas hydrophila*, *A. salmonicida* subsp. *salmonicida*, and *A. salmonicida* subsp. *achromogenes* at various temperatures. *North American Journal of Aquaculture* 73, 456 - 461.

[93] Maiti, B., Shetty, M., Shekar, M., Karunasagar, I. and Karunasagar, I. (2012) Evaluation of two outer membrane proteins, Aha1 and OmpW of *Aeromonas hydrophila* as vaccine candidate for common carp. *Veterinary Immunology and Immunopathology* 149, 298 - 301.

[94] Martin-Carnahan, A. and Joseph, S.W. (2005) Family I. *Aeromononadaceae*. Order XII. *Aeromonadales* ord. nov. In: Garrity, G.M., Brenner, D.J., Krieg, N.R. and Staley, J. R. (eds) *Bergey's Manual of Systematic Bacteriology. Volume 2: The Proteobacteria, Part B: The Gammaproteobacteria*, 2nd edn. Springer, New York, pp. 556 - 587.

[95] Masada, C.L., Lapatra, S.E., Morton, A.W. and Strom, M.S. (2002) An *Aeromonas salmonicida* type IV pilin is required for virulence in rainbow trout *Oncorhynchus mykiss*. *Diseases of Aquatic Organisms* 51, 13 - 25.

[96] Mawdesley-Thomas, L.E. (1969) Furunculosis in the goldfish *Carassius auratus* (L). *Journal of Fish Biology* 1, 19 - 23.

[97] McCarthy, D. H. (1975) Fish furunculosis caused by *Aeromonas salmonicida* var. *achromogenes*. *Journal of Wildlife Diseases* 11, 489 - 493.

[98] McCarthy, D.H. and Roberts, R.J. (1980) Furunculosis of fish the present state of our knowledge. In: Droop, M. A. and Jannasch, H. W. (eds) *Advances in Aquatic Microbiology*. Academic Press, London.

[99] McIntosh, D. and Austin, B. (1991) Atypical characteristics of the salmonid pathogen *Aeromonas salmonicida*. *Journal of General Microbiology* 137, 1341 - 1343.

[100] Menanteau-Ledouble, S., Kattlun, J., Nobauer, K. and ElMatbouli, M. (2014) Protein expression and tran-scription profiles of three strains of *Aeromonas salmonicida*

ssp. *salmonicida* under normal and iron-limited culture conditions. *Proteome Sciences* 12：29.

[101] Merino，S.，Canals，R.，Knirel，Y.A. and Tomas，J.M.（2015）Molecular and chemical analysis of the lipopolysaccharide from *Aeromonas hydrophila* strain AH-1 (Serotype O11). *Marine Drugs* 13，2233－2249.

[102] Mirjalili，M.H.，Moyano，E.，Bonfill，M.，Cusido，R.M. and Palazon，J.（2009）Steroidal lactones from *Withania somnifera*，an ancient plant for novel medicine. *Molecules* 14，2373－2393.

[103] Moki，S.T.，Nomura，T. and Yoshimizu，M.（1995）Effect of incubation temperature for isolation on autoagglutination of *Aeromonas salmonicida*. *Fish Pathology* 30，67－68.

[104] Munn，C.B.，Ishiguro，E.E.，Kay，W.W. and Trust，T.J.（1982）Role of surface components in serum resistance of virulent *Aeromonas salmonicida*. *Infection and Immunity* 36，1069－1075.

[105] Najimi，M.，Lemos，M.L. and Osorio，C.R.（2008）Identification of siderophore biosynthesis genes essential for growth of *Aeromonas salmonicida* under iron limitation conditions. *Applied and Environmental Microbiology* 74，2341－2348.

[106] Najimi，M.，Lemos，M.L. and Osorio，C.R.（2009）Identification of iron regulated genes in the fish pathogen *Aeromonas salmonicida* subsp. *salmonicida*：genetic diversity and evidence of conserved iron uptake systems. *Veterinary Microbiology* 133，377－382.

[107] Nash，J.H.，Findlay，W.A.，Luebbert，C.C.，Mykytczuk，O.L.，Foote，S.J. *et al.*（2006）Comparative genomics profiling of clinical isolates of *Aeromonas salmonicida* using DNA microarrays. *BMC Genomics* 7：43.

[108] Natrah，F.M.I.，Alam，M.I.，Pawar，S.，Harzevili，A.S.，Nevejan，N. *et al.*（2012）The impact of quorum sensing on the virulence of *Aeromonas hydrophila* and *Aeromonas salmonicida* towards burbot（*Lota lota* L.）larvae. *Veterinary Microbiology* 159，77－82.

[109] Nikoskelainen，S.，Ouwehand，A.，Salminen，S. and Bylund，G.（2001）Protection of rainbow trout（*Oncorhynchus mykiss*）from furunculosis by *Lactobacillus rhamnosus*. *Aquaculture* 198，229－236.

[110] Nivaskumar，M. and Francetic，O.（2014）Type II secretion system：a magic beanstalk or a protein escalator. *Biochimica et Biophysica Acta* 1843，1568－1577.

[111] Nomura，T.（1993）The epidemiological study of furunculosis in salmon propagation. *Scientific Reports of the Hokaido Salmon Hatchery* [*Japan*] 47，1－99.

[112] Pang，M.，Jiang，J.，Xie，X.，Wu，Y.，Dong，Y. *et al.*（2015）Novel insights into the pathogenicity of epidemic *Aeromonas hydrophila* ST251 clones from comparative genomics. [*Nature*] *Scientific Reports* 5：9833.

[113] Parasa，L.S.，Tumati，S.R.，Prasad，C. and Kumar，C.A.（2012）*In vitro* antibacterial activity of culinary spices aniseed，star anise and cinnamon against bacterial pathogens of fish. *International Journal of Pharmacy and Pharmaceutical Sciences* 4，667－670.

[114] Paterson，W.D.，Douey，D. and Desautels，D.（1980）Isolation and identification of an atypical *Aeromonas salmonicida* strain causing epizootic losses among Atlantic salmon（*Salmo salar*）reared in a Nova Scotian hatchery. *Canadian Journal of Fisheries and Aquatic Sciences* 37，2236－2241.

[115] Pavan，M.E.，Abbott，S.L.，Zorzopulos，J. and Janda，J.M.（2000）*Aeromonas salmonicida* subsp. *pectinolytica* subsp. nov.，a new pectinase-positive subspecies isolated from a heavily polluted river. *International Journal of Systematic and Evolutionary Microbiology* 50，1119－1124.

[116] Pavan, M.E., Pavan, E.E., Lopez, N.I., Levin, L. and Pettinari, M.J. (2013) Genome sequence of the melaninproducing extremophile *Aeromonas salmonicida* subsp. *pectinolytica* strain 34 melT. *Genome Announcements* 1(5): e00675 - 13.

[117] Pemberton, J.M., Kidd, S.P. and Schmidt, R. (1997) Secreted enzymes of *Aeromonas*. *FEMS Microbiology Letters* 152, 1 - 10.

[118] Phipps, B.M. and Kay, W.W. (1988) Immunoglobulin binding by the regular surface array of *Aeromonas salmonicida*. *The Journal of Biological Chemistry* 263, 9298 - 9303.

[119] Pionnier, N., Falco, A., Miest, J., Frost, P., Irnazarow, I. *et al.* (2013) Dietary betaglucan stimulate complement and C-reactive protein acute phase responses in common carp (*Cyprinus carpio*) during an *Aeromonas salmonicida* infection. *Fish and Shellfish Immunology* 34, 819 - 831.

[120] Piotrowska, M. and Popowska, M. (2015) Insight into the mobilome of *Aeromonas* strains. *Frontiers in Microbiology* 6: 494.

[121] Portaliou, A.G., Tsolis, K.C., Loos, M.S., Zorzini, V. and Economou, A. (2015) Type III secretion: building and operating a remarkable nanomachine. *Trends in Biochemical Sciences* 41, 175 - 189.

[122] Pridgeon, J.W., Klesius, P.H. and Yildirim-Aksoy, M. (2013) Attempt to develop live attenuated bacterial vaccines by selecting resistance to gossypol, proflavine hemisulfate, novobiocin, or ciprofloxacin. *Vaccine* 31, 2222 - 2230.

[123] Rahman, M., Colque-Navarro, P., Kuhn, I., Huys, G., Swings, J. *et al.* (2002) Identification and characterization of pathogenic *Aeromonas veronii* biovar *sobria* associated with epizootic ulcerative syndrome in fish in Bangladesh. *Applied and Environmental Microbiology* 68, 650 - 655.

[124] Rasch, M., Kastbjerg, V.G., Bruhn, J.B., Dalsgaard, I., Givskov, M. and Gram, L. (2007) Quorum sensing signals are produced by *Aeromonas salmonicida* and quorum sensing inhibitors can reduce production of a potential virulence factor. *Diseases of Aquatic Organisms* 78, 105 - 113.

[125] Reith, M.E., Singh, R.K., Curtis, B., Boyd, J.M., Bouevitch, A. *et al.* (2008) The genome of *Aeromonas salmonicida* subsp. *salmonicida* A449: insights into the evolution of a fish pathogen. *BMC Genomics* 9: 427.

[126] Ren, S., Guo, J., Zeng, G. and Sun, G. (2006) Decolorization of triphenylmethane, azo, and anthraquinone dyes by a newly isolated *Aeromonas hydrophila* strain. *Applied Microbiology and Biotechnology* 72, 1316 - 1321.

[127] Rodgers, C.J. and Furones, M.D. (2009) Antimicrobial agents in aquaculture: practice, needs and issues. In: Rogers, C. and Basurco, B. (eds) *The Use of Veterinary Drugs and Vaccines in Mediterranean Aquaculture*. *Options Mediterraneennes: Serie A. Seminaires Mditerraneens* No. 86, CIHEAM (Centre International de Hautes Etudes Agronomiques Mediterraneennes), Montpeller, France/Zaragoza, Spain, pp. 41 - 59.

[128] Rodriguez-Estrada, U., Satoh, S., Haga, Y., Fushimi, H. and Sweetman, J. (2013) Effects of inactivated *Enterococcus faecalis* and mannan oligosaccharide and their combination on growth, immunity, and disease protection in rainbow trout. *North American Journal of Aquaculture* 75, 416 - 428.

[129] Romstad, A.B., Reitan, L.J., Midtlyng, P., Gravningen, K. and Evensen, O. (2013) Antibody responses correlate with antigen dose and *in vivo* protection for oiladjuvanted, experimental furunculosis (*Aeromonas salmonicida* subsp *salmonicida*) vaccines in Atlantic salmon (*Salmo salar* L.) and can be used for batch potency testing of vaccines. *Vaccine* 31, 791 - 796.

[130] Rusin, P.A., Rose, J.B., Haas, C.N. and Gerba, C.P. (1997) Risk assessment of

opportunistic bacterial pathogens in drinking water. *Reviews of Environmental Contamination and Toxicology* 152, 57 - 83.

[131] Saleh, M., Soliman, H., Haenen, O. and El-Matbouli, M. (2011) Antibodycoated gold nanoparticles immunoassay for direct detection of *Aeromonas salmonicida* in fish tissues. *Journal of Fish Diseases* 34, 845 - 852.

[132] Schoenhofen, I.C., Stratilo, C. and Howard, S.P. (1998) An ExeAB complex in the type II secretion pathway of *Aeromonas hydrophila*: effect of ATP-binding cassette mutations on complex formation and function. *Molecular Microbiology* 29, 1237 -1247.

[133] Schwenteit, J., Gram, L., Nielsen, K.F., Fridjonsson, O.H., Bornscheuer, U.T. *et al*. (2011) Quorum sensing in *Aeromonas salmonicida* subsp *achromogenes* and the effect of the autoinducer synthase AsaI on bacterial virulence. *Veterinary Microbiology* 147, 389 - 397.

[134] Schwenteit, J.M., Weber, B., Milton, D.L., Bornscheuer, U.T. and Gudmundsdottir, B.K. (2014) Construction of *Aeromonas salmonicida* subsp. *achromogenes* AsaP1-toxoid strains and study of their ability to induce immunity in Arctic char, *Salvelinus alpinus* L. *Journal of Fish Diseases* 38, 891 - 900.

[135] Scott, C.J.W., Austin, B., Austin, D.A. and Morris, P.C. (2013) Nonadjuvanted flagellin elicits a non-specific protective immune response in rainbow trout (*Oncorhynchus mykiss*, Walbaum) towards bacterial infections. *Vaccine* 31, 3262 - 3267.

[136] Seshadri, R., Joseph, S.W., Chopra, A.K., Sha, J., Shaw, J. *et al*. (2006) Genome sequence of *Aeromonas hydrophila* ATCC 7966T: jack of all trades. *Journal of Bacteriology* 188, 8272 - 8282.

[137] Sha, J., Kozlova, E.V. and Chopra, A.K. (2002) Role of various enterotoxins in *Aeromonas hydrophila*-induced gastroenteritis: generation of enterotoxin gene-deficient mutants and evaluation of their enterotoxic activity. *Infection and Immunity* 70, 1924 -1935.

[138] Sha, J., Rosenzweig, J.A., Kozlova, E.V., Wang, S., Erova, T.E. *et al*. (2013) Evaluation of the roles played by Hcp and VgrG type 6 secretion system effectors in *Aeromonas hydrophila* SSU pathogenesis. *Microbiology* 159, 1120 - 1135.

[139] Sharma, A., Deo, A.D., Riteshkumar, S.T., Chanu, T.I. and Das, A. (2010) Effect of *Withania somnifera* (L. Dunal) root as a feed additive on immunological parameters and disease resistance to *Aeromonas hydrophila* in *Labeo rohita* (Hamilton) fingerlings. *Fish and Shellfish Immunology* 29, 508 - 512.

[140] Sharma, M. and Dixit, A. (2015) Identification and immu-nogenic potential of B cell epitopes of outer membrane protein OmpF of *Aeromonas hydrophila* in translational fusion with a carrier protein. *Applied Microbiology and Biotechnology* 99, 6277 - 6291.

[141] Sierra, J.C., Suarez, G., Sha, J., Baze, W.B., Foltz, S.M. *et al*. (2010) Unraveling the mechanism of action of a new type III secretion system effector AexU from *Aeromonas hydrophila*. *Microbial Pathoglogy* 49, 122 - 134.

[142] Smith, P. and Davey, S. (1993) Evidence for the competitive exclusion of *Aeromonas salmonicida* from fish with stress-inducible furunculosis by a fluorescent pseudomonad. *Journal of Fish Diseases* 16, 521 - 524.

[143] Star, B., Nederbragt, A.J., Jentoft, S., Grimholt, U., Malmstrom, M. *et al*. (2011) The genome sequence of Atlantic cod reveals a unique immune system. *Nature* 477, 207 - 210.

[144] Stintzi, A. and Raymond, K.N. (2000) Amonabactinmediated iron acquisition from transferrin and lactoferrin by *Aeromonas hydrophila*: direct measurement of individual microscopic rate constants. *Journal of Biological Inorganic Chemistry* 5, 57 - 66.

[145] Stratev, D., Stoev, S., Vashin, I. and Daskalov, H. (2015) Some varieties of pathological changes in experimental infection of carps (*Cyprinus carpio*) with *Aeromonas hydrophila*. *Journal of Aquaculture Engineering and Fisheries Research* 1, 191 - 202.
[146] Studer, N., Frey, J. and Vanden Bergh, P. (2013) Clustering subspecies of *Aeromonas salmonicida* using IS630 typing. *BMC Microbiology* 13: 36.
[147] Suarez, G., Sierra, J.C., Erova, T.E., Sha, J., Horneman, A.J. *et al*. (2010a) A type VI secretion system effector protein, VgrG1, from *Aeromonas hydrophila* that induces host cell toxicity by ADP ribosylation of actin. *Journal of Bacteriology* 192, 155 - 168.
[148] Suarez, G., Sierra, J.C., Kirtley, M.L. and Chopra, A.K. (2010b) Role of Hcp, a type 6 secretion system effector, of *Aeromonas hydrophila* in modulating activation of host immune cells. *Microbiology-SGM* 156, 3678 - 3688.
[149] Sun, J.H., Wang, Q.K., Qiao, Z.Y., Bai, D.Q., Sun, J.F. *et al*. (2012) Effect of lipopolysaccharide (LPS) and outer membrane protein (OMP) vaccines on protection of grass carp (*Ctenopharyngodon idella*) against *Aeromonas hydrophila*. *The Israeli Journal of Aquaculture - Bamidgeh* 64, 1 - 8.
[150] Sung, Y.Y. (2014) Heat shock proteins: an alternative to control disease in aquatic organism. *Journal of Marine Science Research and Development* 4, e126.
[151] Swift, S., Karlyshev, A.V., Fish, L., Durant, E.L., Winson, M.K. *et al*. (1997) Quorum sensing in *Aeromonas hydrophila* and *Aeromonas salmonicida*: identification of the LuxRI homologs AhyRI and AsaRI and their cognate *N*-acylhomoserine lactone signal molecules. *Journal of Bacteriology* 179, 5271 - 5281.
[152] Swift, S., Lynch, M.J., Fish, L., Kirke, D.F., Tomas, J.M. *et al*. (1999) Quorum sensing-dependent regulation and blockade of exoprotease production in *Aeromonas hydrophila*. *Infection and Immunity* 67, 5192 - 5199.
[153] Tan, W.S., Yin, W.F. and Chan, K.G. (2015) Insights into the quorum sensing activity in *Aeromonas hydrophila* strain M013 as revealed by whole genome sequencing. *Genome Announcements* 3(1): e01372 - 14.
[154] Tenover, F.C. (2007) Rapid detection and identification of bacterial pathogens using novel molecular technologies: infection control and beyond. *Clinical Infectious Diseases* 44, 418 - 423.
[155] Thanigaivel, S., Hindu, S.V., Vijayakumar, S., Mukherjee, A., Chandrasekaran, N. *et al*. (2015) Differential solvent extraction of two seaweeds and their efficacy in controlling *Aeromonas salmonicida* infection in *Oreochromis mossambicus*: a novel therapeutic approach. *Aquaculture* 443, 56 - 64.
[156] Tomás, J.M. (2012) The main *Aeromonas* pathogenic factors. *ISRN Microbiology* 2012: 256261.
[157] Torgersen, Y. and Hastein, T. (1995) Disinfection in Aquaculture. *Revue Scientifique et Technique* 14, 419 - 434.
[158] Tosi, T., Pflug, A., Discola, K.F., Neves, D. and Dessen, A. (2013) Structural basis of eukaryotic cell targeting by type III secretion system (T3SS) effectors. *Research in Microbiology* 164, 605 - 619.
[159] Trust, T.J., Khouri, A.G., Austen, R.A. and Ashburner, L.D. (1980) First isolation in Australia of atypical *Aeromonas salmonicida*. *FEMS Microbiology Letters* 9, 39 - 42.
[160] Tu, F.P., Chu, W.H., Zhuang, X.Y. and Lu, C.P. (2010) Effect of oral immunization with *Aeromonas hydrophila* ghosts on protection against experimental fish infection. *Letters in Applied Microbiology* 50, 13 - 17.
[161] Turker, H. and Yıldırım, A.B. (2015) Screening for antibacterial activity of some

Turkish plants against fish pathogens: a possible alternative in the treatment of bacterial infections. *Biotechnology and Biotechnological Equipment* 29, 281 - 288.
[162] Udeh, P.J. (2004) *A Guide to Healthy Drinking Water*. iUniverse Inc., Lincoln, Nebraska.
[163] Valera, L. and Esteve, C. (2002) Phenotypic study by numerical taxonomy of strains belonging to the genus *Aeromonas*. *Journal of Applied Microbiology* 93, 77 - 95.
[164] Vanden Bergh, P. and Frey, J. (2014) *Aeromonas salmonicida* subsp. *salmonicida* in the light of its type-three secretion system. *Microbial Biotechnology* 7, 381 - 400.
[165] Ventura, M. T. and Grizzle, J. M. (1988) Lesions associated with natural and experimental infections of *Aeromonas hydrophila* in channel catfish, *Ictalurus punctatus* (Rafinesque). *Journal of Fish Diseases* 11, 397 - 407.
[166] Villumsen, K.R. and Raida, M.K. (2013) Long-lasting protection induced by bath vaccination against *Aeromonas salmonicida* subsp. *salmonicida* in rainbow trout. *Fish and Shellfish Immunology* 35, 1649 - 1653.
[167] Villumsen, K.R., Dalsgaard, I., Holten-Andersen, L. and Raida, M.K. (2012) Potential role of specific antibodies as important vaccine induced protective mechanism against *Aeromonas salmonicida* in rainbow trout. *PLoS ONE* 7(10): e46733.
[168] Villumsen, K.R., Koppang, E.O. and Raida, M.K. (2015) Adverse and long-term protective effects following oiladjuvanted vaccination against *Aeromonas salmonicida* in rainbow trout. *Fish and Shellfish Immunology* 42, 193 - 203.
[169] Vine, N.G., Leukes, W.D. and Kaiser, H. (2004) *In vitro* growth characteristics of five candidate aquaculture probiotics and two fish pathogens grown in fish intestinal mucus. *FEMS Microbiology Letters* 231, 145 - 152.
[170] Vipond, R., Bricknell, I.R., Durant, E., Bowden, T.J., Ellis, A.E. *et al*. (1998) Defined deletion mutants demonstrate that the major secreted toxins are not essential for the virulence of *Aeromonas salmonicida*. *Infection and Immunity* 66, 1990 - 1998.
[171] Wang, N.N., Wu, Y.F., Pang, M.D., Liu, J., Lu, C.P. *et al*. (2015) Protective efficacy of recombinant hemolysin coregulated protein (Hcp) of *Aeromonas hydrophila* in common carp (*Cyprinus carpio*). *Fish and Shellfish Immunology* 46, 297 - 304.
[172] Wang, Z., Liu, X., Dacanay, A., Harrison, B. A., Fast, M. *et al*. (2007) Carbohydrate analysis and serological classification of typical and atypical isolates of *Aeromonas salmonicida*: a rationale for the lipopolysaccharide-based classification of *A. salmonicida*. *Fish and Shellfish Immunology* 23, 1095 - 1106.
[173] Wedermeyer, G.A. and Nelson, N.C. (1977) Survival of two bacterial fish pathogens (*Aeromonas salmonicida* and the enteric redmouth bacterium) in ozonated, chlorinated and untreated water. *Journal of the Fisheries Research Board of Canada* 34, 429 - 432.
[174] Wiklund, T. (1990) Atypical *Aeromonas salmonicida* isolated from ulcers of pike, *Esox lucius* L. *Journal of Fish Diseases* 13, 541 - 544.
[175] Wiklund, T. (1995) Survival of atypical *Aeromonas salmonicida* in water and sediment microcosms of different salinities and temperatures. *Diseases of Aquatic Organisms* 21, 137 - 143.
[176] Wiklund, T. and Bylund, G. (1993) Skin ulcer disease of flounder *Platichthys flesus* in the northern Baltic Sea. *Diseases of Aquatic Organisms* 17, 165 - 174.
[177] Wiklund, T. and Dalsgaard, I. (1998) Occurrence and significance of atypical *Aeromonas salmonicida* in nonsalmonid and salmonid fish species: a review. *Diseases of Aquatic Organisms* 32, 49 - 69.
[178] Wiklund, T., Lonnstrom, L. and Niiranen, H. (1993) *Aeromonas salmonicida* ssp. *salmonicida* lacking pigment production, isolated from farmed salmonids in Finland. *Diseases of Aquatic Organisms* 15, 219 - 223.

[179] Wiklund, T., Dalsgaard, I., Eerola, E. and Olivier, G. (1994) Characteristics of atypical, cytochrome oxi-dasenegative *Aeromonas salmonicida* isolated from ulcerated flounders (*Platichthys flesus* (L). *Journal of Applied Bacteriology* 76, 511 - 520.
[180] Wu, Z. Q., Jiang, C., Ling, F. and Wang, G. X. (2015) Effects of dietary supplementation of intestinal autochthonous bacteria on the innate immunity and disease resistance of grass carp (*Ctenopharyngodon idellus*). *Aquaculture* 438, 105 - 114.
[181] Yang, B., Wang, C., Tu, Y., Hu, H., Han, D. *et al*. (2015) Effects of repeated handling and air exposure on the immune response and the disease resistance of gibel carp (*Carassius auratus gibelio*) over winter. *Fish and Shellfish Immunology* 47, 933 - 941.
[182] Yanong, R. P. E. (2008 /9, rev. 2014) *Use of Vaccines in Finfish Aquaculture*. Publication No. FA156, revised August 2014, Program in Fisheries and Aquatic Sciences, School of Forest Resources and Conservation, Florida Cooperative Extension Service, IFAS (Institute of Food and Agricultural Sciences), University of Florida, Gainesville, Florida. Available at: http://edis.ifas.ufl.edu/pdffiles/FA/FA15600.pdf (accessed 8 November 2016).
[183] Yu, H.B., Zhang, Y.L., Lau, Y.L., Yao, F., Vilches, S. *et al*. (2005) Identification and characterization of putative virulence genes and gene clusters in *Aeromonas hydrophila* PPD134/91. *Applied and Environmental Microbiology* 71, 4469 - 4477.
[184] Zhao, Y., Liu, Q., Wang, X.H., Zhou, L.Y., Wang, Q.Y. *et al*. (2011) Surface display of *Aeromonas hydrophila* GAPDH in attenuated *Vibrio anguillarum* to develop a novel multivalent vector vaccine. *Marine Biotechnology* 13, 963 - 970.

15

爱德华氏菌

Matt J. Griffin* ，Terrence E. Greenway 和 David J. Wise

15.1 引言

爱德华氏菌属[肠杆菌科(Enterobacteriacae)]最初在 20 世纪 60 年代中期被描述定义为肠杆菌科的一个新属。它们主要来自从美国、巴西、厄瓜多尔、以色列和日本等国人和动物的开放性伤口、血液、尿液和粪便中回收分离的 37 种分离株(Ewing et al., 1965)。尽管如此,爱德华氏菌属的种大多被认为是鱼类病原体(Mohanty and Sahoo, 2007; 表 15.1)。迟缓爱德华氏菌(*E. tarda*)最早报道于美国阿肯色州养殖斑点叉尾鮰暴发的病害(Meyer and Bullock,1973),目前已成为全球公认的鱼类病原体之一,影响着世界范围内的野生和养殖鱼类(Park et al., 2012)。

表 15.1　鱼类中的爱德华氏菌

宿　主	参考文献
鳗爱德华氏菌(*Edwardsiella anguillarum*)	
欧洲鳗鲡(*Anguilla anguilla*)	Shao et al., 2015
日本鳗鲡(*A. japonica*)	Shao et al., 2015
花鳗鲡(*A. marmorata*)	Shao et al., 2015
真鲷(*Pagrus major*)	Oguro et al., 2014
日本真鲷(*Evynnis japonica*)	Abayneh et al., 2012
尼罗罗非鱼(*Oreochromis niloticus*)	Griffin et al., 2014
青铜石斑鱼(*Epinephelus aeneus*)	Ucko et al., 2016
鮰爱德华氏菌(*Edwardsiella ictaluri*)	
香鱼(*Plecoglossus altivelis*)	Nagai et al., 2008
叉尾黄颡鱼(*Pelteobagrus nudiceps*)	Sakai et al., 2009b
云斑鮰(*Ameiurus nebulosus*)	Iwanowicz et al., 2006

* 通信作者邮箱: griffin@cvm.msstate.edu。

续 表

宿 主	参 考 文 献
斑点叉尾鮰(*Ictalurus punctatus*)	Hawke et al., 1981
孟加拉斑马鱼(*Danio devario*)	Waltman et al., 1985
玻璃飞刀鱼(*Eigenmannia virescens*)	Kent and Lyons, 1982
虹鳟(*Oncorhynchus mykiss*)	Keskin et al., 2004
玫瑰无须鲃(*Pethia conchonius*)	Humphrey et al., 1986
低眼巨鲇(*Pangasius hypophthalmus*)	Crumlish et al., 2002
蝌蚪石鮰(*Noturus gyrinus*)	Klesius et al., 2003
尼罗罗非鱼(*Oreochromis niloticus*)	Soto et al., 2012
胡鲇(*Clarias batrachus*)	Kasornchandra et al., 1987
黄颡鱼(*Pelteobagrus fulvidraco*)	Ye et al., 2009
斑马鱼(*Danio rerio*)	Hawke et al., 2013
杀鱼爱德华氏菌(*Edwardsiella piscicida*)	
迈氏条尾魟(*Taeniura meyeni*)	Camus et al., 2016
蓝鲇(*Ictalurus furcatus*)	Griffin et al., 2014
斑点叉尾鮰(*I. punctatus*)	Griffin et al., 2014
欧洲鳗鲡(*A. anguilla*)	Abayneh et al., 2013
杂交鲇(*I. furcatus* × *I. punctatus*)	Griffin et al., 2014
鲇鱼(*Silurus asotus*)	Abayneh et al., 2013
大口黑鲈(*Micropterus salmoides*)	Fogleson et al., 2016
牙鲆(*Paralichthys olivaceus*)	Oguro et al., 2014
马鲃脂鲤(*Hyphessobrycon eques*)	Shao et al., 2015
低眼巨鲇(*Pangasianodon hypophthalmus*)	Shetty et al., 2014
尼罗罗非鱼(*O. niloticus*)	Griffin et al., 2014
大菱鲆(*Scophthalmus maximus*)	Abayneh et al., 2013
白鲑(*Coregonus lavaretus*)	Shafiei et al., 2016
迟缓爱德华氏菌(*Edwardsiella tarda*)	
尖齿胡鲇(*Clarias gariepinus*)	Abraham et al., 2015
黄鳝(*Monopterus albus*)	Shao et al., 2016
香鱼(*Plecoglossus altivelis*)	Yamada and Wakabayashi, 1999
澳洲宝石鲈(*Scortum barcoo*)	Ye et al., 2010
尖吻鲈(*Lates calcarifer*)	Humphrey and Langdon, 1986
美洲红点鲑(*Salvelinus fontinalis*)	Uhland et al., 2000
斑点叉尾鮰(*Ictalurus punctatus*)	Meyer and Bullock, 1973
大鳞大麻哈鱼(*Oncorhynchus tshawytscha*)	Amandi et al., 1982
锦鲤(*Cyprinus carpio*)	Sae-oui et al., 1984

续 表

宿　主	参考文献
日本真鲷(*E. japonica*)	Kusuda et al., 1977
欧洲鳗鲡(*A. anguilla*)	Alcaide et al., 2006
虎皮鱼(*Puntius tetrazona*)	Akinbowale et al., 2006
杂交条纹鲈(*Morone chrysops*×*M. saxatilis*)	Griffin et al., 2014
日本鳗鲡(*A. japonica*)	Yamada and Wakabayashi, 1999
褐牙鲆(*Paralichthys olivaceus*)	Yamada and Wakabayashi, 1999
鲇鱼(*Silurus asotus*)	Yu et al., 2009
大口黑鲈(*Micropterus salmoides*)	White et al., 1973
鲻鱼(*Mugil cephalus*)	Kusuda et al., 1976
地图鱼(*Astronotus ocellatus*)	Wang et al., 2011
巨鲇(*Pangasius pangasius*)	Nakhro et al., 2013
红鲷(*Chrysophrus major*)	Yamada and Wakabayashi, 1999
舌齿鲈(*Dicentrarchus labrax*)	Blanch et al., 1990
暹罗斗鱼(*Betta splendens*)	Humphrey et al., 1986
白鲢(*Hypophthalmichthys molitrix*)	Xu and Zhang, 2014
条纹鲈(*Morone saxatilis*)	Herman and Bullock, 1986
尼罗罗非鱼(*O. niloticus*)	Yamada and Wakabayashi, 1999

同样，在20世纪80年代早期，人们从美国东南部养殖叉尾鮰病害中发现了鮰爱德华氏菌。该病原的病害症状不同于最初由迟缓爱德华氏菌导致的脓肿性腐烂病(Meyer and Bullock, 1973)，这种新疾病被认为是鮰肠败血症(enteric septicaemia of catfish, ESC) (Hawke et al., 1981)。第三种报道的爱德华氏菌来自鸟类和爬行动物，被命名为保科爱德华氏菌(*E.hoshinae*)，该菌种与其他爱德华氏菌在表型上和基因型上均存在明显差异(Grimont et al., 1980)。它通常在禽鸟类和爬行动物类宿主中被发现分离，尚不清楚该菌种是否会导致人类、鸟类、爬行动物或鱼类患病(Janda et al., 1991; Yang et al., 2012)。

这种分类一直保持不变，直到后来有了分辨率更高、分辨能力更强的分子生物学方法。种内遗传差异将 *E.tarda* 分离株分为两个不同的多系群，从而产生了鱼类致病性和鱼类非致病性迟缓爱德华氏菌的命名(Yamada and Wakabayashi, 1998; Yamada and Wakabayashi, 1999)。研究人员发现可以根据运动性区分鱼类致病性迟缓爱德华氏菌(Matsuyama et al., 2005)，这也与菌毛基因序列的遗传类群相一致(Sakai et al., 2007, 2009c)，这种分型区分被进一步推进深化。

后来的研究进一步支持了这些发现，因为在欧洲和美国的研究确定 *E.*

tarda 代表了两个或两个以上遗传学截然不同而表型模糊的种(Abayneh et al., 2012; Griffin et al., 2013)。这导致采用杀鱼爱德华氏菌(*E. piscicida*)命名分类了一个新种(Abayneh et al., 2013),并鉴定了第五种遗传学截然不同的 *E. piscicida* 类似种,后来被鉴定命名为鳗爱德华氏菌(*E. anguillarum*) (Shao et al., 2015)。根据最近的研究发现,将存档的核苷酸数据与最近发布的爱德华氏菌基因组进行比较(表 15.2),已经确定 *E. piscicida* 和 *E. anguillarum* 在历史上曾分别被鉴定为"典型、运动性"和"非典型、非运动"的鱼类致病性迟缓爱德华氏菌分离株。

表 15.2 已提供的爱德华氏菌基因组显示谱系和 GenBank(美国国立卫生研究院)登录号

菌 株	接收菌名	谱系 (Shao et al., 2015)	GenBank 登录号	参考文献
全基因组				
080813	*E. anguillarum*	*E. anguillarum*	CP006664	Shao et al., 2015
93-146	*E. ictaluri*	*E. ictalarui*	CP001600	Williams et al., 2012
C07-187	*E. tarda*	*E. piscicida*	CP004141	Tekedar et al., 2013
EA181011	*Edwardsiella* sp.	*E. anguillarum*	CP011364	Reichley et al., 2015b
EIB202	*E. tarda*	*E. piscicida*	CP001135	Wang et al., 2009
FL6-60	*E. tarda*	*E. piscicida*	CP002154	van Soest et al., 2011
FL95-01	*E. tarda*	*E. tarda*	CP011359	Reichley et al., 2015a
LADL 05-105	*Edwardsiella* sp.	*E. anguillarum*	CP011516	Reichley et al., 2015c
基因组草图				
ATCC #15947	*E. tarda*	*E. tarda*	PRJNA39897	Yang et al., 2012
ATCC #23685	*E. tarda*	*E. tarda*	PRJNA28661	Turnbaugh et al., 2007
ATCC #33202	*E. ictaluri*	*E. ictaluri*	PRJNA66365	Yang et al., 2012
ATCC #33379	*E. hoshinae*	*E. hoshinae*	PRJDB228	Nite.go.jp
DT	*E. tarda*	*E. tarda*	PRJNA66369	Yang et al., 2012
ET070829	*E. tarda*	*E. anguillarum*	PRJNA231705	Shao et al., 2015
ET081126R	*E. tarda*	*E. anguillarum*	PRJNA231706	Shao et al., 2015
ET883	*E. tarda*	*E. piscicida*	PRJNA259400	Shao et al., 2015
JF1305	*E. piscicida*	*E. piscicida*	PRJDB1727	Oguro et al., 2014
LADL 11-100	*E. ictaluri*	*E. ictaluri*	PRJ285663	Wang et al., 2015
LADL 11-194	*E. ictaluri*	*E. ictaluri*	PRJNA285852	Wang et al., 2015
RSB1309	*E. piscicida*	*E. anguillarum*	PRJDB1727	Oguro et al., 2014

造成上述分类模糊的一个因素是使用 16S rRNA 序列进行细菌分类，以及错误地将可公开获取的核苷酸数据库作为分类权威性依据。使用 16S rRNA 序列鉴别爱德华氏菌属的局限性有据可查(Griffin et al.，2013，2014，2015)。通常，16S rRNA 不能充分区分爱德华氏菌属，因为前面提到的五个菌种在该基因座上的相似性大于 99%(Griffin et al.，2015)，不完整或碎片化的 16S rRNA 序列可导致错误的分类。其他替代分类标记，如 DNA 促旋酶 B 亚基(*gyrB*)，或铁辅因子超氧化物歧化酶基因(*sodB*)的内部片段，提供了更可靠的鉴定。基于 *gyrB* (Griffin et al.，2013，2014，2015)和 *sodB* (Yamada and Wakabayashi，1999)的分子系统发育与更可靠的分子系统发育推论一致(图 15.1；Yang et al.，2012；Abayneh et al.，2013)，并且基于 *gyrB* 和 *sodB* 的系统发育拓扑图与通过多位点测序和比较系统发育基因组学确定的进化史基本一致(Shao et al.，2015；图 15.2)。

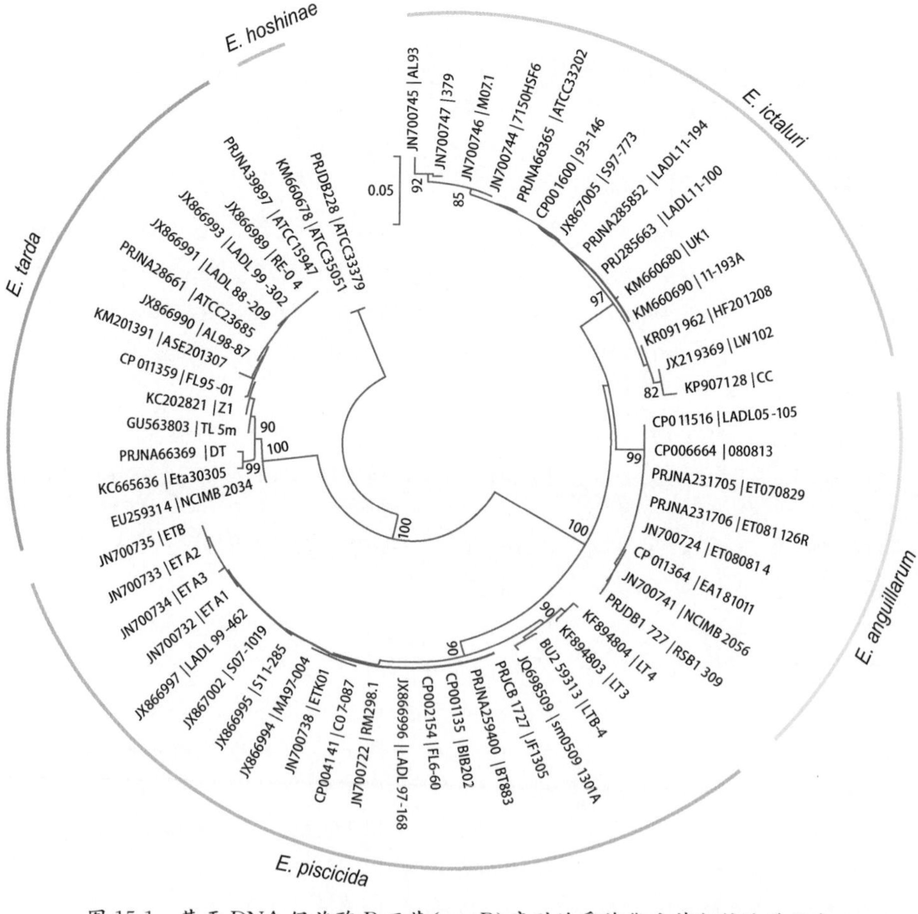

图 15.1　基于 DNA 促旋酶 B 亚基(*gyrB*)序列的爱德华氏菌邻接法进化史

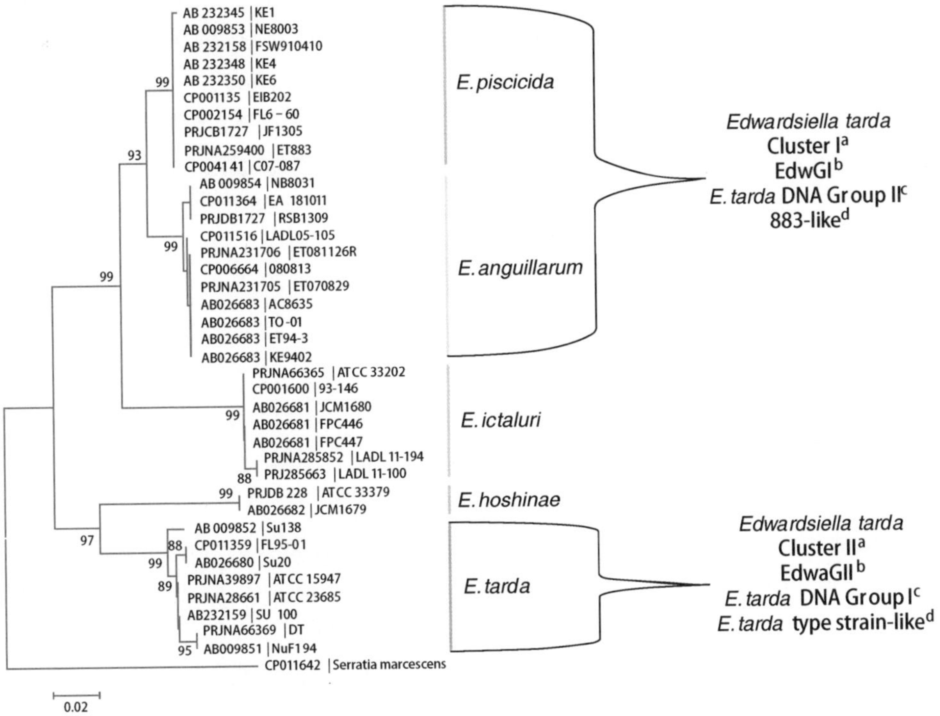

图 15.2　基于铁辅因子超氧化物歧化酶基因(*sodB*)内部片段测序的爱德华氏菌邻接进化史。分类分配(彩色)是根据 Shao 等(2015)所报道的分类。对先前指定的系统群的引用由上标指定：[a] Yamada and Wakabayashi，1999；[b] Yang et al.，2012；[c] Griffin et al.，2013；[d] Abayneh et al.，2013

这些系统发育基因组学研究已将爱德华氏菌属重新划分为五个种(图15.2)。因此，许多关于 *E. tarda* 的历史报道很可能被错误分类了。对已存档的核苷酸序列和历史记录的分子调查和评估表明：*E.ictaluri*、*E.piscicida* 和 *E. anguillarum* 更常与鱼类疾病有关(Griffin et al.，2014；Shao et al.，2015)。因此，任何与鱼类中 *E. tarda* 感染相关的病理生物学和疾病的讨论都可能代表了 *E. tarda*、*E.piscicida* 和 *E. anguillarum* 的混合物。

大量研究报道表明，仅凭表型特征不足以区分 *E. tarda*、*E. piscicida* 和 *E. anguillarum* (Abayneh et al.，2012；Yang et al.，2012；Abayneh et al.，2013；Griffin et al.，2013，2014；Shao et al.，2015)。目前，为准确地分离 *E. tarda* 分类群，需要进行分子水平的分析。建议今后对爱德华氏菌的研究辅以适当的分子分析，以促进不同实验室之间报告的一致性。

最后，迟缓爱德华氏菌被认为是一种人畜共患病原体(Mohanty and Sahoo，2007)。随着近年来水产养殖生产和水产品消费的升级，通过处理或摄入这些产品而引发人畜共患感染的可能性增加。迟缓爱德华氏菌作为来自鱼和贝类的主要病原之一(Haenen et al., 2013)，虽然很少见，但其对人类来说可能是致命的(Hirai et al., 2015)。此外，有报道称人类爱德华氏菌病传染自观赏鱼(Vandepitte et al., 1983)和垂钓运动(Clarridge et al., 1980)。人通过穿刺、开放性伤口或摄食感染 *E. tarda*，可能会发生坏死性皮肤病变和胃肠炎，严重者会引发败血症，导致胆囊炎、骨髓炎、脑膜炎和坏死性筋膜炎(Gilman et al., 1971; Janda and Abbott, 1993; Matshushima et al., 1996)。因此，科学和医学界就爱德华氏菌病的鉴定和报告达成共识是至关重要的。

15.2 迟缓爱德华氏菌

在全球范围内，*E.tarda* 是最具经济性和环境重要性的鱼类病原体之一，所有七大洲均有过报道(Meyer and Bullock, 1973; Van Damme and Vandepitte, 1980; Clavijo et al., 2002; Akinbowale et al., 2006; Alcaide et al., 2006; Leotta et al., 2009; Joh et al., 2011; Xu and Zhang, 2014)。在 28～37 ℃下培养 24～36 h 后，添加 5%羊血的常用细菌培养基(脑心浸液或胰蛋白酶大豆血琼脂)上可见灰白色小菌落，但该菌在 37 ℃下生长得更好。该菌是一种运动性可变的革兰氏阴性短杆菌(0.6 μm×2.0 μm)，细胞色素氧化酶阴性和吲哚阳性，发酵葡萄糖能产酸和产气，并且三糖铁(triple sugar iron, TSI)反应碱性优于酸(K/A)，产生硫化氢(Hawke and Khoo, 2004)，现有的表型诊断检测试剂盒数据库滞后于当前的系统分类。鉴于这种表型模糊性，*E. piscicida* 和 *E. anguillarum* 均视为 *E. tarda*。然而，可以使用种特异性 PCR (Griffin et al., 2014; Reichley et al., 2015d)或相关遗传标记的测序来进行确诊。

15.2.1 病理学

E. tarda 主要被认为是一种暖水性鱼类条件致病菌。环境变化，如高温、水质差和有机物含量高会加大病害严重程度(Meyer and Bullock, 1973; Hawke and Khoo, 2004; Mohanty and Sahoo, 2007; Park et al., 2012)。有研究在 75%的家养鲇(可能是斑点叉尾鮰)池塘水样、64%的鲇池塘泥样和从鲇养殖池塘中采集的 100%的青蛙、乌龟和小龙虾样本中发现了 *E. tarda* (Wyatt et al., 1979)。在中华鳖(*Trionyx sinensis*)中也发现了该病原菌(Pan et al., 2010)。

在斑点叉尾鮰中，这种疾病开始为小的皮肤溃疡，后来发展为肌肉组织中充满气体的大脓肿。在慢性感染中，脓肿充满恶臭气体和坏死组织(Meyer and Bullock, 1973)，但全身性败血症更常见。在某些情况下，疾病表现出类似于鮰爱德华氏菌(*E. ictaluri*)的症状，在颅骨顶部出现出血性溃疡(Hawke and Khoo, 2004)，不过这种溃疡与 *E. ictaluri* 产生的溃疡不同，其边缘更加精致。

感染的临床体征包括肌肉组织和头部出血;突眼(眼球突出);眼混浊;腹部肿胀;血性腹水;肛门出血突出;脾和肾肿大;鳃、肾、肝、脾,偶尔还有肠道有白色小结节(Austin and Austin, 1993; Park et al., 2012)。

最早在阿肯色州、密西西比州、路易斯安那州和得克萨斯州的鲇养殖场,发现报道了 *E. tarda* 引发的疾病暴发(Meyer and Bullock, 1973)。从那以后,这些分离株被分子鉴定为迟缓爱德华氏菌(*E. tarda*)(Griffin,未发表的数据)。每年 7~10 月间,当水温超过 30 ℃且水中含有大量有机物时,疫情主要发生在体重大于 450 g 的鱼中。没有报告过灾难性损失,患病率很少超过 5%。然而,当捕获受感染的鱼并转移到储槽时,死亡率可能接近 50%。实验感染鲇鱼幼鱼(5~10 cm),腹腔(IP)注射剂量为 8×10^5~8×10^7 CFU/尾,27 ℃下 5 尾鱼中有 4 尾在 10 d 内死亡。在 13 ℃下,对褐鳟(*Salmo trutta*)相似剂量攻毒不会造成死亡。

Darwish 等(2000)在模拟自然感染的浸泡之前,对斑点叉尾鮰的皮肤进行了擦伤处理。实验中所观察到的瘀点和皮肤溃疡与自然感染的斑点叉尾鮰(Meyer and Bullock, 1973)和大口黑鲈(Francis-Floyd et al., 1993)的病变损伤一致。可见局灶性坏死性肌炎,但肌肉中未见充满气体的脓肿。在最近的一项比较研究中,Reichley 等(2015d)向鲇腹腔注射了分子鉴定为 *E.tarda*、*E. piscicida* 和 *E.anguillarum* 的分离株。*E.tarda* 分离株 FL95-01 的半数致死剂量比在 *E.piscicida* 中观察到的高了近 100 倍。另外,所有死亡都发生在感染后最初的 24~36 h 内,这与 Darwish 等(2000)描述的死亡率曲线不同。此外,马萨诸塞州斯通维尔(Stoneville)水产研究和诊断实验室的诊断报告显示,自 21 世纪中期以来,美国水产养殖中迟缓爱德华氏菌的发生率有所增加。然而,对这一时期的存档分离株进行的一项调查发现,大部分为 *E. piscicida* 分离株(Griffin et al., 2014),这突出了历史记录与当前分类之间可能存在差异。

E. tarda 还牵涉佛罗里达州一个湖中的一起大口黑鲈死亡事件(Francis-Floyd et al., 1993)。这起为期 6 周的死鱼事件发生在初秋,损失了大约 1 500 尾鱼。近 50%的死鱼样本有深层皮肤恶臭坏死性溃疡,但没有腐烂和气体释放。后肾、脾和肝间质组织中可见多灶性肉芽肿和散在的黑色素巨噬细胞中心。从几条鱼的内脏器官中分离出与 *E. tarda* 表型一致的分离株,与之前报道的从爬行动物、鸟类、鱼类和佛罗里达州湖泊表层水中提取的 *E. tarda* 一致(White et al., 1973)。

Padrós 等(2006)报道了西班牙一个养殖场大菱鲆病害暴发中的 *E. tarda*,当时水温为 15~18 ℃,损失在 3%~10%。由于过多的脓液,病鱼眼部周围出现肿胀;肌肉组织和头部出血,腹部肿胀并伴有腹水;肝出血,脾肾肿大,肠、脾、肾充血。组织学观察,在头肾和躯干肾的造血组织中可见明显的脓肿,这些脓肿周围有明显的炎性细胞浸润。脾中有大量巨噬细胞，但在细胞外空间、血管或巨噬细胞中心观察到的细菌很少。肝未受到大面积损伤，只有一小群含有少量细菌细胞的巨噬细胞嵌入纤维蛋白基质中。

在一次西班牙水产养殖作业中的海鲈（*Dicentrarchus labrax*）死亡事件中也采集到生物化学鉴定为 *E. tarda* 的临床分离株（Blanch et al.，1990）。疾病暴发期间的水温约为 24 ℃，盐度为 37%～38%。以大约 1.6×10^8 CFU/尾的剂量注射采集的分离株时可观察到鱼的死亡。患病鱼表现出尾鳍侵蚀，头部、鳃盖区、鳍和体表有出血点，部分鱼体表溃疡和出现腹水。

据报道，在冷水鱼中也发现了 *E. tarda*，从俄勒冈州秋季濒死和死亡的野生成年大鳞大麻哈鱼中就分离出该病原（Amandi et al.，1982）。在大鳞大麻哈鱼和虹鳟中恢复的分离株感染性相对较低，其 LD_{50} 值比在类似条件下暴露的斑点叉尾鮰高近 10 倍。这些分离株在表型上与来自人源 *E. tarda* 的模式菌（ATCC＃ 15947）和大口黑鲈模式株（FL－77－4）一致。

同样，Amandi 等（1982）和 Uhland 等（2000）报道了加拿大魁北克省两个养殖场的美洲红点鲑在夏季暴发 *E. tarda* 感染的情况。患病鱼表现为眼球突出、鳃瘀点出血、腹部有带血的液体及肠出血。分离株生化鉴定为 *E. tarda*，API20E 谱图为 4544000，与 *E. piscicida*（LTB4，EIB202，TX1）和 *E. tarda*（ATCC＃15947，DT）生成的编码一致（Wang et al.，2011；Shao et al.，2015）。

据报道，捷克共和国集约化养殖的虹鳟中也发现有 *E. tarda*（Řehulka et al.，2012）。患病鱼厌食嗜睡，肝充血和瘀点出血，伴有慢性淋巴细胞门脉型肝炎、局灶性坏死、脂肪变性、血窦扩张和窦状细胞活化。该病原菌表型上与 *E.tarda* 模式菌株 CCM 2238（＝ATCC＃15947）一致。实验表明，该分离株对锦鲤、银鲫（*Carassius gibelio*）或丁鲹（*Tinca tinca*）均无致病性，这使得研究者推测：(1) 这些鱼类不适合进行致病性测定；(2) 分离株在低温储存中丧失毒力；(3) 这是一种非致病性菌株，疾病是由与高密度饲养相关的压力引发的。

最后，在中国暴发疾病的养殖黄鳝中发现并分离出 *E. tarda*（Shao et al.，2016）。患病黄鳝表现出皮肤和内脏出血及肛门组织水肿。这些分离株腹腔注射的半数致死量 LD_{50} 为 1×10^6 CFU，受感染的鱼呈现出与自然疾病相似的临床症状。遗传上，回收分离株（ASE201307）的基因 *gyrB* 与 *E. tarda* 分离株 RE－04 和 ATCC＃15947（Griffin et al.，2014）具有 100%的序列相似性。

15.2.2 防控策略

1. 治疗

对来自美国和中国台湾的分离株进行的一项调查发现，迟缓爱德华氏菌对氨基糖苷类、头孢菌素类、青霉素类、呋喃妥因类、磺胺甲恶唑/甲氧苄啶和喹诺酮类药物敏感。同时,对青霉素G、磺胺嘧啶、黏菌素、新生霉素、壮观霉素、氨苄西林、四环素和氯霉素等有耐药性(Waltman and Shotts, 1986a)。此外,局部施用于表面的常见水产养殖化学品对迟缓爱德华氏菌(USFWS 9.36分离株)的消毒效果发现:酒精(30%、50%或70%)、苄基-4-氯苯酚/苯基苯酚(1%)、次氯酸钠(50 mg/L、100 mg/L、200 mg/L或50 000 mg/L)、*N*-烷基二甲基苄基氯化铵(1∶256)、聚维酮碘(50 mg/L或100 mg/L)、戊二醛(2%)和单过硫酸氢钾复合盐/氯化钠(1%)都是有效的消毒剂,而氯胺-T(15 mg/L)和福尔马林(250 mg/L)则无效(Mainous et al., 2010)。

土霉素、氟苯尼考和噁喹酸已用于防治鱼类中的迟缓爱德华氏菌(Kusuda and Kawai, 1998),在亚洲,四环素经常用于治疗*E. tarda*引起的疾病(Lo et al., 2014)。Meyer和Bullock (1973)的研究报道,在饲料中加入土霉素(55 mg/kg,连续投喂10 d)可在3 d内降低鲇鱼的死亡率。加拿大魁北克省的美洲红点鲑暴发病害时就使用了饲料和植物油混合的土霉素[剂量为100 mg/kg(鱼的体重)]进行治疗(Uhland et al., 2000)。然而,已在斑点叉尾鮰中发现了*E. tarda*的土霉素耐药株(Hilton and Wilson, 1980)。在中国台湾鳗鲡的*E. tarda*中检测到土霉素和多西环素抗性(Lo et al., 2014),在澳大利亚的虎皮鱼(*Puntius tetrazona*)中发现了*E. tarda*对氨苄青霉素、阿莫西林、红霉素、氯霉素、四环素和土霉素耐药(Akinbowale et al., 2006)。对淡水养殖系统中鱼、水和沉积物中*E. tarda*的调查显示,78%的分离株具有多药耐药性(Acharya et al., 2007),并报道了质粒介导的抗生素耐药性(Aoki et al., 1987)。在日本,抗生素耐药性的出现推动了新疗法的开发。采用奥美普林/磺胺间甲氧嘧啶[25 mg/(kg·d)]和噁喹酸[12.5 mg/(kg·d)]1∶3的组合以及米洛沙星以6.2 mg/(kg·d)剂量进行联合用药,成功地治疗了感染耐药*E. tarda*的患病鳗鲡(Aoki et al., 1989)。

2. 预防

迟缓爱德华氏菌对经济的重要性促使研究者们在疫苗开发方面做出了大量研究努力。尽管做出了这些努力,但由于*E. tarda*血清型的变化,广谱疫苗一直难以实现(Kawai et al., 2004; Mohanty and Sahoo, 2007)。因此,Costa等(1998)研究发现,来自海鲷(真鲷和日本真鲷)的非典型、非运动型*E. tarda*与来自日本鳗鲡和牙鲆的典型*E. tarda*具有相同的O血清型和相似的表面抗原。在露斯塔野鲮中,通过浸泡接种迟缓爱德华氏菌疫苗可以获得对*E. tarda*感染的强有力保护,但效力取决于鱼龄和浸泡时间。小于3周龄的鱼对免疫接种没有应答反应(Swain et al., 2002)。在鳗鲡中,当通过肌内注射野生毒株攻毒时,肌内免疫接种似乎提高了存活率(Salati and Kusuda, 1985)。每日饲喂福尔马林灭活的迟缓爱德华氏菌细胞、免疫增强剂可得然胶(细菌

β-1，3-葡聚糖）和皂树皂苷悬浮液连续3周的牙鲆，其存活率优于对照鱼（Ashida et al.，1999）。同样在牙鲆中，通过腹腔注射免疫接种*E. tarda*外膜蛋白可对多种迟缓爱德华氏菌血清型产生强烈保护作用（Kawai et al.，2004）。同样，接种了迟缓爱德华氏菌双基因敲除基因工程疫苗NH1株的牙鲆，在遭受NH1野生株攻毒时可免受感染（Choi and Kim，2011）。此外，在注射接种活的和福尔马林灭活的迟缓爱德华氏菌的大菱鲆中也实现了成功的免疫（Castro et al.，2008）。

为了开发尼罗罗非鱼爱德华氏菌病多价疫苗，将大肠杆菌表达的鮰爱德华氏菌重组GAPDH（甘油醛-3-磷酸脱氢酶）外膜蛋白和鮰爱德华氏菌福尔马林灭活细胞进行了联合注射免疫。这种联合免疫对迟缓爱德华氏菌OT9805-27株攻毒的RPS为71.4%（Trung Cao et al.，2014）。有趣的是，使用基于爱德华氏菌属的环介导等温扩增PCR法（Savan et al.，2004）鉴定了OT9805-27株的铁调节溶血素基因序列，其在GenBank的基因序列为D89876（Hirono et al.，1997），该基因与杀鱼爱德华氏菌EIB202、FL6-60和C07-087具有99%的序列同源性。相比之下，对D89876序列的BlastN（核苷酸本地序列比对搜索工具）分析显示，与*E. tarda* FL95-01株基因片段仅有不到80%的相似性。

15.3 杀鱼爱德华氏菌

在含5%牛血的2号血琼脂培养基（Oxoid，Adelaide，Australia）上，30 ℃条件下培养24 h，*E. piscicida*菌落形态和生化特征与*E. tarda*相似（Abayneh et al.，2013）。在遗传学上，*E. piscicida*等同于已被描述为典型的运动型鱼类致病性*E. tarda*（Griffin et al.，2014）。因此，大多数商业表型诊断试剂盒都将*E. piscicida*鉴定为*E.tarda*。利用种特异性PCR方法（Griffin et al.，2014；Reichley et al.，2015d）或适当遗传标记的序列比较法可以进行确诊。

15.3.1 病理生物学

将存档的核苷酸序列与最近发布的*E. piscicida*基因组数据（Tekedar et al.，2013；Oguro et al.，2014；Shao et al.，2015）进行比较，有助于推断*E. piscicida*的病理生物学。例如，16S rDNA序列比较分析表明，在中国大菱鲆中暴发的疫情（Xiao et al.，2008）是由*E. piscicida*引起的，而不是最初报道的*E. tarda*。随后的攻毒感染显示，剑尾鱼（*Xiphophorus helleri*）肌内注射的半数致死量LD_{50}值的范围为$4\times10^3\sim4\times10^5$ CFU/g。实验感染的鱼有出血症状，伴有皮肤和肌肉坏死及化脓性脓肿。同样，从患病的韩国鲇鱼（*Silurus asotus*）中获得了爱德华氏菌属分离株KE6，患病鱼表现出腹部膨胀和皮肤上的红斑病灶。病鱼体内表现为腹水、肠炎、肝充血、脾肿大、肾肿大和脓肿并伴有脓液(Yu et al., 2009)。*sodB* 测序将 KE6 株归类于 *E. piscicida* 遗传谱系(图 15.2)。

Shafiei 等(2016)报道了芬兰真白鲑(*Coregonus lavaretus*)中的 *E. piscicida*。受感染的白鲑出现全身性败血症,眼球突出,皮肤和鳍基部出血性充血,肛门部肿胀、充血,内脏、鳃和肌肉组织出血,肝斑驳,脾、肾肿大,伴有弥漫性腹膜炎,大部分内脏器官有炎性细胞。感染区腹膜腔中可见革兰氏阴性杆状细菌。肾中也观察到坏死灶,肝中可见伴有小型坏死灶的单细胞坏死。

同样,在圈养大口黑鲈种群的慢性死亡事件中,也分离获得了 *E. piscicida*。心、肝、前肾、后肾和脾中均散布有多灶性坏死区域。许多坏死灶被离散的肉芽肿包裹或替代,并伴随革兰氏阴性菌菌落(Fogelson et al., 2016)。

还从大型展示水族箱中患有脓毒性斑点的迈氏条尾魟(*Taeniura meyeni*)中发现了 *E.piscicida* (Camus et al., 2016)。剖检发现脏器和心脏的腹膜壁、浆膜表面呈白色并有增厚的现象,在心房的心内膜表面多处可见类似物质,腹膜后肌肉组织和肾周结缔组织水肿,左肾尾侧有恶臭的液化区。显微镜检结果与细菌性败血症相符,包括伴有细菌杆状物的坏死性心肌炎和蜂窝织炎、脾血管坏死、鳃和肾血管血栓,以及体腔炎。

在对 *E. piscicida* 最初的描述中,使用了五株鱼类分离株和人源 *E. tarda* 模式菌株(ATCC♯15947)进行致病性研究。有三个分离株来自 *E. piscicida* 遗传世系,其中包括 *E. piscicida* 模式菌株 ET883 (Abayneh et al., 2013)。斑马鱼的实验感染包括肌内注射每个分离株大约 3×10^3 CFU,然后观察 7 d。分离株 ET883 和 LTB4 的致死率分别为 100%和 95%,而 *E. piscicida* 分离株 ETK01 的致病性较低(累计死亡率为 11.1%)。临床体征包括游动不稳定,底栖和食欲不振,背部和注射部位出现溃疡。

分离株 LTB4 的发病机制在大菱鲆中也进行了研究阐释(Lan et al., 2008)。患病鱼表现为食欲不振,游动迟缓乏力,无法保持中性浮力,腹部隆起,眼球突出,鳍基部、头、鳃盖、嘴、下颌和腹部出血。鱼的肝和肾肿大,伴有出血和坏死区域。脾肿大,肠有时从肛门脱出。肝脏可见白色小结节。在 19~20 ℃的海水中维持实验感染的大菱鲆,LTB4 腹腔(IP)注射的半数致死量为 3×10^3 CFU/g(鱼体重)。

Matsuyama 等(2005)研究了典型运动型鱼类致病性迟缓爱德华氏菌(syn. *E. piscicida*)在五条鰤(*Seriola quinqueradiata*)、牙鲆和真鲷中的致病性。25 ℃ 水温下进行 IP 注射时,分离株 NUF806 对黄条鰤、牙鲆和真鲷的 LD_{50} 值分别为 1.6×10^6 CFU、1.4×10^3 CFU 和 2.6×10^9 CFU,表明在不同宿主中该菌的致病力各不相同。

在美国东南部养殖的鲇鱼中,*E. piscicida* 引发了与报道的 *E. tarda* 类似的全身性败血症。同样,一些病例在病鱼颅骨顶部出现出血性溃疡(Griffin,未发表的数据）。最近对从密西西比州养殖鲇鱼中收集的存档细菌标本进行的一项调查显示，从 2007 年到 2012 年分离出的所有菌株，表型上都被鉴定为 *E. tarda*，但实际上是 *E. piscicida*（Griffin et al.，2014）。此外，Reichley 等(2015d)测定了 *E. piscicida*、*E. tarda* 和 *E. anguilluarum* 在斑点叉尾鮰中的 LD_{50} 值。体腔注射 *E. piscicida* 分离株 S11285 的半数致死量为 3.9×10^5 CFU。这是类似条件下 *E. tarda* 分离株 FL95－01 或 *E. anguillarum* 的 LD_{50} 值的 $\frac{1}{1\,000}\sim\frac{1}{100}$，这一数据支持了 *E. piscicida* 是目前比 *E. tarda* 对鲇鱼养殖更大威胁的说法。

15.3.2 防控策略

Takano 等(2010)考察了五种减毒爱德华氏菌活疫苗在牙鲆中预防爱德华氏菌病的效力。存档的核苷酸序列鉴定了属于 *E. tarda* 遗传谱系的四个分离株，和一个与 *E. piscicida* 聚类的分离株(E22)(图 15.1)，使用 *E. piscicida* NUF806 株攻毒并进行免疫效力评价。*E. piscicida* E22 是唯一可显著降低 NUF806 株攻毒死亡率的分离株，表明 *E. tarda* 遗传谱系不能抵御 *E. piscicida*。相反地，暴露于人源 *E. tarda* 模式株(ATCC＃15947)的日本牙鲆对分离株 TX1 的攻毒具有免疫抗性(Cheng et al.，2010)，该分离株为 *E. piscicida* 聚类成员(Yang et al.，2012；Shao et al.，2015)。这表明牙鲆在 *E. piscicida* 和*E. tarda*之间至少存在一定程度的抗原表位保守性。

在大菱鲆的进一步研究工作中，研究者们研究了分离株 TX1 推定表面抗原 Esa1 作为纯化重组亚单位疫苗的应用。假单胞菌表达的重组蛋白通过口服或注射给药，相对免疫保护率(RPS)分别为 52％和 79％(Sun et al.，2010)。Esa1 作为 DNA 疫苗也同样有效，通过注射携带 Esa1 插入片段的质粒免疫鱼，可获得 57％的 RPS (Sun et al.，2011a)。与 Esa1 的研究工作类似，纯化重组亚单位疫苗和 DNA 疫苗也利用了来自 TX1 株的 Eta2 基因产物(Eta2 为另一种表面抗原)的抗原性。使用纯化重组亚单位疫苗或 DNA 疫苗注射免疫接种牙鲆，RPS 分别为 83％和 67％。此外，被动免疫研究表明，重组疫苗接种鱼的血清对幼鱼的保护(RPS＝57％)比 DNA 疫苗免疫鱼的血清更强(RPS＝29％；Sun et al.，2011b)。

同样地，杀鱼爱德华氏菌 *esrB* 活突变株(EsrB 正调节Ⅲ型和Ⅵ型分泌系统 T3SS 和 T6SS)浸泡免疫大菱鲆可对 *E. piscicida* 攻毒产生免疫保护力(Yang et al.，2015)。虽然最初被分类为 *E. tarda*，但在这些研究中使用的菌株 EIB202 已被鉴定为 *E. piscicida* 遗传谱系成员(Griffin et al.，2014；Shao et al.，2015)。使用 EIB202 株的尿苷二磷酸葡萄糖脱氢酶缺失突变株腹腔注射免疫接种，可引起对亲本野生分毒株的显著保护作用，RPS 为 43.3％～76.7％

(Lv et al., 2012)。此外,与进行过相同攻毒的对照鱼相比,腹腔注射免疫 EIB202 株的重组甘油醛-3-磷酸脱氢酶的大菱鲆具有较低的累计死亡率(Liang et al., 2012)。

最后,细菌"菌蜕"也被评估为预防鱼类爱德华氏菌感染的候选疫苗。这些研究中使用的分离株(FSW910410)是从韩国一家养鱼场的垂死褐牙鲆中分离到的。其 *sodB* 序列将该分离株鉴定为 *E. piscicida* 种群的一员。与接种经福尔马林灭活的疫苗或未接种疫苗的对照组群相比,两次(间隔 2 周)腹腔免疫接种约 1×10^6 菌蜕细胞的尼罗罗非鱼存活率显著提高。类似地,与接种福尔马林灭活疫苗和未接种疫苗的对照组相比,口服免疫 FSW910410 细菌菌蜕的牙鲆也提高了存活率(Kwon et al., 2006, 2007)。

15.4 鳗爱德华氏菌

E. anguillarum 代表一组分离自鳗鱼与其他河口鱼类的遗传上截然不同的菌株,这些分离株不属于其他爱德华氏菌谱系。它在表型上与 *E. tarda* 无法区分,因此需要种特异性 PCR(Griffin et al., 2014; Reichley et al., 2015d)或相关遗传标记的测序方法来确认该种。*E. anguillarum* 等同于 Matsuyama 等(2005)和 Sakai 等(2007, 2009c)描述的非典型的非运动型 *E. tarda*。然而,Shao 等(2015)在对 *E. anguillarum* 的描述中指出,该菌通过周生鞭毛运动。同样,Griffin 等(2013)观察到 *E. anguillarum* 分离株 LADL 05-105(Reichley et al., 2015d)在 25 ℃和 37 ℃下的运动性,表明种群细菌的运动性是可变的。

15.4.1 病理生物学

Matsuyama 等(2005)还研究了非典型的非运动型鱼类致病性 *E. tarda* 在鰤鱼、牙鲆和真鲷中的致病性。分离株 FPC503 具有与三个鳗爱德华氏菌基因组(LADL 05-105; 080813; EA181011)98%~100%的同源性(Sakai et al., 2009a; Reichley et al., 2015b, 2015c; Shao et al., 2015)。在 25 ℃下,腹腔注射 FPC503 分离株,对鰤鱼、牙鲆和真鲷的 LD_{50} 值分别为 1.5×10^5 CFU、1.4×10^6 CFU 和 8.9×10^3 CFU,这与典型的可运动鱼类致病性 *E. tarda* 引发的死亡率趋势不同。*E. anguillarum* 在大菱鲆中具有致病性,在 15 ℃下 1×10^2 CFU 剂量的腹腔注射具有很高的死亡率(Shao et al., 2015)。同样地,0.25 g 斑马鱼肌内注射 5×10^2 CFU 剂量该菌,在大约 22 ℃下 3 d 内死亡率可达 100%(Yang et al., 2012)。

Abayneh 等(2013)评估了现在被认为是鳗爱德华氏菌谱系(Shao et al., 2015)成员的两株分离株的致病性。从未知鱼种中分离的 NCIMB 2034 株,肌内注射(3×10^3 CFU)对斑马鱼无致病性。与此同时，同样注射 ET080813 株会导致 87.5%的死亡率。相反地，腹腔注射剂量 1×10^8 CFU 下，*E. anguillarum* LADL 05－105 株在约 30 ℃时对鲇鱼幼鱼仅产生最小的死亡率(Reichley et al.，2015c)，因此这些结果支持先前的研究结论，即在不同宿主中，这些病原菌具有不同程度的致病性。

15.4.2　防控策略

鉴于最近才采用 *E. anguillarum* (Shao et al.，2016)这一病原分类，因此对这种新发现的鱼类病原菌的防治研究很少。尽管如此，利用分离株 FPC503 制备的福尔马林灭活疫苗对真鲷中的鳗爱德华氏菌感染显示出保护作用(Takano et al.，2011)。腹腔注射接种该灭活疫苗，无论是否有佐剂，RPS 均为 85%～100%。

15.5　鮰爱德华氏菌

如第 15.1 节所述，作为鮰肠败血症(ESC)的病原体，*E. ictaluri* 是工业化养殖叉尾鮰和其他温带淡水鱼类的重要病原体(表 15.2)(Hawke et al.，1981)。该疾病是美国东南部大多数商业叉尾鮰养殖业中的地方流行病(Plumb and Vinitnantharat，1993)，可造成灾难性损失，特别对一岁龄幼鱼(Wise et al.，2004)。

初步研究表明，来自不同地理位置的分离株在生化、生物物理和血清学上都具有同源性，并且其外膜蛋白、同工酶和质粒在很大程度上是保守的(Newton et al.，1988；Plumb and Vinitnantharat，1989；Hawke and Khoo，2004)。在遗传学上，尽管在世界范围内已经发现了基因型亚型(Bader et al.，1998；Griffin et al.，2011；Bartie et al.，2012)，*E. ictaluri* 在美国东南部的叉尾鮰养殖区域中大部分是克隆性的。与来自斑点叉尾鮰的分离株相比，来自低眼无齿𩷶(*Pangasianodon hypophthalmus*)的越南分离株具有独特的质粒图谱(Fernandez et al.，2001)，并且与针对 *E. ictaluri* 模式菌株 ATCC ＃33202 的单克隆抗体(Ed9)不发生反应(Rogge et al.，2013)。同样地，最近来自斑马鱼实验室种群中暴发疾病的 *E. ictaluri* 具有独特的质粒图谱，不被 Ed9 识别，并在脑心浸液肉汤(BHI)培养基中自聚集(Hawke et al.，2013)。在日本，分离自野生香鱼的 *E. ictaluri* 与美国或印度尼西亚分离株具有显著的扩增片段长度多态性(AFLP)差异(Sakai et al.，2009b)。最近的研究发现，来自鲇鱼、斑马鱼和尼罗罗非鱼的分离株在基因组排列、质粒确认和含量、毒力因子基因、抗菌药物敏感性和血清学特征方面存在差异(Griffin et al.，2015)。

E. ictaluri 通常通过从肾和脑中分离和培养的细菌来进行确诊。在 25～30 ℃条件下，在添加 5%羊血的常用培养基（胰蛋白酶大豆琼脂培养基，Mueller-Hinton 培养基或脑心浸液琼脂培养基）上培养 36～48 h，可见小的、点状灰白色菌落。*E. ictaluri* 在 37 ℃下生长不良，可以通过在 27 ℃和 37 ℃下的重复培养物来进行推定鉴定。该菌为革兰氏阴性弱运动性杆菌（0.75 μm×1.25 μm），细胞色素氧化酶和吲哚阴性，发酵葡萄糖，TSI 反应为 K/A，不产生 H_2S 气体（Hawke et al.，1981；Hawke and Khoo，2004）。可以通过 PCR（Bilodeau et al.，2003；Panangala et al.，2007）或 LAMP 测定（Yeh et al.，2005）进行分子鉴定。

15.5.1 病理生物学

在美国东南部的叉尾鮰养殖区，ESC 暴发发生在春末和初秋，当时水温在 22～28 ℃（Francis-Floyd et al.，1987；Wise et al.，2004），但死亡可能在这个范围之外发生（Plumb and Shoemaker，1995）。叉尾鮰养殖动物流行病的特点是疾病往往会引起鱼类食欲不振，严重时可能导致鱼类完全停止摄食。鱼在水面附近变得无精打采，沿着背风的池塘在岸边聚集，并且经常绕圈游动。可根据临床体征推定诊断，包括多灶性、针尖状红色和白色溃疡，皮肤瘀点性出血，腹部肿胀，眼球突出和脑膜脑炎。脑膜脑炎可导致覆盖额骨囟门的皮肤糜烂，表现为“头洞病”，这是该疾病的通俗说法。颅内病变有粗糙的侵蚀边缘，相比之下，与 *E. tarda* 和 *E. piscicida* 相关的病变边缘则更平滑、更清晰。腹腔内，腹部充满清晰、黄色或带血的腹水，伴有肾肿大和苍白、坏死、斑驳的肝（Thune et al.，1993）。

初期的研究表明，鮰爱德华氏菌是一种专性病原体，在池塘中存活的时间很短（Plumb and Quinlan，1986）。该病原菌在池塘水中仅存活 10～15 d，在沉积物中存活则长达 3 个月，但存活时间取决于温度。在 5 ℃条件下，*E. ictaluri* 的存活时间小于 15 d，这表明它可能不会在鲇鱼池塘中越冬。用于检测和量化 *E. ictaluri* 的定量 PCR（qPCR）方法的开发（Bilodeau et al.，2003），为评估环境中病原体水平提供了机会（Griffin et al.，2011）。最近的研究发现，在冬末移走鱼类之后和来年夏天重新放养之前，育苗池塘中存在 *E. ictaluri* 的 DNA（Griffin，未发表数据）。然而，从这些数据中不清楚这些微生物是否存活。与先前的发现相反，这些结果表明，*E. ictaluri* 在池塘环境中和除鲇鱼宿主之外的疾病媒介存在下仍然存在并可存活。

应激、营养不良、混合感染和水质不佳增加了 ESC 相关的死亡率。由于 *E. ictaluri* 对斑点叉尾鮰具有高致病性，它可在没有诱发条件下引起严重的流行病（Mqolomba and Plumb，1992；Wise et al.，1993；Ciembor et al.，1995；Plumb and Shoemaker，1995）。没有证据表明疾病通过垂直传播，无论是通过饲料摄入细菌还是通过黏膜表面直接进入，水媒传播被认为是主要的传播方式（Thune et al.，1993；Wise et al.，2004）。插管感染实验表明，*E. ictaluri* 穿过肠上皮并急性发病（Baldwin and Newton，1993）。

更多的研究表明，与不投喂处理相比，饲喂幼龄斑点叉尾鮰的死亡率增加了（Wise et al.，2008）。这种相关性在池塘试验中也得到证实（Wise and Johnson，1998）。此外，*E. ictaluri* 可通过鼻孔进入宿主并在嗅囊内引发疾病发生。一旦感染，病原菌就会通过嗅觉神经迁移至鱼脑从而引发脑膜脑炎（Miyazaki and Plumb，1985；Shotts et al.，1986）。虽然也报道了其他侵染门户（Menanteau-Ledouble et al.，2011），但经口摄入从受感染鱼的粪便或死鱼和腐烂鱼尸体中传播的细菌，被认为在疾病暴发期间推动了感染的快速传播（Klesius，1994；Hawke and Khoo，2004；Wise et al.，2008）。

鮰肠败血症可表现为急性、亚急性和慢性感染（Hawke and Khoo，2004）。急性感染的特征是快速暴发性败血症，死亡前几乎没有临床症状。亚急性疾病的特征表现为有明显的临床症状，包括皮肤上的瘀点和瘀斑出血、肝肾坏死和肠出血。腹腔内常有清澈血液和脾肿大。存活下来的鱼可以恢复并抵抗随后的感染，或者发展为慢性疾病。在大多数慢性病例中，病原菌只能从大脑或偶尔从坏死的肝中分离出来。

有关食鱼鸟类在 *E. ictaluri* 传播中的作用存在相互矛盾的报道。在对雪鹭（*Egretta thula*）、大白鹭（*Casmerodius albus*）、大蓝鹭（*Ardea herodias*）和双冠鸬鹚（*Phalacrocorax auritus*）的肠道和直肠涂片的调查中，从 53%的鸟类样本中检测到 *E. ictaluri*，这其中包括两个有存活细菌的样品（Taylor，1992）。相反地，从喂食了感染 *E. ictaluri* 的鱼的大蓝鹭中没有恢复活菌，这表明细菌不太可能通过大蓝鹭胃肠道传播（Waterstrat et al.，1999）。即便如此，受感染的鱼在池塘间的禽类运输也会传播疾病。

E. ictaluri 也出现在非鮰类鱼中（例如斑马鱼、尼罗罗非鱼、香鱼）。据报道，斑马鱼的实验室种群中疾病的暴发，似乎是由实验过程中处理不当引发的（Hawke et al.，2013）。其临床体征与叉尾鮰类似，以脾、肾、肝、肠和脑部的坏死为全身性疾病特征，并通常在巨噬细胞内可观察到大量细菌。肾、脾中多见炎性细胞浸润，鼻腔伴有弥漫性、严重的炎症和坏死，嗅花结严重损伤。在某些病例中，坏死的白细胞和细菌使中脑室明显扩张（Petrie-Hanson et al.，2007；Hawke et al.，2013）。斑马鱼肌内注射超过 6.0×10^4 CFU/mL 的*E. ictaluri*，在 28 ℃下 6 d 内死亡率超过 75%；即使剂量超过 2.0×10^7 CFU/mL，相对应的浸泡感染导致的死亡率低于 15%（Petrie-Hanson et al.，2007）。

E. ictaluri 也与养殖尼罗罗非鱼的死亡有关。患病幼鱼的脾和头肾有一致的多灶性结节，并伴有肝肿大、肝和腹膜脂肪减少。在组织学上，脾、头肾和肝可见严重的炎性浸润，以及多灶性坏死域和肉芽肿形成。分别采用10^6 CFU/尾鱼

注射或 10^6 CFU/mL 浸泡攻毒，注射后 3 d 和浸泡后 8 d 死亡率可达 100%。低剂量(10^3 CFU/mL)导致死亡率低于 40%，病鱼出现血性腹水，脾、肾肿大及前肾和脾有多灶性白色结节(Soto et al., 2012)。

在日本河流的野生香鱼中观察到与鮰爱德华氏菌相关的夏末初秋死亡率。病鱼有出血性腹水和眼球突出症，体表、肛门或鳍基部发红。在 17～20 ℃下，分离株的实验性腹腔注射感染的 LD_{50} 约为 1.3×10^4 CFU/mL，其临床体征与自然疫情相似(Sakai et al., 2008)。Nagai 和 Nakai (2014)的研究表明，在 28 ℃时，*E. ictaluri* 对香鱼的毒力比在 20 ℃时更强，这与早期对斑点叉尾鮰的研究结果一致(Plumb and Shoemaker, 1995)。

15.5.2 防控策略

1. 控制措施

在 ESC 暴发期间，通过每隔一天或每三天喂食一次的限饲，限制摄入 *E. ictaluri* 和防止粪口传染途径，可以改善叉尾鮰的存活率(Wise and Johnson, 1998; Wise et al., 2004)。在水温允许的情况下限饲，是叉尾鮰养殖中控制 ESC 的常见管理做法。2009 年，近 30%的鱼种场报告称，他们停止投喂以控制 ESC (USDA/APHIS/NAHMS, 2010)，然而，由于饲喂天数减少，预防性限饲导致减产(Wise and Johnson, 1998; Wise et al., 2008)。

与迟缓爱德华氏菌相似，应用乙醇(30%、50%或 70%)、苄基-4-氯苯酚/联苯酚(1%)、次氯酸钠(50 mg/L、100 mg/L、200 mg/L 或 50 000 mg/L)、*N*-烷基二甲基苄基氯化铵(1∶256)、聚维酮碘(50 mg/L 或 100 mg/L)、戊二醛(2%)和单过硫酸氢钾复合盐/氯化钠(1%) 表面消毒处理可在 1 min 内减少或消除可检测的生物数量。相比之下，氯胺 T(15 mg/L)和福尔马林(250 mg/L)则达不到这种效果(Mainous et al., 2010)。此外，*E. ictaluri* 对氨基糖苷类、头孢菌素类、青霉素类、喹诺酮类、四环素类、氯霉素类、呋喃妥因类和增效磺胺类药物敏感(Waltman and Shotts, 1986b)。在美国，Romet® 和 Aquaflor® 被批准用于控制叉尾鮰中由 *E. ictaluri* 引起的疾病。

Romet® (Hoffman LaRoche)是一种由磺胺二甲氧嘧啶和奥美普林以 5∶1 配伍组合的增效磺胺类药物，以 50 mg/kg 的推荐剂量连续施药 5 d 可有效降低斑点叉尾鮰鱼苗的 ESC 死亡率(Plumb et al., 1987)。Aquaflor® (Merck Animal Health)是一种酚类广谱抗生素，已在世界各地被批准用于多种鱼类。它作为饲料预混剂(50%质量分数的氟苯尼考)(Gaunt et al., 2004,2006)以每天每千克体重 10 mg 的用药剂量连续 10 d 进行施药(Gaikowski et al., 2003)。在美国以外，养殖生产者在考虑任何治疗措施之前，必须咨询当地的监管机构。

E. ictaluri 可对四环素、磺胺二甲氧嘧啶/奥普美林和氟苯尼考产生质粒介导的抗生素耐药性(Waltman et al., 1989; Welch et al., 2008)。此外，据报道，*E. ictaluri* 的野生分离株对四环素、磺胺二甲氧嘧啶/奥普美林和氟苯尼考具有耐药性(Starliper et al., 1993; Welch et al., 2008)。

2. 预防措施

为了开发有效的鮰爱德华氏菌疫苗，人们已进行了大量研究工作。最初为斑点叉尾鮰接种疫苗所做的努力主要集中在细胞成分或灭活疫苗给药上，但大部分不成功或不切实际。Saeed 和 Plumb (1986)评价了斑点叉尾鮰对脂多糖(LPS)和全细胞*E. ictaluri*疫苗的免疫应答，他们发现，多次注射弗氏完全佐剂(FCA)的*E. ictaluri*脂多糖免疫的斑点叉尾鮰，对随后的 *E. ictaluri* 攻毒有保护作用。尽管如此，鉴于美国东南部斑点叉尾鮰生产的物流，注射疫苗没有得到行业强有力的支持。

已经投入了大量的研究来开发有效的浸泡型鮰爱德华氏菌疫苗，该疫苗可在鱼从孵化场转移之前进行疫苗接种。鉴于细菌疫苗在鲑鱼养殖业中的成功应用，早期尝试在商业环境中对斑点叉尾鮰免疫接种的目标是通过口服和浸泡方式接种 *E. ictaluri* 疫苗(Thune et al., 1994; Shoemaker and Klesius, 1997)，尽管这些策略并未获得业界认可。

后来尝试使用减毒活疫苗浸泡接种，这些疫苗在实验室试验中非常有效(Lawrence et al., 1997; Thune et al., 1999; Abdelhamed et al., 2013)。然而，按照行业生产时间表进行接种时，这些策略效果有限，因为这些策略的目标是在孵化后7～10 d将鱼从孵化场转移到育苗池之前或转移期间实施接种。效果不佳归因于这个鱼龄段缺乏免疫能力，因为斑点叉尾鮰在孵化后 18～21 d 才具有完全的免疫活性(Petrie-Hanson and Ainsworth, 1999)，远远超出孵化场/育苗场转移窗口。虽然有报道称，在 7 日龄鱼苗(RPS 为 58.4%～77.5%)和斑点叉尾鮰的发眼卵(RPS 为 27.3%～87.9%)中(Shoemaker et al., 1999, 2002, 2007)成功使用了一种市售浸泡型疫苗(Aquavac-ESC, Merck Animal Health)，但由于投资回报率有限，浸泡型疫苗并未得到行业的广泛采用(Bebak and Wagner, 2012)。

为了达到最大的预防效果，疫苗应该给具有免疫活性的较大鱼龄的鱼进行接种。口服疫苗不会产生应激压力，可以在将疫苗放入池塘后接种大鱼龄的鱼。虽然使用灭活疫苗进行口服接种已证实有效(Plumb and Vinitnantharat, 1993; Plumb et al., 1994; Thune et al., 1994)，但接种减毒活疫苗显示出最大的潜力。在实验室攻毒中的鮰爱德华氏菌 Δ*fur* 减毒突变株(缺失铁摄取调节子，Fur 蛋白)有效保护了斑点叉尾鮰(Santander et al., 2012)。同样地，在实验室和田间试验中，口服接种利福平抗性的 *E. ictaluri* 减毒活菌株(340X2)的斑点叉尾鮰对野生型 *E. ictaluri* 攻毒产生了高水平的免疫保护(Wise et al., 2015)。

此外,冷藏 2 年(－74 ℃)后的冻干疫苗仍能提供优异的免疫保护(Greenway et al., 2017)。在实验性池塘试验中,使用这种减毒活疫苗进行口服接种可显著提高鱼类的存活率、饲料喂养率、饲料转化率和养成产量(Wise et al., 2015)。口服疫苗可方便大鱼龄鱼接种,从而避免了许多与浸泡鱼苗有关的限制。虽然疫苗覆盖率受到鱼群内不同摄食率的限制,但人们认为,未接种足量疫苗的鱼仍可受益于池塘鱼群内的群体免疫效应。在群体水平上,与先前的浸泡或注射接种相比,口服接种可能被证明是一种更有效、更实用的针对 *E. ictaluri* 的免疫方法。

15.6　总结与研究展望

对自 2013 年描述杀鱼爱德华氏菌以来发表的文献评估显示,许多迟缓爱德华氏菌研究未能解决错误分类的可能性,进一步增加了科学记录的不一致性。将迟缓爱德华氏菌分类为多个类群意味着大量有关在鱼类中爱德华氏菌致病作用的有效历史研究需要重新评估。将最近的研究发现与存档的序列数据相结合,结果表明许多关于鱼类中的迟缓爱德华氏菌的报道实际上是杀鱼爱德华氏菌或鳗爱德华氏菌。展望未来,对于研究人员和诊断人员来说,按照重新调整的系统分类学采用一致的命名法和鉴定方法是非常重要的。

此外,由于分类学上的模糊性,开发适当的疫苗接种方案以保护鱼类免受迟缓爱德华氏菌的影响变得复杂起来。爱德华氏菌属的多样抗原性被认为是开发有效迟缓爱德华氏菌疫苗的主要障碍。然而,与这种疫苗开发相关的挑战可能在于,直到最近,迟缓爱德华氏菌代表三种遗传上不同的种,它们在不同的鱼类寄主中都具有不同程度的致病性。也就是说,正如从这里总结的工作所表明的那样,缺乏针对不同爱德华氏菌的可行疫苗不是问题所在。有大量有效的候选疫苗,涵盖了广泛的疫苗构建策略,包括重组蛋白疫苗、DNA 疫苗、福尔马林灭活疫苗、减毒疫苗和菌蜕疫苗。相反地,问题在于高效、有效地向鱼类提供这些疫苗。未来的研究应侧重于开发实用的疫苗接种方案,这些方案在操作性上符合水产行业标准并取得生产者更广泛的接受度。

参考文献

[1] Abayneh, T., Colquhoun, D. and Sørum, H. (2012) Multilocus sequence analysis (MLSA) of *Edwardsiella tarda* isolates from fish. *Veterinary Microbiology* 158, 367－375.

[2] Abayneh, T., Colquhoun, D. and Sørum, H. (2013) *Edwardsiella piscicida* sp. nov., a novel species pathogenic to fish. *Journal of Applied Microbiology* 114, 644－654.

[3] Abdelhamed, H., Lu, J., Shaheen, A., Abbass, A., Lawrence, M. and Karsi, A. (2013) Construction and evaluation of an *Edwardsiella ictaluri fhuC* mutant.

Veterinary Microbiology 162, 858 - 865.

[4] Abraham, T., Mallick, P., Adikesavalu, H. and Banerjee, S. (2015) Pathology of *Edwardsiella tarda* infection in African catfish, *Clarias gariepinus* (Burchell 1822), fingerlings. *Archives of Polish Fisheries* 23, 141 - 148.

[5] Acharya, M., Maiti, N. K., Mohanty, S., Mishra, P. and Samanta, M. (2007) Genotyping of *Edwardsiella tarda* isolated from freshwater fish culture system. *Comparative Immunology, Microbiology and Infectious Diseases* 30, 33 - 40.

[6] Akinbowale, O., Peng, H. and Barton, M. (2006) Antimicrobial resistance in bacteria isolated from aquaculture sources in Australia. *Journal of Applied Microbiology* 100, 1103 - 1113.

[7] Alcaide, E., Herraiz, S. and Esteve, C. (2006) Occurrence of *Edwardsiella tarda* in wild European eels *Anguilla anguilla* from Mediterranean Spain. *Diseases of Aquatic Organisms* 73, 77 - 81.

[8] Amandi, A., Hiu, S., Rohovec, J. and Fryer, J. (1982) Isolation and characterization of *Edwardsiella tarda* from fall chinook salmon (*Oncorhynchus tshawytscha*). *Applied and Environmental Microbiology* 43, 1380 - 1384.

[9] Aoki, T. and Takahashi, A. (1987) Class D tetracycline resistance determinants of R plasmids from the fish pathogens *Aeromonas hydrophila*, *Edwardsiella tarda*, and *Pasteurella piscicida*. *Antimicrobial Agents and Chemotherapy* 31, 1278 - 1280.

[10] Aoki, T., Kitao, T. and Fukudome, M. (1989) Chemotherapy against infection with multiple drug resistant strains of *Edwardsiella* tarda in cultured eels. *Fish Pathology* 24, 161 - 168.

[11] Ashida, T., Okimasu, E., Ui, M., Heguri, M., Oyama, Y. and Amemura, A. (1999) Protection of Japanese flounder *Paralichthys olivaceus* against experimental edwardsiellosis by formalin-killed *Edwardsiella tarda* in combination with oral administration of immunostimulants. *Fisheries Science* 65, 527 - 530.

[12] Austin, B. and Austin, D. (1993) *Bacterial Pathogens in Farmed Wild Fish*. Ellis Horwood, New York, pp. 188 - 226.

[13] Bader, J., Shoemaker, C., Klesius, P., Connolly, M. and Barbaree, J. (1998) Genomic subtyping of *Edwardsiella ictaluri* isolated from diseased channel catfish by arbitrarily primed polymerase chain reaction. *Journal of Aquatic Animal Health* 10, 22 - 27.

[14] Baldwin, T. and Newton, J. (1993) Pathogenesis of enteric septicemia of channel catfish, caused by *Edwardsiella ictaluri*: bacteriologic and light and electron microscopic findings. *Journal of Aquatic Animal Health* 5, 189 - 198.

[15] Bartie, K., Austin, F., Diab, A., Dickson, C., Dung, T. *et al*. (2012) Intraspecific diversity of *Edwardsiella ictaluri* isolates from diseased freshwater catfish, *Pangasianodon hypophthalmus* (Sauvage), cultured in the Mekong Delta, Vietnam. *Journal of Fish Diseases* 35, 671 - 682.

[16] Bebak, J. and Wagner, B. (2012) Use of vaccination against enteric septicemia of catfish and columnaris disease by the U.S. catfish industry. *Journal of Aquatic Animal Health* 24, 30 - 36.

[17] Bilodeau, A., Waldbieser, G., Terhune, J., Wise, D. and Wolters, W. (2003) A real-time polymerase chain reaction assay of the bacterium *Edwardsiella ictaluri* in channel catfish. *Journal of Aquatic Animal Health* 15, 80 - 86.

[18] Blanch, A., Pintó, R. and Jofre, J. (1990) Isolation and characterization of an *Edwardsiella* sp. strain, causative agent of mortalities in sea bass (*Dicentrarchus labrax*). *Aquaculture* 88, 213 - 222.

[19] Camus, A., Dill, J., McDermott, A., Hatcher, N. and Griffin, M. (2016) *Edwardsiella piscicida*-associated septicaemia in a blotched fantail stingray *Taeniura meyeni* (Müeller & Henle). *Journal of Fish Diseases* 39, 1125 - 1131.

[20] Castro, N., Toranzo, A., Nunez, S. and Magarinos, B. (2008) Development of an effective *Edwardsiella tarda* vaccine for cultured turbot (*Scophthalmus maximus*). *Fish and Shellfish Immunology* 25, 208-212.
[21] Cheng, S., Hu, Y., Zhang, M. and Sun, L. (2010) Analysis of the vaccine potential of a natural avirulent *Edwardsiella tarda* isolate. *Vaccine* 28, 2716-2721.
[22] Choi, S. and Kim, K. (2011) Generation of two auxotrophic genes knock-out *Edwardsiella tarda* and assessment of its potential as a combined vaccine in olive flounder (*Paralichthys olivaceus*). *Fish and Shellfish Immunology* 31, 58-65.
[23] Ciembor, P., Blazer, V., Dawe, D. and Shotts, E. (1995) Susceptibility of channel catfish to infection with *Edwardsiella ictaluri*: effect of exposure method. *Journal of Aquatic Animal Health* 7, 132-140.
[24] Clarridge, J., Musher, D., Fainstein, V. and Wallace, R. (1980) Extraintestinal human infection caused by *Edwardsiella tarda*. *Journal of Clinical Microbiology* 11, 511-514.
[25] Clavijo, A., Conroy, G., Conroy, D., Santander, J. and Aponte, F. (2002) First report of *Edwardsiella tarda* from tilapias in Venezuela. *Bulletin of the European Association of Fish Pathologists* 22, 280-282.
[26] Costa, A., Kanai, K. and Yoshikoshi, K. (1998) Serological characterization of atypical strains of *Edwardsiella tarda* isolated from sea breams. *Fish Pathology* 33, 265-274.
[27] Crumlish, M., Dung, T., Turnbull, J., Ngoc, N. and Ferguson, H. (2002) Identification of *Edwardsiella ictaluri* from diseased freshwater catfish, *Pangasius hypophthalmus* (Sauvage), cultured in the Mekong Delta, Vietnam. *Journal of Fish Diseases* 25, 733-736.
[28] Darwish, A., Plumb, J. and Newton, J. (2000) Histopathology and pathogenesis of experimental infection with *Edwardsiella tarda* in channel catfish. *Journal of Aquatic Animal Health* 12, 255-266.
[29] Ewing, W., McWhorter, A., Escobar, M. and Lubin, A. (1965) *Edwardsiella*, a new genus of *Enterobacteriaceae* based on a new species, *E. tarda*. *International Bulletin of Bacteriological Nomenclature and Taxonomy* 15, 33-38.
[30] Fernandez, D., Pittman-Cooley, L. and Thune, R. (2001) Sequencing and analysis of the *Edwardsiella ictaluri* plasmids. *Plasmid* 45, 52-56.
[31] Fogelson, S., Petty, B., Reichley, S., Ware, C., Bowser, P. *et al.* (2016) Histologic and molecular characterization of *Edwardsiella piscicida* infection in largemouth bass (*Micropterus salmoides*). *Journal of Veterinary Diagnostic Investigation* 28, 338-344.
[32] Francis-Floyd, R., Beleau, M., Waterstrat, P. and Bowser, P. (1987) Effect of water temperature on the clinical outcome of infection with *Edwardsiella ictaluri* in channel catfish. *Journal of the American Veterinary Medical Association* 191, 1413-1416.
[33] Francis-Floyd, R., Reed, P., Bolon, B., Estes, J. and McKinney, S. (1993) An epizootic of *Edwardsiella tarda* in largemouth bass (*Micropterus salmoides*). *Journal of Wildlife Diseases* 29, 334-336.
[34] Gaikowski, M., Wolf, J., Endris, R. and Gingerich, W. (2003) Safety of Aquaflor (florfenicol, 50% type A medicated article), administered in feed to channel catfish, *Ictalurus punctatus*. *Toxicologic Pathology* 31, 689-697.
[35] Gaunt, P., Endris, R., Khoo, L., Howard, R., McGinnis, A. *et al.* (2004) Determination of dose rate of florfenicol in feed for control of mortality in channel catfish *Ictalurus punctatus* (Rafinesque) infected with *Edwardsiella ictaluri*, etiological agent of enteric septicemia. *Journal of the World Aquaculture Society* 35, 257-267.
[36] Gaunt, P., McGinnis, A., Santucci, T., Cao, J., Waeger, P. and Endris, R. (2006) Field efficacy of florfenicol for control of mortality in channel catfish, *Ictalurus*

punctatus (Rafinesque), caused by infection with *Edwardsiella ictaluri*. *Journal of the World Aquaculture Society*, 37, 1-11.

[37] Gilman, R., Madasamy, M., Gan, E., Mariappan, M., Davis, C. and Kyser, K. (1971) *Edwardsiella tarda* in jungle diarrhea and a possible association with *Entamoeba histolytica*. *Southeast Asian Journal of Tropical Medicine and Public Health* 2, 186-189.

[38] Greenway, T., Byars, T., Elliot, R., Jin, X., Griffin, M. and Wise, D. (2017) Validation of fermentation and processing procedures for the commercial scale production of a live attennated *Edwardsiella ictaluri* vaccine for use in catfish aquaculture. *Journal of Aquatic Animal Health* (in press).

[39] Griffin, M., Mauel, M., Greenway, T., Khoo, L. and Wise, D. (2011) A real-time polymerase chain reaction assay for quantification of *Edwardsiella ictaluri* in catfish pond water and genetic homogeneity of diagnostic case isolates from Mississippi. *Journal of Aquatic Animal Health* 23, 178-188.

[40] Griffin, M., Quiniou, S., Cody, T., Tabuchi, M., Ware, C. *et al*. (2013) Comparative analysis of *Edwardsiella* isolates from fish in the eastern United States identifies two distinct genetic taxa amongst organisms phenotypically classified as *E. tarda*. *Veterinary Microbiology* 165, 358-372.

[41] Griffin, M., Ware, C., Quiniou, S., Steadman, J., Gaunt, P. *et al*. (2014) Edwardsiella piscicida identified in the southeastern USA by *gyrB* sequence, species-specific and repetitive sequence-mediated PCR. *Diseases of Aquatic Organisms*, 108, 23-35.

[42] Griffin, M., Reichley, S., Greenway, T., Quiniou, S., Ware, C. *et al*. (2015) Comparison of *Edwardsiella ictaluri isolates* from different hosts and geographic origins. *Journal of Fish Diseases* 39, 947-969.

[43] Grimont, P., Grimont, F., Richard, C. and Sakazaki, R. (1980) *Edwardsiella hoshinae*, a new species of *Enterobacteriaceae*. *Current Microbiology* 4, 347-351.

[44] Haenen, O., Evans, J. and Berthe, F. (2013) Bacterial infections from aquatic species: potential for and prevention of contact zoonoses. *Revue Scientifique et Technique* 32, 497-507.

[45] Hawke, J. and Khoo, L. (2004) Infectious diseases. In: Tucker, C. and Hargreaves, J. (ed.) *Biology and Culture of the Channel Catfish*, 1st edn. Elsevier, Amsterdam, pp. 347-443.

[46] Hawke, J., McWhorter, A., Steigerwalt, A. and Brenner, D. (1981) *Edwardsiella ictaluri* sp. nov., the causative agent of enteric septicemia of catfish. *International Journal of Systematic Bacteriology* 31, 396-400.

[47] Hawke, J., Kent, M., Rogge, M., Baumgartner, W., Wiles, J. *et al*. (2013) Edwardsiellosis caused by *Edwardsiella ictaluri* in laboratory populations of zebrafish *Danio rerio*. *Journal of Aquatic Animal Health* 25, 171-183.

[48] Herman, R. and Bullock, G. (1986) Pathology caused by the bacterium *Edwardsiella tarda* in striped bass. *Transactions of the American Fisheries Society* 115, 232-235.

[49] Hilton, L. and Wilson, J. (1980) Terramycin-resistant *Edwardsiella tarda* in channel catfish. *The Progressive Fish-Culturist* 42, 159.

[50] Hirai, Y., Ashata-Tago, S., Ainoda, Y., Fujita, T. and Kikuchi, K. (2015) *Edwardsiella tarda* bacteremia. A rare but fatal water-and foodborne infection: review of the literature and clinical cases from a single centre. *Canadian Journal of Infectious Disease and Medical Microbiology* 26, 313-318.

[51] Hirono, I., Tange, N. and Aoki, T. (1997) Iron-regulated haemolysin gene from *Edwardsiella tarda*. *Molecular Microbiology* 24, 851-856.

[52] Humphrey, J. and Langdon, J. (1986) Pathological anatomy and diseases of barramundi

(Lates calcarifer). In: Copland, J. W. and Grey, D. L. (eds) *Management of Wild Cultured Sea Bass /Barramundi* (Lates calcarifer): Proceedings of an International Workshop held at Darwin, N. T., Australia, 24 - 30 September 1986. ACIAR Proceedings No. 20, Australian Centre for International Agricultural Research, Canberra, pp. 198 - 203.

[53] Humphrey, J., Lancaster, C., Gudkovs, N. and McDonald, W. (1986) Exotic bacterial pathogens *Edwardsiella tarda* and *Edwardsiella ictaluri* from imported ornamental fish *Betta splendens* and *Puntius conchonius*, respectively: isolation and quarantine significance. *Australian Veterinary Journal* 63, 369 - 371.

[54] Iwanowicz, L., Griffin, A., Cartwright, D. and Blazer, V. (2006) Mortality and pathology in brown bullheads *Amieurus nebulosus* associated with a spontaneous *Edwardsiella ictaluri* outbreak under tank culture conditions. *Diseases of Aquatic Organisms* 70, 219 - 225.

[55] Janda, J. and Abbott, S. (1993) Infections associated with the genus *Edwardsiella*: the role of *Edwardsiella tarda* in human disease. *Clinical Infectious Diseases* 17, 742 - 748.

[56] Janda, J., Abbott, S., Kroske-Bystrom, S., Cheung, W., Powers, C. *et al*. (1991) Pathogenic properties of *Edwardsiella* species. *Journal of Clinical Microbiology* 29, 1997 - 2001.

[57] Joh, S., Kim, M., Kwon, H., Ahn, E., Jang, H. and Kwon, J. (2011) Characterization of *Edwardsiella tarda* isolated from farm-cultured eels, *Anguilla japonica*, in the Republic of Korea. *Journal of Veterinary Medical Science* 73, 7 - 11.

[58] Kasornchandra, J., Rogers, W. and Plumb, J. (1987) *Edwardsiella ictaluri* from walking catfish, *Clarias batrachus* L., in Thailand. *Journal of Fish Diseases* 10, 137 - 138.

[59] Kawai, K., Liu, Y., Ohnishi, K. and Oshima, S. (2004) A conserved 37 kDa outer membrane protein of *Edwardsiella tarda* is an effective vaccine candidate. *Vaccine* 22, 3411 - 3418.

[60] Kent, M. and Lyons, J. (1982) *Edwardsiella ictaluri in the green knife fish*, *Eigemannia virescens*. *Fish Health News* 2, 2. [Eastern Fish Disease Laboratory, US Department of the Interior, Fish and Wildlife Service, Kearneysville, West Virginia.]

[61] Keskin, O., Seçer, S., I · zgür, M., Türkyilmaz, S. and Mkakosya, R. S. (2004) *Edwardsiella ictaluri* infection in rainbow trout (*Oncorhynchus mykiss*). *Turkish Journal of Veterinary and Animal Sciences* 28, 649 - 653.

[62] Klesius, P. (1994) Transmission of *Edwardsiella ictaluri* from infected, dead to noninfected channel catfish. *Journal of Aquatic Animal Health* 6, 180 - 182.

[63] Klesius, P., Lovy, J., Evans, J., Washuta, E. and Arias, C. (2003) Isolation of *Edwardsiella ictaluri* from tadpole madtom in a southwestern New Jersey River. *Journal of Aquatic Animal Health* 15, 295 - 301.

[64] Kusuda, R. and Kawai, K. (1998) Bacterial diseases of cultured marine fish in Japan. *Fish Pathology* 33, 221 - 227.

[65] Kusuda, R., Toyoshima, T., Iwamura, Y. and Sako, H. (1976) *Edwardsiella tarda* from an epizootic of mullets (*Mugil cephalus*) in Okitsu Bay. *Nippon Suisan Gakkaishi* 42, 271 - 275.

[66] Kusuda, R., Itami, T., Munekiyo, M. and Nakajima, H. (1977) Characteristics of a *Edwardsiella* sp. from an epizootic of cultured sea breams. *Nippon Susan Gakkaishi* 43, 129 - 134.

[67] Kwon, S., Nam, Y., Kim, S. and Kim, K. (2006) Protection of tilapia (*Oreochromis mosambicus*) from edwardsiellosis by vaccination with *Edwardsiella tarda* ghosts. *Fish and Shellfish Immunology* 20, 621 - 626.

[68] Kwon, S., Lee, E., Nam, Y., Kim, S. and Kim, K. (2007) Efficacy of oral

immunization with *Edwardsiella tarda* ghosts against edwardsiellosis in olive flounder (*Paralichthys olivaceus*). *Aquaculture* 269, 84 - 88.
[69] Lan, J., Zhang, X., Wang, Y., Chen, J. and Han, Y. (2008) Isolation of an unusual strain of *Edwardsiella tarda* from turbot and establish a PCR detection technique with the *gyrB* gene. *Journal of Applied Microbiology* 105, 644 - 651.
[70] Lawrence, M., Cooper, R. and Thune, R. (1997) Attenuation, persistence, and vaccine potential of an *Edwardsiella ictaluri purA* mutant. *Infection and Immunity* 65, 4642 -4651.
[71] Leotta, G., Piñeyro, P., Serena, S. and Vigo, G. (2009) Prevalence of *Edwardsiella tarda* in Antarctic wildlife. *Polar Biology* 32, 809 - 812.
[72] Liang, S., Wu, H., Liu, B., Xiao, J., Wang, Q. and Zhang, Y. (2012) Immune response of turbot (*Scophthalmus maximus* L.) to a broad spectrum vaccine candidate, recombinant glyceraldehyde-3-phosphate dehydrogenase of *Edwardsiella tarda*. *Veterinary Immunology and Immunopathology* 150, 198 - 205.
[73] Lo, D., Lee, Y., Wang, J. and Kuo, H. (2014) Antimicrobial susceptibility and genetic characterisation of oxytetracycline-resistant *Edwardsiella tarda* isolated from diseased eels. *Veterinary Record* 175, 203 - 203.
[74] Lv, Y., Zheng, J., Yang, M., Wang, Q. and Zhang, Y. (2012) An *Edwardsiella tarda* mutant lacking UDP-glucose dehydrogenase shows pleiotropic phenotypes, attenuated virulence, and potential as a vaccine candidate. *Veterinary Microbiology* 160, 506 - 512.
[75] Mainous, M., Smith, S. and Kuhn, D. (2010) Effect of common aquaculture chemicals against *Edwardsiella ictaluri* and *E. tarda*. *Journal of Aquatic Animal Health* 22, 224 - 228.
[76] Matshushima, S., Yajima, S., Taguchi, T., Takahashi, A., Shiseki, M. *et al.* (1996) A fulminating case of *Edwardsiella tarda* septicemia with necrotizing fasciitis. *Kansenshogaku Zasshi* 70, 631 - 636.
[77] Matsuyama, T., Kamaishi, T., Ooseko, N., Kurohara, K. and Iida, T. (2005) Pathogenicity of motile and nonmotile *Edwardsiella tarda* to some marine fish. *Fish Pathology* 40, 133 - 135.
[78] Menanteau-Ledouble, S., Karsi, A. and Lawrence, M. (2011) Importance of skin abrasion as a primary site of adhesion for *Edwardsiella ictaluri* and impact on invasion and systematic infection in channel catfish *Ictalurus punctatus*. *Veterinary Microbiology* 148, 425 - 430.
[79] Meyer, F. and Bullock., G. (1973) *Edwardsiella tarda*, a new pathogen of channel catfish (*Ictalurus punctatus*). *Applied Microbiology* 25, 155 - 156.
[80] Miyazaki, T. and Plumb, J. (1985) Histopathology of *Edwardsiella ictaluri* in channel catfish, *Ictalurus punctatus* (Rafinesque). *Journal of Fish Diseases* 8, 389 - 392.
[81] Mohanty, B. and Sahoo, P. (2007) Edwardsiellosis in fish: a brief review. *Journal of Biosciences* 32, 1331 - 1344.
[82] Mqolomba, T. and Plumb, J. (1992) Effect of temperature and dissolved oxygen concentration on *Edwardsiella ictaluri* in experimentally infected channel catfish. *Journal of Aquatic Animal Health* 4, 215 - 217.
[83] Nagai, T. and Nakai, T. (2014) Water temperature effect on *Edwardsiella ictaluri* infection of ayu *Plecoglossus altivelis*. *Fish Pathology* 49, 61 - 63.
[84] Nagai, T., Iwamoto, E., Sakai, T., Arima, T., Tensha, K. *et al.* (2008) Characterization of *Edwardsiella ictaluri* isolated from wild ayu *Plecoglossus altivelis* in Japan. *Fish Pathology* 43, 158 - 163.
[85] Nakhro, K., Devi, T. and Kamilya, D. (2013) *In vitro* immunopathogenesis of *Edwardsiella tarda* in catla *Catla catla* (Hamilton). *Fish and Shellfish Immunology*

35, 175 - 179.
[86] Newton, J. C., Bird, R. C., Blevins, W. T., Wilt, G. R. and Wolfe, L. G. (1988) Isolation, characterization, and molecular cloning of cryptic plasmids isolated from *Edwardsiella ictaluri*. *American Journal of Veterinary Research* 49, 1856 - 1860.
[87] Nite.go.jp (2016) Genome Projects: Whole Genome Shotgun (WGS). (National) Biological Resource Center (NBRC), NITE (National Institute of Technology and Evaluation), Shibuya-ku, Tokyo. Available at: http://www.nite.go.jp/en/nbrc/genome/project/wgs/project_wgs.html (accessed 31 May 2016).
[88] Oguro, K., Tamura, K., Yamane, J., Shimizu, M., Yamamoto, T. *et al*. (2014) Draft genome sequences of two genetic variant strains of *Edwardsiella piscicida*, JF1305 and RSB1309, isolated from olive flounder (*Paralichythys olivaceus*) and red sea bream (*Pagrus major*) cultured in Japan, respectively. *Genome Announcements* 2 (3): e00546 - 14.
[89] Padrós, F., Zarza, C., Dopazo, L., Cuadrado, M. and Crespo, S. (2006) Pathology of *Edwardsiella tarda* infection in turbot, *Scophthalmus maximus* (L.). *Journal of Fish Diseases* 29, 87 - 94.
[90] Pan, X., Hao, G., Yao, J., Xu, Y., Shen, J. and Yin, W. (2010) Identification and pathogenic facts studying for *Edwarsdiella tarda* from edwardsiellosis of Trionyx sinensis [J.]. *Freshwater Fisheries* 6, 40 - 45.
[91] Panangala, V., Shoemaker, C., van Santen, V., Dybvig, K. and Klesius, P. (2007) Multiplex-PCR for simultaneous detection of three bacterial fish pathogens, *Flavobacterium columnare*, *Edwardsiella ictaluri*, and *Aeromonas hydrophila*. *Diseases of Aquatic Organisms* 74, 199 - 208.
[92] Park, S., Aoki, T. and Jung, T. (2012) Pathogenesis of and strategies for preventing *Edwardsiella tarda* infection in fish. *Veterinary Research* 43, 67.
[93] Petrie-Hanson, L. and Ainsworth, J.A. (1999) Humoral immune responses of channel catfish (*Ictalurus punctatus*) fry and fingerlings exposed to *Edwardsiella ictaluri*. *Fish and Shellfish Immunology* 9, 579 - 589.
[94] Petrie-Hanson, L., Romano, C., Mackey, R., Khosravi, P., Hohn, C. and Boyle, C. (2007) Evaluation of zebrafish *Danio rerio* as a model for enteric septicemia of catfish (ESC). *Journal of Aquatic Animal Health* 19, 151 - 158.
[95] Plumb, J. and Quinlan, E. (1986) Survival of *Edwardsiella ictaluri* in pond water and bottom mud. *The Progressive Fish-Culturist* 48, 212 - 214.
[96] Plumb, J. and Shoemaker, C. (1995) Effects of temperature and salt concentration on latent *Edwardsiella ictaluri* infections in channel catfish. *Diseases of Aquatic Organisms* 21, 171 - 175.
[97] Plumb, J. and Vinitnantharat, S. (1989) Biochemical, biophysical, and serological homogeneity of *Edwardsiella ictaluri*. *Journal of Aquatic Animal Health* 1, 51 - 56.
[98] Plumb, J. and Vinitnantharat, S. (1993) Vaccination of channel catfish, *Ictalurus punctatus* (Rafinesque), by immersion and oral booster against *Edwardsiella ictaluri*. *Journal of Fish Diseases* 16, 65 - 71.
[99] Plumb, J., Maestrone, G. and Quinlan, E. (1987) Use of a potentiated sulfonamide to control *Edwardsiella ictaluri* infection in channel catfish (*Ictalurus punctatus*). *Aquaculture* 62, 187 - 194.
[100] Plumb, J., Vinitnantharat, S. and Paterson, W. (1994) Optimum concentration of *Edwardsiella ictaluri* vaccine in feed for oral vaccination of channel catfish. *Journal of Aquatic Animal Health* 6, 118 - 121.
[101] Řehulka, J., Marejková, M. and Petráš, P. (2012) Edwardsiellosis in farmed rainbow trout (*Oncorhynchus mykiss*). *Aquaculture Research* 43, 1628 - 1634.
[102] Reichley, S., Waldbieser, G., Tekedar, H., Lawrence, M. and Griffin, M. (2015a)

Complete genome sequence of *Edwardsiella tarda* isolate FL95－01, recovered from channel catfish. *Genome Announcements* 3(3): e00682－15.
[103] Reichley, S., Waldbieser, G., Ucko, M., Colorni, A., Dubytska, L. *et al*. (2015b) Complete genome sequence of an *Edwardsiella piscicida*-like species isolated from diseased grouper in Israel. *Genome Announcements* 3(4): e00829－15.
[104] Reichley, S., Waldbieser, G., Lawrence, M. and Griffin, M. (2015c) Complete genome sequence of an *Edwardsiella piscicida*-like species, recovered from tilapia in the United States. *Genome Announcements* 3(5): e01004－15.
[105] Reichley, S., Ware, C., Greenway, T., Wise, D. and Griffin, M. (2015d) Real-time polymerase chain reaction assays for the detection and quantification of *Edwardsiella tarda*, *Edwardsiella piscicida*, and *Edwardsiella piscicida*-like species in catfish tissues and pond water. *Journal of Veterinary Diagnostic Investigation* 27, 130－139.
[106] Rogge, M., Dubytska, L., Jung, T., Wiles, J., Elkamel, A. *et al*. (2013) Comparison of Vietnamese and US isolates of *Edwardsiella ictaluri*. *Diseases of Aquatic Organisms* 106, 17－29.
[107] Saeed, M. O. and Plumb, J. A. (1986) Immune response of channel catfish to lipopolysaccharide and whole cell *Edwardsiella ictaluri* vaccines. *Diseases of Aquatic Organisms* 2, 21－26.
[108] Sae-oui, D., Muroga, K. and Nakai, T. (1984) A case of *Edwardsiella tarda* infection in cultured colored carp *Cyprinus carpio*. *Fish Pathology* 19, 197－199.
[109] Sakai, T., Iida, T., Osatomi, K. and Kanai, K. (2007) Detection of Type 1 fimbrial genes in fish pathogenic and non-pathogenic *Edwardsiella tarda* strains by PCR. *Fish Pathology* 42, 115－117.
[110] Sakai, T., Kamaishi, T., Sano, M., Tensha, K., Arima, T. *et al*. (2008) Outbreaks of *Edwardsiella ictaluri* infection in ayu *Plecoglossus altivelis* in Japanese rivers. *Fish Pathology* 43, 152－157.
[111] Sakai, T., Matsuyama, T., Sano, M. and Iida, T. (2009a) Identification of novel putative virulence factors, adhesin AIDA and type VI secretion system, in atypical strains of fish pathogenic *Edwardsiella tarda* by genomic subtractive hybridization. *Microbiology and Immunology* 53, 131－139.
[112] Sakai, T., Yuasa, K., Ozaki, A., Sano, M., Okuda, R. *et al*. (2009b) Genotyping of *Edwardsiella ictaluri* isolates in Japan using amplified-fragment length polymorphism analysis. *Letters in Applied Microbiology* 49, 443－449.
[113] Sakai, T., Yuasa, K., Sano, M. and Iida, T. (2009c) Identification of *Edwardsiella ictaluri* and E. tarda by species-specific polymerase chain reaction targeted to the upstream region of the fimbrial gene. *Journal of Aquatic Animal Health* 21, 124－132.
[114] Salati, F. and Kusuda, R. (1985) Vaccine preparations used for immunization of eel *Anguilla japonica* against *Edwardsiella tarda* infection. *Nippon Suisan Gakkaishi* 51, 1233－1237.
[115] Santander, J., Golden, G., Wanda, S. and Curtiss, R. (2012) Fur-regulated iron uptake system of *Edwardsiella ictaluri* and its influence on pathogenesis and immunogenicity in the catfish host. *Infection and Immunity* 80, 2689－2703.
[116] Savan, R., Igarashi, A., Matsuoka, S. and Sakai, M. (2004) Sensitive and rapid detection of edwardsiellosis in fish by a loop-mediated isothermal amplification method. *Applied and Environmental Microbiology* 70, 621－624.
[117] Shafiei, S., Viljamaa-Dirks, S., Sundell, K., Heinikainen, S., Abayneh, T. and Wiklund, T. (2016) Recovery of *Edwardsiella piscicida* from farmed whitefish, Coregonus lavaretus (L.), in Finland. *Aquaculture*, 454, 19－26.
[118] Shao, S., Lai, Q., Liu, Q., Wu, H., Xiao, J. *et al*. (2015) Phylogenomics

characterization of a highly virulent *Edwardsiella* strain ET080813T encoding two distinct T3SS and three T6SS gene clusters: propose a novel species as *Edwardsiella anguillarum* sp. nov. *Systematic and Applied Microbiology* 38, 36 - 47.

[119] Shao, J., Yuan, J., Shen, Y., Hu, R. and Gu, Z. (2016) First isolation and characterization of *Edwardsiella tarda* from diseased Asian swamp eel, *Monopterus albus* (Zuiew). *Aquaculture Research* 47, 3684 - 3688.

[120] Shetty, M., Maiti, B., Venugopal, M., Karunasagar, I. and Karunasagar, I. (2014) First isolation and characterization of *Edwardsiella tarda* from diseased striped catfish, *Pangasianodon hypophthalmus* (Sauvage). *Journal of Fish Diseases* 37, 265 - 271.

[121] Shoemaker, C. and Klesius, P. (1997) Protective immunity against enteric septicaemia in channel catfish, *Ictalurus punctatus* (Rafinesque), following controlled exposure to *Edwardsiella ictaluri*. *Journal of Fish Diseases* 20, 361 - 368.

[122] Shoemaker, C., Klesius, P. and Bricker, J. (1999) Efficacy of a modified live *Edwardsiella ictaluri* vaccine in channel catfish as young as seven days post hatch. *Aquaculture* 176, 189 - 193.

[123] Shoemaker, C., Klesius, P. and Evans, J. (2002) *In ovo* methods for utilizing the modified live *Edwardsiella ictaluri* vaccine against enteric septicemia in channel catfish. *Aquaculture* 203, 221 - 227.

[124] Shoemaker, C., Klesius, P. and Evans, J. (2007) Immunization of eyed channel catfish, *Ictalurus punctatus*, eggs with monovalent *Flavobacterium columnare* vaccine and bivalent *F. columnare* and *Edwardsiella ictaluri* vaccine. *Vaccine* 25, 1126 - 1131.

[125] Shotts, E., Blazer, V. and Waltman, W. (1986) Pathogenesis of experimental *Edwardsiella ictaluri* infections in channel catfish (*Ictalurus punctatus*). *Canadian Journal of Fisheries and Aquatic Science* 43, 36 - 42.

[126] Soto, E., Griffin, M., Arauz, M., Riofrio, A., Martinez, A. and Cabrejos, M. (2012) *Edwardsiella ictaluri* as the causative agent of mortality in cultured Nile tilapia. *Journal of Aquatic Animal Health* 24, 81 - 90.

[127] Starliper, C., Cooper, R., Shotts, E. and Taylor, P. (1993) Plasmid-mediated Romet resistance of *Edwardsiella ictaluri*. *Journal of Aquatic Animal Health* 5, 1 - 8.

[128] Sun, Y., Liu, C.-S. and Sun, L. (2010) Identification of an *Edwardsiella tarda* surface antigen and analysis of its immunoprotective potential as a purified recombinant subunit vaccine and a surface-anchored subunit vaccine expressed by a fish commensal strain. *Vaccine* 28, 6603 - 6608.

[129] Sun, Y., Liu, C.-S. and Sun, L. (2011a) Construction and analysis of the immune effect of an *Edwardsiella tarda* DNA vaccine encoding a D15-like surface antigen. *Fish and Shellfish Immunology* 30, 273 - 279.

[130] Sun, Y., Liu, C.-S. and Sun, L. (2011b) Comparative study of the immune effect of an *Edwardsiella tarda* antigen in two forms: subunit vaccine vs DNA vaccine. *Vaccine* 29, 2051 - 2057.

[131] Swain, P., Nayak, S., Sahu, A., Mohapatra, B. and Meher, P. (2002) Bath immunisation of spawn, fry and fingerlings of Indian major carps using a particulate bacterial antigen. *Fish and Shellfish Immunology* 13, 133 - 140.

[132] Takano, T., Matsuyama, T., Oseko, N., Sakai, T., Kamaishi, T. et al. (2010) The efficacy of five avirulent *Edwardsiella tarda* strains in a live vaccine against edwardsiellosis in Japanese flounder, *Paralichthys olivaceus*. *Fish and Shellfish Immunology* 29, 687 - 693.

[133] Takano, T., Matsuyama, T., Sakai, T. and Nakayasu, C. (2011) Protective efficacy of a formalin-killed vaccine against atypical *Edwardsiella tarda* infection in red sea bream *Pagrus major*. *Fish Pathology* 46, 120 - 122.

[134] Taylor, P. (1992) Fish-eating birds as potential vectors of *Edwardsiella ictaluri*. *Journal of Aquatic Animal Health* 4, 240 - 243.

[135] Tekedar, H., Karsi, A., Williams, M., Vamenta, S., Banes, M. *et al*. (2013) Complete genome sequence of channel catfish gastrointestinal septicemia isolate *Edwardsiella tarda* C07 - 087. *Genome Announcements* 1(6): e00959 - 13.

[136] Thune, R., Stanley, L. and Cooper, R. (1993) Pathogenesis of Gram-negative bacterial infections in warmwater fish. *Annual Review of Fish Diseases* 3, 37 - 68.

[137] Thune, R., Hawke, J. and Johnson, M. (1994) Studies on vaccination of channel catfish, *Ictalurus punctatus*, against *Edwardsiella ictaluri*. *Journal of Applied Aquaculture* 3, 11 - 24.

[138] Thune, R., Fernandez, D. and Battista, J. (1999) An *aroA* mutant of *Edwardsiella ictaluri* is safe and efficacious as a live, attenuated vaccine. *Journal of Aquatic Animal Health* 11, 358 - 372.

[139] Trung Cao, T., Tsai, M., Yang, C., Wang, P., Kuo, T. *et al*. (2014) Vaccine efficacy of glyceraldehyde-3-phosphate dehydrogenase (GAPDH) from *Edwardsiella ictaluri* against *E. tarda* in tilapia. *The Journal of General and Applied Microbiology* 60, 241 - 250.

[140] Turnbaugh, P., Ley, R., Hamady, M., Fraser-Liggett, C., Knight, R. and Gordon, J. (2007) The human microbiome project: exploring the microbial part of ourselves in a changing world. *Nature* 449, 804.

[141] Ucko, M., Colorni, A., Dubytska, L. and Thune, R. (2016) *Edwardsiella piscicida*-like pathogen in cultured grouper. *Diseases of Aquatic Organisms* 121, 141 - 148.

[142] Uhland, F., Hélie, P. and Higgins, R. (2000) Infections of *Edwardsiella tarda* among brook trout in Quebec. *Journal of Aquatic Animal Health* 12, 74 - 77.

[143] USDA/APHIS/NAHMS (2010) *Catfish 2010. Part I: Reference of Catfish Health and Production Practices in the United States, 2009*. US Department of Agriculture/Animal and Plant Health Inspection Service, National Animal Health Monitoring System, Fort Collins, Colorado.

[144] Van Damme, L. and Vandepitte, J. (1980) Frequent isolation of *Edwardsiella tarda* and *Pleisiomonas shigelloides* from healthy Zairese freshwater fish: a possible source of sporadic diarrhea in the tropics. *Applied and Environmental Microbiology* 39, 475 - 479.

[145] Vandepitte, J., Lemmens, P. and De Swert, L. (1983) Human edwardsiellosis traced to ornamental fish. *Journal of Clinical Microbiology* 17, 165 - 167.

[146] van Soest, J., Stockhammer, O., Ordas, A., Bloemberg, G., Spaink, H. and Meijer, A. (2011) Comparison of static immersion and intravenous injection systems for exposure of zebrafish embryos to the natural pathogen *Edwardsiella tarda*. *BMC Immunology* 12: 58.

[147] Waltman, W. and Shotts, E. (1986a) Antimicrobial susceptibility of *Edwardsiella tarda* from the United States and Taiwan. *Veterinary Microbiology* 12, 277 - 282.

[148] Waltman, W. and Shotts, E. (1986b) Antimicrobial susceptibility of *Edwardsiella ictaluri*. *Journal of Wildlife Diseases* 22, 173 - 177.

[149] Waltman, W., Shotts, E. and Blazer, V. (1985) Recovery of *Edwardsiella ictaluri* from danio (Danio devario). Aquaculture 46, 63 - 66.

[150] Waltman II, W., Shotts, E. and Wooley, R. (1989) Development and transfer of plasmid-mediated antimicrobial resistance in *Edwardsiella ictaluri*. Canadian *Journal of Fisheries and Aquatic Sciences* 46, 1114 - 1117.

[151] Wang, Q., Yang, M., Xiao, J., Wu, H., Wang, X. *et al*. (2009) Genome sequence of the versatile fish pathogen *Edwardsiella tarda* provides insights into its adaptation to broad host ranges and intracellular niches. *PLoS ONE* 4(10): e7646.

[152] Wang, R., Tekedar, H. C., Lawrence, M. L., Chouljenko, V. N., Kim, J. *et al.* (2015) Draft genome sequences of *Edwardsiella ictaluri* strains LADL11 - 100 and LADL11 - 194 isolated from zebrafish *Danio rerio*. *Genome Announcements* 3(6): e01449 - 15.

[153] Wang, Y., Wang, Q., Xiao, J., Liu, Q., Wu, H. and Zhang, Y. (2011) Genetic relationships of *Edwardsiella* strains isolated in China aquaculture revealed by rep-PCR genomic fingerprinting and investigation of *Edwardsiella* virulence genes. *Journal of Applied Microbiology* 111, 1337 - 1348.

[154] Waterstrat, P., Dorr, B., Glahn, J. and Tobin, M. (1999) Recovery and viability of *Edwardsiella ictaluri* from great blue herons *Ardea herodias* fed *E. ictaluri*-infected channel catfish *Ictalurus punctatus* fingerlings. *Journal of the World Aquaculture Society* 30, 115 - 122.

[155] Welch, T., Evenhuis, J., White, D., McDermott, P., Harbottle, H. et al. (2008) IncA/C plasmid-mediated florfenicol resistance in the catfish pathogen *Edwardsiella ictaluri*. *Antimicrobial Agents and Chemotherapy* 53, 845 - 846.

[156] White, F., Simpson, C. and Williams, L. (1973) Isolation of *Edwardsiella tarda* from aquatic animal species and surface waters in Florida. *Journal of Wildlife Diseases* 9, 204 - 208.

[157] Williams, M., Gillaspy, A., Dyer, D., Thune, R., Waldbieser, G. *et al.* (2012) Genome sequence of *Edwardsiella ictaluri* 93 - 146, a strain associated with a natural channel catfish outbreak of enteric septicemia of catfish *Journal of Bacteriology* 194, 740 - 741.

[158] Wise, D. and Johnson, M. (1998) Effect of feeding frequency and Romet-medicated feed on survival, antibody response, and weight gain of fingerling channel catfish Ictalurus punctatus after natural exposure to *Edwardsiella ictaluri*. *Journal of the World Aquaculture Society* 29, 169 - 175.

[159] Wise, D., Schwedler, T. and Otis, D. (1993) Effects of stress on susceptibility of naive channel catfish in immersion challenge with *Edwardsiella ictaluri*. *Journal of Aquatic Animal Health* 5, 92 - 97.

[160] Wise, D., Camus, A., Schwedler, T. and Terhune, J. (2004) Health management. In: Tucker, C. and Hargreaves, J. (eds) *Biology and Culture of the Channel Catfish*, 1st edn. Elsevier, Amsterdam, pp. 444 - 503.

[161] Wise, D., Greenway, T., Li, M., Camus, A. and Robinson, E. (2008) Effects of variable periods of food deprivation on the development of enteric septicemia in channel catfish. *Journal of Aquatic Animal Health* 20, 39 - 44.

[162] Wise, D., Greenway, T., Byars, T., Griffin, M. and Khoo, L. (2015) Oral vaccination of channel catfish against enteric septicemia of catfish using a live attenuated *Edwardsiella ictaluri* isolate. *Journal of Aquatic Animal Health* 27, 135 - 143.

[163] Wyatt, L., Nickelson, R. and Vanderzant, C. (1979) *Edwardsiella tarda* in freshwater catfish and their environment. *Applied and Environmental Microbiology* 38, 710 - 714.

[164] Xiao, J., Wang, Q., Liu, Q., Wang, X., Liu, H. and Zhang, Y. (2008) Isolation and identification of fish pathogen *Edwardsiella tarda* from mariculture in China. *Aquaculture Research* 40, 13 - 17.

[165] Xu, T. and Zhang, X. (2014) *Edwardsiella tarda*: an intriguing problem in aquaculture. Aquaculture 431, 129 - 135.

[166] Yamada, Y. and Wakabayashi, H. (1998) Enzyme electrophoresis, catalase test and PCR-RFLP analysis for the typing of *Edwardsiella tarda*. *Fish Pathology* 33, 1 - 5.

[167] Yamada, Y. and Wakabayashi, H. (1999) Identification of fish-pathogenic strains belonging to the genus *Edwardsiella* by sequence analysis of *sodB*. *Fish Pathology*

34, 145 - 150.

[168] Yang, M., Lv, Y., Xiao, J., Wu, H., Zheng, H. *et al.* (2012) *Edwardsiella* comparative phylogenomics reveal the new intra/inter-species taxonomic relationships, virulence evolution and niche adaptation mechanisms. *PLoS ONE* 7(5): e36987.

[169] Yang, W., Wang, L., Zhang, L., Qu, J., Wang, Q. and Zhang, Y. (2015) An invasive and low virulent *Edwardsiella tarda esrB* mutant promising as live attenuated vaccine in aquaculture. *Applied Microbiology and Biotechnology* 99, 1765 - 1777.

[170] Ye, S., Li, H., Qiao, G. and Li, Z. (2009) First case of *Edwardsiella ictaluri* infection in China farmed yellow catfish *Pelteobagrus fulvidraco*. *Aquaculture* 292, 6 - 10.

[171] Ye, X., Lin, X. and Wang, Y. (2010) Identification and detection of virulence gene of the pathogenic bacteria *Edwardsiella tarda* in cultured Scortum barcoo. *Freshwater Fisheries* 40, 50 - 54.

[172] Yeh, H., Shoemaker, C. and Klesius, P. (2005) Evaluation of a loop-mediated isothermal amplification method for rapid detection of channel catfish Ictalurus punctatus important bacterial pathogen *Edwardsiella ictaluri*. *Journal of Microbiological Methods* 63, 36 - 44.

[173] Yu, J., Han, J., Park, K., Park, K. and Park, S. (2009) *Edwardsiella tarda* infection in Korean catfish, *Silurus asotus*, in a Korean fish farm. *Aquaculture Research* 41, 19 - 26.

16

黄 杆 菌

Thomas P. Loch* 和 Mohamed Faisal

16.1 引言

鱼类黄杆菌病主要是由三种革兰氏阴性产黄色素病原细菌引起的：嗜冷黄杆菌（*Flavobacterium psychrophilum*）、柱状黄杆菌（*Flavobacterium columnare*）、嗜鳃黄杆菌（*Flavobacterium branchiophilum*）。其中嗜冷黄杆菌是细菌性冷水病（bacterial cold water disease，BCWD）和虹鳟鱼苗综合征（rainbow trout fry syndrome，RTFS）的病原菌（Davis，1946；Borg，1948；Holt，1987；Bernardet and Grimont，1989）；柱状黄杆菌是鱼类柱形病（Columnaris disease，CD）的病原菌（Davis，1922；Ordal and Rucker，1944；Bernardet and Grimont，1989）；而嗜鳃黄杆菌则是细菌性鳃病（bacterial gill disease，BGD）的推定病原（Wakabayashi et al.，1989）。

16.1.1 嗜冷黄杆菌

1. 微生物描述

Borg（1948）描述了养殖银大麻哈鱼（*Oncorhynchus kisutch*）鱼苗和幼鱼在 6～10 ℃的水温下由大量杆状细菌（现在被称为嗜冷黄杆菌）引起的动物流行病。该细菌经历多次分类学评估，最终被归入拟杆菌门（Bacteroidetes），黄杆菌纲（Flavobacteria），黄杆菌目（Flavobacteriales），黄杆菌科（Flavobacteriaceae），黄杆菌属（*Flavobacterium*）（Bernardet，2011a），并被命名为嗜冷黄杆菌（*F. psychrophilum*）（Bernardet et al.，1996）。

F. psychrophilum 的最适生长温度为 15～18 ℃，但低至 4 ℃时也可生长（Bernardet and Bowman，2011）。已经报道了用于培养嗜冷黄杆菌的低营养培养基（Bernardet and Bowman，2006；Austin and Austin，2007；Starliper and Schill，2011）。最常见的培养基包括嗜细胞培养基（Anacker and Ordal，1959）、

---

* 通信作者邮箱：lochthom@cvm.msu.edu。

富含胰蛋白胨的嗜细胞培养基(Bernardet and Kerouault，1989)和胰蛋白胨酵母提取物培养基(tryptone yeast extract salts medium，TYES；Holt，1987)。抗生素的添加提高了这些培养基的选择能力(Starliper and Schill，2011)，并有助于从存在其他微生物的病灶中分离出嗜冷黄杆菌。初代培养物需在15～18 ℃下进行连续近7 d的有氧培养(Holt et al.，1993)。尽管菌落形态也会发生变化，细菌产生黄色菌落，中心凸起，边缘薄而稍不规则。该菌长1～5 μm，宽0.3～0.5 μm，呈细杆状(Bernardet and Kerouault，1989)[图16.1(A)]，会滑动运动(Bernardet and Nakagawa，2006)，可耐受不超过1%的NaCl。其他的形态学、生理和生化特征可见其他文献(Bernardet et al.，1996；Bernardet and Bowman，2011)。

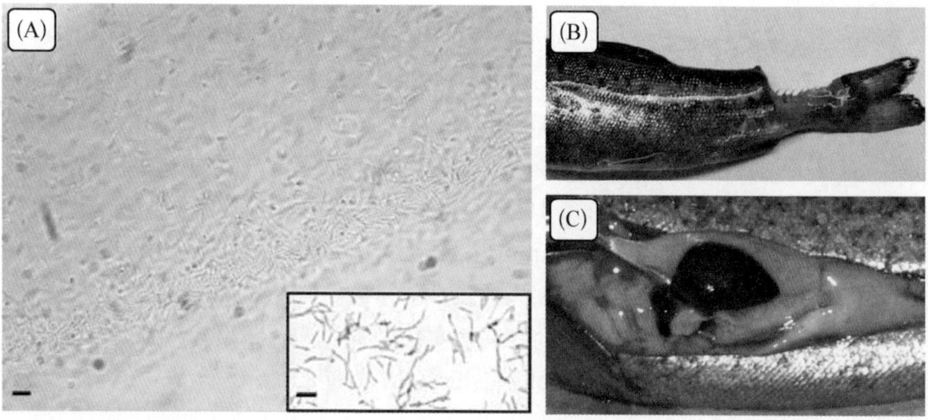

图16.1 (A) 从虹鳟幼鱼脑部分离出的嗜冷黄杆菌镜检湿压片，可见大量细长杆菌(比例尺：5 μm；插图：同一分离物的革兰氏染色)。(B) 感染*F. psychrophilum*的大西洋鲑尾柄深度溃疡并显露脊柱。(C) 感染嗜冷黄杆菌的硬头鳟出现严重的脾肿大

至少存在7种嗜冷黄杆菌血清型(Madsen and Dalsgaard，2000；Mata et al.，2002)，已经使用噬菌体分型(Castillo et al.，2014)、随机扩增多态性DNA(random amplified polymorphic DNA，RAPD)分析(Chakroun et al.，1997)、脉冲场凝胶电泳(pulse-field gel electrophoresis，PFGE)(Arai et al.，2007)、核糖体和持家基因的限制性消化/序列分析(Madsen and Dalsgaard，2000)以及多位点序列分型(multilocus sequence typing，MLST)(Nicolas et al.，2008；Van Vliet et al.，2016)进行了种内遗传异质性描述，并且已有完整的基因组序列数据(Duchaud et al.，2007；Wiens et al.，2014)。

2. 传播方式

水平传播是通过水传播和接触感染(Holt，1987)。濒死鱼和死鱼是养殖场内病原菌的来源(Wiklund et al.，2000；Madetoja et al.，2002)。Vatsos等

(2003)从无菌河水中分离出长达 9 个月存留期的 *F. psychrophilum*，但是 Madetoja 等(2002)发现沉积物增强细菌生存能力。该细菌还形成生物被膜(Alvarez et al.，2006)，出现在底栖硅藻(Izumi et al.，2005)、藻类(Amita et al.，2000)、两栖动物(Brown et al.，1997)和水蛭中(*Myzobdella lugubris*)(Schulz and Faisal，2010)。

F. psychrophilum 是从鲑繁殖液中分离出来的(Holt，1987；Van Vliet et al.，2015)，可以通过配子或卵从亲鱼传给后代(Brown et al.，1997；Cipriano，2015)。如果无嗜冷黄杆菌的未受精鱼卵在含有细菌的水中硬化，也会发生传播(Kumagai et al.，1998，2000)。无论是源自水体还是受污染的孵育产品，嗜冷黄杆菌在鱼卵中的驻留和碘伏消毒后存活的能力都会增强其传播能力(Avendaño-Herrera et al.，2014)。

3. 地理分布和宿主范围

已在北美、南美、欧洲、亚洲和大洋洲发现了 *F. psychrophilum* (Starliper，2011；http://pubmlst.org/fpsychrophilum)。该病原菌会导致养殖鲑科鱼类发生疾病(Starliper，2011)，虹鳟和银大麻哈鱼特别易感(Holt et al.，1993)。在欧洲，这种疾病首先在法国的虹鳟养殖场出现(Bernardet et al.，1988)，不久之后在其他欧洲国家的养殖虹鳟中也被发现(Lorenzen et al.，1991；Evensen and Lorenzen，1996)。由于该病主要在虹鳟鱼苗(0.2～2.0 g)中发现，因此被称为虹鳟鱼苗综合征(RTFS)(Lorenzen et al.，1991)。该病原菌也可从野生鲑科鱼类中分离出来(Loch et al.，2013；Van Vliet et al.，2015)，可以感染香鱼、日本鳗鲡、欧洲鳗鲡、河鲈、虾虎鱼[条尾裸头虾虎鱼(*Chaenogobius urotaenia*)、褐吻虾虎鱼(*Rhinogobius brunneus*)、短棘缟虾虎鱼(*Tridentiger brevispinis*)、裸身虾虎鱼属(*Gymnogobius* sp.)]、高首鲟、胡鲇(*Clarias batrachus*)、海七鳃鳗(*Petromyzon marinus*)、西太公鱼(*Hypomesus nipponensis*)、雷氏叉牙七鳃鳗(*Lethenteron reissneri*)、三刺鱼和欧洲川鲽(Wakabayashi et al.，1994；Elsayed et al.，2006；Starliper，2011；Verma and Prasad，2014；http://pubmlst.org/fpsychrophilum)。

16.1.2 柱状黄杆菌

1. 微生物描述

Davis(1922)在 1917—1919 年夏季观察到鱼类中的多种动物流行病。受感染鱼的体表有"脏白或发黄的区域"，鳍和鳃上有损伤病灶，并最终死亡。它们的鳍常常被侵蚀成"只剩残根"并且鳃也坏死。对受影响组织的湿压片显微镜检查显示有大量长、细且易弯曲的杆状体并形成"柱状团块"。因此，Davis(1922)将这种细菌命名为柱状杆菌(*Bacillus columnaris*)。Ordal 和 Rucker(1944)最初从受类似影响的鱼类中分离出产黄色素的细菌。经过多次分类学变化后，该细菌被归类为柱状黄杆菌 (Bernardet et al., 1996)。

该细菌常用的培养基可能包含抗生素，包括 Shieh 培养基和改良 Shieh 培养基(Song et al., 1988)、Hsu-Shotts 培养基和嗜细胞培养基(Song et al., 1988; Bernardet and Bowman, 2006)。将原代培养物在 20～30 ℃下有氧培养 1～3 d；在实验室中，尽管生长缓慢，*F. columnare* 也可在 15 ℃下生长。*F. columnare*的菌落产黄色素，呈根状，略扁平，并强烈地附着或嵌入琼脂中。*F. columnare* 是可滑动的革兰氏阴性菌[图 16.2(A)]，长 3～10 μm，宽 0.3～0.5 μm。它们在 0～0.5% NaCl 和 15～37 ℃下生长，尽管该菌可在 10～12 ℃下从鱼中分离，但其最适生长温度为 20～25 ℃ (Fujihara and Nakatani, 1971; Loch and Faisal，未发表的结果)。其他特征可见相关文献报道(Bernardet et al., 1996; Bernardet and Bowman, 2011)。

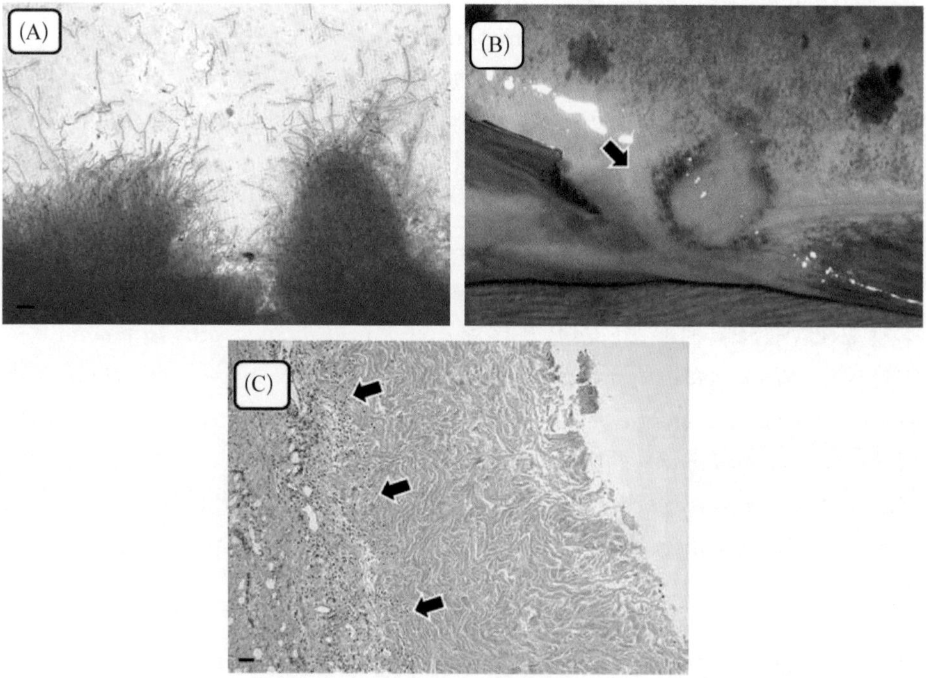

图 16.2 (A) *F. columnare* 革兰氏染色的“柱形”杆菌特征(比例尺：10 μm)；(B) 养殖斑点叉尾鮰躯干上的皮肤溃疡，周围弥漫性出血，其上覆盖着产黄色素的 *F. columnare*(箭头所示)；(C) 感染 *F. columnare* 的皮肤切片，显示其表皮和真皮发生溃疡，暴露出下面的皮下结缔组织(比例尺：50 μm；苏木精-伊红染色切片)，箭头表示所产生的炎症反应

F. columnare 的种内血清学存在异质性(Anacker and Ordal, 1959)，并且该菌有不同的基因组群，Triyanto 和 Wakabayashi (1999)界定划分了三种基因

组型Ⅰ、Ⅱ和Ⅲ。进一步的异质性和其他基因组型(例如Ⅰ,Ⅱ,Ⅱ-B,Ⅲ和Ⅰ/Ⅱ)已被描述,并与毒力和宿主特异性有关(LaFrentz et al., 2014)。已获得完整的注释基因组(菌株 ATCC 49512)(Tekedar et al., 2012)。

2. 传播方式

F. columnare 通过水平方式传播,在感染中存活下来的鱼会成为带菌者,并周期性地脱落细菌(Declercq et al., 2013)。没有证据表明 *F. columnare* 可垂直传播,但这种细菌可以从鲑鱼的生殖液中恢复分离(Loch and Faisal, 2016a)。在实验室条件下,该菌可在蒸馏水中存活超过 5 个月,在湖水中存活 2 年以上(Kunttu, 2010; Kunttu et al., 2012)。不过,多种因素,如水的硬度、碱度、温度、有机物负荷和水质等,都会影响其存活率(Farmer et al., 2011; Declercq et al., 2013)。即使在惰性表面,柱状黄杆菌也可形成生物被膜(Cai et al., 2013)。

3. 地理分布和宿主范围

F. columnare 在世界范围内的淡水中均有分布,几乎所有淡水(以及一些咸水)中的冷水和温水鱼类都对该菌易感(Starliper and Schill, 2011),也包括观赏鱼类(Decostere et al., 1998)。它是美国养殖斑点叉尾鮰的重要病原菌(Wagner et al., 2002)。一些冷水鱼类,特别是鳟鱼和淡水阶段的溯河洄游鲑科鱼类,也易感该病原菌(Avendaño-Herrera et al., 2011; LaFrentz et al., 2012a)。尽管许多疫情都在养殖鱼类中暴发,但野生鱼类也会受到影响(Loch et al., 2013)。

16.1.3 嗜鳃黄杆菌

1. 微生物描述

Davis(1926)记录了幼龄养殖美洲红点鲑和虹鳟的多次疾病暴发,这些鱼的鳃受到一种病原细菌损害,这种细菌在鱼鳃上形成"繁茂的生长"态势。这种定植增加了黏液的产生和鳃上皮的增殖,导致次级鳃片和杵状细胞的融合。Kimura 等(1978)从类似患病鱼的鳃中恢复分离了一种产黄色素细菌,并满足了科赫法则(Wakabayashi et al., 1980)。该细菌被命名为嗜鳃黄杆菌(*Flavobacterium branchiophilum*)(Wakabayashi et al., 1989; Bernardet et al., 1996),被认为是细菌性鳃病(BGD)的病原体(Ostland et al., 1995; Starliper and Schill, 2011)。不过,环境因素和其他细菌(Snieszko, 1981),包括其他黄杆菌(Good et al., 2015; Loch and Faisal, 2016b),也可以产生类似的鳃病理病变。

F. branchiophilum 可从嗜细胞培养基、TYES,特别是酪胨酵母膏琼脂培养基(NCIMB medium no. 218; Bernardet and Bowman, 2006) 上的感染鳃样

本中恢复分离。原代培养物应在 18～25 ℃下通氧培养 12 d(Wakabayashi et al., 1989)。这些革兰氏阴性杆菌产生淡黄色菌落,其表面光滑、凸起并具有完整的边缘。细菌长 5～8 μm、宽 0.5 μm[图 16.3(A)](Wakabayashi et al., 1989),在 0～0.2%的 NaCl 中生长的温度范围为 5～30 ℃(最适生长温度范围为 18～25 ℃),且是非运动性的(Wakabayashi et al., 1989; Bernardet and Bowman, 2011)。其他特征可见相关文献(Wakabayashi et al., 1989; Bernardet and Bowman, 2011)。*F. branchiophilum* 显示出一定的抗原异质性(Huh and Wakabayashi, 1989)。其 FL－15 菌株的完整注释基因组已获得(Touchon et al., 2011)。

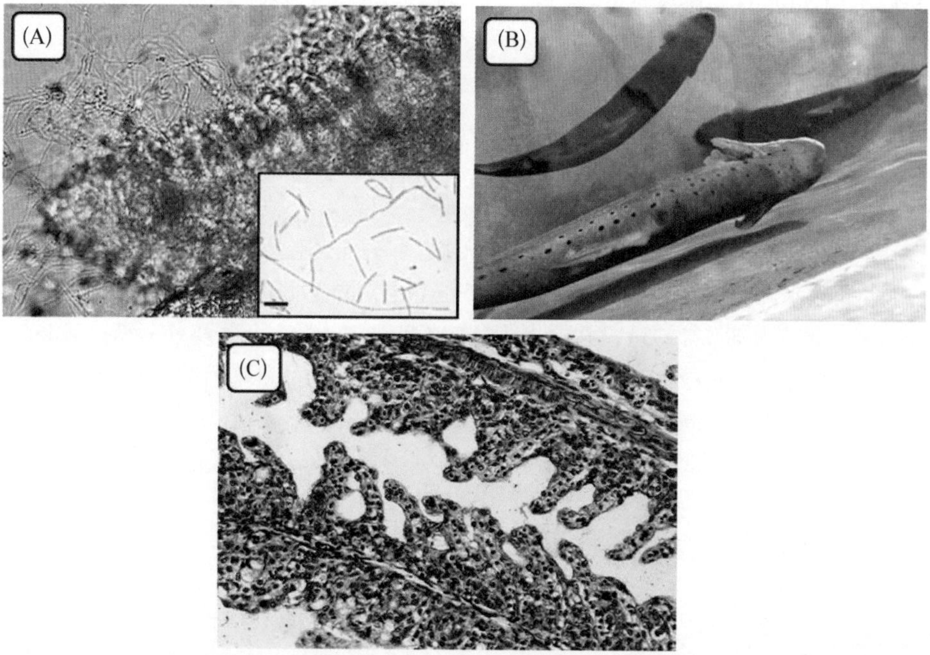

图 16.3 (A) 大量嗜鳃黄杆菌侵染鳃的整装制片(照片由美国国家鱼类健康研究实验室照片档案馆提供;西弗吉尼亚州,Leetown);插图：*F. branchiophilum* 的革兰氏染色;比例尺：10 μm。(B) 在虹鳟中具有 *F. branchiophilum* 感染特征的扩张鳃盖和苍白的鳃;另请注意,鱼在水-空气界面附近的水柱中升起。(C) 感染 *F. branchiophilum* 的鱼中显著的鳃上皮增生,导致许多次级薄片融合(苏木精-伊红染色切片;由美国国家鱼类健康研究实验室照片档案馆提供;西弗吉尼亚州,Leetown)

2. 传播方式

F. branchiophilum 通过水平方式传播,但其传染性因菌株而异(Ostland et al., 1995),环境因素影响其在鳃中的定植(Wakabayashi et al., 1980)。该菌在淡水环境中普遍存在,因此,含有其他鱼的供水设施以及在有 BGD 病史的设

施中养鱼是重要的疾病危险因素(Bebak et al., 1997)。高饲量、低换水率和高密度养殖也会导致 BGD 的暴发(Good et al., 2009, 2010)。

3. 地理分布和寄主范围

F. branchiophilum 在世界范围内都有分布(Snieszko, 1981),并且在鱼类集约化养殖中普遍存在。事实上,BGD 是养殖鱼类的专属问题(Starliper and Schill, 2011)。虽然大多数 BGD 暴发发生在冷水鱼类中,但温水鱼类也很容易受到影响(Bullock, 1990),尽管它们的易感性存在差异。例如,养殖美洲红点鲑、加拿大杂交鳟(*S. namaycush* × *S. fontinalis*)和虹鳟患 BGD 的风险高于其他鲑科鱼类(Good et al., 2008, 2009; Starliper and Schill, 2011)。一般来说,幼龄鲑比成年鲑更易感(Snieszko, 1981)。

16.1.4 三种黄杆菌对鱼类养殖生产的影响

F. psychrophilum 是导致智利水产养殖业经济损失的第二大致病菌(Avendaño-Herrera et al., 2014)。它还造成欧洲(Garcia et al., 2000)、亚洲(Nakayama et al., 2015)和北美洲(Hesami et al., 2008)等地区鱼类养殖场的损失,是密歇根州孵化场饲养鲑鱼中最常见的与病原相关的死亡原因之一(Van Vliet et al., 2015)。虽然它通常会导致鱼苗和幼鱼的损失,但在感染 *F. psychrophilum* 的鱼卵中已观察到发育不良的发眼卵(Cipriano, 2015)。因为存活鱼的生长率低,*F. psychrophilum* 还会造成养殖上的经济困难(Nematollahi et al., 2003)。据估计,在美国斑点叉尾鮰养殖业中,与柱形病相关的病害损失每年可达 3 000 万美元(Wagner et al., 2002; Shoemaker et al., 2011)。在芬兰,CD 的发病率自 20 世纪 90 年代以来不断上升,目前已严重威胁芬兰的鲑鳟养殖生产(Pulkkinen et al., 2010)。这也成为智利鲑鱼养殖(Avendaño-Herrera et al., 2011)、尼罗罗非鱼养殖(Figueiredo et al., 2005)、美国鳟鱼养殖(LaFrentz et al., 2012a)和观赏鱼(Decostere et al., 1998)的主要问题。可能由于细菌性鳃病(BGD)的多因素性质,缺乏对损失的成本估算。不过,在加拿大安大略省,BGD 是一个长期存在的问题(Daoust and Ferguson, 1983),一直制约着渔场的养殖生产(Good et al., 2008)。

16.2 病理学

与三种黄杆菌相关的行为和大体病理变化对于疾病的鉴别诊断很重要。然而,有些迹象并非病征(Loch and Faisal, 2016b),同一种鱼可能感染多个黄杆菌种(Loch and Faisal, 2016a)。

16.2.1 嗜冷黄杆菌

Flavobacterium psychrophilum 感染的银大麻哈鱼在尾鳍前面出现病变，导致脱落并几乎完全丧失组织，直到尾鳍失去作用，只被几条纤维牵住（Borg，1948）。在鱼峡部和背部也有黑变病和病灶。疾病体征通常包括嗜睡、厌食和鳍部发白（Starliper，2011）。外部病理可能仅限于鳍糜烂，但更典型的临床症状包括鳃苍白、出血，头吻部、鱼体和肛门溃疡，眼球突出（伴有或无角膜混浊及眼出血），腹胀，背部和尾柄特征性溃疡[图 16.1(B)]。通常，即使在光线暗淡的情况下，也会出现 *F. psychrophilum* 引起的脂鳍或背鳍溃疡，以及凸眼或角膜混浊和鳍糜烂、脾肿大[图 16.1(C)]。可能有明显的肾肿胀、内脏出血和腹水。在五大湖流域养殖的银大麻哈鱼幼鱼中，经常观察到胸鳍糜烂、单侧眼球突出和并发的角膜混浊和眼部出血，这让人联想起 Ostland 等（1997）报道的与嗜冷黄杆菌相关的双侧坏死性巩膜炎。在慢性 BCWD 中，患病鱼可能出现共济失调、不稳定游动和脊柱异常（如脊柱侧凸、后凸和前凸）。此外，这种患病鱼可能会被纵向压缩成“皱背鱼”或“南瓜籽”（Wood，1974）。当鱼苗受到影响时，覆盖卵黄囊的皮肤经常被侵蚀（Nematollahi et al.，2003）。

RTFS 疫情通常在鱼苗饲喂外源性饵料 6～7 周后开始，如果不及时治疗会导致 60%～90%的死亡率。病鱼浮于水面，表现为嗜睡、食欲不振、黑变病症状、眼球突出和腹水性腹胀。可能存在鳍糜烂，鱼体和头部溃疡，以及鳃、肾、肝和肠发白。脾肿大是最常见的内部病变。实际上，压片制备中脾组织内的细菌及一些特征性组织病理学被认为是特异性病征（Rangdale et al.，1999；Bernardet and Bowman，2006）。此外，Evensen 和 Lorenzen（1996）利用免疫组织化学方法在受感染的虹鳟鱼苗视网膜和脉络膜腺中检测到 *F. psychrophilum*，还观察到炎症反应，这可以解释 5%～10%的 RTFS 存活鱼为什么失明。在虹鳟幼鱼中，RTFS 体征经常与 BCWD 相比较。在香鱼中，*F. psychrophilum* 引起鳃苍白和肉茎/口周溃疡，然而，身体溃疡和鳍糜烂并不常见（Miwa and Nakayasu，2005）。

Wood 和 Yasutake（1956）在银大麻哈鱼和大鳞大麻哈鱼幼鱼的头部、上颚、下颌和鳃盖上发现了大量 *F. psychrophilum*“群集”，它们侵蚀上皮组织并渗透皮下组织和肌肉组织。这些细菌存在于鳃毛细血管内，而不在上皮细胞上。在银大麻哈鱼中，肾有病变，在退化的肾小球中有病菌。心中有大量细菌，伴有轻度炎症和个别小梁坏死。在螺旋游动的鲑中，Kent 等（1989）发现连续的亚急性到慢性骨膜炎、骨炎、脑膜炎和神经节细胞炎，以及脊柱与颅骨连接处的前椎骨膜增生和炎症。他们还发现邻近的神经组织，特别是髓质，被增生性病变压迫，可能导致异常游动。

在 RTFS 中，可见肾小管内的透明液滴变性，以及肝、肾和心局灶性坏死，大部分脏器内可见细菌(Lorenzen et al., 1991)。Bruno (1992)在老龄虹鳟中发现了相似的病变，但也发现了肝空泡变性、固缩和散在坏死的肝细胞，以及肾小管上皮中的嗜酸性粒细胞增多和核固缩。他还发现坏死性淋巴细胞性皮炎和肌炎，许多细菌与嗅囊腔上皮相关。在虹鳟鱼苗中，Rangdale 等(1999)描述了脾肥大、脾边界丧失清晰度、被结构疏松的嗜酸性粒细胞层取代，以及从脾浆膜表面延伸的腹膜炎程度。他们还发现脾中的纤维素性炎症、水肿和许多丝状细菌，同时伴有脾肥大，这些都是 RTFS 的特异性病征。

16.2.2 柱状黄杆菌

一些人认为细菌鱼类柱形病(CD)的临床症状几乎都是特异性病征(Plumb and Hanson, 2010)，尽管病变部位的严重程度和病程可能不同。例如，高毒力的 *F. columnare* 引起暴发性感染，伴有高急性死亡率，但很少有大体病理反应，而低毒力菌株发病缓慢，导致伴有显著病理反应的慢性死亡(Declercq et al., 2013 and references therein)。*F. columnare* 感染从外部的鳍、鳃和皮肤开始。行为变化包括厌食、嗜睡和浮于水面。鳃片的破坏导致白色至棕色病灶的发展，其可能会或不会被黄色"薄膜"覆盖。*F. columnare* 引起的皮肤病变可累及头部和全身，常伴有或没有出血脱色，并可能覆盖有相同的黄色薄膜和坏死碎片[图 16.2(B)]，皮肤损伤可溃烂到下层肌肉。受感染的鳍会磨损并坏死，鳍边缘将出现白色或灰色外观(Plumb and Hanson, 2010)。该菌可在背鳍和脂鳍基部的皮肤上增殖并向外辐射，产生"鞍形"病变(Declercq et al., 2013)。在观赏鱼中，CD 病变通常显现在口腔黏膜中。虽然 CD 通常是一种外部感染，但全身性疾病也可以发生(Hawke and Thune, 1992)，不过很少观察到内部病理(Plumb and Hanson, 2010)。在五大湖区，鲑中的 *F. columnare* 与外部和全身感染有关，这种病原菌经常可从肾中恢复分离(Loch et al., 2013; Loch and Faisal, 2016a)。

F. columnare 引起鳃上皮和结缔组织坏死(Speare and Ferguson, 1989)。Declercq 等(2013)指出，鳃感染通常始于上皮增生和黏液细胞增生，从而阻塞相邻片层之间的间隙。随着疾病的发展，鳃片充血、炎症(即支气管炎)和水肿导致上皮隆起。最终，可能会发生层片融合，导致广泛性出血和伴随的循环衰竭及坏死(Declercq et al., 2013)。Declercq 等(2015)使用低毒力和高毒力 *F. columnare* 分离株感染锦鲤和虹鳟幼鱼。他们发现高毒力分离株引起锦鲤病变，包括鳃上皮脱屑、多灶性片层融合、片层中远端结构缺失，轻度发炎和出血，以及鳃丝弥漫性融合，失去的鳃片被坏死碎片取代。

在感染高毒力 *F. columnare* 的虹鳟中，鳃损伤范围广泛，以层状上皮坏死为特征，随后被嗜酸性基质中的丝状细菌垫所取代，并伴有相邻片层的分支炎和水肿。在一些养殖的斑点叉尾鮰中，真皮和较小程度上的鳃部病变的主要组织病理学特征是坏死，很少或没有炎症反应(Dr Lester Khoo，Mississippi State University，个人交流)。*F. columnare* 引起的微观皮肤或肌肉损伤包括上皮溃疡[图 16.2(C)]和被细菌垫(尤其是背鳍周围)替代，最终侵入真皮和下层肌肉，导致坏死(Morrison et al.，1981)。而 Declercq 等(2015)在感染 *F. columnare* 的鲤鱼皮肤中未检测到微观病变，虹鳟出现了皮肤病变，包括表皮溃疡坏死、鳞片脱落、出血、细菌浸润和炎症。他们还发现细菌渗入皮肤深层和下层肌肉，导致肌炎。

16.2.3 嗜鳃黄杆菌

在细菌性鳃病(BGD)中，鱼变得厌食、在水平面游动，并朝着进水的水流方向游动(Snieszko，1981)。嗜睡、缺乏逃避反射、喘气和双侧鳃盖张开[图 16.3(B)]是常见症状(Speare and Ferguson，1989)。根据鱼龄、大小、放养密度和环境条件的质量，24～48 h 内发病率高达 80%，死亡率达 10%～50%(Snieszko，1981；Speare et al.，1991)。由于 *F. branchiophilum* 是非侵入性的，病理学通常局限于鳃部。明显可见鳃片层充血、黏液、碎屑过多、鳃弓参差不齐或张开(Speare and Ferguson，1989)。与 *F. columnare* 和 *F. psychrophilum* 相反，*F. branchiophilum*引发上皮增生、板层融合(“杵状指”)，严重情况下甚至发生初级鳃片融合[图 16.3(C)](Wood and Yasutake，1957)，导致呼吸表面积显著减少，从而抑制气体和离子交换，以及渗透稳态。Wood 和 Yasutake(1957)发现，BGD 诱导的增生通常始于薄片远尖端的上皮，并最终形成沿着鳃丝不规则散布的“小岛状增生”(多灶性上皮增生)。

16.3 黄杆菌感染的诊断

根据养殖设施、病例史、行为变化、大体病理和组织病理情况等诊断黄杆菌感染，并随后对病原进行分离和表型鉴定。使用分子生物学和血清学分析可以最好地进行确认。但是，不同国家可能遵循类似但不同的诊断准则，特别是对于无须向世界动物卫生组织(OIE)报告的黄杆菌等病原。

16.3.1 设施检查及以往的动物流行病史

如果没有采取适当的生物安全措施，利用含有鱼类的地表水或泉水的水产养殖活动更有可能发生黄杆菌相关疾病的暴发。亲鱼的健康也很重要，因为 *F. psychrophilum* 可以通过配子传播。同样在这种情况下，应该审查有关鱼卵消毒的记录，因为这会影响 *F. psychrophilum* 和 *F. columnare* 的传播。同样，有关养殖设施的历史、水质、孵化场基础设施和休耕过程的信息也很有用。

16.3.2 临床和尸检

如果观察到表明黄杆菌疾病的皮肤或鳃病变时(参见 16.2 节),应对患处的刮屑和整装制片进行光学显微镜检。大量细长杆菌[图 16.1(A)]、杆菌聚集成"柱"或"干草堆"[图 16.2(A)],或与鳃相关的丝状细菌群[图 16.3(A)],分别表明存在 *F. psychrophilum*、*F. columnare* 和 *F. branchiophilum*。整装制片可革兰氏染色以做进一步检查。脾组织的压片制备和细菌检查也可能有助于诊断。

16.3.3 初步分离和推定鉴定

应从外部病灶的前缘采集用于黄杆菌培养的组织样本,以减少次生侵染菌的过度生长。从肾、脾、脑和肝切除的组织,以及任何腹水用于检测全身性感染。虽然 *F. psychrophilum* 以全身性疾病和感染神经组织而闻名(Nematollahi et al., 2003),但 *F. columnare* 也可从脑组织中恢复分离(Loch and Faisal, 2016a)。可以将组织直接接种到培养基上或均质化以增强细菌的恢复(Starliper and Schill, 2011)。对 *F. branchiophilum* 的初步分离及其后续鉴定很少进行,因为该菌的生长非常苛刻(Bernardet and Bowman, 2006),综合设施历史、临床表现和细菌形态[图 16.3(A)],长期以来足以识别各种菌种(Starliper and Schill, 2011)。

革兰氏阴性、无芽孢杆状、产生黄色或橙色菌落(由于产生柔红霉素或类胡萝卜素类型色素)、非运动或滑动运动(Bernardet, 2011a)的细菌被初步确定为黄杆菌科的成员。Bernardet(2011b)综述了黄杆菌科的分离特征。根据作者的经验,金黄色杆菌属(*Chryseobacterium*)是从淡水鱼中分离的黄杆菌科中最常见的"非黄杆菌属",应注意区分这些属。多种表型特征用于区分 *F. psychrophilum*、*F. columnare* 和 *F. branchiophilum* (Bernardet et al., 2002; Bernardet and Bowman, 2011)(表 16.1)。

表 16.1 用于推定区分嗜冷黄杆菌、柱状黄杆菌和嗜鳃黄杆菌的细菌特征[a]

特　征	*F. psychrophilum*	*F. columnare*	*F. branchiophilum*
菌落形态[b]	低凸、圆形、边缘完整;呈"煎蛋"形状	扁平,或具有稍隆起的中心,根状不规则边缘	低凸、圆形,边缘完整
刚果红吸附	—	+	—
柔红霉素类色素	+	+	—
滑行运动	(+)	+	—

续 表

特　征	*F. psychrophilum*	*F. columnare*	*F. branchiophilum*
产生：			
碳水化合物（如葡萄糖、蔗糖、麦芽糖、海藻糖）产酸	—	—	+
β-半乳糖苷酶	—	—	+
硫化氢	—	+	—

[a]数据来源于 Bernardet 和 Bowman (2011 及其引用文献)以及 Loch 和 Faisal 的未发表结果

[b] 形态因培养条件和菌株的不同而不同；"—"表示阴性反应；"+"表示大于 90%的分离株呈阳性反应；"(+)"表示小于等于 90%的分离株呈延迟或弱阳性反应。

16.3.4 血清学检测

已经开发了与全细胞嗜冷黄杆菌抗体反应的基于多克隆抗体的多酶联免疫吸附法(ELISAs)和荧光抗体测试(FATs)(Lindstrom et al., 2009)，然而，与其他黄杆菌的交叉反应限制了它们的使用。

同样，已经开发了基于多克隆血清捕获的 ELISA(Rangdale and Way, 1995; Mata et al., 2002)，但是 *F. psychrophilum* 的血清学异质性使诊断变得复杂化。Lindstrom 等(2009)开发了一种单克隆抗体(Mab FL－43; ImmunoPrecise Antibodies)，对嗜冷黄杆菌外表面的一种蛋白质具有特异性。在 ELISA 设计中，单克隆抗体通过实验检测阈低至 1.6×10^3 CFU/mL。该试验得到了验证(Long et al., 2012)，并证明适用于筛查鲑肾，但不适用于卵巢液。

Mab FL－43 还用于基于过滤的 FAT(Lindstrom et al., 2009)，可检测卵巢液中的 *F. psychrophilum*。免疫组织化学还可用于检测 *F. psychrophilum* 并评估组织定位(Madetoja et al., 2000)。*F. columnare* 的血清学诊断检测方法包括基于单克隆抗体的 IFAT，其可以使鳃表面和胃肠道内的细菌可视化(Speare et al., 1995)。Panangalal 等(2006)开发了一种 IFAT，其敏感性与细菌培养相似，但结果迅速。*F. branchiophilum* 的鳃感染可以使用多克隆抗血清(Bullock et al., 1994)和单克隆抗体(Speare et al., 1995)通过 IFAT 检测。

MacPhee 等(1995)开发了一种 ELISA，可检测大约 500 CFU /mL 的 *F. branchiophilum* 整体制片和大约 1 000 CFU /100 μg 加标鳃组织。与 *F. branchiophilum* 的挑剔性质保持一致，这些研究者仅能够在已知的鳃感染样本中实现 19/54 的细菌分离率，而 ELISA 检测则为 53/54 呈阳性。此外，还有针对 *F. branchiophilum* 的凝集和免疫扩散测定法(Huh and Wakabayashi,

1989；Ko and Heo，1997）。

16.3.5 分子鉴定

PCR 检测，如，常规 PCR、巢式 PCR（nPCR）、多重 PCR、逆转录酶 PCR（RT -PCR）、实时定量 PCR（qPCR），已经开发用于 *F. psychrophilum* 检测，它们可以检测新鲜或冷冻组织、福尔马林固定组织和环境样品中的细菌。它们也可用于确诊。事实上，很多靶向 16S rRNA 和持家基因的常规和巢式 PCR 检测都是这样做的（Starliper and Schill，2011）。简而言之，其中一些 PCR/nPCR 检测法可检测低至 17 CFU/mg 加标虹鳟脑组织（Wiklund et al.，2000），养殖银大麻哈鱼和虹鳟 10 CFU /mg 脾和 5 CFU /mL 卵巢液（Baliarda et al.，2002），110 CFU/mL加标淡水（Wiklund et al.，2000）和 3 CFU/mL 加标非无菌井水（Madetoja and Wiklund，2002）。Suzuki 等（2008）发现，一些针对 *F. psychrophilum* 16S rRNA 基因的 PCR 检测敏感性高于靶向 *gyrA*（DNA 促旋酶 A 亚基）、*gyrB* 和肽基脯氨酰顺反异构酶（ppiC）基因的敏感性，但这增加了假阳性结果。Del Cerro 等（2002a）设计了一种多重 PCR（靶向16S rRNA 基因），用于检测嗜冷黄杆菌、杀鲑气单胞菌和鲁氏耶尔森菌。

Del Cerro 等（2002b）首次报道了使用基于 TaqMan 探针的靶向 *F. psychrophilum* 16S rRNA 基因的 PCR 检测方法。至少有四种其他 qPCR 检测方法被开发出来（Orieux et al.，2011；Marancik and Wiens，2013；Strepparava et al.，2014；Long et al.，2015）。Marancik 和 Wiens（2013）描述了针对 16S rRNA 基因的 qPCR 检测的一些缺陷，如存在 6 个 16S rRNA 基因拷贝（Duchaud et al.，2007），它们可以在序列上变化并且非特异性地扩增其他黄杆菌。因此，他们设计了自己的 qPCR 检测方法，以靶向单拷贝基因（RFPS00910），每个反应可检测 3.1 个基因组当量。即便如此，亚临床感染的鱼有时培养呈阳性，但 qPCR 检测阴性。其他用于 *F. psychrophilum* 的分子检测方法包括原位杂交（Vatsos et al.，2002）和荧光原位杂交，可检测组织和水中的细菌（Strepparava et al.，2012）。Fujiwara-Nagata 和 Eguchi(2009)开发了定量环介导等温扩增（LAMP）测定法，每次反应可在 70 min 内检测到大于等于 20 CFU当量的 *F. psychrophilum*。免疫磁分离结合 PCR 或流式细胞仪也已被用于检测 *F. psychrophilum*，并且可以定量低至 0.4 CFU /mL（Ryumae et al.，2012；Hibi et al.，2012）。

可以使用多种 PCR 检测方法来检测和鉴定 *F. columnare*（Triyanto et al.，1999；Darwish et al.，2004；Welker et al.，2005）。Panangala 等（2007）开发了一种针对软骨素 AC 裂解酶基因的 qPCR，其在纯培养物中检测到大于等于 3 CFU 的 *F. columnare*，在斑点叉尾鮰中的检测限达 3.4 CFU /mL。Yeh 等（2006）描述了一种检测斑点叉尾鮰中细菌的 *F. columnare* 特异性 LAMP 方

法。Suebsing 等(2015)开发了一种用于养殖尼罗罗非鱼的基于钙黄绿素的LAMP方法，可检测性腺、鳃、血液样品中的细菌，其灵敏度比常规PCR(检测限为 2.2×10^{2} CFU/反应)高约10倍。原位杂交检测并可视化受感染鱼中的 *F. columnare* (Tripathi et al., 2005)。*F. branchiophilum* 也可以通过常规PCR(Toyama et al., 1996)和新一代测序技术(Good et al., 2015)检测。Lievens 等(2011)开发的基于DNA阵列的多重PCR可用于检测 *F. psychrophilum*、*F. columnare* 和 *F. branchiophilum*。

16.4 病理生理学

黄杆菌病的病理生理学效应是复杂的。皮肤或鳃上皮和覆盖的黏液是抵御病原的重要屏障，对渗透调节非常重要。*F. psychrophilum* 和 *F. columnare* 对这些组织的损伤破坏了宿主的体内平衡。CD通常伴有"涝渍"症状，这是一种皮肤、肌肉组织水肿的术语(Declercq et al., 2013)。同样地，BGD诱导的鳃上皮增生损害离子交换并干扰渗透稳态(Speare and Ferguson, 1989; Byrne et al., 1995)。在 *F. psychrophilum* 和 *F. columnare* 系统性感染期间，宿主可能通过血管舒张以增加感染部位的血液供应，并增加血管通透性，从而使白细胞、抗体、补体和促炎性细胞因子可以控制感染。宿主对黄杆菌感染的免疫反应是实质性的(Starliper and Schill, 2011; Declercq et al., 2013)，但这种反应可能损害幸存者的生长、性成熟和繁殖。同样地，从 *F. psychrophilum* 系统性感染中恢复的养殖鱼类常常伴有眼部和脊柱异常(LaFrentz and Cain, 2004)。

Marancik 等(2014)用 *F. psychrophilum* 感染虹鳟，结果显示，它们的血细胞比容(Hct)、总蛋白、白蛋白、葡萄糖、胆固醇、氯化物和钙浓度发生改变，导致了与BCWD相关的变化。他们还指出，Hct、总蛋白、胆固醇和钙水平与 *F. psychrophilum* 的脾负荷呈负相关，并且最早观察到的病理生理变化是总蛋白和白蛋白下降，以及电解质失衡。此外，在 *F. psychrophilum* 感染的香鱼中也报道了Hct的减少(Miwa and Nakayasu, 2005)。

Tripathi 等(2005)发现，感染 *F. columnare* 的锦鲤的Hct、红细胞计数、血红蛋白浓度、平均红细胞体积和绝对淋巴细胞计数显著下降，从而导致非再生性贫血(小细胞和正色素性)和白细胞减少症。然而，在有些个体中出现红细胞数量再生。患有大面积皮肤溃疡的鱼类出现严重的血液学变化，血清中出现明显的低钠血症和低氯血症，由于鳃的被动灌注及溃疡造成皮肤屏障的破坏而导致钠离子和氯离子损失。

为了研究感染 *F. branchiophilum* 的虹鳟的血液学变化和酸碱平衡，Byrne 等(1995)用背主动脉导管对鱼进行插管，缝合鼻胃管并在树脂玻璃盒内维持个体。感染细菌后，一些鱼被喂食而另一些则没有。在24 h内，与相似但未饲喂的对照组相比，感染 *F. branchiophilum* 的喂食鱼表现出严重的低氧血症、高碳酸血症、低渗性、低钠血症、低氯血症、血氨升高、鳃呼吸过速和更频繁的“咳嗽”。有趣的是，感染 *F. branchiophilum* 的未饲喂鱼显示出类似的临床症状和血气改变，尽管程度较轻，但没有检测到电解质失衡。研究人员得出结论，*F. branchiophilum*诱导的低氧血症由于摄食行为增加需氧量而恶化。

16.5 防治策略

通过减少应激、降低养殖密度、保持卫生和水质及充足的水流量来优化养殖条件，有助于降低黄杆菌引起的死亡率。例如，亚硝酸盐含量升高和有机物含量增加会增强 *F. columnare* 和 *F. psychrophilum* 对鳃的黏附(Decostere et al.，1999；Nematollahi et al.，2003)，而水质不佳影响虹鳟感染 *F. psychrophilum* 的累计死亡率(Garcia et al.，2000)。低溶解氧和氨氮升高促使鱼更易患细菌性鳃病(Bullock，1990)。相反地，升高的总氨氮干扰了斑点叉尾鮰的 *F. columnare*感染，实际上提高了存活率(Farmer et al.，2011)。不良的水交换是 BGD 暴发的一个风险因素(Good et al.，2009,2010)。在水再利用系统中，对黄杆菌具有较强抗性的养殖品种应置于更易感鱼类的下方。

使用无鱼水源(深井)可减少进入孵化场的黄杆菌数量(Madsen and Dalsgaard，2008)。紫外线(UV)处理进水也可减少或消除黄杆菌的流入。Hedrick 等(2000)发现126 mJ/cm^2 的紫外线照射剂量可杀灭*F. psychrophilum*，而许多其他革兰氏阴性细菌在紫外线照射剂量小于等于42 mJ/cm^{2}时即被杀灭。养殖单元使用单独的工具以及对共用设备消毒可最大限度地减少传染，这一点很重要，因为黄杆菌可以在干燥条件下长期存活(例如塑料带上的 *F. psychrophilum*可存活超过 96 h；Oplinger and Wagner，2010)。

Oplinger 和 Wagner (2010)的研究表明，苯扎氯铵(600 mg/L 溶液)浸泡 10 min后可根除塑料上的 *F. psychrophilum*，他们将含有 *F. psychrophilum* 的条带浸入苯扎氯铵中 10 s 后干燥 1 h，获得了相同的结果。Oplinger 和 Wagner (2013)还发现，*F. psychrophilum* 在暴露于大于等于 55 ℃的温度下被杀灭，但在改变 pH(3.0～11.0，保持 15 min)和渗透压(0～12%的 NaCl，保持 15 min)后仍能存活。Mainous 等(2012)发现，Virkon® Aquatic (0.1%；Western Chemical)、Halamid®(15 ppm；Western Chemical)、戊二醛(2%)、碘，Lysol®(1%)、Roccal®(1：256；Zoetis Services)、Clorox®(50 ppm)，以及 50%和 70%的乙醇可在接触后 1 min 内杀灭 *F. columnare* 和 *F. psychrophilum*，但福尔马林则需要 10～20 min，具体取决于浓度。有趣的是，高锰酸钾在 1 min 内可消除 *F. columnare*，但即使在 1 h 后也没有显著减少 *F. psychrophilum*。然而，Mainous 等(2012)没有研究有机物如何影响消毒效果，也没有研究生物被膜如

何影响生存，这是一个重要问题，因为 *F. psychrophilum* 的生物被膜形成能力在暴露于化疗药物时可增强其存活率（Sundell and Wiklund，2011）。美国国家环境保护局（EPA）仅许可了 Virkon® Aquatic 用于水产养殖设施，而根据美国食品药品监督管理局（FDA）规定，其他化学品（如碘）的监管优先级较低。

F. psychrophilum 和在较小程度上的 *F. columnare* 对配子和生殖液的亲和力（Brown et al.，1997；Loch and Faisal，2016a），以及 *F. psychrophilum* 已通过鱼卵运输传播的证据（Avendaño-Herrera et al.，2014），说明了对鱼卵消毒的重要性。聚维酮碘（50 mg/L 活性 I_2，30 min 或 100 mg/L 活性 I_2，10 min）最常用于鱼卵的消毒，过氧化氢（100 mg/L，10 min）和戊二醛（200 mg/L，20 min）也很有效（Cipriano and Holt，2005）。尽管碘表面消毒可以减少与鱼卵相关的病原传播，但它并不总是能消除黄杆菌（Holt et al.，1993）。事实上，Kumagai 等（1998）的研究表明，用 1 000 ppm 聚维酮碘处理 15 min 和 200 ppm 处理 2 h 后，可从鱼卵中恢复 *F. psychrophilum*。Loch 和 Faisal（未发表数据）使用50 mg/L的聚维酮碘消毒 30 min，然后用 100 mg/L 消毒 10 min，有时也能从这些消毒的鱼卵中分离出 *F. columnare*，他们认为大量的细菌可能会影响消毒效果。

营养是决定黄杆菌病风险和严重程度的一个重要因素。例如，与饲喂对照组相比，禁食 7～10 d 的 *F. columnare* 攻毒的斑点叉尾鮰死亡率显著升高，抗体反应降低（Klesius et al.，1999）。Beck 等（2012）发现，禁食 7 d 的斑点叉尾鮰，其鼠李糖结合凝集素（RBL）显著上调，他们之前已经确定这在易感 CD 的斑点叉尾鮰家系中也显著上调，但在完全抗性的家系中并未上调。有趣的是，当斑点叉尾鮰暴露于 RBL 推定的配体（如 L-鼠李糖和 D-半乳糖）时，它们受到保护免受 *F. columnare* 的攻毒感染，这进一步证明了 RBL 在 CD 易感性中的重要性。Liu 等（2013）的研究发现，*F. columnare* 攻毒的斑点叉尾鮰短期禁食导致免疫适应性显著降低，并标志着参与能量代谢和细胞周期/增殖的相关基因失调。饥饿的鱼下调了 iNOS2b（诱导型一氧化氮合酶 2b）、肽聚糖识别蛋白 6 和溶菌酶-C，它们是先天免疫的关键组成部分。矛盾的是，过度喂食是 BGD 暴发的一个风险因素（Good et al.，2009，2010），这可能与饲料消耗伴随的需氧量增加有关（Byrne et al.，1995）。

用于预防鱼类黄杆菌病的免疫增强剂产生了复杂的效果。Suomalainen 等（2009）评估了在饲料中添加 Ergosan®（Schering-Plough Aquaculture）和 Alkosel®（Lallemand Animal Nutrition）的保护作用，它们分别是海藻酸提取物和含有生物可利用的 L（+）-硒代蛋氨酸的灭活酵母（*Saccharomyces cerevisiae*）。用 Alkosel® 饲喂 7 d 的虹鳟（1.2 g），对于 *F. columnare* 的攻毒，其死亡率显著低于饲喂 Ergosan® 或标准饲料的感染鱼。然而，在喂食免疫增强剂 14 d 的 5 g 鱼中，死亡率并没有显著差异。Kunttu 等（2009）的研究发现，用酵母 β-葡聚糖和 β-羟基-β-甲基丁酸处理并不能保护虹鳟早期生命阶段免受柱形病的侵害。黑种草(*Nigella sativa*)油在体外可强烈抑制 *F. colunmare* 的生长,并保护喂食了该植物油或种子的斑点叉尾鮰和斑马鱼免受随后的细菌攻毒感染(Mohammed and Arias, 2016)。饲料中添加腐殖质提取物可降低香鱼的细菌性冷水病死亡率和皮肤损伤发展(Nakagawa et al., 2009)。将驼背大麻哈鱼(*O. gorbuscha*)睾丸干粉掺入养殖虹鳟鱼苗的饲料中显示出作为免疫增强剂的潜力,添加 β-葡聚糖饮食也是如此(Macrogard®; Biorigin),但 Fehringer 等(2014)的研究发现,长期向养殖鱼类提供睾丸粉或葡聚糖基的膳食可能会对抗 *F. psychrophilum* 特异性抗体的发展产生负面影响。

一些假单胞菌在体外条件下抑制黄杆菌的生长(Strom-Bestor and Wiklund, 2011; Fuente Mde et al., 2015),并为鱼类提供保护。Korkea-Aho 等(2011)对先前喂食 *Pseudomonas* M174 的虹鳟进行 *F. psychrophilum* 攻毒感染,与对照组相比,这导致了呼吸爆发活性增强和死亡率的降低。尽管另一种来自鳟鱼卵的假单胞菌的机制与 *Pseudomonas* M174 不同,但其可保护鱼类免受 RTFS 侵害(Korkea-Aho et al., 2012)。Burbank 等(2012)从虹鳟肠道分离菌中筛选具有 *F. psychrophilum* 体外拮抗作用的细菌,并最终筛选出 16 种益生菌候选株,分别属于柠檬酸杆菌属(*Citrobacter*)、赖氨酸芽孢杆菌属(*Lysinibacillus*)、葡萄球菌属(*Staphylococcus*)、气球菌属(*Aerococcus*)、肠杆菌属(*Enterobacter*)和气单胞菌属(*Aeromonas*)。LaPatra 等(2014)和 Ghosh 等(2016)将其中的一种菌株(*Enterobacter* C6-6)腹腔注射到虹鳟中,提高了它们对 BCWD 的存活率。Boutin 等(2012)从美洲红点鲑皮肤或黏液中分离了多种益生菌菌株,并创建了一种菌株"鸡尾酒",这些菌株在体外可抑制 *F. columnare* 并保护美洲红点鲑幼鱼抵御柱形病。Boutin 等(2013)使用一种红球菌(*Rhodococcus* sp.)在没有破坏美洲红点鲑天然微生物群系(菌群失调)的情况下获得了益生菌效应,它们没有在鱼体中定植,而是在水族箱中形成生物被膜,从而降低了养殖在水族箱中 *F. psychrophilum* 感染鱼的死亡率。Strom-Bestor 和 Wiklund (2011)的研究发现,益生假单胞菌在体外的抑菌程度随 *F. psychrophilum* 血清型的变化而变化。

Greg Wiens 博士的实验室(国家冷水和冷水水产养殖中心,美国农业部农业研究所,利敦,西弗吉尼亚州)培育了一种对 BCWD 具有明显抗性而对生长率没有任何影响的虹鳟家系(Silverstein et al., 2009; Leeds et al., 2010; Wiens et al., 2013)。这些研究已经开始揭示这种抗性背后的机制(Marancik et al., 2014,2015; Wiens et al., 2015),并且还鉴定了可以定位候选抗性基因的单核苷酸多态性(single nucleotide polymorphisms, SNPs)(Liu et al., 2015)。在斑点叉尾鮰(Beck et al., 2012; LaFrentz et al., 2012b)、杂交鲶(Arias et al., 2012)和草鱼(*Ctenopharyngodon idellus*)(Yu et al., 2013)中也报道了对柱形病的差异遗传抗性。Evenhuis 等(2015)发现 BCWD 抗性虹鳟家系对 BCWD 的

抗性与 CD 抗性相关，表明在选育过程中可以获得对多种疾病的抗性。

利用噬菌体的裂解活性来控制黄杆菌病也很有前景。Prasad 等(2011)筛选出可裂解许多 *F. columnare* 分离株的特异性噬菌体。他们感染了胡鲇(*C. batrachus*)，随后通过多种途径（注射、浸泡和口服）将它们暴露于 *F. columnare* 特异性噬菌体，结果表明，经噬菌体处理的鱼存活率为 100%，并且无法恢复 *F. columnare* 培养。Castillo 等(2012)获得了类似的结果，通过噬菌体处理减少了实验诱导的 BCWD 死亡率，然而，对噬菌体的抗性可能会发展(Castillo et al., 2014)。

已经证实黄杆菌有效疫苗的开发存在问题。在实验室条件下使用福尔马林灭活全细菌来预防 BCWD/RTFS 的早期尝试是有希望的(Bernardet, 1997)，但需要对这些进行田间评价(Nematollahi et al., 2003)。此后，人们多次尝试使用被动免疫或开发各种灭活疫苗、亚单位疫苗、野生型疫苗和减毒活疫苗，但这些疫苗都没有在美国商业化(Gómez et al., 2014)。LaFrentz 等(2008)开发了一种减毒活疫苗(CSF259－93B.17)，在实验攻毒的接种疫苗组产生 45%的相对保护率。这些研究者后来的一份报告表明(LaFrentz et al., 2009)，在限铁条件下，几种免疫反应性 *F. psychrophilum* 蛋白过表达，这导致 Long 等(2013)在类似条件下培养 CSF259－93B.17。Long 等发现浸泡接种的银大麻哈鱼的死亡率明显低于模拟免疫的个体，而模拟接种的鱼与富铁条件下生长的 CSF259－93B.17免疫鱼之间的死亡率差异不显著。几个州孵化场进行了田间试验(爱达荷大学鱼类和野生动物科学系 Kenneth Cain 博士，个人交流)。Solís 等(2015)报道称，一种针对 *F. psychrophilum* 的福尔马林灭活疫苗已在智利上市销售(Flavomune®，Veterquímica)。

尽管 *F. columnare* 的疫苗接种取得了相对有希望的结果(Declercq et al., 2013)，但直到最近才有疫苗被批准用于养殖鱼类。一种 *F. columnare* 减毒活疫苗(AQUAVAC－COL™；Merck Animal Health)在美国注册，可用于斑点叉尾鮰和大口黑鲈(Shoemaker et al., 2011)。该疫苗浸泡接种斑点叉尾鮰鱼苗对 *F. columnare* 攻毒可产生 57%～94%的免疫保护率(RPS) (Shoemaker et al., 2011)。相比之下，Kirkland (2010)发现，在田间条件下，与未接种疫苗的群体相比，接种疫苗的杂交鲇的存活率没有显著提高。其他研究人员也质疑 AQUAVAC－COL™在实验室外的疗效，并认为其在这种情况下的疗效降低可能与 *F. columnare* 的种内异质性有关(Mohammed et al., 2013)。然而，在田间试验中，免疫接种 AQUAVAC－COL™的大口黑鲈鱼苗对自然发生的柱形病有着显著保护作用(Bebak et al., 2009)。另一种 *F. columnare* 浸泡型疫苗(FryVacc1; Novartis Animal Health)也已在美国上市并被许可用于鲑科鱼类(≥3 g)(Bowker et al., 2015)。

对于黄杆菌病有许多治疗选择(Bernardet and Bowman, 2006; Austin and

Austin, 2007; Starliper and Schill, 2011; Declercq et al., 2013),但化学品的使用取决于养殖地区的监管法规。例如,Aquaflor®[氟苯尼考;Merck Animal Health;(10~15 mg氟苯尼考/kg 鱼)/d;10 d]和 Terramycin®[土霉素二水合物;(3.75 g/100 磅[①]鱼)/d;10 d]在美国被批准用于治疗淡水养殖鲑科鱼类的 *F. psychrophilum* 感染(Bowker et al., 2015)。过氧化氢(对于幼鱼和成鱼,50~75 mg/L,60 min,每日一次,共三次;对于鱼苗,50 mg/L,60 min,每日一次,共三次)被批准用于治疗淡水养殖冷水鱼类和斑点叉尾鮰的体外 *F. columnare* 感染。氯胺-T 被批准用于治疗大眼狮鲈(10~20 mg/L,60 min,每日一次,共三次)和淡水养殖温水鱼类(20 mg/L,60 min,每日一次,共三次)的体外 *F. columnare* 感染。Aquaflor®[10~15 mg/(kg 鱼·d),10 d]被批准用于治疗淡水养殖鱼类中的全身和体外 *F. columnare* 感染,而 Terramycin®被批准用于淡水养殖虹鳟,剂量如上(Bowker et al., 2015)。过氧化氢(100 mg/L,30 min或50~100 mg/L,60 min,每日一次,共三次)和氯胺-T(按上述剂量)被批准用于治疗淡水养殖鲑科鱼类的 *F. branchiophilum* 感染(Bowker et al., 2015)。5%的 NaCl 浸泡 2 min 可有效防治虹鳟幼鱼中的 *F. branchiophilum* 感染(Kudo and Kimura, 1983)。此外,青霉素和链霉素可减少精液中的 *F. psychrophilum* 传播(Oplinger and Wagner, 2015),并且在水硬化过程中,细菌在由此产生的受精卵中的流行率显著降低(Oplinger et al., 2015)。

已有报道,*F. psychrophilum* 中的抗生素耐药性增加(Smith et al., 2016; Van Vliet et al.,未发表的数据),并且 *F. columnare* 中也存在耐药性(Declercq et al., 2013)。对于经常用于治疗 BCWD/RTFS 的土霉素广泛获得耐药性的发展尤其令人担忧(Smith et al., 2016),智利水产养殖业中氟苯尼考最低抑菌浓度不断增加的报道同样令人关切(Henríquez-Núñez et al., 2012)。因此,从正在调查的疾病事件中恢复和鉴定病原体,并在治疗前进行抗菌药敏试验,对于减少抗生素耐药性的进一步发展至关重要。

16.6 总结与研究展望

嗜冷黄杆菌、柱状黄杆菌和嗜鳃黄杆菌制约了世界范围内的水产养殖生产力,并造成巨大的经济损失。几十年的研究已经开发出灵敏和特异性的诊断方法来检测和识别这些病原,并帮助了解其病理生物学。尽管存在多种预防和控制这些病原的策略,但仍有许多未知之处。

例如,我们需要更好地了解这些细菌的全面的血清学和遗传群体结构,以确保未来的诊断和基于免疫学的防控令人满意。同样,进一步阐明黄杆菌动物流

① 1磅=0.453 6千克。

行病学将有助于改进和靶向防控方法的开发，同时也有助于进一步开发替代控制策略（如免疫增强剂、营养强化、益生菌、噬菌体疗法）的持续研究。为了干扰细菌黏附和感染（可能通过配体干预、益生元和益生菌的使用），以及减轻对宿主组织的损害（可能通过营养强化），需要填补黄杆菌发病机制及其伴随的宿主病理生理变化的知识空白。

不可低估宿主微生物组在预防或助长黄杆菌病方面的重要性。事实上，我们怀疑宿主微生物组的可变性在不同鱼种和实验室之间黄杆菌实验攻毒模型的重现性困难中发挥了重要作用，这可能是通过不同细菌拮抗和协同作用实现的。因此，应用新一代测序技术研究鱼类微生物组构成与致病性黄杆菌的互作关系，以及在不同环境中一些微生物在加剧或预防黄杆菌感染的作用，无疑将使这一假设更加清晰。

未来研究的另一个重要领域是生物被膜在黄杆菌毒力、水产养殖设施中的持久性和化学疗法功效中所起的作用。此外，还迫切需要阐明未被鉴定的黄杆菌在疾病暴发中的作用，特别是因为有些菌类似熟知的黄杆菌疾病症状。Snieszko (1981)指出："似乎很明显细菌性鳃病是一种复杂的疾病，可能涉及一种以上的细菌类型。"这似乎是一个包括黄杆菌病在内的适合多种鱼类疾病的范例，这是由宿主、环境、专性或兼性病原体和共栖或共生微生物之间的复杂相互作用所致。

参考文献

[1] Alvarez, B., Secades, P., Prieto, M., Mcbride, M.J. and Guijarro, J.A. (2006) A mutation in *Flavobacterium psychrophilum tlpB* inhibits gliding motility and induces biofilm formation. *Applied and Environmental Microbiology* 72, 4044 - 4053.

[2] Amita, K., Hoshino, M., Honma, T. and Wakabayashi, H. (2000) An investigation on the distribution of *Flavobacterium psychrophilum* in the Umikawa River. *Fish Pathology* 35, 193 - 197.

[3] Anacker, R.L. and Ordal, E.J. (1959) Studies on the myxobacterium *Chondrococcus columnaris*: I. Serological typing. *Journal of Bacteriology* 78, 25 - 32.

[4] Arai, H., Morita, Y., Izumi, S., Katagiri, T. and Kimura, H. (2007) Molecular typing by pulsed-field gel electrophoresis of *Flavobacterium psychrophilum* isolates derived from Japanese fish. *Journal of Fish Diseases* 30, 345 - 355.

[5] Arias, C.R., Cai, W., Peatman, E. and Bullard, S.A. (2012) Catfish hybrid Ictalurus punctatus. I. furcatus exhibits higher resistance to Columnaris disease than the parental species. *Diseases of Aquatic Organisms* 100, 77 - 81.

[6] Austin, B. and Austin, D.A. (2007) *Bacterial Fish Pathogens: Diseases of Farmed and Wild Fish*, 4th edn. Springer/Praxis Publishing, Chichester, UK.

[7] Avendaño-Herrera, R., Gherardelli, V., Olmos, P., Godoy, M., Heisinger, A. *et al.* (2011) *Flavobacterium columnare* associated with mortality of salmonids farmed in Chile: a case report of two outbreaks. *Bulletin of the European Association of Fish Pathologists* 31, 36 - 44.

[8] Avendaño-Herrera, R., Houel, A., Irgang, R., Bernardet, J.-F., Godoy, M. et al. (2014) Introduction, expansion and coexistence of epidemic *Flavobacterium*

psychrophilum lineages in Chilean fish farms. *Veterinary Microbiology* 170, 298 - 306.
[9] Baliarda, A., Faure, D. and Urdaci, M.C. (2002) Development and application of a nested PCR to monitor brood stock salmonid ovarian fluid and spleen for detection of the fish pathogen *Flavobacterium psychrophilum*. *Journal of Applied Microbiology* 92, 510 - 516.
[10] Bebak, J., Baumgarten, M. and Smith, G. (1997) Risk factors for bacterial gill disease in young rainbow trout (*Oncorhynchus mykiss*) in North America. *Preventive Veterinary Medicine* 32, 23 - 34.
[11] Bebak, J., Matthews, M. and Shoemaker, C. (2009) Survival of vaccinated, feed-trained largemouth bass fry (*Micropterus salmoides floridanus*) during natural exposure to *Flavobacterium columnare*. *Vaccine* 27, 4297 - 4301.
[12] Beck, B.H., Farmer, B.D., Straus, D.L., Li, C. and Peatman, E. (2012) Putative roles for a rhamnose binding lectin in *Flavobacterium columnare* pathogenesis in channel catfish *Ictalurus punctatus*. *Fish and Shellfish Immunology* 33, 1008 - 1015.
[13] Bernardet, J. F. (1997) Immunization with bacterial antigens: *Flavobacterium* and *Flexibacter* infections *Developments in Biological Standardization* 90, 179 - 188.
[14] Bernardet, J.-F. (2011a) Class II. Flavobacteriia class. nov. In: Krieg, N., Staley, J., Brown, D., Hedlund, B., Paster, B., Ward, N., Ludwig, W. and Whitman, W. (eds) *Bergey's Manual of Systematic Bacteriology. Volume 4: The Bacteroidetes, Spirochaetes, Tenericutes (Mollicutes), Acidobacteria, Fibrobacteres*, 2nd edn. Springer, New York /Dordrecht, The Netherlands /Heidelberg, Germany /London, p. 105.
[15] Bernardet, J.-F. (2011b) Family I Flavobacteriaceae Reichenbach 1992b, 327VP. In: Krieg, N., Staley, J., Brown, D., Hedlund, B., Paster, B., Ward, N., Ludwig, W. and Whitman, W. (eds) *Bergey's Manual of Systematic Bacteriology. Volume 4: The Bacteroidetes, Spirochaetes, Tenericutes (Mollicutes), Acidobacteria, Fibrobacteres*, 2nd edn. Springer, New York /Dordrecht, The Netherlands /Heidelberg, Germany / London, pp. 106 - 111.
[16] Bernardet, J.-F. and Bowman, J.P. (2006) The genus Flavobacterium. In: Dworkin, M., Falkow, S., Rosenberg, E., Schleifer, K. and Stackebrandt, E. (eds) *The Prokaryotes: A Handbook on the Biology of Bacteria. Volume 7: Proteobacteria: Delta and Epsilon Subclasses. Deeply Rooting Bacteria*, 3rd edn. Springer-Verlag, New York, pp. 481 - 531.
[17] Bernardet, J. and Bowman, J. (2011) Genus 1. Flavobacterium. In: Krieg, N., Staley, J., Brown, D., Hedlund, B., Paster, B., Ward, N., Ludwig, W. and Whitman, W. (eds) *Bergey's Manual of Systematic Bacteriology. Volume 4: The Bacteroidetes, Spirochaetes, Tenericutes (Mollicutes), Acidobacteria, Fibrobacteres*, 2nd edn. Springer, New York/Dordrecht, The Netherlands/Heidelberg, Germany/London, pp. 112 - 154.
[18] Bernardet, J.-F. and Grimont, P. A. (1989) Deoxyribonucleic acid relatedness and phenotypic characterization of *Flexibacter columnaris* sp. nov., nom. rev., *Flexibacter psychrophilus* sp. nov., nom. rev., and *Flexibacter maritimus* Wakabayashi, Hikida, and Masumura 1986. *International Journal of Systematic Bacteriology* 39, 346 - 354.
[19] Bernardet, J. and Kerouault, B. (1989) Phenotypic and genomic studies of *Cytophaga psychrophila* isolated from diseased rainbow trout (*Oncorhynchus mykiss*) in France. *Applied and Environmental Microbiology* 55, 1796 - 1800.
[20] Bernardet, J.-F. and Nakagawa, Y. (2006) An introduction to the family *Flavobacteriaceae*. In: Dworkin, M., Falkow, S., Rosenberg, E., Schleifer, K. and Stackebrandt, E. (eds) *The Prokaryotes: A Handbook on the Biology of Bacteria. Volume 7: Proteobacteria: Delta and Epsilon Subclasses. Deeply Rooting Bacteria*, 3rd

edn. Springer-Verlag, New York, pp. 455 – 480.

[21] Bernardet, J., Baudin-Laurencin, F. and Tixerant, G. (1988) First identification of *Cytophaga psychrophila* in France. *Bulletin of the European Association of Fish Pathologists* 8, 104 – 105.

[22] Bernardet, J.-F., Segers, P., Vancanneyt, M., Berthe, F., Kersters, K. *et al*. (1996) Cutting a Gordian knot: emended classification and description of the genus *Flavobacterium*, emended description of the family *Flavobacteriaceae*, and proposal of *Flavobacterium hydatis* nom. nov. (basonym, *Cytophaga aquatilis* Strohl and Tait 1978). *International Journal of Systematic Bacteriology* 46, 128 – 148.

[23] Bernardet, J.-F., Nakagawa, Y. and Holmes, B. (2002) Proposed minimal standards for describing new taxa of the family *Flavobacteriaceae* and emended description of the family. *International Journal of Systematic and Evolutionary Microbiology* 52, 1049 – 1070.

[24] Borg, A.F. (1948) *Studies on Myxobacteria Associated with Diseases in Salmonid Fishes*. University of Washington, Seattle, Washington.

[25] Boutin, S., Bernatchez, L., Audet, C. and Derome, N. (2012) Antagonistic effect of indigenous skin bacteria of brook charr (*Salvelinus fontinalis*) against *Flavobacterium columnare* and *F. psychrophilum*. *Veterinary Microbiology* 155, 355 – 361.

[26] Boutin, S., Audet, C. and Derome, N. (2013) Probiotic treatment by indigenous bacteria decreases mortality without disturbing the natural microbiota of *Salvelinus fontinalis*. *Canadian Journal of Microbiology* 59, 662 – 670.

[27] Bowker, J., Trushenski, J., Gaikowski, M. and Straus, D. (2015) *Guide to Using Drugs, Biologics, and Other Chemicals in Aquaculture*. Fish Culture Section, American Fisheries Society. Revised July 2016 version available at: https://drive.google.com/file/d/0B43dblZIJqD3MTZGNHBzR21 mS2c/view (accessed 11 November 2016).

[28] Brown, L., Cox, W. and Levine, R. (1997) Evidence that the causal agent of bacterial cold-water disease *Flavobacterium psychrophilum* is transmitted within salmonid eggs. *Diseases of Aquatic Organisms* 29, 213 – 218.

[29] Bruno, D. (1992) *Cytophaga psychrophila* (='*Flexibacter psychrophilus*') (Borg), histopathology associated with mortalities among farmed rainbow trout, *Oncorhynchus mykiss* (Walbaum) in the UK. *Bulletin of the European Association of Fish Pathologists* 12, 215 – 216.

[30] Bullock, G. (1990) Bacterial Gill Disease of Freshwater Fishes. Fish Disease Leaflet No. 84, US Fish and Wildlife Service, US Department of the Interior, Washington, DC. Available at: http://citeseerx. ist. psu. edu /viewdoc /download? doi = 10. 1. 1. 650. 656&rep=rep1&type=pdf (accessed 11 November 2016).

[31] Bullock, G., Herman, R., Heinen, J., Noble, A., Weber, A. *et al*. (1994) Observations on the occurrence of bacterial gill disease and amoeba gill infestation in rainbow trout cultured in a water recirculation system. *Journal of Aquatic Animal Health* 6, 310 – 317.

[32] Burbank, D., Lapatra, S., Fornshell, G. and Cain, K. (2012) Isolation of bacterial probiotic candidates from the gastrointestinal tract of rainbow trout, *Oncorhynchus mykiss* (Walbaum), and screening for inhibitory activity against *Flavobacterium psychrophilum*. *Journal of Fish Diseases* 35, 809 – 816.

[33] Byrne, P.J., Ostland, V.E., Lumsden, J.S., Macphee, D.D. and Ferguson, H.W. (1995) Blood chemistry and acidbase balance in rainbow trout *Oncorhynchus mykiss* with experimentally-induced acute bacterial gill disease. *Fish Physiology and Biochemistry* 14, 509 – 518.

[34] Cai, W., De La Fuente, L. and Arias, C.R. (2013) Biofilm formation by the fish

pathogen *Flavobacterium columnare*: development and parameters affecting surface attachment. *Applied and Environmental Microbiology* 79, 5633 - 5642.

[35] Castillo, D., Higuera, G., Villa, M., Middelboe, M., Dalsgaard, I. *et al*. (2012) Diversity of *Flavobacterium psychrophilum* and the potential use of its phages for protection against bacterial cold water disease in salmonids. *Journal of Fish Diseases* 35, 193 - 201.

[36] Castillo, D., Christiansen, R.H., Espejo, R. and Middelboe, M. (2014) Diversity and geographical distribution of *Flavobacterium psychrophilum* isolates and their phages: patterns of susceptibility to phage infection and phage host range. *Microbial Ecology* 67, 748 - 757.

[37] Chakroun, C., Urdaci, M.C., Faure, D., Grimont, F. and Bernardet, J.-F. (1997) Random amplified polymorphic DNA analysis provides rapid differentiation among isolates of the fish pathogen *Flavobacterium psychrophilum* and among *Flavobacterium* species. *Diseases of Aquatic Organisms* 31, 187 - 196.

[38] Cipriano, R.C. (2015) Bacterial analysis of fertilized eggs of Atlantic salmon from the Penobscot, Naraguagus, and Machias rivers, Maine. *Journal of Aquatic Animal Health* 27, 172 - 177.

[39] Cipriano, R. C. and Holt, R. A. (2005) *Flavobacterium Psychrophilum, Cause of Bacterial Cold-water Disease and Rainbow Trout Fry Syndrome*. Fish Disease Leaflet No. 86, National Fish Health Research Laboratory, US Geological Survey, US Department of the Interior, Kearneysville, West Virginia. Available at: https://articles.extension.org/sites/default/files/w/8/80/USFWS_Coldwater_disease.pdf (accessed 11 November 2016).

[40] Daoust, P.Y. and Ferguson, H. (1983) Gill diseases of cultured salmonids in Ontario. *Canadian Journal of Comparative Medicine* 47, 358 - 362.

[41] Darwish, A.M., Ismaiel, A.A., Newton, J.C. and Tang, J. (2004) Identification of *Flavobacterium columnare* by a species-specific polymerase chain reaction and renaming of ATCC43622 strain to *Flavobacterium johnsoniae*. *Molecular and Cellular Probes* 18, 421 - 427.

[42] Davis, H.S. (1922) A new bacterial disease of fresh-water fishes. *Bulletin of the United States Bureau of Fisheries* 38, 261 - 280.

[43] Davis, H.[S.] (1926) A new gill disease of trout. *Transactions of the American Fisheries Society* 56, 156 - 160.

[44] Davis, H. S. (1946) *Care and Diseases of Trout*. U. S. Fish and Wildlife Service, Washington, D.C.

[45] Declercq, A.M., Haesebrouck, F., Van den Broeck, W., Bossier, P. and Decostere, A. (2013) Columnaris disease in fish: a review with emphasis on bacterium - host interactions. *Veterinary Research* 44: 27.

[46] Declercq, A.M., Chiers, K., Haesebrouck, F., Van Den Broeck, W., Dewulf, J. *et al*. (2015) Gill infection model for columnaris disease in common carp and rainbow trout. *Journal of Aquatic Animal Health* 27, 1 - 11.

[47] Decostere, A., Haesebrouck, F. and Devriese, L. (1998) Characterization of four *Flavobacterium columnare* (*Flexibacter columnaris*) strains isolated from tropical fish. *Veterinary Microbiology* 62, 35 - 45.

[48] Decostere, A., Haesebrouck, F., Turnbull, J. F. and Charlier, G. (1999) Influence of water quality and temperature on adhesion of high and low virulence *Flavobacterium columnare* strains to isolated gill arches. *Journal of Fish Diseases* 22, 1 - 11.

[49] Del Cerro, A., Marquez, I. and Guijarro, J. A. (2002a) Simultaneous detection of *Aeromonas salmonicida*, *Flavobacterium psychrophilum*, and *Yersinia ruckeri*, three major fish pathogens, by multiplex PCR. *Applied and Environmental Microbiology*

68, 5177-5180.
[50] Del Cerro, A., Mendoza, M.C. and Guijarro, J.A. (2002b) Usefulness of a TaqMan-based polymerase chain reaction assay for the detection of the fish pathogen *Flavobacterium psychrophilum*. *Journal of Applied Microbiology* 93, 149-156.
[51] Duchaud, E., Boussaha, M., Loux, V., Bernardet, J.-F., Michel, C. *et al*. (2007) Complete genome sequence of the fish pathogen *Flavobacterium psychrophilum*. *Nature Biotechnology* 25, 763-769.
[52] Elsayed, E. E., Eissa, A. E. and Faisal, M. (2006) Isolation of *Flavobacterium psychrophilum* from sea lamprey, *Petromyzon marinus* L., with skin lesions in Lake Ontario. *Journal of Fish Diseases* 29, 629-632.
[53] Evensen, O. and Lorenzen, E. (1996) An immunohistochemical study of *Flexibacter psychrophilus* infection in experimentally and naturally infected rainbow trout (*Oncorhynchus mykiss*) fry. *Diseases of Aquatic Organisms* 25, 53-61.
[54] Evenhuis, J.P., Leeds, T.D., Marancik, D.P., Lapatra, S.E. and Wiens, G.D. (2015) Rainbow trout (*Oncorhynchus mykiss*) resistance to columnaris disease is heritable and favorably correlated with bacterial cold water disease resistance. *Journal of Animal Science* 93, 1546-1554.
[55] Farmer, B.D., Mitchell, A.J. and Straus, D.L. (2011) The effect of high total ammonia concentration on the survival of channel catfish experimentally infected with *Flavobacterium columnare*. *Journal of Aquatic Animal Health* 23, 162-168.
[56] Fehringer, T.R., Hardy, R.W. and Cain, K.D. (2014) Dietary inclusion of salmon testes meal from Alaskan seafood processing byproducts: effects on growth and immune function of rainbow trout, *Oncorhynchus mykiss* (Walbaum). *Aquaculture* 433, 34-39.
[57] Figueiredo, H.C.P., Klesius, P.H., Arias, C.R., Evans, J., Shoemaker, C.A. *et al*. (2005) Isolation and characterization of strains of *Flavobacterium columnare* from Brazil. *Journal of Fish Diseases* 28, 199-204.
[58] Fuente Mde, L., Miranda, C.D., Jopia, P., Gonzalez-Rocha, G., Guiliani, N. et al. (2015) Growth inhibition of bacterial fish pathogens and quorum-sensing blocking by bacteria recovered from Chilean salmonid farms. *Journal of Aquatic Animal Health* 27, 112-122.
[59] Fujihara, M.P. and Nakatani, R.E. (1971) Antibody production and immune responses of rainbow trout and coho salmon to *Chondrococcus columnaris*. *Journal of the Fisheries Research Board of Canada* 28, 1253-1258.
[60] Fujiwara-Nagata, E. and Eguchi, M. (2009) Development and evaluation of a loop-mediated isothermal amplification assay for rapid and simple detection of *Flavobacterium psychrophilum*. *Journal of Fish Diseases* 32, 873-881.
[61] Garcia, C., Pozet, F. and Michel, C. (2000) Standardization of experimental infection with *Flavobacterium psychrophilum*, the agent of rainbow trout *Oncorhynchus mykiss* fry syndrome. *Diseases of Aquatic Organisms* 42, 191-197.
[62] Ghosh, B., Cain, K.D., Nowak, B.F. and Bridle, A.R. (2016) Microencapsulation of a putative probiotic *Enterobacter* species, C6-6, to protect rainbow trout, *Oncorhynchus mykiss* (Walbaum), against bacterial coldwater disease. *Journal of Fish Diseases* 39, 1-11.
[63] Gómez, E., Méndez, J., Cascales, D. and Guijarro, J. A. (2014) *Flavobacterium psychrophilum* vaccine development: a difficult task. *Microbial Biotechnology* 7, 414-423.
[64] Good, C.M., Thorburn, M.A. and Stevenson, R.M. (2008) Factors associated with the incidence of bacterial gill disease in salmonid lots reared in Ontario, Canada government hatcheries. *Preventive Veterinary Medicine* 83, 297-307.
[65] Good, C.M., Thorburn, M.A., Ribble, C.S. and Stevenson, R.M. (2009) Rearing unit-

level factors associated with bacterial gill disease treatment in two Ontario, Canada government salmonid hatcheries. *Preventive Veterinary Medicine* 91, 254 - 260.

[66] Good, C. M., Thorburn, M. A., Ribble, C. S. and Stevenson, R. M. (2010) A prospective matched nested case-control study of bacterial gill disease outbreaks in Ontario, Canada government salmonid hatcheries. *Preventive Veterinary Medicine* 95, 152 - 157.

[67] Good, C.[M.], Davidson, J., Wiens, G.D., Welch, T.J. and Summerfelt, S. (2015) *Flavobacterium branchiophilum* and *F. succinicans* associated with bacterial gill disease in rainbow trout *Oncorhynchus mykiss* (Walbaum) in water recirculation aquaculture systems. *Journal of Fish Diseases* 38, 409 - 413.

[68] Hawke, J.P. and Thune, R.L. (1992) Systemic isolation and antimicrobial susceptibility of *Cytophaga columnaris* from commercially reared channel catfish. *Journal of Aquatic Animal Health* 4, 109 - 113.

[69] Hedrick, R.P., Mcdowell, T.S., Marty, G.D., Mukkatira, K., Antonio, D.B. *et al.* (2000) Ultraviolet irradiation inactivates the waterborne infective stages of *Myxobolus cerebralis*: a treatment for hatchery water supplies. *Diseases of Aquatic Organisms* 42, 53 - 59.

[70] Henríquez-Núñez, H., Oscar, E., Goran, K. and Avendaño-Herrera, R. (2012) Antimicrobial susceptibility and plasmid profiles of *Flavobacterium psychrophilum* strains isolated in Chile. *Aquaculture* 354/355, 38 - 44.

[71] Hesami, S., Allen, K.J., Metcalf, D., Ostland, V.E., Macinnes, J.I. *et al.* (2008) Phenotypic and genotypic analysis of *Flavobacterium psychrophilum* isolates from Ontario salmonids with bacterial coldwater disease. *Canadian Journal of Microbiology* 54, 619 - 629.

[72] Hibi, K., Yoshiura, Y., Ushio, H., Ren, H. and Endo, H. (2012) Rapid detection of *Flavobacterium psychrophilum* using fluorescent magnetic beads and flow cytometry. *Sensors and Materials* 24, 311 - 322.

[73] Holt, R.A. (1987) *Cytophaga psychrophila, the causative agent of bacterial cold-water disease in salmonid fish.* PhD thesis, Oregon State University, Corvallis, Oregon.

[74] Holt, R.A., Rohovec, J.S. and Fryer, J.L. (1993) Bacterial cold-water disease. In: Inglis, V., Roberts, R.J. and Bromage, N.R. (eds) *Bacterial Diseases of Fish.* Blackwell Scientific, Oxford, UK.

[75] Huh, G.-J. and Wakabayashi, H. (1989) Serological characteristics of *Flavobacterium branchiophila* isolated from gill diseases of freshwater fishes in Japan, USA, and Hungary. *Journal of Aquatic Animal Health* 1, 142 - 147.

[76] Izumi, S., Fujii, H. and Aranishi, F. (2005) Detection and identification of *Flavobacterium psychrophilum* from gill washings and benthic diatoms by PCR-based sequencing analysis. *Journal of Fish Diseases* 28, 559 - 564.

[77] Kent, L., Groff, J., Morrison, J., Yasutake, W. and Holt, R. (1989) Spiral swimming behavior due to cranial and vertebral lesions associated with *Cytophaga psychrophila* infections in salmonid fishes. *Diseases of Aquatic Organisms* 6, 11 - 16.

[78] Kimura, N., Wakabayashi, H. and Kudo, S. (1978) Studies on bacterial gill disease in salmonids 1: selection of bacterium transmitting gill disease. *Fish Pathology* 12, 233 - 242.

[79] Kirkland, S. (2010) *Evaluation of the live-attenuated vaccine AquaVac-COL® on hybrid channel × blue catfish fingerlings in earthen ponds.* Masters thesis, Auburn University, Auburn, Alabama.

[80] Klesius, P., Lim, C. and Shoemaker, C. (1999) Effect of feed deprivation on innate resistance and antibody response to *Flavobacterium columnare* in channel catfish,

Ictalurus punctatus. *Bulletin of the European Association of Fish Pathologists* 19, 156 - 158.

[81] Ko, Y.-M. and Heo, G.-J. (1997) Characteristics of *Flavobacterium branchiophilum* isolated from rainbow trout in Korea. *Fish Pathology* 32, 97 - 102.

[82] Korkea-Aho, T., Heikkinen, J., Thompson, K., Von Wright, A. and Austin, B. (2011) *Pseudomonas* sp. M174 inhibits the fish pathogen *Flavobacterium psychrophilum*. *Journal of Applied Microbiology* 111, 266 - 277.

[83] Korkea-Aho, T.L., Papadopoulou, A., Heikkinen, J., Von Wright, A., Adams, A. *et al*. (2012) *Pseudomonas* M162 confers protection against rainbow trout fry syndrome by stimulating immunity. *Journal of Applied Microbiology* 113, 24 - 35.

[84] Kudo, S. and Kimura, N. (1983) The recovery from hyperplasia in a natural infection. *Bulletin of the Japanese Society of Scientific Fisheries* 49, 1627 - 1633.

[85] Kumagai, A., Takahashi, K., Yamaoka, S. and Wakabayashi, H. (1998) Ineffectiveness of iodophore treatment in disinfecting salmonid eggs carrying *Cytophaga psychrophila*. *Fish Pathology* 33, 123 - 128.

[86] Kumagai, A., Yamaoka, S., Takahashi, K., Fukuda, H. and Wakabayashi, H. (2000) Waterborne transmission of *Flavobacterium psychrophilum* in coho salmon eggs. *Fish Pathology* 35, 25 - 28.

[87] Kunttu, H. (2010) Characterizing the Bacterial Fish Pathogen *Flavobacterium columnare*, and Some Factors Affecting Its Pathogenicity. Jyväskylä Studies in Biology and Environmental Science 206, University of Jyväskylä, Jyväskylä, Finland. Available at: https://jyx.jyu.fi/dspace/bitstream/handle/123456789/23323/9789513938673.pdf?sequence=1 (accessed 11 November 2016).

[88] Kunttu, H.M., Valtonen, E.T., Suomalainen, L.R., Vielma, J. and Jokinen, I.E. (2009) The efficacy of two immunostimulants against *Flavobacterium columnare* infection in juvenile rainbow trout (*Oncorhynchus mykiss*). *Fish and Shellfish Immunology* 26, 850 - 857.

[89] Kunttu, H. M., Sundberg, L. R., Pulkkinen, K. and Valtonen, E. T. (2012) Environment may be the source of *Flavobacterium columnare* outbreaks at fish farms. *Environmental Microbiology Reports* 4, 398 - 402.

[90] LaFrentz, B. and Cain, K. (2004) *Coldwater Disease*. Extension Bulletin, Western Regional Aquaculture Consortium (WRAC), Washington, DC.

[91] LaFrentz, B. R., Lapatra, S. E., Call, D. R. and Cain, K. D. (2008) Isolation of rifampicin resistant *Flavobacterium psychrophilum* strains and their potential as live attenuated vaccine candidates. *Vaccine* 26, 5582 - 5589.

[92] LaFrentz, B.R., Lapatra, S.E., Call, D.R., Wiens, G.D. and Cain, K.D. (2009) Proteomic analysis of *Flavobacterium psychrophilum* cultured in vivo and in iron-limited media. *Diseases of Aquatic Organisms* 87, 171 - 182.

[93] LaFrentz, B. R., Lapatra, S. E., Shoemaker, C. A. and Klesius, P. H. (2012a) Reproducible challenge model to investigate the virulence of *Flavobacterium columnare* genomovars in rainbow trout *Oncorhynchus mykiss*. *Diseases of Aquatic Organisms* 101, 115 - 122.

[94] LaFrentz, B.R., Shoemaker, C.A., Booth, N.J., Peterson, B.C. and Ourth, D.D. (2012b) Spleen index and mannose-binding lectin levels in four channel catfish families exhibiting different susceptibilities to *Flavobacterium columnare* and *Edwardsiella ictaluri*. *Journal of Aquatic Animal Health* 24, 141 - 147.

[95] LaFrentz, B., Waldbieser, G., Welch, T. and Shoemaker, C. (2014) Intragenomic heterogeneity in the 16S rRNA genes of *Flavobacterium columnare* and standard protocol for genomovar assignment. *Journal of Fish Diseases* 37, 657 - 69.

[96] LaPatra, S.E., Fehringer, T.R. and Cain, K.D. (2014) A probiotic Enterobacter sp.

provides significant protection against *Flavobacterium psychrophilum* in rainbow trout (*Oncorhynchus mykiss*) after injection by two different routes. *Aquaculture* 433, 361－366.

[97] Leeds, T., Silverstein, J., Weber, G., Vallejo, R., Palti, Y. et al. (2010) Response to selection for bacterial cold water disease resistance in rainbow trout. *Journal of Animal Science* 88, 1936－1946.

[98] Lievens, B., Frans, I., Heusdens, C., Justé, A., Jonstrup, S.P. *et al*. (2011) Rapid detection and identification of viral and bacterial fish pathogens using a DNA arraybased multiplex assay. *Journal of Fish Diseases* 34, 861－875.

[99] Lindstrom, N.M., Call, D.R., House, M.L., Moffitt, C.M. and Cain, K.D. (2009) A quantitative enzyme-linked immunosorbent assay and filtration-based fluorescent antibody test as potential tools to screen broodstock for infection with *Flavobacterium psychrophilum*. *Journal of Aquatic Animal Health* 21, 43－56.

[100] Liu, L., Li, C., Su, B., Beck, B. H. and Peatman, E. (2013) Short-term feed deprivation alters immune status of surface mucosa in channel catfish (*Ictalurus punctatus*). *PLoS ONE* 8 (9): e74581.

[101] Liu, S., Vallejo, R. L., Palti, Y., Gao, G., Marancik, D. P. et al. (2015) Identification of single nucleotide polymorphism markers associated with bacterial cold water disease resistance and spleen size in rainbow trout. *Frontiers in Genetics* 6: 298.

[102] Loch, T.P. and Faisal, M. (2016a) *Flavobacteria* isolated from the milt of feral Chinook salmon of the Great Lakes. *North American Journal of Aquaculture* 78, 25－33.

[103] Loch, T.P. and Faisal, M. (2016b) *Flavobacterium spartansii* induces pathological changes and mortality in experimentally challenged Chinook salmon *Oncorhynchus tshawytscha* (Walbaum). *Journal of Fish Diseases* 39, 483－488.

[104] Loch, T. P., Fujimoto, M., Woodiga, S. A., Walker, E. D., Marsh, T. L. *et al*. (2013) Diversity of fish-associated flavobacteria of Michigan. *Journal of Aquatic Animal Health* 25, 149－164.

[105] Long, A., Polinski, M.P., Call, D.R. and Cain, K.D. (2012) Validation of diagnostic assays to screen broodstock for *Flavobacterium psychrophilum* infections. *Journal of Fish Diseases* 35, 407－419.

[106] Long, A., Fehringer, T.R., Swain, M.A., LaFrentz, B.R., Call, D.R. *et al*. (2013) Enhanced efficacy of an attenuated *Flavobacterium psychrophilum* strain cultured under iron-limited conditions. *Fish and Shellfish Immunology* 35, 1477－1482.

[107] Long, A., Call, D. R. and Cain, K. D. (2015) Comparison of quantitative PCR and ELISA for detection and quantification of *Flavobacterium psychrophilum* in salmonid broodstock. *Diseases of Aquatic Organisms* 115, 139－146.

[108] Lorenzen, E., Dalsgaard, I., From, J., Hansen, E., Horlyck, V. *et al*. (1991) Preliminary investigations of fry mortality syndrome in rainbow trout. *Bulletin of the European Association of Fish Pathologists* 11, 77－79.

[109] Macphee, D., Ostland, V., Lumsden, J. and Ferguson, H. (1995) Development of an enzyme-linked immunosorbent assay (ELISA) to estimate the quantity of *Flavobacterium branchiophilum* on the gills of rainbow trout *Oncorhynchus mykiss*. *Diseases of Aquatic Organisms* 21, 13－23.

[110] Madetoja, J. and Wiklund, T. (2002) Detection of the fish pathogen *Flavobacterium psychrophilum* in water from fish farms. *Systematic and Applied Microbiology* 25, 259－266.

[111] Madetoja, J., Nyman, P. and Wiklund, T. (2000) *Flavobacterium psychrophilum*, invasion into and shedding by rainbow trout *Oncorhynchus mykiss*. *Diseases of Aquatic Organisms* 43, 27－38.

[112] Madetoja, J., Dalsgaard, I. and Wiklund, T. (2002) Occurrence of *Flavobacterium psychrophilum* in fishfarming environments. *Diseases of Aquatic Organisms* 52, 109 - 118.

[113] Madsen, L. and Dalsgaard, I. (2000) Comparative studies of Danish *Flavobacterium psychrophilum* isolates: ribotypes, plasmid profiles, serotypes and virulence. *Journal of Fish Diseases* 23, 211 - 218.

[114] Madsen, L. and Dalsgaard, I. (2008) Water recirculation and good management: potential methods to avoid disease outbreaks with *Flavobacterium psychrophilum*. *Journal of Fish Diseases* 31, 799 - 810.

[115] Mainous, M.E., Kuhn, D.D. and Smith, S.A. (2012) Efficacy of common aquaculture compounds for disinfection of *Flavobacterium columnare* and *F. psychrophilum*. *Journal of Applied Aquaculture* 24, 262 - 270.

[116] Marancik, D.P. and Wiens, G.D. (2013) A real-time polymerase chain reaction assay for identification and quantification of *Flavobacterium psychrophilum* and application to disease resistance studies in selectively bred rainbow trout *Oncorhynchus mykiss*. *FEMS Microbiology Letters* 339, 122 - 129.

[117] Marancik, D.P., Camus, M.S., Camus, A.C., Leeds, T.D., Weber, G.M. *et al.* (2014) Biochemical reference intervals and pathophysiological changes in *Flavobacterium psychrophilum*-resistant and -susceptible rainbow trout lines. *Diseases of Aquatic Organisms* 111, 239 - 248.

[118] Marancik, D., Gao, G., Paneru, B., Ma, H., Hernandez, A.G. et al. (2015) Whole-body transcriptome of selectively bred, resistant-, control-, and susceptible-line rainbow trout following experimental challenge with *Flavobacterium psychrophilum*. *Frontiers in Genetics* 5: 453.

[119] Mata, M., Skarmeta, A. and Santos, Y. (2002) A proposed serotyping system for *Flavobacterium psychrophilum*. *Letters in Applied Microbiology* 35, 166 - 170.

[120] Miwa, S. and Nakayasu, C. (2005) Pathogenesis of experimentally induced bacterial cold water disease in ayu *Plecoglossus altivelis*. *Diseases of Aquatic Organisms* 67, 93 - 104.

[121] Mohammed, H.H. and Arias, C.R. (2016) Protective efficacy of *Nigella sativa* seeds and oil against Columnaris disease in fishes. *Journal of Fish Diseases* 39, 693 - 703.

[122] Mohammed, H., Olivares-Fuster, O., LaFrentz, S. and Arias, C.R. (2013) New attenuated vaccine against Columnaris disease in fish: choosing the right parental strain is critical for vaccine efficacy. *Vaccine* 31, 5276 - 5280.

[123] Morrison, C., Cornick, J., Shum, G. and Zwicker, B. (1981) Microbiology and histopathology of 'saddleback' disease of underyearling Atlantic salmon, *Salmo salar* L. *Journal of Fish Diseases* 4, 243 - 258.

[124] Nakagawa, J., Iwasaki, T. and Kodama, H. (2009) Protection against *Flavobacterium psychrophilum* infection (cold water disease) in ayu fish (*Plecoglossus altivelis*) by oral administration of humus extract. *The Journal of Veterinary Medical Science* 71, 1487 - 1491.

[125] Nakayama, H., Tanaka, K., Teramura, N. and Hattori, S. (2015) Expression of collagenase in Flavobacterium psychrophilum isolated from cold-water diseaseaffected ayu (*Plecoglossus altivelis*). *Bioscience, Biotechnology, and Biochemistry* 80, 135 - 144.

[126] Nematollahi, A., Decostere, A., Pasmans, F. and Haesebrouck, F. (2003) *Flavobacterium psychrophilum* infections in salmonid fish. *Journal of Fish Diseases* 26, 563 - 574.

[127] Nicolas, P., Mondot, S., Achaz, G., Bouchenot, C., Bernardet, J.-F. *et al.* (2008) Population structure of the fish-pathogenic bacterium *Flavobacterium psychrophilum*.

Applied and Environmental Microbiology 74, 3702 - 3709.
[128] Oplinger, R. W. and Wagner, E. (2010) Disinfection of contaminated equipment: evaluation of benzalkonium chloride exposure time and solution age and the ability of air-drying to eliminate *Flavobacterium psychrophilum*. *Journal of Aquatic Animal Health* 22, 248 - 253.
[129] Oplinger, R.W. and Wagner, E.J. (2013) Control of *Flavobacterium psychrophilum*: tests of erythromycin, streptomycin, osmotic and thermal shocks, and rapid pH change. *Journal of Aquatic Animal Health* 25, 1 - 8.
[130] Oplinger, R.W. and Wagner, E.J. (2015) Use of penicillin and streptomycin to reduce spread of bacterial coldwater disease I: antibiotics in sperm extenders. *Journal of Aquatic Animal Health* 27, 25 - 31.
[131] Oplinger, R. W., Wagner, E. J. and Cavender, W. (2015) Use of penicillin and streptomycin to reduce spread of bacterial coldwater disease II: efficacy of using antibiotics in diluents and during water hardening. *Journal of Aquatic Animal Health* 27, 32 - 37.
[132] Ordal, E.J. and Rucker, R.R. (1944) Pathogenic myxobacteria. *Proceedings of the Society for Experimental Biology and Medicine* 56, 15 - 18.
[133] Orieux, N., Bourdineaud, J.P., Douet, D.G., Daniel, P. and Le Henaff, M. (2011) Quantification of *Flavobacterium psychrophilum* in rainbow trout, *Oncorhynchus mykiss* (Walbaum), tissues by qPCR. *Journal of Fish Diseases* 34, 811 - 821.
[134] Ostland, V.E., Macphee, D.D., Lumsden, J.S. and Ferguson, H.W. (1995) Virulence of *Flavobacterium branchiophilum* in experimentally infected salmonids. *Journal of Fish Diseases* 18, 249 - 262.
[135] Ostland, V.E., Mcgrogan, D.G. and Ferguson, H.W. (1997) Cephalic osteochondritis and necrotic scleritis in intensively reared salmonids associated with *Flexibacter psychrophilus*. *Journal of Fish Diseases* 20, 443 - 451.
[136] Panangala, V.S., Shelby, R.A., Shoemaker, C.A., Klesius, P.H., Mitra, A. *et al.* (2006) Immunofluorescent test for simultaneous detection of *Edwardsiella ictaluri* and *Flavobacterium columnare*. *Diseases of Aquatic Organisms* 68, 197 - 207.
[137] Panangala, V. S., Shoemaker, C. A. and Klesius, P. H. (2007) TaqMan real-time polymerase chain reaction assay for rapid detection of *Flavobacterium columnare*. *Aquaculture Research* 38, 508 - 517.
[138] Plumb, J.A. and Hanson, L.A. (2010) Catfish bacterial diseases. In: Plumb, J.A. and Hanson, L. A. Health *Maintenance and Principal Microbial Diseases of Cultured Fishes*, 3rd edn. Wiley-Blackwell, Oxford, UK, pp. 275 - 313.
[139] Prasad, Y., Arpana, Kumar, D. and Sharma, A. K. (2011) Lytic bacteriophages specific to *Flavobacterium columnare* rescue catfish, *Clarias batrachus* (Linn.) from columnaris disease. *Journal of Environmental Biology* 32, 161 - 168.
[140] Pulkkinen, K., Suomalainen, L.-R., Read, A.F., Ebert, D., Rintamäki, P. *et al.* (2010) Intensive fish farming and the evolution of pathogen virulence: the case of Columnaris disease in Finland. *Proceedings of the Royal Society of London B: Biological Sciences* 277, 593 - 600.
[141] Rangdale, R. and Way, K. (1995) Rapid identification of *C. psychrophila* from infected spleen tissue using an enzyme-linked immunosorbent assay (ELISA). *Bulletin of the European Association of Fish Pathologists* 15, 213 - 216.
[142] Rangdale, R., Richards, R. and Alderman, D. (1999) Histopathological and electron microscopical observations on rainbow trout fry syndrome. *Veterinary Record* 144, 251 - 254.
[143] Ryumae, U., Hibi, K., Yoshiura, Y., Ren, H. and Endo, H. (2012) Ultra highly sensitive method for detecting *Flavobacterium psychrophilum* using high-gradient

immunomagnetic separation with a polymerase chain reaction. *Aquaculture Research* 43, 929 - 939.
[144] Schulz, C. and Faisal, M. (2010) The bacterial community associated with the leech *Myzobdella lugubris* Leidy 1851 (Hirudinea: Piscicolidae) from Lake Erie, Michigan, USA. *Parasite* 17, 113 - 121.
[145] Shoemaker, C.A., Klesius, P.H., Drennan, J.D. and Evans, J.J. (2011) Efficacy of a modified live *Flavobacterium columnare* vaccine in fish. *Fish and Shellfish Immunology* 30, 304 - 308.
[146] Silverstein, J.T., Vallejo, R.L., Palti, Y., Leeds, T.D., Rexroad, C.E. *et al.* (2009) Rainbow trout resistance to bacterial cold-water disease is moderately heritable and is not adversely correlated with growth. *Journal of Animal Science* 87, 860 - 867.
[147] Smith, P., Endris, R., Kronvall, G., Thomas, V., Verner-Jeffreys, D., Wilhelm, C. and Dalsgaard, I. (2016) Epidemiological cut-off values for *Flavobacterium psychrophilum* MIC data generated by a standard test protocol. *Journal of Fish Diseases* 39,143 - 154.
[148] Snieszko, S.F. (1981) *Bacterial Gill Disease of Freshwater Fishes. Fish Disease Leaflet No. 62*, US Fish Wildlife Service, US Department of the Interior, Washington, DC. Available at: http://digitalcommons.unl.edu/cgi/viewcontent.cgi?article=1146&context=usfwspubs (accessed 11 November 2016).
[149] Solís, C.J., Poblete-Morales, M., Cabral, S., Valdés, J.A., Reyes, A.E. *et al.* (2015) Neutrophil migration in the activation of the innate immune response to different *Flavobacterium psychrophilum* vaccines in zebrafish (*Danio rerio*). *Journal of Immunology Research* 2015: 515187.
[150] Song, Y.L., Fryer, J.L. and Rohovec, J.S. (1988) Comparison of six media for the cultivation of *Flexibacter columnaris*. *Fish Pathology* 23, 91 - 94.
[151] Speare, D.J. and Ferguson, H.W. (1989) Clinical and pathological features of common gill diseases of cultured salmonids in Ontario. *The Canadian Veterinary Journal* 30, 882 - 887.
[152] Speare, D.J., Ferguson, H.W., Beamish, F.W.M., Yager, J.A. and Yamashiro, S. (1991) Pathology of bacterial gill disease: sequential development of lesions during natural outbreaks of disease. *Journal of Fish Diseases* 14, 21 - 32.
[153] Speare, D.J., Markham, R.J., Despres, B., Whitman, K. and Macnair, N. (1995) Examination of gills from salmonids with bacterial gill disease using monoclonal antibody probes for *Flavobacterium branchiophilum* and *Cytophaga columnaris*. *Journal of Veterinary Diagnostic Investigation* 7, 500 - 505.
[154] Starliper, C.E. (2011) Bacterial coldwater disease of fishes caused by *Flavobacterium psychrophilum*. *Journal of Advanced Research* 2, 97 - 108.
[155] Starliper, C. and Schill, W. (2011) Flavobacterial diseases: Columnaris disease, coldwater disease, and bacterial gill disease. In: Woo, P.T.K. and Bruno, D.B. (eds) *Fish Diseases and Disorders. Volume 3: Viral, Bacterial and Fungal Infections*, 2nd edn. CAB International, Wallingford, UK, pp. 606 - 631.
[156] Strepparava, N., Wahli, T., Segner, H., Polli, B. and Petrini, O. (2012) Fluorescent in situ hybridization: a new tool for the direct identification and detection of *F. psychrophilum*. *PLoS ONE* 7 (11): e49280.
[157] Strepparava, N., Wahli, T., Segner, H. and Petrini, O. (2014) Detection and quantification of *Flavobacterium psychrophilum* in water and fish tissue samples by quantitative real time PCR. *BMC Microbiology* 14: 105.
[158] Strom-Bestor, M. and Wiklund, T. (2011) Inhibitory activity of *Pseudomonas* sp. on *Flavobacterium psychrophilum*, in vitro. *Journal of Fish Diseases* 34, 255 - 264.
[159] Suebsing, R., Kampeera, J., Sirithammajak, S., Withyachumnarnkul, B., Turner,

W. *et al*. (2015) Colorimetric method of loop-mediated isothermal amplification with the pre-addition of calcein for detecting *Flavobacterium columnare* and its assessment in tilapia farms. *Journal of Aquatic Animal Health* 27, 38 - 44.
[160] Sundell, K. and Wiklund, T. (2011) Effect of biofilm formation on antimicrobial tolerance of *Flavobacterium psychrophilum*. *Journal of Fish Diseases* 34, 373 - 383.
[161] Suomalainen, L.R., Bandilla, M. and Valtonen, E.T. (2009) Immunostimulants in prevention of Columnaris disease of rainbow trout, *Oncorhynchus mykiss* (Walbaum). *Journal of Fish Diseases* 32, 723 - 726.
[162] Suzuki, K., Arai, H., Kuge, T., Katagiri, T. and Izumi, S. (2008) Reliability of PCR methods for the detection of *Flavobacterium psychrophilum*. *Fish Pathology* 43, 124 - 127.
[163] Tekedar, H.C., Karsi, A., Gillaspy, A.F., Dyer, D.W., Benton, N.R. et al. (2012) Genome sequence of the fish pathogen *Flavobacterium columnare* ATCC 49512. *Journal of Bacteriology* 194, 2763 - 2764.
[164] Touchon, M., Barbier, P., Bernardet, J.-F., Loux, V., Vacherie, B. *et al*. (2011) Complete genome sequence of the fish pathogen *Flavobacterium branchiophilum*. *Applied and Environmental Microbiology* 77, 7656 - 7662.
[165] Toyama, T., Kita-Tsukamoto, K. and Wakabayashi, H. (1996) Identification of Flexibacter maritimus, *Flavobacterium branchiophilum* and *Cytophaga columnaris* by PCR targeted 16S ribosomal DNA. *Fish Pathology* 31, 25 - 31.
[166] Tripathi, N.K., Latimer, K.S., Gregory, C.R., Ritchie, B.W., Wooley, R.E. *et al*. (2005) Development and evaluation of an experimental model of cutaneous columnaris disease in koi *Cyprinus carpio*. *Journal of Veterinary Diagnostic Investigation* 17, 45 - 54.
[167] Triyanto, A. and Wakabayashi, H. (1999) Genotypic diversity of strains of *Flavobacterium columnare* from diseased fishes. *Fish Pathology* 34, 65 - 71.
[168] Triyanto, A., Kumamaru, A. and Wakabayashi, H. (1999) The use of PCR targeted 16S rDNA for identification of genomovars of *Flavobacterium columnare*. *Fish Pathology* 34, 217 - 218.
[169] Van Vliet, D., Loch, T.P. and Faisal, M. (2015) *Flavobacterium psychrophilum* Infections in salmonid broodstock and hatchery-propagated stocks of the Great Lakes basin. *Journal of Aquatic Animal Health* 27, 192 - 202.
[170] Van Vliet, D., Wiens, G., Loch, T., Nicolas, P. and Faisal, M. (2016) Genetic diversity of *Flavobacterium psychrophilum* isolated from three *Oncorhynchus* spp. in the U. S. A. revealed by multilocus sequence typing. *Applied and Environmental Microbiology*. doi: 10.1128/AEM.00411 - 16.
[171] Vatsos, I. N., Thompson, K. D. and Adams, A. (2002) Development of an immunofluorescent antibody technique (IFAT) and in situ hybridization to detect *Flavobacterium psychrophilum* in water samples. *Aquaculture Research* 33, 1087 - 1090.
[172] Vatsos, I.N., Thompson, K.D. and Adams, A. (2003) Starvation of *Flavobacterium psychrophilum* in broth, stream water and distilled water. *Diseases of Aquatic Organisms* 56, 115 - 126.
[173] Verma, V. and Prasad, Y. (2014) Isolation and immunohistochemical identification of *Flavobacterium psychrophilum* from the tissue of catfish, *Clarias batrachus*. *Journal of Environmental Biology* 35, 389 - 393.
[174] Wagner, B.A., Wise, D.J., Khoo, L.H. and Terhune, J.S. (2002) The epidemiology of bacterial diseases in food-size channel catfish. *Journal of Aquatic Animal Health* 14, 263 - 272.
[175] Wakabayashi, H., Egusa, S. and Fryer, J. (1980) Characteristics of filamentous

bacteria isolated from a gill disease of salmonids. *Canadian Journal of Fisheries and Aquatic Sciences* 37, 1499 - 1504.

[176] Wakabayashi, H., Huh, G.J. and Kimura, N. (1989) *Flavobacterium branchiophila* sp. nov., a causative agent of bacterial gill disease of freshwater fishes. *International Journal of Systematic Bacteriology* 39, 213 - 216.

[177] Wakabayashi, H., Toyama, T. and Iida, T. (1994) A study on serotyping of *Cytophaga psychrophila* isolated from fishes in Japan. *Fish Pathology* 29, 101 - 104.

[178] Welker, T.L., Shoemaker, C.A., Arias, C.R. and Klesius, P.H. (2005) Transmission and detection of *Flavobacterium columnare* in channel catfish *Ictalurus punctatus*. *Diseases of Aquatic Organisms* 63, 129 - 138.

[179] Wiens, G.D., Lapatra, S.E., Welch, T.J., Evenhuis, J.P., Rexroad, C.E. III *et al*. (2013) On-farm performance of rainbow trout (*Oncorhynchus mykiss*) selectively bred for resistance to bacterial cold water disase: effect of rearing environment on survival phenotype. *Aquaculture* 388, 128 - 136.

[180] Wiens, G.D., Lapatra, S.E., Welch, T.J., Rexroad, C. 3rd, Call, D.R. *et al*. (2014) Complete genome sequence of *Flavobacterium psychrophilum* strain CSF259 - 93, used to select rainbow trout for increased genetic resistance against bacterial cold water disease. *Genome Announcements* 2(5): e00889 - 14.

[181] Wiens, G.D., Marancik, D.P., Zwollo, P. and Kaattari, S.L. (2015) Reduction of rainbow trout spleen size by splenectomy does not alter resistance against bacterial cold water disease. *Developmental and Comparative Immunology* 49, 31 - 37.

[182] Wiklund, T., Madsen, L., Bruun, M.S. and Dalsgaard, I. (2000) Detection of *Flavobacterium psychrophilum* from fish tissue and water samples by PCR amplification. *Journal of Applied Microbiology* 88, 299 - 307.

[183] Wood, J. (1974) *Diseases of Pacific Salmon, Their Prevention and Treatment*. Washington Department of Fisheries, Olympia, Washington, DC.

[184] Wood, E. and Yasutake, W. (1956) Histopathology of fish: III. Peduncle (cold-water) disease. *The Progressive Fish-Culturist* 18, 58 - 61.

[185] Wood, E. and Yasutake, W. (1957) Histopathology of fish. V. Gill disease. *The Progressive Fish-Culturist* 19, 7 - 13.

[186] Yeh, H.Y., Shoemaker, C. and Klesius, P. (2006) Sensitive and rapid detection of *Flavobacterium columnare* in channel catfish *Ictalurus punctatus* by a loopmediated isothermal amplification method. *Journal of Applied Microbiology* 100, 919 - 925.

[187] Yu, H., Tan, S., Zhao, H. and Li, H. (2013) MH-DAB gene polymorphism and disease resistance to *Flavobacterium columnare* in grass carp (*Ctenopharyngodon idellus*). *Gene* 526, 217 - 222.

17

诺神弗朗西斯菌

Esteban Soto* 和 John P. Hawke

17.1 引言

鲑立克次氏体(*Piscirickettsia salmonis*)是第一种被发现的可以引起鱼类疾病的立克次体样微生物(Fryer et al., 1992),继此之后,在尼罗罗非鱼(Chern and Chao, 1993; Chen et al., 1994)、萨氏巴拉圭鲇(*Panaque suttoni*) (Eigenmann and Eigenmann)和黑点石斑鱼(*Epinephelus melanostigma*) (Khoo et al., 1995; Chen et al., 2000; Mauel et al., 2003)中也发现报道了其他类立克次体微生物。Kamaishi 等(2005)对水产养殖场中患病三线矶鲈(*Parapristipoma trilineatum*)冷冻肾脏中的病原样本进行扩增,并对其16S rDNA进行了测序,与其他真细菌 16S rDNA 序列比对后,发现该序列与弗朗西斯菌属种(*Francisella* spp.)的相似性大于 97%。研究人员从置于添加 1%血红蛋白的半胱氨酸心琼脂(Difco, USA)上的内脏病样中分离出细菌,并通过了科赫法则验证。这是弗朗西斯菌引起养殖鱼类疾病的首次报道。随后,诺神弗朗西斯菌(*Francisella noatunensis*)被确认为是一种世界范围内海水和淡水养殖鱼类的病原菌(表 17.1 和表 17.2)。在某些情况下,其导致的疾病死亡率超过 90%。

表 17.1 感染诺神弗朗西斯菌东方亚种的鱼类

通用名	拉丁学名	栖息地	参考文献
罗非鱼 (Tilapia)	*Oreochromis* spp.	巴西	Leal et al., 2014
		哥斯达黎加	Mikalsen and Colquhoun, 2009; Soto et al., 2009a, 2012c
		印度尼西亚	Ottem et al., 2009
		泰国	Nguyen et al., 2016
		英国	Jeffery et al., 2010
		美国	Soto et al., 2011a
		夏威夷	Soto et al., 2012c, 2013c

* 通信作者邮箱: sotomartinez@ucdavis.edu。

续 表

通用名	拉丁学名	栖息地	参考文献
马拉维湖慈鲷 (Malawi cichlids)	*Nimbochromis venustus*, *Nimbochromis linni*, *Aulonocara stuartgranti*, *Placidochromis* sp., *Protomelas* sp., *Naevochromis chryosogaster*, *Copadichromis mloto*, *Otopharynx tetrastigma*	奥地利	Lewisch，2014
三线矶鲈 (Three-line grunt)	*Parapristipoma trilinineatum*	日本	Fukuda et al.，2002
淡黑镊丽鱼 (Blue-white labido)	*Labidochromis caeruleus*	中国台湾地区	Hsieh et al.，2007
七彩神仙鱼 (Brown discus)	*Symphysodon aequifasciatus*		
紫水晶鱼 (Deep-water hap)	*Haplochromis electra*		
弗氏鬼丽鱼 (Electric blue hap)	*Sciaenochromis fryeri*		
华拟丽鱼 (Elegans)	*Pseudotropheus elegans*		
火鸟鱼 (Firebird)	*Aulonocara rubescens*		
皇冠六间 (Frontosa cichlid)	*Cyphotilapia frontosa*		
马面鲷 (Malawi eyebiter)	*Dimidiochromis compressiceps*		
斑马拟丽鱼 (Maylandia zebra)	*Pseudotropheus zebra*		
帝王鲷 (Rhodes's chilo)	*Chilotilapia rhoadesii*		
蓝绿光鳃鱼 (Blue-green damselfish)	*Chromis viridis*	美国	Camus et al.，2013
凯撒仿石鲈 (Caesar grunt)	*Haemulon carbonarium*		Soto et al.，2014b
丝隆头鱼 (Fairy wrasses)	*Cirrhilabrus*		Camus et al.，2013
法国石鲈 (French grunt)	*Haemulon flavolineatum*		Soto et al.，2014b

续　表

通　用　名	拉　丁　学　名	栖息地	参　考　文　献
杂交条纹鲈 (Hybrid striped bass)	*Morone saxatilis* × *M. chrysops*	美国	Ostland et al., 2006
斑马鱼 (Zebrafish)	*Danio rerio*	美国(试验鱼)	Vojtech et al., 2009

表 17.2　感染诺神弗朗西斯菌诺神亚种的鱼类和其他水生生物

通　用　名	拉　丁　学　名	栖　息　地	参　考　文　献
大西洋鳕,野生鳕 (Atlantic cod, wild cod)	*Gadus morhua*	凯尔特海,丹麦,爱尔兰,挪威,瑞典,英国	Mikalsen et al., 2007 Ruane et al., 2015
大西洋鲑 (Atlantic salmon)	*Salmo salar*	智利,挪威	Birkbeck et al., 2007 Bohle et al., 2009
大西洋鲭 (Atlantic mackerel)	*Scomber scombrus*	挪威	Ottem et al., 2008 Wangen et al., 2012
紫贻贝 (Blue mussel)	*Mytilus edulis*		
黄道蟹 (Crab)	*Cancer pagurus*		
欧洲鲽 (European plaice)	*Pleuronectes platessa*		
帆鳞鲆 (Megrim)	*Lepidorhombus whiffiagonis*		
狭鳕,绿青鳕 (Pollock, saithe)	*Pollachius virens*		
斑马鱼 (Zebrafish)	*Danio rerio*	挪威(实验鱼)	Brudal et al., 2014

弗朗西斯菌属(*Francisella*)是硫发菌目(Thiotrichales)、弗朗西斯菌科(Francisellaceae)中唯一的属。该属菌为多形性、非运动性、专性好氧、无孢子的小革兰氏阴性菌(Foley and Nieto, 2010)。历史上,该属中三个公认的种包括土拉热弗朗西斯菌(*F. tularensis*)、新凶手弗朗西斯菌(*F. novicida*)和蜃楼弗朗西斯菌(*F. philomiragia*)。它们都可以感染人,其中 *F. tularensis* 最具致病性,是土拉菌病(tularaemia)的病原菌,该病是一种潜在致命性的多系统人畜共患疾病(Petersen et al., 2009; Foley and Nieto, 2010)。

温水性或暖水性鱼类[如养殖罗非鱼、三线矶鲈和杂交条纹鲈(*Morone saxatilis*×*M. chrysops*)]与冷水性鱼类[如挪威和智利的大西洋鳕(*Gadus morhua*)和大西洋鲑(*Salmo salar*)]中的 *Francisella* 分离株存在遗传、表型和生化差异（Mikalsen et al.，2007；Mikalsen and Colquhoun，2009；Ottem et al.，2009）。由于从尼罗罗非鱼中分离出的弗朗西斯氏菌种与 *F. tularensis* 种株脂多糖(LPS)的组成存在差异，Kay 等(2006)将尼罗罗非鱼分离株命名为维多利亚弗朗西斯菌(*F. Victoria*)新种。Ottem 等(2007a,b)利用分子方法、脂肪酸分析、生化和表型特性对挪威鳕鱼分离株进行了鉴定。基于遗传差异，他们提出了一个新种并命名为杀鱼弗朗西斯菌(*F. piscicida*)。与此同时，挪威的其他研究人员对源自呈现慢性肉芽肿性疾病的养殖大西洋鳕分离株进行了表型和分子特征鉴定(Olsen et al.，2006)。这些细菌也被确认为弗朗西斯菌，该种虽与蜃楼弗朗西斯菌(*F. philomiragia*)密切相关但却有区别(Olsen et al.，2006)。该大西洋鳕分离株被命名为蜃楼弗朗西斯菌诺神新亚种(*F. philomiragia* subsp. *noatunensis* subsp. nov.)(Mikalsen et al.，2007)。

随后，Mikalsen 和 Colquhoun(2009)利用表型和分子分类方法，对患病的哥斯达黎加养殖罗非鱼(*Oreochromis* sp.)、智利大西洋鲑和日本三线矶鲈的分离株进行了鉴定。基于分子遗传学，他们提出将来自哥斯达黎加和日本的分离株确认为亚洲弗朗西斯菌(*F. asiatica* sp. nov.)。由于 *F. piscicida* 和 *F. philomiragia* subsp. *noatunensis* 之间没有差异，并且有越来越多的证据表明 *F. philomiragia* 存在生态差异和特定的鱼类致病分支，他们提议把 *F. philomiragia* subsp. *noatunensis* 提升到种级，即诺神弗朗西斯菌合并新种(*F. noatunensis* comb. nov. sp.)。如同 *F. noatunensis* comb. nov 一样，Ottem 等(2009)采用日本三线矶鲈和挪威大西洋鳕分离株的若干持家基因序列和表型特征将 *F. philomiragia* subsp. *noatunensis* 提升至种级。*F. piscicida* 被认为是 *F. noatunensis* comb. nov. 的异型同名。这些研究者还提出日本株 *Francisella* sp. Ehime－1 代表一个新的亚种，即诺神弗朗西斯菌东方亚种(*F. noatunensis* subsp. *orientalis* subsp. nov.)。

现在鱼类致病性弗朗西斯菌被分为两个亚种，与它们存在于冷水或温水性鱼类中有关。冷水中的诺神弗朗西斯菌诺神亚种 *F. noatunensis* subsp. *noatunensis* (*Fnn*)和温水中的 *F. noatunensis* subsp. *orientalis* (*Fno*)是养殖鱼类和野生鱼类的病原菌。它们为非运动性、高度多形性的革兰氏阴性球杆菌，宽 0.2～0.4 μm，长 0.4～1.9 μm。这些细菌是过氧化氢酶阳性和氧化酶阴性的需氧微生物，需要半胱氨酸才能生长(Soto et al.，2009a；Birkbeck et al.，2011；Colquhoun and Duodu，2011)。

F. noatunensis 主要通过水平途径传播(Kamaishi et al.，2005；Nylund et al.，2006；Ostland et al.，2006)。然而，已经在鱼的性腺中观察到这种细菌(Soto et al.，2009a)，鱼苗中也有疾病暴发的报道(Soto et al.，2009a；Duodu and Colquhoun，2010)，所以垂直传播仍然是可能的。此外，目前还不知道 *Fnn* 是否可以在宿主之外繁殖。有研究表明，在淡水或盐水中数周后，*Fnn* 和 *Fno*

会进入存活不可培养(viable butnonculturable, VBNC)状态(Duodu and Colquhoun, 2010;Duodu et al., 2012a; Soto and Revan, 2012)。VBNC状态的微生物不能在培养基上生长,但具有代谢活性。此外,*Fno* 可以在24 h内产生生物被膜,细菌在30 ℃以下的淡水、半咸水和海水中可以在生物被膜中存活5 d以上。而且,这种生物被膜使 *Fno* 对过氧化氢、漂白剂和 Virkon® 具有更强的抵抗力,从而证明了其在环境中生存的重要性(Soto et al., 2015b)。

F. philomiragia 和 *F. tularensis* 能感染棘阿米巴原虫(*Acanthamoeba*)并在其中进行复制(Abd et al., 2003;Verhoeven et al., 2010),弗朗西斯菌内纤毛虫亚种(*F. noatunensis* subsp. *endociliophora*)是纤毛游仆虫(*Euplotes*)的体内共生菌(Schrallhammer et al., 2011)。*F. noatunensis* 在其他水生生物中的存活情况尚不清楚,但在受污染水产养殖场附近的贻贝和蟹类中已检测到 *Fnn* 的DNA(Ottem et al., 2008; Duodu and Colquhoun, 2010)。这些无脊椎动物可能在营养性动物流行病学中发挥作用。紫贻贝(*Mytilus edulis*)是滤食性动物,可从水中、鱼尸体或感染鱼类的上皮细胞、黏液等中获得细菌。黄道蟹(*Cancer pagurus*)可在水中或食用受感染的贻贝后被感染(Ottem et al., 2008)。不过,由于研究人员只对 *Fnn* 的DNA进行了测试,所以这种细菌在这些水生动物体中的生存能力尚不清楚。有研究表明,非临床带菌动物和观赏鱼类(表17.1和表17.2)的养殖可以传播该病原菌(Camus et al., 2013);同样,进口和运输受感染商品鱼类也会将该病原菌传播到新的地区(Ottem et al., 2008; Birkbeck et al., 2011)。

17.2 感染诊断

弗朗西斯菌病(francisellosis)的临床症状是非特定性的,并呈现多样化。在不同大小和鱼龄的鱼中存在广泛的死亡率(从10%到90%以上),并经常观察到异常泳动和厌食症状(Soto et al., 2009a, 2012b)。尼罗罗非鱼养殖场病害死亡率的巨大差异(5%~90%)与水温、水质、混合感染和养殖操作压力有关(Soto et al., 2009a, 2012b)。温水性或暖水性鱼类的疾病大多暴发于20~25 ℃(Soto et al., 2009a, 2011a, 2014b)。养殖鳕鱼的死亡率在5%~40%,挪威野生鳕鱼的感染率可以达到13%(Ottem et al., 2008)。挪威鳕鱼养殖场疾病暴发时的水温为10~15 ℃(Nylund et al., 2006; Olsen et al., 2006),而智利鲑鱼养殖场观测的数据为8~10 ℃(Birkbeck et al., 2007)。

对弗朗西斯菌病的初步诊断基于临床病史、死亡模式、大体组织学检查,通常附带非特异性的吉姆萨和革兰氏染色(图17.1~图17.3)。鉴别诊断包括分枝杆菌病(mycobacteriosis)、诺卡氏菌病(nocardiosis)、立克次氏体病(piscirickettsiosis)、肾杆菌病(renibacteriosis)、巴氏杆菌病(pasteurellosis)和爱德华氏菌病(edwardsiellosis)。可以通过病原分离、分子诊断、血清学方法来实现明确诊断，理论上来说应至少结合使用两种诊断方法。

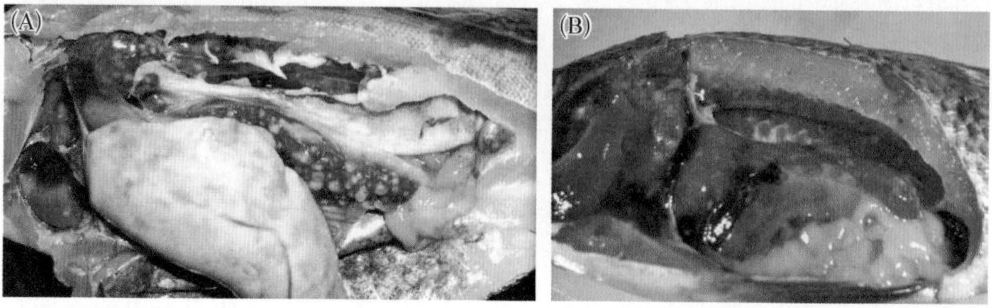

图 17.1 鱼类弗朗西斯菌病的大体病理学。(A) 感染 *Fnn* 的大西洋鳕的脾肿大和肾肿大，伴多灶性白色结节。由挪威国家兽医研究所的 Duncan Colquhoun 博士提供。(B)感染 *Fno* 的尼罗罗非鱼。由哥斯达黎加国立大学的 Juan A. Morales 博士提供

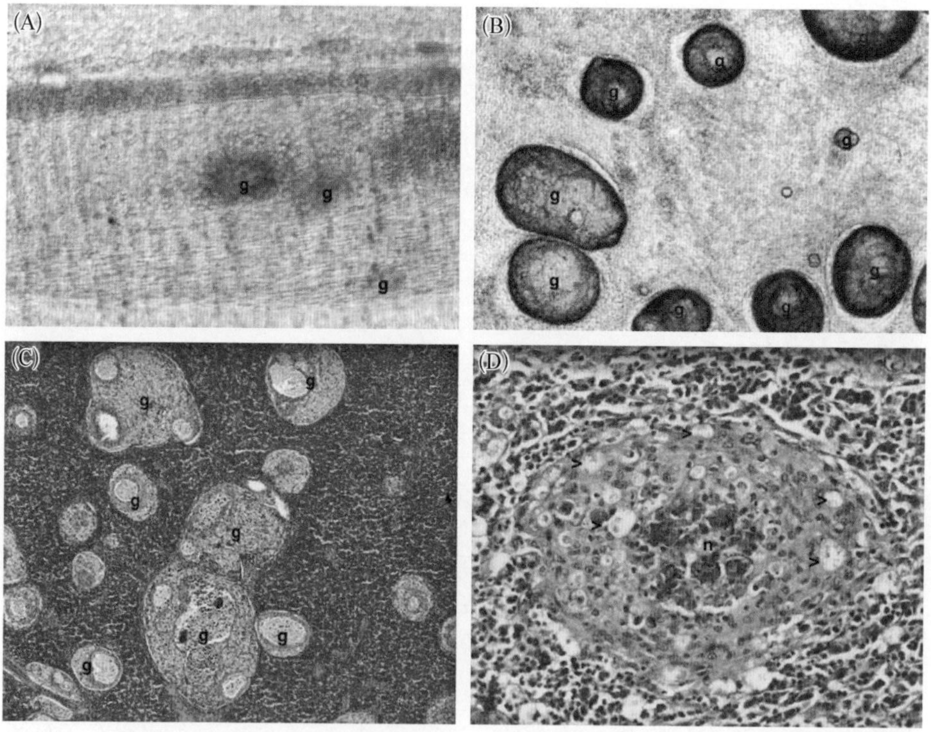

图 17.2 尼罗罗非鱼感染 *Fno* 的显微镜检。(A) 鳃肉芽肿湿压片，"g"表示肉芽肿(苏木精-伊红染色，40 倍放大)，由哥斯达黎加国立大学的 Juan A. Morales 博士提供；(B) 脾肉芽肿湿压片，"g"表示肉芽肿(苏木精-伊红染色，40 倍放大)，由哥斯达黎加国立大学的 Juan A. Morales 博士提供；(C) 广泛的多灶性肉芽肿病变，伴有脾混合炎性浸润，"g"表示肉芽肿(苏木精-伊红染色，10 倍放大)；(D) 典型头肾肉芽肿，呈现上皮样巨噬细胞、淋巴细胞和坏死细胞围绕的坏死的深染核(n)，箭头表示头肾吞噬细胞中的细胞内球状细菌(苏木精-伊红染色，40 倍放大)

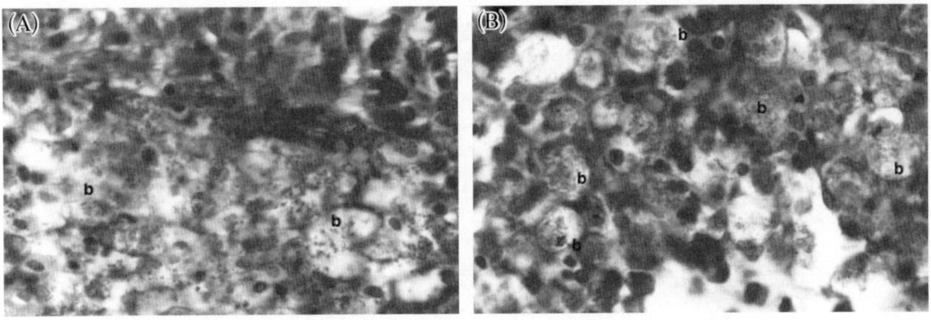

图 17.3 自然感染 *Fno* 的尼罗罗非鱼。鳃(A)和脾(B)中的肉芽肿性炎症显微照片。注意细胞内和细胞外广泛存在的阳性染色细菌，以"b"表示(吉姆萨染色，放大 100 倍)。由哥斯达黎加国立大学的 Juan A. Morales 博士提供

培养苛养微生物的培养基已被用于分离 *F. noatunensis* (Birkbeck et al., 2011; Colquhoun and Duodu, 2011)。此外，成功恢复分离 *Fnn* 的细胞系包括三文鱼头肾细胞、大西洋鲑肾细胞和虹鳟性腺细胞系，而 *Fno* 则从大鳞大麻哈鱼胚胎细胞中被分离出(Hsieh et al., 2006; Nylund et al., 2006)。

识别和鉴定 *F. noatunensis* 的分子诊断方法包括属特异性常规 PCR (Forsman et al., 1994)、检测 *Fnn* 的多重实时 PCR(Ottem et al., 2008)、检测 *Fno* 和 *Fnn* 的实时定量 PCR(Soto et al., 2010a; Duodu et al., 2012b)和环介导等温扩增(LAMP) (Caipang et al., 2010)。

17.3 病理学

患病鱼的临床症状包括皮肤苍白或暗沉、出血(从瘀点到瘀斑)、脱鳞溃疡伴有鳞片脱落、鱼鳍磨损、眼球突出和鳃苍白伴白色结节，还可能产生肉芽肿性或脓性肉芽肿性炎症、腹部肿大伴结肠炎、腹水及广泛分布的多灶性浅棕色或奶油色结节，且结节主要分布于心、肝、性腺、头肾、后肾和脾，并伴有明显的脾肿大和肾肿大(图 17.1)。白色结节也出现在鳃、肝(尤其在鳕鱼中)、眼部，偶尔也可见于胃肠壁和肠系膜脂肪中。

聚合肉芽肿性炎症可以导致高达 90%的组织实质替代(Soto et al., 2009a; Birkbeck et al., 2011)。在组织(特别是鳃、脾和肾)湿压片显微镜检中，重症病例可以观察到形态良好的肉芽肿[图 17.2(A)(B)]。肉芽肿的特征是坏死的中心被一层巨噬细胞包围，通常含有大量直径为 0.5～1.0 μm 的球状细菌，外层为成纤维细胞、空泡细胞、淋巴细胞和小血管[图 17.2(C)(D)；图 17.3]。在慢性感染的鱼中，含黑色素巨噬细胞的结构良好的肉芽肿在脾和头肾中特别明显。肝有小的坏死灶和充满细菌的巨噬细胞。躯干肾的肉芽肿性炎症相对较轻，很大程度上保留了肾元。鳃有中度的上皮增生并伴有少量小肉芽肿，以及充

满了菌体的巨噬细胞的鳃片层融合。心脏可见心房内皮肥大。重症病例中，在心包和心肌中可见广泛的细胞浸润和肉芽肿。肉芽肿在脑部并不常见，有报道称，在严重感染的鱼中存在大量的单核炎性浸润症状（Soto et al., 2009a）。特别是在急性和亚急性感染中，在细胞内外可以检测到小的多形性球杆菌（图 17.3）。

17.4 病理生理学、致病机制和毒力

Soto 等（2013a）研究了野生型 *Fno* 浸泡感染尼罗罗非鱼后的弗朗西斯菌病。感染 3 h 后收集的体表黏液中含有最多的 *Fno* 基因组当量。感染 96 h 后，鱼出现败血症，并且在脾、头肾、后肾、鳃、心、肝、脑、性腺和胃肠道中呈现显著增加的 *Fno* 基因组当量。主要见于脾、前肾和后肾中的典型弗朗西斯菌病肉芽肿也与基因组当量的检出相关联（Soto et al., 2013a）。Gjessing 等（2011）研究了实验室感染 *Fnn* 的鳕脾脏中的组织和细胞动态变化。他们发现了脾病灶内的巨噬细胞样细胞和粒细胞样细胞的过氧化物酶和溶菌酶染色。由此，研究者认为，至少在鳕鱼中弗朗西斯菌病会引起化脓性肉芽肿炎症反应。

F. noatunensis 不仅在胞外环境中，也能在真核细胞（特别是单核细胞/巨噬细胞系）内存在和复制（Soto et al., 2010b; Bakkemo et al., 2011; Furevik et al., 2011; Vestvik et al., 2013; Brudal et al., 2014）。在罗非鱼头肾源巨噬细胞中，*Fno* 能在 72 h 内完成有效复制，并最终形成细胞毒性。在鳕鱼中，*Fnn* 能黏附、穿透单核细胞和巨噬细胞系细胞并在其中复制，而在中性粒细胞和 B 细胞中则较少（Furevik et al., 2011）。Brudal 等（2014）的研究证实，在斑马鱼胚胎中，中性粒细胞和巨噬细胞中均携带 *F. noatunensis*，其中巨噬细胞是细菌复制的主要场所。

Francisella 致病岛 FPIs（*Francisella* pathogenicity islands）（一个在大多数已测序 *Francisella* 基因组中重复的由 16～19 个基因组成的簇）中的基因似乎编码介导细菌胞内存活的毒力因子（Sridhar et al., 2012; Sjödin et al., 2012）。*iglABCD* 操纵子对 *Fno* 的毒力测定具有重要意义，对罗非鱼和斑马鱼的疾病诱发和病原的巨噬细胞内存活是必需的（Soto et al., 2009b, 2010b; Hansen 个人交流）。关于诺神弗朗西斯菌致病岛 FPI 中与毒力相关的基因突变仍知之甚少（Soto et al., 2009b, 2010b; Hansen et al., 2013）。Soto 等（2009b, 2010b）描述了通过 *iglC* 基因突变对 *Fno* 的致弱作用，此外在早期的两项研究中，报道了罗非鱼腹腔注射约 1.0×10^8 CFU/mL 野生型 *Fno* 后的死亡率为 100%。感染野生型菌株（LADL 07 - 285A）的鱼仅 48 h 后死亡，而腹腔注射 *iglC* 突变株，攻毒后 30 d 内只有一条鱼死亡。用野生型菌株（1.8×10^7～3.7×10^7 CFU/mL）浸泡攻毒时，约 50%的鱼存活，而减毒株浸泡攻毒的存活率为

100%。Soto 等(2010b)研究发现 *iglC* 基因突变对 *Fno* 的补体裂解敏感性没有影响。然而,*iglC* 基因被认为对病原菌在巨噬细胞和内皮细胞内的复制很重要(Soto et al., 未发表)。虽然巨噬细胞和上皮细胞内化了野生型和突变型细菌,但突变株没有复制。*iglC* 基因突变也减少了对巨噬细胞的细胞毒性和细胞凋亡(Soto et al., 2010b)。由于在胞内生长过程中高表达,*iglC* 基因是 FPI 中研究最广泛的基因之一,从而证明了其对致病性和毒力的重要性(Nano and Schmerk,2007)。

Hansen 等(2013)研究表明,*Fno* 中 *pdpA* 基因(编码致病性决定蛋白 A,PdpA)的突变会导致斑马鱼模型中毒力的显著减弱。该基因全长约 2.4 kbp,是 FPI 中最大的可读框之一(Nano and Schmerk, 2007)。它位于一个包含 *pdpB*、*vgrG* 和 *dotU* 基因的假定操纵子的起始端。人们对 *pdpA* 的作用和功能知之甚少,相关研究仍在继续。我们对该基因的大部分理解来自对新凶手弗朗西斯菌(*F. novicida*)和土拉热弗朗西斯菌(*F. tularensis*)的研究(Nano et al., 2004; Schmerk et al., 2009a, b; Chou et al., 2013)。该基因在后一个种中是保守的,同源性超过 97%(Schmerk et al., 2009a, b)。尽管没有怀疑它是Ⅵ型分泌系统中一个组成装配蛋白,但 PdpA 可能作为伴侣蛋白或效应蛋白发挥作用(Schmerk et al., 2009b; Bröms et al., 2012a, b)。Nano 等(2004)的实验表明,高达 10^7 CFU/mL 剂量的新凶手弗朗西斯菌 *pdpA* 突变株在皮下注射小鼠后并未产生死亡,而野生型菌株的 LD_{50} 则为 $2.0\times10^2\sim5.0\times10^2$ CFU/mL。此外,Schmerk 等(2009b)的研究表明,*F. novicida* 的 *pdpA* 突变株未能逃脱受感染巨噬细胞的吞噬体,并且在鸡胚和小鼠中表现出高度减弱的毒力。

Fno 的 *pdpA* 基因编码 *F. tularensis* 致病性决定蛋白 A 的同源蛋白。在 *F. tularensis* 中,PdpA 是细胞内生长和毒力所必需的,然而,*Fno* 中的 *pdpA* 基因在鱼类弗朗西斯菌病致病机制中的作用尚不清楚。Ⅵ型分泌系统在细菌致病性中的作用已被公认为细菌毒力的重要机制。通过类似于其他分泌系统针状复合物的保守基质穿过细菌和真核细胞的细胞膜,Ⅵ型分泌系统将大分子从细菌细胞质输送到宿主细胞中(de Bruin et al., 2007; Nano and Schmerk, 2007)。弗朗西斯菌属的Ⅵ型分泌系统在致病岛 FPI 中编码,而系统核心组件的缺失会导致细菌无法逃脱吞噬体并在细胞内复制(Bröms et al., 2012a, b)。此外,尽管弗朗西斯菌Ⅵ型分泌系统的核心组件与典型的Ⅵ型分泌蛋白没有结构上的相似性,但它们的遗传位置和分泌模式表明它们可能具有相似的功能(Ludu et al., 2008; Bröms et al., 2012a, b)。

最近,与野生型亲本菌株一起,Farrell (2015)研究了两种标记不同的 *Fno* 的 *pdpA* 突变株(Δ*pdpA* - 1 和 Δ*pdpA* - 2)浸泡感染杂交红罗非鱼(*Oreochromis* sp.)的毒力情况。两种突变株均呈现高度弱毒性,在水中它们的

LD_{50}大于 2.0×10^6 CFU /mL，而野生型菌株的 LD_{50} 为 891 CFU /mL。与 Δ*pdpA* 突变株相比，野生型菌株对过氧化氢的氧化杀伤更具抗性，但所有菌株对次氯酸钠和血清介导的溶菌作用都具有相似的敏感性。该研究进一步证实了 *pdpA* 基因产物在 *Fno* 中作为重要毒力因子的作用，并证实了该基因突变导致减毒的假设。

目前人们对鱼的免疫病理反应知之甚少。在攻击鳕吞噬细胞时，*Fnn* 的 LPS 似乎会刺激一个温和的免疫反应(Bakkemo et al., 2011)。此外，鳕中高水平的 IL－10 应答表明 *Fnn* 可以促使更强的免疫应答向非完全有效的 Th2(二型辅助性 T 细胞)反应转变(Bakkemo et al., 2011)。鳕巨噬细胞感染 *Fnn* 后，白介素 IL－1b 和 IL－8 的表达水平略有增加(5～10 倍)(Bakkemo et al., 2011)。此外，*Fnn* 能抑制鳕白细胞中的呼吸暴发，这是细胞内生物体在吞噬细胞(如巨噬细胞)的溶酶体中生存的一个特性(Vestvik et al., 2013)。

这些研究表明，*F.noatunensis* 可以通过调节宿主免疫力而在细胞内存活，尽管这种反应可能因种而异。对免疫相关基因表达的体内分析表明，斑马鱼在感染 *Fno* 后 6 h 内通过上调 IL－1β、γ 干扰素(IFNγ)和肿瘤坏死因子 α(TNFα) mRNA 表达，呈现出显著的组织特异性促炎反应，并且反应可持续 7 d(Vojtech et al., 2009)。

17.5 防控策略

预防和控制鱼类弗朗西斯菌病的疫苗、治疗方法和其他策略的开发必须基于对诺神弗朗西斯菌动物流行病学和鱼类弗朗西斯菌病致病机制的充分考虑。在预防和治疗成功的背景下，需要关注病原菌兼性胞内性质，及形成生物被膜和进入 VBNC 状态的能力。尽管实验室研究中已取得可激发鱼类保护性免疫反应的应用前景，但目前尚未有商业化的疫苗上市，并且也很少有关于鱼类弗朗西斯菌病治疗的报道。在鳕鱼和罗非鱼中已实现了 *F. noatunensis* 抗体的产生(Schrøder et al., 2009; Soto et al., 2011b)，但是 *Fnn* 的 LPS 在攻击鳕吞噬细胞时可产生较为温和的免疫应答，这可能会取代高效的保护性免疫反应而削弱免疫效力(Bakkemo et al., 2011)。最近的研究结果表明，黏膜和血清抗体通过与病原体上的特异性黏附素结合，使之凝集从而阻止附着于靶细胞，发挥重要的保护作用(Soto et al., 2015a)。这些抗体调理吞噬细胞的病原体，也激活补体的经典途径和增强抗体介导的细胞杀伤。被动免疫后的成年罗非鱼体内的抗体面对 10^4 CFU/尾和 10^5 CFU/尾剂量的野生型 *Fno* 攻毒提供了显著的保护效力($P<0.001$)，但无法抵御 10^6 CFU/尾的攻毒剂量(Soto et al., 2011b)。

强烈的细胞免疫反应可预防脊椎动物感染弗朗西斯菌属病原(Kirimanjeswara et al., 2008; Chou et al., 2013)。因此，灭活或亚单位疫苗可能需要佐剂来激发细胞免疫(Soto et al., 2011b; Brudal et al., 2015)。另一种可能的策略是使用减毒活疫苗,它能持续存在于组织中并激发保护性免疫(Soto et al., 2011b, 2014a; Brudal et al., 2015)。Soto 等(2009b, 2011b, 2014a)研究表明,*Fno* 的 *iglC* 基因插入突变能使细菌减毒,接种免疫该减毒株的罗非鱼显现出对野生型 *Fno* 浸泡攻毒的显著保护作用。Δ*iglC* 减毒株保护罗非鱼免受 *Fno* 野生型菌株高剂量(LD_{80})的攻毒。浸泡免疫 Δ*iglC* 减毒株的天然罗非鱼对致死剂量 *Fno* 的攻毒表现出平均存活率为 90%的保护力(Soto et al., 2011b)。尽管这是一项实验室阶段的研究,但它表明减毒活疫苗具有潜在的应用前景。最近,Brudal 等(2014)开发了一种基于"外膜囊泡"的疫苗,该疫苗可保护斑马鱼胚胎免受致病性 *Fnn* 的侵害。

将养殖水温升高至 30 ℃可缓解实验室感染的尼罗罗非鱼的临床症状和死亡率(Soto et al., 2012b),并已成功应用于观赏鱼疾病的预防(Soto et al., 2014b)。温度对罗非鱼弗朗西斯菌病的发生有显著影响,但盐度对其无影响。这一方法可应用于能操作控制升温的室内养殖设施中。

F. noatunensis 的抗菌药敏试验基于为 *F. tularensis* 设计的方法(Baker et al., 1985)。肉汤微量稀释试验是在改良 Mueller-Hinton Ⅱ 肉汤(阳离子调节)[Becton, Dickinson and Company (BD), Sparks, Maryland]培养基上进行的,补加了 2%的 IsoVitalex (BD BBL 培养基)和 0.1%的葡萄糖。两个亚种对氟苯尼考和噁喹酸具有较低的最小抑菌浓度(minimal inhibitory concentration, MIC),而对红霉素和甲氧苄啶/磺胺甲噁唑具有高 MIC 值。此外,*Fno* 对恩诺沙星、庆大霉素、新霉素、土霉素、四环素、链霉素和呋喃妥因的 MIC 较低,但对青霉素、氨苄西林、阿莫西林、头孢噻呋、磺胺二甲氧嘧啶、磺胺噻唑、新生霉素、酒石酸泰乐菌素和克林霉素的 MIC 较高(Soto et al., 2012a)。Isachsen 等(2012)发现 *Fnn* 对土霉素、环丙沙星和链霉素具有高的 MIC,但对氟甲喹和利福平敏感。

美国目前仅批准了三种抗菌药物用于食用鱼。其中,氟苯尼考和土霉素是治疗弗朗西斯菌病的潜在药物。在饲料中添加土霉素和氟苯尼考投喂处理的罗非鱼死亡率(自然感染和实验感染)均有所下降(Mauel et al., 2003; Soto et al., 2010c, 2013b),不过,快速治疗是十分必要的,因为受感染的鱼很快变得厌食。抗生素治疗在急性疾病中特别成功,但对亚急性或慢性感染则不那么有效(Soto et al., 2010c, 2013b)。我们猜测抗菌剂无法穿透亚急性和慢性感染期间形成的肉芽肿。另外,在推荐治疗方案之前必须确定药物进入真核细胞的能力。

按每天每千克鱼饲服 15 mg 氟苯尼考和 20 mg 氟苯尼考的剂量,连续投喂 10 d,可显著降低尼罗罗非鱼弗朗西斯菌病的死亡率(Soto et al., 2013b)。Mauel 等(2003)报道称,以每 0.45 kg 鱼投喂 4.0 g 土霉素强化的 Rangen™饲料来饲喂与病菌共栖的鱼,存活率可达 83%。相比之下,非药物治疗组只有 17%

的鱼存活。

最近，一种涉及温度控制和抗菌治疗双管齐下的方法被推荐(Soto et al.，2014b)。在2周的时间内将养殖水温逐渐升至30 ℃并维持3个月，一旦水温升至30 ℃，就以每日每4.5 kg鱼饲服3 g土霉素，连续10 d的总剂量给鱼投喂药饵饲料(Soto et al.，2014b)。

一旦病害暴发，明智的做法是降低养殖密度并对养殖设施进行消毒。三种常用消毒剂对水产设施中浮游和生物被膜形式的 *Fno* 均有效，其中包括1%的Virkon®(10 min)、10%的家用漂白剂(约6 000 ppm有效氯，10 min)和3%过氧化氢(10 min)的体外消毒(Soto et al.，2015b)。

17.6 研究展望

鱼类弗朗西斯菌病被认为是养殖鱼类最重要的新发疾病之一。其广泛的宿主范围、苛养特性、在多种环境下生存的能力及遍布全球的存在凸显了这一引人关注病原的重要性，然而我们对其知之甚少。鱼类弗朗西斯菌病的环境持久性和发病机理需要进一步的研究，同时需要建立预防、治疗和根除方案，以有效对抗浮游、生物被膜和VBNC状态下的病原菌。

鱼类弗朗西斯菌病疫苗的研制，应基于选择可能刺激鱼类黏膜和全身免疫应答的抗原。最近的研究表明，通过分泌系统释放的效应蛋白显示出不错的前景，不过，灭活疫苗应当激发强烈的细胞免疫。这对于其他需要佐剂和注射接种的病原体也是一个挑战。如果要将减毒活疫苗用于接种罗非鱼或鳕鱼，还应调查研究该疫苗在其他鱼类中的减毒效果。虽然不同国家和地区的疫苗安全法规不同，但研究人员和提供疫苗的制药公司应该实施严格的措施，以便为鱼类、人和环境提供最安全的产品。疫苗应高效且易于接种，例如通过浸泡或口服方法。理想的抗菌剂应该具有杀菌作用，并能够穿透肉芽肿和活体生物被膜。基于“有机”产品进行治疗和预防的研究是必要的，特别像“鱼菜共生”这样的“有机”实践都很重要。

为应对这一病害，需要更好的生物安全和疾病早期诊断措施。在运输前对亚临床感染、带菌或潜伏感染的鱼类进行病原检测，有助于预防新的疫情暴发和环境污染。同样，由于弗朗西斯菌病可能存在垂直传播，开发非致死性诊断方法对亲鱼进行检测是可取的。消毒和消除的方案应考虑细菌的生物学特性。选育抗病鱼种是必要的，这对重要鱼类养殖品种尤其必要。应明确弗朗西斯菌的宿主范围，包括爬行动物和两栖动物。由于混合感染的报道很常见，应调查慢性感染、接种疫苗和天然鱼类对机会性和原发性病原菌(如链球菌、弧菌和气单胞菌)混合感染的易感性。最后，更好地了解鱼类不同生命阶段和受影响种类中弗朗西斯菌病的总体影响后果，为将来研究的必要性提供有力依据。

参考文献

[1] Abd, H., Johansson, T., Golovliov, I., Sandström, G. and Forsman, M. (2003) Survival and growth of *Francisella tularensis* in *Acanthamoeba castellanii*. *Applied and Environmental Microbiology* 69, 600 - 606.

[2] Baker, C.N., Hollis, D.G. and Thornsberry, C. (1985) Antimicrobial susceptibility testing of *Francisella tularensis* with a modified Mueller-Hinton Broth. *Journal of Clinical Microbiology* 22, 212 - 215.

[3] Bakkemo, K.R., Mikkelsen, H., Bordevik, M., Torgersen, J., Winther-Larsen, H.C. *et al.* (2011) Intracellular localization and innate immune responses following *Francisella noatunensis* infection of Atlantic cod (*Gadus morhua*) macrophages. *Fish and Shellfish Immunology* 31, 993 - 1004.

[4] Birkbeck, T.H., Bordevik, M., Frøystad, M.K. and Baklien, A. (2007) Identification of *Francisella* sp. from Atlantic salmon, *Salmo salar* L., in Chile. *Journal of Fish Diseases* 30, 505 - 507.

[5] Birkbeck, T.H, Feist, S.W. and Verner-Jeffreys, D.W. (2011) *Francisella* infections in fish and shellfish. *Journal of Fish Diseases* 34, 173 - 187.

[6] Bohle, H., Tapia, E., Martínez, A., Rozas, M. Figueroa, A. *et al.* (2009) Francisella philomiragia, bacteria asociada con altas mortalidades en salmones del Atlántico (Salmo salar) cultivados en balsas-jaulas en el Lago Llanquihue. *Archivos de Medicina Veterinaria* 41, 237 - 244.

[7] Bröms, J.E., Meyer, L., Lavander, M., Larsson, P., Sjöstedt, A. *et al.* (2012a) DotU and VgrG, core components of type VI secretion systems, are essential for *Francisella* LVS pathogenicity. *PLoS One* 7(4): e34639.

[8] Bröms, J.E., Meyer, L., Sun, K., Lavander, M. and Sjöstedt, A. (2012b) Unique substrates secreted by the type VI secretion system of *Francisella tularensis* during intramacrophage infection. *PLoS One* 7(11): e50473.

[9] Brudal, E., Ulanova, L.S., O Lampe, E., Rishovd, A.L., Griffiths, G. *et al.* (2014) Establishment of three *Francisella* infections in zebrafish embryos at different temperatures. *Infection and Immunity* 82, 2180 - 2194.

[10] Brudal, E., Lampe, E.O., Reubsaet, L., Roos, N., Hegna, I.K. *et al.* (2015) Vaccination with outer membrane vesicles from *Francisella noatunensis* reduces development of francisellosis in a zebrafish model. *Fish and Shellfish Immunology* 42, 50 - 57.

[11] Caipang, C.M., Kulkarni, A., Brinchmann, M.F., Korsnes, K. and Kiron, V. (2010) Detection of *Francisella piscicida* in Atlantic cod (*Gadus morhua* L.) by the loop-mediated isothermal amplification (LAMP) reaction. *Veterinary Journal* 184, 357 - 361.

[12] Camus, A.C., Dill, J.A., McDermott, A.J., Clauss, T.M., Berliner, A.L. et al. (2013) *Francisella noatunensis* subsp. *orientalis* infection in Indo-Pacific reef fish entering the United States through the ornamental fish trade. *Journal of Fish Diseases* 36, 681 - 684.

[13] Chen, S.-C., Tung, M.-C., Chen, S.-P., Tsai, J.-F. and Wang, P.-C. (1994) Systematic granulomas caused by a *rickettsia*-like organism in Nile tilapia, *Oreochromis niloticus* (L.), from southern Taiwan. *Journal of Fish Diseases* 17, 591 - 599.

[14] Chen, S.-C., Wang, P.-C., Tung, M.-C., Thompson, K.D. and Adams, A. (2000) A *Piscirickettsia salmonis*-like organism in grouper, *Epinephelus melanostigma* in Taiwan. *Journal of Fish Diseases* 23, 415 - 418.

[15] Chern, R.S. and Chao, C.B. (1993) Outbreaks of a disease caused by rickettsia-like

organism in cultured tilapias in Taiwan. *Fish Pathology* 29, 61－71.

[16] Chou, A.Y., Kennett, N.J., Nix, E.B., Schmerk, C.L., Nano, F.E. *et al.* (2013) Generation of protection against *Francisella novicida* in mice depends on the pathogenicity protein PdpA, but not PdpC or PdpD. *Microbes and Infection* 15, 816－827.

[17] Colquhoun, D.J. and Duodu, S. (2011) *Francisella* infections in farmed and wild aquatic organisms. *Veterinary Research* 42: 47.

[18] de Bruin, O.M., Ludu, J.S. and Nano, F.E. (2007) The *Francisella* pathogenicity island protein IglA localizes to the bacterial cytoplasm and is needed for intracellular growth. *BMC Microbiology* 7: 1.

[19] Duodu, S. and Colquhoun, D. (2010) Monitoring the survival of fish-pathogenic *Francisella* in water microcosms. *FEMS Microbiology Ecology* 74, 534－541.

[20] Duodu, S., Larsson, P., Sjödin, A., Forsman, M. and Colquhoun, D.J. (2012a) The distribution of *Francisella*-like bacteria associated with coastal waters in Norway. *Microbial Ecology* 64, 370－377.

[21] Duodu, S., Larsson, P., Sjödin, A., Soto, E., Forsman, M. *et al.* (2012b) Real-time PCR assays targeting unique DNA sequences of fish-pathogenic *Francisella noatunensis* subspecies *noatunensis* and *orientalis*. *Diseases of Aquatic Organisms* 101, 225－234.

[22] Farrell, F. (2015) *Disruption of the pathogenicity determinant protein A gene (pdpA) in Francisella noatunensis subsp. orientalis results in attenuation and greater susceptibility to oxidative stress.* MSc thesis, Ross University School of Veterinary Medicine, Bassaterre, Saint Kitts and Nevis, West Indies.

[23] Foley, J. E. and Nieto, N. C. (2010) Tularemia. *Veterinary Microbiology* 140, 332－338.

[24] Forsman, M., Sandstrom, G. and Sjostedt, A. (1994) Analysis of 16S ribosomal DNA sequences of *Francisella* strains and utilization for determination of the phylogeny of the genus and for identification of strains by PCR. *International Journal of Systematic Bacteriology* 44, 38－46.

[25] Fryer, J.L., Lannan, C.N., Giovannoni, S.J. and Wood, N.D. (1992) *Piscirickettsia salmonis* gen. nov., sp. nov., the causative agent of an epizootic disease in salmonid fishes. *International Journal of Systematic Bacteriology* 42, 120－126.

[26] Fukuda, Y., Okamura, A., Nishiyama, M., Kawakami, H., Kamaishi, T. *et al.* (2002) Granulomatosis of cultured three-line grunt *Parapristipoma trilineatum* caused by an intracellular bacterium. *Fish Pathology* 37, 119－124.

[27] Furevik, A., Pettersen, E. F., Colquhoun, D. and Wergeland, H. I. (2011) The intracellular lifestyle of *Francisella noatunensis* in Atlantic cod (*Gadus morhua* L.) leucocytes. *Fish and Shellfish Immunology* 30, 488－494.

[28] Gjessing, M. C., Inami, M., Weli, S. C., Ellingsen, T., Falk, K. *et al.* (2011) Presence and interaction of inflammatory cells in the spleen of Atlantic cod, *Gadus morhua* L., infected with *Francisella noatunensis*. *Journal of Fish Diseases* 34, 687－699.

[29] Hansen, J.D., Ray, K., Woodson, J.C., Soto, E. and Welch, T.J. (2013) Disruption of the *Francisella noatunensis pdpA* gene results in virulence attenuation. *Fish and Shellfish Immunology* 34, 1655.

[30] Hsieh, C.Y., Tung, M.C., Tu, C., Chang, C.D. and Tsai, S.S. (2006) Enzootics of visceral granulomas associated with *Francisella*-like organism infection in tilapia (*Oreochromis* sp.) *Aquaculture* 254, 129－138.

[31] Hsieh, C. Y., Wu, Z. B., Tung, M. C. and Tsai, S. S. (2007) PCR and *in situ* hybridization for the detection and localization of a new pathogen *Francisella*-like bacterium (FLB) in ornamental cichlids. *Diseases of Aquatic Organisms* 75, 29－36.

[32] Isachsen, C. H., Vågnes, O., Jakobsen, R. A. and Samuelsen, O. B. (2012) Antimicrobial susceptibility of *Francisella noatunensis* subsp. *noatunensis* strains isolated from Atlantic cod *Gadus morhua* in Norway. *Diseases of Aquatic Organisms* 98, 57 - 62.

[33] Jeffery, K.R., Stone, D., Feist, S.W. and Verner-Jeffreys, D.W. (2010) An outbreak of disease caused by *Francisella* sp. in Nile tilapia *Oreochromis niloticus* at a recirculation fish farm in the UK. *Diseases of Aquatic Organisms* 91, 161 - 165.

[34] Kamaishi, T., Fukuda, Y. and Nishiyama, M. (2005) Identification and pathogenicity of intracellular *Francisella* bacterium in three-line grunt *Parapristipoma trilineatum*. *Fish Pathology* 40, 67 - 71.

[35] Kay, W., Petersen, B. O., Duus, J. Ø., Perry, M. B. and Vinogradov, E. (2006) Characterization of the lipopolysaccharide and beta-glucan of the fish pathogen *Francisella victoria*. *FEBS Journal* 273, 3002 - 3013.

[36] Khoo, L., Dennis, P.M. and Lewbart, G.A. (1995) *Rickettsia*-like organisms in the blue-eyed plecostomus, *Panaque suttoni* (Eigenmann & Eigenmann). *Journal of Fish Diseases* 18, 157 - 164.

[37] Kirimanjeswara, G.S., Olmos, S., Bakshi, C.S. and Metzger, D.W. (2008) Humoral and cell-mediated immunity to the intracellular pathogen *Francisella tularensis*. *Immunological Reviews* 225, 244 - 255.

[38] Leal, C.A.G., Tavares, G.C. and Figueiredo, H.C.P. (2014) Outbreaks and genetic diversity of *Francisella noatunensis* subsp *orientalis* isolated from farm-raised Nile tilapia (*Oreochromis niloticus*) in Brazil. *Genetics and Molecular Research* 13, 5704 - 5712.

[39] Lewisch, E. (2014) Francisellosis in ornamental African cichlids in Austria. *Bulletin of the European Association of Fish Pathologists* 34, 63 - 70.

[40] Ludu, J.S., de Bruin, O.M., Duplantis, B.N., Schmerk, C.L., Chou, A.Y. *et al.* (2008) The *Francisella* pathogenicity island protein PdpD is required for full virulence and associates with homologues of the type VI secretion system. *Journal of Bacteriology* 190, 4584 - 4595.

[41] Mauel, M. J., Miller, D. L. and Frazier, K. (2003) Characterization of a Piscirickettsiosis-like disease in Hawaiian tilapia. *Diseases of Aquatic Organisms* 53, 249 - 255.

[42] Mikalsen, J. and Colquhoun, D.J. (2009) *Francisella asiatica* sp. nov. isolated from farmed tilapia (*Oreochromis* sp.) and elevation of *Francisella philomiragia* subsp. *noatunensis* to species rank as *Francisella noatunensis* comb. nov., sp. nov. *International Journal of Systematic and Evolutionary Microbiology*. Available as abstract only at: doi: 10.1099/ijs.0.002139 - 0 (accessed 14 November 2016).

[43] Mikalsen, J., Olsen, A. B., Tengs, T. and Colquhoun, D. J. (2007) *Francisella philomiragia* subsp. *noatunensis* subsp. nov., isolated from farmed Atlantic cod (*Gadus morhua* L.). *International Journal of Systematic and Evolutionary Microbiology* 57, 1960 - 1965.

[44] Nano, F.E. and Schmerk, C. (2007) The *Francisella* pathogenicity island. *Annals of the New York Academy of Sciences* 1105, 122 - 137.

[45] Nano, F.E., Zhang, N., Cowley, S.C., Klose, K.E., Cheung, K.K. *et al.* (2004) A *Francisella tularensis* pathogenicity island required for intramacrophage growth. *Journal of Bacteriology* 186, 6430 - 6436.

[46] Nguyen, V. V., Dong, H. T., Senapin, S., Pirarat, N. and Rodkhum, C. (2016) *Francisella noatunensis* subsp. *orientalis*, an emerging bacterial pathogen affecting cultured red tilapia (*Oreochromis* sp.) in Thailand. *Aquaculture Research* 47, 3697 - 3702.

[47] Nylund, A., Ottem, K.F, Watanabe, K., Karlsbakk, E. and Krossøy, B. (2006) Francisella sp. (Family *Francisellaceae*) causing mortality in Norwegian cod (*Gadus morhua*) farming. *Archives of Microbiology* 185, 383 - 392.

[48] Olsen, A.B., Mikalsen, J., Rode, M., Alfjorden, A., Hoel, E. *et al*. (2006) A novel systemic granulomatous inflammatory disease in farmed Atlantic cod, *Gadus morhua* L., associated with a bacterium belonging to the genus *Francisella*. *Journal of Fish Diseases* 29, 307 - 311.

[49] Ostland, V.E., Stannard, J.A., Creek, J.J., Hedrick, R.P. and Ferguson, H.W. (2006) Aquatic *Francisella*-like bacterium associated with mortality of intensively cultured hybrid striped bass *Morone chrysops* × *M. saxatilis*. *Diseases of Aquatic Organisms* 72, 135 - 145.

[50] Ottem, K.F., Nylund, A., Karlsbakk, E., Friis-Møller, A. and Krossøy, B. (2007a) Characterization of *Francisella* sp., GM2212, the first *Francisella* isolate from marine fish, Atlantic cod (*Gadus morhua*). *Archives of Microbiology* 187, 343 - 350.

[51] Ottem, K.F., Nylund, A., Karlsbakk, E., Friis-Møller, A., Krossøy, B. *et al*. (2007b), New species in the genus *Francisella* (*Gammaproteobacteria*; *Francisellaceae*); *Francisella piscicida* sp. nov. isolated from cod (*Gadus morhua*). *Archives of Microbiology* 188, 547 - 550.

[52] Ottem, K.F., Nylund, A., Isaksen, T.E., Karlsbakk, E. and Bergh, Ø. (2008) Occurrence of *Francisella piscicida* in farmed and wild Atlantic cod, *Gadus morhua* L., in Norway. *Journal of Fish Diseases* 31, 525 - 534.

[53] Ottem, K.F., Nylund, A., Karlsbakk, E., Friis-Møller, A. and Kamaishi, T. (2009) Elevation of *Francisella philomiragia* subsp. *noatunensis* Mikalsen *et al*. (2007) to *Francisella noatunensis* comb. nov. (syn. *Francisella piscicida* Ottem *et al*. (2008) syn. nov.) and characterization of *Francisella noatunensis* subsp. *orientalis* subsp. nov., two important fish pathogens. *Journal of Applied Microbiology* 106, 1231 - 1243.

[54] Petersen, J.M., Carlson, J., Yockey, B., Pillai, S., Kuske, C. *et al*. (2009) Direct isolation of *Francisella* spp. From environmental samples. *Letters in Applied Microbiology* 48, 663 - 667.

[55] Ruane, N.M., Bolton-Warberg, M., Rodger, H.D., Colquhoun, D.J., Geary, M. *et al*. (2015) An outbreak of francisellosis in wild-caught Celtic Sea Atlantic cod, *Gadus morhua* L., juveniles reared in captivity. *Journal of Fish Diseases* 38, 97 - 102.

[56] Schmerk, C.L., Duplantis, B.N., Wang, D., Burke, R.D., Chou, A.Y. *et al*. (2009a) Characterization of the pathogenicity island protein PdpA and its role in the virulence of *Francisella novicida*. *Microbiology* 155, 1489 - 1497.

[57] Schmerk, C.L., Duplantis, B.N., Howard, P.L. and Nano, F.E. (2009b) A *Francisella novicida pdpA* mutant exhibits limited intracellular replication and remains associated with the lysosomal marker LAMP-1. *Microbiology* 155, 1498 - 1504.

[58] Schrallhammer, M., Schweikert, M., Vallesi, A., Verni, F. and Petroni, G. (2011) Detection of a novel subspecies of *Francisella noatunensis* as an endosymbiont of the ciliate *Euplotes raikovi*. *Microbial Ecology* 61, 455 - 464.

[59] Schrøder, M.B., Ellingsen, T., Mikkelsen, H., Norderhus, E.A. and Lund, V. (2009) Comparison of antibody responses in Atlantic cod (*Gadus morhua* L.) to *Vibrio anguillarum*, *Aeromonas salmonicida* and *Francisella* sp. *Fish and Shellfish Immunology* 27, 112 - 119.

[60] Sjödin, A., Svensson, K., Ohrman, C., Ahlinder, J., Lindgren, P. *et al*. (2012) Genome characterisation of the genus *Francisella* reveals insight into similar evolutionary paths in pathogens of mammals and fish. *BMC Genomics* 13: 268.

[61] Soto, E. and Revan, F. (2012) Culturability and persistence of *Francisella noatunensis* subsp. *orientalis* (syn. *Francisella asiatica*) in sea-and freshwater microcosms.

Microbial Ecology 63, 398-404.
[62] Soto, E., Hawke, J.P., Fernandez, D. and Morales, J.A. (2009a) *Francisella* sp., an emerging pathogen of tilapia, *Oreochromis niloticus* (L.) in Costa Rica. *Journal of Fish Diseases* 32, 713-722.
[63] Soto, E., Fernandez, D. and Hawke, J.P. (2009b) Attenuation of the fish pathogen *Francisella* sp. By mutation of the *iglC* gene. *Journal of Aquatic Animal Health* 21, 140-149.
[64] Soto, E., Bowles, K., Fernandez, D. and Hawke, J.P. (2010a) Development of a real-time PCR assay for identification and quantification of the fish pathogen *Francisella noatunensis* subsp. *orientalis*. *Diseases of Aquatic Organisms* 89, 199-207.
[65] Soto, E., Fernandez, D., Thune, R. and Hawke, J. P. (2010b) Interaction of *Francisella asiatica* with tilapia (Oreochromis niloticus) innate immunity. *Infection and Immunity* 78, 2070-2078.
[66] Soto, E., Endris, R.G. and Hawke, J.P. (2010c) *In vitro* and *in vivo* efficacy of florfenicol for treatment of *Francisella asiatica* infection in tilapia. *Antimicrobial Agents and Chemotherapy* 54, 4664-4670.
[67] Soto, E., Baumgartner, W., Wiles, J. and Hawke, J.P. (2011a) *Francisella asiatica* as the causative agent of piscine francisellosis in cultured tilapia (*Oreochromis* sp.) in the United States. *Journal of Veterinary Diagnostic Investigation* 23, 821-825.
[68] Soto, E., Wiles, J., Elzer, P., Macaluso, K. and Hawke, J.P. (2011b) Attenuated *Francisella asiatica iglC* mutant induces protective immunity to francisellosis in tilapia. *Vaccine* 29, 593-598.
[69] Soto, E., Griffin, M., Wiles, J. and Hawke, J. P. (2012a) Genetic analysis and antimicrobial susceptibility of *Francisella noatunensis* subsp. *orientalis* (syn. *F. asiatica*) isolates from fish. *Veterinary Microbiology* 154, 407-412.
[70] Soto, E., Abrams, S. B. and Revan, F. (2012b) Effects of temperature and salt concentration on *Francisella noatunensis* subsp. *orientalis* infections in Nile tilapia *Oreochromis niloticus*. *Diseases of Aquatic Organisms* 101, 217-223.
[71] Soto, E., Illanes, O., Hilchie, D., Morales, J.A., Sunyakumthorn, P. *et al*. (2012c) Molecular and immunohistochemical diagnosis of *Francisella noatunensis* subsp. *orientalis* from formalin-fixed, paraffin-embedded tissues. *Journal of Veterinary Diagnostic Investigation* 24, 840-845.
[72] Soto, E., Kidd, S., Mendez, S., Marancik, D., Revan, F. *et al*. (2013a) *Francisella noatunensis* subsp. *orientalis* pathogenesis analyzed by experimental immersion challenge in Nile tilapia, *Oreochromis niloticus* (L.). *Veterinary Microbiology* 164, 77-84.
[73] Soto, E., Kidd, S., Gaunt, P.S. and Endris, R. (2013b) Efficacy of florfenicol for control of mortality associated with *Francisella noatunensis* subsp. *orientalis* in Nile tilapia, *Oreochromis niloticus* (L.). *Journal of Fish Diseases* 36, 411-418.
[74] Soto, E., McGovern-Hopkins, K., Klinger-Bowen, R., Fox, B.K., Brock, J. *et al*. (2013c) Prevalence of *Francisella noatunensis* subsp. *orientalis* in cultured tilapia on the island of Oahu, Hawaii. *Journal of Aquatic Animal Health* 25, 104-109.
[75] Soto, E., Brown, N., Gardenfors, Z.O., Yount, S., Revan, F. *et al*. (2014a) Effect of size and temperature at vaccination on immunization and protection conferred by a live attenuated *Francisella noatunensis* immersion vaccine in red hybrid tilapia. *Fish and Shellfish Immunology* 41, 593-599.
[76] Soto, E., Primus, A.E., Pouder, D.B., George, R.H., Gerlach, T.J. *et al*. (2014b) Identification of *Francisella noatunensis* in novel host species French grunt (*Haemulon flavolineatum*) and Caesar grunt (*Haemulon carbonarium*). *Journal of Zoo and Wildlife Medicine* 45, 727-731.
[77] Soto, E., Tobar, J. and Griffin, M. (2015a). Mucosal vaccines. In: Beck, B.H. and

Peatman, E. (eds) *Mucosal Health in Aquaculture*. Academic Press/Elsevier, London/San Diego, California/Waltham, Massachusetts/Kidlington, UK, pp. 297 - 323.

[78] Soto, E., Halliday-Simmonds, I., Francis, S., Kearney, M. T. and Hansen, J. D. (2015b) Biofilm formation of *Francisella noatunensis* subsp. *orientalis*. *Veterinary Microbiology* 181, 313 - 317.

[79] Sridhar, S., Sharma, A., Kongshaug, H., Nilsen, F. and Jonassen, I. (2012) Whole genome sequencing of the fish pathogen *Francisella noatunensis* subsp. *orientalis* Toba04 gives novel insights into *Francisella* evolution and pathogenicity. *BMC Genomics* 13: 598.

[80] Verhoeven, A. B., Durham-Colleran, M. W., Pierson, T., Boswell, W. T. and Van Hoek, M.L. (2010) *Francisella philomiragia* biofilm formation and interaction with the aquatic rotest *Acanthamoeba castellanii*. *The Biological Bulletin* 219, 178 - 188.

[81] Vestvik, N., Rønneseth, A., Kalgraff, C.A., Winther-Larsen, H.C., Wergeland, H.I. *et al*. (2013) *Francisella noatunensis* subsp. *noatunensis* replicates within Atlantic cod (*Gadus morhua* L.) leucocytes and inhibits respiratory burst activity. *Fish and Shellfish Immunology* 35, 725 - 733.

[82] Vojtech, L.N., Sanders, G.E., Conway, C., Ostland, V. and Hansen, J.D. (2009) Host immune response and acute disease in a zebrafish model of *Francisella* pathogenesis. *Infection and Immunity* 77, 914 - 925.

[83] Wangen, I.H., Karlsbakk, E., Einen, A.C., Ottem, K.F., Nylund, A. et al. (2012) Fate of *Francisella noatunensis*, a pathogen of Atlantic cod *Gadus morhua*, in blue mussels *Mytilus edulis*. *Diseases of Aquatic Organisms* 98, 63 - 72.

18

分 枝 杆 菌

David T. Gauthier* 和 Martha W. Rhodes

18.1 引言

18.1.1 分枝杆菌属(*Mycobacterium* spp.)

分枝杆菌属,属放线菌目(Actinomycetales)、分枝杆菌科(Mycobacteriaceae),由一类抗酸、好氧至微需氧、非运动型的革兰氏阳性杆菌组成。除某些不可培养菌种[如麻风分枝杆菌(*M. leprae*)]外,分枝杆菌通常根据细菌的生长速度和色素沉着的表型特征进行分类,称之为鲁尼恩分类法(Runyon, 1959)。鲁尼恩Ⅰ~Ⅲ型的分枝杆菌较难培养,在固体培养基中需要培养5 d以上才能产生菌落。Ⅰ型分枝杆菌,如海分枝杆菌(*M. marinum*)、假肖茨分枝杆菌(*M. pseudoshottsii*)和堪萨斯分枝杆菌(*M. kansasii*),具有光照产色性,在光照下产生橙黄色素(图18.1)。Ⅱ型分枝杆菌具有黑暗产色性,不论有无光照都能产生色素,这类菌包括戈登分枝杆菌(*M. gordonae*)和瘰疬分枝杆菌(*M. scrofulaceum*)。Ⅲ型分枝杆菌不产色素,属中许多重要的病原菌,如结核分枝杆菌(*M. tuberculosis*)、鸟分枝杆菌(*M. avium*)、溃疡分枝杆菌(*M. ulcerans*)和鱼类病原肖茨分枝杆菌(*M. shottsii*),均属于该型。Ⅳ型分枝杆菌生长速度较快,不到5 d便能在琼脂培养基上形成菌落,该型菌不产色素或生长后期产色素,包括偶发分枝杆菌(*M. fortuitum*)、龟分枝杆菌(*M. chelonae*)、脓肿分枝杆菌(*M. abscessus*)和外来分枝杆菌(*M. peregrinum*)等病原菌。

尽管鲁尼恩分类法目前仍在使用,但与最新的分子分类法相比则存在一些缺陷。例如,根据生长温度的不同,苏尔加分枝杆菌(*M. szulgai*)可表现为光致产色或暗致产色;溃疡分枝杆菌不产色素,而与其密切相关的拟议生态变种假肖茨分枝杆菌则产色素(Doig et al., 2012)。此外,生长速度与系统发育之间的关系也更为密切。Ⅰ~Ⅲ型分枝杆菌与Ⅳ型分枝杆菌形成不同的进化枝,而后者的分类特征是16S rRNA基因有明显缺失以及存在两个rRNA操纵子(Bercovier et al., 1986; Rogall et al., 1990; Menendez et al., 2002)。采用多

* 通信作者邮箱:dgauthie@odu.edu。

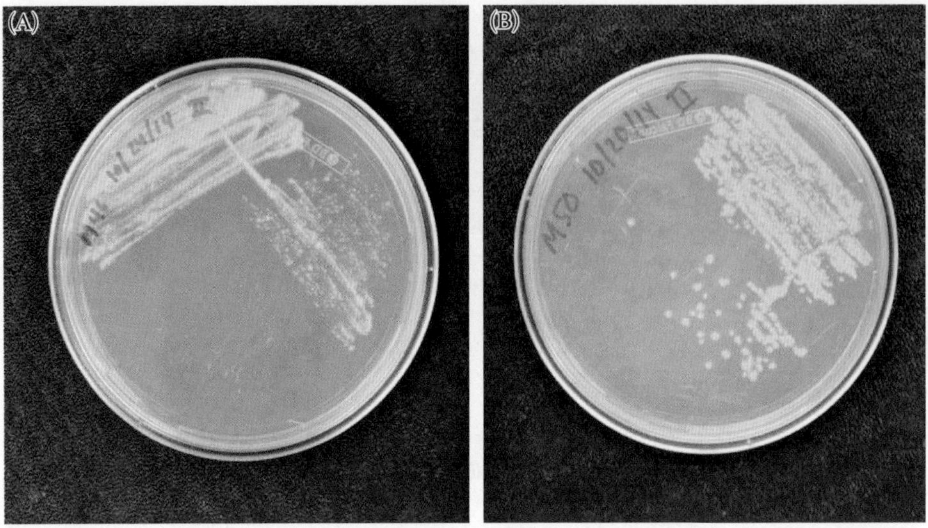

图 18.1 (A) 生长快速、无色素的龟分枝杆菌；(B) 生长缓慢、光产色的海分枝杆菌

位点测序，特别是 16S rRNA、*hsp65*（热休克蛋白，65 kDa）、*rpoB*（RNA 聚合酶β亚基）和 *erp*（外部重复蛋白）等基因位点，是鉴定已知菌种的有效方法（Devulder et al.，2005；Gauthier et al.，2011）。

分枝杆菌包括几种重要的人类病原菌：结核分枝杆菌和牛分枝杆菌（*M. bovis*）引起结核病；麻风分枝杆菌（*M. leprae*）为汉森氏病（Hansen's disease）或麻风的病原；溃疡分枝杆菌则是西非和澳大利亚新发疾病布鲁里溃疡的病因（Van Der Werf et al.，2005）。分枝杆菌还感染其他动物宿主并存在于多种环境生态位中。如范巴伦氏分枝杆菌（*M. vanbaalenii*）和霍德勒分枝杆菌（*M. hodleri*）等种属与烃类化合物污染土壤相关，并可降解多环芳烃物质（Kleespies et al.，1996；Khan et al.，2002）。环境中的其他分枝杆菌，如土分枝杆菌（*M. terrae*）和戈登分枝杆菌，偶尔涉及动物和人类免疫功能缺陷的疾病（Bonnet et al.，1996；Carbonara et al.，2000）。特别是鸟分枝杆菌-胞内分枝杆菌-瘰疬分枝杆菌（MAIS）复合群成员的分枝杆菌，则是艾滋病患者和其他免疫功能不全人群的重要机会致病菌（Kirschner et al.，1992）。

18.1.2 鱼类中的分枝杆菌：病因

分枝杆菌病是世界范围内鱼类发病和死亡的一个主要病因。尽管近年来的研究表明，引发鱼类疾病的分枝杆菌种类日益增多，但 *M. marinum*、*M. fortuitum* 和 *M. chelonae* 是最常报道的病原菌（Gauthier and Rhodes，2009；Jacobs et al.，2009a）。

作为最常见的鱼类分枝杆菌——海分枝杆菌，1926 年从几种海洋观赏鱼中被分离出来（Aronson，1926）。世界范围内的淡水和咸水以及温水和冷水鱼类都有海分枝杆菌的感染病例报道。海分枝杆菌是一种生长苛刻的光产色菌(鲁尼恩Ⅰ型),初次分离时只限在35 ℃以下生长,在之后培养中可以适应较高的温度。它在固体琼脂上产生粗糙至光滑的菌落,通过硝酸盐还原阴性、脲酶产生阳性和Tween-20水解阳性等生化反应可以与其他生长缓慢的光致产色分枝杆菌区分开来(Leão et al., 2004)。

偶发分枝杆菌于1953年首次从一种淡水观赏鱼[红灯鱼(*Hyphessobrycon innesi*)]中被分离出来(Ross and Brancato, 1959),类似于海分枝杆菌,该菌具有广泛的宿主、温度和盐度范围。偶发分枝杆菌不产色,生长快速,通过芳基硫酸酯酶反应阳性,硝酸盐还原、NaCl耐受性和甘露醇利用试验阴性等特性可与其他鲁尼恩Ⅳ型分枝杆菌区分开来(Leão et al., 2004)。外来分枝杆菌,一种与偶发分枝杆菌密切相关并有时被认为是其亚种的分枝杆菌,是斑马鱼的病原菌(Kent et al., 2004)。

龟分枝杆菌(*M. chelonae*),最初于1903年从蠵龟(*Caretta caretta*)中被分离出来,是世界范围内鲑科鱼类的致病菌(Ashburner, 1977; Grange, 1981; Bruno et al., 1998)。包括实验室饲养的斑马鱼在内,该菌对其他鱼类也有感染的相关报道(Whipps et al., 2008)。*M. chelonae* 是鲁尼恩Ⅳ型分枝杆菌,通过芳基硫酸酯酶反应阳性、硝酸盐还原阴性和对NaCl缺乏耐受性等生化特性与该群中的其他分枝杆菌相区别(Leão et al., 2004)。在被上升认定为一个种之前,脓肿分枝杆菌一直被认为是龟分枝杆菌的一个亚种(Kusunoki and Ezaki, 1992),其宿主范围与龟分枝杆菌相似。嗜鲑分枝杆菌(*M. salmoniphilum*) (ex Ross, 1960)作为龟分枝杆菌的一个亚种,最近被重新提升为种的地位(Whipps et al., 2007a)。主要作为鲑科鱼类的病原菌,*M. salmoniphilum* 也感染江鳕(*Lota lota*)和鲟鱼(*Acipenser gueldenstaedtii*) (Berg et al., 2013; Righetti et al., 2014)。

另外两种分枝杆菌,*M. shottsii* 和 *M. pseudoshottsii*,发现于美国大西洋中部东海岸内陆河口的切萨皮克湾。1997年,在切萨皮克湾的条纹鲈中首次观察到含有抗酸细菌的真皮和内部肉芽肿病变,与鱼类分枝杆菌病特征一致(Vogelbein et al., 1998; Pieper, 2006)。最初尝试从含抗酸菌条纹鲈中分离培养分枝杆菌没有成功,但随后对脾脏匀浆进行了长期培养。2001年,Rhodes等描述了一种新的生长非常缓慢的分离菌株,后来被命名为 *M. shottsii*(Rhodes et al., 2003)。差不多在同一时间,从切萨皮克湾条纹鲈中也分离到另一种分枝杆菌,并暂时命名为"*M. chesapeaki*" (Heckert et al., 2001),这两个分离株很可能来自同一个种(Gauthier and Rhodes, 2009)。*M. pseudoshottsii* 是2005年发现的与切萨皮克湾条纹鲈相关的第二种生长缓慢的分枝杆菌(Rhodes et al., 2005)。基于分离培养和PCR检测技术,*M. pseudoshottsii* 和 *M. shottsii* 在切萨皮克湾条纹鲈中都具有较高的患病率,似乎是该区域中分枝杆菌病的主要病原(Rhodes et al., 2004; Gauthier et al., 2008b)。

在美国东海岸其他地区的条纹鲈及相关的美洲狼鲈(*Morone americana*)中也检测到了 *M. shottsii* 和 *M. pseudoshottsii*(Ottinger et al., 2007; Stine et al., 2009)。在红海海水鱼类(Colorni, 1992; Ranger et al., 2006)及五条鰤(*Seriola quinqueradiata*)(Kurokawa et al., 2013)中也发现了与 *M. pseudoshottsii* 相似的分离株。基于 PCR 的研究还在切萨皮克湾的鲱(*Brevoortia tyrannus*)和湾鳀(*Anchoa mitchilli*)中检测到了这种细菌(Gauthier et al., 2010)。

M. shottsii 和 *M. pseudoshottsii* 的 16S rRNA 基因与 *M. marinum* 相似(>99%)，除了 *M. shottsii* 的耐热性较低(<30 ℃)，它们在生化性质上非常相似，且两种菌都产烟酸(Rhodes et al., 2005)。*M. shottsii* 不产色素(Ⅲ型)，而 *M. pseudoshottsii* 是光产色菌(Ⅰ型)。基因组分析表明 *M. pseudoshottsii* 是溃疡分枝杆菌进化枝的基群，并且具有溃疡分枝杆菌的基因组特征，包括高拷贝数的插入序列 IS*2404* 和存在一个大于 170 kb 编码霉内酯毒素的大质粒。尽管在目前的命名法中 *M. pseudoshottsii* 保留其菌种地位，这些相似性表明 *M. pseudoshottsii* 应被认为是溃疡分枝杆菌的生态变种(Doig et al., 2012)。据报道，溃疡分枝杆菌在路易斯安那州东南部广泛分布(Hennigan et al., 2013)，尽管该研究使用了 IS*2404* 和一组有限的附加基因座，这些基因座无法区分人类溃疡分枝杆菌病原和从变温动物分离出来的拟议生态变种(如 *M. pseudoshottsii*)，并且这些变种对人类具有不确定的毒力。实际上，IS*2404* 序列变种存在于生化和遗传上确认的海分枝杆菌中(Gauthier et al., 2010)，因此其检测不应被视为人类致病性溃疡分枝杆菌的证据。溃疡分枝杆菌和 *M. pseudoshottsii* 产生不同形式的霉内酯毒素，*M. pseudoshottsii* 霉内酯 F 比溃疡分枝杆菌霉内酯 A/B 在小鼠组织培养中产生明显更少的致细胞病变效应(Mve-Obiang et al., 2003; Mve-Obiang et al., 2005)。此外，没有证据表明 *M. pseudoshottsii* 会感染人类，并且从人类分离出的溃疡分枝杆菌对鱼类几乎没有毒力(Mosi et al., 2012)。

对 *M. shottsii*(模式菌株 M175)的初步基因组分析表明，它是海分枝杆菌的高度衍生变种，与 *M. pseudoshottsii* 没有什么相似之处，与一般海分枝杆菌菌株的差异较小。*M. shottsii* 不具有 IS*2404*、霉内酯质粒和其他溃疡分枝杆菌进化枝基因组变化，拥有使其与海分枝杆菌区别开来的几个独特插入序列。*M. shottsii* 和 *M. pseudoshottsii* 在生态学上也有所不同：*M. pseudoshottsii* 广泛分布于切萨皮克湾的水体和沉积物中，而 *M. shottsii* 仅在狼鲈科鱼类中被分离出来(Gauthier et al., 2010)。

18.1.3 传播

Nigrelli 和 Vogel(1963)列举了 151 种对分枝杆菌属易感的鱼类，这表明大多数硬骨鱼类即使不患病也容易受这些细菌的感染。然而，人们对于分枝杆菌的传播方式仍知之甚少。早期关于鲑鳟鱼类孵化场的记录表明，幼鱼摄食受感染成鱼内脏后经口传播而患病(Wood and Ordal, 1958; Ross et al., 1959)。此外,水族馆鱼类感染与食用受污染的颤蚓有关(Nenoff and Uhlemann, 2006)。Gauthier 等(2003)提出海分枝杆菌通过水媒(或颗粒携带)传播感染条纹鲈,斑马鱼胚胎则可能通过水接触感染海分枝杆菌(Davis et al., 2002)。还有研究报道了在含有感染和未感染海分枝杆菌的杂交条纹鲈(*Morone saxatilis* × *M. chrysops*)系统之间的推测水传播(Li and Gatlin, 2005)。

Hariff 等(2007)证实了海分枝杆菌和外来分枝杆菌通过水体和摄食传播感染斑马鱼的可能性。研究证实斑马鱼可以经口传播感染海分枝杆菌和龟分枝杆菌,投喂感染的大草履虫(*Paramecium caudatum*)会增加斑马鱼感染率(Peterson et al., 2013)。除了可能的胎生鱼经卵巢传播外,分枝杆菌的垂直传播尚未得到明确证实(Conroy, 1966)。Ross 和 Johnson (1962)的实验没有证实大鳞大麻哈鱼的垂直传播,但 Ashburner(1977)的研究指出,大鳞大麻哈鱼能从亲本将分枝杆菌传播给后代。鱼配子中分枝杆菌含量不明显,不能排除表面污染的精液或卵巢液引起水传染的可能性。虽然已从条纹鲈的配子和配子液中培养出分枝杆菌,但尚不清楚这些细菌是否真的存在于配子中(Stine et al., 2006)。

18.1.4 影响

在水产养殖中经常观察到与分枝杆菌病相关的发病率和死亡率(Hedrick et al., 1987; Bruno et al., 1998),特别是在有应激胁迫或水质差的情况下。分枝杆菌感染鱼引发的死亡通常是慢性的,偶尔会发生高的死亡率(Bruno et al., 1998)。急性、高死亡率主要发生在鱼类养殖密度高的温水养殖场中(Whipps et al., 2007b)。野生鱼类中的严重疾病比较少见,相关病原也很少得到确认。Lund 和 Abernethy (1978)报道,来自华盛顿州亚基马河的山地柱白鲑(*Prosopium williamsoni*)的抗酸杆菌引起的肉芽肿病变发病率为 8%。在东北部的大西洋鲭(*Scomber scombrus*)中观察到可能的分枝杆菌病(Mackenzie, 1988),其患病率随鱼龄缓慢增长,在鱼龄较大的鱼中可达 90%以上。据报道,太平洋海岸条纹鲈内脏分枝杆菌病患病率很高(25%~80%)(Sakanari et al., 1983)。在加拿大艾伯塔省的野生黄金鲈(*Perca flavescens*)中也观察到与龟分枝杆菌有关的疾病暴发,患病鱼 80%以上为内脏感染,5%~25%为皮肤损伤(Daoust et al., 1989)。

在切萨皮克湾条纹鲈中一直可以检测到高发的分枝杆菌病(患病率高于 50%)(Cardinal, 2001; Overton et al., 2003; Gauthier et al., 2008a)。在切萨皮克湾主干中的条纹鲈,5、6 龄前的雄鱼和雌鱼的内脏分枝杆菌病随着鱼龄而增加。7 龄以上的雄鱼患病率仍然很高(>70%),但雌鱼患病率显著下降(20%~30%)(Gauthier et al., 2008a)。标记重捕法数据分析表明,1999 年后湾区中条纹鲈的自然(非捕捞)死亡率增加(Jiang et al., 2007),并且表观流行率数据的建模支持与疾病相关的死亡率(Gauthier et al.，2008a)。一项检测有和没有皮肤损伤鱼的存活率的标记重捕研究有相似的结果(Sadler et al.，2012)，表明分枝杆菌病可能对该野生种群的生存情况产生了影响。

18.2 诊断

鱼类分枝杆菌病的临床特征既有明显菌血症的急性暴发性(Wolf and Smith，1999；Whipps et al.，2007b)，也有更常见以低细菌载量和宿主肉芽肿性炎症为特征的慢性感染形式。外部临床体征(如果存在)是非特异性的，包括消瘦、腹水、眼球突出、脊柱侧凸和鳞片脱落。皮肤溃疡可能发生在病程晚期，是切萨皮克湾野生条纹鲈分枝杆菌感染的特征(图 18.2)。行为改变包括嗜睡、失去平衡或浮力控制及厌食。内脏器官和肌肉中可能出现呈灰白色至米黄色结节状的肉芽肿性炎症(图 18.3)。病原感染主要靶向前肾、脾和肝，但所有鱼体组织都有可能受到影响。

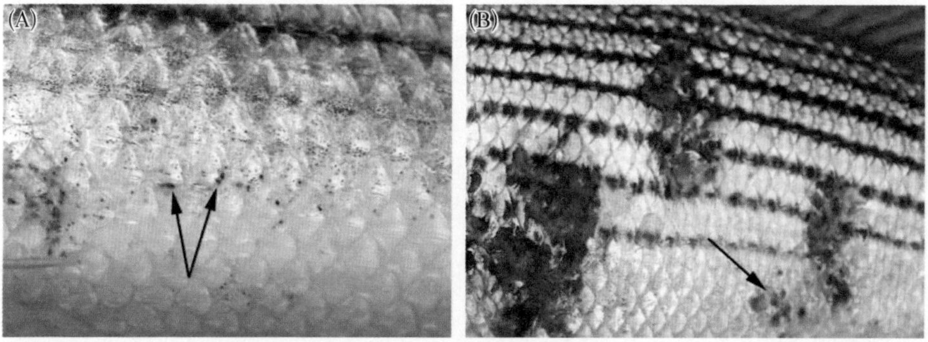

图 18.2　切萨皮克湾条纹鲈的皮肤分枝杆菌病：(A) 早期皮肤损伤表现为色素沉着病灶(箭头所示)，局部鳞片侵蚀；(B) 晚期表现为多灶性溃疡和色素病灶，伴随局部溃疡(箭头所示)

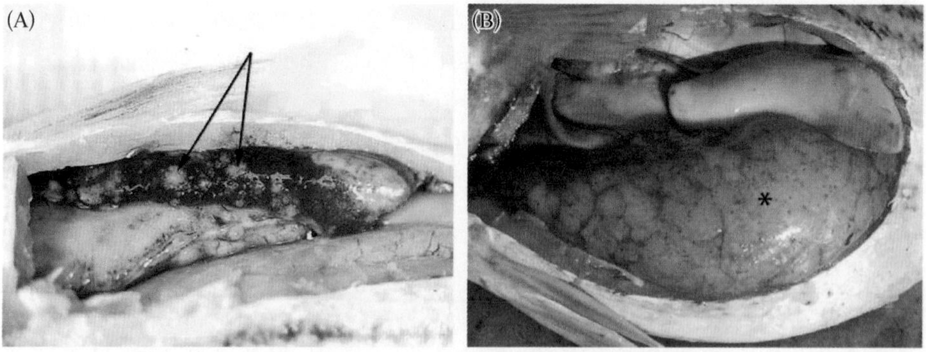

图 18.3　切萨皮克湾条纹鲈中的脾分枝杆菌病：(A) 脾中度肉芽肿性炎症和多灶性灰色结节(箭头所示)；(B) 一种重度肉芽肿性炎症，其中大部分脾实质(星号)被肉芽肿组织取代。照片来自 Wolfgang Vogelbein

许多方法被用于检测和鉴别鱼组织中的分枝杆菌，包括组织学、培养和分子生物学方法(Kaattari et al., 2006; Gauthier and Rhodes, 2009)。组织学是最传统的方法，通常通过 Ziehl-Neelsen 染色法或其改良法检测抗酸杆菌。尽管在实验感染中常报道不含可见抗酸杆菌的肉芽肿，但肉芽肿中存在抗酸的非分枝杆菌通常被认定为分枝杆菌感染的标志性特征(Colorni et al., 1998; Gauthier et al., 2003; Watral and Kent, 2007)。

分枝杆菌通常在 Middlebrook 7H9 肉汤培养基或 7H10 琼脂培养基上培养，前者使用 Tween－20 作为碳源和去垢剂以防止结块。Löwenstein-Jensen 琼脂培养基也常用于分枝杆菌培养(Kent and Kubica, 1985)。因为许多鱼类分枝杆菌在 35 ℃以上被抑制，在某些情况下 30 ℃时便受到抑制，建议在22～24 ℃和所研究鱼类的环境温度下培养。一些分离株在人工培养基上(如 *M. shottsii*)也表现出生长异常(生长缓慢且相对较差)，因此应延长培养期(>60 d)以获得最大的敏感性。利用生物化学特性可作为鱼类分枝杆菌的诊断要点，通常集中在海分枝杆菌、偶发分枝杆菌和龟分枝杆菌的诊断中(Frerichs, 1993; Chinabut, 1999)。尽管从鱼类中分离出的菌株常常不完全符合命名种的生化定义，但其他常用的关键特征也被使用(Lévy-Frébault and Portaels, 1992; Tortoli, 2003; Leão et al., 2004)。除非呈现出优势的生物化学和多基因位点遗传信息，尤其是如果从鱼体中发现人类病原体，否则，谨慎的做法应是将鱼类或其他变温动物的新分离株描述为“类似”的种(例如“*M. triplex*-like”)，这在文献中有先例，比如将分离自海鳗(*Gymnothorax funebris*)的一个病原新种描述为蒙特非奥分枝杆菌(*M. montefiorense*)(Levi et al., 2003)，而在此之前基于 16S rRNA 基因的相似性将该分离株最初命名为“*M. triplex*-like” (Herbst et al., 2001)。

采用无菌技术从鱼体内部培养鱼类分枝杆菌是最好的，不过，也可使用对外部样本进行“净化”的技术。抗酸细胞壁使分枝杆菌具有高度疏水性，对杀菌剂和抗生素具有相对抗性。因此，当无法进行无菌操作时，诸如次氯酸盐和苯扎氯铵等化合物可以限制其他环境细菌的过度生长，从而更好地分离分枝杆菌(Brooks et al., 1984; Rhodes et al., 2004)。

分子生物学技术在鱼类分枝杆菌检测的应用日益广泛，包括 PCR、PCR/RFLP(限制性片段长度多态性)、qPCR(定量 PCR)、环介导扩增(LAMP)等多种方法(Kaattari et al., 2006)。这些方法大多靶向 16S rRNA(小核糖体亚基)、热休克蛋白(*hsp*65, 65 kDa)、RNA 聚合酶 β 亚基(*rpo*B)和外部重复蛋白(*erp*)的基因。尽管插入序列(细菌转座因子)具有高度的种或菌株特异性，被广泛用于人致病性分枝杆菌的检测和分型(如结核分枝杆菌中的 IS*6110*)，但很少用于鱼类分离菌的检测，不过 IS*2404* 是个例外，它被用于检测溃疡分枝杆菌进化枝成员，也包括 *M. pseudoshottsii*。IS*901* /IS*1245* 也被用于鉴定热带鱼鸟分枝杆菌相关细菌(Lescenko et al., 2003)。与生物化学特性鉴定一样,许多来自鱼类的分枝杆菌扩增子可能与已公布的具有命名地位种的序列不完全匹配,并且基于一个或有限数量的扩增子序列来确定一个已命名种的鉴定可能会产生误导。

人和动物分枝杆菌感染的检测通常通过皮肤试验进行,即皮下注射纯化的分枝杆菌蛋白衍生物(PPD),随后检测Ⅳ型超敏反应中硬结的存在和大小。该测试还补充了干扰素捕获分析,如 QuantiFERON-Gold®,用外周血白细胞体外检测 T 细胞致敏性。与人类和具有兽医重要性动物中分枝杆菌血清学检测的良好发展状况相反,鱼类的分枝杆菌血清学检测很少使用。针对分枝杆菌细胞外产物(Chen et al., 1996)、全细胞超声处理(Chen et al., 1997)和活体感染(Colorni et al., 1998),鱼类均产生抗体反应。此外,虹鳟对结核分枝杆菌抗原产生迟发型超敏反应(Bartos and Sommer, 1981),在用弗氏佐剂免疫后注射灭活结核分枝杆菌或嗜鲑分枝杆菌,则呈现典型的硬结。因此,鱼体中分枝杆菌血清学诊断的研究开发似乎很容易,但尚未取得太多进展。

18.3 病理学

表面上看,鱼类分枝杆菌病的内脏肉芽肿性炎症与人类结核病相似。过去曾使用"鱼结核病(fish /piscine tuberculosis)"这个术语,但鉴于病因学上的差异,这种疾病称为"鱼分枝杆菌病(fish /piscine mycobacteriosis)"更为恰当。鱼肉芽肿具有典型的组织病变性状,由上皮样巨噬细胞组成,周围有炎性白细胞层(图 18.4)。

肉芽肿可表现为细胞性或坏死性,也可能呈现为干酪样病变。通常,坏死核心周围有一层高度压缩的纺锤形上皮细胞。某些鱼类的分枝杆菌肉芽肿中存在巨型多核细胞,而另一些鱼类中则不存在。肉芽肿性炎症可以是轻微的,仅可以通过组织学才能检测到,也存在大量多灶性或汇聚性肉芽肿性炎症,并伴有正常组织结构的明显位移和破坏。此类病例中灰色至灰白色结节可能很明显,但应注意与其他肉芽肿病症区别开来[如弗朗西斯菌、发光杆菌(*Photobacterium*)或诺卡氏菌(*Nocardia*)感染或包囊蠕虫寄生感染]。严重病例中也可见内脏粘连。急性分枝杆菌病的特征是肉芽肿形成不完整或不存在,细胞内和细胞外存在大量抗酸细菌,以及广泛的组织坏死(Talaat et al., 1998)。

真皮损伤也呈肉芽肿性,并可延伸到皮下组织。相关炎症导致上皮组织和鳞片脱落,并可形成合并浅溃疡。条纹鲈早期的真皮分枝杆菌病表现为单一鳞片的局灶性糜烂,并伴有底层色素沉着病灶(图 18.4)。肉芽肿性炎症可能是这些局灶性糜烂的基础。

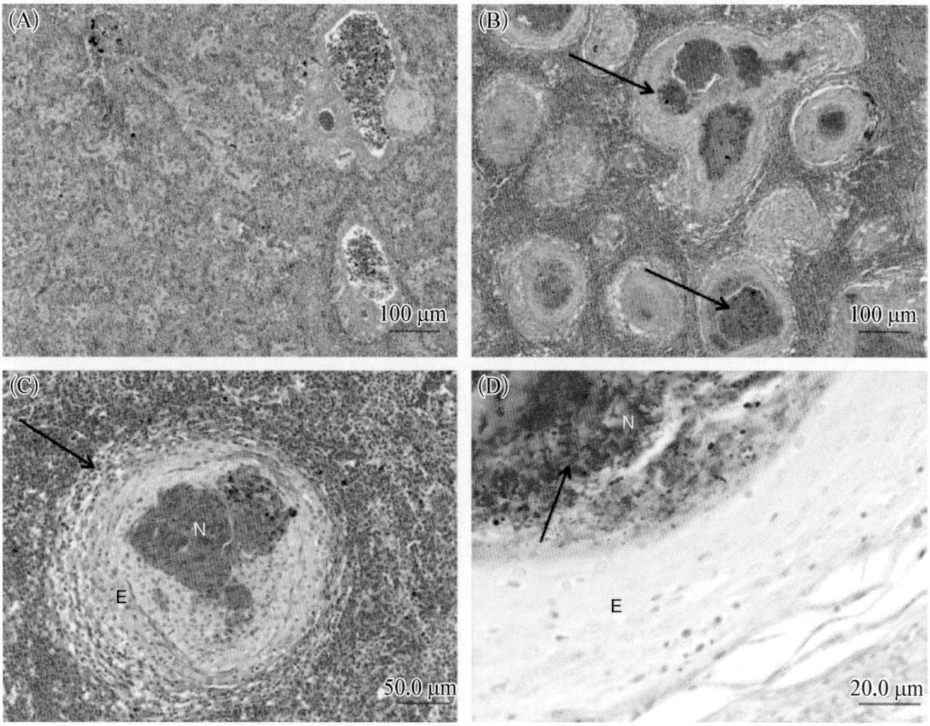

图 18.4 鱼类脾脏分枝杆菌病的组织病理学：(A) 健康的脾(HE 染色)；(B) 严重受损脾表现出多个肉芽肿(箭头所示)(HE 染色)；(C) 单个脾肉芽肿呈现出中央坏死(N)，由上皮样细胞(E)和白细胞不明显的外囊(箭头所示)组成的周围细胞层(HE 染色)；(D) 表现出中央坏死(N)的单个脾肉芽肿、抗酸分枝杆菌(箭头所示)和包裹状上皮样细胞(E) (Ziehl-Neelsen 染色)

18.4 病理生理学

Swanson 等(2002)研究发现感染了分枝杆菌的三角洲胡瓜鱼(*Hypomesus transpacificus*)的游动性能(临界游泳速度)降低(Brett, 1964)。Lapointe 等(2014)研究了患病和健康条纹鲈在不同氧浓度和温度下的代谢参数，并得出结论：高温、缺氧和疾病协同作用，能够显著减少患病鱼的需氧范围。重度脾脏疾病降低了条纹鲈的缺氧耐受性，特别是水温升高的情况下。然而，在最佳温度和正常氧浓度条件下，重度脾脏疾病并不影响最大代谢率或需氧范围。此外，血细胞比容和血红蛋白浓度均未受到影响。同样，患皮肤病的鱼鳃和肠的钠钾 ATP 酶活性也均未受到影响，但推测它们的渗透调节受损。这些研究发现表明野生条纹鲈能够耐受显著病理变化，对氧运输和渗透调节的影响很小，尽管疾病确实会降低非适宜环境条件下的代谢恢复能力(如温度升高、缺氧)。

生长减缓和发育迟缓在鱼分枝杆菌感染中很常见，这很可能是由与慢性疾病长期免疫反应相关的代谢耗损增加导致的。在患病严重的条纹鲈中观察到鱼体消瘦和各种器官功能退化(Lapointe et al., 2014)，Latour 等(2012)发现，相对于未患病条纹鲈，患内脏分枝杆菌病的鱼生长迟缓(鱼龄体长)。

18.5 防控策略

目前已经尝试用抗生素治疗鱼类中的分枝杆菌感染，包括四环素(Boos et al., 1995)、红霉素(Kawakami and Kusuda, 1990)和链霉素(Colorni et al., 1998)，但没有一种能成功清除感染。据报道，用水溶性硫酸卡那霉素(50 ppm)治疗孔雀鱼(*Lebistes reticulatus*)取得了一些成效，但未证实该抗生素是否成功清除了感染(Conroy and Conroy, 1999)。利福平和乙胺丁醇也曾用于治疗鱼的分枝杆菌感染，但疗效有限(Colorni, 1992; Chinabut, 1999)。人用结核病药物异烟肼和吡嗪酰胺并不适用于治疗鱼类分枝杆菌病，因为大多数非结核菌株都具有遗传抗药性(Rastogi et al., 1992)。

目前尚未有针对鱼类分枝杆菌病的商业化疫苗。基于纤连蛋白结合蛋白 Ag85A 的重组疫苗和 DNA 疫苗可以诱导产生特异性和非特异性免疫应答，在某些情况下，可对杂交条纹鲈提供短期免疫保护(Pasnik et al., 2003; Pasnik and Smith, 2005, 2006)，但长期保护尚未实现。

由于治疗方案有限，水产养殖中分枝杆菌病的控制主要基于严格的检疫和筛查措施(Astrofsky et al., 2000)。Li 和 Gatlin (2005)报道在饲料中补加益生菌剂(GroBiotic®- A)后，杂交条纹鲈受海分枝杆菌感染后的存活率略有改善。相较于对照组，接受益生菌(GroBiotic®- A，自溶酿造酵母)的鱼生长速度和饲料系数也有所增加，因此患病鱼的存活率增加可能与鱼状态改善有关，这与 Jacobs 等(2009a)的研究是一致的：营养不良的条纹鲈表现出更加严重的分枝杆菌病。

在水产养殖中根除分枝杆菌病的唯一选择是销毁受影响的种群并对养殖设施进行消毒(Noga, 2000; Roberts, 2001)。水产养殖设施的彻底消毒是比较困难的，因为常用的消毒剂如 *N*-烷基二甲基苄基氯化铵(Roccal-D, Micronex)和过氧单硫酸钾(Virkon - S)对海分枝杆菌无效。乙醇和苄基-4-氯苯酚/苯基苯酚(Lysol)和次氯酸钠(漂白剂)一样有效，只是消毒时间要延长(Mainous and Smith, 2005)。

18.6 总结与研究展望

鱼类中三大分枝杆菌(海分枝杆菌、偶发分枝杆菌和龟分枝杆菌)的范式已经发生变化，目前感染鱼类的分枝杆菌种类多样性可能仍未得到充分认识。仅在切萨皮克湾内，Stine 等(2010)采用脂肪酸和甲酯气相色谱法(FAME)，发现了 29 组感染鱼类的分枝杆菌，Rhodes 等(2004)报道了至少八个表型不同的群，其中大多数可能包含不同的种或菌株。从迁徙条纹鲈的福尔马林保存组织中扩增出分枝杆菌 DNA 序列，进一步表明分枝杆菌存在的多样性(Matsche et al.，2010)。对这些分枝杆菌的系统发生以及它们在不同条件下对鱼类的致病性，几乎没有什么研究。由于这种多样性，准确识别并专注于研究与疾病相关的数量优势分离株非常重要。此外，由于分枝杆菌的种类涵盖非致病性腐生菌和专性病原体，因此有必要将培养和利用 PCR 检测到的分枝杆菌种属鉴定区分到尽可能低的分类水平，包括菌株的分化。如果没有这种区分，就很有可能得出关于单一分枝杆菌致病性的错误结论。

虽然分枝杆菌病在水产养殖和水族馆中的影响严重且易于评估，但对野生鱼类种群的影响尚不清楚。鉴于该疾病的慢性特征以及现场检测隐性死亡率的困难，确定群体水平的影响具有挑战性。表观流行数据和标记重捕法研究的数学建模可以提供帮助，然而，将这些影响纳入管理模型对于了解分枝杆菌病对生物量和群体健康的影响是必要的。此外，重要的是检查分枝杆菌的亚致死效应，如代谢障碍和生长抑制，以获得更完整的疾病影响信息。由于在不良条件下的水产养殖设施中分枝杆菌病最常见，野生鱼种群中的地方病可能表明环境压力源很重要。野生鱼种群中这些环境压力源的流行病学鉴定是未来研究的重要领域，可以增强我们对疾病生态学的理解。

基于细菌培养技术、组织学和分子技术的鱼类分枝杆菌感染诊断已经相当成熟，并且有足够的分子和生化指标参考资源来确定新分离株的亲缘关系。这些目前都是基于致死性的诊断方法，可能会限制它们在珍贵鱼类或大量个体中的实际应用。基于血液的非致死性诊断，无论是血清学的还是细胞学的，目前仍没有得到很好的研究，这些方法的开发建立将促进对亲鱼或其他鱼类在引入养殖设施之前的病原携带筛查。

显然，由于分枝杆菌固有的抗生素抗性及难治性，以及普遍缺乏成功的药物应用，疫苗的开发对于鱼类病害防控将是非常重要的。如果将人类结核分枝杆菌疫苗开发的困难作为指标，开发有效的鱼用分枝杆菌疫苗可能具有挑战性。

参考文献

[1] Aronson, J.D. (1926) Spontaneous tuberculosis in saltwater fish. *Journal of Infectious Diseases* 39, 315 - 320.

[2] Ashburner, L. D. (1977) Mycobacteriosis in hatchery-confined chinook salmon (*Oncorhynchus tshawytscua* Walbaum) in Australia. *Journal of Fish Biology* 10, 523 - 528.

[3] Astrofsky, K.M., Schrenzel, M.D., Bullis, R.A., Smolowitz, R.M. and Fox, J.G. (2000) Diagnosis and management of atypical *Mycobacterium* spp. infections in established laboratory zebrafish (*Brachydanio rerio*) facilities. *Comparative Medicine* 50, 666 - 672.
[4] Bartos, J.M. and Sommer, C.V. (1981) *In vivo* cell-mediated immune response to *M. tuberculosis* and *M. salmoniphilum* in rainbow trout (*Salmo gairdneri*). *Developmental and Comparative Immunology* 5, 75 - 83.
[5] Bercovier, H., Kafri, O. and Sela, S. (1986) Mycobacteria possess a surprisingly small number of ribosomal RNA genes in relation to the size of their genome. *Biochemical and Biophysical Research Communications* 136, 1136 - 1141.
[6] Berg, V., Zerihun, M.A., Jørgensen, A., Lie, E., Dale, O.B. *et al.* (2013) High prevalence of infections and pathological changes in burbot (*Lota lota*) from a polluted lake (Lake Mjøsa, Norway). *Chemosphere* 90, 1711 - 1718.
[7] Bonnet, E., Massip, P., Bauriaud, R., Alric, L. and Auvergnat, J.-C. (1996) Disseminated *Mycobacterium gordonae* infection in a patient infected with human immunodeficiency virus. *Clinical Infectious Diseases* 23, 644 - 645.
[8] Boos, S., Schmidt, H., Ritter, G. and Manz, D. (1995) Untersuchungen zur oralen Wirksamkeit von Rifampicin gegen die Mykobakteriose der Zierfische [Effectiveness of oral rifampicin against mycobacteriosis in tropical fish]. *Berliner und Münchener* Tierärztliche Wochenschrift 108, 253 - 255.
[9] Brett, J.R. (1964) The respiratory metabolism and swimming performance of young sockeye salmon. *Journal of the Fisheries Research Board of Canada* 21, 1183 - 1226.
[10] Brooks, R.W., George, K.L., Parker, B.C. and Falkinham, J.O. III (1984) Recovery and survival of nontuberculous mycobacteria under various growth and decontamination conditions. *Canadian Journal of Microbiology* 30, 1112 - 1117.
[11] Bruno, D.W., Griffiths, J., Mitchell, C.G., Wood, B.P., Fletcher, Z.J. *et al.* (1998) Pathology attributed to *Mycobacterium chelonae* infection among farmed and laboratory-infected Atlantic salmon *Salmo salar*. *Diseases of Aquatic Organisms* 33, 101 - 109.
[12] Carbonara, S., Tortoli, E., Costa, D., Monno, L., Fiorentino, G. *et al.* (2000) Disseminated *Mycobacterium terrae* infection in a patient with advanced human immunodeficiency virus disease. *Clinical Infectious Diseases* 30, 831 - 835.
[13] Cardinal, J.L. (2001) Mycobacteriosis in striped bass, *Morone saxatilis*, from Virginia waters of Chesapeake Bay. Master's thesis MSc., College of William and Mary, Virginia Institute of Marine Science, Gloucester Point, Virginia.
[14] Chen, S.-C., Yoshida, T., Adams, A., Thompson, K.D. and Richards, R.H. (1996) Immune response of rainbow trout to extracellular products of *Mycobacterium* spp. *Journal of Aquatic Animal Health* 8, 216 - 222.
[15] Chen, S.-C., Adams, A., Thompson, K.D. and Richards, R.H. (1997) A comparison of the antigenicity of the extracellular products and whole-cell sonicates from *Mycobacterium* spp. in rabbits, mice and fish by immunoblotting and enzyme-linked immunosorbent assay. *Journal of Fish Diseases* 20, 427 - 442.
[16] Chinabut, S. (1999) Mycobacteriosis and nocardiosis. In: Woo, P.T.K. and Bruno, D. W. (eds) *Fish Diseases and Disorders. Volume 3: Viral, Bacterial and Fungal Infections*. CAB International, Wallingford, UK, pp. 319 - 340.
[17] Colorni, A. (1992) A systemic mycobacteriosis in the European sea bass *Dicentrarchus labrax* cultured in Eilat (Red Sea). *The Israeli Journal of Aquaculture-Bamidgeh* 44, 75 - 81.
[18] Colorni, A., Avtalion, R., Knibb, W., Berger, E., Colorni, B. and Timan, B. (1998) Histopathology of sea bass (*Dicentrarchus labrax*) experimentally infected with *Mycobacterium marinum* and treated with streptomycin and garlic (*Allium sativum*)

extract. *Aquaculture* 160, 1 - 17.
[19] Conroy, D. (1966) A report on the problems of bacterial fish diseases in the Argentine Republic. *Bulletin, Office International des Epizooties* 65, 755 - 768.
[20] Conroy, G. and Conroy, D.A. (1999) Acid-fast bacterial infection and its control in guppies (*Lebistes reticulatus*) reared on an ornamental fish farm in Venezuela. *Veterinary Record* 144, 177 - 178.
[21] Daoust, P.-Y., Larson, B.E. and Johnson, G.R. (1989) Mycobacteriosis in yellow perch (*Perca flavescens*) from two lakes in Alberta. *Journal of Wildlife Diseases* 25, 31 - 37.
[22] Davis, J.M., Clay, H., Lewis, J.L., Ghori, N., Herbomel, P. and Ramakrishnan, L. (2002) Realtime visualization of *Mycobacterium*-macrophage interactions leading to initiation of granuloma formation in zebrafish embryos. *Immunity* 17, 693 - 702.
[23] Devulder, G., De Montclos, M.P. and Flandrois, J.P. (2005) A multigene approach to phylogenetic analysis using the genus *Mycobacterium* as a model. *International Journal of Systematic and Evolutionary Microbiology* 55, 293 - 302.
[24] Doig, K.D., Holt, K.E., Fyfe, J.A.M., Lavender, C.J., Eddyani, M. *et al*. (2012) On the origin of *Mycobacterium ulcerans*, the causative agent of Buruliulcer. *BMC Genomics* 13: 258.
[25] Frerichs, G.N. (1993) Mycobacteriosis: nocardiosis. In: Inglis, V., Roberts, R.J. and Bromage, N.R. (eds) *Bacterial Diseases of Fish*. Blackwell Scientific, Oxford, UK, pp. 219 - 235.
[26] Gauthier, D.T. and Rhodes, M.W. (2009) Mycobacteriosis in fishes: a review. *The Veterinary Journal* 180, 33 - 47.
[27] Gauthier, D.T., Rhodes, M.W., Vogelbein, W.K., Kator, H. and Ottinger, C.A. (2003) Experimental mycobacteriosis in striped bass (*Morone saxatilis*). *Diseases of Aquatic Organisms* 54, 105 - 117.
[28] Gauthier, D.T., Latour, R.J., Heisey, D.M., Bonzek, C.F., Gartland, J. *et al*. (2008a) Mycobacteriosis-associated mortality in wild striped bass (*Morone saxatilis*) from Chesapeake Bay, USA. *Ecological Applications* 18, 1718 - 1727.
[29] Gauthier, D.T., Vogelbein, W.K., Rhodes, M.W. and Reece, K. (2008b) Nested PCR assay for detection of *Mycobacterium shottsii* and *Mycobacterium pseudoshottsii* in striped bass (*Morone saxatilis*). *Journal of Aquatic Animal Health* 20, 192 - 201.
[30] Gauthier, D.T., Reece, K.S., Xiao, J., Rhodes, M.W., Kator, H.I. *et al*. (2010) Quantitative PCR assay for *Mycobacterium pseudoshottsii* and *Mycobacterium shottsii* and application to environmental samples and fishes from the Chespeake Bay. *Applied and Environmental Microbiology* 76, 6171 - 6179.
[31] Gauthier, D.T., Helenthal, A.M., Rhodes, M.W., Vogelbein, W.K. and Kator, H.I. (2011) Characterization of photochromogenic *Mycobacterium* spp. from Chesapeake Bay striped bass (*Morone saxatilis*). *Diseases of Aquatic Organisms* 95, 113 - 124.
[32] Grange, J.M. (1981) *Mycobacterium chelonae*. *Tubercle* 62, 273 - 276.
[33] Hariff, M.J., Bermudez, L.E. and Kent, M.L. (2007) Experimental exposure of zebrafish, *Danio rerio* (Hamilton) to *Mycobacterium marinum* and *Mycobacterium peregrinum* reveals the gastrointestinal tract as the primary route of infection: a potential model for environmental mycobacterial infection. *Journal of Fish Diseases* 30, 587 - 600.
[34] Heckert, R.A., Elankumaran, S., Milani, A. and Baya, A. (2001) Detection of a new *Mycobacterium* species in wild striped bass in the Chesapeake Bay. *Journal of Clinical Microbiology* 39, 710 - 715.
[35] Hedrick, R.P., Mcdowell, T. and Groff, J. (1987) Mycobacteriosis in cultured striped bass from California. *Journal of Wildlife Diseases* 23, 391 - 395.
[36] Hennigan, C.E., Myers, L. and Ferris, M.J. (2013) Environmental distribution and

seasonal prevalence of *Mycobacterium ulcerans* in southern Louisiana. *Antimicrobial Agents and Chemotherapy* 79, 2648 - 2656.

[37] Herbst, L.H., Costa, S.F., Weiss, L.M., Johnson, L.K., Bartell, J. *et al*. (2001) Granulomatous skin lesions in moray eels caused by a novel *Mycobacterium* species related to *Mycobacterium triplex*. *Infection and Immunity* 69, 4639 - 4646.

[38] Jacobs, J. M., Rhodes, M. R., Baya, A., Reimschuessel, R., Townsend, H. and Harrell, R.M. (2009a) Influence of nutritional state on the progression and severity of mycobacteriosis in striped bass *Morone saxatilis*. *Diseases of Aquatic Organisms* 87, 183 - 197.

[39] Jacobs, J. M., Stine, C. B., Baya, A. M. and Kent, M. L. (2009b) A review of mycobacteriosis in marine fish. *Journal of Fish Diseases* 32, 119 - 130.

[40] Jiang, H., Pollock, J.M., Brownie, C., Hoenig, J.M., Latour, R.J. *et al*. (2007) Tag return models allowing for harvest and catch and release: evidence of environmental and management impacts on striped bass fishing and natural mortality rates. *North American Journal of Fisheries Management* 27, 387 - 396.

[41] Kaattari, I., Rhodes, M.W., Kaattari, S.L. and Shotts, E.B. (2006) The evolving story of *Mycobacterium tuberculosis* clade members detected in fish. *Journal of Fish Diseases* 29, 509 - 520.

[42] Kawakami, K. and Kusuda, R. (1990) Efficacy of rifampicin, streptomycin, and erythromycin against experimental *Mycobacterium* infection in cultured yellowtail. *Nippon Suisan Gakkaishi* 56, 51 - 53.

[43] Kent, M.L., Whipps, C.M., Matthews, J.L., Florio, D., Watral, V. *et al*. (2004) Mycobacteriosis in zebrafish (*Danio rerio*) research facilities. *Comparative Biochemistry and Physiology-Part C: Toxicology and Pharmacology* 138, 383 - 390.

[44] Kent, P.T. and Kubica, G.P. (1985) *Public Health Mycobacteriology.A Guide for the Level III Laboratory*. US Department of Health and Human Services Publication No. (CDC) 86 - 8230, Centers for Disease Control, Atlanta, Georgia.

[45] Khan, A.A., Kim, S.-J., Paine, D.D. and Cerniglia, C.E. (2002) Classification of a polycyclic aromatic hydro-carbon-metabolizing bacterium, *Mycobacterium* spp. strain PYR - 1, as *Mycobacterium vanbaalenii* spp. nov. *International Journal of Systematic and Evolutionary Microbiology* 52, 1997 - 2002.

[46] Kirschner, R. A., Parker, B. C. and Falkinham, J. O. III (1992) Epidemiology of infection by nontuberculous mycobacteria: *Mycobacterium avium*, *Mycobacterium intracellulare*, and *Mycobacterium scrofulaceum* in acid, brown-water swamps of the southeastern United States and their association with environmental variables. *American Review of Respiratory Disease* 145, 271 - 275.

[47] Kleespies, M., Kroppenstedt, R.M., Rainey, F.A., Webb, L.E. and Stackebrandt, E. (1996) *Mycobacterium hodleri* sp. nov., a new member of the fast-growing mycobacteria capable of degrading polycyclic aromatic hydrocarbons. *International Journal of Systematic Bacteriology* 46, 683 - 687.

[48] Kurokawa, S., Kabayama, J., Fukuyasu, T., Hwang, S.D., Park, C.I. *et al*. (2013) Bacterial classification of fish-pathogenic *Mycobacterium* species by multigene phylogenetic analyses and MALDI Biotyper identification system. *Marine Biotechnology* 15, 340 - 348.

[49] Kusunoki, S. and Ezaki, T. (1992) Proposal of *Mycobacterium peregrinum* sp. nov., nom. rev., and elevation of *Mycobacterium chelonae* subsp. *abscessus* (Kubica *et al*.) to species status: *Mycobacterium abscessus* comb. nov. *International Journal of Systematic Bacteriology* 42, 240 - 245.

[50] Lapointe, D., Vogelbein, W.K., Fabrizio, M.C., Gauthier, D.T. and Brill, R.W. (2014) Temperature, hypoxia, and mycobacteriosis: effects on adult striped bass

Moronesaxatilis metabolic performance. *Diseases of Aquatic Organisms* 108, 113 - 127.
[51] Latour, R. J., Gauthier, D. T., Gartland, J., Bonzek, C. F., Mcnamee, K. and Vogelbein, W. K. (2012) Impacts of mycobacteriosis on the growth of striped bass (*Morone saxatilis*) in Chesapeake Bay. *Canadian Journal of Fisheries and Aquatic Sciences* 69, 247 - 258.
[52] Leão, S. C., Martin, A., Mejia, G. I., Palomino, J. C., Robledo, J. *et al.* (2004) *Practical Handbook for the Phenotypic and Genotypic Identification of Mycobacteria*. Vanden Broelle, Brugges, Belgium.
[53] Lescenko, P., Matlova, L., Dvorska, L., Bartos, M., Vavra, O. *et al.* (2003) Mycobacterial infection in aquarium fish. *Veterinární Medicína* 48, 71 - 78.
[54] Levi, M. H., Bartell, J., Gandolfo, L., Smole, S. C., Costa, S. F. *et al.* (2003) Characterization of *Mycobacterium montefiorense* sp. nov., a novel pathogenic mycobacterium from moray eels that is related to *Mycobacterium triplex*. *Journal of Clinical Microbiology* 41, 2147 - 2152.
[55] Lévy-Frébault, V. V. and Portaels, F. (1992) Proposed minimal standards for the genus *Mycobacterium* and for description of new slowly growing *Mycobacterium* species. *International Journal of Systematic Bacteriology* 42, 315 - 323.
[56] Li, P. and Gatlin, D.M.I. (2005) Evaluation of the prebiotic GroBiotic®-A and brewers yeast as dietary supplements for subadult hybrid striped bass (*Morone chrysops* × *M. saxatilis*) challenged *in situ* with *Mycobacterium marinum*. *Aquaculture* 248, 197 - 205.
[57] Lund, J.E. and Abernethy, C.S. (1978) Lesions of tuberculosis in mountain whitefish (*Prosopium williamsoni*). *Journal of Wildlife Diseases* 14, 222 - 228.
[58] Mackenzie, K. (1988) Presumptive mycobacteriosis in northeast Atlantic mackerel, *Scomber scombrus* L. *Journal of Fish Biology* 32, 263 - 275.
[59] Mainous, M. E. and Smith, S. A. (2005) Efficacy of common disinfectants against *Mycobacterium marinum*. *Journal of Aquatic Animal Health* 17, 284 - 288.
[60] Matsche, M., Overton, A., Jacobs, J., Rhodes, M.R. and Rosemary, K.M. (2010) Low prevalence of splenic mycobacteriosis in migratory striped bass *Morone saxatilis* from North Carolina and Chesapeake Bay, USA. *Diseases of Aquatic Organisms* 90, 181 - 189.
[61] Menendez, M.C., Garcia, M.J., Navarro, M.C., Gonzalez-Y-Merchand, J.A., Rivera-Gutierrez, S. *et al.* (2002) Characterization of an rRNA operon (rrnB) of *Mycobacterium fortuitum* and other mycobacterial species: implications for the classification of mycobacteria. *Journal of Bacteriology* 184, 1078 - 1088.
[62] Mosi, L., Mutoji, N. K., Basile, F. A., Donnell, R., Jackson, K. L. *et al.* (2012) *Mycobacterium ulcerans* causes minimal pathogenesis and colonization in medaka (*Oryzias latipes*): an experimental fish model of disease transmission. *Microbes and Infection* 14, 719 - 729.
[63] Mve-Obiang, A., Lee, R.E., Portaels, F. and Small, P.L.C. (2003) Heterogeneity of mycolactones produced by clinical isolates of *Mycobacterium ulcerans*: implications for virulence. *Infection and Immunity* 71, 774 - 783.
[64] Mve-Obiang, A., Lee, R. E., Umstot, E. S., Trott, K. A., Grammer, T. C. *et al.* (2005) A newly discovered mycobacterial pathogen isolated from laboratory colonies of *Xenopus* species with lethal infections produces a novel form of mycolactone, the *Mycobacterium ulcerans* macrolide toxin. *Infection and Immunity* 73, 3307 - 3312.
[65] Nenoff, P. and Uhlemann, R. (2006) Mycobacteriosis in mangrove killifish (*Rivulus magdalanae*) caused by living fish food (*Tubifex tubifex*) infected with *Mycobacterium marinum*. *Deutsche Tierärztliche Wochenschrift* 113, 209 - 248.
[66] Nigrelli, R.F. and Vogel, H. (1963) Spontaneous tuberculosis in fishes and in other

cold-blooded vertebrates with special reference to *Mycobacterium fortuitum* Cruz from fish and human lesions. *Zoologica: Scientific Contributions of the New York Zoological Society* 48, 131 - 144.

[67] Noga, E.J. (2000) *Fish Disease: Diagnosis and Treatment*. Iowa State University Press, Ames, Iowa.

[68] Ottinger, C.A., Brown, J.J., Densmore, C., Starliper, C.E., Blazer, V. *et al*. (2007) Mycobacterial infections in striped bass from Delaware Bay. *Journal of Aquatic Animal Health* 19, 99 - 108.

[69] Overton, A.S., Margraf, F.J., Weedon, C.A., Pieper, L.H. and May, E.B. (2003) The prevalence of mycobacterial infections in striped bass in Chesapeake Bay. *Fisheries Management and Ecology* 10, 301 - 308.

[70] Pasnik, D.J. and Smith, S.A. (2005) Immunogenic and protective effects of a DNA vaccine for *Mycobacterium marinum* in fish. *Veterinary Immunology and Immunopathology* 103, 195 - 206.

[71] Pasnik, D.J. and Smith, S.A. (2006) Immune and histopathologic responses of DNA-vaccinated hybrid striped bass *Morone saxatilis* × *M. chrysops* after acute *Mycobacterium marinum* infection. *Diseases of Aquatic Organisms* 73, 33 - 41.

[72] Pasnik, D.J., Vemulapalli, R., Smith, S.A. and Schurig, G.G. (2003) A recombinant vaccine expressing a mammalian *Mycobacterium* sp. antigen is immunostimulatory but not protective in striped bass. *Veterinary Immunology and Immunopathology* 95, 43 - 52.

[73] Peterson, T.S., Ferguson, J.A., Watral, V.G., Mutoji, K.N., Ennis, D.G. and Kent, M. L. (2013) *Paramecium caudatum* enhances transmission and infectivity of *Mycobacterium marinum* and *Mycobacterium chelonae* in zebrafish (*Danio rerio*). *Diseases of Aquatic Organisms* 106, 229 - 239.

[74] Pieper, L. (2006) Striped bass disease overview for the past ten years plus. In: Ottinger, C.A. and Jacobs, J. (eds) *USGS /NOAA Workshop on Mycobacteriosis in Striped Bass, May 7 - 10, 2006, Annapolis, Maryland*. USGS Scientific Investigations Report 2006 - 5214 /NOAA Technical Memorandum NOS NCCOS 41, US Geological Survey /National Oceanic and Atmospheric Administration, Reston, Virginia, pp. 10 - 11. Available at: http: //aquaticcommons. org /2233 /1 /SIR2006 - 5214-complete.pdf (accessed 15 November 2016).

[75] Ranger, B.S., Mahrous, E.A., Mosi, L., Adusumilli, S., Lee, R.E. *et al*. (2006) Globally distributed mycobacterial fish pathogens produce a novel plasmid-encoded toxic macrolide, mycolactone F. *Infection and Immunity* 74, 6037 - 6045.

[76] Rastogi, N., Goh, K.S., Guillou, N. and Labrousse, V. (1992) Spectrum of drugs against atypical mycobac-teria: how valid is the current practice of drug susceptibility testing and the choice of drugs? *Zentralblatt für Bakteriologie* 277, 474 - 84.

[77] Rhodes, M.W., Kator, H., Kotob, S., Van Berkum, P., Kaattari, I. *et al*. (2001) A unique *Mycobacterium* species isolated from an epizootic of striped bass (*Morone saxatilis*). *Emerging Infectious Diseases* 7, 1 - 3.

[78] Rhodes, M.W., Kator, H., Kotob, S., Van Berkum, P., Kaattari, I. *et al*. (2003) *Mycobacterium shottsii* sp. nov., a slowly growing species isolated from Chesapeake Bay striped bass. *International Journal of Systematic and Evolutionary Microbiology* 53, 421 - 424.

[79] Rhodes, M.W., Kator, H., Kaattari, I., Gauthier, D., Vogelbein, W.K. and Ottinger, C.A. (2004) Isolation and characterization of mycobacteria from striped bass *Morone saxatilis* from the Chesapeake Bay. *Diseases of Aquatic Organisms* 61, 41 - 51.

[80] Rhodes, M.W., Kator, H., McNabb, A., Deshayes, C., Reyrat, J.-M. *et al*. (2005) *Mycobacterium pseudoshottsii* sp. nov., a slowly growing chromogenic species isolated

from Chesapeake Bay striped bass (*Morone saxatilis*). *International Journal of Systematic and Evolutionary Microbiology* 55, 1139 - 1147.

[81] Righetti, M., Favaro, L., Antuofermo, E., Caffara, M., Nuvoli, S. *et al*. (2014) *Mycobacterium salmoniphilum* infection in a farmed Russian sturgeon, *Acipenser gueldenstaedtii* (Brandt & Ratzeburg). *Journal of Fish Diseases* 37, 671 - 674.

[82] Roberts, R.J. (2001) *Fish Pathology*, 3rd edn. W.B. Saunders, London.

[83] Rogall, T., Wolters, J., Flohr, T. and Böttger, E.C. (1990) Towards a phylogeny and definition of species at the molecular level within the genus *Mycobacterium*. *International Journal of Systematic Bacteriology* 40, 323 - 330.

[84] Ross, A. J. (1960) *Mycobacterium salmoniphilum* spp. nov. from salmonid fishes. *American Review of Respiratory Disease* 81, 241 - 250.

[85] Ross, A.J. and Brancato, F.P. (1959) *Mycobacterium fortuitum* Cruz from the tropical fish, *Hyphessobrycon innessi*. *Journal of Bacteriology* 78, 392 - 395.

[86] Ross, A. J. and Johnson, H. E. (1962) Studies of transmission of mycobacterial infections in Chinook salmon. *The Progressive Fish-Culturist* 24, 147 - 149.

[87] Ross, A.J., Earp, B.J. and Wood, J.W. (1959) *Mycobacterial Infections in Adult Salmon and Steelhead Trout Returning to the Columbia River basin and Other Areas in 1957*. Special Scientific Report-Fisheries No. 332, US Department of the Interior. Available at: http://spo.nmfs.noaa.gov/SSRF/SSRF332.pdf (accessed 15 November 2016).

[88] Runyon, E.H. (1959) Anonymous mycobacteria in pulmonary disease. *The Medical Clinics of North America* 43, 273 - 290.

[89] Sadler, P., Smith, M.W., Sullivan, S.E., Hoenig, J.M., Harris, R.E. Jr and Goins, L. (2012) *Evaluation of Striped Bass Stocks in Virginia: Monitoring and Tagging Studies, 2010 - 2014*. Progress Report F - 77 - R - 24 to Virginia Marine Resources Commission, Newport News, Virginia, 26 January 2012. Department of Fisheries Science, School of Marine Science, Virginia Institute of Marine Science, The College of William and Mary, Gloucester Point, Virginia. Available at: http://fluke.vims.edu/hoenig/pdfs/Final_report_2011.pdf (accessed 15 November 2016).

[90] Sakanari, J.A., Reilly, C.A. and Moser, M. (1983) Tubercular lesions in Pacific coast populations of striped bass. *Transactions of the American Fisheries Society* 112, 565 - 566.

[91] Stine, C.B., Kane, A.S., Matsche, M., Pieper, L., Rosemary, K.M. *et al*. (2006) Microbiology of gametes and age 0 - 3 striped bass (*Morone saxatilis*). In: Ottinger, C.A. and Jacobs, J. (eds) *USGS/NOAA Workshop on Mycobacteriosis in Striped Bass, May 7 - 10, 2006, Annapolis, Maryland*. USGS Scientific Investigations Report 2006-5214/NOAA Technical Memorandum NOS NCCOS 41, US Geological Survey/National Oceanic and Atmospheric Administration, Reston, Virginia, pp. 13 - 14. Available at: http://aquaticcommons.org/2233/1/SIR2006 - 5214-complete.pdf (accessed 15 November 2016).

[92] Stine, C.B., Jacobs, J.M., Rhodes, M.R., Overton, A., Fast, M. and Baya, A.M. (2009) Expanded range and new host species of *Mycobacterium shottsii* and *M. pseudoshottsii*. *Journal of Aquatic Animal Health* 21, 179 - 183.

[93] Stine, C.B., Kane, A.S. and Baya, A.M. (2010) Mycobacteria isolated from Chesapeake Bay fish. *Journal of Fish Diseases* 33, 39 - 46.

[94] Swanson, C., Baxa, D.V., Young, P.S., Cech, J.J.J. and Hedrick, R.P. (2002) Reduced swimming performance in delta smelt infected with *Mycobacterium* spp. *Journal of Fish Biology* 61, 1012 - 1020.

[95] Talaat, A.M., Reimschuessel, R., Wasserman, S.S. and Trucksis, M. (1998) Goldfish, *Carassius auratus*, a novel animal model for the study of *Mycobacterium*

marinum pathogenesis. *Infection and Immunity* 66, 2938–2942.

[96] Tortoli, E. (2003) Impact of genotypic studies on mycobacterial taxonomy: the new mycobacteria of the 1990s. *Clinical Microbiology Reviews* 16, 319–354.

[97] Van Der Werf, T. S., Steinstra, Y., Johnson, R. C., Phillips, R., Adjei, O. *et al.* (2005) *Mycobacterium ulcerans* disease. *Bulletin of the World Health Organization* 83, 785–791.

[98] Vogelbein, W.K., Zwerner, D., Kator, H., Rhodes, M.W., Kotob, S.I. and Faisal, M. (1998) Mycobacteriosis in the striped bass, *Morone saxatilis*, from Chesapeake Bay. In: Kane, A.S. and Poynton, S.L. (eds) *3rd Symposium on Aquatic Animal Health*. APC Press, Baltimore, Maryland.

[99] Watral, V. and Kent, M.L. (2007) Pathogenesis of *Mycobacterium* spp. in zebrafish (*Danio rerio*) from research facilities. *Comparative Biochemistry and Physiology – Part C: Toxicology and Pharmacology* 145, 55–60.

[100] Whipps, C.M., Butler, W.R., Pourahmad, F., Watral, V. and Kent, M.L. (2007a) Molecular systematics support the revival of *Mycobacterium salmoniphilum* (*ex* Ross 1960) sp. nov., nom. rev., a species closely related to *Mycobacterium chelonae*. *International Journal of Systematic and Evolutionary Microbiology* 57, 2525–2531.

[101] Whipps, C.M., Dougan, S.T. and Kent, M.L. (2007b) *Mycobacterium haemophilum* infections of zebrafish (*Danio rerio*) in research facilities. *FEMS Microbiology Letters* 270, 21–26.

[102] Whipps, C.M., Matthews, J. L. and Kent, M. L. (2008) Distribution and genetic characterization of *Mycobacterium chelonae* in laboratory zebrafish *Danio rerio*. *Diseases of Aquatic Organisms* 82, 45–54.

[103] Wolf, J.C. and Smith, S.A. (1999) Comparative severity of experimentally induced mycobacteriosis in striped bass *Morone saxatilis* and hybrid tilapia *Oreochromis* spp. *Diseases of Aquatic Organisms* 38, 191–200.

[104] Wood, J.W. and Ordal, E.J. (1958) Tuberculosis in *Pacific Salmon* and *Steelhead Trout*. Contribution No. 25, Fish Commission of Oregon, Portland, Oregon. Available at: http://library.state.or.us/repository/2013/201310171436594/index.pdf (accessed 15 November 2016).

19

美人鱼发光杆菌

John P. Hawke*

19.1 引言

美人鱼发光杆菌(*Photobacterium damselae*)是发光杆菌病的病原菌。它是一种革兰氏阴性嗜盐细菌,属 γ-变形杆菌纲、弧菌目、弧菌科。该菌属包括在深海鱼类发光器官中定植的许多发光菌群,故命名为发光杆菌(*Photobacterium*)。鱼类中该菌的两个致病性亚种为美人鱼发光杆菌杀鱼亚种(*P. damselae* subsp. *piscicida*)和美人鱼发光杆菌美人鱼亚种(*P. damselae* subsp. *damselae*),两者均不表现生物发光性。由这两个亚种引起的疾病综合征及其临床症状有很大不同,它们自身的表型特性也是迥然不同的(表 19.2)。

19.2 美人鱼发光杆菌杀鱼亚种

P. damselae subsp. *piscicida* 曾被称为"鱼巴氏杆菌病(fish pasteurellosis)"和"假结核病(pseudotuberculosis)"的病原菌。"鱼巴氏杆菌病"在 1963 年马里兰州切萨皮克湾大规模鱼病暴发后被报道,这场疫情导致大约 50%的本地美洲狼鲈和条纹鲈死亡(Snieszko et al., 1964)。此次疫情中的一株细菌分离株(ATCC 17911)被保存在弗吉尼亚州马纳萨斯美国典型培养物保藏中心。根据其生理生化特征,将该菌暂时归类于巴氏杆菌属(*Pasteurella*)。Janssen 和 Surgalla(1968)根据其形态学、生理学及血清学研究认为该细菌是一个新种,他们提议将其命名为杀鱼巴斯德氏菌(*Pasteurella piscicida*)。然而,由于与巴斯德氏菌所描述的生理学特性不一致,该命名并没有得到细菌分类学家的认可。这些生理特性差异包括缺少硝酸还原酶、pH 超出正常范围的耐受性、嗜盐性、较低的最适生长温度及特殊的宿主范围。因此,该细菌的命名在《伯杰氏系统细菌学手册》(Mannheim, 1984)和《核准的细菌名录》(Skerman et al., 1989)中均未被录入。尽管如此,这个命名一直沿用至 1995 年,后来基于 16S 核糖体 RNA 测序(Gauthier et al., 1995),该菌被重新命名为

* 作者邮箱:jhawke1@lsu.edu。

美人鱼发光杆菌杀鱼亚种(Truper and Dc'Clari，1997)。

在美国有多起关于 *P. damselae* subsp. *piscicida* 导致野生和养殖鱼类急性败血症的记录。在最初发生在切萨皮克湾的疫情暴发之后，这种病原菌又导致了该地区(Paperna and Zwerner，1976)及纽约长岛海湾西部条纹鲈幼鱼的死亡(Robohm，1983)。在亚拉巴马州海洋资源局位于墨西哥湾海岸的克劳德·皮特海水养殖中心的微咸水土塘中，用于扩充种群的养殖条纹鲈鱼苗中也发现了该疾病(Hawke et al.，1987)。该病原也曾导致路易斯安州咸水湖商业化网箱养殖杂交条纹鲈[HSB，条纹狼鲈(*Morone saxatilis*)× 金眼狼鲈(*M. chrysops*)]的严重死亡(Hawke et al.，2003)。表 19.1 汇总了美人鱼发光杆菌杀鱼亚种在野生和养殖鱼类病害暴发中的宿主、分布及参考文献。

表 19.1 美人鱼发光杆菌杀鱼亚种在野生和养殖鱼类病害暴发中的宿主、分布及参考文献

鱼宿主	栖息地	参考文献
香鱼(*Plecoglossus altevelis*)	日本	Kusuda and Miura，1972
黑鲷(*Mylio macrocephalus*)	日本	Muroga et al.，1977
军曹鱼(*Rachycentron canadum*)	中国台湾地区，巴西	Lopez et al.，2002；Moraes et al.，2015
金头鲷(*Sparus aurata*)	西班牙，葡萄牙，马耳他，意大利	Toranzo et al.，1991；Magariños et al.，1992；Baptista et al.，1996；Bakopoulos et al.，1997
卵形鲳鲹(*Trachinotus ovatus*)	中国	Wang et al.，2013
杂交条纹鲈(*Morone saxatilis*×*M. chrysops*)	亚拉巴马州和路易斯安那墨西哥湾，以色列	Hawke et al.，1987，2003；Nitzan et al.，2001
绿鳍马面鲀(*Navodon modestus*)	日本	Yasunaga et al.，1984
盖斑斗鱼(*Macropodus opercularis*)	中国台湾地区	Liu et al.，2011
真鲷(*Pagrus major*)	日本	Yasunaga et al.，1983
赤点石斑鱼 (*Epinephelus akarra*)	日本	Ueki et al.，1990
舌齿鲈(*Dicentrarchus labrax*)	法国，土耳其，希腊	Magariños et al.，1992；Bakopoulos et al.，1995；Candan et al.，1996
斑鳢(Channa maculata)	中国台湾地区	Tung et al.，1985
条纹狼鲈(*Morone saxatilis*)	马里兰州切萨皮克湾，纽约长岛海湾，日本	Paperna and Zwerner，1976；Robohm，1983
黄带拟鲹 (*Pseudocaranx dentex*)	日本	Nakai et al.，1992
美洲狼鲈(*Morone americana*)	马里兰州切萨皮克湾	Snieszko et al. 1964
副鳚(*Pictiblennius yatabei*)	日本	Hamaguchi et al.，1991
五条鰤(*Seriola quinqueradiata*)	日本	Kubota et al.，1970

正如一些综述文献强调的那样，*P. damselae* subsp. *piscicida* 作为世界范围内野生和养殖鱼类病原菌的重要性已得到确认（Austin and Austin，1993；Magariños et al.，1996；Romalde，2002；Plumb and Hanson，2007；Andreoni and Magnani，2014）。

19.2.1 细菌描述

P. damselae subsp. *piscicida* 是革兰氏阴性、多形性、杆状细菌[(0.5～0.8) μm× (0.7～2.6) μm]。该菌无运动性，无鞭毛，通常表现为革兰氏和吉姆萨染色的两极染色。中度嗜盐、氧化酶阳性（60 s 显色）、过氧化氢酶阳性，对弧菌抑制剂 O/129 敏感。其在含 5%绵羊血的胰蛋白酶大豆琼脂（TSAB）或补加 2%盐的脑心浸液琼脂（BHIA）上生长缓慢（Hawke et al.，2003）。该菌不能在硫代硫酸盐-柠檬酸盐-胆盐-蔗糖琼脂（TCBS）上生长，但在 TSAB 上，28 ℃培养 48 h 后可产生 1 mm 大小非溶血性、灰白色菌落。最适生长条件为 22.5～30 ℃、1.0%～2.5%的 NaCl 和 pH 为 6.47～7.24（Hashimoto et al.，1985）。该菌为兼性厌氧菌，发酵葡萄糖、甘露糖、果糖和半乳糖产酸，但不产气。该菌在生物梅里埃 API 20E 鉴定系统中生成一个唯一的代码（2005004）。该亚种来自不同宿主和地理区域的所有菌株都具有相似的生化和生理特性（Magarinõs et al.，1992）。

19.2.2 动物流行病学

由 *P. damselae* subsp. *piscicida* 引发的发光杆菌病暴发于水温在 14～29 ℃，盐度在 3～21 ppt① 的范围内，而急性发病的最佳温度和盐度范围是 18～25 ℃和 5～15 ppt。从切萨皮克湾分离的菌株能够在无菌半咸水中存活 3 d（Jassen and Surgalla，1968）。然而，有证据表明，其存活但不可培养状态（viable but non-culturable，VBNC）在海水和沉积物中可能长期存在（Magariños et al.，1994a）。有研究指出，某种养殖鱼类或无脊椎动物易感种群可能成为疾病传播的带菌者和传染源（Robohm，1983）。路易斯安那墨西哥湾分离株与切萨皮克湾、希腊、日本及以色列的分离株在生化表型和酶活性上几乎完全相同，但它们的质粒图谱和抗菌药敏性不同。路易斯安那分离株具有独特的质粒带谱，它们通常产生两个大于 30 kb 的大质粒条带及分别为 8.0 kb 和 5.0 kb 的两个小质粒条带。希腊和以色列的分离株有 10 kb 和 8 kb 的质粒条带，日本分离株的质粒条带为 5 kb 和 3.5 kb（Hawke，1996）。一些路易斯安那菌株所具有的 Romet® 和土霉素抗药性来自其获得的 R -质粒（Hawke et al.，2003；Kim et al.，2008）。利用随机扩增多态性 DNA（random amplified polymorphic DNA，

① ppt=10^{-3}。

RAPD)法对路易斯安那墨西哥湾菌株的分析，表明其属于克隆谱系2群，与日本菌株相似(Magariños et al., 2000; Hawke et al., 2003)。

19.2.3 毒力因子

P. damselae subsp. *piscicida* 利用多种毒力机制来克服鱼类宿主免疫防御机制。金头鲷(*Sparus aurata*)和舌齿鲈(*Dicentarchus labrax*)皮肤黏液的正常抗菌活性对这一病原菌无效(Magariños et al., 1995)。该病原菌通过未知途径经由鳃进入杂交条纹鲈宿主体内，并通过血液循环迁移扩散至靶细胞和器官(Hawke, 1996)。除了鳃，对鱼宿主肠黏膜的黏附也是重要的感染途径(Magariños et al., 1996)。该过程由一种蛋白或糖蛋白受体介导，并通过肌动蛋白微丝依赖性机制来实现细菌肠黏膜内化。细菌外膜上的多糖荚膜可抵抗宿主补体及血清诱导的免疫杀伤。血清抗性使得该菌能够在血液中持续存在并迁移扩散至靶细胞和组织，这还使得细菌在杂交条纹鲈血液内达到约 1.0×10^6 CFU/mL的极高密度(Hawke, 1996)。该病原为兼性胞内病原菌，一旦被巨噬细胞吞噬，其会在吞噬体内复制，最终杀死巨噬细胞并释放大量细菌细胞至周围的组织和血液中。在对杂交条纹鲈的侵染过程中，病原菌的胞内复制过程涉及巨噬细胞内吞噬体-溶酶体融合的抑制(Elkamel et al., 2003)。上述相关研究均开展了杂交条纹鲈体内和头肾源巨噬细胞体外实验。

任何鱼龄的条纹鲈和杂交条纹鲈都对 *P. damselae* subsp. *piscicida* 易感，而对于其他鱼种，当体重超过50 g时则表现出增强的抗感染和抗病能力(Andreoni and Magnani, 2014)。Noya等(1995)发现该病原菌可在金头鲷幼鱼(0.5 g)的腹膜渗出(peritoneal exudate, PE)细胞中不受抑制地复制，而在较大鱼(20～30 g)的PE细胞中被抑制。受感染的金头鲷幼鱼死亡率为100%，大鱼中没有死亡。关键的毒力因子是一个质粒编码的56 kDa大小的凋亡诱导蛋白(AIP56)，该蛋白诱导海鲈巨噬细胞和中性粒细胞凋亡(do Vale et al., 2005)。

最后，*P. damselae* subsp. *piscicida* 最重要的毒力机制之一是利用铁载体从宿主体内获取铁元素。铁在宿主体内极其有限，细菌分泌的铁载体可摄取与宿主蛋白(如转铁蛋白)结合的铁。这使得细菌可以通过特异性外膜受体获取铁元素。铁载体合成基因发生突变的菌株对杂交条纹鲈无毒力(Hawke, 1996)。*P. damselae* subsp. *piscicida* DI21株(来自金头鲷)中的铁载体合成基因簇已被鉴定，其与耶尔森菌(*Yersinia*)的高致病性岛相似(Osorio et al., 2006)。这表明DI21株的铁载体可能在结构与功能上与人致病菌小肠结肠炎耶尔森菌(*Y. enterocolitica*)中的重要铁载体耶尔森菌素(yersiniabactin)存在关联。美人鱼发光杆菌杀鱼亚种DI21株的 *irp1* 基因突变会导致限铁条件下的生长受损，无法合成铁载体，并在大菱鲆幼鱼体内的毒力下降到原来的$\frac{1}{100}$(Osorio et al., 2006)。

19.2.4 诊断程序

1. 推定诊断

对于推定诊断，应从宿主体内分离细菌并进行纯化培养，并按19.2.1节中概述的程序进行验证。*P. damselae* subsp. *piscicida* 的表型特征见表19.2。

表19.2 *P. damselae* subsp. *piscicida* 与 *P. damselae* subsp. *damselae* 典型菌株的表型比较

检测科目	*P. damselae* subsp. *piscicida*	*P. damselae* subsp. *damselae*
O/129	S	S
API 20E 编号	2005004	2015004
		6015004
		2011004
双极性染色	+	—
过氧化氢酶	+	+
革兰氏染色	—	—
35 ℃下生长	—	+
TCBS下生长	—	+G
溶血性	—	+
精氨酸	—	75%+
脱羧酶		
运动性	—	+
硝酸盐生成	—	+
氧化酶	W+	+
多形性	+	—
脲酶	—	+
Voges - Proskauer 检测	+	10%+

O/129为弧菌抑制剂；API 20E编号在bioMérieux系统中；Voges-Proskauer检测为产丙酮测试。"+"表示阳性结果；"−"表示阴性结果；"+G"表示绿色菌落的阳性结果；"S"表示敏感；"W+"表示弱或延迟阳性结果；"10%+"表示只有10%的菌株是阳性的；"75%+"表示75%的菌株是阳性的。

2. 确诊

(1) 分子学方法

① 16S rRNA 测序　16S rRNA 的通用引物可用于PCR扩增。PCR产物(完整或部分序列)测序后，利用NCBI的BLAST(美国国家生物技术信息中心基于局部比对算法搜索工具，https://blast.ncbi.nlm.nih.gov/Blast.cgi)与GenBank(美国国立卫生研究院基因序列数据库，https://www.ncbi.nlm.nih.

gov/genbank/)中的序列进行比对分析，可以确定是否为 *P. damsela*。遗憾的是，这种方法无法区分美人鱼发光杆菌的两个亚种。

② 种特异性 PCR　Aoki 等(1997)利用种特异性质粒 pZP1 的序列开发了首个种特异性 PCR 方法。然而，由于欧洲菌株并不具有相同的质粒谱，这种鉴定方法并不完备。随后，利用设计的荚膜多糖基因扩增引物，开发建立了一种快速鉴定和区分两种 *P. damsela* 亚种的方法。该特异性引物对用于扩增 *P. damselae* subsp. *piscicida* 荚膜多糖基因中 410 bp 大小的片段(GenBank 登录号为 AB074290)。正向引物 CPSF 为 5′- AGGGGATCCGATTATTACTG - 3′，对应荚膜多糖基因的 531～550 碱基位，反向引物 CPSR 为 5′- TCCCATTGAGAAGATTTGAT - 3′，对应 921～940 碱基位。由于两个亚种都使用该方法进行检测，还应通过在 TCBS 培养基上的生长状态进一步区分——*P. damselae* subsp. *damselae* 在 TCBS 培养基上生长形成绿色菌落，而 *P. damselae* subsp. *piscicida* 无法生长(Rajan et al., 2003)。

③ 亚种特异性多重 PCR　利用针对 *16S rRNA* 和 *ureC* 基因内部区域的两对引物，开发建立了一种可以识别和区分美人鱼发光杆菌两个亚种的多重 PCR 检测法(Osorio et al., 2000a)。通过该法，*P. damselae* subsp. *damselae* 菌株产生两个扩增产物，一个大小为 267 bp，另一个大小为 448 bp，分别对应于 *16S rRNA* 和 *ureC* 基因的内部片段。相反地，*P. damselae* subsp. *piscicida* 仅会产生来自 *16S rRNA* 的 267 bp 大小的 PCR 片段，表明其缺少脲酶基因。从 GenBank 中检索到 *P. damselae* subsp. *damselae* 的 *ureC* 基因的部分序列，登录号为 U40071。正向引物 Ure - 5′(20 - mer 5′- TCCGGAATAGGTAAAGCGGG -3′)和反向引物 Ure - 3′(22 - mer 5′- CTTGAATATCCATCTCATCTGC -3′)被设计用来与 *ureC* 基因的上一段 448 bp 大小的片段配对。一对匹配于美人鱼发光杆菌杀鱼亚种 ATCC 29690 株 *16S rRNA* 基因 267 bp 片段(GenBank 登录号为 Y18496)的正向引物(118 - mer 5′- GCTTGAAGAGATTCGAGT - 3′；位于大肠杆菌 *16S rRNA* 基因的 1016～1033 碱基位)和反向引物(18 - mer 5′- CACCTCGCGGTCTTGCTG - 3′；位于 1266～1283 的碱基位)，被设计用来与 Ure - 5′和 Ure - 3′一同用于多重 PCR(Osorio et al., 1999)。最近，另一种靶向青霉素结合蛋白 1A 编码基因和 *ureC* 基因的多重 PCR 检测法被开发出来(Amagliani et al., 2009)。这一优化的多重 PCR 检测法可鉴定和区分两个亚种，检出限为 500 fg DNA(100 个基因当量单位)。

(2) 血清学方法

使用 BIONOR™ AQUARAPID(BIONOR，希恩，挪威)检测试剂盒可检出微小临床症状的急性感染或亚临床感染。该试剂盒使用针对 *P. damselae* subsp. *piscicida* 抗原的特异性抗体，对组织匀浆样本进行酶联免疫吸附试验(ELISA)(Romalde et al., 1995a)。从病鱼中鉴定疑似 *P. damselae* subsp.

piscicida 的细菌时，可以使用 BIONOR™ Mono Aqua 试剂盒（Romalde et al., 1995b）。该检测法依赖于一种应用于细菌纯培养物悬浮液的颗粒凝集试验法。

19.2.5 疾病临床症状

1. 大体病理学

根据鱼的种类、环境条件和药物干预，受感染的鱼可能表现出不同的临床症状和病理。一般来说，感染了发光杆菌病的鱼变得无精打采，在水面附近缓慢游动，最终在死亡前沉入水底，可能出现明显的呼吸频次增加和失去平衡。在多种鱼类中都描述报道了急性和慢性发光杆菌病（Thune et al., 1993）。急性发病时，能够观察到的大体病理症状很少（Bullock, 1978; Toranzo et al., 1991）。在慢性疾病中，脾脏和肾脏中可见类似肉芽肿的白色小病灶，因此这种疾病被称为“假结核病（pseudotuberculosis）”。

患病的养殖条纹鲈和杂交条纹鲈可能是嗜睡的，表现为鳃苍白，鳃盖区有瘀斑，体色较正常色素沉着深（Hawke et al., 1987, 2003）。在体内，脾肿胀易碎是唯一明显的大体临床症状（图 19.1），但肝脏可能有轻微斑点，肾脏可能有出血。美洲狼鲈只表现出鳃盖和鳍基部轻微出血。在野生条纹鲈和美洲狼鲈中，在肿胀的脾和肾中可见小的白色“粟粒样”病变（Wolke, 1975; Bullock, 1978）。慢性形式的疾病可能因受影响的鱼种类和是否饲喂抗生素饲料而有所不同。抗生素治疗后，在养殖杂交条纹鲈的脾和肾中可能可以看到类似的白色小肉芽肿样病变（图 19.2）（Hawke et al., 2003）。这表明，抗生素减缓了感染的进程，从而使肉芽肿性炎症变得肉眼可见。

图 19.1 由 *P. damselae* subsp. *piscicida* 引起急性发光杆菌病而处于濒死状态的杂交条纹鲈，其脾肿胀易碎，其他临床症状不明显。照片由 Joe Newton 博士提供

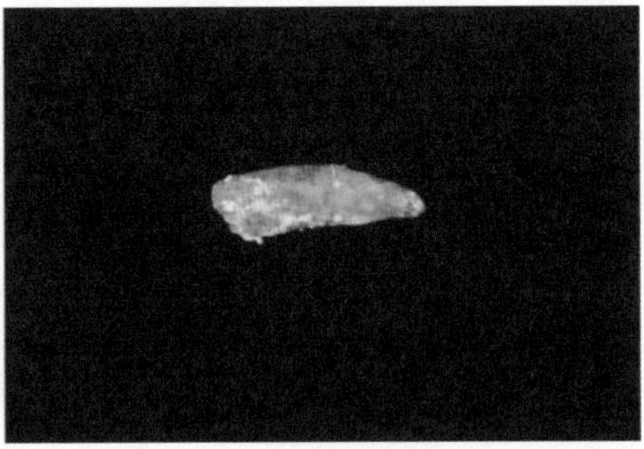

图 19.2 抗生素治疗后杂交条纹鲈中由 *P. damselae* subsp. *piscicida* 引起的慢性发光杆菌感染。脾以脾实质中的坏死灶、细菌菌落和肉芽肿样病变为临床特征，可见白色“粟粒样”病变。照片由 Al Camus 博士提供

受感染的养殖黄条鰤和五条鰤(*Seriola* spp.)出现水肿并且无法调节色素沉着。在黄条鰤中，慢性病变的典型特征是脾、肾和心中呈现1～2 mm的肉芽肿样病变，这是由大量的病原菌、上皮细胞和成纤维细胞组成的。这些病变非常类似于肉芽肿(Kubota et al., 1970)。除罕见个体头盖周围出血外，患病金头鲷没有明显的外部临床症状。养殖军曹鱼则在肾、肝和脾中有白色肉芽肿样病变。濒死的军曹鱼嗜睡，体色比正常鱼更暗(Liu et al., 2003)。

2. 组织病理学

Wolke(1975)首次报道了切萨皮克湾自然感染的美洲狼鲈和条纹鲈的组织病理学。在一种明显的慢性疾病中，可从脾中观察到坏死淋巴和外周血细胞的聚集。肝脏内，明显可见肝细胞凝固性坏死和明显缺乏炎性细胞反应的病灶区。在条纹鲈和HSB的急性疾病中，脾淋巴组织发生急性多灶性坏死，表现为细胞损失、凝固性坏死、核破裂和细菌大菌落(图19.3)。肝常见伴有明显核破裂的急性多灶性坏死，无炎性细胞聚集(Hawke et al., 1987)。在金头鲷中也报道了类似的显微病变(Toranzo et al., 1991)。在急性疾病中，每毫升血液中细菌数量接近100万个(图19.4；Hawke, 1996)。黄条鰤和军曹鱼的反应不同，脾脏中的细菌菌落形成肉芽肿样结构——“假结节”含有嗜酸性粒细胞并被上皮细胞包围。

图19.3 与大量*P. damselae* subsp. *piscicida*菌落有关的杂交条纹鲈脾局灶性坏死。缺乏炎性细胞积聚。照片由John Hawke博士提供

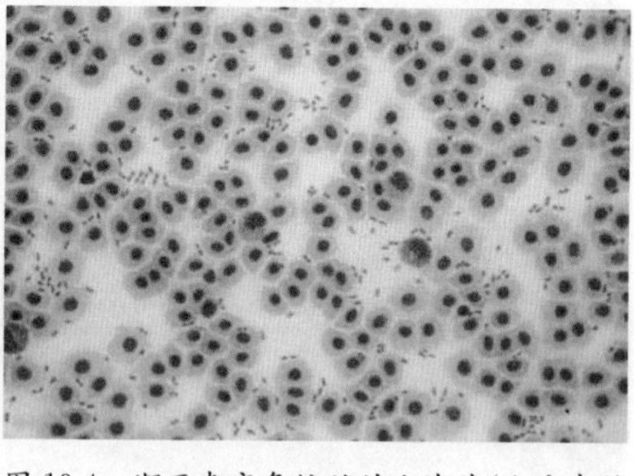

图19.4 濒死杂交条纹鲈的血涂片(血液中约含1×10^6 CFU/mL的*P. damselae* subsp. *piscicida*)。照片由John Hawke博士提供

在所有种类的患病鱼中，毛细血管中均会形成细菌团(细菌栓子)，并阻塞内脏器官间质中的血液流动。在实验感染的杂交条纹鲈中，死亡归因于鳃毛细血管中的细菌栓塞和由于无法实现血流和气体交换而导致的窒息(Hawke, 1996)。

19.2.6 疾病治疗

早期发现对于疾病的有效治疗和控制至关重要。最常见的措施是在感染早期施用添加抗生素的饲料(Andreoni and Magnani, 2014)。如前所述,感染和死亡的进程受鱼的大小、鱼龄和种类的影响。药用饲料投喂时间和方法的选择受到养殖方式(池塘、海水网箱、围网室内系统)以及水温、死亡率和季节等因素的影响。在美国,获得批准的抗菌药很少,可能需要兽医处方或临床新兽药(investigational new animal drug,INAD)许可。使用 Romet® 和土霉素饲料治疗网箱和围网中感染的杂交条纹鲈最初成功地降低了死亡率甚至停止死亡(Hawke et al., 2003)。不幸的是,疾病常常复发,并且继发病害的分离株由于获得 R-质粒通常产生多重耐药性(Hawke et al., 2003)。来自日本病原株的 R-质粒赋予菌株对卡那霉素、氯霉素、四环素和磺胺类药物的耐药性,而美国菌株则对四环素、甲氧苄啶和磺胺类药物具有抗药性(Kime et al., 2008)。

19.2.7 发光杆菌病的预防

抗生素饲料缺乏一致的功效以及 *P. damselae* subsp. *piscicida* 抗生素耐药菌株的出现,促进了疫苗、益生菌剂和预防性膳食补充剂等防控手段的开发。一种替代抗生素使用的方案是在摄食中掺加植物提取物[例如在饲喂中添加大蒜(*Allium sativum*)或其活性成分大蒜素]。大蒜具有悠久的药用历史,并具有广泛的抗菌性能(Guo et al., 2015)。在军曹鱼的实验性试验中,将大蒜粉掺入饲料中并以每天 1.2 g/kg 的剂量投喂 21 d,实验鱼对 *P. damselae* subsp. *piscicida* 的攻毒具有显著的抵抗力。来自海水鱼类肠道的弧菌菌株 NM10 可产生一种热不稳定蛋白类物质(<5 kDa),具有抑菌活性。该菌株可能具有抗发光杆菌病益生菌的潜力(Sugita et al., 1997)。

早期针对日本五条鰤(*S. quinqueradiata*)发光杆菌病的疫苗,无论是福尔马林灭活疫苗还是胞外产物亚单位疫苗[ECPs-脂多糖(LPS)和核糖体组分],都是无效的(Fukuda and Kusuda, 1981)。在培养上清液中一种富含 97 kDa 外膜蛋白和 52 kDa 蛋白的菌苗,能在鳃和黏膜中引发抗体反应,并具有保护作用(Barnes et al., 2005)。一种在欧洲商业化使用的富含 ECP 的菌苗,通过浸泡在海鲈和五条鰤中显示出良好的免疫效果(Magariños et al., 1994b)。

P. damselae subsp. *piscicida* 的两种减毒活菌株 LSU-P1 和 LSU-P2,由分离自杂交条纹鲈的亲本株 LADL 91～197 的突变体产生。铁载体生物合成突变株 LSU-P1 由亲本株 LADL 91～197 使用电转化方法导入自杀质粒 pEIS,并通过自杀质粒上携带卡那霉素抗性基因的 *tn10* 微型转座子随机插入失活铁载体生物合成基因转化产生。在含卡那霉素和铬天青 S 的培

养基上进行突变株的筛选，以检测铁载体的产生（Hawke，1996）。减毒株 LSU - P2 在芳香族氨基酸生物合成基因（*aroA*）上发生突变。该突变是通过同源重组和插入卡那霉素抗性基因在 *aroA* 基因中产生移码突变而实现的（Thune et al.，2003）。在实验室试验中，杂交条纹鲈浸泡接种这两种减毒株后对毒株 LADL 91～197 的攻毒都表现出显著的免疫保护作用。然而，由于美国墨西哥湾沿岸的杂交条纹鲈养殖业的衰退和最终消亡，该疫苗从未商业化。

在中国台湾地区，探索开发预防军曹鱼发光杆菌病亚单位疫苗的研究工作已经开展（Ho et al.，2011）。在这些研究中，免疫蛋白质组学被用于鉴定 *P. damselae* subsp. *piscicida* 的抗原，以克隆和生产用于开发潜在的疫苗重组蛋白。一种被称为"反向疫苗学"的方法利用生物信息学来识别重要的候选疫苗蛋白。已经在体外确认了参与细菌黏附和内化的抗原。用该方法鉴定的重组抗原对海鲈进行免疫接种后进行的攻毒实验诱导了特异性抗体和保护（Andreoni et al.，2013）。

19.3 美人鱼发光杆菌美人鱼亚种

P. damselae subsp. *damselae* 最初被命名为美人鱼弧菌（*Vibrio damsela*），可导致海水鱼类溃疡病（ulcer disease）和出血性败血症（haemorrhagic septicaemia）。Love 等（1981）报道引起栖息于南加州沿海水域雀鲷（*Chromis punctipinnis*）皮肤溃疡的病原为 *V. damsela*，并指出已从水中和两处人体伤口中分离出来该细菌，其可能是造成人体伤口感染和疾病的病原。该细菌与弧菌有许多共同的特征和表型特性。它曾被重命名为美人鱼利斯顿氏菌（*Listonella damsela*）（MacDonell and Colwell，1985）、美人鱼发光杆菌（*Photobacterium damsela*）（Smith et al.，1991）、美人鱼发光杆菌美人鱼亚种（*P. damselae* subsp. *damselae*）（Gauthier et al.，1995），并最终命名为美人鱼发光杆菌美人鱼亚种（*P. damselae* subsp. *damselae*）（Truper and De'Clari，1997）。

该菌对许多海洋鱼类、甲壳类动物、软体动物和鲸类具有致病性。对于人类来说，它会引起机会性感染，进而发展为坏死性筋膜炎，导致肢体残缺甚至死亡（Rivas et al.，2013）。特定的鱼类宿主为斑鳍光鳃鱼（*Chromis punctipinnis*）（Love et al.，1981）、多种鲨鱼（Grimes et al.，1984，1985；Han et al.，2009），五条鰤（*S. quinqueradiata*）（Sakata et al.，1989）、金头鲷（Vera et al.，1991）、欧洲海鲈（Botella et al.，2002），大菱鲆（Fouz et al.，1992）、尖吻鲈（Renault et al.，1994）及虹鳟（Pedersen et al.，2009）。从西班牙新养殖的海水鱼类[三长棘赤鲷、赤鲷、重牙鲷和大西洋白姑鱼（*Argyrosomus regius*）]中也分离出该菌（Labella et al.，2011）。除了鱼类，美人鱼发光杆菌美人鱼亚种也在

患病虾(*Penaeus monodon*)(Wang and Chen, 2006)、章鱼(*Octopus joubini*)(Hanlon et al., 1984)、海龟(*Dermochelys coriacea*)(Obendorf et al., 1987)、海豚(*Tursiops truncatus*)(Fujioka et al., 1988)及人类伤口感染(Love et al., 1981; Clarridge and Zighelboim-Daum, 1985)中被分离出来。

19.3.1 菌株描述

P. damselae subsp. *damselae* 是革兰氏阴性杆菌或球杆菌,大小为(0.5~0.8) μm×(0.7~2.0) μm。该细菌通过单个或多个无鞘极性鞭毛运动,表现为典型的氧化酶、过氧化氢酶和脲酶阳性(Smith et al., 1991)。具有嗜盐性,可在1%~6%的 NaCl 范围内生长,在含 2%盐的 BHIA 培养基上的生长温度范围为20~37 ℃,35 ℃下生长最适宜。在含 5%绵羊血的 TSAB 培养基上,菌落呈现典型的溶血性,但这一特性随菌株和产生的溶血素而变化。35 ℃下该菌在 TCBS 上生长 24 h 可产生绿色菌落,在 2%盐的 TSAB 或 BHIA 上产生1~2 mm的灰白色菌落。该菌是一种兼性厌氧菌,可发酵葡萄糖产生酸和气体,发酵甘露糖、蜜二糖、麦芽糖、纤维二糖和 D-半乳糖产酸,对弧菌抑制剂 O/129 敏感(Fouz et al., 1992)。该菌典型的鱼类致病株在 API 20E 系统(bioMérieux)中生成唯一的代码(2015004),但是在一些菌株中,Voges-Proskauer 检测(用于检测丙酮生成)和赖氨酸脱羧酶检测的结果可变,导致产生不同的代码(表19.2)。一些研究表明,来自不同宿主的菌株之间存在着其他生化异质性,在葡萄糖产气、脲酶产生以及对某些糖的利用测试中结果不一(Labella et al., 2011)。

19.3.2 动物流行病学

美人鱼发光杆菌美人鱼亚种通常栖息在海水和海洋沉积物中,偏好温暖水域(20~30 ℃)。毒株可以在 14~22 ℃的海水微环境中以可培养的状态长期存活,并保持其对鱼类的感染力(Fouz et al., 1998)。在土塘养鱼场、网栏和海水网箱以及室内水槽和水族箱中都暴发过由该病原菌引发的病害。该病原菌通过水传播给高度易感的宿主生物。感染通常是机会性的,而易感性较低的鱼类需要一些胁迫或损伤作为诱发因素。在丹麦海水养殖虹鳟中,这种病原菌可在水温较高时引发疾病。在水温为 20 ℃时对虹鳟的致病力比 13 ℃时高 1 000 倍(Pedersen et al., 2009)。当水温从 18 ℃升高到 25 ℃时,实验大菱鲆的死亡率会出现急剧上升(Fouz et al., 1992)。这种病原菌在西班牙南部养殖的海水鱼类中造成过多次疾病暴发,如三长棘赤鲷、赤鲷和重牙鲷(Labella et al., 2011)。西班牙养鱼场的死亡率从 12 月份的 5%到 8 月份的 94%不等。由于其偏好较高的水温,一些研究者预测,随着全球气候变暖,由该病原引起的疾病问题将会增加(Pedersen et al., 2009)。

19.3.3 毒力因子

P. damselae subsp. *damselae* 的胞外产物对不同鱼类和哺乳动物细胞系都具有细胞毒性。只有致病株能够产生有毒性的胞外产物并且这些细胞毒性物质不耐热(Labella et al., 2010)。其中主要的毒力因子是一种磷脂酶 D,传统上被称为豆蔻蛋白酶。其他磷脂酶也可能是毒力因子,因为一些毒株缺乏磷脂酶 D 基因 *dly*,但它们的胞外产物中仍具有磷脂酶活性(Osorio et al., 2000b)。最近,一个 150 kb 的质粒 pPHDD1 被鉴定携带 Dly 及成孔毒素家族 Hly 毒素的编码基因(Rivas et al., 2011)。只有一小部分菌株携带 pPHDD1 质粒,这些菌株比没有该质粒的菌株形成更大的 β-溶血区(Rivas et al., 2013)。这两种溶血素都很重要,编码基因突变会导致菌株对鱼类无致病力,并降低对小鼠的毒力(Rivas et al., 2013)。*hlyA* 基因在一些菌株中也可以被染色体编码。毒力研究证实,无论是质粒还是染色体编码,Dly 都与 HlyA 协同作用(Rivas et al., 2013)。此外,在胞外产物中检测到的酶活性可能与细菌体内或体外的毒性有关(Labella et al., 2010)。大多数菌株都没有可检测到的蛋白酶(Fouz et al., 1992),这表明其他未知的分子在毒性中发挥了作用。对于 *P. damselae* subsp. *damselae* 全基因组的测序预计将有助于弥补我们目前对其认知上存在的空白。

19.3.4 诊断程序

1. 推定诊断

对于推定诊断,应该从宿主体内分离病菌并进行纯化培养,并使用 19.3.1 节中概述的程序进行鉴定。*P. damselae* subsp. *damselae* 与 *P. damselae* subsp. *piscicida* 的表型特征一起列于表 19.2。

2. 确诊

使用 19.2.4 节中所述的鉴定 *P. damselae* subsp. *damselae* 的分子学方法可以实现对该病原的确诊。如下所述,不同的 *P. damselae* subsp. *damselae* 菌株也可以根据它们的血清学差异和遗传变异进行检测。

血清学与菌株异质性　*P. damselae* subsp. *damselae* 菌株在血清学上存在异质性(Smith et al., 1991; Fouz et al., 1992; Pedersen et al., 2009)。从大菱鲆中鉴定出四个可识别的基于脂多糖的血清群(A～D)(Fouz et al., 1992)。通过扩增片段长度多态性(amplified fragment length polymorphism, AFLP)确定的遗传变异是显著的:来自欧洲海鲈和金头鲷的 33 个菌株中的 24 个表现出不同的染色体带型,这与分离株或宿主鱼类的地理来源存在关联(Botella et al., 2002)。

19.3.5 疾病临床症状与大体病理学

根据宿主种类，美人鱼发光杆菌美人鱼亚种引发的临床症状可能会略有不同，且大多数宿主的行为症状尚未被描述。在澳洲金赤鲷(*Pagrus auratus*)中，濒死鱼在死亡前以侧卧的姿势在水面上漂浮数小时(Stephens et al., 2006)。在雀鲷中，溃疡性病变在胸鳍和尾柄附近形成，死亡前溃疡大小可增大至 20 mm(Love et al., 1981)。在患病虹鳟中可观察到肛门周围的皮肤、壁层腹膜和肠道的大面积出血。皮肤出血在鳍基部和沿腹部腹侧中线最明显。瘀斑常见于肝脏，并伴有腹腔出血性腹水(Pedersen et al., 2009)。海鲈和海鲷的临床症状与弧菌病相似。主要的外部症状是眼球突出、皮肤色素沉着发暗、鳃苍白和鱼鳍溃烂，鱼体表常见出血区和溃疡；内部可见肝苍白、脾肿大和血性腹水(Labella et al., 2011)。在患病大菱鲆中，最显著的临床症状是腹胀、眼和嘴(包括上颚和下颚)以及肛门周围的出血(Fouz et al., 1992)。墨西哥湾野外捕获的患病红鲷(*Lutjanus campechanus*)有浅表皮肤溃疡(Hawke and Baumgartner, 2011)。

19.3.6 组织病理学

关于这种疾病的组织病理学和显微病变信息非常少。受感染的澳洲金赤鲷患有粒细胞性肠炎，伴有水肿和嗜酸性粒细胞浸润。这种相同的细胞状态在鳃中很明显。受感染鱼的脾、肾和肝中的黑色素巨噬细胞中心和血铁黄素沉积物数量增加(Stephens et al., 2006)。皮肤损伤的红鲷呈现出溃疡性皮炎，伴有中度真皮纤维化和炎性细胞浸润(Hawke and Baumgartner, 2011)。

19.3.7 疾病治疗

对于美人鱼发光杆菌美人鱼亚种引起的发光杆菌病暴发通常采用药物饲料进行治疗，并且有研究报道了养鱼场典型分离株的抗生素药敏性(Pedersen et al., 2009; Labella et al., 2011)。西班牙养鱼场典型菌株对甲氧苄氨嘧啶-磺胺甲噁唑、氯霉素、恩诺沙星、氟甲喹、萘啶酸、噁喹酸和呋喃妥因敏感。目前已观察到该菌株对链霉素、红霉素、四环素、土霉素、阿莫西林和氨苄西林产生的耐药性(Labella et al., 2011)。由于不同分离株之间的抗原异质性，尚未有相应疫苗投入商业化应用(Botella et al., 2002)。

19.4 总结与研究展望

美人鱼发光杆菌的两个亚种都是重要的致病菌，但在表型和流行病学上两者存在差异。*P. damselae* subsp. *piscicida* 是一种专性鱼类病原菌，在环境(水和沉积物)中无法长期存活，而 *P. damselae* subsp. *damselae* 在环境(水和沉积物)中可长期保持感染性，是一种鱼类机会致病菌，也是包括人类在内的恒温动物的机会致病菌。*P. damselae* subsp. *piscicida* 作为鱼类病原菌出现于温带气候下 18～28 ℃的水域范围中，引起的疾病很少伴随临床症状，而 *P. damselae* subsp. *damselae* 偏好更暖的水域，引起的疾病具有明显临床症状，包括皮肤多灶性出血和溃疡。灭活疫苗对发光杆菌病的预防效果不明显，而预防 *P. damselae* subsp. *piscicida* 感染的减毒活疫苗为今后水产养殖中该病害的管控提供了前景。相比之下，由于存在血清学多样性，使得针对 *P. damselae* subsp. *damselae* 的免疫疗法的建立变得复杂，对于该亚种的管控可能更加依赖于养殖方式的改进和有效的抗生素疗法。

参考文献

[1] Amagliani, G., Omiccioli, E., Andreoni, F., Boiani, R., Bianconi, I. *et al*. (2009) Development of a multiplex PCR assay for *Photobacterium damselae* subsp. *piscicida* identification in fish samples. *Journal of Fish Diseases* 32, 645 - 653.

[2] Andreoni, F. and Magnani, M. (2014) Photobacteriosis: prevention and diagnosis. *Journal of Immunology Research* 2014: Article ID 793817.

[3] Andreoni, F., Boiani, R., Serafini, G., Amagliani, G., Dominici, S. *et al*. (2013) Isolation of a novel gene from *Photobacterium damselae* subsp. *piscicida* and analysis of the recombinant antigen as a promising vaccine candidate. *Vaccine* 31, 820 - 826.

[4] Aoki, T., Ikerda, D., Katagiri, T. and Hirono, I. (1997) Rapid detection of the fish pathogenic bacterium *Pasteurella piscicida* by polymerase chain reaction targeting nucleotide sequences of the species specific plasmid pZP1. *Fish Pathology* 32, 143 - 151.

[5] Austin, B. and Austin, D.A. (1993) *Bacterial Fish Pathogens. Disease of Farmed and Wild Fish*, 2nd edn. Ellis Horwood, Chichester, UK, pp. 23 - 42.

[6] Bakopoulos, V., Adams, A. and Richards, R.H. (1995) Some biochemical properties and antibiotic sensitivities of *Pasteurella piscicida* isolated in Greece and comparison with strains from Japan, France and Italy. *Journal of Fish Diseases* 18, 1 - 7.

[7] Bakopoulos, V., Peric, Z., Rodger, H., Adams, A. and Richards, R. (1997) First report of fish pasteurellosis from Malta. *Journal of Aquatic Animal Health* 9, 26 - 33.

[8] Baptista, T., Romalde, J. L. and Toranzo, A. E. (1996) First occurrence of pasteurellosis in Portugal affecting cultured gilthead seabream (*Sparus aurata*). *Bulletin of the European Association of Fish Pathologists* 16, 92 - 95.

[9] Barnes, A.C., dos Santos, N.M.S. and Ellis, A.E. (2005) Update on bacterial vaccines: *Photobacterium damselae* subsp. *piscicida*. *Developments in Biologicals* 121, 75 - 84.

[10] Botella, S., Pujalta, M.-J., Macian, M.-C., Ferrus, M.-A., Hernandez, J. *et al*. (2002) Amplified fragment length polymorphism (AFLP) and biochemical typing of *Photobacterium damselae* subsp. *damselae*. *Journal of Applied Microbiology* 93, 681 - 688.

[11] Bullock, G. L. (1978) *Pasteurellosis of Fishes*. *Fish Disease Leaflet* No. 54, US Department of the Interior, Fish and Wildlife Service, Washington, DC.

[12] Candan, A., Kucuker, M.A. and Karatas, S. (1996) Pasteurellosis in cultured sea bass (*Dicentrarchus labrax*) in Turkey. *Bulletin of the European Association of Fish Pathologists* 16, 150 - 153.

[13] Clarridge, J.E. and Zighelboim-Daum, S. (1985) Isolation and characterization of two haemolytic phenotypes of *Vibrio damsela* associated with a fatal wound infection. *Journal of Clinical Microbiology* 21, 302 - 306.

[14] doVale, A., Silva, M.T., dosSantos, N.M.S., Nascimento, D.S., Reis-Rodrigues, P. *et al*. (2005) AIP56, a novel plasmid-encoded virulence factor of *Photobacterium damselae* subsp. *piscicida* with apoptotic activity against sea bass macrophages and neutrophils. *Molecular Microbiology* 58, 1025 - 1038.

[15] Elkamel, A., Hawke, J.P., Henk, W.G. and Thune, R.L. (2003) *Photobacterium damselae* subsp. *piscicida* is capable of replicating in hybrid striped bass macrophages. *Journal of Aquatic Animal Health* 15, 175 - 183.

[16] Fujioka, R.S., Greco, S.B., Cates, M.B. and Schroeder, J.P. (1988) *Vibrio damsela* from wounds in bottlenose dolphins Tursiops truncatus. *Diseases of Aquatic Organisms* 4, 1 - 8.

[17] Fouz, B., Larsen, J. L., Nielsen, B, Barja, J. and Toranzo, A. E. (1992) Characterization of *Vibrio damsela* strains isolated from turbot *Scophthalmus maximus* in Spain. *Diseases of Aquatic Organisms* 12, 155 - 166.

[18] Fouz, B., Toranzo, A.E., Marco-Noales, E. and Amaro, C. (1998) Survival of fish-virulent strains of *Photobacterium damselae* subsp. *damselae* in seawater under starvation conditions. *FEMS Microbiology Letters* 168, 181 - 186.

[19] Fukuda, Y. and Kusuda, R. (1981) Efficacy of vaccination for pseudotuberculosis in cultured yellowtail by various routes of administration. *Bulletin of the Japanese Society for Scientific Fisheries* 47, 147 - 150.

[20] Gauthier, G., LaFay, B., Ruimy, R., Breittmayer, V., Nicolas, J.L. *et al*. (1995) Small-subunit rRNA sequences and whole DNA relatedness concur for the reassignment of *Pasteurella piscicida* (Snieszko *et al*.) (Janssen and Surgalla) to the genus *Photobacterium* as *Photobacterium damsela* subsp. *piscicida* comb. nov. *International Journal of Systematic Bacteriology* 45, 139 - 144.

[21] Grimes, D.J., Colwell, R.R., Stemmler, J., Hada, H., Maneval, D. *et al*. (1984) *Vibrio* species as agents of elasmobranch disease. *Helgoländer Meeresunters* 37, 309 - 315.

[22] Grimes, D.J., Brayton, P., Colwell, R.R. and Gruber, S.H. (1985) Vibrios as authocthonous flora of neritic sharks. *Systematic and Applied Microbiology* 6, 221 - 226.

[23] Guo, J.J, Kuo, C.M., Hong, J.W., Chou, R.L., Lee, Y.H. *et al*. (2015) The effects of garlic supplemented diets on antibacterial activities against *Photobacterium damselae* subsp. *piscicida* and *Streptococcus iniae* and on growth in cobia, *Rachycentron canadum*. *Aquaculture* 435, 111 - 115.

[24] Hamaguchi, M., Usui, H. and Kusuda, R. (1991) *Pasteurella piscicida* infection in yatabe blenny (*Pictiblennius yatabei*). *Gyobio Kenkyu* 26, 93 - 94.

[25] Han, J.E., Gomez, D.K., Kim, J.H., Choresca, C.H., Shin, S.P. *et al*. (2009) Isolation of *Photobacterium damselae* subsp. *damselae* from zebra shark *Stegostoma fasciatum*. *Korean Journal of Veterinary Research* 49, 35 - 38.

[26] Hanlon, R.T., Forsythe, J.W., Cooper, K.M., Dinuzzo, A.R., Folse, D.S. *et al*. (1984) Fatal penetrating skin ulcers in laboratory reared octopuses. *Journal of Invertebrate Pathology* 44, 67 - 83.

[27] Hashimoto, S., Muraoka, A., Mihara, S. and Kusuda, R. (1985) Effects of cultivation temperature, NaCl concentration, and pH on the growth of *Pasteurella piscicida*. *Bulletin of the Japanese Society for Scientific Fisheries* 51, 63 - 67.

[28] Hawke, J.P. (1996) Importance of a siderophore in the pathogenesis and virulence of *Photobacterium damsela* subsp. *piscicida* in hybrid striped bass (*Morone saxatilis* ×

Morone chrysops). Ph. D. dissertation, Louisiana State University, Baton Rouge, Louisiana.

[29] Hawke, J. P. and Baumgartner, W. A. (2011) *Unpublished Case Reports*. Louisiana Aquatic Diagnostic Laboratory, Louisiana State University, Baton Rouge, Louisiana.

[30] Hawke, J.P., Plakas, S.M., Minton, R.V., McPhearson, R.M., Snider, T.G. *et al*. (1987) Fish pasteurellosis of cultured striped bass (*Morone saxatilis*) in coastal Alabama. *Aquaculture* 65, 193 - 204.

[31] Hawke, J.P., Thune, R.L., Cooper, R.K., Judice, E. and Kelly-Smith, M. (2003) Molecular and phenotypic characterization of strains of *Photobacterium damselae* subsp. *piscicida* from hybrid striped bass cultured in Louisiana, USA. *Journal of Aquatic Animal Health* 15, 189 - 201.

[32] Ho, L.-P., Lin, J. H.-Y., Liu, H.-C., Chen, H.-E., Chen, T.-Y. *et al*. (2011) Identification of antigens for the development of a subunit vaccine against *Photobacterium damselae* subsp. *piscicida*. *Fish and Shellfish Immunology* 30, 412 - 419.

[33] Janssen, W.A. and Surgalla, M.J. (1968) Morphology, physiology, and serology of a *Pasteurella* species pathogenic for white perch (*Roccus americanus*). *Journal of Bacteriology* 96, 1606 - 1610.

[34] Kim, M., Hirono, I., Kurokawa, K., Maki, T., Hawke, J. *et al*. (2008) Complete DNA sequence and analysis of the transferable multiple drug resistance plasmids (R-plasmids) from *Photobacterium damselae* subsp. *piscicida* isolated in Japan and USA. *Antimicrobial Agents and Chemotherapeutics* 52, 606 - 611.

[35] Kubota, S.S., Kimura, M. and Egusa, S. (1970) Studies of a bacterial tuberculoidosis of the yellowtail. I. Symptomatology and histopathology. *Fish Pathology* 4, 111 - 118.

[36] Kusuda, R. and Miura, W. (1972) Characteristics of a *Pasteurella* sp. pathogenic for pond cultured ayu. *Fish Pathology* 7, 51 - 57.

[37] Labella, A.C., Sanchez-Montes, N., Berbel, C., Aparicio, M. Castro, D. *et al*. (2010) Toxicity of *Photobacterium damselae* subsp. *damselae* isolated from new cultured marine fish. *Diseases of Aquatic Organisms* 92, 31 - 40.

[38] Labella, A., Berbel, C., Manchado, M., Castro, D. and Borrego, J.J. (2011) Chapter 9. *Photobacterium damselae* subsp. *damselae*, an emerging pathogen affecting new cultured marine fish species in southern Spain. In: Aral, F. and Doğu, Z. (eds) *Recent Advances in Fish Farms*. InTech Open Science, Rijeka, Croatia. Available at: http://www.intechopen.com/articles/show/title/photobacterium-damselaesubsp-damselae-an-emerging-pathogen-affectingnew-cultured-marine-fish-speci (accessed 16 November 2016).

[39] Liu, P.-C., Lin, J.-Y. and Lee, K.-K. (2003) Virulence of *Photobacterium damselae* subsp. piscicida in cultured cobia *Rachycentron canadum*. *Journal of Basic Microbiology* 43, 499 - 507.

[40] Liu, P.-C., Cheng, C.-F., Chang, C.-H., Lin, S.-L., Wang, W.-S. *et al*. (2011) Highly virulent *Photobacterium damselae* subsp. *piscicida* isolated from Taiwan paradise fish, *Macropodus opercularis* (L.) in Taiwan. *African Journal of Microbiology Research* 5, 2107 - 2113.

[41] Lopez, C., Rajan, P.R., Lin, J.H.-Y. and Yang, H.-L. (2002) Disease outbreak in sea farmed cobia, *Rachycentron canadum* associated with *Vibrio* spp., *Photobacterium damselae* subsp. *piscicida*, monogenean, and myxosporean parasites. *Bulletin of the European Association of Fish Pathologists* 23, 206 - 211.

[42] Love, M., Teebkin-Fisher, D., Hose, J. E., Farmer, J. J., Hickman, F. W. *et al*. (1981) *Vibrio damsela*, a marine bacterium, causes skin ulcers on the damselfish *Chromis punctipinnis*. *Science* 214, 1139 - 1140.

[43] MacDonell, M. T. and Colwell, R. R. (1985) Phylogeny of the *Vibrionaceae*, and recommendation of two new genera, *Listonella* and *Shewanella*. *Systematic and Applied Microbiology* 6, 171 - 182.
[44] Magariños, B., Romalde, J. L., Bandin, I., Fouz, B. and Toranzo, A. E. (1992) Phenotypic, antigenic, and molecular characterization of *Pasteurella piscicida* strains isolated from fish. *Applied and Environmental Microbiology* 58, 3316 - 3322.
[45] Magariños, B., Romalde, J.L., Barja, J.L. and Toranzo, A.E. (1994a) Evidence of a dormant but infective state of the fish pathogen *Pasteurella piscicida* in seawater and sediment. *Applied and Environmental Bacteriology* 60, 180 - 186.
[46] Magariños, B., Noya, M., Romalde, J. L., Perez, G. and Toranzo, A. E. (1994b) Influence of fish size and vaccine formulation on the protection of gilthead seabream against *Pasteurella piscicida*. *Bulletin of the European Association of Fish Pathologists* 14, 120 - 122.
[47] Magariños, B., Pazos, F., Santos, Y., Romalde, J. L. and Toranzo, A. E. (1995) Response of *Pasteurella piscicida* and *Flexibacter maritimus* to skin mucus of marine fish. *Diseases of Aquatic Organisms* 21, 103 - 108.
[48] Magariños, B, Toranzo, A.E. and Romalde, J.L. (1996) Phenotypic and pathobiological characteristics of *Pasteurella piscicida*. *Annual Review of Fish Diseases* 6, 41 - 64.
[49] Magariños, B., Toranzo, A.E., Barja, J.L. and Romalde, J.L. (2000) Existence of two geographically-linked clonal lineages in the bacterial fish pathogen *Photobacterium damselae* subsp. *piscicida* evidenced by random amplified polymorphic DNA analysis. *Epidemiology and Infection* 125, 213 - 219.
[50] Mannheim, W. (1984) Family III. Pasteurellaceae Pohl 1981a, 382. In: Krieg, N.R. and Holt, J. G. (eds) *Bergey's Manual of Systematic Bacteriology*, 9th edn, Vol. 1. Williams and Wilkins, Baltimore, Maryland, pp. 550 - 552.
[51] Moraes, J.R.E., Shimada, M.T., Yunis, J., Claudiano, G.S., Filho, J.R.E. *et al.* (2015) *Photobacteriosis outbreak in cage reared Rachycentron canadum: predisposing conditions*. In: Abstracts of the Annual Meeting of the European Aquaculture Society, Aquaculture Europe 15, Rotterdam, Netherlands. World Aquaculture Society, Ostend, Belgium, Session abstracts 119. Available at: https: //www. was. org /easOnline / AbstractDetail.aspx? i=4421 (accessed 16 November 2016).
[52] Muroga, K., Sugiyama, T. and Ueki, N. (1977) Pasteurellosis in cultured black sea bream *Mylio macrocephalus*. *Journal of the Faculty of Fisheries and Animal Husbandry, Hiroshima University* 16, 17 - 21.
[53] Nakai, T., Fujiie, N., Muroga, K., Arimoto, M., Mizuta, Y. *et al.* (1992) *Pasteurella piscicida* in hatchery-reared striped jack. *Gyobio Kenkyu* 27, 103 - 108.
[54] Nitzan, S., Shwartsburd, B., Vaiman, R. and Heller, E.D. (2001) Some characteristics of *Photobacterium damselae* subsp. *piscicida* isolated in Israel during outbreaks of pasteurellosis in hybrid striped bass (*Morone saxatilis* × *M. chrysops*). *Bulletin of the European Association of Fish Pathologists* 21, 77 - 80.
[55] Noya, M., Magariños, B., Toranzo, A.E. and Lamas, J, (1995) Sequential pathology of experimental pasteurellosis in gilthead seabream *Sparus aurata*. A lightand electron-microscopic study. *Diseases of Aquatic Organisms* 21, 177 - 186.
[56] Obendorf, D.L., Carson, J. and McManus, T.J. (1987) *Vibrio damsela* infection in a stranded leatherback turtle (*Dermochelys coriacea*). *Journal of Wildlife Diseases* 23, 666 - 668.
[57] Osorio, C.R., Collins, M.D., Toranzo, A.E., Barja, J.L. and Romalde, J.L. (1999) 16S rRNA gene sequence analysis of *Photobacterium damselae* and nested PCR method for rapid detection of the causative agent of fish pasteurellosis. *Applied and Environmental Microbiology* 65, 2942 - 2946.

[58] Osorio, C.R., Toranzo, A.E., Romalde, J.L. and Barja, J.L. (2000a) Multiplex PCR assay for ureC and 16s rRNA genes clearly discriminates between both subspecies of *Photobacterium damselae*. *Diseases of Aquatic Organisms* 40, 177 - 183.

[59] Osorio, C.R., Romalde, J.L., Barja, J.L. and Toranzo, A.E. (2000b) Presence of phospholipase-D (*dly*) gene encoding for damselysin production is not a pre-requisite for pathogenicity in *Photobacterium damselae* subsp. *damselae*. *Microbial Pathogenesis* 28, 119 - 126.

[60] Osorio, C.R., Juiz-Río, S. and Lemos, M.L. (2006) A siderophore biosynthesis gene cluster from the fish pathogen *Photobacterium damselae* subsp. *piscicida* is structurally and functionally related to the Yersinia high-pathogenicity island. *Microbiology* 152, 3327 - 3341.

[61] Paperna, I. and Zwerner, D.E. (1976) Parasites and diseases of striped bass, *Morone saxatilis* (Walbaum) from the lower Chesapeake Bay. *Journal of Fish Biology* 9, 267 - 287.

[62] Pedersen, K., Skall, H.F., Lassen-Nielsen, H.M., Bjerrum, L. and Olesen, N.J. (2009) *Photobacterium damselae* subsp. *damselae*, an emerging pathogen in Danish rainbow trout *Oncorhrynchus mykiss* (Walbaum) mariculture. *Journal of Fish Diseases* 32, 465 - 472.

[63] Plumb, J.A. and Hanson, L.A. (2007) Striped bass bacterial diseases. In: Plumb, J.A. and Hanson, L.A. *Health Maintenance and Principal Microbial Diseases of Cultured Fishes*, 3rd edn. Wiley-Blackwell, Ames, Iowa, pp. 429 - 433.

[64] Rajan, P.R., Lin, J.H.-Y., Ho, M.-S. and Yang, H.-L. (2003) Simple and rapid detection of *Photobacterium damselae* ssp. *piscicida* by a PCR technique and plating method. *Journal of Applied Microbiology* 95, 1375 - 1380.

[65] Renault, T., Haffner, P., Malfondet, C. and Weppe, M. (1994) *Vibrio damsela* as a pathogenic agent causing mortalities in cultured sea bass (*Lates calcarifer*). *Bulletin of the European Association of Fish Pathologists* 14, 117 - 119.

[66] Rivas, A.J., Balado, M., Lemos, M.L. and Osorio, C.R. (2011) The *Photobacterium damselae* subsp. *damselae* hemolysins damselyn and HlyA are encoded within a new virulence plasmid. *Infection and Immunity* 79, 4617 - 4627.

[67] Rivas, A.J., Lemos, M.L. and Osorio, C.R. (2013) *Photobacterium damselae* subsp. *damselae*, a bacterium pathogenic for marine animals and humans. *Frontiers in Microbiology* 4: 283.

[68] Robohm, R.A. (1983) *Pasteurella piscicida*. In: Anderson, D.P., Dorson, M. and Dubourget, P. (eds) *Antigens of Fish Pathogens: Development and Production for Vaccines and Serodiagnostics*. Collection Foundation Marcel Merieux, Association Corporative des Etudiants en Médecine de Lyon, Lyon, France, pp. 161 - 175.

[69] Romalde, J.L. (2002) *Photobacterium damselae* subsp. *piscicida*: an integrated view of a bacterial fish pathogen. *International Microbiology* 5, 3 - 9.

[70] Romalde, J.L., LeBreton, A., Magariños, B. and Toranzo, A.E. (1995a) Use of BIONOR Aquarapid-Pp kit for the diagnosis of *Pasteurella piscicida* infections. *Bulletin of the European Association of Fish Pathologists* 15, 64 - 66.

[71] Romalde, J.L., Magariños, B., Fouz, D.B., Bandin, I., Nunez, S. *et al*. (1995b) Evaluation of BIONOR Mono-kits for rapid detection of bacterial fish pathogens. *Diseases of Aquatic Organisms* 21, 25 - 34.

[72] Sakata, T., Matsuura, M. and Shimkawwa, Y. (1989) Characteristics of *Vibrio damsela* isolated from diseased yellowtail *Seriola quinqueradiata*. *Nippon Suisan Gakkashi* 55, 135 - 141.

[73] Skerman, V.B.D., McGowan, V. and Sneath, P.H.A. (eds) (1989) *Approved Lists of Bacterial Names*, amended edn. American Society for Microbiology, Washington, DC,

p. 72.
[74] Smith, S.K., Sutton, D.C., Fuerst, J.A. and Reichelt, J.L. (1991) Evaluation of the genus Listonella and reassignment of *Listonella damsela* (Love *et al.*) MacDonell and Colwell to the genus *Photobacterium* as *Photobacterium damsela* comb. nov. with an emended description. *International Journal of Systematic Bacteriology* 41, 529 - 534.
[75] Snieszko, S.F., Bullock, G.L., Hollis, E. and Boone, J.G. (1964) *Pasteurella* sp. from an epizootic of white perch (*Roccus americanus*) in Chesapeake Bay tidewater areas. *Journal of Bacteriology* 88, 1814 - 1815.
[76] Stephens, F. J., Raidal, S. R., Buller, N. and Jones, B. (2006) Infection with *Photobacterium damselae* subsp. *damselae* and *Vibrio harveyi* in snapper, *Pagrus auratus* with bloat. *Australian Veterinary Journal* 84, 173 - 177.
[77] Sugita, H., Matsuo, N., Hirose, Y., Iwato, M. and Deguchi, Y. (1997) *Vibrio* strain NM 10, isolated from the intestine of a Japanese coastal fish, has an inhibitory effect against *Pasteurella piscicida*. *Applied and Environmental Microbiology* 63, 4986 -4989.
[78] Thune, R.L., Stanley, L.A. and Cooper, R.K. (1993) Pathogenesis of Gram negative bacterial infections in warmwater fish. *Annual Review of Fish Diseases* 3, 37 - 68.
[79] Thune, R.L., Fernandez, D.H., Hawke, J.P. and Miller, R. (2003) Construction of a safe, stable, efficacious vaccine against *Photobacterium damselae* ssp. *piscicida*. *Diseases of Aquatic Organisms* 57, 51 - 58.
[80] Toranzo, A.E., Barreiro, S., Casal, J.F., Figueras, A., Magariños, B. *et al.* (1991) Pasteurellosis in cultured gilthead seabream (*Sparus aurata*): first report in Spain. *Aquaculture* 99, 1 - 15.
[81] Truper, H.G. and De'Clari, L. (1997) Taxonomic note: necessary correction of epithets formed as substantives (nouns) "in apposition". *International Journal of Systematic Bacteriology* 47, 908 - 909.
[82] Tung, M.C., Tsai, S.S., Ho, L.F., Huang, S.T. and Chen, S.C. (1985) An acute septicemic infection of *Pasteurella* organism in pond cultured Formosa snake-head fish (*Channa maculata*) in Taiwan. *Fish Pathology* 20, 143 - 148.
[83] Ueki, N., Kayano, Y. and Muroga, K. (1990) *Pasteurella piscicida* in juvenile red grouper. *Fish Pathology* 25, 43 - 44.
[84] Vera, P., Navas, J.I. and Fouz, B. (1991) First isolation of *Vibrio damsela* from sea bream *Sparus aurata*. *Bulletin of the European Association of Fish Pathologists* 11, 112 - 113.
[85] Wang, F.-I. and Chen, J.-C. (2006) Effect of salinity on the immune response of tiger shrimp *Penaeus monodon* and its susceptibility to *Photobacterium damselae* subsp. *damselae*. *Fish and Shellfish Immunity* 20, 671 - 681.
[86] Wang, R., Feng, J., Su, Y., Ye, L. and Wang, J. (2013) Studies on the isolation of *Photobacterium damselae* subsp. *piscicida* from diseased golden pompano (*Trachinotus ovatus*) and antibacterial agents sensitivity. *Veterinary Microbiology* 162, 957 - 963.
[87] Wolke, R. E. (1975) Pathology of bacterial and fungal diseases affecting fish. In: Ribelin, W. E. and Migaki, G. (eds) *The Pathology of Fishes*. The University of Wisconsin Press, Madison, Wisconsin, pp. 33 - 116.
[88] Yasunaga, N., Hatai, K and Tsukahara, J. (1983) *Pasteurella piscicida* from an epizootic of cultured red seabream. *Fish Pathology* 18, 107 - 110.
[89] Yasunaga, N., Yasumoto, S., Hirakawa, E. and Tsukahara, J. (1984) On a massive mortality of oval file fish (*Navodan modestus*) caused by *Pasteurella piscicida*. *Fish Pathology* 19, 51 - 55.

20

鲑立克次氏体

Jerri Bartholomew*, Kristen D. Arkush 和 Esteban Soto

20.1 引言

20.1.1 细菌描述

鲑立克次氏体(*Piscirickettsia salmonis*, Fryer et al., 1992)是一种革兰氏阴性、非运动型兼性胞内细菌。从智利海水网箱养殖银大麻哈鱼的流行病中分离了模式菌株 LF-89(ATCC VR 1361) (Fryer et al., 1990)。在其他海水和淡水鱼中也发现过并分离出了该病原菌(表 20.1)。尽管从非鲑科鱼类中分离出来的分离株在形态上是相似的,但仅有部分分离株被鉴定证实(Lannan et al., 1991; Alday-Sanz et al., 1994)。

表 20.1 鲑立克次氏体的地理分布与宿主范围

地理区域	鱼　种	微生物观察(O) 微生物分离(I) DNA 检测(DNA)	参考文献
澳大利亚塔斯马尼亚州	大西洋鲑	O/DNA	Corbeil et al., 2005
加拿大不列颠哥伦比亚省	驼背大麻哈鱼、银大麻哈鱼、大鳞大麻哈鱼、大西洋鲑	O/I/DNA	Brocklebank et al., 1993; Evelyn et al., 1998
加拿大新斯科舍省	大西洋鲑	O/I/DNA	Jones et al., 1998; Cusack et al., 2002
智利	驼背大麻哈鱼、银大麻哈鱼、马苏大麻哈鱼、虹鳟、大鳞大麻哈鱼、大西洋鲑	O/I/DNA	Bravo and Campos, 1989; Fryer et al., 1990; Branson and Nieto Díaz-Munoz, 1991; Cvitanich et al., 1991; Garcés et al., 1991; Bravo, 1994; Gaggero et al., 1995; Smith et al., 1995

* 通信作者邮箱: bartholj@science.oregonstate.edu。

续 表

地理区域	鱼 种	微生物观察(O) 微生物分离(I) DNA检测(DNA)	参 考 文 献
智利	智利油南极鱼、银河鱼、澳洲犁齿鳕、南非平鲉	DNA	Contreras-Lynch et al., 2015
欧洲	舌齿鲈	O/I/DNA	Comps et al., 1996; Steiropoulos et al., 2002; Athanassopoulou et al., 2004
爱尔兰	大西洋鲑	O/I/DNA	Rodger and Drinan, 1993
挪威	大西洋鲑	O/I/DNA	Olsen et al., 1997
美国加利福尼亚州	有名锤形石首鱼	O/I/DNA	Chen et al., 2000
美国俄勒冈州	海岸水样	DNA	Mauel and Fryer, 2001

P. salmonis 的形态主要为球状,可能出现成对的弯曲杆状菌或环状菌(Fryer et al., 1990)。单个细胞直径为0.5～1.5 μm。细菌复制发生在宿主细胞内的膜结合细胞质空泡内(图20.1)。细菌可以在细胞系中培养(Fryer et al., 1990),也可在强化琼脂和肉汤中培养(Mikalsen et al., 2008)。细菌吉姆萨(Giemsa)染液呈深蓝色,Giménez法染色阴性(Giménez, 1964),当采用立克次氏体和衣原体Pinkerton法(US Surgeon-General's Office, 1994)染色时,细胞

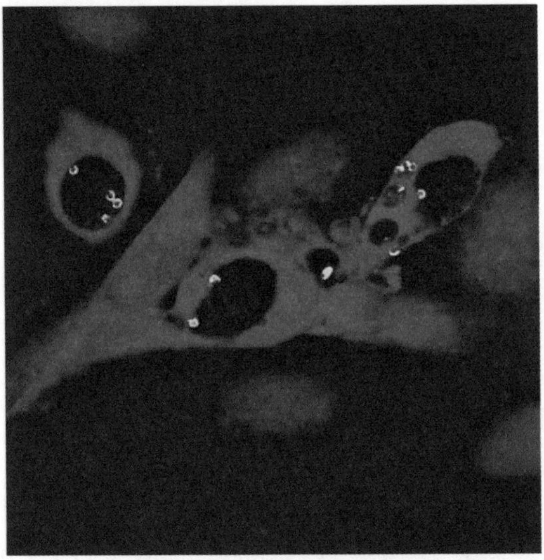

图20.1 *Piscirickettsia salmonis* 在CHSE-214(大鳞大麻哈鱼,胚胎)细胞胞质液泡中复制的共聚焦图像。比例尺:5 μm。图片经许可转载

保持碱性品红(Fryer et al., 1990)。

16S、内转录间隔区(internal transcribed spacer, ITS)和23S核糖体DNA序列的比较表明，来自智利、挪威、爱尔兰、苏格兰和加拿大的 *P. salmonis* 具有相似性并且聚类为一个单系群，但智利分离株 EM-90 则不同(Mauel et al., 1999; Heath et al., 2000)。这些菌株在抗生素敏感性、致细胞病变效应(CPE)、组织培养(Smith et al., 1996b)、宿主范围(Cvitanich et al., 1991)和毒力(Smith et al., 1996a; House et al., 1999)方面的差异已有报道。

20.1.2 传播方式

在陆生环境中，节肢动物作为中间宿主或媒介传播立克次氏体(Weiss and Moulder, 1984)，而在寄生等足目动物高氏触角鞘(*Ceratothoa gaudichaudii*, Garcés et al., 1994)中检测到 *P. salmonis* 提示外寄生物可能会促进*P. salmonis*的传播。

然而，实验研究显示，水平传播可以在无媒介的情况下在淡水和海水中发生(Cvitanich et al., 1991; Almendras et al., 1997a)。海洋鱼类也是该病原的潜在宿主库，被感染的有名锤形石首鱼(*Atractoscion nobilis*)通过共栖可将感染传播给大鳞大麻哈鱼和银大麻哈鱼(Arkush et al., 2005)。实验中，*P. salmonis*可以通过皮肤、鳃以及肠和胃的插管侵入鱼体(Smith et al., 1999, 2004, 2015)，表明传播也可通过食用被感染的猎物而发生。在被感染鱼的肠道和肾脏中检测到细菌以及黏膜上皮和肾小管坏死和脱落，暗示了病原菌随脱落物质的潜在传播途径(Cvitanich et al., 1991; Arkush et al., 2005; Smith et al., 2015)。鳃丝中被感染细胞的脱落，甚至同类相食，都可能导致病原体的进一步传播(Arkush et al., 2005)。实验证明，*P. salmonis* 可以在盐水中存活14 d，但在淡水中几乎立即失活(Lannan and Fryer, 1994)。*P. salmonis* 在海水中的长期存活也在一项实地研究中得到证实，研究发现，在该病害暴发40 d后仍可在商业养殖场周围的水域中检测到 *P. salmonis* (Olivares and Marshall, 2010)。虽然没有观察到孢子期，但是从老化的组织培养物和自然感染的鱼中发现了一种小的感染性 *P. salmonis*变体，这表明当条件限制增殖时该病菌仍能存活(Rojas et al., 2008)。

通过在虹鳟亲鱼腹腔中接种该菌，证实了 *P. salmonis* 的垂直传播(Larenas et al., 1996, 2003)。后来在接种鱼的性腺、体腔液、精液及卵中检测到这些菌。Larenas 等(2003)发现在接种虹鳟亲本的一方或双方后均可在子代鱼苗中检测到该菌，即使在一定比例的子代中都检测到该菌(孵化稚鱼中为16%～24%，幼鱼中为12%～16%)，但是它们并未发展出鱼立克次氏体病(piscirickettsiosis)。在受精过程中，也可通过在含有 *P. salmonis* 的媒介中孵化鱼卵而感染后代。Larenas 等(2003)利用扫描电子显微镜检证实，*P. salmonis* 在接触卵子1 min后就通过绒毛膜穿入卵子。在淡水中，幼鱼自然发生鱼立克次氏体病的病例报告很少(Bravo, 1994; Gaggero et al., 1995)，表明垂直传播很罕见。然而，由于 *P. salmonis* 在淡水中会迅速失活(Lannan and Fryer, 1994),除非有宿主细胞膜或其他生物材料的保护,否则也不大可能发生水平传播。

20.1.3 地理和宿主分布

被鉴定为鲑立克次氏体、鱼立克次氏体样微生物(*Piscirickettsia*-like organisms, PLOs)或立克次氏体样微生物(*Rickettsia*-like organisms, RLOs)的细菌在全球范围内的淡水和海水鱼类以及无脊椎动物中普遍存在(Fryer and Lannan, 1994; Mauel and Miller, 2002; Rozas and Enríquez, 2014)。其中许多PLOs或RLOs随后被证明与弗朗西斯菌属有亲缘性。

智利、爱尔兰、挪威、苏格兰和加拿大都已报道过与鲑科鱼类立克次氏体病相关的*P. salmonis* (表20.1),随着进一步的研究,*P. salmonis* 的宿主范围和分布可能会扩大。可能最早于20世纪70年代早期在加拿大不列颠哥伦比亚省太平洋生物站海水饲养的实验用驼背大麻哈鱼中发现了该疾病(Evelyn et al., 1998)。尽管对驼背大麻哈鱼、大鳞大麻哈鱼和银大麻哈鱼进行了时断时续的观察,但直到1992年,当细菌培养方法(Lannan and Fryer, 1994)和基于荧光抗体的检测技术(Lannan et al., 1991)得到运用时,才从海水围网养殖的患病大西洋鲑和大鳞大麻哈鱼中分离出*P. salmonis* (Brocklebank et al., 1993)。

1989年在智利南部蒙特港附近养殖的鲑科鱼类暴发疾病,但有证据表明这种疾病自20世纪80年代初就存在了(Bravo and Campos, 1989)。在银大麻哈鱼、大西洋鲑、大鳞大麻哈鱼、虹鳟(Cvitanich et al., 1991)和马苏大麻哈鱼中都检测到该病原(Bravo, 1994)。虽然感染主要发生在海水和微咸水鱼中,但该疾病也发生在淡水鱼中(Cvitanich et al., 1991)。在20世纪90年代,从挪威(Olsen et al., 1997)、爱尔兰(Rodger and Drinan, 1993)、苏格兰(Grant et al., 1996)和加拿大(Jones et al., 1998)海水网箱养殖的大西洋鲑中都分离出了*P. salmonis*。随后在塔斯马尼亚州东南部养殖的大西洋鲑中发现了一种遗传学上截然不同的*P. salmonis* 菌株,该菌株与来自智利大西洋鲑的EM-90株最密切相关(Corbeil et al., 2005)。

P. salmonis 也存在于非鲑科鱼类中。从美国加利福尼亚州孵化养殖的白海鲈中获得的一种分离株与模式菌株LF-89有99%的同源性(Arkush et al., 2005),该分离株在实验感染的白海鲈幼鱼中诱发了疾病和死亡(Arkush et al., 2005)。一株从地中海患有严重脑炎的欧洲海鲈中分离出的RLO也与LF-89菌株有关(McCarthy et al., 2005)。利用聚合酶链反应(PCR)在来自智利的智利油南极鱼(*Eleginops maclovinus*)、南非平鲉(*Sebastes capensis*)和澳洲犁齿鳕(*Salilota australis*)中也发现了该病原菌(Contreras-Lynch et al., 2015)。

20.1.4 病害影响

智利是仅次于挪威的世界第二大鲑鳟鱼类出口国,每年因鱼立克次氏体病造成的损失已经影响了智利经济(Wilhelm et al., 2006)。1989 年,超过 150 万尾银大麻哈鱼死亡,经济损失达 1 000 万美元(Larenas et al., 2000)。1995 年,超过 1 000 万尾银大麻哈鱼在海水养殖阶段死亡,大部分死于鱼立克次氏体病,损失达 4 900 万美元(Smith et al., 1997)。在 21 世纪初,预防和控制疾病方法的改进,以及从银大麻哈鱼到更具抗性的大西洋鲑的养殖品种转变,降低了该疾病的严重影响(Fryer and Hedrick, 2003)。2005 年左右,传染性鲑贫血症危机在鱼类中造成的死亡率超过了鱼立克次氏体病造成的损失。然而,鱼立克次氏体病再度成为智利鱼类健康的主要挑战(Rozas and Enríquez, 2014),据报道该病害导致了海水中 90%的鱼类因病死亡(Ibieta et al., 2011)。2014 年,鱼立克次氏体病导致的养殖大西洋鲑、银大麻哈鱼和虹鳟的死亡率分别为 74.1%、36.7% 和 73.5%(Sernapesca, 2014)。2010 年至 2014 年期间,智利鲑鱼养殖业使用了超过 1 700 t 的抗菌剂。这些抗生素大部分用于海水养殖(96%的病例),其中 90%用于防治鱼立克次氏体病(Sernapesca, 2015)。

在世界其他地区,鱼立克次氏体病对鲑鱼养殖的影响没有被很好地记录下来。在不列颠哥伦比亚省,鱼立克次氏体感染通常与其他传染病同时发生,不过 *P. salmonis* 是导致鱼类死亡的主要病因(Evelyn et al., 1998)。在苏格兰,尽管大西洋鲑养殖场报告的死亡率很高,但该疾病的发生率和总体影响被认为很低(Rozas and Enríquez, 2014)。在挪威,死亡率与恶劣的环境条件(特别是藻华)、过高的网箱养殖密度和初次入海幼鲑的不良状况有关。这些条件的改善减少了疾病的影响(Olsen et al., 1997)。

20.2 感染诊断

20.2.1 临诊疾病体征

鲑科鱼类中的 *P. salmonis* 感染曾被称为“银大麻哈鱼综合征”或“Hitra 疾病”(Branson and Nieto Díaz-Munoz, 1991)、“鲑立克次体败血症”(Cvitanich et al., 1991)和“鱼立克次氏体病”(Fryer et al., 1992)。

外部体征各不相同,但都包括食欲不振和体色发黑。受感染的鱼可能嗜睡,并在水面附近或网箱围栏周边游动。发现被感染的银大麻哈鱼游动方向不定,并且从具有临床症状的鱼脑部分离出该病原菌(Skarmeta et al., 2000)。同样,感染的海鲈(*D. labrax*)幼鱼也表现出异常的游动和旋转行为(McCarthy et al., 2005)。受感染鱼可能有皮肤病变,范围从小区域的竖鳞到浅层出血性溃疡(Branson and Nieto Díaz-Munoz, 1991)。尽管死亡可能在没有疾病症状的情况下发生,但贫血症始终存在,苍白的鳃可作为证明(Arkush et al., 2005)。

该疾病通常在鲑科鱼类转移入海后 6～12 周发生(Bravo and Campos, 1989; Fryer et al., 1990; Cvitanich et al., 1991)。尽管淡水中的自然感染很少

见，但 Smith 等(2015)报道了虹鳟鱼苗在淡水中的实验感染死亡率超过 90%，且临床症状与那些急性立克次氏体病期间自然感染的鲑科鱼类的症状一致。

20.2.2 诊断

推定诊断基于对染色组织印迹或涂片中细菌的显微镜观察，或使用细胞培养(Fryer et al., 1990)或琼脂培养基分离病菌(Mauel et al., 2008; Mikalsen et al., 2008; Yañez et al., 2013)。使用免疫荧光抗体测试(IFAT; Lannan et al., 1991; OIE, 2006)、免疫组织化学(Alday-Sanz et al., 1994; OIE, 2006)或 PCR 检测(House and Fryer, 2002; OIE, 2006)可确诊 *P. salmonis*。

将感染组织(如肾、肝、脑和血液等)与易感细胞系共培养。除了在大多数鲑科鱼细胞系中复制外(Fryer et al., 1990)，还可使用来自胖头鲹(Fryer et al., 1990)、云斑鮰(*Ameiurus nebulosus*) (Almendras et al., 1997b)和草地贪夜蛾(*Spodoptera frugiperda*, Sf21 细胞系)(Birkbeck et al., 2004a)的细胞系分离该病原菌。因为该菌对细胞培养中常用的大多数抗生素都很敏感(Lannan and Fryer, 1991)，所以所有培养基，包括组织转运或处理缓冲液，都应该不含抗生素。该病原细菌对高温和冷冻也很敏感，因此无菌取出的组织应保持在 4 ℃。将接种的培养物在 15～18 ℃下培养至最佳生长状态，并进行 28 d 的致细胞病变效应(CPE)检测(Fryer et al., 1990)，CPE 表现为具有大空泡的圆形细胞的斑块样簇。在初次分离时，任何 CPE(图 20.2)可能需要 21 d 或更长时间才能出现。在随后的传代中，CPE 在接种后 4～7 d 变得明显。在鲑细胞系中，该病原菌的复制在 20 ℃以上或 10 ℃以下时受到抑制。在−70 ℃，经过一次冻融循环

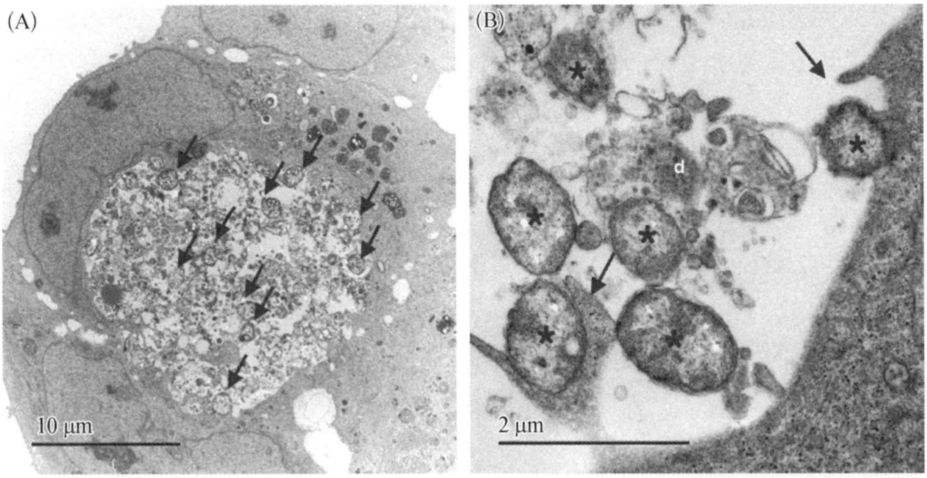

图 20.2 *Piscirickettsia salmonis* 感染 CHSE-214(大鳞大麻哈鱼，胚胎)细胞的超薄切片：(A) 在直径约为 20 μm 的液泡内的球形细菌(箭头所示)和相关的坏死细胞碎片；(B) CHSE-214 细胞最近释放的细菌(星号)的内化(箭头所示)，来自最近裂解细胞的碎片由“d”标识

后，*P. salmonis*的滴度可降低99%，然而，添加10%的二甲基亚砜作为冷冻保护剂可将存活率提高10倍(Fryer et al., 1990)。在培养的鱼细胞中，该菌可以产生10^6～10^7的$TCID_{50}$/mL的滴度，而在Sf21细胞中的产量提高约100倍(Birkbeck et al., 2004a)。

该菌也可以在添加有羊血和半胱氨酸的琼脂和肉汤中培养(Mauel et al., 2008; Mikalsen et al., 2008)。形成的菌落呈凸状、灰白色、有光泽、中央不透明，边缘半透明、略微起伏(图20.3)。最近报道了两种无血琼脂培养基(Yañez et al., 2013)。

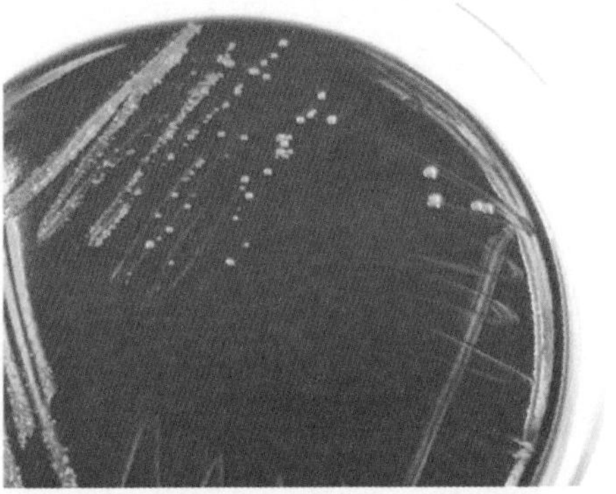

图20.3　在改良的Thayer-Martin (MTMII)琼脂(Becton Dickenson BBL, Sparks, Maryland)上20 ℃下放置10 d的*Piscirickettsia salmonis*菌落

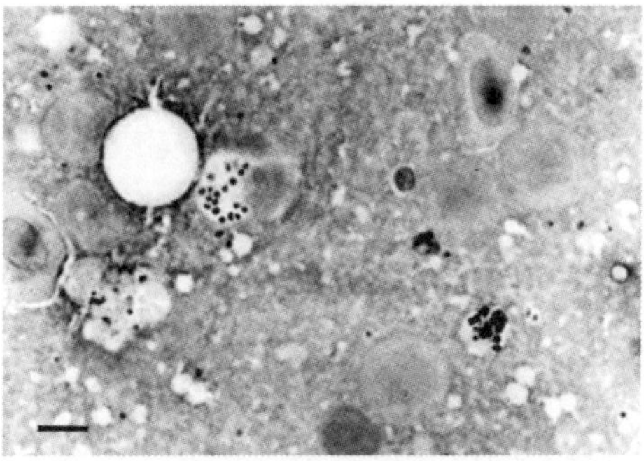

图20.4　濒死银大麻哈鱼的肾脏印迹显示宿主细胞内和细胞间的*Piscirickettsia salmonis*生物体。吉姆萨染色。比例尺：10 μm。照片由智利蒙特港的Sandra Bravo提供。图像经许可转载

一种不太灵敏但更快速的推定诊断方法对革兰氏(Gram)、吉姆萨(Giemsa)或亚甲基蓝溶液染色的肾、肝或脾组织涂片、印模或组织切片进行显微镜检(Lannan and Fryer,1991)。Giemsa染色标本中，在宿主细胞的细胞质液泡中，出现的多形性细菌呈现为被染成深色的球状或环状，通常成对出现，直径为0.5～1.5 μm (图20.4)。

在初步检测后，可以使用免疫荧光(Lannan et al., 1991; Jamett et al., 2001)、免疫组织化学(Alday-Sanz et al., 1994; Steiropoulos et al., 2002)或酶联免疫吸附试验(ELISA) (Aguayo et al., 2002)等血清学确诊*P. salmonis*。或者，可以使用分子技术确定该病原菌的存在。PCR检测可扩增小亚基核糖体(16S)基因(Mauel et al., 1996)或ITS区域的靶序列(Marshall et al., 1998)。斑点DNA杂交可以检测鱼组织和细胞培养上清液中的*P. salmonis*，而原位杂交可以观察组织中感染的细胞(Venegas et al., 2004)。Warsen等(2004)描述了一种DNA微阵列，通过该微阵列可以用来检测PCR产物或基因组DNA样品中是否存在*P. salmonis*和其他14种鱼类病原体。最近，有研究者设计了一

种多重PCR用于检测鲑立克次氏体(*P. salmonis*)、海豹链球菌(*Streptococcus phocae*)、杀鲑气单胞菌和鳗弧菌(*Vibrio anguillarum*)(Tapia-Cammas et al., 2011)。

20.3 病理学

被*P. salmonis*感染的鱼通常有皮肤病变,从小范围的竖鳞到浅层溃疡(图20.5)(Cvitanich et al., 1990; Fryer et al., 1990),并可能演变成全身弥漫性皮肤溃疡,包括鳃盖和尾柄部位(Rozas and Enríquez, 2014)。

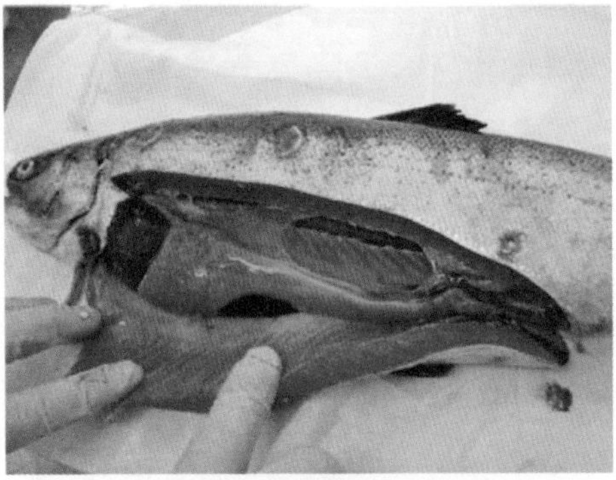

图20.5 感染*Piscirickettsia salmonis*的虹鳟表现出皮肤浅层出血性溃疡以及肝脏和体腔的点状出血。照片由智利圣地亚哥Centrovet公司的Jaime Tobar提供

在体内,鱼立克次氏体全身扩散,导致肾脏和脾脏肿胀。虽然仅存在于5%~10%的受感染鱼中,但最具诊断性的病变发生在肝脏中,它们呈现出直径为5~6 mm的白色或奶油色圆形不透明结节或环形病灶(Lannan and Fryer, 1993)。败血性感染(经常出现腹水)的特征,表现为肌肉组织和器官(包括胃、肠、幽门盲囊、鱼鳔和脂肪组织)的点状出血(图20.5)(Cvitanich et al., 1991; Brocklebank et al., 1993; Rodger and Drinan, 1993)。在骨骼肌中,带有血清血液渗出物的空洞越来越多地被报道,尤其是在鳟鱼中。这些独特的病变通常与*P. salmonis*的小变种有关(Rojas et al., 2008)。

已有多篇与*P. salmonis*相关的组织病理学综述报道(Almendras and Fuentealba, 1997; Fryer and Hedrick, 2003; Rozas and Enríquez, 2014)。肾和脾中的正常造血和淋巴组织被慢性炎性细胞和宿主细胞碎片取代。肝脏病变(图20.6)严重,鱼立克次氏体经常出现在退化变性的肝细胞胞质和巨噬细胞中,或游离于血液中。据报道,肝组织有肉芽肿,在急性感染时肝脏呈斑驳状(Arkush et al., 2005)。肝、肾和脾可能出现血管和血管周围坏死,血管内凝血导致小血管内的纤维蛋白血栓和炎性细胞浸润(Branson and Nieto Díaz-Munoz, 1991; Cvitanich et al., 1991)。大肠肉芽肿性炎症很常见,常导致黏膜上皮坏死和脱落。鳃上皮的肉芽肿性炎症导致鳃瓣融合(Branson and Nieto Díaz-Munoz, 1991)。在真皮和表皮中,可观察到肉芽肿性炎症和坏死,以及皮下肌肉组织的退化变性(Brocklebank et al., 1993; Chen et al., 1994)。病鱼也可能存在脑膜炎、心内膜炎和胰腺炎。

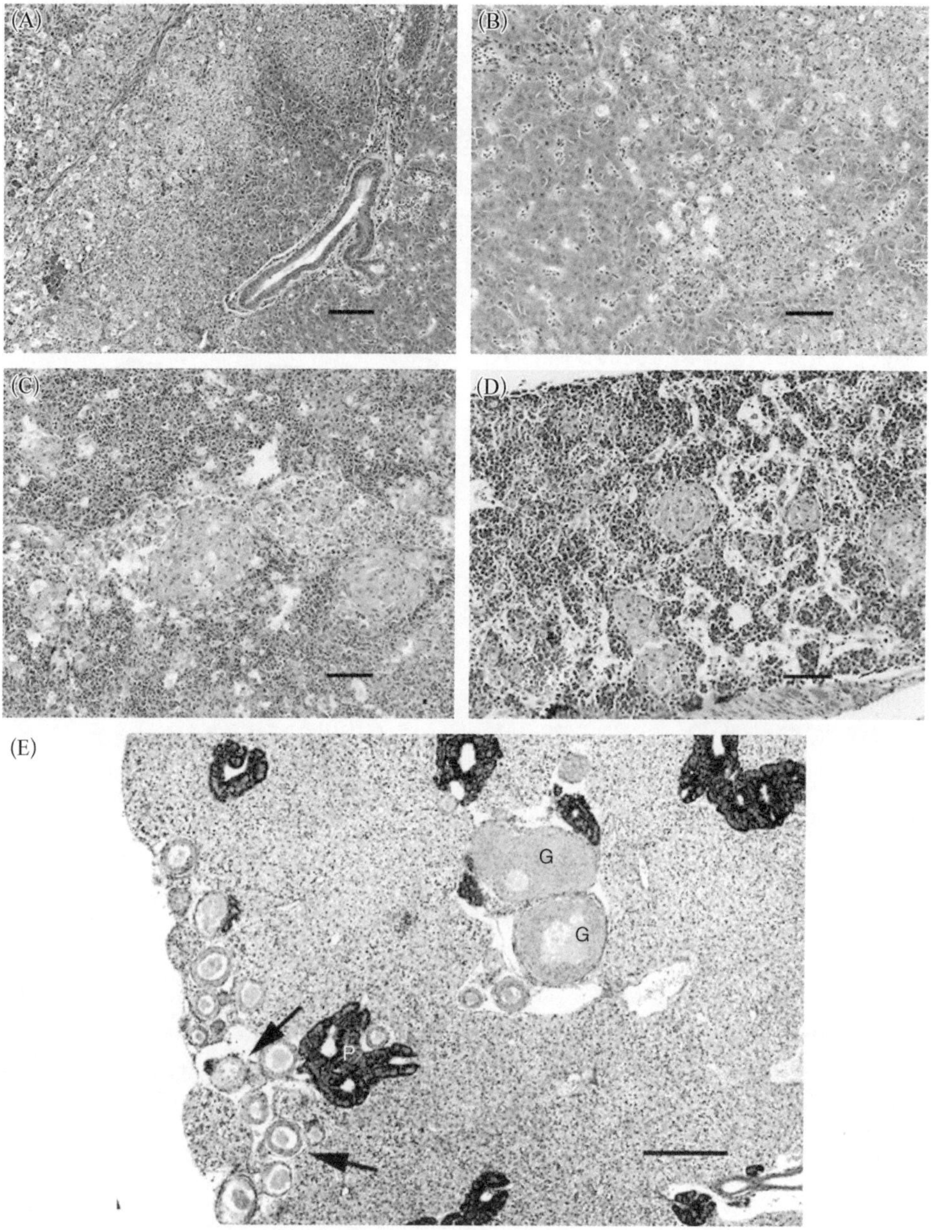

图 20.6 *Piscirickettsia salmonis* 实验感染白海鲈(*Atractoscion nobilis*)后，苏木精和伊红(H&E)染色的组织切片微观病变：(A) 局灶性肝病变，以坏死和单核细胞浸润为特征，比例尺为 100 μm；(B) 肝坏死病灶，比例尺为 50 μm；(C) 脾坏死病灶，比例尺为 100 μm；(D) 头肾间质中巨噬细胞的多发性和局灶性聚集，比例尺为 100 μm；(E) 腹腔注射细菌后 123 d，肝脏中外分泌胰腺(P)胰岛附近的大而发达的肉芽肿(G)和小而众多的肉芽肿(箭头所示)，比例尺为 200 μm

Almendras 等(1997a)通过腹腔注射、口服和鳃接触等途径以及与受感染的鱼共栖感染大西洋鲑。在腹腔感染的鱼中,肝脏和脾脏出现包膜(浆膜)状,并扩散到脾、肝和肾的白细胞和其他实质细胞。通过口服和鳃途径接触病原菌的鱼及共栖间接感染的鱼,感染的特征是细菌从血管侵入器官到周围组织的血原性模式。血管病变似乎是由 *P. salmonis* 对内皮细胞的嗜性引起的,而肝病变是缺血性坏死和胞质内细菌直接损伤的结果(Almendras et al., 2000)。

采用免疫组织化学方法,Smith 等(2015)在浸泡后 5 min 内检测到 *P. salmonis*附着于虹鳟鱼苗皮肤和鳃外上皮细胞层表面,15 min 后出现在食道中,18 h 后出现在血液中。皮肤和鳃也是实验性暴露研究中银大麻哈鱼的侵染入口(Smith et al., 2004)。无论是初始的浆膜传播还是血原性传播,在感染后期,这些鱼类的内部大体和微观病理变化都是相似的,可能是败血症所致。

在白海鲈中,组织炎症反应和巨噬细胞组织成离散的聚集体促进了病原菌的血管侵入和通过被感染的巨噬细胞传播到其他组织(Arkush et al., 2005)。*P. salmonis* 在细胞内的存活和持留,特别是在肉芽肿病变内,能够逃避宿主的免疫监视,并允许病原菌在宿主中持续存在。这些病变的再激活尚未被实验证实,因此还没有建立起真正的带菌状态。

实验感染的银大麻哈鱼和虹鳟的特异性抗体反应较弱,很可能针对脂多糖的脂寡糖成分(Kuzyk et al., 1996)。在暴露病菌实验中存活下来的白海鲈中,大部分鱼产生了抗 *P. salmonis* 血清抗体(81.4%),但这些反应的强度各不相同(Arkush et al., 2006)。这些抗体反应的变化在鱼类中并不罕见,并且细胞介导的免疫对 *P. salmonis* 的作用有待进一步研究。

20.3.1 毒力因子

目前关于 *P. salmonis* 毒力基因的文献描述很少,预测蛋白的表达和功能也没有得到很好的研究。许多研究已将 *P. salmonis* 菌株间的表型差异与遗传多样性联系起来。脂多糖(LPS)基因的多态性与黏液状表型多样性相关(Bohle et al., 2014);该研究还发现了 *P. salmonis* 中Ⅳ型分泌系统(type Ⅳ secretion systems, T4SSs)的多个同源基因,这些同源基因显示 T4SS 相关基因数量的多样性。有趣的是,*P. salmonis* 中某些 T4SS 同源基因在 pH<5 时发生过表达,这表明在吞噬后它们在细菌存活中的潜在作用(Gómez et al., 2013)。利用感染了 *P. salmonis* 的 SHK-1(大西洋鲑头肾)细胞系,Isla 等(2014)发现 *P. salmonis*蛋白 ClpB 和 BipA 的表达显著增加。这些蛋白是军团菌(*Legionella*)和弗朗西斯菌(*Francisella*)等哺乳动物兼性胞内病原菌的重要毒力因子,可能在 *P. salmonis* 的细胞内存活中发挥类似作用(Meibom et al., 2008; Isla et al., 2014)。全基因组测序还揭示了假定的 Fur 表达盒,Fur(铁摄取调节子)蛋白以高亲和力与之结合,这些基因在缺铁条件下的表达可允许细菌

应对宿主中的铁剥夺机制(Pulgar et al., 2015)。最后,*gyrA*(DNA促旋酶A亚基)序列的差异与对喹诺酮类药物的不同易感性相关(Henríquez et al., 2015)。遗憾的是,缺乏用于对*P. salmonis*进行遗传操作的分子方法阻碍了这些候选毒力因子的功能解析。

20.4 防控策略

20.4.1 饲养管理

简单的饲养管理措施,包括养殖场休耕、降低饲养密度和在同一养殖区域养殖单一年份的同种鱼以减少水平传播,可能有助于避免鱼立克次氏体病的发生。虽然垂直传播仅在实验中得到证实,但对成年亲鱼的筛选可能进一步降低感染风险。在智利,实施了一项监测计划,即通过早期检测和适当控制来减少疾病的影响(Sernapesca, 2013)。

20.4.2 抗菌疗法

在体外条件下,*P. salmonis*对多种抗生素敏感,包括庆大霉素、链霉素、红霉素、四环素和氯霉素(Fryer et al., 1990; Cvitanich et al., 1991)。在智利,噁喹酸、阿莫西林、红霉素、氟苯尼考、氟甲喹和土霉素被用于治疗鱼立克次氏体病,不过由于噁喹酸和氟甲喹在人类医学中的重要性,它们的使用已经减少(Rozas and Enríquez, 2014)。此外,抗生素治疗已被证实在田间的疗效并不一致(Smith et al., 1996b, 1997)。例如,用噁喹酸(每天10～30 mg/kg,连续10～15 d)和土霉素(每天55～100 mg/kg,连续10～15 d)药饵饲喂是最有效的,但被感染的鱼对药效响应缓慢,需要反复使用(Palmer et al., 1997)。对某些抗生素的耐药性已有报道(Rodger and Drinan, 1993),可变的药敏性增加了治疗次数和剂量(Rozas and Enríquez, 2014)。

20.4.3 疫苗

智利有大量商业化疫苗(33种,Rozas and Enríquez, 2014),其中29种是注射用细菌灭活疫苗,4种为亚单位疫苗,可与其他细菌或病毒疫苗联合使用。基于*P. salmonis*菌苗的免疫效果在不同情况下效果不同。在银大麻哈鱼中,福尔马林灭活疫苗引起剂量依赖性反应,这在某些情况下加重了疾病(Kuzyk et al., 2001)。Birkbeck等(2004b)的研究表明,*P. salmonis*热灭活或福尔马林灭活疫苗具有显著的保护作用。通过使用专有的MicroMatrix®技术封装的灭活菌苗,连续10 d口服接种,无论是作为初次接种还是作为初次IP接种的加强免疫,都激发了免疫保护(Tobar et al., 2011)。

使用兔多克隆抗血清检测的*P. salmonis*蛋白和糖类抗原被认为是亚单位

疫苗的潜在成分(Kuzyk et al., 1996; Barnes et al., 1998),尽管使用一些相同的纯化抗原检测恢复期的银大麻哈鱼和虹鳟血清时,仅检测到微弱反应(Kuzyk et al., 1996)。由于 *P. salmonis* 是一种胞内病原菌,细胞介导的免疫刺激,包括增强吞噬和细胞内杀伤,可能对疫苗的成功免疫至关重要(Barnes et al., 1998)。

已经测试了几种重组疫苗。使用大肠杆菌(*Escherichia coli*)重组产生的 *P. salmonis* 外表面脂蛋白 OspA 接种银大麻哈鱼,免疫保护率(RPS)可达 59%(Kuzyk et al., 2001)。将破伤风毒素和麻疹病毒融合蛋白(两者在哺乳动物中均具有免疫原性)的 T 细胞表位整合到 OspA 融合蛋白中,疫苗免疫保护率可达 83.0%,免疫效力几乎是单一 OspA 蛋白疫苗(RPS 为 30.2%)的三倍(Kuzyk et al., 2001)。

使用一种以上的重组抗原可以增强免疫反应。用热休克蛋白(Hsp)60 和 Hsp70 加上 *P. salmonis* 鞭毛蛋白制备的重组疫苗在大西洋鲑中的 RPS 高达 95%(Wilhelm et al., 2006)。在接种后至少 8 个月的鱼血清中检测到反应性抗体。来自智利南部自然感染的银大麻哈鱼的一种高免疫原性蛋白 ChaPs 是疫苗开发的另一种潜在候选抗原(Marshall et al., 2007)。ChaPs 的序列分析表明,它也是一种热休克蛋白,并且这些分子先前已在重组疫苗中被开发利用(Wilhelm et al., 2006)。

在水产养殖中,市售灭活疫苗和亚单位疫苗的效力在第一个冬季通常很高,但在早春的已接种组和未接种组鱼中的效力均有所下降(Leal and Woywood, 2007)。注射疫苗对预防鲑鱼初次转移入海后发生的疫情相对有效,但对随后的疫情却没有效果(Tobar et al., 2011)。这些后来的疫情影响较大的鱼,并导致更大的经济损失。追加免疫接种在技术上和经济上都不可行。不过,口服加强免疫接种可能是一种有吸引力的选择。口服注射型疫苗溶液后饲喂 300 度日产生了针对 *P. salmonis* 的抗体,并且无论作为初次接种还是加强接种施用都具有保护性(Rozas and Enríquez, 2014)。尽管实验性疫苗接种取得了这些进展,但水产养殖场的死亡率并未显著下降(Rozas and Enríquez, 2014)。

通过鉴定分子生物标记物,治疗方法也正在发生变革。基于微阵列的实验已被用于鉴定大西洋鲑巨噬细胞和头肾细胞中响应感染的差异表达基因(Rise et al., 2004)。选择转录产物作为 *P. salmonis* 感染的分子发病机制以及疫苗和治疗剂评估中潜在效用的生物标记物。例如,检测到感染的巨噬细胞的氧化还原状态变化,这些变化可能使这些细胞耐受 *P. salmonis* 感染,那么,使用抗氧化剂来减少造血组织损伤是有可能的。

20.5 总结与研究展望

近年来,鱼立克次氏体病重组疫苗的研究开发取得了重大进展。据推测,这些疫苗接种措施将进行田间试验，随后进行商业化。然而，必须更充分地探索揭示该病原的生物学关键因素，例如恢复鱼中可能存在的带菌状态、溯河鲑类和海洋鱼种相互间的传播以及可能导致疾病暴发的环境或宿主因素。需要通过分子分析对 *P. salmonis* 和 PLO 病原进行更广泛的地理和宿主比较，以描述系统发育关系。

随着 *P. salmonis* 培养方法的改进，对该病原菌的早期检测将有助于快速诊断并减少该病原菌造成的损失。今后最大的研发工作挑战将来自鲑立克次氏体发病机制的阐明、毒力表征和针对这种苛养病原菌的疫苗开发。

参考文献

[1] Aguayo, J., Miquel, A., Aranki, N., Jamett, A., Valenzuela, P.D.T. *et al*. (2002) Detection of *Piscirickettsia salmonis* in fish tissue by an enzyme-linked immunosorbent assay using specific monoclonal antibodies. *Diseases of Aquatic Organisms* 49, 33–38.

[2] Alday-Sanz, V., Rodger, H., Turnbull, T., Adams, A. and Richards, R.H. (1994) An immunohistochemical diagnostic test for rickettsial disease. *Journal of Fish Diseases* 17, 189–191.

[3] Almendras, F.E. and Fuentealba, I.C. (1997) Salmonid rickettsial septicaemia caused by *Piscirickettsia salmonis*: a review. *Diseases of Aquatic Organisms* 29, 137–144.

[4] Almendras, F.E., Fuentealba, I.C., Jones, S.R.M., Markham, F. and Spangler, E. (1997a) Experimental infection and horizontal transmission of *Piscirickettsia salmonis* in freshwater-raised Atlantic salmon, *Salmo salar* L. *Journal of Fish Diseases* 20, 409–418.

[5] Almendras, F.E., Jones, S.R.M., Fuentealba, C. and Wright, G.M. (1997b) In vitro infection of a cell line from *Ictalurus nebulosus* with *Piscirickettsia salmonis*. *Canadian Journal of Veterinary Research* 61, 66–68.

[6] Almendras, F.E., Fuentealba, I.C., Frederic Markham, R.F. and Speare, D.J. (2000) Pathogenesis of liver lesions caused by experimental infection with *Piscirickettsia salmonis* in juvenile Atlantic salmon, *Salmo salar* L. *Journal of Diagnostic Investigation* 12, 552–557.

[7] Arkush, K.D., McBride, A.M., Mendonca, H.L., Okihiro, M.S., Andree, K.B. *et al*. (2005) Genetic characterization and experimental pathogenesis of *Piscirickettsia salmonis* isolated from white seabass *Atractoscion nobilis*. *Diseases of Aquatic Organisms* 73, 131–139.

[8] Arkush, K.D., Edes, H.L., McBride, A.M., Adkinson, M.A. and Hedrick, R.P. (2006) Persistence of *Piscirickettsia salmonis* and the detection of serum antibodies to the bacterium in white seabass *Atractoscion nobilis* following experimental exposure. *Diseases of Aquatic Organisms* 73, 139–149.

[9] Athanassopoulou, F., Groman, D., Prapas, T. and Sabatakou, O. (2004) Pathological and epidemiological observations on rickettsiosis in cultured sea bass (*Dicentrarchus labrax* L.) from Greece. *Journal of Applied Ichthyology* 20, 525–529.

[10] Barnes, M.N., Landolt, M.L., Powell, D.B. and Winston, J.R. (1998) Purification of *Piscirickettsia salmonis* and partial characterization of antigens. *Disease of Aquatic Organisms* 33, 33–41.

[11] Birkbeck, T.H., Griffen, A.A., Reid, H.I., Laidler, L.A. and Wadsworth, S. (2004a) Growth of *Piscirickettsia salmonis* to high titers in insect tissue culture cells. *Infection*

and Immunity 72, 3693－3694.

[12] Birkbeck, T.H., Rennie, S., Hunter, D., Laidler, L.A. and Wadsworth, S. (2004b) Infectivity 400 of a Scottish isolate of *Piscirickettsia salmonis* for Atlantic salmon *Salmo salar* and 401 immune response of salmon to this agent. *Diseases of Aquatic Organisms* 60, 97－103.

[13] Bohle, H., Henríquez, P., Grothusen, H., Navas, E., Sandoval, A. *et al.* (2014) Comparative genome analysis of two isolates of the fish pathogen *Piscirickettsia salmonis* from different hosts reveals major differences in virulence-associated secretion systems. *Genome Announcements* 2(6): e01219－14.

[14] Branson, E.J. and Nieto Díaz-Munoz, D. (1991) Description of a new disease condition occurring in farmed coho salmon, *Oncorhynchus kisutch* (Walbaum), in South America. *Journal of Fish Diseases* 14, 147－156.

[15] Bravo, S. (1994) Piscirickettsiosis in freshwater. *Bulletin of the European Association of Fish Pathologists* 14, 137－138.

[16] Bravo, S. and Campos, M. (1989) Coho salmon syndrome in Chile. *American Fisheries Society/Fish Health Section Newsletter* 17(3), 3.

[17] Brocklebank, J.R., Evelyn, T.P., Speare, D.J. and Armstrong, R.D. (1993) Rickettsial septicemia in farmed Atlantic and Chinook salmon in British Columbia: clinical presentation and experimental transmission. *The Canadian Veterinary Journal* 34, 745－748.

[18] Chen, M.F., Yun, S., Marty, G.D., McDowell, T.S., House, M.L. *et al.* (2000) A *Piscirickettsia salmonis*-like bacterium associated with mortality of white seabass *Atractoscion nobilis*. *Diseases of Aquatic Organisms* 43, 117－126.

[19] Chen, S.C., Tung, M.C., Chen, S.P., Tsai, J.F., Wang, P.C. *et al.* (1994) System granulomas caused by a rickettsialike organism in Nile tilapia, *Oreochromis niloticus* (L.), from southern Taiwan. *Journal of Fish Diseases* 17, 591－599.

[20] Comps, M., Raymond, J. C. and Plassiart, G. N. (1996) Rickettsia-like organism infecting juvenile sea-bass *Dicentrarchus labrax*. *Bulletin of the European Association of Fish Pathologists* 16, 30－33.

[21] Contreras-Lynch, S., Olmos, P., Vargas, A., Figueroa, J., González-Stegmaier, R. *et al.* (2015) Identification and genetic characterization of *Piscirickettsia salmonis* in native fish from southern Chile. *Diseases of Aquatic Organisms* 115, 233－244.

[22] Corbeil, S., Hyatt, A.D. and Crane, M.S.J. (2005) Characterisation of an emerging rickettsia-like organism in Tasmanian farmed Atlantic salmon *Salmo salar*. *Diseases of Aquatic Organisms* 64, 37－44.

[23] Cusack, R.R., Groman, D.B. and Jones, S.R.M. (2002) The first case of rickettsial infections in farmed Atlantic salmon in eastern North America. *The Canadian Veterinary Journal* 43, 435－440.

[24] Cvitanich, J.D., Garate, N.O. and Smith, C.E. (1990) Etiological agent in Chilean Coho disease isolated and confirmed by Koch's postulates. *American Fisheries Society/Fish Health Section Newsletter* 18(1), 1－2.

[25] Cvitanich, J.D., Garate, N.O. and Smith, C.E. (1991) The isolation of a rickettsia-like organism causing disease and mortality in Chilean salmonids and its confirmation by Koch's postulate [sic]. *Journal of Fish Diseases* 14, 121－145.

[26] Evelyn, T.P.T., Kent, M.L., Poppe, T.T. and Bustos, P. (1998) Bacterial diseases. In: Kent, M.L. and Poppe, T.T (eds) *Diseases of Seawater Netpen-reared Salmonid Fishes*. Pacific Biological Station, Nanaimo, British Columbia, Canada, pp. 17－35.

[27] Fryer, J. L. and Hedrick, R. P. (2003) *Piscirickettsia salmonis*: a Gram-negative intracellular bacterial pathogen of fish. *Journal of Fish Diseases* 26, 251－262.

[28] Fryer, J.L. and Lannan, C.N. (1994) Rickettsial and chlamydial infections of freshwater

and marine fishes, bivalves, and crustaceans. *Zoological Studies* 33, 95 - 107.

[29] Fryer, J.L., Lannan, C.N., Garcés, L.H., Larenas, J.J. and Smith, P.A. (1990) Isolation of a Rickettsiales-like organism from diseased coho salmon *Oncorhynchus kisutch* in Chile. *Fish Pathology* 25, 107 - 114.

[30] Fryer, J.L., Lannan, C.N., Giovannoni, S.J. and Wood, N.D. (1992) *Piscirickettsia salmonis* gen. nov., sp. nov., the causative agent of an epizootic disease in salmonid fishes. *International Journal of Systematic Bacteriology* 42, 120 - 126.

[31] Gaggero, A., Castro, H. and Sandino, A.M. (1995) First isolation of *Piscirickettsia salmonis* from coho salmon, *Oncorhynchus kisutch* (Walbaum), and rainbow trout, *Oncorhynchus mykiss* (Walbaum), during the freshwater stage of their life cycle. *Journal of Fish Diseases* 18, 277 - 279.

[32] Garcés, L.H., Larenas, J.J., Smith, P.A., Sandino, S., Lannan, C.N. and Fryer, J.L. (1991) Infectivity of a rickettsia isolated from coho salmon *Oncorhynchus kisutch*. *Diseases of Aquatic Organisms* 11, 93 - 97.

[33] Garcés, L.H., Correal, P., Larenas, J.J., Contreras, J., Oyandel, S. *et al.* (1994) Finding Piscirickettsia salmonis on Cerathothoa gaudichaudii. In: Hedrick, R.P. and Winton, J.R. (organizers) *International Symposium on Aquatic Fish Health: Program and Abstracts*, Sheraton Hotel, Seattle, Washington, September 4 - 8, 1994, abstract, p. 109.

[34] Giménez, D.F. (1964) Staining rickettsiae in yolk-sac cultures. *Stain Technology* [now *Biotechnic and Histochemistry*] 39, 139 - 140.

[35] Gómez, F., Tobar, J.A., Henríquez, V., Sola, M., Altamirano, C. and Marshall, S. (2013) Evidence of the presence of a functional Dot/Icm type IV-B secretion system in the fish bacterial pathogen *Piscirickettsia salmonis*. *PLoS ONE* 8(1): e54934.

[36] Grant, A.N., Brown, A.C., Cox, D.I., Birkbeck, T.H. and Griffen, A.A. (1996) Rickettsia-like organism in farmed salmon. *Veterinary Record* 138, 423.

[37] Heath, S., Pak, S., Marshall, S., Prager, E.M. and Orrego, C. (2000) Monitoring *Piscirickettsia salmonis* by denaturant gel electrophoresis and competitive PCR. *Diseases of Aquatic Organisms* 41, 19 - 29.

[38] Henríquez, P., Bohle, H., Bustamante, F., Bustos, P. and Mancilla, M. (2015) Polymorphism in *gyrA* is associated to quinolones resistance in Chilean *Piscirickettsia salmonis* field isolates. *Journal of Fish Diseases* 38, 415 - 418.

[39] House, M. L. and Fryer, J. L. (2002) The biology and molecular detection of *Piscirickettsia salmonis*. In: Cunningham, C. O. (ed.) *Molecular Diagnosis of Salmonid Diseases*. Kluwer, Dordrecht, The Netherlands, pp. 141 - 155.

[40] House, M.L., Bartholomew, J.L., Winton, J.R. and Fryer, J.L. (1999) Relative virulence of three isolates of *Piscirickettsia salmonis* for coho salmon *Oncorhynchus kisutch*. *Diseases of Aquatic Organisms* 35, 107 - 113.

[41] Ibieta, P., Venegas, C., Takle, H., Hausdorf, M. and Tapia, V. (2011) *Chilean Salmon Farming on the Horizon of Sustainability: Review of the Development of a Highly Intensive Production, the ISA Crisis and Implemented Actions to Reconstruct a more Sustainable Aquaculture Industry*. InTech Open Science, Rijeka, Croatia.

[42] Isla, A., Haussmann, D., Vera, T., Kausel, G. and Figueroa, J. (2014) Identification of the *clpB* and *bipA* genes and an evaluation of their expression as related to intracellular survival for the bacterial pathogen *Piscirickettsia salmonis*. *Veterinary Microbiology* 173, 390 - 394.

[43] Jamett, A., Aguayo, J., Miquel, A., Muller, I., Arriagada, R. *et al.* (2001) Characteristics of monoclonal antibodies against *Piscirickettsia salmonis*. *Journal of Fish Diseases* 24, 205 - 215.

[44] Jones, S.R.M., Markham, R.J.F., Groman, D.B. and Cusack, R.R. (1998) Virulence

and antigenic characteristics of a cultured *Rickettsiales*-like organism isolated from farmed Atlantic Salmon *Salmo salar* in eastern Canada. *Diseases of Aquatic Organisms* 33, 25 - 31.

[45] Kuzyk, M.A., Thornton, J.C. and Kay, W.W. (1996) Antigenic characterization of the salmonid pathogen *Piscirickettsia salmonis*. *Infection and Immunity* 64, 5205 - 5210.

[46] Kuzyk, M.A., Burian, J., Machander, D., Dolhaine, D., Cameron, S. *et al.* (2001) An efficacious recombinant subunit vaccine against the salmonid rickettsial pathogen *Piscirickettsia salmonis*. *Vaccine* 19, 2337 - 2344.

[47] Lannan, C.N. and Fryer, J.L. (1991) Recommended methods for inspection of fish for the salmonid rickettsia. *Bulletin of the European Association of Fish Pathologists* 11, 135 - 136.

[48] Lannan, C.N. and Fryer, J.L. (1993) *Piscirickettsia salmonis*, a major pathogen of salmonid fish in Chile. *Fisheries Research* 17, 115 - 121.

[49] Lannan, C.N. and Fryer, J.L. (1994) Extracellular survival of *Piscirickettsia salmonis*. *Journal of Fish Diseases* 17, 545 - 548.

[50] Lannan, C.N., Ewing, S.A. and Fryer, J.L. (1991) A fluorescent antibody test for detection of the rickettsia causing disease in Chilean salmonids. *Journal of Aquatic Animal Health* 3, 229 - 234.

[51] Larenas, J.J., Astorga, C., Contreras, J. and Smith, P. (1996) *Piscirickettsia salmonis* in ova obtained from rainbow trout (*Oncorhynchus mykiss*) experimentally inoculated. *Archivos de Medicina Veterinaria* 28, 161 - 166.

[52] Larenas, J., Contreras, J. and Smith, P. (2000) Piscirickettsiosis: uno de los principales problemas en cultivos de salmones en Chile. *Revista de Extension Tecno Vet* 6, 28 - 30.

[53] Larenas, J.J., Bartholomew, J.L., Troncoso, O., Fernandez, S., Ledezma, H. *et al.* (2003) Experimental vertical transmission of *Piscirickettsia salmonis* and in vitro study of attachment and mode of entrance into the fish ovum. *Diseases of Aquatic Organisms* 56, 25 - 30.

[54] Leal, J. and Woywood, D. (2007) Piscirickettsiosis en Chile: avances y perspectivas para su control. *Salmociencia* 2, 34 - 42.

[55] Marshall, S., Heath, S., Henríquez, V. and Orrego, C. (1998) Minimally invasive detection of *Piscirickettsia salmonis* in cultivated salmonids via the PCR. *Applied and Environmental Microbiology* 64, 3066 - 3069.

[56] Marshall, S.H., Conejeros, P., Zahr, M., Olivares, J., Gómez, F. *et al.* (2007) Immunological characterization of a bacterial protein isolated from salmonid fish naturally infected with *Piscirickettsia salmonis*. *Vaccine* 25, 2095 - 2102.

[57] Mauel, M.J. and Fryer, J.L. (2001) Amplification of a *Piscirickettsia salmonis*-like 16S rDNA product from bacterioplankton DNA collected from the coastal waters of Oregon, USA. *Journal of Aquatic Animal Health* 13, 280 - 284.

[58] Mauel, M.J. and Miller, D.L. (2002) Piscirickettsiosis and piscirickettsiosis-like infections in fish: a review. *Veterinary Microbiology* 87, 279 - 289.

[59] Mauel, M.J., Giovannoni, S.J. and Fryer, J.L. (1996) Development of polymerase chain reaction assays for detection, identification, and differentiation of *Piscirickettsia salmonis*. *Diseases of Aquatic Organisms* 26, 189 - 195.

[60] Mauel, M.J., Giovannoni, S.J. and Fryer, J.L. (1999) Phylogenetic analysis of *Piscirickettsia salmonis* by 16S, internal transcribed spacer (ITS) and 23S ribosomal DNA sequencing. *Diseases of Aquatic Organisms* 35, 115 - 123.

[61] Mauel, M.J., Ware, C. and Smith, P.A. (2008) Culture of *Piscirickettsia salmonis* on enriched blood agar. *Journal of Veterinary Diagnostic Investigation* 20, 213 - 214.

[62] McCarthy, U., Steiropoulos, N.A., Thompson, K.D., Adams, A., Ellis, A.E. *et al.*

(2005) Confirmation of *Piscirickettsia salmonis* as a pathogen in European sea bass *Dicentrarchus labrax* and phylogenetic comparison with salmonid strains. *Diseases of Aquatic Organisms* 64, 107 - 119.

[63] Meibom, K.L., Dubail, I., Dupuis, M., Barel, M., Lenco, J., Stulik, J. *et al.* (2008) The heat-shock protein ClpB of *Francisella tularensis* is involved in stress tolerance and is required for multiplication in target organs of infected mice. *Molecular Microbiology* 67, 1384 - 1401.

[64] Mikalsen, J., Skjaervik, O., Wiik-Nielsen, J., Wasmuth, M.A. and Colquhoun, D.J. (2008) Agar culture of *Piscirickettsia salmonis*, a serious pathogen of farmed salmonid and marine fish. *FEMS Microbiology Letters* 278, 43 - 47.

[65] OIE (2006) Chapter 2.1.13. Piscirickettsiosis (*Piscirickettsia salmonis*). In: *Manual of Diagnostic Tests for Aquatic Animals 2006*, 5th edn. World Organisation for Animal Health, Paris, pp. 236 - 241. Available at: http://www.oie.int/doc/ged/D6510.PDF (accessed 17 November 2016).

[66] Olivares, J. and Marshall, S. H. (2010) Determination of minimal concentration of *Piscirickettsia salmonis* in water columns to establish a fallowing period in salmon farms. *Journal of Fish Diseases* 33, 261 - 266.

[67] Olsen, A.B., Melby, H.P., Speilberg, L., Evensen, Ø. And Håstein, T. (1997) *Piscirickettsia salmonis* infection in Atlantic salmon *Salmo salar* in Norway - epidemiological, pathological and microbiological findings. *Diseases of Aquatic Organisms* 31, 35 - 48.

[68] Palmer, R., Rutledge, M., Callanan, K. and Drinan, E. (1997) A piscirickettsiosis-like disease in farmed Atlantic salmon in Ireland - isolation of the agent. *Bulletin of the European Association of Fish Pathologists* 17, 68 - 72.

[69] Pulgar, R., Hödar, C., Travisany, D., Zuñiga, A., Domínguez, C. *et al.* (2015) Transcriptional response of Atlantic salmon families to *Piscirickettsia salmonis* infection highlights the relevance of the iron-deprivation defense system. *BMC Genomics* 16: 495.

[70] Rodger, H.D. and Drinan, E.M. (1993) Observation of a rickettsia-like organism in Atlantic salmon, *Salmo salar* L., in Ireland. *Journal of Fish Diseases* 16, 361 - 369.

[71] Rojas, V., Olivares, J., del Río, R. and Marshall, S.H. (2008) Characterization of a novel and genetically different small infective variant of *Piscirickettsia salmonis*. *Microbial Pathogenesis* 44, 370 - 378.

[72] Rozas, M. and Enríquez, R. (2014) Piscirickettsiosis and *Piscirickettsia salmonis* in fish: a review. *Journal of Fish Diseases* 37, 163 - 188.

[73] Rise, M.L., Jones, S.R.M., Brown, G.D., von Schalburg, K.R., Davidson, W.S. *et al.* (2004) Microarray analyses identify molecular biomarkers of Atlantic salmon macrophage and hematopoietic kidney response to *Piscirickettsia salmonis* infection. *Physiological Genomics* 20, 21 - 35.

[74] Sernapesca (2013) *Establece Programa Sanitario Específico de Vigilancia y Control de Piscirickettsiosis (PSEVC-PISCIRICKETTSIOSIS)*. Servicio Nacional de Pesca y Acuicultura, Valparaíso, Chile. Available at: http://www.sernapesca.cl/index.php?option=com_remository&Itemid=246&func=startdown&id=6726 (accessed 1 February 2016).

[75] Sernapesca (2014) *Informe Sanitario de Salmonicultura en Centros Marinos - Año 2014*. Servicio Nacional de Pesca y Acuicultura, Valparaíso, Chile. Available at: http://www.sernapesca.cl/index.php?option=com_remository&Itemid=246&func=fileinfo&id=11083 (accessed 17 November 2016).

[76] Sernapesca (2015) *Informe Sobre el Uso de Antimicrobianos en Salmonicultura Nacional 2015*. Servicio Nacional de Pesca y Acuicultura, Valparaíso, Chile. Available at: http://www.sernapesca.cl/presentaciones/Comunicaciones/Informe_Sobre_Uso_

de_Antimicrobianos_2015.pdf (accessed 15 February 2016).
[77] Skarmeta, A.M., Henriquez, V. Zahr, M., Orrego, C. and Marshall, S.H. (2000) Isolation of a virulent *Piscirickettsia salmonis* from the brain of naturally infected coho salmon (*Oncorhynchus kisutch*). *Bulletin of the European Association of Fish Pathologists* 20, 261-264.
[78] Smith, P.A., Lannan, C.N., Garcés, L.H., Jarpa, M., Larenas, J. *et al.* (1995) Piscirickettsiosis: a bacterin field trial in coho salmon (*Oncorhynchus kisutch*). *Bulletin of the European Association of Fish Pathologists* 15, 137-141.
[79] Smith, P.A., Contreras, J.R., Garcés, L.H., Larenas, J., Oyanedel, S. *et al.* (1996a) Experimental challenge of coho salmon and rainbow trout with *Piscirickettsia salmonis*. *Journal of Aquatic Animal Health* 8, 130-134.
[80] Smith, P.A., Vecchiolla, I.M., Oyanedel, S., Garcés, L.H., Larenas, J. *et al.* (1996b) Antimicrobial sensitivity of four isolates of *Piscirickettsia salmonis*. *Bulletin of the European Association of Fish Pathologists* 16, 164-168.
[81] Smith, P.A., Contreras, J.R., Larenas, J., Aguillon, J.C., Garcés *et al.* (1997) Immunization with bacterial antigens: piscirickettsiosis. In: Gudding, R., Lillehaug, A., Midtlyng, P.J. and Brown, F. (eds) *Fish Vaccinology*. Development of Biological Standards, Vol. 90, Karger, Basel, Switzerland, pp. 161-166.
[82] Smith, P.A., Pizarro, P., Ojeda, P., Contreras, J.R., Oyanedel, S. et al. (1999) Routes of entry of *Piscirickettsia salmonis* in rainbow trout *Oncorhynchus mykiss*. *Diseases of Aquatic Organisms* 37, 165-172.
[83] Smith, P.A., Pizarro, P., Ojeda, P., Contreras, J.R., Oyanedel, S. *et al.* (2004) Experimental infection of coho salmon *Oncorhynchus mykiss* by exposure of skin, gills and intestine with *Piscirickettsia salmonis*. *Diseases of Aquatic Organisms* 61, 53-57.
[84] Smith, P.A., Díaz, F.E., Rojas, M.E., Díaz, S., Galleguillos, M. *et al.* (2015) Effect of *Piscirickettsia salmonis* inoculation on the ASK continuous cell line. *Journal of Fish Diseases* 38, 321-324.
[85] Steiropoulos, N.A., Yuksel, S.A., Thompson, K.D., Adams, A. and Ferguson, H.W. (2002) Detection of *Rickettsia*-like organisms (RLOs) in European sea bass (*Dicentrarchus labrax*) by immunohistochemistry. *Bulletin of the European Association of Fish Pathologists* 22, 260-267.
[86] Tapia-Cammas, D., Yañez, A., Arancibia, G., Toranzo, A.E. and Avendaño-Herrera, R. (2011) Multiplex PCR for the detection of *Piscirickettsia salmonis*, *Vibrio anguillarum*, *Aeromonas salmonicida* and *Streptococcus phocae* in Chilean marine farms. *Diseases of Aquatic Organisms* 97, 135-142.
[87] Tobar, J.A., Jerez, S., Caruffo, M., Bravo, C., Contreras, F. *et al.* (2011) Oral vaccination of Atlantic salmon (*Salmo salar*) against salmonid rickettsial septicaemia. *Vaccine* 29, 2336-2340.
[88] US Surgeon-General's Office (1994) Laboratory methods of the United States Army. In: Simmons, J.S. and Gentzkow, C.J. (eds) *Laboratory Methods in the United States Army*, 5th edn. Lea and Febiger, Philadelphia, Pennsylvania, p. 572.
[89] Venegas, C.A., Contreras, J.R., Larenas, J.J. and Smith, P.A. (2004) DNA hybridization assay for the detection of *Piscirickettsia salmonis* in salmonid fish. *Journal of Fish Diseases* 27, 431-433.
[90] Warsen, A.E., Krug, M.J., LaFrentz, S., Stanek, D.R., Loge, F.J. *et al.* (2004) Simultaneous discrimination between 15 fish pathogens by using 16S ribosomal DNA PCR and DNA microarrays. *Applied and Environmental Microbiology* 70, 4216-4221.
[91] Weiss, E. and Moulder, J.W. (1984) Order I. Rickettsiales Gieszczkiewicz 1939, 25AL. In: Kreig, N.R. (ed.) *Bergey's Manual of Systematic Bacteriology*, Vol. 1. Williams and Wilkins, Baltimore/London, pp. 687-729.

[92] Wilhelm, V., Miquel, A., Burzio, L.O., Rosemblatt, M., Engel, E., Valenzuela, S. et al. (2006) A vaccine against the salmonid pathogen *Piscirickettsia salmonis* based on recombinant proteins. *Vaccine* 24, 5083 - 5091.
[93] Yañez, A.J., Silva, H., Valenzuela, K., Pontigo, J.P., Godoy, M. *et al*. (2013) Two novel blood-free solid media for the culture of the salmonid pathogen *Piscirickettsia salmonis*. *Journal of Fish Diseases* 36, 587 - 591.

21

鲑肾杆菌

Diane G. Elliott*

21.1 引言

自20世纪30年代早期对苏格兰野生大西洋鲑的细菌性肾脏病(bacterial kidney disease, BKD)的初始描述以来,为了更好地认识和控制这一严重疾病,人们付出了大量的努力,并对BKD及其病原鲑肾杆菌(*Renibacterium salmoninarum*)进行数次全面综述(Fryer and Sanders, 1981; Austin and Austin, 1987; Elliott et al., 1989; Evelyn, 1993; Evenden et al., 1993; Fryer and Lannan, 1993; Pascho et al., 2002; Wiens, 2011)。

BKD的地理分布区域几乎覆盖了世界范围内野生或养殖鲑科鱼类存在的淡水和海水栖息地,包括北美、南美、欧洲及亚洲的许多国家(Wiens, 2011; Kristmundsson et al., 2016)。虽然所有鲑科鱼类都易患BKD,但太平洋鲑[太平洋鲑属(*Oncorhynchus*)]通常更易受鲑肾杆菌感染。BKD的自然暴发仅限于鲑科鱼类(*Salmonidae*),但从鲑鱼养殖场及周边水域或有鲑科鱼类的水域中采集的非鲑科鱼类和双壳类软体动物都曾检测出鲑肾杆菌(US Fish and Wildlife Service, 2011; Elliott et al., 2014)。一些非鲑科鱼类也被实验性地感染(Wiens, 2011),这就增加了其他物种成为该病原的宿主或媒介的可能性。

鲑肾杆菌的水平传播和垂直传播均增强了其在鱼类种群体中的持久性(Evelyn, 1993),其中水平传播在淡水及海水环境中均可发生(Pascho et al., 2002)。该菌能从受感染鱼的粪便(Balfry et al., 1996)及表皮组织剥落的细胞中脱落(Elliott et al., 2015)。鲑肾杆菌在环境中的生存能力可能受到正常水生菌群竞争的限制,有研究(Pascho et al., 2002)报道了鲑肾杆菌在环境样品中存活的时间相对较短(10～18 ℃下存活4～21 d)。因此,鲑肾杆菌的水平传播可能只在短距离内发生(Murray et al., 2012)。水平传播的侵入位点包括摄入受感染鱼的尸体或粪便后的胃肠道,或受损伤部位的表皮组织(Elliott et al.,

* 作者邮箱:dgelliott@usgs.gov。

2015)。

鲑肾杆菌与鱼卵一起垂直传播给子代(Pascho et al., 2002),有些卵细胞的卵黄中携带细菌。含有高浓度细菌的卵巢(体腔)液可能造成鱼卵感染,排卵后感染可能通过卵孔发生。在卵母细胞发育早期,鱼卵感染也可直接发生于卵巢组织(Elliott et al., 2014)。

鲑肾杆菌是一种很小(长 1.0～1.5 μm,宽 0.3～1.0 μm)、非运动性、不产孢子、不耐酸、繁殖慢的杆状革兰氏阳性菌,通常成对出现(Sanders and Fryer, 1980)。在微球菌科(*Micrococcaceae*)中,鲑肾杆菌与非致病性土壤微生物节杆菌(*Arthrobacter* spp.)的亲缘关系最近(Wiens et al., 2008)。鲑肾杆菌的基因组由一条大小为 3.15 Mb 的环状染色体构成,不含质粒或噬菌体元件(Wiens et al., 2008)。尽管鲑肾杆菌分布于广大区域的鲑科鱼类中,但不同分离株在生化和血清学特征上显示出有限的差异(Pascho et al., 2002; Elliott et al., 2014)以及较低的遗传多样性(Pascho et al., 2002; Brynildsrud et al., 2014)。然而,近来使用新一代测序技术构建了 *R. salmoninarum* 不同分离株全基因组范围内的单核苷酸多态性(single-nucleotide polymorphism, SNP) 数据,鉴别了该菌的不同亚型并用于流行病学研究(Brynildsrud et al., 2014)。

BKD 是一种鲑科鱼类流行病,由于慢性感染的特性和时常发生的混合感染(Evenden et al., 1993),很难准确估计该疾病的死亡率和 *R. salmoninarum* 对鱼类种群的总体影响,但在一些大西洋鲑和太平洋鲑种群中的疾病死亡率分别高达 70%和 80% (Evenden et al., 1993; Wiens, 2011)。自然放养鱼群的死亡率评估尤其成问题,因为患病鱼很容易被捕食(Mesa et al., 1998)。尽管如此,已在野生鲑(包括无孵化场鱼补充历史的野生产卵鱼群)中观察到临床细菌性肾病(Pascho et al., 2002; Elliott et al., 2014)。在养殖鱼类中报告的疫情最频繁,鲑科鱼类养殖规模的扩张以及受感染的鱼和鱼卵的相关流通也促进了 *R. salmoninarum* 的扩散(Evenden et al., 1993; Murray et al., 2012)。此外,*R. salmoninarum* 感染的免疫抑制效应(见 21.4 节)可能会导致由继发病原引起的死亡(Munson et al., 2010)。

21.2 临床症状与诊断

21.2.1 临床症状

R. salmoninarum 感染通常发病较慢,一般在鱼龄为 6～12 月时才出现明显病症(Evelyn, 1993)。严重感染的鱼可能没有外部病症,或可能出现以下一种或多种症状:嗜睡,皮肤发暗或斑驳,腹水性腹胀,鳃贫血苍白,眼球突起,含清澈、混浊或血性液体的皮肤水泡,鱼体瘀斑性出血,鱼鳍基部和肛门附近出血,皮肤浅表溃疡以及延伸至骨骼肌的大囊性空洞或脓肿(Pascho et al., 2002; Bruno et al., 2013)。在一些成熟鲑中,特别是虹鳟中,一种被称作"产卵疹"的浅表感染可能表现为覆盖大片皮肤的脓疱性皮炎,伴有表皮出血性结节和许多小水疱(Bruno et al., 2013)。这些体征都不是BKD的病理病症,确认 *R. salmoninarum* 感染仍需准确的诊断。*R. salmoninarum* 也可能以亚临床感染方式在鱼类种群中长期存在(Pascho et al., 2002; Elliott et al., 2013)。

21.2.2 诊断

许多不同的方法已被开发用于检测 *R. salmoninarum*,但尚未确定出一种理想的诊断检测方法(Pascho et al., 2002; Elliott et al., 2013)。推荐使用那些灵敏度、特异性和可靠性已知的标准化和经过验证的诊断方法(Purcell et al., 2011; Elliott et al., 2013)。

通过疾病的特征性体征和病变以及随后的病变组织涂片革兰氏染色或荧光抗体染色,通常可以快速诊断出临床BKD。革兰氏染色检测亚临床感染的可靠性受其灵敏度低(1.0×10^7~1.0×10^9 cells/g组织,Wiens, 2011)的限制。组织中黑色素颗粒的存在也会掩盖少量细菌(Pascho et al., 2002),此外革兰氏染色无法提供特异性鉴定。

培养是验证 *R. salmoninarum* 活性的基准分析方法,为此已开发出多种培养基(Pascho et al., 2002; Wiens, 2011)。其中大多数是对KDM2(kidney disease medium-2)培养基的改良,该培养基包含1%(质量浓度)的蛋白胨、0.05%(质量浓度)的酵母提取物、0.1%(质量浓度)的半胱氨酸以及7%~20%(体积分数)的血清。*R. salmoninarum* 的培养生长是需氧的,最适温度为15~18 ℃(Sanders and Fryer, 1980)。在接种后5~7 d内会出现可见菌落(Faisal et al., 2010a),但在亚临床病例中,可能需要长达19周的孵育时间(Pascho et al., 2002)。报道的肾脏匀浆培养物中 *R. salmoninarum* 检出限为8.0×10^1~5.0×10^2 cells/g(Elliott et al., 2013)。然而,即便使用了含抗生素的选择性培养基,培养物也可能会被快速生长的异养细菌污染(Pascho et al., 2002),这将阻碍对 *R. salmoninarum* 的检测。需使用生化检测(Austin and Austin, 1987)或血清学和分子生物学方法(OIE, 2003)来确诊病原。

由于 *R. salmoninarum* 增殖缓慢且需求苛刻,已经开发了许多免疫学和分子生物学方法对其进行检测(Pascho et al., 2002)。免疫学方法包括广泛使用和值得推荐的荧光抗体技术(FAT)和酶联免疫吸附试验(ELISA)(OIE, 2003; AFS-FHS, 2014)。使用单克隆或多克隆抗体的直接FAT(DFAT)或间接FAT(IFAT)均可用于组织涂片或病理切片中的 *R. salmoninarum* 检测(Pascho et al., 2002)。膜过滤荧光抗体技术(Membrane-filtration-FAT, MF-FAT)被用于卵巢液和肾组织匀浆稀释液样本的检测(Pascho et al., 2002)。FAT在组织涂片中的*R. salmoninarum*检出限为10^3～10^7 cells/g组织(Pascho et al., 2002; Elliott et al., 2013)，而过滤样品的MF-FAT检出限小于10^2 cells/mL (Wiens, 2011; Elliott et al., 2013)。FAT通过形态学和特异性染色特征鉴别*R. salmoninarum*，但它无法检测细菌的活力，同时耗费人工。

利用多克隆或单克隆抗体开发的ELISA检测组织中特定的*R. salmoninarum*抗原(Pascho et al., 2002; Wiens, 2011)。有限的研究显示，与单抗ELISA相比，由于多克隆可识别额外的抗原表位(Pascho et al., 2002)，多抗ELISA具有更高的灵敏度(Jansson et al., 1996; Elliott，未发表的数据)。建议对抗血清进行亲和纯化，以确保多抗ELISA足够的特异性。产生足以被ELISA检测出的可溶性抗原水平所需的最小*R. salmoninarum*数量很难确定(Pascho et al., 2002; Elliott et al., 2013)，但多抗ELISA的估计检测限在10^3 cells/g肾组织(Jansson et al., 1996)。ELISA是一种半定量技术(Pascho et al., 2002; Elliott et al., 2013)，常被用于筛查亲鱼(见21.5节)或在BKD流行地区监测种群中的感染状况(Maule et al., 1996; Elliott et al., 1997; Meyers et al., 2003; Faisal et al., 2010b, 2012; Kristmundsson et al., 2016)。ELISA的优点之一是它能够检测从局部组织感染传送的可溶性*R. salmoninarum*抗原(Elliott et al., 2015)。由于在无活菌的情况下可溶性抗原也能存在，因此ELISA阳性结果并不一定表明活性感染，通常还需另外的检测来证实(Wiens, 2011)。因此ELISA更适合用于筛查，而不是用作确证检测(Elliott et al., 2013)。

分子检验，特别是聚合酶链反应(PCR)，因其特异性而在*R. salmoninarum*检测中获得了广泛应用(Pascho et al., 2002; Wiens, 2011)，一些PCR技术被建议作为确证检测法(OIE, 2003; AFS-FHS, 2014)。常规PCR(cPCR)、巢式PCR (nPCR)和实时定量PCR (qPCR)已被开发用于检测*R. salmoninarum*的DNA序列(Pascho et al., 2002; Wiens, 2011)。这些PCR的相对检测灵敏度具有检测特定性，取决于诸如引物序列和反应条件等因素(Purcell et al., 2011)。PCR的一个优势就是它能从样本中存在的少量微生物中扩增目标核酸序列。在实际应用中，由于亚临床感染鱼中*R. salmoninarum*的集中或不均匀分布，加上DNA抽提后每次PCR中实际检测的组织比例很小，PCR检测的灵敏度通常会降低(Purcell et al., 2011; Elliott et al., 2013)。nPCR或qPCR对肾组织的*R. salmoninarum*检测限大于等于10^3 cells/g组织(Elliott et al., 2013)。用于检测DNA序列的PCR无法检测细菌活力。然而，反转录PCR (RT-PCR)已被开发用于检测信使RNA(messenger RNA, mRNA)，mRNA半衰期短，可以表明活菌存在(Pascho et al., 2002; Elliott et al., 2013)。不过，在活跃和潜伏感染期状态下的细菌基因表达的差异性可能会影响mRNA表达，因此RT-PCR检测mRNA的灵敏度可能会低于PCR检测鲑肾杆菌DNA的灵敏度(Elliott et al., 2013)。

21.3 病理学

21.3.1 体内大体病变

在剖检中,患有 BKD 鱼的内部检查通常会显示肾(图 21.1)以及心、肝或脾等其他器官的局灶性或多灶性结节性病变。肾和脾常见肿大(图 21.1)。其他肉眼观察病症可能包括积存于腹腔和心包腔的混浊或血清血液状液体、肠道内存在黄色黏稠或带血液体、脏器及腹壁出血、一个或多个内部器官表面有弥漫性白色膜层(假膜)(Fryer and Sanders, 1981; Evelyn, 1993; Evenden et al., 1993)。

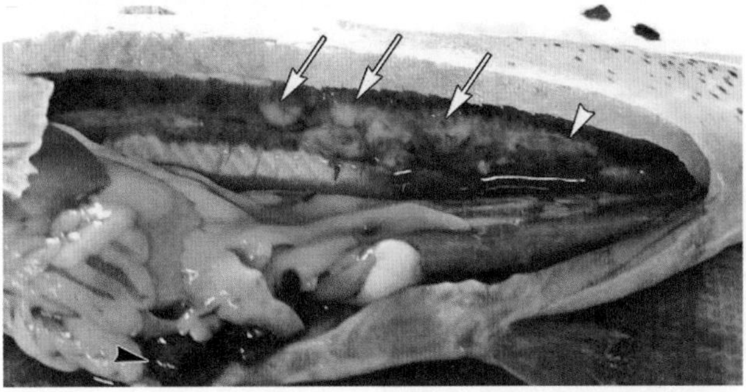

图 21.1 自然感染鲑肾杆菌的人工饲养成年雄性切喉鳟(*Oncorhynchus clarkii*)(长 305 mm)。在肾脏(如光标所示)处有特征性肉眼可见灰白色肉芽肿性病变,此外脾脏(以黑箭头标出)及后肾(如箭头所示)明显浮肿。图片由 John Drennan 博士(Aquatic Animal Health Laboratory, Colorado Parks and Wildlife, Brush, Colorado)提供

21.3.2 组织病理学

BKD 常被描述成以弥漫性全身慢性肉芽肿性炎症为特征的菌血症(Bruno, 1986)。虽然可能主要是肾间质组织受影响(Ferguson, 1989),但肉芽肿性炎症通常存在于大多数感染组织中。太平洋鲑中的肉芽肿通常呈弥漫性,边界模糊[图 21.2(A)],而大西洋鲑中的肉芽肿更多被包裹,并含有大量上皮样巨噬细胞(Evelyn, 1993)。较老的肉芽肿可能会出现干酪样坏死的中心区域,周围环绕上皮样细胞、纤维化和浸润性淋巴细胞[图 21.2(B)]。由纤维蛋白和胶原蛋白薄层以及捕获的吞噬细胞和细菌构成的假膜,可能在肾、脾及肝等器官囊上形成。

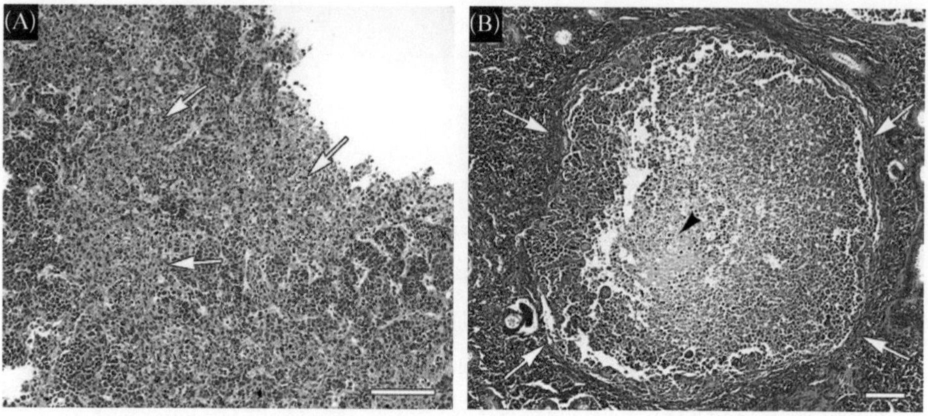

图 21.2　(A) 在公共水域自然感染鲑肾杆菌的幼年银大麻哈鱼的前肾组织切片。可见带有模糊边界的弥散状肉芽肿性炎症(如光标所示)。(B) 自然感染鲑肾杆菌的人工饲养幼年大鳞大麻哈鱼的后肾组织切片。已形成的 BKD 肉芽肿中有被纤维(如光标所示)包裹的中央坏死区(如黑箭头所示)。本图使用苏木精-伊红染色法，比例尺为 100 μm。图片由 Carla Conway (US Geological Survey, Western Fisheries Research Center, Seattle, Washington)提供

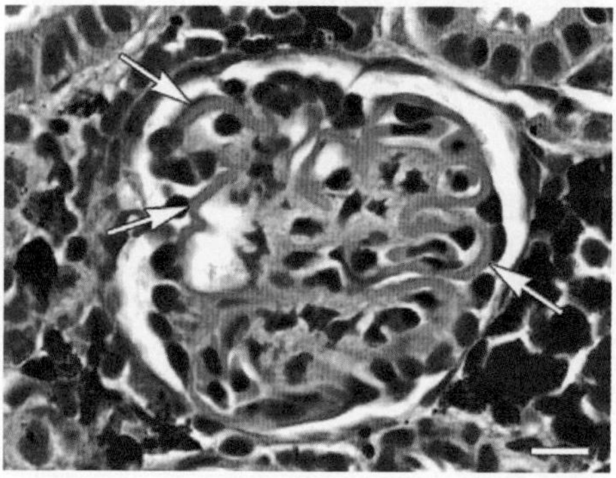

图 21.3　实验感染鲑肾杆菌的大鳞大麻哈鱼后肾组织切片中可见膜性肾小球肾炎。肾小球内皮下抗原-抗体复合物堆积在肾小球毛细血管中导致肾小球基膜透明层加厚(如光标所示)。本图使用苏木精-伊红染色法，比例尺为 10 μm。图片由 Carla Conway (US Geological Survey, Western Fisheries Research Center, Seattle, Washington)提供

在严重的 BKD 病例中，血管病变区域的纤维蛋白渗出尤为常见，其特征可能是伴有少量出血或血栓形成的血管中膜或内皮坏死(Ferguson, 1989)。心脏是 BKD 的常见病灶部位，在致密和海绵状心肌的界面表现为白喉性心外膜炎和肉芽肿性心肌炎。在肾脏中，膜性肾小球肾炎或血管球性肾炎很常见(图 21.3)，并与肾小球中抗原-抗体复合物的皮下沉积有关(Elliott et al., 2014)，这导致肾小球基底膜增厚。

虽然 BKD 通常表现为全身性疾病，但据报道局部感染的病变限定在诸如中枢神经系统(CNS)、眼睛或眶后组织、皮肤和鳍等组织(Pascho et al., 2002)。与全身性感染相似，*R. salmoninarum* 与感染组织的肉芽肿或化脓性肉芽肿病变有关。例如，在没有发生全身性感染的情况下，该菌也可能引起脑膜炎和脑炎(Speare et al., 1993)。CNS 感染可能是由于抗生素治疗清除了其他器官而非脑部的感染(Speare, 1997)，或者从眼后葡萄膜(脉络膜)经视神经结缔组织鞘(神经外膜和神经束膜)逆行延伸至间脑底部引起(Speare et al.,

1993)。在受鲑肾杆菌相关产卵疹影响的鱼中,肉芽肿性炎症可侵入邻近鳞囊,并沿真皮纤维层横向延伸。在鱼体其他部位可能检测不到该菌的存在,而皮损在产卵后消失(Ferguson, 1989)。一些局部感染可能代表了 *R. salmoninarum* 通过皮肤或眼睛等表皮组织破损后引入进行的增殖(Elliott et al., 2015)。

鲑肾杆菌是一种兼性胞内病原菌,在被感染鱼的巨噬细胞内外均可存在。胞内 *R. salmoninarum* 存在于上皮细胞、内皮细胞或中性粒细胞的细胞质中(Bruno, 1986; Ferguson, 1989),以及肾、脾造血组织的网状细胞、窦细胞或成纤维母细胞屏障细胞中(Flaño et al., 1996)。通常巨噬细胞内的细菌数量最多(图 21.4)。*R. salmoninarum* 通过一种宿主防御逃逸机制能够在巨噬细胞内存活并可能在细胞内进行复制(Pascho et al., 2002)。上皮样吞噬细胞可能显示有限的吞噬活性,而黑色素巨噬细胞则吞噬性活跃(Flaño et al., 1996)。色素在组织中扩散是 BKD 的组织病理学特征之一(图 21.5),这可能是黑色素巨噬细胞破裂和溶解造成的(Bruno, 1986; Flaño et al., 1996)。

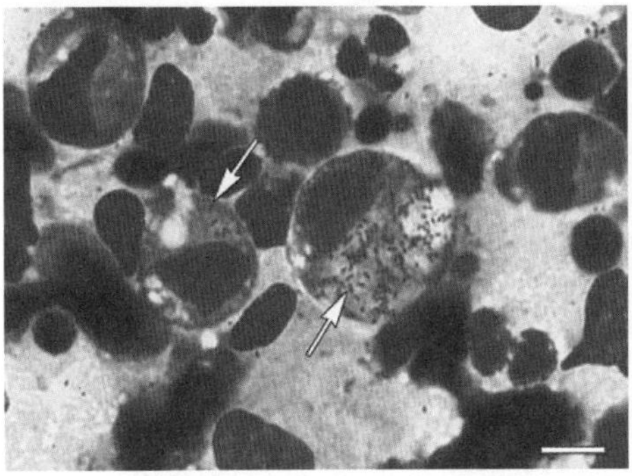

图 21.4 养殖场中自然感染鲑肾杆菌的成年切喉鳟脾印片。巨噬细胞胞质内可见细菌存在(如光标所示)。Dip Quick[1] (Jorgensen 实验室)专利 Romanowski 染色,比例尺为 10 μm。图片由 John Drennan 博士(Aquatic Animal Health Laboratory, Colorado Parks and Wildlife, Brush, Colorado)提供

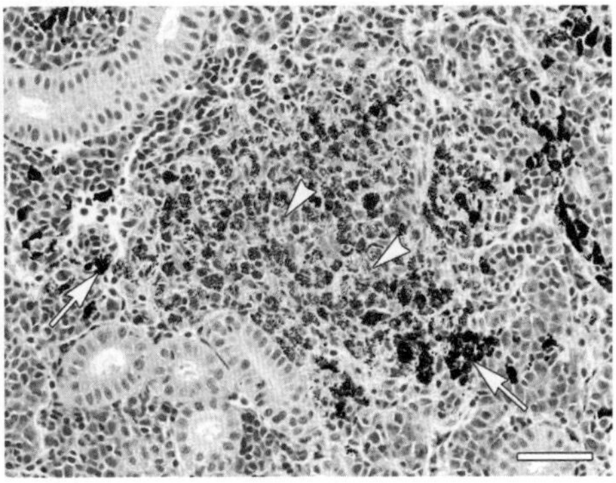

图 21.5 实验感染鲑肾杆菌的幼年大鳞大麻哈鱼后肾组织中局灶性肉芽肿组织切片。组织病变处的边界十分模糊,裂解的黑色素巨噬细胞处可见标志性色素扩散(如箭头所示)。在组织病变处附近可见含高密度胞质色素(如光标所示)的完好黑色素巨噬细胞。苏木精-伊红染色法,比例尺为 50 μm。图片由 Carla Conway (US Geological Survey, Western Fisheries Research Center, Seattle, Washington)提供

21.4 病理生理学

对与 *R. salmoninarum* 的自然或实验感染相关的血液学参数进行研究

后发现（Wiens, 2011），与红细胞相关的一些变化包括血细胞比容和循环红细胞计数下降，以及血红蛋白、红细胞直径和成熟与未成熟红细胞比率降低，红细胞沉降率上升。循环红细胞的减少与这些细胞在脾脏中的潴留有关，这导致感染 *R. salmoninarum* 的鱼脾脏肿大。实验感染鱼还表现出单核细胞增加以及血小板和中性粒细胞短暂增加，但循环的大小淋巴细胞数量没有变化。BKD的病情发展与总血清蛋白（特别是电泳迁移速度较快的部分）、胆固醇和钠含量降低以及血清胆红素、血尿素氮和钾含量升高有关。

某些生理变化可能会降低鱼的生产性能和存活率。例如，Mesa 等（1998, 2000）发现，当随着 *R. salmoninarum* 攻毒感染的大鳞大麻哈鱼幼鱼中 BKD 的病情发展，血浆皮质醇和乳酸水平显著上升，而葡萄糖水平则明显下降。皮质醇含量升高表明 *R. salmoninarum* 感染（特别是在后期阶段）造成应激压力，乳酸水平升高则提示低氧血症，进而会导致氧输送不良，这与严重的血液学变化一致。Mesa 等（1998）推测血糖下降是抵御 *R. salmoninarum* 感染时过度消耗这一能量底物造成的。研究者还认为，虽然在受感染鱼和健康鱼之间无明显的行为差异，但遭受中度至高度 *R. salmoninarum* 感染的大鳞大麻哈鱼幼鱼明显更容易受到捕食。他们推测，慢性感染的生物能量需求降低了可用于逃避捕食者等活动的能量。更严重感染的大鳞大麻哈鱼代谢范围缩减也是这些鱼对诸如溶解气体过饱和等应激源更敏感性的潜在机制（Wiens, 2011）。然而，Mesa 等（2000）的研究结果显示，与未处理的受感染鱼相比，经缺氧和轻度搅动多次处理的大鳞大麻哈鱼的死亡率或 *R. salmoninarum*感染进程没有差异。不过，与经多种应激源处理的未感染鱼不同，感染了 *R. salmoninarum* 的鱼无法持续产生明显的高血糖以响应应激源。

R. salmoninarum 感染与宿主生理过程之间的相互作用可能会产生其他的后果。有限的研究表明，鲑科鱼类的慢性感染对生长有不利影响（Sandell et al., 2015）。此外，银化可显著增加鲑中 BKD 的严重程度（Mesa et al., 1999），并可能导致鲑进入海水后不久的死亡率升高（Pascho et al., 2002; Wiens, 2011）。与 BKD 相关的肾小球肾炎在疾病恢复中的大西洋鲑成鱼中可能很严重，并可能因渗透调节失败而发病（Bruno et al., 2013）。

最后，由 *R. salmoninarum* 引起的免疫抑制可能会降低对次生病原的抵抗力。*R. salmoninarum* 的体外和体内免疫抑制能力已得到充分验证。*R. salmoninarum*丰富的 57 kDa 的胞外主要可溶性抗原（p57 或 MSA）已被确认为主要毒力因子，是免疫抑制的关键媒介，例如降低抗体产生和吞噬细胞杀菌活性（Pascho et al., 2002; Wiens, 2011）。一种 22 kDa 的 *R. salmoninarum* 表面蛋白 p22 也与抑制抗体产生有关（Fredriksen et al., 1997）。

21.5 预防和控制

BKD是较难控制的鱼类细菌疾病之一(Elliott et al., 1989)。已对多种BKD管控策略进行了综述(Elliott et al., 1989, 2014; Pascho et al., 2002; Wiens, 2011),包括化疗药物和疫苗的应用、垂直传播阻断和卫生状况改变、饲养和生物安全措施等。

对*R. salmoninarum*具有抗菌活性的化疗药物曾取得部分成功应用(Pascho et al., 2002; Elliott et al., 2014)。大环内酯类抗生素红霉素是测试最广泛的抗菌剂,在美国仅作为一种美国食品药品监督管理局批准的临床新兽药(investigational new animal drug, INAD)使用(Wiens, 2011)。口服红霉素可使药物的组织浓度超过*R. salmoninarum*的最低抑菌浓度,从而降低BKD的死亡率(Pascho et al., 2002)。以每千克体重100 mg的剂量伴饲投喂硫氰酸红霉素连续28 d,是治疗严重*R. salmoninarum*腹腔注射感染的最有效方法(Wiens, 2011)。然而,抗生素无法完全消除病原体(Pascho et al., 2002),病菌有可能出现抗生素耐药性(Elliott et al., 2014)。

在大鳞大麻哈鱼成鱼产卵前,每隔21~30 d注射一次磷酸红霉素(每千克体重11 mg),可减少产卵前死亡率,但较高剂量可能引起黄疸和死亡(Wiens, 2011)。在产卵前9~60 d对雌鱼注射红霉素可使成熟鱼卵内达到治疗性抗生素水平(Haukenes and Moffitt, 2002; Pascho et al., 2002),并能预防实验条件下*R. salmoninarum*的垂直传播(Pascho et al., 2002),但此法在鱼类生产规模下的有效性尚未得到确认。鱼卵在红霉素中发生水硬化,无法持续降低处理后鱼卵孵化的幼苗中*R. salmoninarum*流行率(Pascho et al., 2002)。在这种情况下,快速浸沥出鱼卵中的抗生素可能会阻止与细菌的充分接触。

最近Elliott等(2014)对疫苗接种进行了综述。目前只有一种商业许可的BKD疫苗可用。该疫苗的商品名为Renogen®(Elanco Animal Health,礼来动保),是一种含非致病性环境细菌*Arthrobacter davidanieli*(拟命名)的冻干制剂,已在美国、加拿大和智利获准销售,但欧洲和日本未获许可。在对幼鲑腹腔注射前,需将疫苗培养物在无菌生理盐水中重新悬浮,疫苗制造商建议鱼群接种至少400度日后才能暴露于*R. salmoninarum*。*R. salmoninarum*和节杆菌之间的紧密亲缘关系很可能是Renogen®展现跨种保护的基础(Wiens et al., 2008)。该疫苗的实验室和田间试验表明,接种了节杆菌活疫苗的大西洋鲑幼鱼具有显著的抵御BKD能力,但在实验中受到*R. salmoninarum*感染的大鳞大麻哈鱼幼鱼的抵御能力有限或没有抵御能力(Elliott et al., 2014)。Renogen®疫苗接种后抵御BKD的持续时间尚不清楚,但据报道,在两家自然暴发过BKD的商业渔场中,接种Renogen®疫苗的大西洋鲑在23和27个月后的存活率明显高于对照鱼群。不过，这些鱼可能分别早在 BKD 造成死亡之前，即接种疫苗 21 和 19 个月后就已暴露于 *R. salmoninarum*（Elliott et al., 2014）。

大多数其他 BKD 疫苗的开发集中于热失活或甲醛灭活的鲑肾杆菌全细胞或裂解菌苗，或含有减毒 *R. salmoninarum* 分离株的活疫苗（Elliott et al., 2014）。这其中，通过热处理 *R. salmoninarum* 细胞或筛选低水平表达 MSA 的菌株实现细菌细胞表面 MSA 抗原减少的疫苗更有前景。一种减少 MSA 的疫苗显示了较低但可检测到的治疗效果，使已感染 *R. salmoninarum* 的大鳞大麻哈鱼在接种后的存活率高于未接种的感染鱼。与之相反，在腹腔注射接种的大西洋鲑中，具有正常细胞相关 MSA 的 *R. salmoninarum* 减毒活疫苗比 MSA 减少的减毒株提供了更好（但不完全）的保护力。

由于缺少完全有效的疫苗和化学治疗药物，已经实施了其他的 BKD 治理手段。在 BKD 流行地区，淘汰或隔离受感染鱼类的卵用于减少垂直感染的影响。成功的扑杀或隔离程序被普遍应用于养殖中（Gudmundsdóttir et al., 2000），或者最近一种多抗 ELISA（Pascho et al., 1991; Gudmundsdóttir et al., 2000; Meyers et al., 2003; Munson et al., 2010; Faisal et al., 2012）也被用于筛查产卵期雌鱼的肾组织。*R. salmoninarum* 检测阳性的雌鱼所产鱼卵通常被淘汰。然而，在产卵鱼数量少或者 *R. salmoninarum* 患病率高的种群中，来自阳性雌鱼（或者 ELISA 检测显示中高水平的抗原水平的雌鱼）的鱼卵批可能会被隔离孵化，而不会被淘汰。或者，可以上调淘汰卵批的 ELISA 吸光度（光密度）阈值，以包括被判断为垂直传播风险较低的阳性雌鱼（Munson et al., 2010）。

持续的亲鱼筛查和淘汰计划已大大降低了 BKD 的流行率和严重性，以及孵育期间鱼的总体死亡率，并降低洄游期成年鲑子代中 *R. salmoninarum* 感染的患病率和水平（Gudmundsdóttir et al., 2000; Meyers et al., 2003; Munson et al., 2010; Faisal et al., 2012）。有证据表明，与中高 *R. salmoninarum* 含量的成鱼后代相比，低或检测不到 *R. salmoninarum* 的鲑后代在下游迁徙和入海过程中的存活率更高，同时成鱼的回归数也更多（Wiens, 2011）。此外，这类计划可以降低对抗生素化疗的需求，并减少为弥补预期损失所必需的产卵成鱼和幼鱼的数量（Munson et al., 2010）。

虽然亲鱼淘汰计划成功实施，但对受感染雌鱼的选择性剔除可能导致一些不良的遗传变化，例如后代对 BKD 更易感（Munson et al., 2010）。Hard 等（2006）报道说，大鳞大麻哈鱼子代对 *R. salmoninarum* 感染的抗性与亲鱼体内 *R. salmoninarum* 抗原水平没有遗传相关性，这表明产卵时成鱼体内抗原水平反映的是环境而非遗传来源的变异。然而，无法确定 Hard 等（2006）的研究结果是否可以扩展到水产养殖的其他鲑鱼种类和环境背景。

养殖人员采用包括改进生物安全措施、卫生状况和饲养方法等其他策略以减少病菌的垂直和水平传播，并减少可能加剧 BKD 恶化的胁迫性饲育条件。一般实施的措施包括对预产卵期成鱼注射抗生素、对未受精鱼卵表面消毒、每个饲育养殖单元使用单独的网和刷、养殖单元或养殖设施之间设置足底消毒池或鞋底消毒垫、定期对养殖单元进行清洁消毒、应激事件时审慎使用预防性口服抗生素、隔离不同鱼龄鱼群以及养殖单元休耕、使用单通水而不是循环水、在室内或有顶盖的单元内养鱼（Maule et al.，1996；Gudmundsdóttir et al.，2000；Meyers et al.，2003；Munson et al.，2010；Faisal et al.，2012；Murray et al.，2012）。

由于病原传播的双重模式、野生或逃脱的养殖鲑鱼感染的存在，以及难以长期检测的亚临床感染，从鱼类养殖设施中根除 *R. salmoninarum* 是困难的（Murray et al.，2012；Peeler and Otte，2016）。从水源中清除野生鱼类是值得考虑的策略（Meyers et al.，2003；Murray et al.，2012）。由于 *R. salmoninarum* 在池塘或网箱之间传播的可能性，生产周期之间实施养殖场休耕是最基本的要求（Murray et al.，2012）。高度生物安全的陆基养殖是个例外，鱼类被饲养在室内水槽中，而独立水槽可降低养殖密度和进行消毒（Murray et al.，2012）。尽管如此，如果在恢复养殖时引进的鱼卵或鱼导致 *R. salmoninarum* 再次引入或者野生鱼传播源仍然存在，即使是养殖级别的休耕也可能失效（Murray et al.，2012）。在英国，动物流行病学和成本效益分析已引导 BKD 防控管理计划的目标从根除病原体向控制病原传播转变（Murray et al.，2012；Hall et al.，2014；Peeler and Otte，2016）。Murray 等（2012）的研究结论是，根据鱼类种类和地理区域划分 BKD 治理策略可达到更有效的 BKD 防控目标。Hall 等（2014）的研究表明，在大西洋鲑养殖中，通过检测 *R. salmoninarum* 感染的养殖场来限制扩散，以及除非 *R. salmoninarum* 被成功消除，否则限制从这些养殖场转移鱼类，这样的 BKD 控制策略比其他更严格或更宽松的策略更具成本效益。

21.6 总结与研究展望

BKD 仍是导致鲑科鱼类死亡和发病的重要原因。对多种鱼类完全有效的预防和治疗方法仍未商业化。为实现对 BKD 管控策略的进一步改进，需要更深入地了解鲑肾杆菌的传播、发病机制和宿主反应。

基因组测序和分析的最新进展为致病机制提供了新观点，这可能对疫苗开发有益（Elliott et al.，2014）。例如，*R. salmoninarum* 有两个铁摄取系统，其中一个涉及铁载体的产生（Bethke et al.，2016）。相关商业化疫苗已被开发出来以减少病原菌在哺乳动物中定植（Cull et al.，2012）；其设计策略就是基于铁载体的免疫进而限制铁营养的获取；鲑肾杆菌疫苗的开发也可考虑采用类似的策略。

另有研究已鉴定了 *R. salmoninarum* 中推定编码转肽酶和转肽酶底物的基因(Wiens et al., 2008)。在革兰氏阳性菌中,转肽酶及其表面蛋白底物对黏附宿主细胞、定植及逃避宿主免疫反应至关重要。实验证实,接种体外生产的重组转肽酶或转肽酶底物的小鼠对链球菌属(*Streptococcus*)和葡萄球菌属(*Staphylococcus*)致病菌株的感染可产生免疫保护作用(Elliott et al., 2014)。药物研发工作发现,*R. salmoninarum* 转肽酶也可能成为转肽酶抑制剂的治疗靶点,这可能被证实是通过防止体内细菌黏附和侵入宿主细胞而发挥作用(Wiens, 2011)。

新一代基因组测序的结果可用于 BKD 区域化管理(Murray et al., 2012),以限制疾病的蔓延。来自不同 *R. salmoninarum* 分离株的 SNP 分析数据可以推测全球 *R. salmoniarum* 传播模式,以及分离株在养殖场内和养殖场间的种内和种间传播方式(Brynildsrud et al., 2014)。这些数据,以及今后大量来自更多不同地理区域分离株的数据,可用于评估 *R. salmoninarum* 在不同 BKD 管理方案下传播的潜在源头和风险。

新一代测序技术的另一个潜在应用是研究宿主应答,以帮助开发改良疫苗和其他 BKD 管控策略。全转录组鸟枪法测序(RNA-seq)用于检测和量化指定时间内细胞内的整套基因转录物,这一技术经常揭示出不同条件下基因表达谱的变化(Wang et al., 2009)。之前的研究已经发现了某些可能与 *R. salmoninarum* 免疫有关的宿主基因表达的变化(Wiens, 2011)。然而,在涉及不同 BKD 疫苗、强毒株和减毒株,或鲑种群 BKD 易感性差异等多方面的研究中,RNA-seq 技术评价可以提供对 BKD 抗性十分重要的更完整的宿主基因定位。

最后,为了更准确地监测鱼类种群中的亚临床 *R. salmoninarum* 感染,需要进一步改进诊断方法,从而更好地确定治疗时机或其他 BKD 管控策略,并评估这些策略的成效。非致死取样方法是监测濒危鲑种群或珍贵亲鱼的首选方法。例如,Elliott 等(2015)证实,在大鳞大麻哈鱼幼鱼的 *R. salmoninarum* 检测中使用表面黏液样本进行 qPCR 分析是取代致死性肾脏采样的最适选择。Shulz (2014)在另一项实验研究中称,当联合使用 nPCR 和多克隆 ELISA 检测时,尿液或粪便样本显示出非致死性评价 *R. salmoninarum* 流行率和严重性的良好潜力。虽然这些研究结果很有前景,我们仍需对鱼的自然感染种群及其他鲑科鱼种类做进一步的研究。

注　意

书中提及的商标、公司或组织名称仅供参考和方便读者了解相关信息,并不构成美国政府对任何产品或服务的官方认可或批准,而排除其他可能适用产品或服务。

参考文献

[1] SMITH I W. The occurrence and pathology of Dee disease [J]. Freshwater Salmon Fish Res, 1964, 34: 1-12.

[2] FRYER J, SANDERS J. Bacterial kidney disease of salmonid fish [J]. Annual Reviews in Microbiology, 1981, 35(1): 273-298.

[3] AUSTIN B, AUSTIN D A. Bacterial fish pathogens: disease in farmed and wild fish [M]. Chichester PO19 1EB: Ellis Horwood Ltd, 1987.

[4] ELLIOTT D G, PASCHO R J, BULLOCK G L. Developments in the control of bacterial kidney disease of salmonid fishes [J]. Diseases of Aquatic Organisms, 1989, 6 (3): 201-215.

[5] EVELYN T. Bacterial kidney disease - BKD [M] //INGLIS V, ROBERTS R J, BROMAGE N R. Bacterial diseases of fish. New York: Halsted Press. 1993: 177-195.

[6] EVENDEN A, GRAYSON T, GILPIN M, *et al*. *Renibacterium salmoninarum* and bacterial kidney disease—the unfinished jigsaw [J]. Annual Review of Fish Diseases, 1993, 3: 87-104.

[7] FRYER J, LANNAN C. The history and current status of *Renibacterium salmoninarum*, the causative agent of bacterial kidney disease in Pacific salmon [J]. Fisheries Research, 1993, 17(1-2): 15-33.

[8] PASCHO R J, ELLIOTT D G, CHASE D M. Comparison of traditional and molecular methods for detection of *Renibacterium salmoninarum* [M] //CUNNINGHAM C O. Molecular diagnosis of salmonid diseases. Springer. 2002: 157-209.

[9] WIENS G D. Bacterial kidney disease (*Renibacterium salmoninarum*) [J]. Fish diseases and disorders, 2011, 3: 338-374.

[10] KRISTMUNDSSON Á, ÁRNASON F, GUDMUNDSDÓTTIR S, *et al*. Levels of *Renibacterium salmoninarum* antigens in resident and anadromous salmonids in the River Ellidaár system in Iceland [J]. Journal of fish diseases, 2016, 39(6): 681-692.

[11] ELLIOTT D G, WIENS G D, HAMMELL K L, *et al*. Vaccination against bacterial kidney disease [M]//GUDDING R, LILLEHAUG A, EVENSEN O. Fish vaccination. Chichester/Oxford: Wiley Blackwell. 2014: 255-272.

[12] SERVICE U F A W. National Wild Fish Health Survey Database [M]. Washington, DC: US Fish and Wildlife Service 2011.

[13] BALFRY S K, ALBRIGHT L J, TPT E. Horizontal transfer of *Renibacterium salmoninarum* among farmed salmonids via the fecal-oral route [J]. Diseases of Aquatic Organisms, 1996, 25(1-2): 63-69.

[14] ELLIOTT D G, MCKIBBEN C L, CONWAY C M, *et al*. Testing of candidate non-lethal sampling methods for detection of *Renibacterium salmoninarum* in juvenile Chinook salmon *Oncorhynchus tshawytscha* [J]. Diseases of aquatic organisms, 2015, 114(1): 21-43.

[15] MURRAY A G, MUNRO L A, WALLACE I S, *et al*. Epidemiology of *Renibacterium salmoninarum* in Scotland and the potential for compartmentalised management of salmon and trout farming areas [J]. Aquaculture, 2012, 324: 1-13.

[16] SANDERS J, FRYER J L. Renibacterium *salmoninarum* gen. nov., sp. nov., the causative agent of bacterial kidney disease in salmonid fishes [J]. International Journal of Systematic and Evolutionary Microbiology, 1980, 30(2): 496-502.

[17] WIENS G D, ROCKEY D D, ZAINING W, *et al*. Genome sequence of the fish pathogen *Renibacterium salmoninarum* suggests reductive evolution away from an environmental *Arthrobacter* ancestor [J]. Journal of Bacteriology, 2008, 190(21): 6970.

[18] BRYNILDSRUD O, FEIL E J, BOHLIN J, *et al.* Microevolution of *Renibacterium salmoninarum*: evidence for intercontinental dissemination associated with fish movements [J]. Isme Journal, 2014, 8(4): 746.
[19] MESA M G, POE T P, MAULE A G, *et al.* Vulnerability to predation and physiological stress responses in juvenile chinook salmon (*Oncorhynchus* tshawytscha) experimentally infected with Renibacterium salmoninarum [J]. Canadian Journal of Fisheries & Aquatic Sciences, 1998, 55(55): 1599 - 1606.
[20] MUNSON A, ELLIOTT D, KEITHJOHNSON. Management of Bacterial Kidney Disease in Chinook Salmon Hatcheries Based on Broodstock Testing by Enzyme-Linked Immunosorbent Assay: A Multiyear Study [J]. North American Journal of Fisheries Management, 2010, 30(4): 940 - 955.
[21] BRUNO D W, NOGUERA P A, POPPE T T. A colour atlas of salmonid diseases [M]. Springer Science & Business Media, 2013.
[22] ELLIOTT D, APPLEGATE L, MURRAY A, *et al.* Bench-top validation testing of selected immunological and molecular *Renibacterium salmoninarum* diagnostic assays by comparison with quantitative bacteriological culture [J]. Journal of fish diseases, 2013, 36(9): 779 - 809.
[23] PURCELL M K, GETCHELL R G, MCCLURE1 C A, *et al.* Quantitative polymerase chain reaction (PCR) for detection of aquatic animal pathogens in a diagnostic laboratory setting [J]. Journal of Aquatic Animal Health, 2011, 23(3): 148 - 161.
[24] FAISAL M, EISSA A E, STARLIPER C E. Recovery of *Renibacterium salmoninarum* from naturally infected salmonine stocks in Michigan using a modified culture protocol [J]. Journal of Advanced Research, 2010, 1(1): 95 - 102.
[25] OIE. Chapter 2.1.11. Bacterial kidney disease (*Renibacterium salmoninarum*).
[26] [M]. Manual of Diagnostic Tests for Aquatic Animals. Paris, France; World Organisation for Animal Health. 2003: 167 - 184.
[27] AFS - FHS. Blue Book: Suggested Procedures for the Detection and Identification of Certain Finfish and Shellfish Pathogens, 2014 edn. American Fisheries Society-Fish Health Section, Bethesda, Maryland. Available at: http://www.afs-fhs.org/bluebook/bluebook-index.php (accessed 5 January 2016). [M]. 2014.
[28] JANSSON E, HONGSLO T, HöGLUND J, *et al.* Comparative evaluation of bacterial culture and two ELISA techniques for the detection of *Renibacterium* salmoninarum antigens in salmonid kidney tissues [J]. Diseases of aquatic organisms, 1996, 27(3): 197 - 206.
[29] ELLIOTT D G. Unpublised data [M].
[30] MAULE A G, RONDORF D W, BEEMAN J, *et al.* Incidence of *Renibacterium salmoninarum* infections in juvenile hatchery spring chinook salmon in the Columbia and Snake Rivers [J]. Journal of Aquatic Animal Health, 1996, 8(1): 37 - 46.
[31] ELLIOTT D G, PASCHO R J, JACKSON L M, *et al. Renibaeterium salmoninarum* in Spring - Summer Chinook Salmon Smolts at Dams on the Columbia and Snake Rivers [J]. Journal of Aquatic Animal Health, 1997, 9(2): 114 - 126.
[32] MEYERS T R, KORN D, GLASS K, *et al.* Retrospective Analysis of Antigen Prevalences of *Renibacterium salmoninarum* (Rs) Detected by Enzyme-Linked Immunosorbent Assay in Alaskan Pacific Salmon and Trout from 1988 to 2000 and Management of Rs in Hatchery Chinook and Coho Salmon [J]. Journal of Aquatic Animal Health, 2003, 15(2): 101 - 110.
[33] FAISAL M, LOCH T P, BRENDEN T O, *et al.* Assessment of *Renibacterium salmoninarum* infections in four lake whitefish (*Coregonus clupeaformis*) stocks from northern Lakes Huron and Michigan [J]. Journal of Great Lakes Research, 2010, 36(Jul 2010): 29 - 37.

[34] FAISAL M, SCHULZ C, EISSA A, *et al*. Epidemiological investigation of *Renibacterium salmoninarum* in three *Oncorhynchus* spp. in Michigan from 2001 to 2010 [J]. Preventive Veterinary Medicine, 2012, 107(3-4): 260-274.
[35] BRUNO D. Histopathology of bacterial kidney disease in laboratory infected rainbow trout, *Salmo gairdneri* Richardson, and Atlantic salmon, *Salmo* salar L., with reference to naturally infected fish [J]. Journal of fish diseases, 1986, 9(6): 523-537.
[36] FERGUSON H W. Systemic pathology of fish. A text and atlas of comparative tissue responses in diseases of teleosts [M]. Iowa State University Press, 1989.
[37] SPEARE D J, OSTLAND V, FERGUSON H. Pathology associated with meningoencephalitis during bacterial kidney disease of salmonids [J]. Research in veterinary science, 1993, 54(1): 25-31.
[38] SPEARE D J. Differences in patterns of meningoencephalitis due to bacterial kidney disease in farmed Atlantic and chinook salmon [J]. Research in veterinary science, 1997, 62(1): 79-80.
[39] FLAñO E, LóPEZ-FIERRO P, RAZQUIN B, *et al*. Histopathology of the renal and splenic haemopoietic tissues of coho salmon *Oncorhynchus kisutch* experimentally infected with *Renibacterium salmoninarum* [J]. Diseases of aquatic organisms, 1996, 24(2): 107-115.
[40] MESA M G, MAULE A G, SCHRECK C B. Interaction of infection with *Renibacterium salmoninarum* and physical stress in juvenile chinook salmon: physiological responses, disease progression, and mortality [J]. Transactions of the American Fisheries Society, 2000, 129(1): 158-173.
[41] SANDELL T, TEEL D J, FISHER J, *et al*. Infections by *Renibacterium salmoninarum* and *Nanophyetus salmincola* Chapin are associated with reduced growth of juvenile Chinook salmon, *Oncorhynchus tshawytscha* (Walbaum), in the Northeast Pacific Ocean [J]. Journal of fish diseases, 2015, 38(4): 365-378.
[42] MESA M G, MAULE A G, POE T P, *et al*. Influence of bacterial kidney disease on smoltification in salmonids: is it a case of double jeopardy? [J]. Aquaculture, 1999, 174(1-2): 25-41.
[43] FREDRIKSEN Å, ENDRESEN C, WERGELAND H I. Immunosuppressive effect of a low molecular weight surface protein from *Renibacterium salmoninarum* on lymphocytes from Atlantic salmon (*Salmo salar* L.) [J]. Fish & Shellfish Immunology, 1997, 7(4): 273-282.
[44] HAUKENES A H, MOFFITT C M. Hatchery evaluation of erythromycin phosphate injections in prespawning spring Chinook salmon [J]. North American Journal of Aquaculture, 2002, 64(3): 167-174.
[45] GUðMUNDSDóTTIR S U, HELGASON S U, SIGURJóNSDóTTIR H S, *et al*. Measures applied to control *Renibacterium salmoninarum* infection in Atlantic salmon: a retrospective study of two sea ranches in Iceland [J]. Aquaculture, 2000, 186(3): 193-203.
[46] PASCHO R J, ELLIOTT D G, STREUFERT J M. Brood stock segregation of spring chinook salmon *Oncorhynchus tshawytscha* by use of the enzyme-linked immunosorbent assay (ELISA) and the fluorescent antibody technique (FAT) affects the prevalence and levels of *Renibacterium salmoninarum* infection in proogeny [J]. Diseases of Aquatic Organisms, 1991, 12(1): 25-40.
[47] HARD J J, ELLIOTT D G, PASCHO R J, *et al*. Genetic effects of ELISA-based segregation for control of bacterial kidney disease in Chinook salmon (*Oncorhynchus tshawytscha*) [J]. Canadian Journal of Fisheries & Aquatic Sciences, 2006, 63(12): 2793-2808.
[48] PEELER E, OTTE M. Epidemiology and economics support decisions about freedom

from aquatic animal disease [J]. Transboundary and emerging diseases, 2016, 63(3): 266 - 277.
[49] HALL M, SOJE J, KILBURN R, *et al*. Cost-effectiveness of alternative disease management policies for Bacterial kidney disease in Atlantic salmon aquaculture [J]. Aquaculture, 2014, 434: 88 - 92.
[50] BETHKE J, POBLETE-MORALES M, IRGANG R, *et al*. Iron acquisition and siderophore production in the fish pathogen *Renibacterium salmoninarum* [J]. Journal of Fish Diseases, 2016, 39(11): 1275 - 1283.
[51] CULL C A, PADDOCK Z D, NAGARAJA T G, *et al*. Efficacy of a vaccine and a direct-fed microbial against fecal shedding of *Escherichia coli* O157: H7 in a randomized pen-level field trial of commercial feedlot cattle [J]. Vaccine, 2012, 30(43): 6210 -6215.
[52] WANG Z, GERSTEIN M, SNYDER M. RNA-Seq: a revolutionary tool for transcriptomics [J]. Nature reviews genetics, 2009, 10(1): 57.
[53] SCHULZ C A. Factors and pitfalls influencing the detection of bacterial kidney disease [M]. Michigan State University, 2014.

22

海豚链球菌和无乳链球菌

Craig A. Shoemaker*，De-Hai Xu 和 Esteban Soto

22.1 引言

海豚链球菌（*Streptococcus iniae*）和无乳链球菌（*Streptococcus agalactiae*）是养殖鱼类和野生鱼类的革兰氏阳性致病菌，已有多篇文献进行了专门综述（Agnew and Barnes，2007；Klesius et al.，2008；Plumb and Hanson，2010；Salati，2011）。两种病原菌都有人畜共患的风险（Weinstein et al.，1997；Delannoy et al.，2013）。海豚链球菌可感染接触或处理过活鱼的免疫缺陷病人（Gauthier，2015）。对鱼类 *S. agalactiae* 分离株的比较基因组分析表明，在鱼类、青蛙和水生哺乳动物中也存在人类 *S. agalactiae*，从而构成人类疾病的潜在风险（Delannoy et al.，2013；Liu et al.，2013）。在 20 世纪 90 年代末到 21 世纪的头十年，*S. iniae* 是养殖鱼类的主要致病菌（Shoemaker et al.，2001；Agnew and Barnes，2007）。现在，*S. agalactiae* 已成为亚洲、拉丁美洲和南美洲养殖罗非鱼（*Oreochromis* spp.）的主要病原菌（Suanyuk et al.，2008；Mian et al.，2009；Sheehan et al.，2009；Chen et al.，2012；Zamri-Saad et al.，2014）。每年这些病菌在全世界造成的经济损失最初估计至少有 1 亿美元（Shoemaker et al.，2001）。仅中国就占全球罗非鱼产量的 40%（约 30 亿美元），而中国生产者报告称因 *S. agalactiae* 造成了 30%～80%的养殖损失（Ye et al.，2011；Chen et al.，2012）。假设年平均损失 40%，仅在中国就要损失约 10 亿美元的收入。

22.1.1 细菌概述

S. iniae 是一种兰氏血清分类法无法分群的链球菌，Pier 和 Madin（1976）最早在亚马逊河豚（*Inia geoffrensis*）中发现并命名该菌。*S. agalactiae* 是一种兰氏 B 群链球菌，最早在野生鱼类中被发现报道（Robinson and Meyer，1966），随后被称为难辨链球菌（*S. difficile*）（Eldar et al.，1994）。Vandamme 等

* 通信作者邮箱：craig.shoemaker@ars.usda.gov。

(1997)将 *S. difficile* 重新分类为非溶血性Ⅰb型无乳链球菌。根据表型特征和荚膜分型,*S. agalactiae* 被分为两种生物型(Sheehan et al., 2009)。根据基于多位点序列分析(Evans et al., 2008; Delannoy et al., 2013)和基因组测序确定的鱼类特异和相关基因(Rosinski-Chupin et al., 2013; Delannoy et al., 2016),证实了*S. agalactiae*包含两种谱系组。使用 Imperi 等(2010)所述引物,Shoemaker 等(未发表的数据)证实了 *S. agalactiae* 有三种荚膜分子类型(图22.1)。表 22.1 列出了 *S. iniae* 和 2 种 *S. agalactiae* 谱系组的典型生化特征、荚膜类型和常见多位点序列类型。

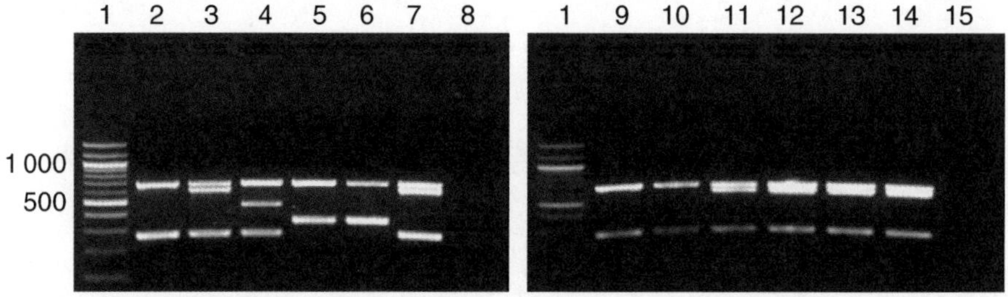

图 22.1 使用改进后 Imperi 所述多重 PCR 法(Imperi et al., 2010)获得凝胶图谱确定鱼类无乳链球菌荚膜分型。美国菌种保藏中心(ATCC)的菌株为实验对照。泳道编号标注在图片顶部,不同 *cps* (荚膜多糖)基因的碱基对数量标注在图片左侧。泳道 1: 100 bp 的分子标记;泳道 2: ATCC 12400,Ⅰa 型(688 bp *cpsL* 和 272 bp *cpsG*);泳道 3: ATCC 51487,Ⅰb 型(688 bp *cpsL*,621 bp *cpsJ* 和 272 bp *cpsG*);泳道 4: ATCC 13813,Ⅱ型(688bp *cpsL*, 465 bp *cpsJ* 和 272 bp *cpsG*);泳道 5: ATCC 31475,Ⅲ型(688 bp *cpsL* 和 352 bp *cpsG*);泳道 6: 马来西亚分离株(Ⅲ型);泳道 7: 墨西哥分离株(John Plumb,Ⅰb 型);泳道 8: 阴性对照;泳道 9: 科威特分离株(KU-MU-11B,Ⅰa 型);泳道 10: 巴西分离株(04-ARS-BZ-TN-002,Ⅰa 型);泳道 11: 洪都拉斯分离株(LADL-05-108A,Ⅰb 型);泳道 12: 巴西分离株(ARS-BZ-TN-004,Ⅰb 型);泳道 13: 洪都拉斯分离株(TN-HON-08A,Ⅰb 型);泳道 14: 厄瓜多尔分离株(8 Br,Ⅰb 型);泳道 15: 海豚链球菌(ARS-98-60)。*cpsL* 的 688 bp 条带可用于诊断无乳链球菌

表 22.1 鱼源海豚链球菌和无乳链球菌特性

特　　性	*S. iniae*	菌种/菌株	
		S. agalactiae (Ⅰb)	*S. agalactiae* (Ⅰa 或Ⅲ)
精氨酸双水解酶(ADH)	+(可变)	+	+
荚膜血清型	Ⅰ, Ⅱ[a]	Ⅰb[b]	Ⅰa[b], Ⅲ[b]
谱系组	NA[c]	552[d]	7[d, e, f]
37 ℃下生长状况	+(可变)	+(弱)	+
溶血性	β[g]	无	β
马尿酸盐水解	—	+	+
兰氏分群	无	B群	B群

续 表

特 性	*S. iniae*	菌种/菌株	
		S. agalactiae（Ⅰb）	*S. agalactiae*（Ⅰa 或Ⅲ）
多位点序列类型	NA	257[h]，258[h]，259[h]，260[h, j]，261[h, j]	7[e, h, j]，283[i]，500[i]，491[i]
吡咯烷酮芳基酰胺酶(PYR)试验	+	—	—
淀粉水解能力	+	—	—

[a]已描述了两种血清型(Barnes et al.，2003)。
[b]根据使用 Imperi 等(2010)所述引物，以及荚膜分型抗血清(日本 Denka Seiken Co.，Ltd.)(Shoemaker and Xu，未发表)进行的荚膜分子分型。
[c]NA 表示不适用。
[d]见 Delannoy et al.，2016。
[e]见 Kayansamruaj et al.，2015。
[f]见 Rosinski-Chupin et al.，2013。
[g]血琼脂类型可影响血溶性，添加 5%羊血的胰酶大豆琼脂可见 α、β、γ 三种血溶性(Chou et al.，2014)。
[h]见 Evans et al.，2008。
[i]见 Delannoy et al.，2013。

22.1.2 传播方式

链球菌通过水进行水平传播，而水中新引进的带菌鱼是传染源(Eldar et al.，1994；Shoemaker et al.，2001；Tavares et al.，2016)。这些病原菌可在养殖场附近的水和沉积物中持续存活超过一年(Nguyen et al.，2002)。当用感染的死鱼投喂健康鱼时会发生粪口传播(Robinson and Meyer，1966；Kim et al.，2007)。Iregui 等(2016)研究证实，灌服 *S. agalactiae* 可通过胃肠道上皮侵入罗非鱼体内，导致败血症。有趣的是，鞍带石斑鱼(*Epinephelus lanceolatus*)在摄食掺有*S. agalactiae* 的饲料后变得厌食和嗜睡，并可在鱼的内脏中分离出 *S. agalactiae*，但未发生鱼的死亡(Delamare-Deboutteville et al.，2015)。Shoemaker 等(2000)观察到，蚕食受感染同类的内脏和眼球的健康鱼会传播 *S. iniae* 并引起死亡。但是不能排除鼻孔、皮肤和鱼鳃介导的其他病原传播途径(McNulty et al.，2003)。不管怎样，由于死鱼和濒死鱼会脱落散播病原菌，应优先清除这些鱼。

海豚链球菌和无乳链球菌的实验感染是经水体细菌暴露的方式诱发的，但实验结果有时并不稳定：实验感染死亡率通常较低(6%～20%)，而且在高密度下饲养的鱼需擦伤才能被感染(Rasheed and Plumb，1984；Chang and Plumb，1996；Shoemaker et al.，2000；Mian et al.，2009；Soto et al.，2015)。大多数研究采用腹腔注射来实现科赫法则和实验室条件下的疾病再现。注射感染重现性好，产生的临床疾病与野外条件下观察到的相似。此外，IP 注射Ⅰb 型*S. agalactiae*分离株对罗非鱼(Mian et al.，2009；Evans et al.，2015；Delannoy et al.，2016)和鞍带石斑鱼(Delamare-Deboutteville et al.，2015)的半数致死量(LD_{50})极低(小于 1×10^2 CFU/mL)。注射(IP 注射或 IM 注射)可绕过先天免疫系统，特别是皮肤和黏膜隔室。LaFrentz 等(2016)最近将未感染的尼罗罗非鱼和腹腔注射了 *S. iniae* 的罗非鱼养殖在一起，实验导致 8%～10%的未感染组尼罗罗非鱼死亡，而 IP 注射组的鱼有 60%～70%的死亡率。在一个饲养超过 2 300 尾鱼(176 g±50 g)的大型鱼槽中重复了该实验，结果显示 6%的同栖鱼死亡，而注射鱼死亡率为 60%。分别将三种剂量(2×10^3 CFU/mL，2×10^4 CFU/mL，2×10^5 CFU/mL)的Ⅰb 型*S. agalactiae* 通过肌肉各注射入每组 25 尾剪过鱼鳍的尼罗罗非鱼，随后将这些鱼和 25 尾未感染处理的鱼混养在一起(每种剂量对应一槽鱼，每槽 50 条鱼)(Shoemaker et al.，未发表的数据)，得到了与海豚链球菌实验相似的结果，同栖鱼的死亡率为 12%～20%，而肌内注射鱼的死亡率为 76%～88%(图 22.2)。

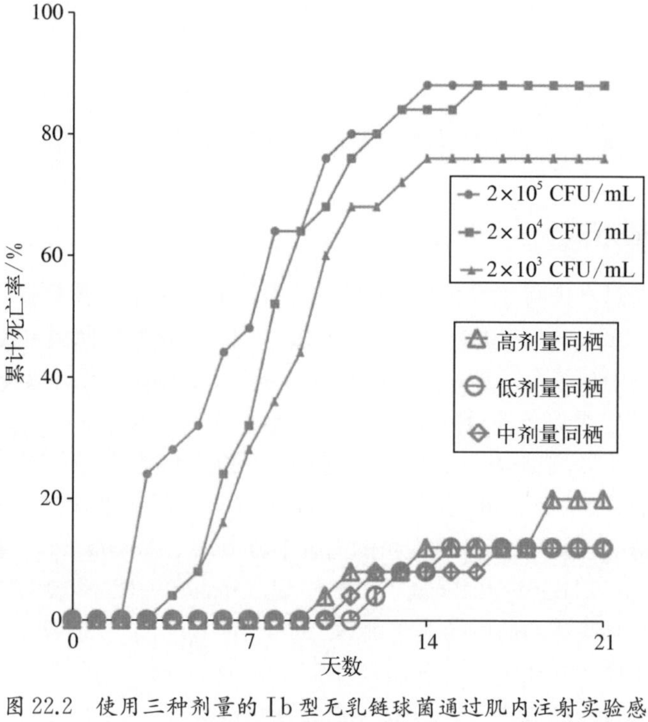

图 22.2 使用三种剂量的Ⅰb 型无乳链球菌通过肌内注射实验感染的尼罗罗非鱼与同栖幼鱼的死亡率。同栖鱼死亡但死亡率较低，说明注射感染鱼体内的病原菌通过水体和粪口途径传播至同栖幼鱼

有研究认为 *S. iniae* 和 *S. agalactiae* 应该均可进行垂直传播，因为使用环介导等温扩增(LAMP)法在罗非鱼受精卵和子代中均检测到这两种病原菌(Suebseing et al.，2013)。人工催生罗非鱼亲鱼后，使用 LAMP 法可在未受精的卵子(70%)、鱼精液(90%)和幼体(40%)中检测出这两种病原菌(Pradeep et al., 2016)。垂直传播的可能性使 *S. iniae* 和 *S. agalactiae* 的防控变得困难。

22.1.3 地理分布

S. iniae 和 *S. agalactiae* 分布于世界各地，可感染包括罗非鱼在内的 27 种以上鱼类(Klesius et al., 2008; Sheehan et al., 2009)。这两种病原菌可影响淡水、半咸水和海水中的野生或养殖鱼类(Bowater et al., 2012; Chou et al., 2014; Keirstead et al., 2014)。

22.2 诊断

22.2.1 临床症状

临床症状(图 22.3)包括食欲减退、皮肤变黑，聚集在槽(池)底、嗜睡、鱼体弯曲以及鳍基部和鳃盖出血(Shoemaker et al., 2006b)。有些受感染鱼会在水面下不规律地进行原地打转或旋转游动。最明显的临床体征为单侧或双侧眼球突出[图 22.3(A)(B)]、腹胀(Amal and Zamri-Saad, 2011)及体腔炎[图 22.3

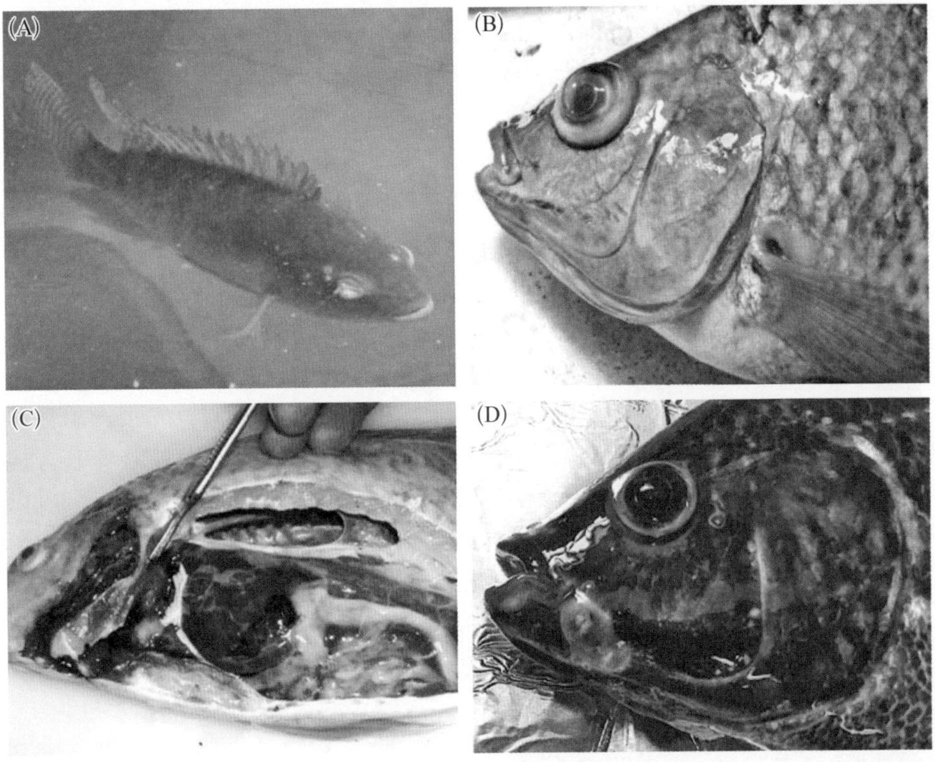

图 22.3 自然感染(A)(B)(C)及实验感染(D)链球菌的性成熟罗非鱼的大体病理表现：(A)(B) 眼球突起及眼混浊；(C) 严重的体腔炎伴组织粘连；(D) 嘴底部出血性皮肤脓疱。图(A)(B)(C)中的照片由 Juan A. Morales 博士提供

(C)],然而,这些症状不能作为链球菌病的确诊病征。通常,眼球突出及相关的眼混浊在感染后期出现,这表明与慢性疾病存在关联(Shoemaker et al., 2006b)。LaFrentz 等(2016)研究发现,在死亡和幸存的感染 *S. iniae* 的尼罗罗非鱼下颌和尾部出现脓疱[图 22.3(D)]。在 *S. agalactiae* 感染中也可见相似的病症,伴有鱼颊侧麻痹(Tavares et al., 2016)和肛门处挤压出粪管或粪便串(Pasnik et al., 2009)。尸检 *S. iniae* 感染的鱼,发现体腔内有血性液体,脾肿大发红,肝苍白肿大,以及心和肾发炎(Amal and Zamri-Saad, 2011)。在某些情况下,感染鱼死前并未表现出明显临床症状,死亡归因于脑和神经系统感染的败血症(Barham et al., 1979)。

22.2.2 感染诊断

链球菌的感染诊断依赖于 5%羊血琼脂平板(Remel, Lenexa, Kansas)的细菌培养。新鲜鱼,特别是其肾脏和大脑,是培养病原菌的最好样本材料。研究人员也会使用干拭子法(Shoemaker et al., 2001)和商用拭子(如 Remel BactiSwab)(Keirstead et al., 2014)进行样本采集。在纯化培养中分离出过氧化氢酶阴性的革兰氏阳性菌之后可以做出推定诊断。如分子技术不可用,利用细菌生化特性(Pier and Madin, 1976; Vandamme et al., 1997)和使用商用抗血清(Denka Seiken, Japan)进行荚膜分型(如 *S. agalactiae*)可以确诊。小型快速系统通常很有效,使用 API 20 Strep 和 API rapid ID 32 Strep 鉴定测试条(bioMérieux, Durham, North Carolina)可以很容易鉴定出 *S. agalactiae*。许多研究者都报道了鉴定 *S. iniae* 的典型 API 20 Strep 特征谱(Chou et al., 2014)。使用基于扩增基因重复区域而开发的引物(Chou et al., 2014),分子生物学可进一步强化鉴定过程。在 *S. agalactiae* 的鉴定中,就使用了多种技术完成荚膜分子分型(Imperi et al., 2010;图 22.1)。22.2.3 节中有对近期 PCR 技术的更详尽介绍。

使用选择性培养基有助于链球菌的分离。Ellner 等(1966)首先报道了含动物源蛋白胨、酪蛋白酶解物和去纤维蛋白绵羊血的哥伦比亚血琼脂培养基。这种培养基通过添加萘啶酸和多黏菌素(哥伦比亚 CNA 琼脂)被进一步改进,以抑制大多数革兰氏阴性细菌的生长,提高链球菌恢复生长的能力。另有两种用于 *S. iniae* 的选择性培养基,分别是乙酸铊-噁喹酸血琼脂(Nguyen et al., 2002)和添加有乙酸铊、黏杆菌素和噁喹酸的 Todd-Hewitt 肉汤培养基。Nguyen 等(2002)证实了以上两种培养基从水中和沉积物中分离 *S. iniae* 的有效性。

22.2.3 分子诊断

S. iniae 和 *S. agalactiae* 的分子鉴定主要基于 PCR 扩增 16S rRNA 基因

或 16S－23 S 基因间隔序列（intergenic spacer regions，IGS），并进行序列分析确认（Berridge et al.，1998；Chou et al.，2014；Soto et al.，2015）。Mata 等（2004）开发了一种具有诊断潜力的基于乳酸氧化酶基因（*lctO*）的种特异性 PCR 方法，可检测 32～61 个 *S. iniae* 细胞。生化测试、常规 PCR 和相关 16S rRNA 基因测序等鉴定过程都非常耗时，且依赖于细菌分离。实时定量 PCR（qPCR）可定量检测鱼类样本中的 *S. agalactiae*。Sebastião 等（2015）设计了一种基于 *cfb* 基因的 qPCR 方法，*cfb* 基因编码 *S. agalactiae* 特异性 CAMP 因子（一种细胞外蛋白），但 *S. iniae* 和 *S. agalactiae* 中均有扩增，因此，这两种病原菌在同一样品中存在的可能性使这种检测方法的应用存在问题。针对 *S. agalactiae*，Su 等（2016）以 *cpsE*（荚膜多糖编码基因）、*sip*（表面免疫原性蛋白编码基因）和 16S－23S rRNA 的 IGS 为模板，开发了一种使用 3 对不同基因引物的 qPCR 检测方法。基于 *cpsE* 和 *sip* 的引物可能与其他细菌或鱼类 DNA 发生交叉反应，但扩增 16S－23S IGS 的引物可特异性检测约 9 个拷贝的无乳链球菌 DNA。有研究对感染鱼组织内的细菌进行定量分析，并评估了痊愈期罗非鱼体内细菌的持留状况（Su et al.，2016）。Tavares 等（2016）在罗非鱼上采用一种非致死取样技术来验证 Su 等（2016）开发的检测方法；抽吸肾样本和静脉穿刺结合 qPCR 技术检测 *S. agalactiae* 非常有效。

LAMP 法在低技术条件下非常实用。使用对应 *S. iniae* 和 *S. agalactiae* 的三组引物和嗜热脂肪芽孢杆菌（*Bacillus stearothermophilus*，*Bst*）DNA 聚合酶，LAMP 可在恒温下（60～65 ℃）扩增靶序列。如果没有电泳设备，在反应管中添加钙黄绿素等显色试剂，可使阳性反应可视化。Suebsing 等（2013）证实了基于超氧化物歧化酶基因（*sodA*）引物的种特异性 LAMP 检测方法可用于 *S. iniae* 和 *S. agalactiae* 的检测。该技术已被用于评估亲鱼、配子和鱼苗中 *S. iniae* 和 *S. agalactiae* 携带状态（Pradeep et al.，2016）。

22.3 病理学

图 22.4（A）～（D）和图 22.5（A）～（D）展示了 *S. iniae* 感染敏尾笛鲷（*Ocyurus chrysurus*）和高首鲟以及 Ⅰb 型 *S. agalactiae* 感染尼罗罗非鱼和大底鳉（*Fundulus grandis*）的相关组织病理学变化。罗非鱼感染链球菌会导致脑膜炎、多浆膜炎、脾炎、卵巢炎和心肌炎等病症（Chang and Plumb，1996）。心脏心包炎的病理特征在于巨噬细胞、淋巴细胞和红细胞的浸润。巨噬细胞也会浸润动脉球和心室。病菌存在于心脏心外膜和动脉球外膜（Miyazaki et al.，1984）。组织切片显示充满细菌的坏死巨噬细胞浸润和纤维蛋白沉淀（Miyazaki et al.，1984）。

感染肝组织的病理组织学展现为肝实质结构丧失、肝坏死、嗜酸性粒状炎性

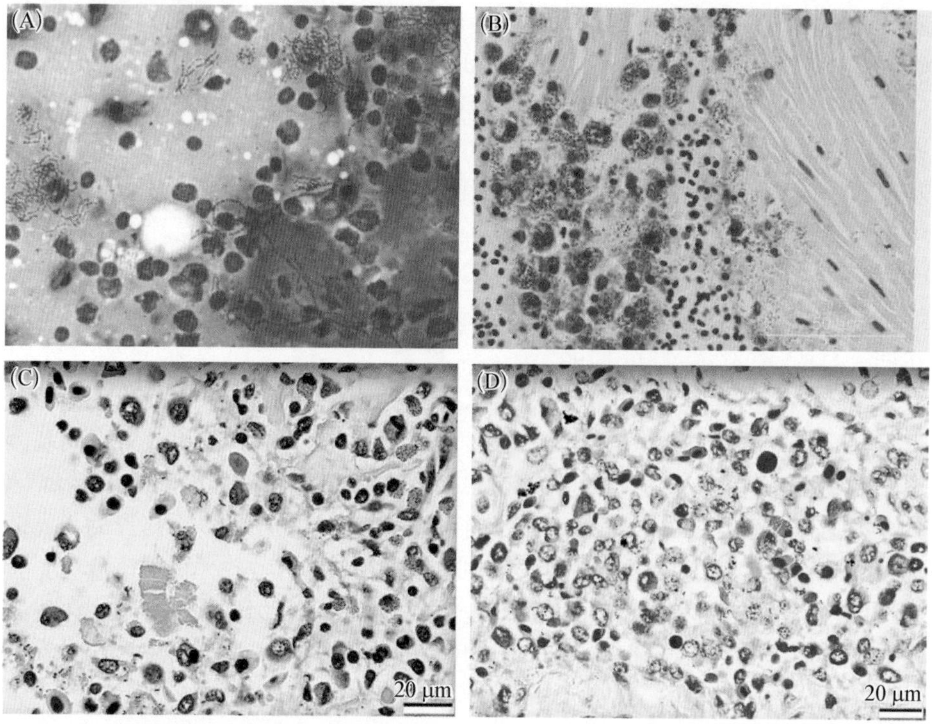

图 22.4　鱼组织中海豚链球菌感染引起的组织病理变化：(A) 细菌感染的敏尾笛鲷肝组织伴有溶解白细胞，说明曾发生过吞噬作用。本图使用瑞氏染色，100 倍放大；(B) 感染敏尾笛鲷的心肌纤维呈分散状，大量革兰氏阳性球菌和双球菌存活于巨噬细胞中或形成胞外克隆，与其他多种物质构成了炎性渗透液使心肌组织扭曲变形。本图使用革兰氏 Brown-Brenn 染色使组织切片中细菌差异染色；(C) (D) 高首鲟感染肾脏组织的病变主要被限制在间质造血区，包括分散的单个到小群细胞的坏死和多见于动脉周的更大片的坏死区的产生(较少见)、不定性、灰白嗜酸性、纤维样物质的堆积物，以及少数革兰氏阳性球菌成对或成链状广泛分散在间质区域。图(C) 使用革兰氏染色法，图(D) 使用吉姆萨染色法。图(A)由 MaryAnna Thrall 博士提供，图(B)由 Natalie Keirstead 博士提供，图(C)(D)由 Al Camus 博士提供

细胞浸润及严重淤血(Abuseliana et al.，2011)。*S. agalactiae* 感染的红罗非鱼脾脏则表现出肉芽肿、单核细胞浸润、红髓变性及含铁血黄素沉着等组织病理症状(Abuseliana et al.，2011)。此外，脾血管明显充血。利用透射电子显微镜(TEM)检测，Iregui 等(2016)发现经口感染 *S. agalactiae* 的罗非鱼肠上皮细胞顶面附着有分裂的细菌。这些细菌在上皮细胞细胞质内大量增殖。

感染鱼脑血管充血、血管壁扩张脱落。在端脑和小脑脑膜出血区域，巨噬细胞与淋巴细胞和成纤维细胞混杂在一起。眼部感染导致晶状体纤维断裂分离，形成空泡或者晶体皮质成纤维细胞增生(Chang and Plumb，1996)。Mayazaki 等(1984)也报道了细菌在角膜、眼眶脂肪和动眼肌中的播散，导致渗出性或肉芽肿性炎症、坏死和出血。

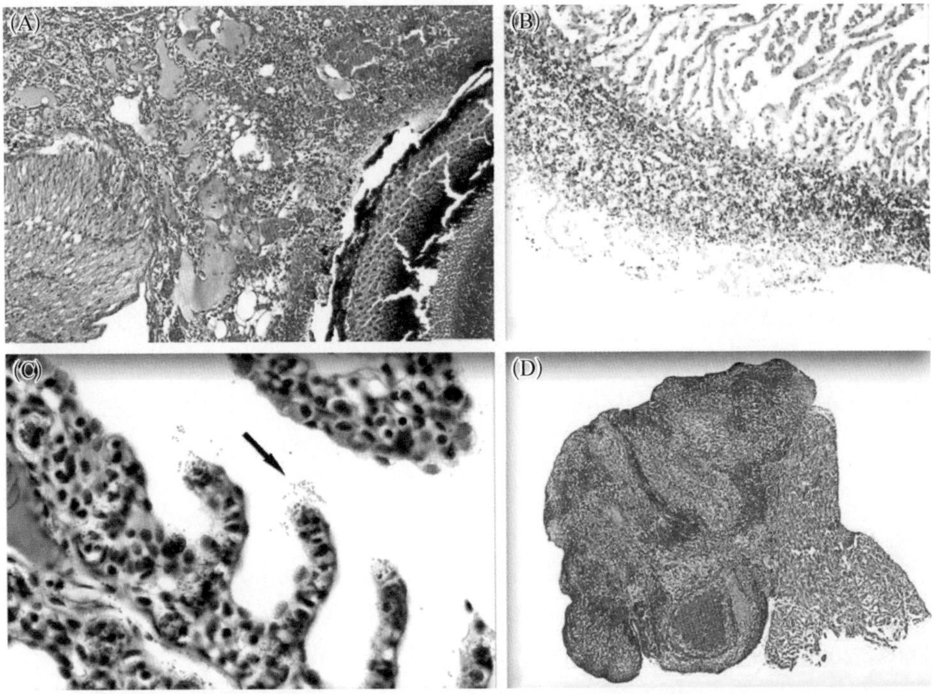

图 22.5 鱼组织中海豚链球菌感染引起的组织病理变化：(A) 体腔注射处理后的实验感染尼罗罗非鱼出现严重的肉芽肿脉络膜炎(本图使用苏木素-伊红染色法,20 倍放大)；(B) 自然感染尼罗罗非鱼出现严重的心包炎(本图使用苏木素-伊红染色法,20 倍放大)；(C) 体腔注射无乳链球菌后大底鳉的层状毛细血管和纤维状间质中的球状细菌导致层状顶端坏死及细菌释放(箭头所示)(本图使用苏木素-伊红染色法,100 倍放大)；(D) 体腔注射无乳链球菌后尼罗罗非鱼动脉球出现肉芽肿心外膜炎(本图使用苏木素-伊红染色法,10 倍放大)。图(A)由 Michele Dennis 博士提供,图(B)由 Juan A. Morales 博士提供,图(C)由 Wes Baumgartner 博士提供,图(D)由 Oscar Illanes 博士提供

随着感染的进程,感染的精巢和卵巢会出现渗出性或肉芽肿性炎症(Miyazaki et al., 1984)。精巢渗出性炎症表现出严重的细菌增殖、坏死、毛细血管充血、水肿以及满载细菌的巨噬细胞和中性粒细胞的浸润。

Suanyuk 等(2010)表示,在尖吻鲈的肝、胰腺、心、眼和脑部发现的组织病理学变化比红罗非鱼更严重。肝组织呈现出肝窦扩张及淋巴细胞浸润。肝细胞出现严重的空泡化、变性及局灶性坏死。胰腺组织则会发生腺泡细胞变性和酶原颗粒丢失。脑部感染区域周围有淋巴细胞浸润。

S. iniae 也会在养殖虹鳟中引发疾病(Akhlaghi and Mahjor, 2004; Lahav et al., 2004)。Lahav 等(2004)从以色列北部鳟鱼养殖场收集了因遭受Ⅱ型 *S. iniae*感染而大批死亡的鱼样本。Ⅱ型 *S. iniae* 造成的组织病理学与Ⅰ型 *S. iniae*中常见的不同。感染了Ⅰ型 *S. iniae* 的鱼骨骼肌未见损伤,但Ⅱ型

S. iniae 感染可呈现多灶性至弥散性变性和坏死，并伴有异嗜性组织细胞浸润。

在中国广西多个斑点叉尾鮰养殖场发生的一起 *S. iniae* 病害暴发造成了急性败血症和鱼类死亡(Chen et al.，2011)。对内脏的组织学检查发现有重型肝炎以及充血的血管周围存在坏死肝细胞，而其他病理变化与罗非鱼相似。

罗非鱼感染 *S. iniae* 和 *S. agalactiae* 后的病理变化包括心包炎、心外膜炎、心肌炎、心内膜炎和脑膜炎(Chen et al.，2007)。肾脏中发生淋巴组织增生，伴有肾小管细胞胞质中嗜酸性物质聚集，导致细胞核被挤至一侧。*S. agalactiae*感染中，肾小管解体，链球菌环绕肾小管并包含在间质细胞和肾小球中。此外，病菌遍布脾脏和头肾。在 *S. agalactiae* 感染期间，大量的球菌出现在病鱼的组织和体循环中，但这并非 *S. iniae* 感染的显著特征。

22.4 病理生理学

养殖密度高、管理不良、极端水温、高氨、低溶氧、pH 大于 8、盐度(Zamri-Saad et al.，2014)和体外寄生虫(如车轮虫、三代虫、小瓜虫感染)(Xu et al.，2007，2009)等环境胁迫可导致病理生理改变和链球菌病加重。

在低(5.6 g/L)、中(11.2 g/L)、高(22.4 g/L)三种养殖密度下，有研究评估了养殖密度对罗非鱼感染 *S. iniae* 的死亡率影响(Shoemaker et al.，2000)。当养殖密度大于 11.2 g/L 时，*S. iniae* 的感染死亡率显著提高。相较于低密度养殖组 4.8%的死亡率，中、高密度养殖组的死亡率分别为 28.4%和 25.6%。

Rodkhum 等(2011)将Ⅰa 型 *S. agalactiae* 以 10^8 CFU/mL 的浸泡浓度感染罗非鱼，并在 25 ℃、30 ℃或 33 ℃下养殖一周以评估温度对病死率的影响。研究发现，33 ℃下养殖的鱼死亡率最高(60%)，而 30 ℃实验组的死亡率为 40%，25 ℃下养殖的鱼存活下来并没有出现临床症状。这些结果表明，高温(>30 ℃)养殖的罗非鱼更容易感染Ⅰa 型 *S. agalactiae*。养殖场反馈的报告也显示，超过 30 ℃的养殖温度会导致 *S. agalactiae* 相关的商品规格罗非鱼的严重死亡(Mian et al.，2009；Noraini et al.，2013)。

Ndong 等(2007)评估了莫桑比克罗非鱼(*O. mossambicus*)在不同温度下对 *S. iniae* 的易感性。从 27 ℃的水温移至 19 ℃和 35 ℃ 时，注射感染 *S. iniae* 的鱼的累计死亡率分别为 57%和 50%，明显高于转移并保持在 23 ℃ (20%)、27 ℃(10%)和 31 ℃ (17%) 下鱼的累计死亡率。此外从 27 ℃转移到 19 ℃或 35 ℃时，鱼的白细胞水平和呼吸爆发下降，吞噬活性与指数也会降低。这些结果表明，27 ℃上下浮动 8 ℃时会造成免疫抑制并引发 *S. iniae* 的病死率上升。

Evans 等(2003)比较了亚致死溶氧(*DO*≤1 mg/L)和正常溶氧(*DO*≥4 mg/L)条件下罗非鱼的血糖和死亡率情况。亚致死溶氧条件下，罗非鱼经 IP 注射 *S. agalactiae*后的血糖水平(>100 mg/dL)和死亡率(80%)显著升高。在养殖鱼类或野生鱼类中，有害藻华、藻华死亡、水体富营养化或物理分层可能造成亚致死溶氧状况（Evans et al.，2003；Amal et al.，2013）。

Amal 等（2013）研究发现，水体氨浓度与河流湖泊养殖红罗非鱼中 *S. agalactiae* 的存在呈正相关。*S. agalactiae* 引起的野生鲻鱼（*Liza klunzingeri*）大规模死亡与氨氮浓度从 0.11 mg/L 上升至 0.33 mg/L 有关（Glibert et al.，2002）。通过血糖水平升高的检测，Evans 等（2006）指出，暴露于 0.37 mg/L 的未解离氨 24 h 会引起尼罗罗非鱼快速且短暂的应激反应，但不会增加它们对 *S. agalactiae* 的易感性。

Perera 等（1997）证实，在碱性水中（pH＞8），杂交罗非鱼（*O. nilotica* × *O. aurea*）在感染 *S. iniae* 后比养殖在酸性或中性（pH 为 6～7）水中的鱼的死亡率更高。Milud 等（2013）发现，与水体盐度为 5 ppt 和 10 ppt 相比，水体盐度为 15 ppt 时红罗非鱼对 *S. agalactiae* 的易感性增加。

体外寄生虫也增加了鱼类对 *S. iniae* 和 *S. agalactiae* 的易感性。破损皮肤、损伤和溃疡是细菌入侵和随后疾病表现的公认感染途径（Xu et al.，2007）。Xu 等（2007）的研究证明，被尼罗三代虫（*Gyrodactylus niloticus*）寄生的罗非鱼感染 *S. iniae* 后的死亡率（42.2%）明显高于未被寄生的罗非鱼（6.7%）。同时，Xu 等（2009）还报道了小瓜虫（*Ichthyophthirius multifiliis*，Ich）和海豚链球菌合并感染尼罗罗非鱼后的死亡率增加。无乳链球菌或小瓜虫单独感染大致可造成 20%的死亡率，但合并感染的鱼死亡率高达 88%。

22.5 防控策略

对海豚链球菌和无乳链球菌的防控应纳入鱼类健康管理计划（Plumb and Hanson，2010），这些计划依赖健全的鱼类饲养，包括生物安全、水质调控、合理营养等健康措施。

通过成活率和生产性能平衡养殖密度可以实现罗非鱼养殖的高产（Zamri-Saad et al.，2014）。当死亡率增加时，减少养殖密度可减轻鱼群应激压力和病原载量，养殖者必须平衡放养率以限制水质差和疾病传播加剧而引起的疾病风险，从而最大限度地提高产量（Shoemaker et al.，2000）。

在将亲鱼或鱼卵引入新的或现有的育苗设施时，应采取严格的预防措施。消毒感染了 *S. iniae* 和 *S. agalactiae* 的鱼卵十分困难。可用于食用鱼鱼卵的化学制剂只有一些表面消毒剂，可减少卵壳上的病菌含量，但对鱼卵内病菌的清除效果有限（Salvesen and Vadstein，1995）。因此，鱼卵和鱼苗应从无病原种源获取。

22.5.1 益生菌、益生元和共生学

Shelby 等（2006）发现商用益生菌（Biomate SF－20、Bioplus 2B、Bactocell

PA10 MD、Levucell SB 20)不能增强鱼群先天免疫力或对 *S. iniae* 感染的存活率。有研究单独或混合使用枯草芽孢杆菌(*Bacillus subtilis*)和嗜酸乳酸杆菌(*Lacotobacillus acidophilus*)两种益生菌饲喂罗非鱼(Aly et al., 2008)。饲喂益生菌后，鱼的先天免疫参数增加，但对 *S. iniae* 的抗性在连续饲喂嗜酸乳酸杆菌 2 个月后才显现出来。Iwashita 等(2015)以每千克饲料混有 10 g 混合益生菌[枯草芽孢杆菌、米曲霉(*Aspergillus oryzae*)和酿酒酵母(*Saccharomyces cerevisiae*)]的剂量饲喂尼罗罗非鱼，对 *S. iniae* 攻毒的存活率从 4%提升至 34%。饲喂乳酸链球菌(*Lactococcus lactis*) (BFE920)、芽孢杆菌(*Bacillus* sp.) (IS－2)和其他芽孢杆菌(*Bacillus* spp.)可提升牙鲆的先天免疫力和对 *S. iniae* 的抵抗力(Cha et al., 2013; Jang et al., 2013; Kim et al., 2013)。

Widanarni 和 Tanbiyaskur(2015)用芽孢杆菌 NP5、甘薯寡糖或者它们的组合物饲喂尼罗罗非鱼，*S. agalactiae* 感染后的存活率分别为 74%、74%和 85%，而对照组存活率仅为 18.5%。Agung 等分别将芽孢杆菌 NP5、甘露寡糖及两者组合微胶囊化，连续饲喂罗非鱼上述微胶囊 40 d 后采用 *S. agalactiae* 攻毒并评估了防病效果。益生菌饲喂组的存活率(55%～60%)明显好于对照组(33%)(Agung et al., 2015)。其他可使用的益生菌包括鼠李糖乳杆菌 GG(*Lactobacillus rhamnosus*, LGG)、枯草芽孢杆菌、地衣芽孢杆菌(*B. licheniformis*)以及芽孢杆菌与小球菌(*Pediococcus* sp.)的组合菌(Ng et al., 2014; Pirarat et al., 2015)。对 LGG 的研究显示，藻酸盐基质改善了乳杆菌在刺激性胃条件下的体外存活率以及在尼罗罗非鱼体内的存活率(胶囊化乳杆菌存活率为 50%，而对照组仅为 12%)(Pirarat et al., 2015)。Ng 等(2014)使用商用或原型益生菌开展了为期 8 周的饲喂实验研究，益生菌包括 *B. subtilis*、*B. licheniformis* 以及 *Bacillus* sp.和 *Pediococcus* sp.的组合菌。除了 *B. licheniformis*，饲喂组鱼对 *S. agalactiae* 的存活率都得到了提升，并认为 *B. subtilis* 的效果最好(Ng et al., 2014)。

22.5.2 疫苗与接种

目前针对 *S. iniae* 和 *S. agalactiae* 的免疫预防主要依靠灭活疫苗产品(Klesius et al., 2006; Sheehan et al., 2009)。Merck Animal Health 生产两种 *S. iniae* 甲醛灭活疫苗产品(www.merck-animal-health.com)。Norvax® Strep Si 是一款甲醛灭活疫苗，已通过 Merck Animal Health 对亚洲海鲈(尖吻鲈)的安全性和效力测试。在 Merck 公司的实验室试验中，将 50 尾 5 g 的尖吻鲈按 Norvax® Strep Si 推荐剂量浸泡接种，并在 3 周后用 *S. iniae* 浸泡攻毒，测试结果显示接种鱼的 RPS 为 68%。在养殖场中，体重为 20 g 的鱼注射接种 6 周后感染 *S. iniae*，其 RPS 达到了 100%。Aquavac-Garvetil™ 是一种包含 *S. iniae* 和 *L. garviae*(格氏乳球菌)抗原的灭活疫苗，通过浸泡免疫或口服加强免疫。

Merck 公司的田间试验显示，在链球菌病自然暴发后罗非鱼的 RPS 分别为 15%（仅一次浸泡免疫）和 50%（浸泡免疫追加口服免疫）。这些疫苗并未在所有国家获得使用许可，因此渔业养殖者应咨询当地所辖监管机构。疫苗的保护作用归因于相应抗体的产生（LaFrentz et al., 2011），但也涉及其他免疫机制（Allen and Neely, 2012; Aviles et al., 2013）。以色列（Eyngor et al., 2008）和澳大利亚（Millard et al., 2012）出现的新血清型 *S. iniae* 降低了疫苗效力，因此亟须研发包含不同血清型的多价疫苗。

美国已开发出针对海豚链球菌的改性胞外产物（ECP）疫苗（Klesius et al., 2000）。Shoemaker 等（2010）的研究指出，IP 注射接种 ECP 菌苗的性反转罗非鱼对多株异源分离株的攻毒产生了 RPS 为 79%～100%的保护效果。免疫印迹（western blot）分析证实的血清学反应表明，各分离株的抗原相对保守。此外，被动免疫和免疫蛋白组学进一步证实了疫苗接种诱发的抗体保护性（LaFrentz et al., 2011）。与其他病菌（Shoemaker et al., 2012）和寄生虫（Martins et al., 2011）的合并感染削弱了疫苗在养殖应用中的表现。

在一些地区已商业许可了两种无乳链球菌 IP 注射疫苗。Merck 公司生产的 AQUAVAC® Strep Sa 是一种油包水剂型的生物 2 型 *S. agalactiae* 灭活疫苗，以单剂量（0.05 mL）为体重大于 15 g 的鱼免疫接种。在实验室研究中，该疫苗的 RPS 为 70%～90%，在两家养殖场的田间试验中存活率分别提升了 13% 和 1%（www.merck-animal-health.com）。AlphaJect® micro1 TiLa 是 Pharmaq 公司生产的类似疫苗产品，据产品说明书所述其可产生大于等于 60%的 RPS。该疫苗仅限在哥斯达黎加、巴拿马和洪都拉斯出售（www.pharmaq.no）。Intervet/Schering-Plough 公司生产了针对 *S. agalactiae* 生物 1 型（β-溶血性）和生物 2 型（非溶血性）的灭活疫苗，可产生生物型特异免疫保护（Sheehan et al., 2009）。Evans 等（2004）利用 β-溶血性 *S. agalactiae* Ⅰa 分离株（Evans et al., 2008）开发了一种 ECP 疫苗，IP 免疫后可有效保护罗非鱼。基于 10 种不同脉冲场凝胶电泳（PFGE）基因型分离株的不同组合，Chen 等（2012）开发了一种 *S. agalactiae* 多价灭活疫苗，针对不同野生分离株，该多价疫苗具有提供广泛保护作用的应用潜力。

一些研究报告了海豚链球菌（Shoemaker et al., 2006a）和无乳链球菌（Firdaus-Nawi et al., 2013; Nur-Nazifah et al., 2014）灭活疫苗的浸泡和口服接种有效性，但注射仍是最有效的接种途径。针对其他鱼类病原菌开发的活疫苗已通过浸泡和口服接种方式获得成功应用（Shoemaker and Klesius, 2014）。*S. iniae* 减毒活疫苗是通过转座子突变或等位基因交换缺失相关毒力基因而研制开发的（Buchanan et al., 2005; Lowe et al., 2007）。Locke 等（2010）采用注射和浸泡接种方式，在杂交条纹鲈（*Morone chrysops*×*M. saxatilis*）上比较了减毒疫苗菌株和 *S. iniae* 全细胞灭活疫苗的效力。结果显示，尽管减毒株仍保留一定毒力，但 Δ*simA*（表面 M 样蛋白）减毒突变株浸泡接种后产生的免疫效力最好(100%保护，但有一定的疫苗相关死亡)。其他浸泡评价的候选疫苗完全减毒，但 Δ*cpsD* 突变株（不会发生疫苗相关死亡）的 RPS 降低了 51%，而 Δ*pgmA*（葡糖磷酸变位酶）突变株的 RPS 降低了 26%，但仍优于灭活疫苗的 RPS(19%)。

通过在胰蛋白酶大豆肉汤培养基中的体外连续传代，筛选制备了一株 *S. agalactiae*减毒株(Li et al., 2015)。以荚膜Ⅰa 型 *S. agalactiae* HN016 (Chen et al., 2012)为出发菌株，Li 等(2015)将其改造成 YM001 菌株，该菌株具有更长的细胞链长(超过 5 个细胞)，血琼脂上为非溶血性表型，PFGE 特征与亲本菌株有差异。传代后菌株毒力降低且该菌株经 IP 注射、浸泡和口服接种后，对亲本毒株攻毒的保护效力得到证实(RPS 为 53%～96%)(Li et al., 2015)。

最近，有研究报道了一种编码 *S. agalactiae* 表面免疫原蛋白(sip)的 DNA 口服疫苗，可通过减毒鼠伤寒沙门氏菌给药(Huang et al., 2014)。研究者设计了重组质粒 pVAX1-sip，并将其转化进减毒沙门氏菌活菌载体中对罗非鱼进行口服免疫和保护。尽管该质粒没有整合到罗非鱼染色体中(Huang et al., 2014)，但该疫苗仍然归属于转基因生物(genetically modified organism, GMO)，在欧洲和美国，此类产品的公众接受度较低。

其他疫苗开发策略包括通过注射和口服接种的海豚链球菌(Aviles et al., 2013)和无乳链球菌(Nur-Nazifah et al., 2014)重组蛋白疫苗。Aviles 等基于 *S. iniae* 的保守表面 M 蛋白开发应用了一种 *S. iniae* 重组疫苗。该重组蛋白可诱导产生特异性抗体，但对亚洲海鲈(尖吻鲈)无保护作用。还有研究利用大肠杆菌表达的 *S. agalactiae* 细胞壁表面锚蛋白开发了一种饲喂剂型重组疫苗。罗非鱼饲喂该重组疫苗后的存活率可达 70%，但其他治疗方案中的鱼全部死亡(Nur-Nazifah et al., 2014)。

22.5.3 抗菌药物疗法

当其他策略无法维持鱼类健康时，谨慎的抗菌剂治疗是水产养殖者的一个重要工具。在美国，唯一批准用于控制 *S. iniae* 死亡率的抗生素是 AQUAFLOR® (氟苯尼考)。使用剂量为每日每千克鱼服用 15 mg，连续用药10 d，休药期为 15 d (Gaunt et al., 2010; Gaikowski et al., 2013)。AQUAFLOR®必须按照兽医指导进行药饵饲喂。据报道，马来西亚的养殖者使用红霉素和土霉素作为治疗和预防链球菌病的药物(Zamri-Saad et al., 2014)。许多国家(如美国)禁止预防性使用抗生素，全世界都在倡导抗菌剂的审慎使用。几乎没有使用抗生素控制 *S. agalactiae*死亡率的相关信息。

22.5.4 选择育种

促进鱼类生长和抗病能力的选择育种是有效鱼类健康管理的重要组成部分。LaFrentz 等(2016)首次开展了尼罗罗非鱼对 *S. iniae* 的抗性大规模测试，试验中对父系半同胞家系和母系全同胞家系进行攻毒感染，以确定鱼群 *S. iniae* 抗性的加性遗传变异潜力。结果发现罗非鱼具有较高的遗传力评估值(0.42±0.07)，这表明 *S. iniae* 抗性是可遗传的，可以通过家系选择进行遗传改良。

22.6 总结与研究展望

海豚链球菌和无乳链球菌是世界范围内野生和养殖鱼类的重要经济病原菌。在科学出版物中应鼓励对无乳链球菌的荚膜分型、生物型和基因型进行谨慎的研究报道。近期许多报道描述了无乳链球菌的研究，但未能对所使用菌株进行界定，这种界定能更好地对不同研究进行比较，还可加快免疫机制和潜在生物控制策略的科研发现。

目前缺乏对 *S. iniae*，Ⅰa 型、Ⅲ型和Ⅰb 型 *S. agalactiae*(不同生物型或基因型的无乳链球菌)之间组织病理学和发病机制的比较研究。由于 *S. agalactiae* 的鱼类特异性或鱼类相关基因型(Delannoy et al., 2016)已被确定，这一信息应该被纳入上述比较中。此外，马来西亚和泰国的最新研究报道了渔场同时发生 *S. iniae* 和 *S. agalactiae* 感染情况。因此有必要针对多种链球菌感染进行宿主-病原互作机制的实验研究。这些实验研究可能对设计鱼类链球菌病的有效控制和疫苗接种策略有直接意义。

致谢

本项工作得到了 USDA-ARS (CRIS Project No. 6010－32000－026－00D) 以及 Spring Genetics, Homestead, Florida, USA (MTA－CRADA No. 58－6010－6－005)项目的支持。本文提及的相关商标或商业产品仅为了提供具体信息，不代表美国农业部的推荐或认可。

参考文献

[1] AGNEW W, BARNES A C. *Streptococcus iniae*: an aquatic pathogen of global veterinary significance and a challenging candidate for reliable vaccination [J]. Veterinary microbiology, 2007, 122(1－2): 1－15.

[2] KLESIUS P, SHOEMAKER C, EVANS J. *Streptococcus*: a worldwide fish health problem; proceedings of the 8th international symposium on tilapia in aquaculture Cairo, F, 2008 [C].

[3] PLUMB J A, HANSON L A. Tilapia bacterial diseases [M].//PLUMB J A, HANSON

L A. Health maintenance and principal microbial diseases of cultured fishes. John Wiley & Sons. 2010：445－463.

[4] SALATI F. 10 *Enterococcus seriolicida* and *Streptococcus* spp.(*S. iniae*, *S. agalactiae* and *S. dysgalactiae*) [J]. Fish diseases and disorders, 2011, 3: 375.

[5] WEINSTEIN M R, LITT M, KERTESZ D A, *et al*. Invasive infections due to a fish pathogen, *Streptococcus iniae* [J]. New England Journal of Medicine, 1997, 337(9): 589－594.

[6] DELANNOY C M, CRUMLISH M, FONTAINE M C, *et al*. Human *Streptococcus agalactiae* strains in aquatic mammals and fish [J]. BMC microbiology, 2013, 13(1): 41.

[7] GAUTHIER D T. Bacterial zoonoses of fishes: a review and appraisal of evidence for linkages between fish and human infections [J]. The Veterinary Journal, 2015, 203(1): 27－35.

[8] LIU G, ZHANG W, LU C. Comparative genomics analysis of *Streptococcus agalactiae* reveals that isolates from cultured tilapia in China are closely related to the human strain A909 [J]. BMC genomics, 2013, 14(1): 775.

[9] SHOEMAKER C A, KLESIUS P H, EVANS J J. Prevalence of *Streptococcus iniae* in tilapia, hybrid striped bass, and channel catfish on commercial fish farms in the United States [J]. American journal of veterinary research, 2001, 62(2): 174－177.

[10] SUANYUK N, KONG F, KO D, *et al*. Occurrence of rare genotypes of *Streptococcus agalactiae* in cultured red tilapia *Oreochromis* sp. and Nile tilapia *O. niloticus* in Thailand—relationship to human isolates? [J]. Aquaculture, 2008, 284(1－4): 35－40.

[11] MIAN G, GODOY D, LEAL C, *et al*. Aspects of the natural history and virulence of *S. agalactiae* infection in Nile tilapia [J]. Veterinary microbiology, 2009, 136(1－2): 180－183.

[12] SHEEHAN B, LABRIE L, LEE Y, *et al*. Streptococcosis in tilapia-vaccination effective against main strep species [J]. Global aquaculture advocate, 2009.

[13] CHEN M, WANG R, LI L-P, *et al*. Screening vaccine candidate strains against *Streptococcus agalactiae* of tilapia based on PFGE genotype [J]. Vaccine, 2012, 30(42): 6088－6092.

[14] ZAMRI-SAAD M, AMAL M, SITI-ZAHRAH A, *et al*. Control and Prevention of Streptococcosis in Cultured Tilapia in Malaysia: A Review [J]. Pertanika Journal of Tropical Agricultural Science, 2014, 37(4).

[15] YE X, LI J, LU M, *et al*. Identification and molecular typing of *Streptococcus agalactiae* isolated from pond-cultured tilapia in China [J]. Fisheries Science, 2011, 77(4): 623－632.

[16] PIER G B, MADIN S H. *Streptococcus iniae* sp. nov., a beta-hemolytic streptococcus isolated from an Amazon freshwater dolphin, *Inia geoffrensis* [J]. International Journal of Systematic and Evolutionary Microbiology, 1976, 26(4): 545－553.

[17] ROBINSON J A, MEYER F P. Streptococcal fish pathogen [J]. Journal of Bacteriology, 1966, 92(2): 512.

[18] ELDAR A, BEJERANO Y, BERCOVIER H. *Streptococcus shiloi* and *streptococcus difficile*: Two new streptococcal species causing a meningoencephalitis in fish [J]. Current Microbiology, 1994, 28(3): 139－143.

[19] VANDAMME P, DEVRIESE L, POT B, *et al*. *Streptococcus difficile* is a nonhemolytic group B, type Ⅰb *Streptococcus* [J]. International Journal of Systematic and Evolutionary Microbiology, 1997, 47(1): 81－85.

[20] EVANS J J, BOHNSACK J F, KLESIUS P H, *et al*. Phylogenetic relationships among *Streptococcus agalactiae* isolated from piscine, dolphin, bovine and human sources: a dolphin and piscine lineage associated with a fish epidemic in Kuwait is also associated

with human neonatal infections in Japan [J]. Journal of medical microbiology, 2008, 57 (11): 1369 - 1376.
[21] ROSINSKI-CHUPIN I, SAUVAGE E, MAIREY B, *et al*. Reductive evolution in *Streptococcus agalactiae* and the emergence of a host adapted lineage [J]. BMC genomics, 2013, 14(1): 252.
[22] DELANNOY C M, ZADOKS R N, CRUMLISH M, *et al*. Genomic comparison of virulent and non-virulent *Streptococcus agalactiae* in fish [J]. Journal of fish diseases, 2016, 39(1): 13 - 29.
[23] IMPERI M, PATARACCHIA M, ALFARONE G, *et al*. A multiplex PCR assay for the direct identification of the capsular type (Ⅰa to Ⅸ) of *Streptococcus agalactiae* [J]. Journal of microbiological methods, 2010, 80(2): 212 - 214.
[24] SHOEMAKER C A. Unpublised data [M].
[25] BARNES A C, YOUNG F M, HORNE M T, *et al*. *Streptococcus iniae*: serological differences, presence of capsule and resistance to immune serum killing [J]. Diseases of aquatic organisms, 2003, 53(3): 241 - 247.
[26] SHOEMAKER C, XU D. unpublished [M].
[27] KAYANSAMRUAJ P, PIRARAT N, KONDO H, *et al*. Genomic comparison between pathogenic *Streptococcus agalactiae* isolated from Nile tilapia in Thailand and fish-derived ST7 strains [J]. Infection, Genetics and Evolution, 2015, 36: 307 - 314.
[28] CHOU L, GRIFFIN M J, FRAITES T, *et al*. Phenotypic and genotypic heterogeneity among *Streptococcus iniae* isolates recovered from cultured and wild fish in North America, Central America and the Caribbean islands [J]. Journal of aquatic animal health, 2014, 26(4): 263 - 271.
[29] TAVARES G C, DE ALCâNTARA COSTA F A, SANTOS R R D, *et al*. Nonlethal sampling methods for diagnosis of *Streptococcus agalactiae* infection in Nile tilapia, *Oreochromis niloticus* (L.) [J]. Aquaculture, 2016, 454: 237 - 242.
[30] NGUYEN H T, KANAI K, YOSHIKOSHI K. Ecological investigation of *Streptococcus iniae* in cultured Japanese flounder (*Paralichthys olivaceus*) using selective isolation procedures [J]. Aquaculture, 2002, 205(1 - 2): 7 - 17.
[31] KIM J H, GOMEZ D K, CHORESCA JR C H, *et al*. Detection of major bacterial and viral pathogens in trash fish used to feed cultured flounder in Korea [J]. Aquaculture, 2007, 272(1 - 4): 105 - 110.
[32] IREGUI C, COMAS J, VÁSQUEZ G, *et al*. Experimental early pathogenesis of *Streptococcus agalactiae* infection in red tilapia *Oreochromis* spp [J]. Journal of fish diseases, 2016, 39(2): 205 - 215.
[33] DELAMARE-DEBOUTTEVILLE J, BOWATER R, CONDON K, *et al*. Infection and pathology in Queensland grouper, *Epinephelus lanceolatus*, (Bloch), caused by exposure to *Streptococcus agalactiae* via different routes [J]. Journal of fish diseases, 2015, 38(12): 1021 - 1035.
[34] SHOEMAKER C A, EVANS J J, KLESIUS P H. Density and dose: factors affecting mortality of *Streptococcus iniae* infected tilapia (*Oreochromis niloticus*) [J]. Aquaculture, 2000, 188(3 - 4): 229 - 235.
[35] MCNULTY S T, KLESIUS P H, SHOEMAKER C A, *et al*. *Streptococcus iniae* infection and tissue distribution in hybrid striped bass (*Morone chrysops* × *Morone saxatilis*) following inoculation of the gills [J]. Aquaculture, 2003, 220 (1 - 4): 165 - 173.
[36] RASHEED V, PLUMB J A. Pathogenicity of a non-haemolytic group B *Streptococcus* sp. in gulf killifish (*Fundulus grandis* Baird and Girard) [J]. Aquaculture, 1984, 37 (2): 97 - 105.
[37] CHANG P A, PLUMB J. Histopathology of experimental *Streptococcus* sp. infection in

tilapia, *Oroochromis niloticus* (L.), and channel catfish, *Ictafurus punctatus* (Ratinesque) [J]. Journal of Fish Diseases, 1996, 19(3): 235 - 241.

[38] SOTO E, WANG R, WILES J, *et al*. Characterization of isolates of *Streptococcus agalactiae* from diseased farmed and wild marine fish from the US Gulf Coast, Latin America, and Thailand [J]. Journal of aquatic animal health, 2015, 27(2): 123 - 134.

[39] EVANS J J, PASNIK D J, KLESIUS P H. Differential pathogenicity of five *Streptococcus agalactiae* isolates of diverse geographic origin in N ile tilapia (*Oreochromis niloticus* L.) [J]. Aquaculture research, 2015, 46(10): 2374 - 2381.

[40] LAFRENTZ B R, LOZANO C A, SHOEMAKER C A, *et al*. Controlled challenge experiment demonstrates substantial additive genetic variation in resistance of Nile tilapia (*Oreochromis niloticus*) to *Streptococcus iniae* [J]. Aquaculture, 2016, 458: 134 - 139.

[41] SUEBSING R, KAMPEERA J, TOOKDEE B, *et al*. Evaluation of colorimetric loop-mediated isothermal amplification assay for visual detection of *Streptococcus agalactiae* and *Streptococcus iniae* in tilapia [J]. Letters in applied microbiology, 2013, 57(4): 317 - 324.

[42] PRADEEP P J, SUEBSING R, SIRTHAMMAJAK S, *et al*. Evidence of vertical transmission and tissue tropism of Streptococcosis from naturally infected red tilapia (*Oreochromis* spp.) [J]. Aquaculture Reports, 2016, 3: 58 - 66.

[43] BOWATER R, FORBES-FAULKNER J, ANDERSON I, *et al*. Natural outbreak of *Streptococcus agalactiae* (GBS) infection in wild giant Queensland grouper, *Epinephelus lanceolatus* (Bloch), and other wild fish in northern Queensland, Australia [J]. Journal of fish diseases, 2012, 35(3): 173 - 186.

[44] KEIRSTEAD N, BRAKE J, GRIFFIN M, *et al*. Fatal septicemia caused by the zoonotic bacterium *Streptococcus iniae* during an outbreak in Caribbean reef fish [J]. Veterinary pathology, 2014, 51(5): 1035 - 1041.

[45] SHOEMAKER C, XU D, EVANS J, *et al*. Parasites and diseases [J]. Tilapia: Biology, Culture and Nutrition, 2006: 561 - 582.

[46] AMAL M, ZAMRI-SAAD M. Streptococcosis in tilapia (*Oreochromis niloticus*): a review [J]. Pertanika J Trop Agric Sci, 2011, 34(2): 195 - 206.

[47] PASNIK D J, EVANS J J, KLESIUS P H. Fecal strings associated with *Streptococcus agalactiae* infection in Nile tilapia, *Oreochromis niloticus* [J]. Open veterinary science journal, 2009, 3: 6 - 8.

[48] BARHAM W, SCHOONBEE H, SMIT G. The occurrence of *Aeromonas* and *Streptococcus* in rainbow trout, *Salmo gairdneri* Richardson [J]. Journal of Fish Biology, 1979, 15(4): 457 - 460.

[49] ELLNER P D, STOESSEL C J, DRAKEFORD E, *et al*. New Culture Medium for Medical Bacteriology [J]. American journal of clinical pathology, 1966, 45(4): 502 - 504.

[50] BERRIDGE B R, FULLER J D, DE AZAVEDO J, *et al*. Development of specific nested oligonucleotide PCR primers for the *Streptococcus iniae* 16S - 23S ribosomal DNA intergenic spacer [J]. Journal of Clinical Microbiology, 1998, 36(9): 2778 - 2781.

[51] MATA A I, BLANCO M M, DOMíNGUEZ L, *et al*. Development of a PCR assay for *Streptococcus iniae* based on the lactate oxidase (*lctO*) gene with potential diagnostic value [J]. Veterinary microbiology, 2004, 101(2): 109 - 116.

[52] SEBASTIÃO F D A, LEMOS E G, PILARSKI F. Validation of absolute quantitative real-time PCR for the diagnosis of *Streptococcus agalactiae* in fish [J]. Journal of microbiological methods, 2015, 119: 168 - 175.

[53] SU Y L, FENG J, LI Y W, *et al*. Development of a quantitative PCR assay for monitoring *Streptococcus agalactiae* colonization and tissue tropism in experimentally infected tilapia [J]. Journal of fish diseases, 2016, 39(2): 229 - 238.

[54] MIYAZAKI T, KUBOTA S S, KAIGE N, *et al*. A histopathological study of streptococcal disease in tilapia [J]. Fish pathology, 1984, 19(3): 167 - 172.
[55] ABUSELIANA A F, DAUD H M, AZIZ S A, *et al*. Pathogenicity of *Streptococcus agalactiae* isolated from a fish farm in Selangor to juvenile red tilapia (*Oreochromis* sp.) [J]. Journal of Animal and Veterinary Advances, 2011, 10(7): 914 - 919.
[56] SUANYUK N, SUKKASAME N, TANMARK N, *et al*. *Streptococcus iniae* infection in cultured Asian sea bass (*Lates calcarifer*) and red tilapia (*Oreochromis* sp.) in southern Thailand [J]. Songklanakarin Journal of Science & Technology, 2010, 32(4).
[57] AKHLAGHI M, MAHJOR A. Some histopathological aspects of streptococcosis in cultured rainbow trout (*Oncorhynchus mykiss*) [J]. BULLETIN-EUROPEAN ASSOCIATION OF FISH PATHOLOGISTS, 2004, 24(3): 132 - 136.
[58] LAHAV D, EYNGOR M, HURVITZ A, *et al*. *Streptococcus iniae* type Ⅱ infections in rainbow trout *Oncorhynchus mykiss* [J]. Diseases of aquatic organisms, 2004, 62(1 - 2): 177 - 180.
[59] CHEN D-F, WANG K-Y, GENG Y, *et al*. Pathological changes in cultured channel catfish *Ictalurus punctatus* spontaneously infected with *Streptococcus iniae* [J]. Diseases of aquatic organisms, 2011, 95(3): 203 - 208.
[60] CHEN C, CHAO C, BOWSER P. Comparative histopathology of *Streptococcus iniae* and *Streptococcus agalactiae*-infected tilapia [J]. Bulletin-European Association of Fish Pathologists, 2007, 27(1): 2.
[61] XU D H, SHOEMAKER C, KLESIUS P. Evaluation of the link between gyrodactylosis and streptococcosis of Nile tilapia, *Oreochromis niloticus* (L.) [J]. Journal of Fish Diseases, 2007, 30(4): 233 - 238.
[62] RODKHUM C, KAYANSAMRUAJ P, PIRARAT N. Effect of water temperature on susceptibility to *Streptococcus agalactiae* serotype Ⅰa infection in Nile tilapia (*Oreochromis niloticus*) [J]. Thai J Vet Med, 2011, 41(3): 309 - 314.
[63] NORAINI O, JAHWARHAR N, SABRI M, *et al*. The effect of heat stress on clinicopathological changes and immunolocalization of antigens in experimental *Streptococcus agalactiae* infection in Red hybrid tilapia (*Oreochromis* spp.) [J]. Veterinary World, 2013, 6(12): 997.
[64] NDONG D, CHEN Y-Y, LIN Y-H, *et al*. The immune response of tilapia *Oreochromis mossambicus* and its susceptibility to *Streptococcus iniae* under stress in low and high temperatures [J]. Fish & Shellfish Immunology, 2007, 22(6): 686 - 694.
[65] EVANS J J, SHOEMAKER C A, KLESIUS P H. Effects of sublethal dissolved oxygen stress on blood glucose and susceptibility to *Streptococcus agalactiae* in Nile tilapia *Oreochromis niloticus* [J]. Journal of Aquatic Animal Health, 2003, 15(3): 202 - 208.
[66] AMAL M N A, SAAD M Z, ZAHRAH A S, *et al*. Water quality influences the presence of *Streptococcus agalactiae* in cage cultured red hybrid tilapia, *Oreochromis niloticus* × *Oreochromis mossambicus* [J]. Aquaculture Research, 2015, 46 (2): 313 - 323.
[67] GLIBERT P M, LANDSBERG J H, EVANS J J, *et al*. A fish kill of massive proportion in Kuwait Bay, Arabian Gulf, 2001: the roles of bacterial disease, harmful algae, and eutrophication [J]. Harmful Algae, 2002, 1(2): 215 - 231.
[68] EVANS J J, PASNIK D J, BRILL G C, *et al*. Un-ionized ammonia exposure in Nile tilapia: toxicity, stress response, and susceptibility to *Streptococcus agalactiae* [J]. North American Journal of Aquaculture, 2006, 68(1): 23 - 33.
[69] PERERA R P, JOHNSON S K, LEWIS D H. Epizootiological aspects of *Streptococcus iniae* affecting tilapia in Texas [J]. Aquaculture, 1997, 152(1 - 4): 25 - 33.
[70] ALSAID M, DAUD H, MOHAMED N, *et al*. Environmental factors influencing the susceptibility of red hybrid tilapia (*Orechromis* sp.) to *Streptococcus agalactiae* infection

[J]. Advanced Science Letters, 2013, 19(12): 3600 - 3604.
[71] XU D-H, SHOEMAKER C A, KLESIUS P H. Enhanced mortality in Nile tilapia *Oreochromis niloticus* following coinfections with ichthyophthiriasis and streptococcosis [J]. Diseases of aquatic organisms, 2009, 85(3): 187 - 192.
[72] SALVESEN I, VADSTEIN O. Surface disinfection of eggs from marine fish: evaluation of four chemicals [J]. Aquaculture International, 1995, 3(3): 155 - 171.
[73] SHELBY R A, LIM C, YILDIRIM-AKSOY M, *et al*. Effects of probiotic diet supplements on disease resistance and immune response of young Nile tilapia, *Oreochromis niloticus* [J]. Journal of Applied Aquaculture, 2006, 18(2): 23 - 34.
[74] ALY S M, AHMED Y A-G, GHAREEB A A-A, *et al*. Studies on *Bacillus subtilis* and *Lactobacillus acidophilus*, as potential probiotics, on the immune response and resistance of *Tilapia nilotica* (*Oreochromis niloticus*) to challenge infections [J]. Fish & shellfish immunology, 2008, 25(1 - 2): 128 - 136.
[75] IWASHITA M K P, NAKANDAKARE I B, TERHUNE J S, *et al*. Dietary supplementation with *Bacillus subtilis*, *Saccharomyces cerevisiae* and *Aspergillus oryzae* enhance immunity and disease resistance against *Aeromonas hydrophila* and *Streptococcus iniae* infection in juvenile tilapia *Oreochromis niloticus* [J]. Fish & shellfish immunology, 2015, 43(1): 60 - 66.
[76] CHA J-H, RAHIMNEJAD S, YANG S-Y, *et al*. Evaluations of *Bacillus* spp. as dietary additives on growth performance, innate immunity and disease resistance of olive flounder (*Paralichthys olivaceus*) against *Streptococcus iniae* and as water additives [J]. Aquaculture, 2013, 402: 50 - 57.
[77] JANG I-S, KIM D-H, HEO M-S. Dietary administration of probiotics, *bacillus* sp. IS-2, Enhance the innate immune response and disease resistance of *Paralichthys olivaceus* against *streptococcus iniae* [J]. The Korean Journal of Microbiology, 2013, 49(2): 172 - 178.
[78] KIM D, BECK B R, HEO S-B, *et al*. *Lactococcus lactis* BFE920 activates the innate immune system of olive flounder (*Paralichthys olivaceus*), resulting in protection against *Streptococcus iniae* infection and enhancing feed efficiency and weight gain in large-scale field studies [J]. Fish & shellfish immunology, 2013, 35(5): 1585 - 1590.
[79] WIDANARNI W, TANBIYASKUR. Application of Probiotic, Prebiotic and Synbiotic for the Control of Streptococcosis in Tilapia *Oreochromis niloticus* [M]. 2015.
[80] AGUNG L A, YUHANA M. Application of micro-encapsulated probiotic *Bacillus* NP5 and prebiotic mannan oligosaccharide (MOS) to prevent streptococcosis on tilapia *Oreochromis niloticus* [J]. Research Journal of Microbiology, 2015, 10(12): 571.
[81] NG W-K, KIM Y-C, ROMANO N, *et al*. Effects of dietary probiotics on the growth and feeding efficiency of red hybrid tilapia, *Oreochromis* sp., and subsequent resistance to *Streptococcus agalactiae* [J]. Journal of Applied Aquaculture, 2014, 26(1): 22 - 31.
[82] PIRARAT N, PINPIMAI K, RODKHUM C, *et al*. Viability and morphological evaluation of alginate-encapsulated *Lactobacillus rhamnosus* GG under simulated tilapia gastrointestinal conditions and its effect on growth performance, intestinal morphology and protection against *Streptococcus agalactiae* [J]. Animal Feed Science and Technology, 2015, 207: 93 - 103.
[83] KLESIUS P, EVANS J, SHOEMAKER C, *et al*. Streptococcal vaccinology in aquaculture [J]. Tilapia: biology, culture, and nutrition Haworth Press, New York, 2006: 583 - 605.
[84] LAFRENTZ B R, SHOEMAKER C A, KLESIUS P H. Immunoproteomic analysis of the antibody response obtained in Nile tilapia following vaccination with a *Streptococcus iniae* vaccine [J]. Veterinary microbiology, 2011, 152(3 - 4): 346 - 352.
[85] ALLEN J P, NEELY M N. CpsY influences *Streptococcus iniae* cell wall adaptations

important for neutrophil intracellular survival [J]. Infection and immunity, 2012, 80 (5): 1707 - 1715.
[86] AVILES F, ZHANG M M, CHAN J, *et al*. The conserved surface M-protein SiMA of *Streptococcus iniae* is not effective as a cross-protective vaccine against differing capsular serotypes in farmed fish [J]. Veterinary microbiology, 2013, 162(1): 151 - 159.
[87] EYNGOR M, TEKOAH Y, SHAPIRA R, *et al*. Emergence of novel *Streptococcus iniae* exopolysaccharide-producing strains following vaccination with nonproducing strains [J]. Appl Environ Microbiol, 2008, 74(22): 6892 - 6897.
[88] MILLARD C M, BAIANO J C, CHAN C, *et al*. Evolution of the capsular operon of *Streptococcus iniae* in response to vaccination [J]. Appl Environ Microbiol, 2012, 78 (23): 8219 - 8226.
[89] KLESIUS P H, SHOEMAKER C A, EVANS J J. Efficacy of single and combined *Streptococcus iniae* isolate vaccine administered by intraperitoneal and intramuscular routes in tilapia (*Oreochromis niloticus*) [J]. Aquaculture, 2000, 188(3 - 4): 237 - 246.
[90] SHOEMAKER C A, LAFRENTZ B R, KLESIUS P H, *et al*. Protection against heterologous *Streptococcus iniae* isolates using a modified bacterin vaccine in Nile tilapia, *Oreochromis niloticus* (L.) [J]. Journal of fish diseases, 2010, 33 (7): 537 - 544.
[91] SHOEMAKER C A, LAFRENTZ B R, KLESIUS P H. Bivalent vaccination of sex reversed hybrid tilapia against *Streptococcus iniae* and *Vibrio vulnificus* [J]. Aquaculture, 2012, 354: 45 - 49.
[92] MARTINS M L, SHOEMAKER C A, XU D, *et al*. Effect of parasitism on vaccine efficacy against *Streptococcus iniae* in Nile tilapia [J]. Aquaculture, 2011, 314(1 - 4): 18 - 23.
[93] EVANS J J, KLESIUS P H, SHOEMAKER C A. Efficacy of *Streptococcus agalactiae* (group B) vaccine in tilapia (*Oreochromis niloticus*) by intraperitoneal and bath immersion administration [J]. Vaccine, 2004, 22(27 - 28): 3769 - 3773.
[94] SHOEMAKER C A, VANDENBERG G W, DéSORMEAUX A, *et al*. Efficacy of a *Streptococcus iniae* modified bacterin delivered using Oralject™ technology in Nile tilapia (*Oreochromis niloticus*) [J]. Aquaculture, 2006, 255(1 - 4): 151 - 156.
[95] FIRDAUS-NAWI M, YUSOFF S M, YUSOF H, *et al*. Efficacy of feed-based adjuvant vaccine against S treptococcus agalactiae in O reochromis spp. in M alaysia [J]. Aquaculture Research, 2013, 45(1): 87 - 96.
[96] NUR-NAZIFAH M, SABRI M, SITI-ZAHRAH A. Development and efficacy of feed-based recombinant vaccine encoding the cell wall surface anchor family protein of *Streptococcus agalactiae* against streptococcosis in *Oreochromis* sp [J]. Fish & shellfish immunology, 2014, 37(1): 193 - 200.
[97] SHOEMAKER C A, KLESIUS P H. Replicating vaccines [J]. Fish Vaccination, 2014: 33 - 46.
[98] BUCHANAN J T, STANNARD J A, LAUTH X, *et al*. *Streptococcus iniae* phosphoglucomutase is a virulence factor and a target for vaccine development [J]. Infection and immunity, 2005, 73(10): 6935 - 6944.
[99] LOWE B A, MILLER J D, NEELY M N. Analysis of the polysaccharide capsule of the systemic pathogen *Streptococcus iniae* and its implications in virulence [J]. Infection and immunity, 2007, 75(3): 1255 - 1264.
[100] LOCKE J B, VICKNAIR M R, OSTLAND V E, *et al*. Evaluation of *Streptococcus iniae* killed bacterin and live attenuated vaccines in hybrid striped bass through injection and bath immersion [J]. Diseases of aquatic organisms, 2010, 89(2): 117 - 123.
[101] LI L, WANG R, LIANG W, *et al*. Development of live attenuated *Streptococcus agalactiae* vaccine for tilapia via continuous passage *in vitro* [J]. Fish & shellfish

immunology, 2015, 45(2): 955 - 963.

[102] HUANG L Y, WANG K Y, XIAO D, *et al*. Safety and immunogenicity of an oral DNA vaccine encoding Sip of *Streptococcus agalactiae* from Nile tilapia *Oreochromis niloticus* delivered by live attenuated *Salmonella typhimurium* [J]. Fish Shellfish Immunol, 2014, 38(1): 34 - 41.

[103] GAUNT P S, ENDRIS R, MCGINNIS A, *et al*. Determination of florfenicol dose rate in feed for control of mortality in Nile tilapia infected with *Streptococcus iniae* [J]. Journal of Aquatic Animal Health, 2010, 22(3): 158 - 166.

[104] GAIKOWSKI M P, WOLF J C, SCHLEIS S M, *et al*. Safety of florfenicol administered in feed to tilapia (*Oreochromis* sp.) [J]. Toxicologic pathology, 2013, 41(4): 639 - 652.

23

弧菌病：鳗弧菌，奥氏弧菌和杀鲑别弧菌

Alicia E. Toranzo*, Beatriz Magariños 和 Ruben Avendaño-Herrera

23.1 引言

弧菌病是一类由曾经归类为弧菌属的不同细菌种引起的疾病，不过该属的一些种已被移入弧菌科的新属中，这其中包括了黏放线菌(*Moritella viscosa*)[曾名黏弧菌(*Vibrio viscosus*)](Benediktsdóttir et al., 2000)、杀鲑别弧菌(*Aliivibrio salmonicida*)[曾名杀鲑弧菌(*Vibrio salmonicida*)](Urbanczyk et al., 2007)及美人鱼发光杆菌(*Photobacterium damselae*)[曾名海鱼弧菌(*Vibrio damsela*)](Smith et al., 1991)。

本章重点介绍鳗弧菌(*Vibrio anguillarum*)、奥氏弧菌(*V. ordalii*)和杀鲑别弧菌(*Aliivibrio salmonicida*)这三种菌的检测/诊断、抗原/遗传特征、疾病机制和控制/预防。它们在世界范围内造成海水和咸淡水鱼类的严重出血性败血病。它们均为嗜盐的革兰氏阴性运动杆状菌，兼性厌氧，氧化酶和过氧化氢酶阳性。*V. anguillarum*的地理分布和寄主范围广泛，而*V. ordalii*和*A. salmonicida*主要局限于特定地理区域的鲑科鱼类。它们之间的主要差异特征包括在弧菌特异性培养基上形成菌落的能力、生长温度的宽广范围、精氨酸双水解酶的存在以及从几种糖(如蔗糖、麦芽糖、甘露糖、海藻糖、苦杏仁苷和甘露醇)中的产酸能力。

23.2 鳗弧菌

23.2.1 菌种描述

V. anguillarum 是经典弧菌病的病原菌，能够在重要经济性温水和冷水鱼类中引起典型的出血性败血症，主要感染的鱼类包括太平洋鲑和大西洋鲑、虹鳟、大菱鲆、舌齿鲈、金头鲷、条纹鲈、大西洋鳕、日本鳗和欧洲鳗、香鱼(Toranzo and Barja, 1990, 1993; Actis et al., 1999; Toranzo et al., 2005; Samuelsen

* 通信作者邮箱：alicia.estevez.toranzo@usc.es。

et al.，2006；Austin and Austin，2007；Angelidis，2014）。它是弧菌科(Vibrionaceae)革兰氏阴性杆状菌，具运动性、嗜盐性和兼性厌氧，20～30 ℃下，在含1%～2% NaCl的培养基中快速生长。鳗弧菌的基因组约为4.2～4.3 Mbp，鸟嘌呤-胞嘧啶(GC)含量为43%～46%(Naka et al.，2011；Li et al.，2013；Holm et al.，2015)，细胞中包含两个环形染色体，这是弧菌种的一个共同特征。尽管*V. anguillarum*分离株中共出现了23种O血清型(Sørensen and Larsen，1986；Pedersen et al.，1999)，但只有O1和O2血清型以及较小程度的O3血清型菌株才会引起鱼类的死亡(Tajima et al.，1985；Toranzo and Barja，1990，1993；Larsen et al.，1994；Toranzo et al.，1997，2005)。O1和O2血清型鳗弧菌的宿主种类和地理区域分布广泛，而O3血清型则主要影响日本养殖的鳗鲡和香鱼，以及智利养殖的大西洋鲑(Silva-Rubio et al.，2008a)。

与抗原同质性的O1血清型不同，O2和O3血清型是异质性的，每种血清型中的不同亚群被命名为O2a、O2b、O2c以及O3A和O3B(Olsen and Larsen，1993；Santos et al.，1995；Mikkelsen et al.，2007；Silva-Rubio et al.，2008a)。O1和O2a血清型存在于鲑科鱼类和非鲑科鱼类中，而O2b和O2c血清型仅在海水鱼类中检测到。O3A血清型可从病鱼中分离，但O3B亚群仅在环境中发现。

已对鳗弧菌主要致病血清型(O1、O2和O3)的种内变异进行了遗传研究(Pedersen and Larsen，1993；Skov et al.，1995；Tiainen et al.，1995；Toranzo et al.，1997；Mikkelsen et al.，2007)。无论采用核糖分型、扩增片段长度多态性(AFLP)和脉冲场凝胶电泳(PFGE)中的哪种方法，O1型菌株都是最具遗传基因同源性的血清型。不过，使用PFGE法在三种血清型中可以检测到具有流行病学价值的不同克隆谱系(Skov et al.，1995；Toranzo et al.，2011)。

*V. anguillarum*的传播方式存在争议。在大多病例中，感染是通过皮肤黏液丧失区域的细菌渗透引发的(Muroga and de la Cruz，1987；Kanno et al.，1989；Svendsen and Bogwald，1997；Spanggaard et al.，2000；O'Toole et al.，2004；Croxatto et al.，2007；Weber et al.，2010)。事实上，皮肤黏液对这种微生物具有很高的抗菌活性(Harrel et al.，1976a；Fouz et al.，1990)。通过水或食物摄入鳗弧菌也会引发弧菌病，这主要发生在仔鱼中，因为它们的胃肠道酸性不足以使细菌失活(Grisez et al.，1996；Olsson et al.，1998；O'Toole et al.，1999，2004；Engelsen et al.，2008)。*V. anguillarum*可在盐水中以可培养形式存活超过一年(Hoff，1989a)，这有利于其在水体中的广泛传播。

23.2.2 感染诊断

1. 疾病临床体征

患有典型弧菌病的鱼呈现全身性败血症，并在鳍基部、腹部和体侧区域、嘴部、鳃盖和眼部出血(图 23.1)。常可见眼球突出和角膜混浊。濒死鱼常厌食，伴有鳃苍白，这反映了严重的贫血症状。经常可见主要集中在皮下组织的水肿性病变。在病鱼体内，可见脾肿大、肾肿胀，肠可能出现胀大并充满黏性清液。肾实质中有出血，肾小管上皮细胞空泡化。在肾脏的黑色素巨噬细胞中心，可见铁沉积物堆积。在脾、肝和肾中，有出血和坏死区域，并且肌肉中可能存在大的坏死病变（Ransom et al.，1984；Lamas et al.，1994；Toranzo et al.，2005；Austin and Austin，2007；Angelidis，2014）。

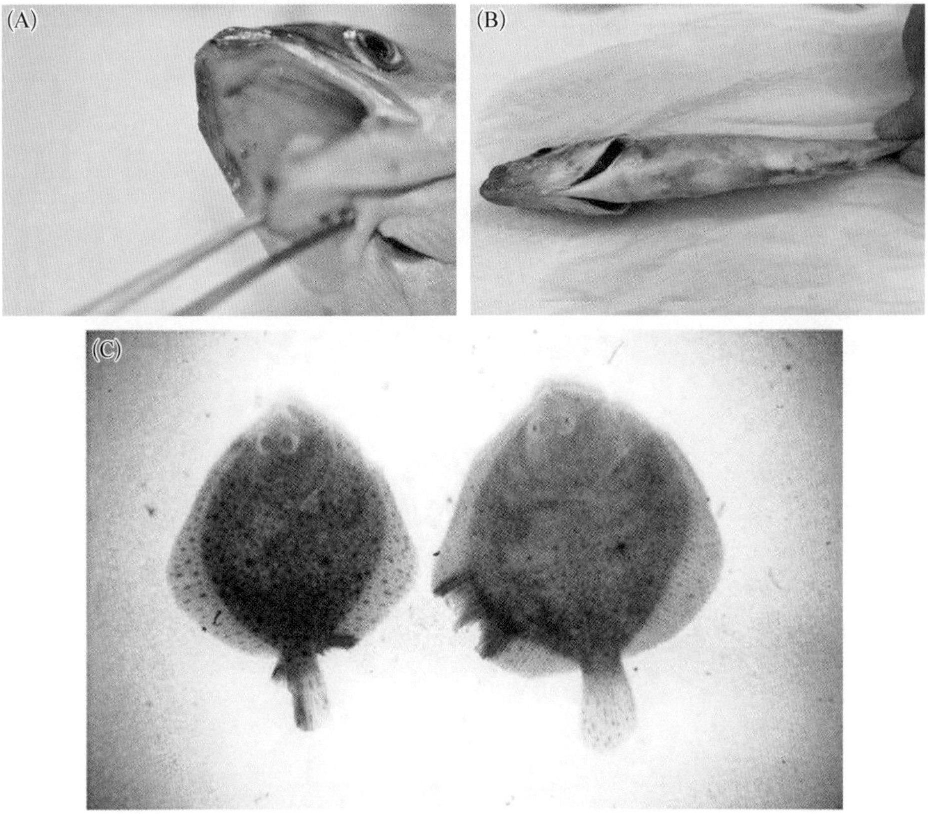

图 23.1 鳗弧菌引起的弧菌病临床表现：(A) 大西洋鲑嘴部和眼睛周围出血(体重 500 g)；(B) 海鲈腹侧和体侧出血区域(体重 25 g)；(C) 大菱鲆鳍出血(体重 10 g)

2. 检测

V. anguillarum 是一种嗜盐菌，在添加 1% NaCl 的普通培养基中易生长。此外，同时使用硫代硫酸盐-柠檬酸盐-胆盐-蔗糖琼脂(TCBS)(Oxoid)培养基有助于鉴别其典型的黄色(蔗糖阳性)菌落。

使用标准生化检测可以推定诊断 *V. anguillarum*（表 23.1）。该菌为革兰氏阴性、运动型杆状细菌，氧化酶、过氧化氢酶、伏-波试验（Voges-Proskauer）（乙酰甲基甲醇）、β-半乳糖苷酶和精氨酸双水解酶阳性，但赖氨酸和鸟氨酸脱羧阴性。该菌水解明胶和淀粉，发酵葡萄糖、蔗糖、麦芽糖、甘露糖、海藻糖、苦杏仁苷和甘露醇（Toranzo and Barja，1990）。需使用血清型特异性抗血清进行血清学确认（Toranzo et al.，1987）。虽可以使用基于载玻片凝集和 ELISA 的商业诊断试剂盒进行诊断，但它们无法区分血清型（Romalde et al.，1995），因此其流行病学价值有限。

表 23.1 *Vibrio anguillarum*、*V. ordalii* 和 *A. salmonicida* 的表型特征[a]

特　征	*Vibrio anguillarum*	*V. ordalii*	*A. salmonicida*
产酸：			
苦杏仁苷	+	—	—
葡萄糖	+	+	+
肌醇	—	—	—
麦芽糖	+	—	+
甘露醇	+	V	+
甘露糖	+	—	—
蔗糖	+	—	—
海藻糖	+	V	+
精氨酸双水解酶	+	—	—
过氧化氢酶	+	+	+
柠檬酸盐利用	+	—	—
细胞色素氧化酶	+	+	+
β-半乳糖苷酶	+	—	—
明胶水解	+	V	—
革兰氏染色	—	—	—
生长：			
25 ℃	+	+	—
37 ℃	V	—	—
0% NaCl	—	—	—
3% NaCl	+	+	+
5% NaCl	—	—	—
TCBS	+（Y）	—	—
吲哚产生	+	—	—
赖氨酸脱羧酶	—	—	—
运动性	+	+	+
硝酸盐还原	+	V	—

续 表

特 征	*Vibrio anguillarum*	*V. ordalii*	*A. salmonicida*
O/F(氧化-发酵)Leifson 试验	F	F	F
鸟氨酸脱羧酶	—	—	—
淀粉水解	+	—	—
伏-波试验	+	—	—

[a] F 为发酵；V 为菌株间的可变反应；Y 为黄色菌落。

灵敏的聚合酶链反应(PCR)方法已被开发用于从纯培养物和鱼组织中快速、特异性地鉴定 *V. anguillarum*。σ^{54} 因子编码基因 *rpoN*、σ^{38} 因子编码基因 *rpoS*、细胞外锌金属蛋白酶 EmpA 编码基因 *empA*、肽聚糖水解酶编码基因 *amiB* 或细菌伴侣蛋白编码基因 *groEL* 可用作鉴定的靶标基因(Osorio and Toranzo，2002；Hong et al.，2007；Xiao et al.，2009；Kim et al.，2010)。这些检测方法对于区分 *V. anguillarum* 和 *V. ordalii* 特别有用，因为这两种菌的 16S rDNA 序列同一性接近 99%。另外，*rpoN* 基因可作为多重 PCR 的靶基因，用于同时检测鳗弧菌、杀鲑气单胞菌、鲑立克次氏体和海豹链球菌(*Streptococcus phocae*)(Tapia-Cammas et al.，2011)。基于 *amiB* 或 *empA* 基因的环介导等温扩增(LAMP)是一种经济有效的 PCR 替代方法，可快速、特异和灵敏地检测鳗弧菌(Kulkarni et al.，2009；Gao et al.，2010)。

23.2.3 致病机制

尽管对鳗弧菌的致病机制尚未完全了解，但其毒力因子包括黏附、定植和侵袭因子，外毒素，细胞表面成分和铁摄取系统(Frans et al.，2011)。此外，这其中的一些毒力因子受群体感应系统的调控。

1. 黏附、定植和侵袭因子

在进入宿主的过程中，细菌的鞭毛促进病原体附着并穿透宿主黏液(O'Toole et al.，1996)。*V. anguillarum* 对鱼的鳃、皮肤和肠黏液表现出趋化作用(Bordas et al.，1998；O'Toole et al.，1999；Larsen et al.，2001)，并黏附和定植于肠道(Bordas et al.，1998)。此外，当在鲑鱼肠道黏液中培养时，*V. anguillarum* 会特异性表达各种蛋白，包括金属蛋白酶 EmpA，其在毒力中的作用由 Milton 等(1992)提出，并被 Denkin 和 Nelson(2004)进一步证实。另外，还鉴定了三种与蛋白水解酶相关的基因，并认为它们连同金属蛋白酶是侵入宿主组织所必需的(Rodkhum et al.，2006)。

2. 外毒素

V. anguillarum 产生胞外酶（溶血素和 RTX 毒素），可能导致组织损伤和促进细菌增殖。已经发现描述了鳗弧菌中五个编码溶血素的不同基因（*vah*1，Hirono et al.，1996；*vah*2，*vah*3，*vah*4，*vah*5，Rodkhum et al.，2005）。*vah*2 至 *vah*5 的单缺陷突变株对虹鳟的毒力低于亲本菌株，表明每一种溶血素都有助于病原菌的毒力。该病原菌还产生一种成孔毒素 RTX（Rodkhum et al.，2006），但这种毒素在 *V. anguillarum* 毒力中的作用尚不清楚。

3. 细胞表面成分

V. anguillarum 可抵抗鱼血清的杀菌作用，这种抗性是由 O-抗原多糖长链的合成介导的（Boesen et al.，1999）。而且，*V. anguillarum* 的 O-抗原似乎参与铁载体介导的铁摄取过程，因为 O-抗原生物合成操纵子的突变导致铁载体摄取缺陷（Welch and Crosa，2005）。

此外，*V. anguillarum* 还可以通过五种盐响应性外膜蛋白的有效渗透调节，帮助细菌适应宿主和海水的不同盐度（Kao et al.，2009）。另外两种主要外膜蛋白调节对胆汁的抗性，增强了该菌在鱼体内的存活和肠定植（Wang et al.，2009）。

4. 铁摄取系统

V. anguillarum 利用铁载体从宿主组织或环境中猎取其代谢所需的铁元素。已经在 O1 和 O2 型鳗弧菌中描述了两种不同铁载体介导的系统（Lemos et al.，2010）。在大多数 O1 血清型致病株中，一种 65 kb 大小的 pJM1 质粒含有铁载体鳗弧菌素 anguibactin 的编码基因（Crosa，1980；Stork et al.，2002），尽管其生物合成也需要染色体编码的酶以及质粒基因编码的酶（Alice et al.，2005）。

O2 血清型菌株和一些无质粒的 O1 血清型菌株产生一种与 pJM1 介导系统无关的儿茶酚类铁载体（Lemos et al.，1988；Conchas et al.，1991）。这种染色体编码的系统合成铁载体 vanchrobactin（Balado et al.，2006，2008；Soengas et al.，2006）。

铁载体在鳗弧菌毒力中的重要性已在质粒消除株以及参与铁载体生物合成的基因或铁载体转运至细菌细胞所需基因的缺陷突变株中得到证实。这种 *V. anguillarum*缺陷突变株严重削弱了对鱼类的感染力（Wolf and Crosa，1986；Stork et al.，2002，2004）。

其他铁同化系统，例如利用血红素、血红蛋白和血红蛋白-触珠蛋白作为铁源，并通过血红素转运的特异性机制，同样存在于 *V. anguillarum* 中（Mouriño et al.，2004）。血红素转运系统在细菌生物学中的发生率尚不确定，但它是弧菌中的保守系统（Mazoy et al.，2003；Lemos and Osorio，2007）。虽然该系统对毒力不是必需的，但当铁作为血红素基团可用时，它确实有助于提高细菌在宿主中定植的能力（Mazoy et al.，2003）。

5. 毒力因子的调节

群体感应（quorum sensing，QS）系统参与了包括 *V. anguillarum* 在内的几种革兰氏阴性鱼类病原菌的毒力基因调控（Bruhn et al.，2005）。研究最深入的 QS 系统依赖于 *N*-酰基高丝氨酸内酯（*N*-acylhomoserine lactones，AHLs）的产生，它是细胞间通信系统中的信号分子，能够监测细菌种群密度以协调基因表达。已在 O1 血清型鳗弧菌中鉴定出 3 种 QS 系统，它们与生物发光弧菌费氏弧菌（*V. fischeri*）和哈维氏弧菌（*V. harveyi*）中存在的系统相似（Milton et al.，1997，2001；Croxatto et al.，2004；Milton，2006；Weber et al.，2008）。研究结果显示，QS 调控转录激活因子 VanT 的鳗弧菌突变株缺乏金属蛋白酶 EmpA 和生物被膜形成，而这是病原菌侵染和存活所必需的（Croxatto et al.，2002）。

23.2.4 预防和控制

大多数商业鳗弧菌疫苗是通过浸泡或腹腔（IP）注射的全细胞灭活疫苗，作为多组分无佐剂或油佐剂疫苗的一部分（Toranzo et al.，1997，2005；Håstein et al.，2005；Angelidis，2014；Colquhoun and Lillehaug，2014）。海水鱼类，如大菱鲆或海鲈，当鱼的平均体重为 1～2 g 时，浸泡接种液体 *V. anguillarum* 疫苗，1 个月后进行加强免疫接种。在海鲈中，通常在转移到网箱之前（20～25 g）通过 IP 注射接种进行加强免疫。对于鲑科鱼类，在银化前 1 个月 IP 接种多价油佐剂疫苗，从而提供免疫保护直至收获（约 24 个月）。非矿物油佐剂的使用减少了使用矿物油佐剂疫苗经常报道的生长迟缓和腹内损伤的发展（Midtlyng and Lillehaug，1998；Poppe and Koppang，2014）。

可通过浸泡和口服的鳗弧菌减毒活疫苗的开发代表了研究的一个重要趋势（Yu et al.，2012；Zhang et al.，2013）。然而，这些活疫苗赋予的免疫保护作用仅在疫苗接种后 1 个月进行了评估，因此，尚无有关保护作用持续时间的数据。

虽然 *V. anguillarum* 疫苗是有效的，但一些疫情仍需要应用抗菌药物来进行控制。土霉素、噁喹酸、氟甲喹、增效磺胺类药物、氟苯尼考和氟喹诺酮类药物被用于治疗由 *V. anguillarum* 感染引起的疾病（Lillehaug et al.，2003；Samuelsen and Bergh，2004；Samuelsen et al.，2006；Angelidis，2014），不过已产生对噁喹酸的抗药性（Colquhoun et al.，2007）。目前，选择的药物和给药方案为土霉素（每日每千克体重 75～100 mg，10 d）、氟苯尼考（每日每千克体重

10 mg，7 d）和恩诺沙星（每日每千克体重 10 mg，7 d）（Rigos and Troisi，2005）。

此外还研究了弧菌病防控中的益生菌应用。浸泡在含荧光假单胞菌、温和气单胞菌（*Aeromonas sobria*）或芽孢杆菌等的抗菌菌株溶液后，可抑制虹鳟中的鳗弧菌感染（Gram et al.，1999；Brunt et al.，2007）。德氏乳杆菌德氏亚种（*Lactobacillus delbruekii* subsp. *delbrueckii*）和河流漫游球菌（*Vagococcus fluvialis*）这两种益生菌在水中施用（Carnevali et al.，2006）或伴饲摄入（Sorroza et al.，2012）后可保护海鲈免受 *V. anguillarum* 的攻毒感染。Chabrillon 等（2006）将分离的弧菌科细菌与其商品饲料混合施用时，显著保护了受到 *V. anguillarum* 攻毒的海鲷。使用轮虫生物包埋的褐杆菌属（*Phaeobacter*）、玫瑰杆菌属（*Roseobacter*）和鲁杰氏菌属（*Ruegeria*）益生菌在大菱鲆仔鱼中也显示出良好的抗 *V. anguillarum* 感染的应用前景（Planas et al.，2006；Porsby et al.，2008）。

由于 *V. anguillarum* 中普遍存在群体感应系统，因此抑制高丝氨酸内酯（AHL）介导的 QS 过程也成为一种防控方法（Defoirdt et al.，2004）。事实上，研究发现，使用特异性 QS 抑制剂，如呋喃酮 C－30，可以降低感染 *V. anguillarum* 的虹鳟死亡率（Rasch et al.，2004），并且 Brackman 等（2009）还证实核苷类似物 LMC－21（一种腺苷衍生物）会影响鳗弧菌的生物被膜形成和蛋白酶产生。此外，分离和鉴定可通过酶法或产生抑制剂或拮抗剂抑制 QS 的海洋细菌菌株，有助于开发水产养殖中的新生物技术方法。为了利用这些有效工具，需要考虑实用性，例如如何将 QS 干扰剂或微生物递送到特定的感染部位，以及治疗成本等。

23.3 奥氏弧菌

23.3.1 菌种描述

V. ordalii 最初被称为非典型弧菌 1669（*Vibrio* sp. 1669）（Harrell et al.，1976b）、*Vibrio* sp. RT（Ohnishi and Muroga，1976）、*V. anguillarum* 生物Ⅱ型（Schiewe et al.，1977）及 *V. anguillarum* 同型种Ⅱ（Ezura et al.，1980）等。Schiewe 等（1981）重新阐明了该病原菌的分类地位，并将其命名为奥氏弧菌。

该种最初是作为弧菌病病原菌从美国太平洋西北沿海水域的野生和养殖银大麻哈鱼（*O. kisutch*）中分离出来的（Harrell et al.，1976b）。自那以后，日本、澳大利亚和新西兰都有相关病例报道，其主要感染养殖鲑科鱼类（Ransom et al.，1984；Toranzo et al.，1997）。2004 年，智利南部报道了该病原菌导致养殖大西洋鲑、太平洋鲑和虹鳟死亡的情况（Colquhoun et al.，2004；Silva-Rubio et al.，2008b）。其他鱼类也有该菌感染的报道，如日本的香鱼和许氏平鲉（Muroga et al.，1986），以及土耳其的金头鲷（Akayli et al.，2010）。

V. ordalii 是一种运动型革兰氏阴性杆状细菌，可发酵葡萄糖，过氧化氢酶和氧化酶阳性，对弧菌抑制剂 O/129 敏感（Farmer et al.，2005）。该菌可在补充有 1%～2% NaCl 的常规培养基中生长。

V. ordalii 的总基因组大小约为 3.4 Mbp，GC 含量为 43%～44%（Schiewe et al.，1981；Naka et al.，2011）。其基因组由两条环形染色体组成，这是弧菌种的共同特征。然而，与 *V. anguillarum* 相比，*V. ordalii* 存在较大的染色体缺失（约 600 kb）。此外，所有 *V. ordalii* 中的另一个共同特征是存在一个称为 pMJ101 的 30 kb 隐蔽性质粒（Schiewe and Crosa，1981）。有趣的是，这种染色体外元件与大多数 O1 血清型鳗弧菌菌株中存在的 pJM1 毒力质粒无关（Crosa，1980）。

V. ordalii 具有抗原同质性，并且与鳗弧菌 O2a 血清亚群的抗原相同（Schiewe et al.，1981）。这两种微生物都具有 O-抗原多糖，它们的区别只是聚合度不同（Sadovskaya et al.，1998）。*V. ordalii* 和 *V. anguillarum* O2a 都含有特异性的分化抗原决定簇（Chart and Trust，1984；Mutharia et al.，1993），但在 *V. ordalii* 中也存在抗原差异（Bohle et al.，2007；Silva-Rubio et al.，2008b）。事实上，Silva-Rubio 等（2008b）描述了智利 *V. ordalii* 脂多糖类似的电泳图谱，具有低相对分子质量 O-抗原梯式谱带，但与美国银大麻哈鱼的模式分离株不同，其在低相对分子质量区域和高相对分子质量区域都显示出抗原反应。这表明智利分离株可能构成一个新的血清学亚群。

可通过核糖分型、PFGE、随机扩增多态性 DNA（RAPD）、重复序列回文 PCR（REP-PCR）和肠杆菌基因间重复一致序列 PCR（ERIC-PCR）等方法分析确定 *V. ordalii* 中存在的种内遗传多样性程度。结果表明，分离株之间的同源性不受宿主和地理来源的影响，从而证明了该种的克隆性（Wards and Patel，1991；Tiainen et al.，1995；Silva-Rubio et al.，2008b）。然而，Silva-Rubio 等（2008b）检测到银大麻哈鱼模式株和智利分离株之间的微小遗传差异，从而表明从智利大西洋鲑中获得的 *V. ordalii* 可构成一个独特的克隆群。

关于水生环境中 *V. ordalii* 特性的信息有限，因此，尚不清楚水是否可以作为疾病的天然病原储库和传播途径。Ruiz 等（2016b）的研究表明，*V. ordalii* 菌株在不添加营养物的无菌海水中存活了一年，且长期保存不影响其生化或遗传特性。此外，*V. ordalii* 在无菌海水中对虹鳟的感染性仍能保持 60 d，不过在非无菌条件下，可培养的 *V. ordalii* 数量迅速下降，在实验开始 2 d 后已无可生化检测到的 *V. ordalii* 菌落。

该观察结果得到了对微环境中 *V. ordalii* DNA 定量 PCR（qPCR）分析结果的支持，其中 qPCR 仅在实验前 3 d 给出了阳性反应。拮抗试验表明，原生菌群降低了非无菌微环境中的 *V. ordalii* 存活率，表明原生菌群影响了其生存（Ruiz et al.，2016b）。在水环境中，*V. ordalii* 也可采用其他不利环境条件下存活的替代策略，例如Naka等(2011)所述的生物被膜形成，其中除了负调控因子 SypE 外，*V. ordalii* 基因组携带整个生物被膜形成簇(Syp 系统)。有趣的是，Syp 系统对另一种费氏弧菌在乌贼器官中的定植是必不可少的(Yip et al., 2006)。*V. ordalii*形成生物被膜，黏附于涂有鱼黏液的玻璃表面，并具有疏水特性(Ruiz et al., 2015)，有利于其对鱼的定植。

23.3.2 感染诊断

1. 疾病临床症状

V. ordalii 引起的弧菌病通常伴有感染部位周围组织的坏死和出血病变，包括腹鳍和肛孔[图 23.2(A)]。有些病例则出现脓肿并伴有皮肤出血(Ransom et al., 1984)。Bohle 等(2007)描述了 14 例智利大西洋鲑临床病例，临床症状包括外侧和侧背有复发性疖疮、广泛的体侧出血性溃疡、脂鳍出血性水肿、胸鳍基部溃疡和出血及腹部瘀点。剖检可见感染的鱼主要在肝脏和脾脏出现白点，脾肿大，在一些病例中，可见带瘀斑的灰白肝[图 23.2(B)]。

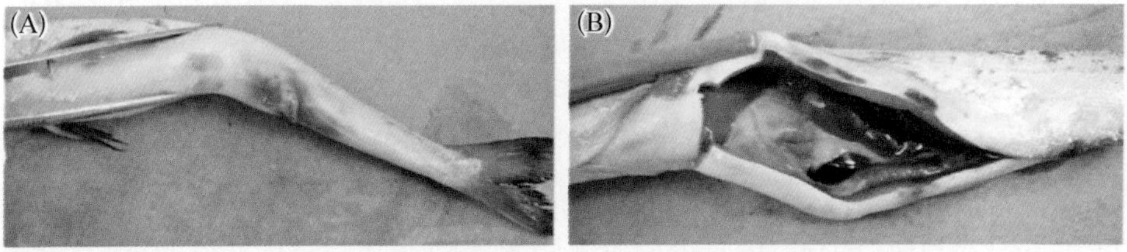

图 23.2　大西洋鲑(平均体重 50 g)中由 *V. ordalii* 引起的弧菌病临床体征：(A) 肛孔周围组织的出血性病变；(B) 脾肿大和伴有瘀斑的灰白肝

V. ordalii 实验感染的组织病理学表明，该菌以聚集体或菌落的形式大量集中在肌肉组织(骨骼肌、心肌和平滑肌)中而非血液中(Ransom et al., 1984)。此外，黏膜病理显示，显性感染开始于降肠和直肠，这些部位的病变最为严重，而前胃肠道的病变不太明显。另外，皮肤上存在 *V. ordalii* 表明细菌通过皮层侵入鱼宿主(Ransom, 1978)，下消化道组织的脱落表明这可能是病原菌排出和传播的方式(Ransom et al., 1984)。

2. *V. ordalii* 的检测

V. ordalii 是一种嗜盐、运动型革兰氏阴性杆状细菌，不能在 TCBS 或 MacConkey 琼脂上生长(Bohle et al., 2007)。在补充添加 1%～2% NaCl 的适宜培养基上分离细菌菌落，然后使用常规管和平板测试程序进行生化测试，可以对该菌进行推定诊断(表 23.1)。*V. ordalii* 在 18 ℃下孵育 2～3 d 后产生半透明、圆形、凸起的 1～2 mm 大小菌落(Poblete-Morales et al., 2013)。该菌对精

氨酸双水解酶、β-半乳糖苷酶和吲哚试验均呈阴性反应，很容易将其与*V. anguillarum*区分开来（Schiewe et al.，1981）。除此之外，BIOLOG - GN（Biolog Inc.，加利福尼亚）指纹图谱和 API - 20E （BioMérieux）分析可进一步揭示*V. ordalii* 菌株之间的差异（Austin et al.，1997）。这种种内差异与分离株从海藻糖和甘露醇产酸以及水解明胶的能力有关（Colquhoun et al.，2004；Bohle et al.，2007；Silva-Rubio et al.，2008b）。虽然 *V. ordalii* 只有一种血清型，但由于其与 O2 型 *V. anguillarum* 的交叉反应（Chart and Trust，1984），使用商业诊断试剂盒的血清学确认对该种的鉴别能力较差（Bohle et al.，2007）。

染色体 DNA - DNA 配对显示，*V. ordalii* 和 *V. anguillarum* 的总体 DNA 序列高度相似（Schiewe et al.，1981）。此外，5S rRNA 基因序列的 DNA 比较揭示了 *V. anguillarum* 和 *V. ordalii* 之间的密切关系，因为它们在 120 bp 中仅存在 7 bp 的差异（MacDonell and Colwell，1984；Pillidge and Colwell，1988；Ito et al.，1995）。类似地，*V. ordalii* 与 *V. anguillarum* 的 16S rRNA 基因序列难以区分（Wiik et al.，1995）。相比之下，编码溶血素产生的 *vohB* 基因是 PCR 和 qPCR 法检测和鉴别 *V. ordalii* 的合适靶基因（Avendaño-Herrera et al.，2014）。该方法快速、具有特异性且灵敏，足以检测和定量感染组织中的*V. ordalii*。

23.3.3 致病机制

V. ordalii 有许多基因是黏附、定植和侵袭因子的潜在毒力基因，此外还包括外毒素、细胞表面组分和铁摄取系统等毒力因子（Naka et al.，2011）。

1. 黏附、定植和侵袭因子

成功定植的先决条件是能够黏附宿主组织，因为这在细菌渗透到靶细胞之前会促进细菌毒素的传递（Montgomery and Kirchman，1994）。Ruiz 等（2015）认为，细菌在鱼体表面生长的能力可以促进病原菌的定植和随后的侵染。体外研究表明，*V. ordalii* 的致病性与大西洋鲑红细胞的血细胞凝集或生物被膜特征无关，但与疏水性有关（Ruiz et al.，2015）。疏水性和生物被膜的产生是黏附过程和病原菌在细胞内存活的主要因素（Daly and Stevenson，1987）。特异性黏附是通过细菌表面上的化合物介导的，这些化合物通过刚性立体化学键与它们所黏附载体上的特定分子结合。相反地，非特异性黏附取决于细菌表面和支撑物上的某些结构之间的疏水或离子相互作用（Ofek and Doyle，1994）。Naka 等（2011）在 *V. ordalii* 中鉴定了金属蛋白酶基因 *empA*，该基因已被认为是 *V. anguillarum* 毒力的一个因素（Milton et al.，1992；Denkin and Nelson，2004）。然而，尚不清楚该基因在 *V. ordalii* 中是否具有相似的功能。

2. 外毒素

在 *V. ordalii* 中，一种白细胞溶解因子的产生可能对致病机制很重要，因为患病鱼的循环血液中白细胞计数减少了 80%～95%(Ransom et al., 1984)。细菌溶血素是细胞溶解毒素，其通常被认为是毒力因子，因为它们能够影响红细胞和其他细胞(Rowe and Welch, 1994)。尽管如此，*V. ordalii* 中溶血素的产生与致病性之间的关系仍存在争议。Kodama 等(1984)报道，该菌不产生溶血素，但 Naka 等(2011)分析 *V. ordalii* ATCC 33509^{T}的全基因组序列发现，除了 *vah*4 基因之外，几乎所有编码溶血素(一种假定的毒力因子)生物合成的基因都存在。

3. 细胞表面组分

V. ordalii 对非免疫虹鳟血清的杀菌作用具有抗性并凝集不同的真核细胞(Trust et al., 1981)。这些研究表明，该病原菌附着于宿主细胞并与宿主细胞相互作用(Larsen and Mellergaard, 1984)。Ruiz 等(2015)发现大西洋鲑黏液有利于 *V. ordalli* 智利分离株的生长，但不利于银大麻哈鱼模式菌株的生长。这表明在 *V. ordalii* 智利分离株中存在一种增强细菌抗性的细胞结构(如荚膜物质)。Sadovskaya 等(1998)证实，*V. ordalii* 能够产生荚膜多糖，但它们对发病机制的意义尚不清楚。

4. 铁摄取系统

另一种推定的 *V. ordalii* 毒力因子是其表达与宿主铁结合蛋白竞争的高亲和力铁摄取机制的能力。最初，Pybus 等(1994)提出鳗弧菌铁载体抑制 *V. ordalii*，这与 *V. ordalii* 无法使用鳗弧菌 anguibactin -铁复合物，共同表明了这两种微生物中存在不同的铁摄取系统。最近，Naka 等(2011)描述了涉及 *V. ordalii*模式株铁摄取机制的基因，包括铁载体 anguibactin 生物合成必需的基因和铁转运基因，包括 *vabA*～*vabE*，*tonB*2，*exbB*，*exbD*1 和 *fur*。

Ruiz 等(2016a)的研究证实 *V. ordalii* 具有不同的铁摄取系统，一种涉及铁载体合成，而另一种直接结合血红素。对来自大西洋鲑的 *V. ordalii* 分离株和来自银大麻哈鱼的 *V. ordalii* 模式株的研究发现，所有菌株能在螯合剂 2,2′-联吡啶存在下生长，并在固体和液体培养基中产生铁载体。

V. ordalii 菌株之间的交互饲养试验表明，所有菌株被模式菌株及一种智利分离株交互饲养，但其余的智利分离株无法促进任何其他 *V. ordalii* 菌株的生长，这表明在该菌产生的铁载体中存在种内变异性(Ruiz et al., 2016a)。根据基于基因组数据的生物信息学搜索和 Naka 等(2011)报道的基因组数据，*V. ordalii* ATCC 33509^T 含有 *vab* 簇基因，其编码 vanchrobactin 的合成，

vanchrobactin 是一种在鳗弧菌中鉴定的典型儿茶酚类铁载体（Balado et al.，2006）。此外，模式菌株含有一个同源基因簇，该基因簇与美人鱼发光杆菌杀鱼亚种中描述的 piscibactin（一种混合型铁载体）生物合成基因同源（Osorio et al.，2006；Souto et al.，2012）。当细胞在低可用铁下生长时，参与 piscibactin 合成（*irp*1 和 *irp*2）、转运（*frpA*）和调控（*araC*1 和 *araC*2）的基因表达显著上调（通过 RT－PCR 检测）。应该注意的是，即使获得了这样的 PCR 结果，也并不意味着用于合成和转运 vanchrobactin 和 piscibactin 的所有必需基因都存在于*V. ordalii* 中，或者它们都表达了。不过，Ruiz 等（2016a）使用纯铁载体的生物测定法证明，没有一种 *V. ordalii* 菌株可以利用 vanchrobactin，但它们都可以使用 piscibactin 作为铁载体。

所有的 *V. ordalii* 菌株都可以使用氯化血红素和血红蛋白作为唯一的铁源。此外，*V. ordalii* 细胞表现出铁调控的血红素结合活性，并且在测试的菌株中没有观察到血红蛋白获取的差异，这表明可能存在位于 *V. ordalii* 细胞表面的组成性结合分子（Ruiz et al.，2016a）。最后，使用虹鳟作为感染模型的毒力测试揭示铁摄取能力与 *V. ordalii* 的致病性之间存在明确关联。事实上，攻毒研究的结果表明，在铁限制条件下培养时，模式菌株和智利大西洋鲑代表性菌株的毒力增强了。

23.3.4 预防和控制

由 *V. ordalii* 引起的弧菌病目前是智利鲑鱼养殖业中的一个主要健康问题。在智利，有 9 种已注册的奥氏弧菌商业疫苗，其中 7 种为注射型乳剂疫苗，1 种为口服疫苗，1 种为浸泡型疫苗。这些生产的商业疫苗有 3 种是单价疫苗，1 种是二价疫苗，另外 3 种是三价疫苗，还有 2 种是四价疫苗。因此，这些制剂含有 1～4 种抗原，包括来自其他鱼类病原体的抗原，例如非典型杀鲑气单胞菌、鲑立克次氏体以及传染性胰腺坏死病和传染性鲑贫血症的病毒。

在智利，用于治疗 *V. ordalii* 感染的最常用抗生素是土霉素（每天每千克体重 55～100 mg，持续 10～15 d）和氟苯尼考（每天每千克体重 10 mg，持续 7～10 d）（Poblete-Morales et al.，2013）。这两种都是广谱抑菌剂。然而，智利的一些养殖户报告说，这些抗菌药物已在很大程度上被噁喹酸所取代（每天每千克体重 10～30 mg，持续 10～15 d）（San Martín et al.，2010），虽然人们对其有效性普遍缺乏了解。

为了监测抗生素的正确应用和预防抗药性，必须对评估抗菌药物敏感性的方案进行标准化和验证。值得注意的是，抗药性的概念涵盖了流行病学、临床和药理学领域，并需要利用抗生素浓度值或临界点来区分菌株类别（Turnidge and Paterson，2007）。临床临界点是指将具有高治疗概率的菌株与治疗可能失败的菌株区分开的阈值。流行病学临界点解决了耐药机制的问题，并将野生型（wild type，WT)细菌从那些已获得或选择了一种抗性机制的非野生型(non-wild types，NWTs)细菌中分开，而这需要对很长一段时间内收集的来自不同地理位置和宿主的大量分离株进行研究(Henríquez et al.，2016)。使用临床与实验室标准化协会(CLSI，2006)发布指南进行的 *V. ordalii* 对土霉素、氟苯尼考和噁喹酸敏感性的体外研究表明，大多数 *V. ordalii* 智利分离株为 NWT 菌株(Poblete-Morales et al.，2013)。

23.4 杀鲑别弧菌

23.4.1 菌种描述

A. salmonicida 以前归类为杀鲑弧菌(*V. salmonicida*)(Urbanczyk et al.，2007)，是“希特拉病(Hitra disease)”或“冷水弧菌病(cold water vibriosis)”的病原，能够对加拿大、挪威和英国的养殖鲑科鱼类和鳕造成影响(Egidius et al.，1981，1986；Bruno et al.，1986；Sørum et al.，1990)。该病主要发生在深秋、冬季和早春海水温度低于 15 ℃的情况下。

该菌为弯曲和运动型革兰氏阴性杆状菌[0.5 μm×(2～3) μm]，极生有鞘鞭毛多达 9 根，但无侧生鞭毛。菌落小、呈灰色、表面光滑、边缘完整。可以在1～22 ℃下生长，但最适生长温度为 15 ℃。*A. salmonicida* 是嗜盐性的，可在添加 0.5%～4% NaCl 的培养基中生长，最适盐浓度为 1.5% NaCl。

A. salmonicida 的基因组大小为 4.6 Mbp，GC 含量为 39.6%，含有 3.3 Mbp 和1.2 Mbp大小的 2 条环状染色体、4 个环状质粒和 111 个蛋白编码序列(Hjerde et al.，2008)。

在血清学上，*A. salmonicida* 是异质性的。尽管存在于表面层中的一种称为VS－P1 的疏水蛋白是所有菌株中的显性抗原(Espelid et al.，1987；Hjelmeland et al.，1988)，但基于它们的脂多糖结构已描述了两种不同的血清型：第一种血清型全部分离自鲑科鱼类，大部分来自鳕；第二种血清型包括来自挪威北部鳕鱼养殖场的一些分离株(Schrøder et al.，1992)。

希特拉病出现于 1977 年，但首次大规模疫情发生在 1979 年的特隆赫姆南部希特拉岛周围的挪威鲑鱼养殖场内(Egidius et al.，1981，1986)。这次疫情在希特拉和北方一直持续到 1983 年的深秋，后来据报道在更南部的斯塔万格附近也有暴发。鱼类养殖场最密集的卑尔根地区受到严重影响。苏格兰(Bruno et al.，1985，1986)、法罗群岛(Dalsgaard et al.，1988)及加拿大的新不伦瑞克和新斯科舍(Sørum et al.，1992)也有疫情报道。

A. salmonicida 的传播方式和感染途径尚不清楚。该微生物在海洋环境具有进入活的非可培养(viable but non-culturable，VBNC)状态的高潜力(Enger et al.，1990)。此外，还认为该病原可作为浮游细菌或在颗粒物表面通过海水传

播(Holm et al., 1985; Sørum et al., 1990; Nordmo et al., 1997)。

在持续流行冷水弧菌病的养殖场沉积物中可分离出*A. salmonicida* (Hoff, 1989a,b)。在实验感染鱼的粪便中也检测到了该病原菌。因此，*A. salmonicida* 可以通过污染的粪便在海洋沉积物中持续存在(Enger et al., 1989)。在此前没有希特拉病史的养殖场沉积物中也检测到该细菌(Enger et al., 1990, 1991)。基于这些观察结果，无症状带菌鱼可能存在，并有助于希特拉病在海洋环境中的传播(Actis et al., 1999)。

23.4.2 感染诊断

1. 疫病临床症状

希特拉病的临床症状是非特异性的，但通常包括嗜睡和停止摄食。受影响的鱼体色发暗，呈现眼球突出，肛门肿胀，沿腹部、胸鳍、腹鳍和臀鳍基部针尖状出血等病征。鳃通常苍白。在体内，该疾病的特征是贫血和出血，并伴有全身性败血症，在垂死或新近死鱼的血液中有大量细菌。出血主要见于盲肠、腹部脂肪和肾脏等内脏器官周围的包膜内(Poppe et al., 1985; Egidius et al., 1986; Kent and Poppe, 2002)。在鳕鱼苗中进行的实验研究表明，可能发生白内障、颅内出血和脾肿大(Sørum et al., 1990)。

组织病理学上，在大西洋鲑组织中检测到的细菌数量与形态学损伤程度之间存在密切关系；最严重的细胞损伤发生在血液供应丰富的地方(Totland et al., 1988)。在毛细血管内皮细胞的腔侧细胞衣和细胞膜中可见最早的损伤迹象。后来，在内皮细胞和白细胞中也发现了细菌。内皮细胞可能完全崩解，并且可以在周围组织的血管外腔中检测到活跃增殖的细菌。

2. *A. salmonicida* 的检测

希特拉病的推定诊断基于培养和随后使用标准生化测试鉴定致病菌(表23.1)。*A. salmonicida* 是革兰氏阴性菌，过氧化氢酶和细胞色素氧化酶阳性，在TCBS上不生长。该菌兼性厌氧，并对弧菌抑菌剂 O/129 非常敏感。该菌无法产生吲哚和硫化氢；柠檬酸盐不能用作唯一碳源，不能还原硝酸盐。该菌从果糖、半乳糖、葡萄糖、麦芽糖、核糖、海藻糖和甘露醇中产酸。它对人或绵羊红细胞不产生溶血。对 *A. salmonicida* 的血清学确认是基于使用特异性商用多克隆抗血清的玻片凝集试验，而荧光抗体试验通常用于常规检测(Toranzo et al., 2005)。据我们所知，目前还没有开发出基于 PCR 的方法来鉴定该病原菌。

23.4.3 致病机制

虽然 *A. salmonicida* 在挪威水产养殖中已闻名超过 25 年，但只有少数研究确定了该病原菌的潜在毒力因子。运动性与一些菌的定植和毒力有关。*A. salmonicida*依靠其运动性来侵入鱼体(Bjelland et al.，2012)。Karlsen 等(2008)指出，*A. salmonicida* 的运动性受环境因素(如渗透压和温度)的调节，而 Raeder 等(2007)的研究则表明，合成培养基中鱼类皮肤黏液的存在增加了与运动有关的蛋白水平(鞭毛蛋白 FlaC、FlaD 和 FlaE)。这三种鞭毛蛋白分子聚合形成鞭毛丝，虽然它们单独存在时对于鞭毛合成和运动性不是必需的，但在鳗弧菌中，FlaD 和 FlaE 的缺失显著降低了毒力。因此，观察到的 *A. salmonicida* 鞭毛蛋白的诱导使其成为可能的毒力因子。

A. salmonicida 具有神奇的生物发光现象(Fidopiastis et al.，1999)。它的细胞在培养中会发光，但只有当它们接触脂肪醛和主要的费氏弧菌自诱导物 *N*－(3－羰基己酰基)－1－高丝氨酸内酯(发光基因 *lux* 的转录激活剂)时才会发光。对 *A. salmonicida* 的 *lux* 操纵子的克隆和测序表明，发光所需的所有基因存在同源性：*luxAB*(荧光素酶)和 *luxCDE*(脂肪醛合成)。此外，*A. salmonicida*菌株具有减少生物发光的群体感应调节基因 *luxR* 和 *luxI* 的新排列及同源基因数量。*luxA* 基因的突变导致了大西洋鲑死亡率的显著延迟(Nelson et al.，2007)。

在患有希特拉病的鱼中观察到的组织损伤表明，*A. salmonicida* 在感染期间会分泌蛋白质。已经在其基因组中鉴定出几种蛋白分泌系统，包括 3 种Ⅰ型分泌系统(T1SS)、1 种Ⅱ型分泌系统(T2SS)、2 种Ⅵ型分泌系统(T6SSI 和 T6SS2)和 1 种 Flp 型菌毛系统。这些系统的存在表明，该菌可以分泌细胞外毒素和酶(Hjerde et al.，2008；Bjelland et al.，2013)。

据推测，在希特拉病感染期间观察到的大量瘀点出血是由细胞外毒素引起的(Holm et al.，1985；Totland et al.，1988)。虽然基因组分析已经确定了三种推定的类似于鳗弧菌和创伤弧菌中的溶血素，但在活跃感染期间这些溶血素的表达水平较低。因此，在疾病期间发生的广泛性溶血可能是鱼类免疫系统活动的结果(Bjelland et al.，2013)。实际上，这些研究者描述了大西洋鲑中的补体系统如何对自身组织造成极大损害的可能性，并且 C3 因子可以回答在希特拉病中观察到的病理体征。

Colquhoun 和 Sørum (2001)证实，在 *A. salmonicida* 中存在一种基于铁载体的铁螯合机制，其包含一种氧肟酸盐型的单一铁螯合分子和一系列铁调控外膜蛋白。他们还报告说，这种铁载体在 10 ℃左右及以下温度时产生最多，这与希特拉病发生时的温度密切相关。Winkelmann 等(2002)的研究进一步证实，该分子为 bisucaberin，是一种先前从海洋细菌河豚毒素交替单胞菌(*Alteromonas haloplanktis*)中分离出的环状二异羟肟酸盐。此外，对 *A. salmonicida*基因组序列的分析表明，该菌存在一种与其他弧菌科细菌具有高序列相似性和共通性的血红素摄取系统。此外，还描述了一种有助于血红素复合物和铁-铁载体跨外膜转运的功能性 TonB 系统(Hjerde et al., 2008)。

A. salmonicida 可在鱼血中增殖并抵抗血清的杀菌活性(Bjelland et al., 2013)。在鲑鱼中进行的研究表明，该菌可被宿主识别并诱导一种快速、强烈但短暂的免疫反应。这表明该菌对重要的鲑鱼细胞因子具有抗性，并可能还具有在感染期间抑制宿主免疫应答的机制。也有证据表明，该病原菌已经发展出抵抗补体杀菌作用的抗性机制。

23.4.4 预防和控制

冷水弧菌病的暴发与鱼类分级、运输、不良养殖管理和环境条件差、养殖密度高、水质差等环境因素有关。当确认疾病暴发时，可使用化学药物治疗，如土霉素(每天每千克体重 75～100 mg，持续 10 d)、噁喹酸(每天每千克体重 10～30 mg，持续 10 d)和磺胺类药物(每天每千克体重 25 mg，持续 15～10 d) (Rigos and Troisi, 2005)。

已经对免疫接种预防冷水弧菌病的有效性进行了大量研究。大多数被测试的疫苗为福尔马林灭活疫苗，并通过不同给药途径对这些疫苗进行了评估(Holm and Jørgensen, 1987; Hjeltnes et al., 1989; Lillehaug et al., 1990; Schrøder et al., 1992)。结果表明，这些疫苗制剂显示出良好的应用前景(提供 90%～100%的保护)，可用于预防大西洋鲑和鳕中 *A. salmonicida* 引起的疾病。此外，含有至少两种致病性弧菌(*V. anguillarum* 和 *A. salmonicida*)的油佐剂灭活疫苗被系统地用于北欧国家鲑科鱼类的免疫接种(Toranzo et al., 1997)。

23.5 总结与研究展望

本章所介绍的弧菌科中 3 种不同致病菌所引起的弧菌病是海水养殖业中具有严重经济危害性的疾病。目前，人们在弧菌病病原诊断、预防控制及相关毒力因子研究等方面都做了大量工作，以保护鱼类免受这些病原的侵害。

虽然目前对这些病原与宿主相互作用的研究取得了重要进展，但其发病机制尚未完全阐明。新研究策略的发展，如"双转录组测序技术(dual RNA-seq)"，可以在感染期间同时分析病原体和宿主中的 RNA 表达，从而可以揭示正在发生的基因功能，并可以进一步有效预防疾病。

最后，气候变化引起的水温升高和海洋酸化会改变养殖鱼类的生理机能以及弧菌病病原菌种群的丰富和多样性。因此，气候变化对海洋鱼类弧菌病流行的影响需要进一步评估。

致谢

A. E. Toranzo 和 B. Magariños 感谢迅塔 · 德加利西亚(西班牙)项目

GRC-2014/007所提供的经费资助。这项工作得到智利国家科学技术研究委员会(CONICYT)以及智利安德烈斯·贝洛大学基金项目FONDAP No. 15110027和FONDECYT No. 1150695的部分资助。

参考文献

[1] Actis, L.A., Tolmasky, M.E. and Crosa, J.H. (1999) Vibriosis. In: Woo, P.T.K. and Bruno, D.W. (eds) *Fish Diseases and Disorders, Volume. 3: Viral, Bacterial and Fungal Infections*. CAB International, Wallingford, UK, pp. 523-558.

[2] Akayli, T., Timur, G., Albayrak, G. and Aydemir, B. (2010) Identification and genotyping of *Vibrio ordalii*: a comparison of different methods. *Israeli Journal of Aquaculture Bamidgeh* 62, 9-18.

[3] Alice, A.F., López, C.S. and Crosa, J.H. (2005) Plasmidand chromosome-encoded redundant and specific functions are involved in biosynthesis of the siderophore anguibactin in *Vibrio anguillarum* 775: a case of chance and necessity? *Journal of Bacteriology* 187, 2209-2214.

[4] Angelidis, P. (2014) *Vibrio anguillarum*-associated vibriosis in the Mediterranean aquaculture. In: Angelidis, P. (ed.) *Aspects of Mediterranean Marine Aquaculture*. Blue Crab PC, Thessaloniki, Greece, pp. 243-264.

[5] Austin, B. and Austin, D.A. (2007) *Bacterial Fish Pathogens: Diseases of Farmed and Wild Fish*, 4th edn. Springer and Praxis Publishing, Dordrecht, The Netherlands, Berlin/Heidelberg, New York and Chichester, UK.

[6] Austin, B., Austin, D.A., Blanch, A.R., Cerda, M., Grimont, F. *et al*. (1997) A comparison of methods for the typing of fish-pathogenic *Vibrio* spp. *Systematic and Applied Microbiology* 20, 89-101.

[7] Avendaño-Herrera, R., Maldonado, J.P., Tapia-Cammas, D., Feijoó, C.G., Calleja, F. *et al*. (2014) PCR protocol for the detection of *Vibrio ordalii* by amplification of the *vohB* (hemolysin) gene. *Diseases of Aquatic Organisms* 107, 223-234.

[8] Balado, M., Osorio, C.R. and Lemos, M.L. (2006) A gene cluster involved in the biosynthesis of vanchrobactin, a chromosome-encoded siderophore produced by *Vibrio anguillarum*. *Microbiology* 152, 3517-3528.

[9] Balado, M., Osorio, C.R. and Lemos, M.L. (2008) Biosynthetic and regulatory elements involved in the production of the siderophore vanchrobactin in *Vibrio anguillarum*. *Microbiology* 154, 1400-1413.

[10] Benediktsdóttir, E., Verdonk, L., Spröer, C., Helgason, S. and Swings, J. (2000) Characterization of *Vibrio viscosus* and *Vibrio wodanis* isolated at different geographical locations: a proposal for reclassification of *Vibrio viscosus* as *Moritella viscosa* comb. nov. *International Journal of Systematic and Evolutionary Microbiology* 50, 479-488.

[11] Bjelland, A.M., Johansen, R., Brudal, E., Hansen, H., Winther-Larsen, H. *et al*. (2012) *Vibrio salmonicida* pathogenesis analyzed by experimental challenge of Atlantic salmon (*Salmo salar*). *Microbial Pathogenesis* 52, 77-84.

[12] Bjelland, A.M., Fauske, A.K., Nguyen, A., Orlien, I.E., Østgaard, I.M. *et al*. (2013) Expression of *Vibrio salmonicida* virulence genes and immune response parameters in experimentally challenged Atlantic salmon (*Salmo salar* L.). *Frontiers in Microbiology* 4: 401.

[13] Boesen, H.T., Pedersen, K., Larsen, J.L., Koch, C. and Ellis, A.E. (1999) *Vibrio anguillarum* resistance to rainbow trout (*Oncorhynchus mykiss*) serum: role of O-

antigen structure of lipopolysaccharide. *Infection and Immunity* 67, 294 - 301.
[14] Bohle, H., Kjetil, F., Bustos, P., Riofrío, A. and Peters, C. (2007) Fenotipo atípico de Vibrio ordalii, bacteria altamente patogénica aislada desde salmón del Atlántico cultivado en las costas marinas del sur de Chile. *Archivos de Medicina Veterinaria* 39, 43 - 52.
[15] Bordas, M.A., Balebona, M.C., Rodríguez-Maroto, J.M., Borrego, J.J. and Moriñigo, M.A. (1998) Chemotaxis of pathogenic *Vibrio* strains towards mucus surfaces of gilt-head sea bream (*Sparus aurata* L.). *Applied and Environmental Microbiology* 64, 1573 - 1575.
[16] Brackman, G., Celen, S., Baruah, K., Bossier, P., Van Calenbergh, S. *et al.* (2009) AI - 2 quorum-sensing inhibitors affect the starvation response and reduce virulence in several *Vibrio* species, most likely by interfering with Lux-PQ. *Microbiology* 155, 4114 -4122.
[17] Bruhn, J.B., Dalsgaard, I., Nielsen, K.F., Buchholtz, C., Larsen, J.L. *et al.* (2005) Quorum sensing signal molecules (acylated homoserine lactones) in Gram-negative fish pathogenic bacteria. *Diseases of Aquatic Organisms* 65, 43 - 52.
[18] Bruno, D.W., Hasting, T.S., Ellis, A.E. and Wotten, R. (1985) Outbreak of a cold water vibriosis in Atlantic salmon in Scotland. *Bulletin of the European Association of Fish Pathologists* 5, 62 - 63.
[19] Bruno, D.W., Hastings, T.S. and Ellis, A.E. (1986) Histopathology, bacteriology and experimental transmission of a cold water vibriosis in Atlantic salmon *Salmo salar*. *Diseases of Aquatic Organisms* 1, 163 - 168.
[20] Brunt, J., Newaj-Fyzul, A. and Austin, B. (2007) The development of probiotics for the control of multiple bacterial diseases of rainbow trout, *Oncorhynchus mykiss* (Walbaum). *Journal of Fish Diseases* 30, 573 - 579.
[21] Carnevali, O., de Vivo, L., Sulpizio, R., Gioacchini, G., Olivotto, I. *et al.* (2006) Growth improvement by probiotic in European sea bass juveniles (*Dicentrarchus labrax*, L.), with particular attention to IGF-1, myostatin and cortisol gene expression. *Aquaculture* 258, 430 - 438.
[22] Chabrillon, M., Arijo, S., Díaz-Rosales, P., Balebona, M. C. and Moriñigo, M. A. (2006) Interference of *Listonella anguillarum* with potential probiotic microorganisms isolated from farmed gilthead sea bream (*Sparus aurata* L.). *Aquaculture Research* 37, 78 - 86.
[23] Chart, H. and Trust, T.J. (1984) Characterization of surface antigens of the marine fish pathogens *Vibrio anguillarum* and *Vibrio ordalii*. *Canadian Journal of Microbiology* 30, 703 - 710.
[24] CLSI (2006) *Methods for Antimicrobial Disk Susceptibility Testing of Bacteria Isolated from Aquatic Animals. Approved Guidelines*. CLSI Document M42 - A, Clinical and Laboratory Standards Institute, Wayne, Pennsylvania.
[25] Colquhoun, D.J. and Lillehaug, A. (2014) Vaccination against vibriosis. In: Gudding, R., Lillehaug, A. and Evensen, Ø. (eds) *Fish Vaccination*. Wiley-Blackwell, Chichester/Oxford, UK, pp. 172 - 184.
[26] Colquhoun, D.J. and Sørum, H. (2001) Temperature dependent siderophore production in *Vibrio salmonicida*. *Microbial Pathogenesis* 31, 213 - 219.
[27] Colquhoun, D.J., Aase, I.L., Wallace, C., Baklien, Å. And Gravningen, K. (2004) First description of *Vibrio ordalii* from Chile. *Bulletin of the European Association of Fish Pathologists* 24, 185 - 188.
[28] Colquhoun, D.J., Aarflot, L. and Melvold, C.F. (2007) *gyrA* and *parC* mutations and associated quinolone resistance in *Vibrio anguillarum* serotype O2b strains isolated from farmed Atlantic cod (*Gadus morhua*) in Norway. *Antimicrobial Agents and*

Chemotherapy 51, 2597 - 2599.

[29] Conchas, R.F., Lemos, M.L., Barja, J.L. and Toranzo, A.E. (1991) Distribution of plasmid-and chromosomemediated iron uptake systems in *Vibrio anguillarum* strains of different origins. *Applied and Environmental Microbiology* 57, 2956 - 2962.

[30] Crosa, J.H. (1980) A plasmid associated with virulence in the marine fish pathogen *Vibrio anguillarum* specifies an iron-sequestering system. *Nature* 284, 566 - 568.

[31] Croxatto, A., Chalker, V.J., Lauritz, J., Jass, J., Hardman, A. *et al.* (2002) Van T, a homologue of *Vibrio harveyi* LuxR, regulates serine, metalloprotease, pigment, and biofilm production in *Vibrio anguillarum*. *Journal of Bacteriology* 184, 1616 - 1629.

[32] Croxatto, A., Pride, J., Hardman, A., Williams, P., Camara, M. *et al.* (2004) A distinctive dual-channel quorum-sensing system operates in *Vibrio anguillarum*. *Molecular Microbiology* 52, 1677 - 1689.

[33] Croxatto, A., Lauritz, J., Chen, C. and Milton, D.L. (2007) *Vibrio anguillarum* colonization of rainbow trout integument requires a DNA locus involved in exopolysaccharide transport and biosynthesis. *Environmental Microbiology* 9, 370 - 383.

[34] Dalsgaard, I., Jürgens, O. and Mortensen, A. (1988) *Vibrio salmonicida* isolated from farmed Atlantic salmon in Faroe Islands. *Bulletin of the European Association of Fish Pathologists* 8, 53 - 54.

[35] Daly, J. and Stevenson, R. (1987) Hydrophobic and haemagglutinating properties of *Renibacterium salmoninarum*. *Journal of General Microbiology* 133, 3575 - 3580.

[36] Defoirdt, T., Boon, N., Bossier, P. and Verstraete, W. (2004) Disruption of bacterial quorum sensing: an unexplored strategy to fight infections in aquaculture. *Aquaculture* 240, 69 - 88.

[37] Denkin, S.M. and Nelson, D.R. (2004) Regulation of *Vibrio anguillarum empA* metalloprotease expression and its role in virulence. *Applied and Environmental Microbiology* 70, 4193 - 4204.

[38] Egidius, E., Andersen, K., Claussen, E. and Raa, J. (1981) Cold-water vibriosis or Hitra-disease in Norwegian salmonid farming. *Journal of Fish Diseases* 4, 353 - 354.

[39] Egidius, E., Wiik, R., Andersen, K., Holff, K.A. and Hjeltness, B. (1986) *Vibrio salmonicida* sp. nov., a new fish pathogen. *International Journal of Systematic Bacteriology* 36, 518 - 520.

[40] Engelsen, A.R., Sandlund, N., Fiksdal, I.U. and Bergh, Ø. (2008) Immunohistochemistry of Atlantic cod larvae *Gadus morhua* experimentally challenged with *Vibrio anguillarum*. *Diseases of Aquatic Organisms* 80, 13 - 20.

[41] Enger, Ø., Husevag, B. and Goksøyr, J. (1989) Presence of *Vibrio salmonicida* in fish farm sediments. *Applied and Environmental Microbiology* 55, 2815 - 2818.

[42] Enger, Ø., Hoff, K.A., Schei, G.H. and Dundas, I. (1990) Starvation survival of the fish pathogenic bacteria *Vibrio anguillarum* and *Vibrio salmonicida* in marine environments. *FEMS Microbiology Letters* 74, 215 - 220.

[43] Enger, Ø., Husevag, B. and Goksøyr, J. (1991) Seasonal variation in presence of *Vibrio salmonicida* and total bacterial counts in Norwegian fish-farm water. *Canadian Journal of Microbiology* 37, 618 - 623.

[44] Espelid, S., Hjelmeland, K. and Jørgensen, T. (1987) The specificity of Atlantic salmon antibodies made against the fish pathogen *Vibrio salmonicida*, establishing the surface protein VS-P1 as the predominant antigen. *Developmental and Comparative Immunology* 11, 529 - 537.

[45] Ezura, Y., Tajima, K., Yoshimizu, M. and Kimura, T. (1980) Studies on the taxonomy and serology of causative organisms of fish vibriosis. *Fish Pathology* 14, 167 - 179.

[46] Farmer, J.J., Janda, M., Brenner, F.W., Cameron, D.N. and Birkhead, K.M. (2005)

Genus I. Vibrio Pacini 1854. In: Brenner, D.J., Krieg, N.R. and Staley, J.T. (eds) *Bergeys's Manual of Systematic Bacteriology, Volume 2: The Proteobacteria, Part B: The Gammaproteobacteria*, 2nd edn. Springer, New York, pp. 495 - 545.

[47] Fidopiastis, P.M., Sørum, H. and Ruby, E.G. (1999) Cryptic luminescence in the coldwater fish pathogen *Vibrio salmonicida*. *Archives in Microbiology* 171, 205 - 209.

[48] Fouz, B., Devesa, S., Gravningen, K., Barja, J. L. and Toranzo, A. E. (1990) Antibacterial action of the mucus of turbot. *Bulletin of the European Association of Fish Pathologists* 10, 56 - 59.

[49] Frans, I., Michiels, C.W., Bossier, P., Willems, K.A., Lievens, B. *et al.* (2011) *Vibrio anguillarum* as a fish pathogen: virulence factors, diagnosis and prevention. *Journal of Fish Diseases* 34, 643 - 661.

[50] Gao, H., Li, F., Zhang, X., Wang, B. and Xiang, J. (2010) Rapid, sensitive detection of *Vibrio anguillarum* using loop mediated isothermal amplification. *Chinese Journal of Oceanology and Limnology* 28, 62 - 66.

[51] Gram, L., Melchiorsen, J., Spanggard, B., Huber, I. and Nielsen, T. F. (1999) Inhibition of *Vibrio anguillarum* by *Pseudomonas fuorescens* AH2, a possible treatment of fish. *Applied and Environmental Microbiology* 65, 963 - 973.

[52] Grisez, L., Chair, M., Zorruelos, P. and Ollevier, F. (1996) Mode of infection and spread of *Vibrio anguillarum* in turbot *Scophthalmus maximus* larvae after oral challenge through live feed. *Diseases of Aquatic Organisms* 26, 181 - 187.

[53] Harrel, L.W., Etlinger, H.M. and Hodgins, H.O. (1976a) Humoral factors important in resistance of salmonid fish to bacterial disease. II. Anti-*Vibrio anguillarum* activity in mucus and observations on complement. *Aquaculture* 7, 363 - 370.

[54] Harrell, L.W., Novotny, A.J., Schiewe, M.J. and Hodgins, H.O. (1976b) Isolation and description of two vibrios pathogenic to Pacific salmon in Puget Sound, Washington. *Fishery Bulletin* 74, 447 - 449.

[55] Håstein, T., Gudding, R. and Evensen, Ø. (2005) Bacterial vaccines for fish — an update of the current situation worldwide. *Developmental Biology* 121, 55 - 74.

[56] Henríquez, P., Kaiser, M., Bohle, H., Bustos, P. and Mancilla, M. (2016) Comprehensive antibiotic susceptibility profiling of Chilean *Piscirickettsia salmonis* field isolates. *Journal of Fish Diseases* 39, 441 - 448.

[57] Hirono, I., Masuda, T. and Aoki, T. (1996) Cloning and detection of the hemolysin gene of *Vibrio anguillarum*. *Microbial Pathogenesis* 21, 173 - 182.

[58] Hjelmeland, K., Stensvåg, K., Jørgensen, T. and Espelid, S. (1988) Isolation and characterization of a surface layer antigen from *Vibrio salmonicida*. *Journal of Fish Diseases* 11, 197 - 205.

[59] Hjeltnes, B., Andersen, K. and Ellingsen, H. (1989) Vaccination against *Vibrio salmonicida*. The effect of different routes of administration and of revaccination. *Aquaculture* 83, 1 - 2.

[60] Hjerde, E., Lorentzen, M.S., Holden, M.T.G., Seeger, K., Paulsen, S. *et al.* (2008) The genome sequence of the fish pathogen *Aliivibrio salmonicida* strain LFI1238 shows extensive evidence of gene decay. *BMC Genomics* 9: 616.

[61] Hoff, K.A (1989a) Survival of *Vibro anguillarum* and *Vibrio salmonicida* at different salinities. *Applied and Environmental Microbiology* 55, 1775 - 1786.

[62] Hoff, K.A. (1989b) Occurrence of the fish pathogen *Vibrio salmonicida* in faeces from Atlantic salmon, *Salmo salar* L., after experimental infection. *Journal of Fish Diseases* 12, 595 - 597.

[63] Holm, K.O. and Jørgensen, T. (1987) A successful vaccination of Atlantic salmon, *Salmo salar* L. against "Hitra disease" or coldwater vibriosis. *Journal of Fish Diseases* 10, 85 - 90.

[64] Holm, K. O., Strøm, E., Stensvaag, K., Raa, J. and Jørgense, T. Ø. (1985) Characteristics of a *Vibrio* sp. associated with the "Hitra disease" of Atlantic salmon in Norwegian fish farms. *Fish Pathology* 20, 125 - 129.
[65] Holm, K. O., Nilsson, K., Hjerde, E., Willassen, N.-P. and Milton, D. (2015) Complete genome sequence of *Vibrio anguillarum* NB10, a virulent isolate from the Gulf of Bothnia. *Standards in Genomic Sciences* 10: 60.
[66] Hong, G.-E., Kim, D.-G., Bae, J.-Y., Ahn, S.-H., Bai, S.C. *et al.* (2007) Species-specific PCR detection of the fish pathogen, *Vibrio anguillarum*, using the *amiB* gene, which encodes N-acetylmuramoyl-L-alanine amidase. *FEMS Microbiology Letters* 269, 201 - 206.
[67] Ito, H., Uchida, I., Sekizaki, T. and Terakado, N. (1995) A specific oligonucleotide probe based on 5S rRNA sequences for identification of *Vibrio anguillarum* and *Vibrio ordalii*. *Veterinary Microbiology* 43, 167 - 171.
[68] Kanno, T., Nakai, T and Muroga, K. (1989) Mode of transmission of vibriosis among ayu, *Plecoglossus altivelis*. *Journal of Aquatic Animal Health* 1, 2 - 6.
[69] Kao, D. Y., Cheng, Y. C., Kuo, T. Y., Lin, S. B., Lin, C. C. *et al.* (2009) Salt-responsive outer membrane proteins of *Vibrio anguillarum* serotype O1 as revealed by comparative proteome. *Journal of Applied Microbiology* 106, 2079 - 2085.
[70] Karlsen, C., Paulsen, S.M., Tunsjø, H.S., Krinner, S., Sørum, H. *et al.* (2008) Motility and flagellin gene expression in the fish pathogen *Vibrio salmonicida*: effects of salinity and temperature. *Microbial Pathogenesis* 45, 258 - 264.
[71] Kent, M.L. and Poppe, T.T. (2002) Infectious diseases coldwater fish in marine and brackish water. In: Woo, P.T.K., Bruno, D.W. and Lim, H.S. (eds) *Diseases and Disorders of Finfish in Cage Culture*, 1st edn. CAB International, Wallingford, UK, pp. 61 - 105.
[72] Kim, D.-G., Kim, Y.-R., Kim, E.-Y., Cho, H.M., Ahn, S.-H. *et al.* (2010) Isolation of the *groESL* cluster from *Vibrio anguillarum* and PCR detection targeting *groEL* gene. *Fisheries Science* 76, 803 - 810.
[73] Kodama, H., Moustafa, M., Ishiguro, S., Mikami, T. and Izawa, H. (1984) Extracellular virulence factors of fish *Vibrio*: relationships between toxic material, hemolysin, and proteolytic enzyme. *American Journal of Veterinary Research* 45, 2203 -2207.
[74] Kulkarni, A., Caipang, C. M. A., Brinchmann, M. F., Korsnes, K. and Kiron, V. (2009) Use of loop-mediated isothermal amplification (LAMP) assay for the detection of *Vibrio anguillarum* O2b, the causative agent of vibriosis in Atlantic cod, *Gadus morhua*. *Journal of Rapid Methods and Automation in Microbiology* 17, 503 - 518.
[75] Lamas, J., Santos, Y., Bruno, D., Toranzo, A. E. and Anadon, R. (1994) A comparison of pathological changes caused by *Vibrio anguillarum* and its extracellular products in rainbow trout (*Oncorhynchus mykiss*). *Fish Pathology* 29, 79 - 89.
[76] Larsen, J.L. and Mellergaard, S. (1984) Agglutination typing of *Vibrio anguillarum* isolates from diseased fish and from the environment. *Applied and Environmental Microbiology* 47, 1261 - 1265.
[77] Larsen, J. L., Pedersen, K. and Dalsgaard, I. (1994) *Vibrio anguillarum* serovars associated with vibriosis in fish. *Journal of Fish Diseases* 17, 259 - 267.
[78] Larsen, M.H., Larsen, J.L. and Olsen, J.E. (2001) Chemotaxis of *Vibrio anguillarum* to fish mucus: role of the origin of the fish mucus, the fish species and the serogroup of the pathogen. *FEMS Microbiology Ecology* 38, 77 - 80.
[79] Lemos, M.L. and Osorio, C.R. (2007) Heme, an iron supply for vibrios pathogenic for fish. *BioMetals* 20, 615 - 626.
[80] Lemos, M. L., Salinas, P., Toranzo, A. E., Barja, J. L. and Crosa, J. H. (1988)

Chromosome-mediated iron uptake systems in pathogenic strains of *Vibrio anguillarum*. *Journal of Bacteriology* 170, 1920 – 1925.

[81] Lemos, M.L., Balado, M. and Osorio, C.R. (2010) Anguibactin versus vanchrobactin-mediated iron uptake in *Vibrio anguillarum*: evolution and ecology of a fish pathogen. *Environmental Microbiology Reports* 2, 19 – 26.

[82] Li, G., Mo, Z., Li, J., Xiao, P. and Hao, B. (2013) Complete genome sequence of *Vibrio anguillarum* M3, a serotype O1 strain isolated from Japanese flounder in china. *Genome Announcement* 1(5): e00769 – 13.

[83] Lillehaug, A., Sørum, R. H. and Ramstad, A. (1990) Cross-protection after immunization of Atlantic salmon *Salmo salar* L., against different strains of *Vibrio salmonicida*. *Journal of Fish Diseases* 13, 519 – 523.

[84] Lillehaug, A., Lunestad, B.T. and Grave, K. (2003) Epidemiology of bacterial diseases in Norwegian aquaculture – a description based on antibiotic prescription data for the ten-year period 1991 to 2000. *Diseases of Aquatic Organisms* 53, 115 – 125.

[85] MacDonell, M.T. and Colwell, R.R. (1984) Nucleotide base sequence of *Vibrionaceae* 5S rRNA. *FEBS Letters* 175, 183 – 188.

[86] Mazoy, R., Osorio, C. R., Toranzo, A. E. and Lemos, M. L. (2003) Isolation of mutants of *Vibrio anguillarum* defective in haeme utilization and cloning of *huvA*, a gene coding for an outer membrane protein involved in the use of haeme as iron source. *Archives of Microbiology* 179, 329 – 338.

[87] Midtlyng, P.J. and Lillehaug, A. (1998) Growth of Atlantic salmon *Salmo salar* after intraperitoneal administration of vaccines containing adjuvants. *Diseases of Aquatic Organisms* 32, 91 – 97.

[88] Mikkelsen, H., Lund, V., Martinsen, L.C., Gravningen, K. and Schrøeder, M.B. (2007) Variability among *Vibrio anguillarum* O2 isolates from Atlantic cod (*Gadus morhua* L.): characterization and vaccination studies. *Aquaculture* 266, 16 – 25.

[89] Milton, D. L. (2006) Quorum sensing in vibrios: complexity for diversification. *International Journal of Medical Microbiology* 296, 61 – 71.

[90] Milton, D.L., Norqvist, A. and Wolfwatz, H. (1992) Cloning of a metalloprotease gene involved in the virulence mechanism of *Vibrio anguillarum*. *Journal of Bacteriology* 174, 7235 – 7244.

[91] Milton, D. L., Hardman, A., Camara, M., Chhabra, S. R., Bycroft, B. W. *et al.* (1997) Quorum sensing in *Vibrio anguillarum*: characterization of the *vanI*/*vanR* locus and identification of the autoinducer N-(3-oxodecanoyl)-L homoserine lactone. *Journal of Bacteriology* 179, 3004 – 3012.

[92] Milton, D.L., Chalker, V.J., Kirke, D., Hardman, A., Camara, M. and Williams, P. (2001) The LuxM homologue VanM from *Vibrio anguillarum* directs the synthesis of N-(3-hydroxyhexanoyl) homoserine lactone and N-hexanoylhomoserine lactone. *Journal of Bacteriology* 183, 3537 – 3547.

[93] Montgomery, M.T. and Kirchman, D.L. (1994) Induction of chitin-binding proteins during the specific attachment of the marine bacterium *Vibrio harveyi* to chitin. *Applied and Environmental Microbiology* 60, 4284 – 4288.

[94] Mouriño, S., Osorio, C.R. and Lemos, M.L. (2004) Characterization of heme uptake cluster genes in the fish pathogen *Vibrio anguillarum*. *Journal of Bacteriology* 186, 6159 – 6167.

[95] Muroga, K. and de la Cruz, M.C. (1987) Fate and localization of *Vibrio anguillarum* in tissues of artificially infected ayu (*Plecoglossus altivelis*). *Fish Pathology* 22, 99 – 103.

[96] Muroga, K., Yasuhiko, J. and Masumura, K. (1986) *Vibrio ordalii* isolated from diseased ayu (*Plecoglossus altivelis*) and rockfish (*Sebastes schlegeli*). *Fish Pathology* 21, 239 – 243.

[97] Mutharia, L.W., Raymond, B.T., Dekievit, T.R. and Stevenson, R.M.W. (1993) Antibody specificities of polyclonal rabbit and rainbow trout antisera against *Vibrio ordalii* and serotype O2 strains of *Vibrio anguillarum*. *Canadian Journal of Microbiology* 39, 492 - 499.
[98] Naka, H., Dias, G.M., Thompson, C.C., Dubay, C., Thompson, F.L. *et al*. (2011) Complete genome sequence of the marine fish pathogen *Vibrio anguillarum* harboring the pJM1 virulence plasmid and genomic comparison with other virulent strains of *V. anguillarum* and *V. ordalii*. *Infection and Immunity* 79, 2889 - 2900.
[99] Nelson, E.J., Tunsjø, H.S., Fidopiastis, P.M., Sørum, H. and Ruby, E.G. (2007) A novel lux operon in the cryptically bioluminescent fish pathogen *Vibrio salmonicida* is associated with virulence. *Applied and Environmental Microbiology* 73, 1825 - 1833.
[100] Nordmo, R., Sevatdal, S. and Ramstad, A. (1997) Experimental infection with *Vibrio salmonicida* in Atlantic salmon (*Salmo salar* L.): an evaluation of three different challenge methods. *Aquaculture* 158, 23 - 32.
[101] Ofek, I. and Doyle, R.J. (1994) *Bacterial Adhesion to Cells and Tissues*. Chapman and Hall, New York.
[102] Ohnishi, K. and Muroga, K. (1976) *Vibrio* sp. as a cause of disease in rainbow trout cultured in Japan. I. *Biochemical characteristics*. *Fish Pathology* 11, 159 - 165.
[103] Olsen, J. E. and Larsen, J. L. (1993) Ribotypes and plasmid contents of *Vibrio anguillarum* strains in relation to serovars. *Applied and Environmental Microbiology* 59, 3863 - 3870.
[104] Olsson, J. C., Joborn, A., Westerdahl, A., Blomberg, L., Kjelleberg, S. *et al*. (1998) Survival, persistence and proliferation of *Vibrio anguillarum* in juvenile turbot, *Scophthalmus maximus* (L.), in intestine and faeces. *Journal of Fish Diseases* 21, 1 - 9.
[105] Osorio, C. and Toranzo, A.E. (2002) DNA-based diagnostics in sea farming. In: Fingerman, M. and Nagabhushanam, R. (eds) *Recent Advances in Marine Biotechnology*, *Volume 7: Seafood Safety and Human Health*. Science Publishers, Plymouth, UK, pp. 253 - 310.
[106] Osorio, C.R., Juiz-Río, S. and Lemos, M.L. (2006) A siderophore biosynthesis gene cluster from the fish pathogen *Photobacterium damselae* subsp. *piscicida* is structurally and functionally related to the *Yersinia* high-pathogenicity island. *Microbiology* 152, 3327 - 3341.
[107] O'Toole, R., Milton, D.L. and Wolf-Watz, H. (1996) Chemotactic motility is required for invasion of the host by the fish pathogen *Vibrio anguillarum*. *Molecular Microbiology* 19, 625 - 637.
[108] O'Toole, R., Lundberg, S., Fredricksson, S.A., Jansson, A., Nilsson, B. *et al*. (1999) The chemotactic response of *Vibrio anguillarum* to fish intestinal mucus is mediated by a combination of multiple mucus components. *Journal of Bacteriology* 181, 4308 - 4317.
[109] O'Toole, R., von Hofsten, J., Rosqvist, R., Olsson, P.E. and Wolf-Watz, H. (2004) Visualisation of zebrafish infection by GFP labelled *Vibrio anguillarum*. *Microbial Pathogenesis* 37, 41 - 46.
[110] Pedersen, K. and Larsen, J. L. (1993) rRNA gene restriction patterns of *Vibrio anguillarum* serogroup O1. *Diseases of Aquatic Organisms* 16, 121 - 126.
[111] Pedersen, K., Grisez, L., van Houdt, R., Tiainen, T., Ollevier, F. *et al*. (1999) Extended serotyping scheme for *Vibrio anguillarum* with the definition of seven provisional O-serogroups. *Current Microbiology* 38, 183 - 189.
[112] Pillidge, C.J. and Colwell, R.R. (1988) Nucleotide sequence of the 5S rRNA from *Listonella* (*Vibrio*) *ordalii* ATCC 33509 and *Listonella* (*Vibrio*) *tubiashii* ATCC

19105. *Nucleic Acids Research* 16, 3111.

[113] Planas, M., Pérez-Lorenzo, M., Hjelm, M., Gram, L., Fiksdal, I.U. et al. (2006) Probiotic effect *in vivo* of *Roseobacter* strain 27 - 4 against *Vibrio* (*Listonella*) *anguillarum* infections in turbot (*Scophthalmus maximus* L.) larvae. *Aquaculture* 255, 323 - 333.

[114] Poblete-Morales, M., Irgang, R., Henríquez-Núñez, H., Toranzo, A.E., Kronvall, G. *et al.* (2013) *Vibrio ordalii* antimicrobial susceptibility testing — modified culture conditions required and laboratory-specific epidemiological cut-off values. *Veterinary Microbiology* 165, 434 - 442.

[115] Poppe, T.T. and Koppang, E. (2014) Vaccination against vibriosis. In: Gudding, R., Lillehaug, A. and Evensen, Ø. (eds) *Fish Vaccination*, Wiley-Blackwell, Chichester/Oxford, UK, pp. 153 - 161.

[116] Poppe, T. T., Håstein, T. and Salte, R. (1985) Hitra disease (haemorrhagic syndrome) in Norwegian salmon farming: present status. In: Ellis, A.E. (ed.) *Fish and Shellfish Pathology. First International Conference of the European Association of Fish Pathologists, Plymouth, England, Sept. 20 - 23*, 1983. Academic Press, London/Orlando, Florida, pp. 223 - 229.

[117] Porsby, C.H., Nielsen, K.F. and Gram, L. (2008) *Phaeobacter* and *Ruegeria* species of the *Roseobacter clade* colonize separate niches in a Danish turbot (*Scophthalmus maximus*)-rearing farm and antagonize *Vibrio anguillarum* under different growth conditions. *Applied and Environmental Microbiology* 74, 7356 - 7364.

[118] Pybus, V., Loutit, M.W., Lamont, I.L and Tagg, J.R. (1994) Growth inhibition of the salmon pathogen *Vibrio ordalii* by a siderophore produced by *Vibrio anguillarum* strain VL4355. *Journal of Fish Diseases* 17, 311 - 324.

[119] Raeder, I.L., Paulsen, S.M., Smalås, A.O. and Willassen, N.P. (2007) Effect of skin mucus on the soluble proteome of *Vibrio salmonicida* analysed by 2 - D gel electrophoresis and tandem mass spectrometry. *Microbial Pathogenesis* 42, 36 - 45.

[120] Ransom, D.P. (1978) Bacteriologic, immunologic and pathologic studies of *Vibrio* sp. pathogenic to salmonids. PhD thesis, Oregon State University, Corvallis, Oregon.

[121] Ransom, D.P., Lannan, C.N., Rohovec, J.S. and Fryer, J.L. (1984) Comparison of histopathology caused by *Vibrio anguillarum* and *Vibrio ordalii* in three species of Pacific salmon. *Journal of Fish Diseases* 7, 107 - 115.

[122] Rasch, M., Buch, C., Austin, B., Slierendrecht, W.J., Ekmann, K.S. *et al.* (2004) An inhibitor of bacterial quorum sensing reduces mortalities caused by vibriosis in rainbow trout (*Oncorhynchus mykiss*, Walbaum). *Systematic and Applied Microbiology* 27, 350 - 359.

[123] Rigos, G. and Troisi, G. M. (2005) Antibacterial agents in Mediterranean finfish farming: a synopsis of pharmacokinetics in important euryhaline fish species and possible environmental implications. *Reviews in Fish Biology and Fisheries* 15, 53 - 73.

[124] Rodkhum, C., Hirono, I., Crosa, J. H. and Aoki, T. (2005) Four novel hemolysin genes of *Vibrio anguillarum* and their virulence to rainbow trout. *Microbial Pathogenesis* 39, 109 - 119.

[125] Rodkhum, C., Hirono, I., Stork, M., DiLorenzo, M., Crosa, J. H. and Aoki, T. (2006) Putative virulence-related genes in *Vibrio anguillarum* identified by random genome sequencing. *Journal of Fish Diseases* 29, 157 - 166.

[126] Romalde, J. L., Magariños, B., Fouz, B., Bandín, I., Nuñez, S. *et al.* (1995) Evaluation of Bionor mono-kits for rapid detection of bacterial fish pathogens. *Diseases of Aquatic Organisms* 21, 25 - 34.

[127] Rowe, G. E. and Welch, R. A. (1994) Assays of hemolytic toxins. *Methods in*

Enzymology 235, 657-667.
[128] Ruiz, P., Poblete, M., Yáñez, A.J., Irgang, R., Toranzo, A.E. *et al.* (2015) Cell surface properties of *Vibrio ordalii* strains isolated from Atlantic salmon *Salmo salar* in Chilean farms. *Diseases of Aquatic Organisms* 113, 9-23.
[129] Ruiz, P., Balado, M., Toranzo, A.E., Poblete-Morales, M., Lemos, M.L. *et al.* (2016a) Iron assimilation and siderophore production by *Vibrio ordalii* strains isolated from diseased Atlantic salmon (*Salmo salar*) in Chile. *Diseases of Aquatic Organisms* 118, 217-226.
[130] Ruiz, P., Poblete-Morales, M., Irgang, R., Toranzo, A.E. and Avendaño-Herrera, R. (2016b) Survival behavior and virulence of the fish pathogen *Vibrio ordalii* in seawater microcosms. *Diseases of Aquatic Organisms* 120, 27-38.
[131] Sadovskaya, I., Brisson, J.R., Khieu, N.H., Mutharia, L.M. and Altman, E. (1998) Structural characterization of the lipopolysaccharide O-antigen and capsular polysaccharide of *Vibrio ordalii* serotype O: 2. *European Journal of Biochemistry* 253, 319-327.
[132] Samuelsen, O.B. and Bergh, Ø. (2004) Efficacy of orally administered florfenicol and oxolinic acid for the treatment of vibriosis in cod (*Gadus morhua*). *Aquaculture* 235, 27-35.
[133] Samuelsen, O.B., Nerland, A.H., Jørgensen, T., Schøder, M.B., Svåsand, T. *et al.* (2006) Viral and bacterial diseases of Atlantic cod *Gadus morhua*, their prophylaxis and treatment: a review. *Diseases of Aquatic Organisms* 71, 239-254.
[134] San Martín N., B., Yatabe, T., Gallardo, A. and Medina, P. (2010) *Manual de Buenas Prácticas en el Uso de Antibióticos y Antiparasitarios en la Salmonicultura Chilena*. Universidad de Chile and Servicio Nacional de Pesca (SERNAPESCA), Santiago, Chile.
[135] Santos, Y., Pazos, F., Bandín, I. and Toranzo, A.E. (1995) Analysis of antigens present in the extracellular products and cell surface of *Vibrio anguillarum* O1, O2 and O3. *Applied and Environmental Microbiology* 61, 2493-2498.
[136] Schiewe, M.H. and Crosa, J.H. (1981) Molecular characterization of *Vibrio anguillarum* biotype 2. *Canadian Journal of Microbiology* 27, 1011-1018.
[137] Schiewe, M.H., Crosa, J.H. and Ordal, E.J. (1977) Deoxyribonucleic acid relationships among marine vibrios pathogenic to fish. *Canadian Journal of Microbiology* 23, 954-958.
[138] Schiewe, M.H., Trust, T.J. and Crosa, J.H. (1981) *Vibrio ordalii* sp. nov.: a causative agent of vibriosis in fish. *Current Microbiology* 6, 343-348.
[139] Schrøder, M.B., Espelid, S. and Jørgensen, T.O. (1992) Two serotypes of *Vibrio salmonicida* isolated from diseased cod (*Gadus morhua* L.): virulence, immunological studies and vaccination experiments. *Fish and Shellfish Immunology* 2, 211-221.
[140] Silva-Rubio, A., Avendaño-Herrera, R., Jaureguiberry, B., Toranzo, A.E. and Magariños, B. (2008a) First description of serotype O3 in *Vibrio anguillarum* strains isolated from salmonids in Chile. *Journal of Fish Diseases* 31, 235-239.
[141] Silva-Rubio, A., Acevedo, C, Magariños, B., Jaureguiberry, B, Toranzo, A.E. *et al.* (2008b) Antigenic and molecular characterization of *Vibrio ordalii* strains isolated from Atlantic salmon *Salmo salar* in Chile. *Diseases of Aquatic Organisms* 79, 27-35.
[142] Skov, M.N., Pedersen, K. and Larsen, J.L. (1995) Comparison of pulse-field gel electrophoresis, ribotyping and plasmid profiling for typing of *Vibrio anguillarum* serovar O1. *Applied and Environmental Microbiology* 61, 1540-1545.
[143] Smith, S.K., Sutton, D.C., Fuerst, J.A. and Reichelt, J.L. (1991) Evaluation of the genus *Listonella* and reassignment of *Listonella damsela* (Love *et al.*) MacDonell and Colwell to the genus *Photobacterium* as *Photobacterium damsela* comb. nov. with an

emended description. *International Journal of Systematic and Evolutionary Microbiology* 41, 529 – 534.

[144] Soengas, R.G., Anta, C., Espada, A., Paz, V., Ares, I.R. *et al*. (2006) Structural characterization of vanchrobactin, a new catechol siderophore produced by the fish pathogen *Vibrio anguillarum* serotype O2. *Tetrahedron Letters* 47, 7113 – 7116.

[145] Sørensen, U.B.S. and Larsen, J.L. (1986) Serotyping of *Vibrio anguillarum*. *Applied and Environmental Microbiology* 51, 593 – 597.

[146] Sørum, H., Hvaal, A.B., Heum, M., Daae, F.L. and Wiik, R. (1990) Plasmid profiling of *Vibrio salmonicida* for epidemiological studies of cold-water vibriosis in Atlantic salmon (*Salmo salar*) and cod (*Gadus morhua*). *Applied and Environmental Microbiology* 56, 1033 – 1037.

[147] Sørum, H., Roberts, M.C. and Crosa, J.H. (1992) Identification and cloning of a tetracycline resistance gene from the fish pathogen *Vibrio salmonicida*. *Antimicrobial Agents and Chemotherapy* 36, 611 – 615.

[148] Sorroza, L., Padilla, D., Acosta, F., Román, L., Grasso, V., Vega, J. and Real, F. (2012) Characterization of the probiotic strain *Vagococcus fluvialis* in the protection of European sea bass (*Dicentrarchus labrax*) against vibriosis by Vibrio anguillarum. *Veterinary Microbiology* 155, 369 – 373.

[149] Souto A., Montaos, M.A., Rivas, A.J., Balado, M., Osorio, C. *et al*. (2012) Structure and biosynthetic assembly of piscibactin, a siderophore from *Photobacterium damselae* subsp. piscicida, predicted from genome analysis. *European Journal of Organic Chemistry* 29, 5693 – 5700.

[150] Spanggaard, B., Huber, I., Nielsen, J., Nielsen, T. and Gram, L. (2000) Proliferation and location of *Vibrio anguillarum* during infection of rainbow trout, *Oncorhynchus mykiss* (Walbaum). *Journal of Fish Diseases* 23, 423 – 427.

[151] Stork, M., Di Lorenzo, M., Welch, T.J., Crosa, L.M. and Crosa, J.H. (2002) Plasmid-mediated iron uptake and virulence in *Vibrio anguillarum*. *Plasmid* 48, 222 – 228.

[152] Stork, M., Di Lorenzo, M., Mouriño, S., Osorio, C.R., Lemos, M.L. and Crosa, J. H. (2004) Two *tonB* systems function in iron transport in *Vibrio anguillarum*, but only one is essential for virulence. *Infection and Immunity* 72, 7326 – 7329.

[153] Svendsen, Y.S. and Bogwald, J. (1997) Influence of artificial wound and non-intact mucus layer on mortality of Atlantic salmon (*Salmo salar* L.) following a bath challenge with *Vibrio anguillarum* and *Aeromonas salmonicida*. *Fish and Shellfish Immunology* 7, 317 – 325.

[154] Tajima, K., Ezura, Y. and Kimura, T. (1985) Studies on the taxonomy and serology of causative organisms of fish vibriosis. *Fish Pathology* 20, 131 – 142.

[155] Tapia-Cammas, D., Yañez, A., Arancibia, G., Toranzo, A.E. and Avendaño-Herrera, R. (2011) Multiplex PCR for the detection of *Piscirickettsia salmonis*, *Vibrio anguillarum*, *Aeromonas salmonicida* and *Streptococcus phocae* in Chilean marine farms. *Diseases of Aquatic Organisms* 97, 135 – 142.

[156] Tiainen, R., Pedersen, K. and Larsen, J.L. (1995) Ribotyping and plasmid profiling of *Vibrio anguillarum* serovar O2 and *Vibrio ordalii*. *Journal of Applied Bacteriology* 79, 384 – 392.

[157] Toranzo, A.E. and Barja, J.L. (1990) A review of the taxonomy and seroepizootiology of *Vibrio anguillarum*, with special reference to aquaculture in the northwest of Spain. *Diseases of Aquatic Organisms* 9, 73 – 82.

[158] Toranzo, A.E. and Barja, J.L. (1993) Virulence factors of bacteria pathogenic for cold water fish. *Annual Review of Fish Diseases* 3, 5 – 36.

[159] Toranzo, A.E., Baya, A., Roberson, B.S., Barja, J.L., Grimes, D.J. *et al*. (1987)

Specificity of slide agglutination test for detecting bacterial fish pathogens. *Aquaculture* 61, 81 - 97.

[160] Toranzo, A. E., Santos, Y. and Barja, J. L. (1997) Immunization with bacterial antigens: Vibrio infections. In: Gudding, R., Lillehaug, A., Midtlyng, P. J. and Brown, F. (eds) *Fish Vaccinology*. Karger, Basel, Switzerland, pp. 93 - 105.

[161] Toranzo, A. E., Magariños, B. and Romalde, J. L. (2005) A review of the main bacterial fish diseases in mariculture systems. *Aquaculture* 246, 37 - 61.

[162] Toranzo, A. E., Avendaño-Herrera, R., Lemos, M. L. and Osorio, C. R. (2011) Vibriosis. In: Avendaño-Herrera, R. (ed.) *Enfermedades Infecciosas del Cultivo de Salmónidos en Chile y el Mundo*. NIVA, Puerto Varas, Chile, pp. 133 - 159.

[163] Totland, G. K., Nylund, A. and Holm, K. O. (1988) An ultrastructural study of morphological changes in Atlantic salmon, *Salmo salar* L., during the development of cold water vibriosis. *Journal of Fish Diseases* 11, 1 - 13.

[164] Trust, T.J., Courtice, I.D., Khouri, A.G., Crosa, J.H. and Schiewe, M.H. (1981) Serum resistance and hemagglutination ability of marine vibrios pathogenic to fish. *Infection and Immunity* 34, 702 - 707.

[165] Turnidge, J. and Paterson, D.L. (2007) Setting and revising antibacterial susceptibility breakpoints. *Clinical Microbiology Reviews* 20, 391 - 408.

[166] Urbanczyk, H., Ast, J. C., Higgins, M. J., Carson, J. and Dunlap, P. V. (2007) Reclassification of *Vibrio fischeri*, *Vibrio logei*, *Vibrio salmonicida* and *Vibrio wodanis* as *Aliivibrio fischeri* gen. nov., comb. nov., *Aliivibrio salmonicida* comb nov. and *Aliivibrio wodanis* comb. nov. *International Journal of Systematic and Evolutionary Microbiology* 57, 2823 - 2829.

[167] Wang, Y., Xu, Z., Jia, A., Chen, J., Mo, Z. and Zhang, X. (2009) Genetic diversity between two *Vibrio anguillarum* strains exhibiting different virulence by suppression subtractive hybridization. *Wei Sheng Wu Xue Bao* (Acta Microbiologica Sinica) 49, 363 - 371.

[168] Wards, B. J. and Patel, H. H. (1991) Characterization by restriction endonuclease analysis and plasmid profiling of *Vibrio ordalii* strains from salmon (*Oncorhynchus tshawytscha* and *Oncorhynchus nerka*) with vibriosis in New Zealand. *Journal of Marine and Freshwater Research* 25, 345 - 350.

[169] Weber, B., Croxatto, A., Chen, C. and Milton, D.L. (2008) RpoS induces expression of the *Vibrio anguillarum* quorum-sensing regulator VanT. *Microbiology* 154, 767 - 780.

[170] Weber, B., Chen, C. and Milton, D. L. (2010) Colonization of fish skin is vital for *Vibrio anguillarum* to cause disease. *Environmental Microbiology Reports* 21, 133 - 139.

[171] Welch, T.J. and Crosa, J.H. (2005) Novel role of lipopolysaccharide O1 side chain in ferric siderophore transport and virulence of *Vibrio anguillarum*. *Infection and Immunity* 73, 5864 - 5872.

[172] Wiik, R., Stackebrandt, E., Valle, O., Daae, F. L., Rødseth, O. M. et al. (1995) Classification of fish-pathogenic vibrios based on comparative 16S rRNA analysis. *International Journal of Systematic Bacteriology* 45, 421 - 428.

[173] Winkelmann, G., Schmid, D. G., Nicholson, G., Jung, G. and Colquhoun, D. J. (2002) Bisucaberin: a dihydroxamate siderophore isolated from *Vibrio salmonicida*, an important pathogen of farmed Atlantic salmon (*Salmo salar*). Biometals 15, 153 - 160.

[174] Wolf, M. K. and Crosa, J. H. (1986) Evidence for the role of a siderophore in promoting *Vibrio anguillarum* infections. *Journal of General Microbiology* 132, 2949 -2952.

[175] Xiao, P., Mo, Z.L., Mao, Y.X., Wang, C.L., Zou, Y.X. and Li, J. (2009) Detection

of *Vibrio anguillarum* by PCR amplification of the *empA* gene. *Journal of Fish Diseases* 32, 293 - 296.

[176] Yip, E. S., Geszvain, K., DeLoney-Marino, C. R. and Visick, K. L. (2006) The symbiosis regulator *rscS* controls the syp gene locus, biofilm formation and symbiotic aggregation by *Vibrio fischeri*. *Molecular Microbiology* 62, 1586 - 1600.

[177] Yu, L.P., Hu, Y.H., Sun, B.G. and Sun, L. (2012) C312M: an attenuated *Vibrio anguillarum* strain that induces immunoprotection as an oral and immersion vaccine. *Diseases of Aquatic Organisms* 102, 33 - 42.

[178] Zhang, Z., Wu, H., Xiao, J., Wang, Q., Liu, Q. *et al*. (2013) Immune response evoked by infection with *Vibrio anguillarum* in zebrafish bath-vaccinated with a live attenuated strain. *Veterinary Immunology and Immunopathology* 154, 138 - 144.

24

鲸魏斯氏菌

Timothy J. Welch* , David P. Marancik 和 Christopher M. Good

24.1 引言

魏斯氏菌(*Weissella*)通常与疾病无关,然而,最近发现的新菌株鲸魏斯氏菌(*Weissella ceti*)被确认为虹鳟的病原菌。2007 年,在中国的一个商业虹鳟养殖场中分离鉴定出 *W. ceti* (Liu et al., 2009),随后,在巴西(Figueiredo et al., 2012; Costa et al., 2015)与美国北卡罗来纳州(Welch and Good, 2013)的养殖虹鳟中也发现该菌。来源于美国与巴西的代表性菌株的基因组序列表明,尽管相关养殖场和国家之间缺乏动物流行病的联系,但它们的序列显示出高度的遗传相似性(Figueiredo et al., 2015)。与这些鱼病暴发相关的细菌起源尚不清楚,但在三大洲中 *W. ceti* 的出现表明它是一种新发病原。

W. ceti 是一种棒状革兰氏阳性菌,过氧化氢酶和氧化酶阴性,长约 1.5 μm,宽 0.30 μm,非运动性(图 24.1)。该菌在添加 5%羊血的 TSA 上形成小(0.25 mm)的白色 α-溶血菌落(图 24.1)。迄今为止,魏斯氏菌病(Weissellosis)仅在集约化养殖的虹鳟中有过报道,不过由于 *W. ceti* 生长条件苛刻且具有易感鱼脑的倾向,而脑组织是鱼类健康调查中通常不会采样的组织,因此这种病原菌有可能在野生或养殖鱼类的健康调查中被遗漏。魏斯氏菌病似乎主要影响体重为 0.25~1 kg 的鱼(Figueiredo et al., 2012; Welch and Good, 2013),因此该病可造成严重的经济损失。实验感染研究证实这种病菌具有感染较大鱼类的倾向,并已建立了体重与死亡率之间的相关性(Marancik et al., 2013)。在北卡罗来纳州,温度的升高是疫情暴发的关键诱因(Welch and Good, 2013),因为仅在温度超过 16 ℃时才能检测到魏斯氏菌病。该病在秋季气温下降时消退,在冬季消失。在中国和巴西的疫情中,水温升高也至关重要(Figueiredo et al., 2012; Liu et al., 2009)。其具体感染途径和感染源正在调查中。

---

* 通信作者邮箱: tim.welch@ars.usda.gov。

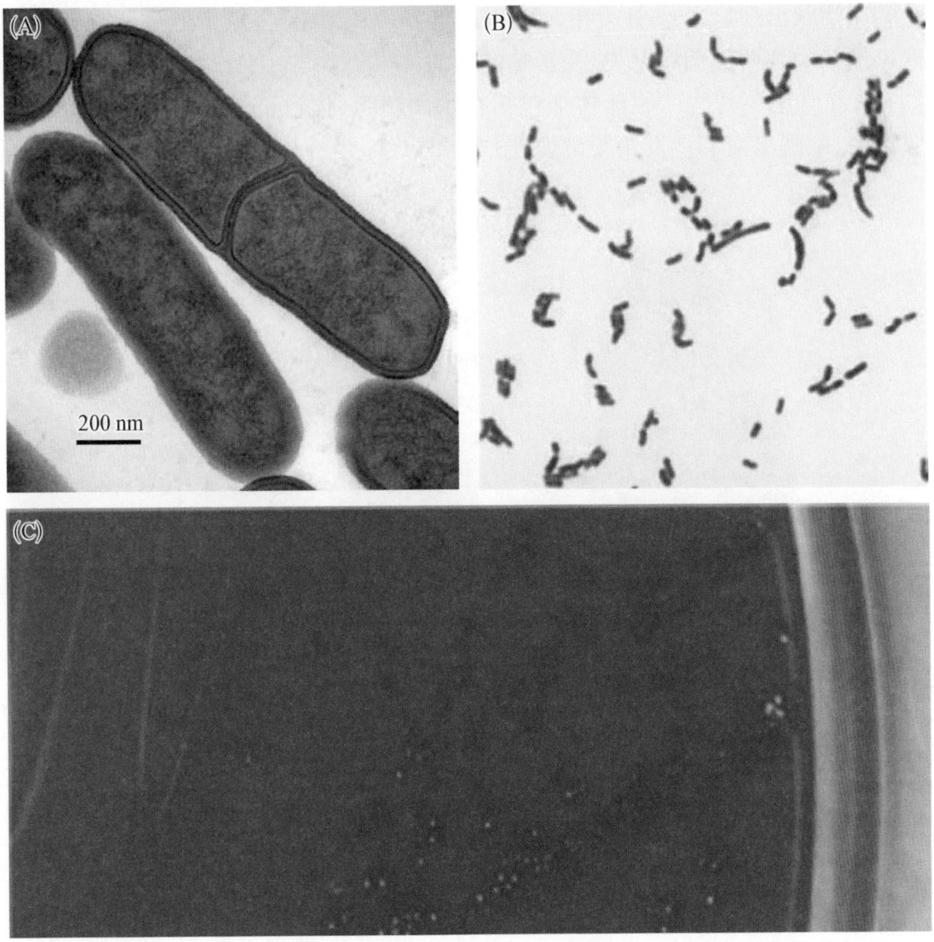

图 24.1 *Weissella ceti* NC36 株的透射电子显微镜照片(A)和革兰氏染色菌细胞的光学显微镜照片(B);(C) 添加 5%羊血的 TSA 上 30 ℃培养 18 h 后,NC36 株菌落呈现出 α-溶血活性

24.2 临床症状与诊断

Liu 等(2009)在中国的一个商业虹鳟养殖场首次发现报道了魏斯氏菌病。患病虹鳟表现为败血症,症状包括眼睛和肛门出血;剖检证实有广泛的组织受累,包括肠道和腹膜壁出血,以及肝脏瘀斑。该疫情持续了两个多月,累计死亡率约为 40%。随后,Figueiredo 等(2012)在巴西五个渔场的虹鳟中观察到魏斯氏菌病的暴发,临床体征包括厌食,之后变得嗜睡,表现为眼球突出、有腹水、眼睛和嘴严重出血。大多数受影响的鱼都接近上市大小(平均体重为250 g),只有一家渔场受感染的鱼从幼鱼到亲鱼不等。鱼的死亡率很高(50%～80%),大约

在临床症状出现后 4～5 d 发生。Welch 和 Good(2013)报道了北卡罗来纳州的两家渔场的魏斯氏菌病暴发，受影响的鱼的体重为 0.5～1 kg，死亡率很高，每天约有 2 000 尾鱼死亡。临床体征包括在水面嗜睡、体色变深、双侧眼球突出、角膜混浊、眼睛出血和偶发角膜破裂(图 24.2)。

图 24.2 虹鳟表现出典型魏斯氏菌病中的眼部大体症状，包括：(A) 双侧眼球突出，(B) 角膜破裂和眼内出血

W. ceti 的实验研究在很大程度上重现了自然疫情暴发期间的临床症状，包括虹鳟、尼罗罗非鱼(Figueiredo et al.，2012)和长丰鲫(*Carassius auratus gibelio*)(Liu et al.，2009)感染时的高死亡率。经腹腔感染虹鳟，表现出厌食、嗜睡、腹水、眼球突出、游动异常、直肠脱垂以及眼、鳍和肛门出血(Figueiredo et al.，2012)。Marancik 等(2013)在实验感染的鳟鱼中报道了相似的临床症状。总之，这些研究表明，在自然和实验感染中，魏斯氏菌病的临床症状类似。

出血性败血症发生在很多细菌感染中，并不是魏斯氏菌病的病征。初步分离该菌的培养基为添加 5%羊血的 TSA 和脑心浸液肉汤(BHI)(Liu et al.，2009；Figueiredo et al.，2012；Welch and Good，2013)。可以从鱼的脾、前肾和脑部分离出 *W. ceti*。该菌通常在脑部的载量最高，在某些情况下，病原菌只能从脑中分离出来。30 ℃下，在 TSA 血琼脂上培养 15～18 h 就可检测到 *W. ceti* 的生长和 α-溶血。*W. ceti* 的这种快速生长在鲑科鱼类细菌中是很罕见的。在最初的报道中，必须要通过表型特征和 16S rRNA 基因测序来确诊，但现在已经开发出基于 PCR 的检测方法，已经能够从细菌培养物或是直接从感染的组织中快速鉴定鲸魏斯氏菌(Snyder et al., 2015)。这些较新的鉴定方法(常规和定量 PCR)是基于对美国鳟鱼菌株 NC36 基因组序列(Ladner et al., 2013)中推定的血小板和胶原黏附基因的 PCR 扩增而开发的。在北卡罗来纳州存在病原菌的渔场的现场评估证实了这些检测方法的有效性。接下来需要进一步证实这些基于 PCR 的方法对其他地理区域 *W. ceti* 分离株的特异性。

24.3 病理学和病理生理学

组织病理学病变仅在实验感染虹鳟中描述过(Figueiredo et al., 2012; Marancik et al., 2013)。与该病的大体症状相似，显微镜检中的病变与细菌性败血症基本一致。病鱼脏器广泛出血与血管炎有关(图 24.3)，表现为巨噬细胞、淋巴细胞及中性粒细胞透壁浸润血管，血管膜和弹性膜的变性退化(Marancik et al., 2013)。眼球后区(恶化为肉芽肿性全眼炎和角膜溃疡)、心外膜、心肌和脑中也可见肉芽肿性炎症和坏死(Figueiredo et al., 2012; Marancik et al., 2013)。

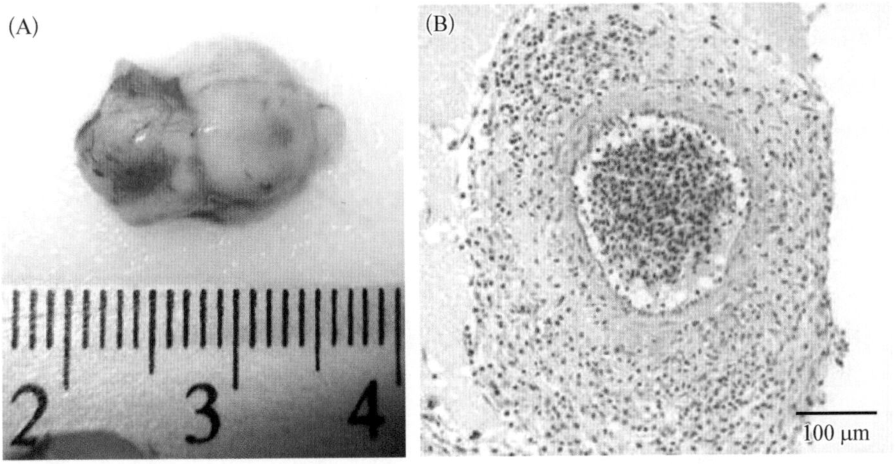

图 24.3 (A) *W. ceti* 造成的脑出血；(B) 很大可能与血管炎相关，如组织病理学表现出的中等动脉壁内的单核细胞和中性粒细胞浸润(20 倍放大，苏木精-伊红染色)

病理生理作用尚未得到充分研究。轶闻观察表明眼球突出症和随后失明影响鱼的进食。对巴西和美国分离菌株的基因组测序表明，*W. ceti* 具有潜在影响患病鱼健康和病程的毒力因子(Ladner et al., 2013; Figueiredo et al., 2014a, b)。例如，检测到与溶血素、胶原结合蛋白、血小板相关黏附蛋白和黏膜结合蛋白的同源基因，这些可能有助于细菌黏附宿主组织并侵入。然而，这些因子对疾病进程的影响，包括血管炎以及大脑和心脏等重要器官的损害，尚未确定。

24.4 防控策略

在体外条件下，*W. ceti* 中国分离株对四环素和氯霉素敏感，但是对新诺明/三甲氧苄啶和诺氟沙星耐药。在这一药敏测试之前，用新诺明/三甲氧苄啶和诺氟沙星治疗病鱼未获成功。当换用土霉素治疗时，死亡率降低（Liu et al.，2009）。巴西分离株对磺胺类药物有抗性，但在体外条件下大多数对土霉素和氟苯尼考敏感（Figueiredo et al.，2012）。美国分离株抗生素药敏性尚未确定，并且也没有尝试过治疗该疾病（Welch and Good，2013）。巴西和美国的分离株含有几个分别编码推定的抗生素抗性蛋白和多药外排泵蛋白的基因（Figueiredo et al.，2015）。目前，没有药物被批准用于食用鱼中魏斯氏菌病的治疗。

通过腹腔注射福尔马林灭活的 *W. ceti* 水基疫苗，实验鱼对北卡罗来纳州分离株的攻毒产生显著的免疫保护（Welch and Good，2013）。其在接种后 38 d（608 度日）的 RPS 为 87.5%，72 d（1 152 度日）的 RPS 为 85%。当与商用鲁氏耶尔森菌（*Yersinia ruckeri*）疫苗（Novartis Animal Health）联合使用时，这种 *W. ceti* 菌苗同样有效，并且在实验室条件下，二价疫苗给药不会改变 *Y. ruckeri* 组分的效力（Welch and Good，2013）。这一点很重要，因为在北卡罗来纳州的鳟鱼养殖场通常通过接种疫苗来控制*Y. ruckeri*，同时接种 *W. ceti* 和*Y. ruckeri* 两种抗原既不会增加生产成本，也不会增加与多次接种相关的操作负担。之后通过以未接种疫苗的同龄鱼作对照，对接种疫苗组约 12 个月后进行攻毒，证实了该二价疫苗的有效性（Welch and Hinshaw，未发表）。目前，北卡罗来纳州每年约有 400 万尾鱼接种二价疫苗，自 2012 年启动该疫苗的接种计划以来，在最初暴发疫情的养殖场就再未发现该病原菌（Welch，未发表）。这些疫苗由一家定制疫苗制造商生产，并经兽医处方使用。相比之下，根据 Costa 等（2015）的报道，灭活的全细胞水基疫苗对巴西*W. ceti*株无效（RPS 为 58%），尽管油基佐剂配方提供了高水平的免疫保护（RPS 为 92%）。

24.5 总结与研究展望

W. ceti 是养殖虹鳟的一种新发病原菌，人们对这种病原菌的生物学及其所致疾病知之甚少。魏斯氏菌病可能成为水产养殖业的一个主要问题。今后的研究应该涵盖：(1) 可以亚临床携带病原菌或是发展为临床魏斯氏菌病的鱼种类范围；(2) 在较暖季节可再次感染鱼类之前，病原菌如何以及是在何种环境中持续存在并越冬；(3) 病原菌的起源和促进病菌迅速传播至高度分散地区的流行病学途径；(4) 阐明病原菌的传播和发病机制；(5) 在孵化场预防和控制疾病的方法；(6) 鳟鱼对于病原菌和疫苗的体液与细胞免疫应答。由于 *W. ceti* 在美国

的传播范围并不广泛，鳟鱼养殖业应限制受感染鱼群在受污染设施外的转运和流通。

参考文献

[1] Costa, F.A., Leal, C.A., Schuenker, N.D., Leite, R.C. and Figueiredo, H.C. (2015) Characterization of *Weissella ceti* infections in Brazilian rainbow trout, *Oncorhynchus mykiss* (Walbaum), farms and development of an oil adjuvanted vaccine. *Journal of Fish Disease* 38, 295 - 302.

[2] Figueiredo, H.C., Costa, F.A., Leal, C.A., Carvalho-Castro, G.A. and Leite, R.C. (2012) *Weissella* sp. Outbreaks in commercial rainbow trout (*Oncorhynchus mykiss*) farms in Brazil. *Veterinary Microbiology* 156, 359 - 366.

[3] Figueiredo, H.C., Leal, G., Pereira, F.L., Soares, S.C., Dorella, F.A. *et al.* (2014a) Whole-genome sequence of *Weissella ceti* strain WS08, isolated from diseased rainbow trout in Brazil. *Genome Announcement* 2(4): e00851 - 14.

[4] Figueiredo, H.C., Leal, C.A., Dorella, F.A., Carvalho, A.F., Soares, S.C. *et al.* (2014b) Complete genome sequences of fish pathogenic *Weissella ceti* strains WS74 and WS105. *Genome Announcement* 2(5): e01014 - 14.

[5] Figueiredo, H.C., Soares, S.C., Pereira, F.L., Dorella, F.A., Carvalho, A.F. *et al.* (2015) Comparative genome analysis of *Weissella ceti*, an emerging pathogen of farm-raised rainbow trout. *BMC Genomics* 16: 1095.

[6] Ladner, J.T., Welch, T.J., Whitehouse, C.A. and Palacios, G.F. (2013) Genome sequence of *Weissella ceti* NC36, an emerging pathogen of farmed rainbow trout in the United States. *Genome Announcement* 1(1): e00187 - 12.

[7] Liu, J.Y., Li, A.H., Ji, C. and Yang, W.M. (2009) First description of a novel *Weissella* species as an opportunistic pathogen for rainbow trout *Oncorhynchus mykiss* (Walbaum) in China. *Veterinary Microbiology* 136, 314 - 320.

[8] Marancik, D.P., Welch, T.J., Leeds, T.D. and Wiens, G.D. (2013) Acute mortality, bacterial load, and pathology of select lines of adult rainbow trout challenged with *Weissella* sp. NC36. *Journal of Aquatic Animal Health* 25, 230 - 236.

[9] Snyder, A.K., Hinshaw, J.M. and Welch, T.J. (2015) Diagnostic tools for rapid detection and quantification of *Weissella ceti* NC36 infections in rainbow trout. *Letters in Applied Microbiology* 60, 103 - 110.

[10] Welch, T.J. and Good, C.M. (2013) Mortality associated with weissellosis (*Weissella* sp.) in USA farmed rainbow trout: potential for control by vaccination. *Aquaculture* 388, 122 - 127.

25

鲁氏耶尔森菌

Michael Ormsby 和 Robert Davies*

25.1 引言

鲁氏耶尔森菌(*Yersinia ruckeri*)属肠杆菌科(Enterobacteriaceae)革兰氏阴性菌,可引起鲑科鱼类肠红嘴病(enteric redmouth disease, ERM)或耶尔森菌病(yersiniosis)。自从在美国和加拿大分离出该菌以来(Ross et al., 1966; Bullock et al., 1978; Busch, 1978; Stevenson and Daly, 1982),在欧洲、南美洲、非洲、亚洲和澳大拉西亚也相继发现了该病原菌(Horne and Barnes, 1999)。该病原菌最初是在虹鳟中被分离出来的,后来在鲑科和非鲑科鱼类中都发现了该疾病(Horne and Barnes, 1999; Carson and Wilson, 2009)。*Y. ruckeri*逐渐成为澳大利亚(Carson and Wilson, 2009)、智利(Bastardo et al., 2011)、挪威(Shah et al., 2012)和苏格兰(Ormsby et al., 2016)大西洋鲑的重要病原菌,在这些地区,鲑鱼生产具有重要的经济意义。该病原菌也会感染其他养殖鱼类,如斑点叉尾鮰(Danley et al., 1999)、鲟(*Acipenser baeri* 和 *A. schrenckii*)(Vuillaume et al., 1987; Shaowu et al., 2013)和高白鲑(*Coregonus peled*)(Rintamaki et al., 1986)。

25.2 肠红嘴病

已有文献对虹鳟和大西洋鲑 ERM 的临床表现、病理学和组织病理学进行了综述(Tobback et al., 2007; Carson and Wilson, 2009; Kumar et al., 2015)。简要地说,该疾病主要为虹鳟的一种急性感染疾病,其典型特征为口腔内部及周围[图 25.1(A)]、胸鳍和腹鳍基部及沿腹侧[图 25.1(B)]皮下出血,也可能会出现鳃丝出血。耶尔森菌病或"鲑血斑病"被认为是大西洋鲑中一种不太严重的疾病(Carson and Wilson, 2009; Costa et al., 2011)。除了口腔内部和周围及鳍基部皮下出血外,耶尔森菌病的病征还包括眼睛虹膜上的出血性充血斑块("血

---

* 通信作者邮箱: Robert.Davies@glasgow.ac.uk。

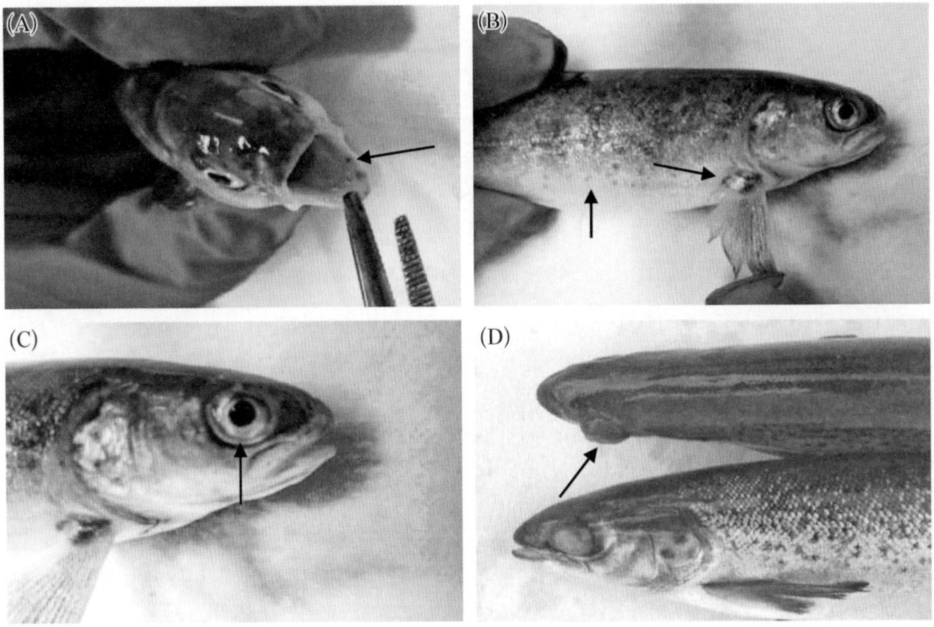

图 25.1 实验感染大西洋鲑患肠红嘴病/耶尔森菌病的临床特征(箭头所示):(A) 口腔出血;(B) 胸鳍基部和体侧出血;(C) 眼出血;(D) 眼球突出

斑”)[图 25.1(C)]和明显的单侧或双侧眼球突出[图 25.1(D)]。

该病原菌会引起全身性败血症并伴有炎症反应,细菌常定植于肾、脾、心、肝和鳃以及瘀点出血区域(Rucker, 1966)。鳃中发生的病理变化包括次级鳃瓣的充血、水肿和上皮细胞脱落(Horne and Barnes, 1999; Tobback et al., 2007)。肝、肾和脾中的局灶性坏死也很常见。在大西洋鲑中,典型败血症是其组织病理学特征,很容易在血液和循环巨噬细胞中检测到细菌,也可在组织出血部位检测到(Carson and Wilson, 2009)。在大西洋鲑二龄鲑适应性应激期间被 *Y. ruckeri* 感染的特征表现为细菌侵入血液中,伴有充血、出血和组织定位(Carson and Wilson, 2009)。

病原菌通过直接接触被感染的鱼或者受感染鱼或带菌鱼通过水进行传播。Rucker (1966)最早发现 *Y. ruckeri* 的病原携带状态。周期性的病原体肠道脱落导致反复感染和死亡,而周期性脱落对于疾病的传播至关重要(Rucker, 1966; Busch and Lingg, 1975)。在 6 ℃和 18 ℃的河水、湖水、河口水和沉积物中,该菌可存活 100 d 以上(Romalde et al., 1994),但在较高盐度(35‰)下存活率大大降低(Thorsen et al., 1992)。生物被膜可能是养殖场发生反复感染的根源(Coquet et al., 2002)。鳃是该病原菌传播到其他器官的重要侵入口(Tobback et al., 2009; Ohtani et al., 2014)。然而,鳃和肠道作为侵入口并且在 *Y. ruckeri* 与宿主的初始相互作用中同等重要(Tobback et al., 2010a)。

25.3 诊断

诊断鉴定包括基于培养诊断、血清学诊断和分子生物学诊断等技术。通常从头肾、脾或出血部位分离 *Y. ruckeri* (Carson and Wilson，2009)，但也可以从后肠分离以确定病原携带状态(Busch and Lingg，1975；Rodgers，1992)。选择性培养基，如 Waltman-Shotts (Waltman and Shotts，1984)和 ROD(核糖鸟氨酸脱氧胆酸盐)琼脂(Rodgers，1992)适用于鉴别。可以通过生化特征确定 *Y. ruckeri*，Davies 和 Frerichs (1989)、Austin 和 Austin (2012)对其关键特性进行了描述。用于快速诊断 *Y. ruckeri* 的血清学检测技术包括酶联免疫吸附试验(ELISA) (Cossarini-Dunier，1985)、免疫荧光抗体技术(IFAT) (Smith et al.，1987)和商用凝集试剂盒(Romalde et al.，1995)。各种基于聚合酶链反应(PCR)的检测方法已被开发出来，用于从受感染的鱼和亚临床携带者的血液或其他组织中直接检测病原菌(Gibello et al.，1999；Altinok et al.，2001；Temprano et al.，2001；Saleh et al.，2008；Seker et al.，2012)。特别地，实时荧光定量 PCR 已被开发用于快速检测和确诊(Bastardo et al.，2012a；Keeling et al.，2012)。

25.4 菌株鉴别

鲁氏耶尔森菌的菌株鉴别对于疾病的监测和控制具有重要意义。有关菌株也对病理生物学产生影响，因为不同菌株的宿主特异性和毒力不同(Davies，1991c；Haig et al.，2011；Calvez et al.，2014)。例如，O1 血清型分离株是导致世界范围内虹鳟疾病暴发的主要原因，而非 O1 血清型通常与其他鱼类的疾病相关。生物分型、血清学分型和外膜蛋白(OMP)分型等表型分型方法常用于鉴定和区分 *Y. ruckeri* 菌株(Tobback et al.，2007；Bastardo et al.，2012b；Kumar et al.，2015)。这些分型方法利用的是细菌细胞表面结构或分子的变化：生物分型评估鞭毛存在与否的变异；血清学分型评估表面脂多糖(LPS)的变化；OMP 分型评估特定“主要”外膜蛋白的相对分子质量的变化。细菌表面是病原和宿主的交界面，在发病机制中发挥着重要作用。鞭毛、LPS 和 OMPs 等细胞表面结构和分子变化是由包括宿主免疫应答在内的选择性压力驱动的，这些结构和分子是菌株鉴别的重要标志物。分子方法也被用于评估 *Y. ruckeri* 内的遗传变异，这些方法为该病原的种群生物学和动物流行病学研究提供了重要依据。

25.4.1 生物分型

Y. ruckeri 有两种生物型：生物 1 型株可运动(有鞭毛)，脂肪酶阳性(水解

吐温);而生物 2 型株对这两种特性均为阴性(图 25.2)。自从在英国分离鉴定出生物 2 型株以来(Davies and Frerichs, 1989),非运动型变种已在欧洲被广泛发现(Horne and Barnes, 1999; Wheeler et al., 2009),并且也在美国(Arias et al., 2007; Welch et al., 2011)和澳大利亚(Carson and Wilson, 2009)被分离鉴定。生物 2 型株引起越来越多的关注,是因为它们可导致接种了生物 1 型株疫苗的鱼类致病(Tinsley et al., 2011a)。

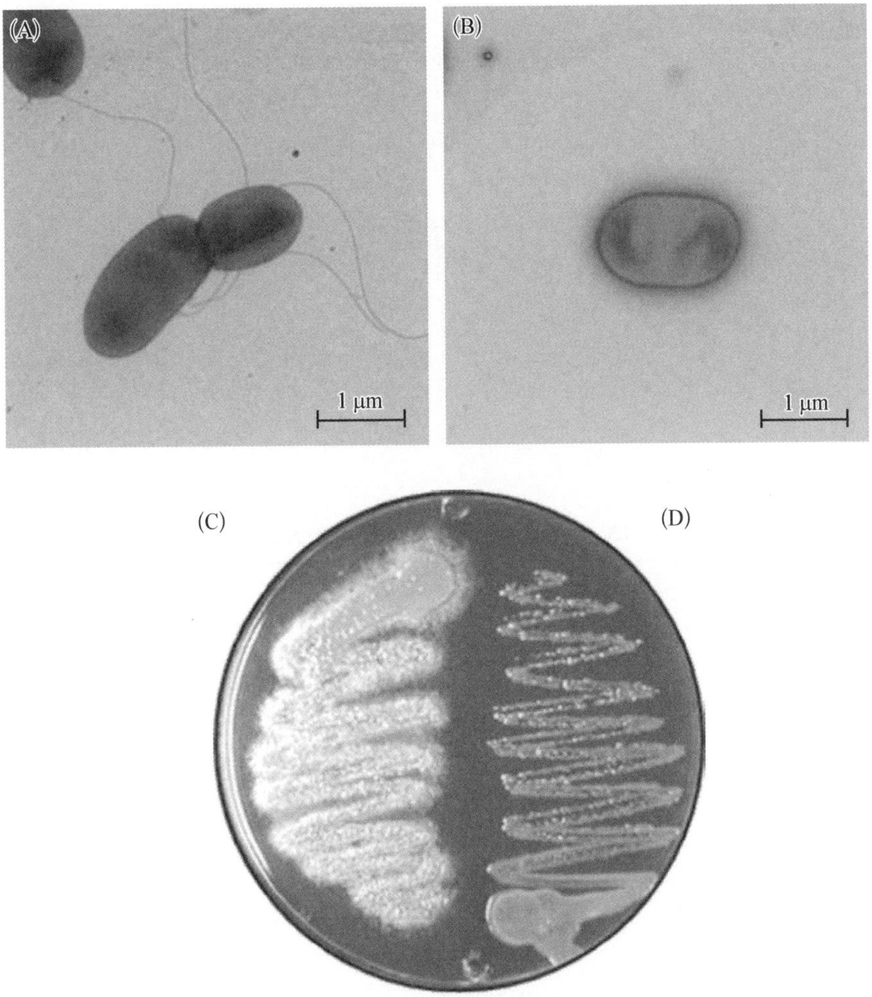

图 25.2　*Y. ruckeri* 代表性生物 1 型和 2 型分离株中存在鞭毛和吐温水解。(A) 运动型 *Y. ruckeri* 分离株具有鞭毛;(B) 透射电子显微镜显示非运动型分离株不具有鞭毛;相对应的是,在添加吐温 80 的胰蛋白酶大豆琼脂培养基上呈吐温水解阳性(C)和吐温水解阴性(D)。运动型和吐温水解阳性分离株[(A)和(C)]代表了生物 1 型;非运动型和吐温水解阴性分离株[(B)和(D)]代表了生物 2 型(Applied and Environmental Microbiology 82, 5785 - 5794, 2016)

25.4.2 血清学分型

血清学分类对于菌株的鉴别很重要。在首次鉴定出Ⅰ型和Ⅱ型血清型后(O'Leary，1977)，随后又确定了其他血清型(Bullock et al.，1978；Stevenson and Airdrie，1984；Pyle and Schill，1985；Daly et al.，1986；Pyle et al.，1987)。可惜，由于在大多数情况下分配的血清型命名与其他的命名方案不匹配，因而造成混淆，如表 25.1 所示。为了解决这种混乱，Davies (1990)开发了一种简单、快速的载玻片凝集测定方法，该方法基于 O-抗原，关键是使用了与之前发布方案共同的关键参考菌株。他提出了一种血清学分型方案，该方案可识别 O1、O2、O5、O6 和 O7(表 25.1)5 种 O 血清型，该方案已被广泛采用(Strom-Bestor et al.，2010；Tinsley et al.，2011a；Shah et al.，2012；Calvez et al.，2014)。随后，Romalde 等(1993)提出了另一种识别 4 种 O 血清型的方案。由于利用膜蛋白质谱和胞外产物(ECPs)中的蛋白质，该方案相对更加复杂；该方案也未能充分区分 O2 和 O7 血清型，因为它们被归类在同一 O2 血清型体系下(表 25.1)。然而，这种方法鉴定出了澳大利亚(Carson and Wilson，2009)和智利(Bastardo et al.，2011)大西洋鲑中发现的 O1 血清型亚型(O1b)。最近，从苏格兰大西洋鲑中鉴定出一种新的 O 血清型，称为 O8 血清型(Ormsby et al.，2016)。新的 LPS 型具有与 O1 血清型相似的多糖核心区域，但具有独特的 O-抗原侧链(图 25.3)。

表 25.1 鲁氏耶尔森菌血清学分型方案比较

参考文献							
O'Leary，1977[a]	Bullock et al.，1978[a]	Stevenson and Airdrie，1984[a]	Pyle and Schill，1985[a]	Daly et al.，1986[a]	Pyle et al.，1987[a]	Davies，1990[b]	Romalde et al.，1993[c]
Ⅰ	Ⅰ	Ⅰ	—	Ⅰ	1	O1	O1a
Ⅱ	Ⅱ	Ⅱ	2	Ⅱ	2	O2	O2a，O2b，O2c
—	Ⅲ	Ⅲ	—	Ⅲ	6	O1	O1b
—	—	Ⅳ	—	—	—	—	—
—	—	Ⅴ	6	Ⅴ	5	O5	O3
—	—	Ⅵ	5	Ⅵ	4	O6	O4
—	—	—	4	—	3	O7	—

注：所示的血清学分型方案和命名基于：[a]全细胞抗原；[b] O-抗原；[c] O-抗原和胞外产物。

25.4.3 外膜蛋白分型

根据主要外膜蛋白(OmpA、OmpC 和 OmpF)相对分子质量的变化，建立了

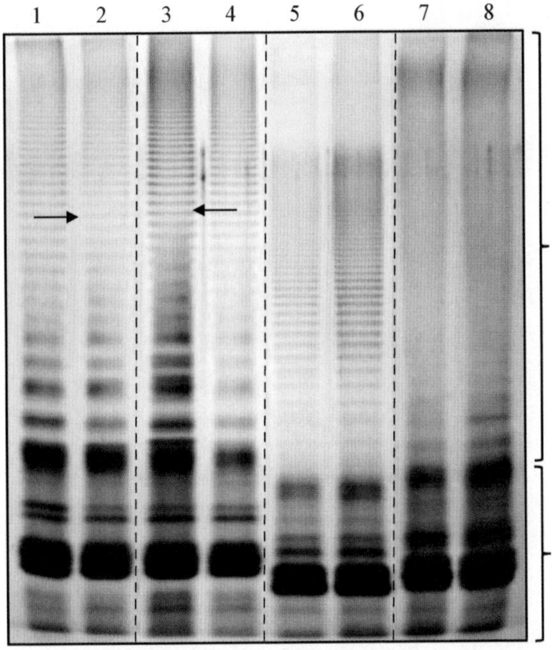

图 25.3 *Y. ruckeri* O1、O2、O5 和 O8 血清型分离株的代表性脂多糖 (LPS) 谱。图中显示了代表性 O1 血清型(泳道 1 和 2)、O8 血清型(泳道 3 和 4)、O2 血清型(泳道 5 和 6)和 O5 血清型(泳道 7 和 8)的菌株对。LPS 分子的核心多糖和 O-抗原区域被标记。O1 血清型(泳道 1 和 2)和 O8 血清型(泳道 3 和 4)分离株的 LPS 共享一个核心多糖区域,但 O-抗原梯度模式略有不同。箭头表示 O1 血清型和 O8 血清型的 LPS 中 O-抗原单位的不同迁移率(Applied and Environmental Microbiology 82, 5785-5794, 2016)

一种 OMP 分型方案,以进一步区分 *Y. ruckeri* 菌株(Davies, 1989, 1991a)。OmpA 与 OmpC 和 OmpF 的区别在于其独特的热变性特性,而 OmpC 与 OmpF 的区别在于 OmpC 在厌氧而非需氧生长条件下表达(图 25.4)。外膜蛋白比较分析在各种动物流行病学研究中发挥了重要作用(Davies, 1991a; Romalde et al., 1993; Sousa et al., 2001; Wheeler et al., 2009; Bastardo et al., 2011, 2012b; Tinsley et al., 2011a)。Bastardo 等(2012b)总结发现,当采用三种或更多分型方法,特别是 API 20E 系统生物分型(来自 bioMérieux 的肠杆菌科分析谱指标)、LPS 血清分型和 OMP 分析时,可获得最大的区分度。根据生物型、血清型和 OMP 型,鉴定出不同的克隆群(Davies, 1991b),它们的毒力和宿主特异性不同(Davies, 1991c)。这些克隆群被认为是不同的"致病型",它们的病理生物学可能有所不同。

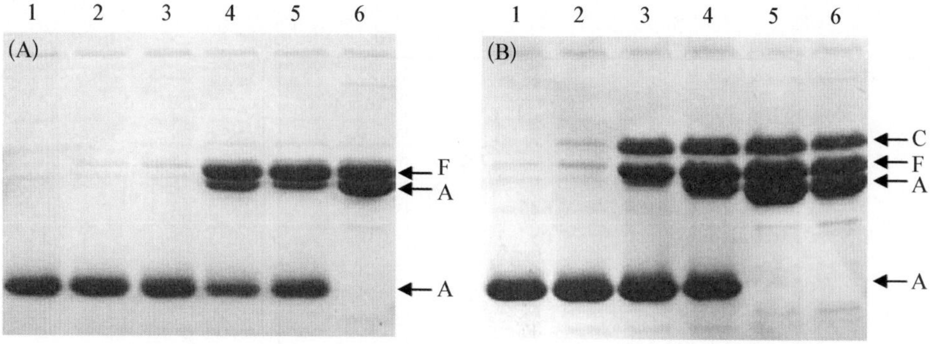

图 25.4 利用 SDS-PAGE 对不同溶解温度和通气条件下的 *Y. ruckeri* 外膜蛋白 OmpA、OmpC 和 OmpF 进行鉴定。*Y. ruckeri* 在有氧(A)和厌氧(B)条件下生长。每个图谱右侧的标签是：A 为 OmpA 蛋白，C 为 OmpC 蛋白，F 为 OmpF 蛋白。泳道 1～6 分别代表 50 ℃、60 ℃、70 ℃、80 ℃、90 ℃和 100 ℃的溶解温度。热变性蛋白 OmpA 可以与 OmpC 和 OmpF 区分开来，因为 OmpA 在 90 ℃或 100 ℃加热时，相对分子质量发生了典型变化，从大约 30 kDa 变化到 40 kDa。OmpC 与 OmpF 的区别在于其在不同通气条件下的表现，它在厌氧生长条件下表达(B)但在有氧生长条件下不表达(A)(Applied and Environmental Microbiology 82，5785-5794，2016)

25.4.4 分子分型

质粒分析(De Grandis and Stevenson，1982；Garcia et al.，1998)、核糖分型(Lucangeli et al.，2000)、重复序列 PCR (Huang et al.，2013)、脉冲场凝胶电泳(PFGE)(Wheeler et al.，2009；Strom-Bestor et al.，2010；Huang et al.，2013；Calvez et al.，2015)、多位点酶电泳(multilocus enzyme electrophoresis，MLEE)(Schill et al.，1984)和多位点序列分型(MLST)(Bastardo et al.，2012c；Calvez et al.，2015)都已被用于评估 *Y. ruckeri* 的遗传变异。Schill 等(1984)证明 *Y. ruckeri* 具有高度同源性。最近，Bastardo 等(2012a)使用 MLST 研究证实，尽管分析的 103 株菌株来自不同的地理区域和宿主物种，但病原菌的遗传多样性相对较低。重组而非变异是造成这种遗传变异的主要原因，这使得研究者认为*Y. ruckeri*具有动物流行病的种群结构。该研究者还确定了特定序列类型(specific sequence types，STs)和宿主鱼种之间的关系，并提出这可能是由生态位适应性特化造成的。Calvez 等(2015)证实了 *Y. ruckeri* 具有普遍的同源性，也证实了来自不同宿主物种菌株的遗传差异化。他们得出结论，养殖鱼类的迁移、地理起源和生态位对 *Y. ruckeri* 的传播和进化有一定影响。

25.5 毒力因子和病理生物学

人们对于 *Y. ruckeri* 感染的病理生物学和在疾病过程中特定毒力因子的作

用还知之甚少。推测的 *Y. ruckeri* 毒力因子与细菌黏附和侵染、铁摄取、胞内存活和免疫逃逸或胞外产物有关。

25.5.1 黏附和侵染

Y. ruckeri 的黏附和侵染已在细胞系(Romalde and Toranzo, 1993; Kawula et al., 1996; Tobback et al., 2010b)、鳃和肠组织中(Tobback et al., 2010a,b)得到证实。然而,参与这些过程的具体分子和机制尚未被发现和揭示。

25.5.2 铁摄取

致病菌已经进化出一系列获取铁元素的机制。其中一种机制涉及极高亲和力铁螯合的铁载体分泌(Faraldo-Gómez and Sansom, 2003)。Davies (1991d) 研究证实,在限铁生长条件下,大小为 66 kDa、68 kDa、69.5 kDa 和 72 kDa 的 4 种高相对分子质量外膜蛋白的表达上调,尽管在此过程中没有检测到铁载体的产生。Romalde 等(1991)发现了三种受铁调节的外膜蛋白,相反地,能够在铁限制培养基中检测到铁载体。随后,鉴定了一种称为鲁氏耶尔森菌素的儿茶酚类铁载体 ruckerbactin,在体外显示其参与铁摄取(Fernández et al., 2004),ruckerbactin 也被证实在 *Y. ruckeri* 致病机制中具有潜在作用。

25.5.3 胞内存活和免疫逃逸

Y. ruckeri 胞内存活和免疫逃逸机制尚不清楚,但在 ERM 发病机制中具有潜在的重要作用(Tobback et al., 2009)。Ryckaert 等(2010)体外研究显示,*Y. ruckeri*能在鳟巨噬细胞中存活和复制。然而,Tobback 等(2010)证实,该菌不会在培养细胞系中繁殖或存活。此外,在组织切片中,细胞外细菌比胞内细菌数量更多,这使得这些研究者认为,胞内存活在 *Y. ruckeri* 致病机理中并不重要。

Tobback 等(2009)的研究表明,鳃对于毒株和无毒株来说均是重要侵入门户,但随后毒株而非无毒株在内脏器官中的持续存在推论免疫逃逸是该病原菌的主要毒力因子。Davies (1991c)证实了*Y. ruckeri*血清抗性与其在虹鳟中的毒力之间的关系。LPS 分型是对补体介导的裂解产生抗性的决定因素,但也可能涉及其他因素。Tobback 等(2010b)证实血清抗性对病原菌的胞外存活至关重要。

25.5.4 胞外产物

Romalde 和 Toranzo(1993)研究证实,将 *Y. ruckeri* 胞外产物(ECPs)注入虹鳟体内会引起 ERM 的一些特征性病征,并认为 ECPs 可能在该疾病的致病机制中发挥重要作用。涉及胞外产物毒力的组分分子包括金属蛋白酶 Yrp1

(Secades and Guijarro, 1999; Fernández et al., 2002, 2003)、溶血素 YhlA (Fernández et al., 2007)和偶氮酪蛋白酶(Secades and Guijarro, 1999)。Yrp1 是在对数生长期末期产生的一种 47k Da 的蛋白酶(Secades and Guijarro, 1999)。它能消化各种细胞外基质和肌肉蛋白,被认为有助于提高 *Y. ruckeri* 毒力,例如,它可能导致毛细血管渗漏,引起口腔周围典型性出血(Fernández et al., 2003)。*Y. ruckeri* 溶血素 YhlA 被认为在致病性中发挥积极作用(Fernández et al., 2007)。当细菌在限铁条件下生长时,YhlA 的产生增加,这表明 YhlA 可能参与宿主铁营养的获取(Fernández et al., 2007; Tobback, 2009)。

25.6 防控策略

25.6.1 抗生素治疗

尽管使用的抗生素仅限于阿莫西林、噁喹酸、土霉素和磺胺嘧啶,并与甲氧苄啶和氟苯尼考合用(Michel et al., 2003),但抗菌化合物常用于治疗鲁氏耶尔森菌感染引起的疾病;这种限制可能导致抗生素耐药性的发展。在英国,仅有的完全许可和可用的抗生素是氟苯尼考、土霉素和阿莫西林;鲑类养殖业中普遍使用氟苯尼考和土霉素。抗生素噁喹酸(每天每千克体重 10 mg,连续 10 d)也可在英国获得特别进口许可时使用(Verner-Jeffreys and Taylor, 2015),因为该药物被认为最有效和最具成本效益(Richard Hopewell, Dawnfresh Seafoods,个人通信)。

一般而言,不同国家有关允许抗生素使用的法规有所不同,需要咨询相应的监管机构。大多数 *Y. ruckeri* 欧洲分离株仍对抗生素治疗具有广泛的敏感性,但美国的一些分离株对治疗水平的磺胺甲嘧啶和土霉素完全耐药(Post, 1987),并且对四环素和磺胺类药物的耐药也有过报道(De Grandis and Stevenson, 1985)。在 *Y. ruckeri* 中发现了一种 β-内酰胺酶基因(Mammeri et al., 2006),它可以抵抗 β-内酰胺类抗生素的使用,尽管有研究表明该基因不太可能高水平表达(Stock et al., 2002)。Rodgers(2001)利用体外试验证实了 *Y. ruckeri* 对噁喹酸、土霉素和增效磺胺类药物耐药性的快速发展。这已在保加利亚鳟鱼养殖场的 *Y. ruckeri* 分离株中得到证实(Orozova et al., 2014)。

25.6.2 益生菌

抗生素耐药性的出现促使人们寻求替代控制措施,而使用益生菌是其中一种选择。益生菌是培养产物或活微生物饲料添加剂,通过改善肠道微生物平衡对宿主产生有益作用(Fuller, 1989)。Irianto 和 Austin(2002)及 Newaj-Fyzul 等(2014)已对它们在水产养殖中的应用进行了综述。益生菌的保护机制通常是未知的，不过最可能的作用方式包括刺激免疫应答、改变微生物代谢、通过产生抑制性化合物或竞争营养物、空间或氧气而产生竞争性排斥（Fuller, 1989; Newaj-Fyzul et al., 2014）。

许多益生菌可以提高对 *Y. ruckeri* 感染的抵抗力，这种方法具有控制 ERM 的巨大潜力（Raida et al., 2003; Kim and Austin, 2006; Brunt and Austin, 2007; Balcazar et al., 2008; Capkin and Altinok, 2009; Sica et al., 2012）。温和气单胞菌和枯草芽孢杆菌的亚细胞成分，包括细胞壁蛋白、OMPs、LPS 和全细胞蛋白，也能提高鱼类抗 *Y. ruckeri* 感染的存活率（Abbass et al., 2010）。然而，益生菌的使用也有潜在风险（Irianto and Austin, 2002; Brunt and Austin, 2007），因为某些具有益生菌潜能的细菌（如 *A. sobria*）也可能对鱼类具有致病性（Austin and Austin, 2012）。将细菌掺入动物饲料中也可能导致耐药性，这将对人类和动物健康产生更广泛的影响。此外，益生菌和宿主天然菌群之间可能发生毒力因子的遗传交换，导致新病原菌的出现。

25.6.3 疫苗接种

ERM 疫苗被广泛使用，这是一个成功的鱼类疫苗接种故事。基于灭活菌苗的 ERM 疫苗已经在市场上销售了 30 多年（Ross and Klontz, 1965），不过对其确切的保护机制仍知之甚少。疫苗可通过浸泡、喷雾、注射或口服方式接种，它们提供了良好的疾病预防策略（Kumar et al., 2015）。目前的疫苗接种策略包括在鱼苗阶段浸泡或注射（腹腔或肌内注射）。浸泡疫苗接种后，通常在接种后 6 个月进行一次加强免疫，以提供持续保护（Tatner and Horne, 1985）。首次接种时饲喂大西洋鲑鱼苗微胶囊口服疫苗比浸泡接种能够提供更有效的保护（Ghosh et al., 2016）。

ERM 疫苗通常是 O1 血清型、生物 1 型分离株的灭活全细胞悬液单价疫苗（Kumar et al., 2015），但一种含有生物 1 型和 2 型分离株的二价疫苗（AquaVac RELERA™）可用于虹鳟（Tinsley et al., 2011b; Deshmukh et al., 2012），而另一种 O1b 血清型的单价疫苗（Yersinivac-B）用于大西洋鲑（Costa et al., 2011）。基于 O1 血清型的疫苗主要用于预防虹鳟中的 ERM，但它们也广泛用于鲑鱼养殖业。基于 O1b 血清型的疫苗专门针对澳大拉西亚养殖的大西洋鲑（Costa et al., 2011）。

全细胞疫苗的替代产品包括亚单位疫苗和 DNA 疫苗。Fernández 等（2003）的研究表明，肌内（IM）注射 Yrp1 类毒素在虹鳟中产生高水平的保护作用（RPS 为 79%）。一种 ECP 制剂也为腹腔注射攻毒的虹鳟提供了保护（RPS 为 74%～81%）（Ispir and Dorucu, 2010）。使用无佐剂鞭毛蛋白作为亚单位疫苗可为虹鳟提供针对生物 1 型株（有鞭毛）和生物 2 型株（无鞭毛）的非特异性保护（Scott et al., 2013）。腹腔注射和浸泡接种 LPS 在腹腔注射攻毒后提供了极好的免疫保护作用(RPS 为 77%～84%)(Ispir and Dorucu，2014)。纯化的LPS是一种高效的保护性抗原(Welch and LaPatra，2016)，有人认为LPS是全细胞疫苗激发免疫保护的唯一成分。这些实验结果证实了 Tinsley 等(2011b)的结论，即O-抗原是主要的免疫原。因此，这些发现意味着O1血清型疫苗可能对其他血清型提供有限的交叉保护，因为根据定义，每种血清型都是高度免疫特异性的。

一般来说，减毒活疫苗比全细胞灭活疫苗能够激发更强的细胞免疫。一种高度减毒的O1血清型 *aroA*(5-烯醇丙酮酰莽草酸-3-磷酸合酶基因)突变株被开发为ERM活疫苗(Temprano et al.，2005)。接种该类疫苗的虹鳟对生物1型 *Y. ruckeri* 的IP攻毒提供了显著保护作用(RPS 为 90%)。然而，对转基因生物环境安全性的担忧可能会限制这种减毒活疫苗的商业用途。

随着抗生素耐药性的增加，疫苗接种可能会成为一种更重要的疾病控制策略，尽管其广泛和长期使用也需要付出一定的代价。这是因为疫苗接种对细菌种群施加了选择性压力，这可能会通过血清型替换或转换而改变宿主物种内循环的主要血清型。由于荚膜生物合成操纵子中有限数量基因的专一性突变，这种血清型转换已在海豚链球菌中发生(Millard et al.，2012)。那么在 *Y. ruckeri* 中是否存在血清型取代和转换的证据？1995年智利引入基于O1血清型的疫苗以控制大西洋鲑中的耶尔森菌病(Bravo and Midtlyng，2007)，但在2008年出现了O1b血清型毒株，而 Bastardo 等(2011)认为疫苗接种可能为这种血清型的出现提供了选择压力。同样，近年来苏格兰大西洋鲑中O8血清型的出现可能是鲑鱼养殖业广泛使用基于O1血清型疫苗的结果(Ormsby et al.，2016)。也有人提出，使用生物1型和O1血清型疫苗驱动了生物2型 *Y. ruckeri* 的出现(Fouz et al.，2006；Welch et al.，2011；Scott et al.，2013)。显然，未来对于监测可能由疫苗接种引起的主要致病血清型的潜在变化将是十分重要的。

25.7 总结与研究展望

由于水产养殖的重要性日益增加以及 *Y. ruckeri* 对鱼类养殖业的潜在经济影响，在ERM快速诊断、预防和治疗方面的进展应成为未来的主要优先事项之一。实时荧光定量PCR等能够快速确认临床样本的细菌种属、生物型和血清型的诊断工具在疾病诊断方面将极为有益。由于抗生素耐药性的增加，疫苗和益生菌可能在疾病控制中承担更重要的意义。最终，对宿主-病原相互作用和分子致病机制的深入了解将使我们能够开发出更可靠和有效的控制策略。疫苗的改进开发可能利用一些新方法，如cDNA微阵列(Bridle et al.，2012)和“组学”技术，这些方法可以监测特定宿主基因的个体反应，作为疫苗成功的预测指标。这样的方法还可以鉴定参与刺激保护性免疫应答的特定细菌抗原。适当体外模型的开发将减少对昂贵、耗时和不道德的活体实验需求。

参考文献

[1] Abbass, A., Sharifuzzaman, S. M. and Austin, B. (2010) Cellular components of probiotics control *Yersinia ruckeri* infection in rainbow trout, *Oncorhynchus mykiss* (Walbaum). *Journal of Fish Diseases* 33, 31－37.

[2] Altinok, I., Grizzle, J.M. and Liu, Z. (2001) Detection of *Yersinia ruckeri* in rainbow trout blood by use of the polymerase chain reaction. *Diseases of Aquatic Organisms* 44, 29－34.

[3] Arias, C.R., Olivares-Fuster, O., Hayden, K., Shoemaker, C.A., Grizzle, J.M. and Klesius, P.H. (2007) First report of *Yersinia ruckeri* biotype 2 in the USA. *Journal of Aquatic Animal Health* 19, 35－40.

[4] Austin, B. and Austin, D.A. (2012) *Bacterial Fish Pathogens: Disease in Farmed and Wild Fish*, 5th edn. Springer/Praxis Publishing, Chichester, UK.

[5] Bachrach, G., Zlotkin, A., Hurvitz, A., Evans, D.L. and Eldar, A. (2001) Recovery of *Streptococcus iniae* from diseased fish previously vaccinated with a *Streptococcus* vaccine. *Applied and Environmental Microbiology* 67, 3756－3758.

[6] Balcazar, J. L., Vendrell, D., de Blas, I., Ruiz-Zarzuela, I., Muzquiz, J. L. and Girones, O. (2008) Characterization of probiotic properties of lactic acid bacteria isolated from intestinal microbiota of fish. *Aquaculture* 278, 188－191.

[7] Bastardo, A., Bohle, H., Ravelo, C., Toranzo, A. E. and Romalde, J. L. (2011) Serological and molecular heterogeneity among *Yersinia ruckeri* strains isolated from farmed Atlantic salmon (*Salmo salar*) in Chile. *Diseases of Aquatic Organisms* 93, 207－214.

[8] Bastardo, A., Ravelo, C. and Romalde, J. L. (2012a) Highly sensitive detection and quantification of the pathogen *Yersinia ruckeri* in fish tissues by using realtime PCR. *Applied Microbiology and Biotechnology* 96, 511－520.

[9] Bastardo, A., Ravelo, C. and Romalde, J.L. (2012b) A polyphasic approach to study the intraspecific diversity of *Yersinia ruckeri* strains isolated from recent outbreaks in salmonid culture. *Veterinary Microbiology* 160, 176－182.

[10] Bastardo, A., Ravelo, C. and Romalde, J.L. (2012c) Multilocus sequence typing reveals high genetic diversity and epidemic population structure for the fish pathogen Yersinia ruckeri. *Environmental Microbiology* 14, 1888－1897.

[11] Bravo, S. and Midtlyng, P.J. (2007) The use of fish vaccines in the Chilean salmon industry 1999－2003. *Aquaculture* 270, 36－42.

[12] Bridle, A.R., Koop, B.F. and Nowak, B.F. (2012) Identification of surrogates of protection against yersiniosis in immersion vaccinated Atlantic salmon. *PLoS One* 7 (7): e40841.

[13] Brunt, J. and Austin, B. (2007) The development of probiotics for the control of multiple bacterial diseases of rainbow trout, *Oncorhynchus mykiss* (Walbaum). *Journal of Fish Diseases* 30, 573－579.

[14] Bullock, G.L., Stuckley, H.M. and Shotts, E.B. (1978) Enteric redmouth bacterium: comparison of isolates from different geographic areas. Journal of Fish Diseases 1, 351－356.

[15] Busch, R.A. (1978) Enteric redmouth disease (Hagerman strain). *Marine Fisheries Review* 40, 42－51.

[16] Busch, R.A. and Lingg, A.J. (1975) Establishment of an asymptomatic carrier strate

infection of enteric redmouth disease (Hagerman Strain) in rainbow trout (*Salmo giardneri*). *Journal of the Fisheries Research Board of Canada* 32, 2429 - 2432.

[17] Calvez, S., Gantelet, H., Blanc, G., Douet, D.G. and Daniel, P. (2014) *Yersinia ruckeri* biotypes 1 and 2 in France: presence and antibiotic susceptibility. *Diseases of Aquatic Organisms* 109, 117 - 126.

[18] Calvez, S., Mangion, C., Douet, D. G. and Daniel, P. (2015) Pulsed-field gel electrophoresis and multi locus sequence typing for characterizing genotype variability of *Yersinia ruckeri* isolated from farmed fish in France. *Veterinary Research* 46, 1 - 13.

[19] Capkin, E. and Altinok, I. (2009) Effects of dietary probiotic supplementations on prevention/treatment of yersiniosis disease. *Journal of Applied Microbiology* 106, 1147 - 1153.

[20] Carson, J. and Wilson, T. (2009) *Yersiniosis in Fish*. Australia and New Zealand Standard Diagnostic Procedure, Committee on Animal Health Laboratory Standards (SCAHLS) [of Australia and New Zealand], Animal Health Committee, Australia. Available at: http://www.scahls.org.au/procedures/documents/aqanzsdp/yersiniosis.pdf (accessed 23 November 2016).

[21] Coquet, L., Cosette, P., Quillet, L., Petit, F., Junter, G.A. and Jouenne, T. (2002) Occurrence and phenotypic characterization of *Yersinia ruckeri* strains with biofilm-forming capacity in a rainbow trout farm. *Applied and Environmental Microbiology* 68, 470 - 475.

[22] Cossarini-Dunier, M. (1985) Indirect enzyme-linked immunosorbent assay (ELISA) to titrate rainbow trout serum antibodies against two pathogens: *Yersinia ruckeri* and Egtved virus. *Aquaculture* 49, 197 - 208.

[23] Costa, A.A., Leef, M.J., Bridle, A.R., Carson, J. and Nowak, B.F. (2011) Effect of vaccination against yersiniosis on the relative percent survival, bactericidal and lysozyme response of Atlantic salmon, *Salmo salar*. *Aquaculture* 315, 201 - 206.

[24] Daly, J.G., Lindvik, B. and Stevenson, R.M.W. (1986) Serological heterogeneity of recent isolates of *Yersinia ruckeri* from Ontario and British Columbia. *Diseases of Aquatic Organisms* 1, 151 - 153.

[25] Danley, M.L., Goodwin, A.E. and Killian, H.S. (1999) Epizootics in farm-raised channel catfish, *Ictalurus punctatus* (Rafinesque), caused by the enteric redmouth bacterium *Yersinia ruckeri*. *Journal of Fish Diseases* 22, 451 - 456.

[26] Davies, R.L. (1989) Biochemical and cell-surface characteristics of *Yersinia ruckeri* in relation to the epizootology and pathogenesis of infections in fish. PhD thesis, University of Stirling, Stirling, UK.

[27] Davies, R.L. (1990) O-serotyping of *Yersinia ruckeri* with special emphasis on European isolates. *Veterinary Microbiology* 22, 299 - 307.

[28] Davies, R.L. (1991a) Outer membrane protein profiles of *Yersinia ruckeri*. *Veterinary Microbiology* 26, 125 - 140.

[29] Davies, R.L. (1991b) Clonal analysis of *Yersinia ruckeri* based on biotypes, serotypes and outer membrane protein types. *Journal of Fish Diseases* 14, 221 - 228.

[30] Davies, R.L. (1991c) Virulence and serum-resistance in different clonal groups and serotypes of *Yersinia ruckeri*. *Veterinary Microbiology* 29, 289 - 297.

[31] Davies, R.L. (1991d) *Yersinia ruckeri* produces four ironregulated outer membrane proteins but does not produce detectable siderophores. *Journal of Fish Diseases* 14, 563 - 570.

[32] Davies, R.L. and Frerichs, G.N. (1989) Morphological and biochemical differences among isolates of *Yersinia ruckeri* obtained from wide geographical areas. *Journal of Fish Diseases* 12, 357 - 365.

[33] De Grandis, S.A. and Stevenson, R.M.W. (1982) Variations in plasmid profiles and

growth characteristics of *Yersinia ruckeri* strains. *FEMS Microbiology Letters* 15, 199－202.

[34] De Grandis, S.A. and Stevenson, R.M.W. (1985) Antimicrobial susceptibility patterns and R plasmid mediated resistance of the fish pathogen *Yersinia ruckeri*. *Antimicrobial Agents and Chemotherapy* 27, 938－942.

[35] Deshmukh, S., Raida, M. K., Dalsgaard, I., Chettri, J. K., Kania, P. W. and Buchmann, K. (2012) Comparative protection of two different commercial vaccines against *Yersinia ruckeri* serotype O1 and biotype 2 in rainbow trout (*Oncorhynchus mykiss*). *Veterinary Immunology and Immunopathology* 145, 379－385.

[36] Faraldo-Gómez, J.D. and Sansom, M.S.P. (2003) Acquisition of siderophores in Gram-negative bacteria. *Nature Reviews Molecular Cell Biology* 4, 105－116.

[37] Fernández, L., Secades, P., Lopez, J. R., Márquez, I. and Guijarro, J. A. (2002) Isolation and analysis of a protease gene with an ABC transport system in the fish pathogen *Yersinia ruckeri*: insertional mutagenesis and involvement in virulence. Microbiology 148, 2233－2243.

[38] Fernández, L., Lopez, J.R., Secades, P., Menendez, A. and Guijarro, J.A. (2003) In vitro and in vivo studies of the Yrp1 protease from *Yersinia ruckeri* and its role in protective immunity against enteric redmouth disease of salmonids. *Applied and Environmental Microbiology* 69, 7328－7335.

[39] Fernández, L., Marquez, I., Guijarro, J.A. and Ma, I. (2004) Identification of specific in vivo-induced (ivi) genes in *Yersinia ruckeri* and analysis of ruckerbactin, a catecholate siderophore iron acquisition system. *Applied and Environmental Microbiology* 70, 5199－5207.

[40] Fernández, L., Prieto, M. and Guijarro, J. A. (2007) The iron-and temperature-regulated haemolysin YhlA is a virulence factor of *Yersinia ruckeri*. *Microbiology* 153, 483－489.

[41] Fouz, B., Zarza, C. and Amaro, C. (2006) First description of non-motile *Yersinia ruckeri* serovar I strains causing disease in rainbow trout, *Oncorhynchus mykiss* (Walbaum), cultured in Spain. *Journal of Fish Diseases* 29, 339－346.

[42] Fuller, R. (1989) Probiotics in man and animals. *Journal of Applied Bacteriology* 66, 365－378.

[43] Garcia, J.A., Dominguez, L., Larsen, J.L. and Pedersen, K. (1998) Ribotyping and plasmid profiling of *Yersinia ruckeri*. *Journal of Applied Microbiology* 85, 949－955.

[44] Ghosh, B., Nguyen, T.D., Crosbie, P.B.B., Nowak, B.F. and Bridle, A.R. (2016) Oral vaccination of first feeding Atlantic salmon, *Salmo salar* L., confers greater protection against yersiniosis than immersion vaccination. *Vaccine* 34, 1－10.

[45] Gibello, A., Blanco, M.M., Moreno, M.A., Cutuli, M.T. and Domínguez, L. (1999) Development of a PCR assay for detection of *Yersinia ruckeri* in tissues of inoculated and naturally infected trout. *Applied and Environmental Microbiology* 65, 346－350.

[46] Haig, S.J., Davies, R.L., Welch, T.J., Reese, R.A. and Verner-Jeffreys, D.W. (2011) Comparative susceptibility of Atlantic salmon and rainbow trout to *Yersinia ruckeri*: relationship to O antigen serotype and resistance to serum killing. *Veterinary Microbiology* 147, 155－161.

[47] Horne, M.T. and Barnes, A.C. (1999) Enteric redmouth disease (*Yersinia ruckeri*). In: Woo, P.K.T. and Bruno, D.W. (eds) *Fish Diseases and Disorders, Volume 3: Viral, Bacterial and Fungal Infections*. CAB International, Wallingford, UK, pp. 455－477.

[48] Huang, Y., Runge, M., Michael, G.B., Schwarz, S., Jung, A. and Steinhagen, D. (2013) Biochemical and molecular heterogeneity among isolates of *Yersinia ruckeri* from rainbow trout (*Oncorhynchus mykiss*, Walbaum) in north west Germany. *BMC Veterinary Research* 9: 215.

[49] Irianto, A. and Austin, B. (2002) Probiotics in aquaculture. *Journal of Fish Diseases* 25, 633 - 642.
[50] Ispir, U. and Dorucu, M. (2010) Effect of immersion booster vaccination with *Yersinia ruckeri* extracellular products (ECP) on rainbow trout *Oncorhynchus mykiss*. *International Aquatic Research* 2, 127 - 130.
[51] Ispir, U. and Dorucu, M. (2014) Efficacy of lipopolysaccharide antigen of *Yersinia ruckeri* in rainbow trout by intraperitoneal and bath immersion administration. *Research in Veterinary Science* 97, 271 - 273.
[52] Kawula, T.H., Lelivelt, M.J. and Orndorff, P.E. (1996) Using a new inbred fish model and cultured fish tissue cells to study *Aeromonas hydrophila* and *Yersinia ruckeri* pathogenesis. *Microbial Pathogenesis* 20, 119 - 125.
[53] Keeling, S. E., Johnston, C., Wallis, R., Brosnahan, C. L., Gudkovs, N. and McDonald, W.L. (2012) Development and validation of real-time PCR for the detection of *Yersinia ruckeri*. *Journal of Fish Diseases* 35, 119 - 125.
[54] Kim, D.-H. and Austin, B. (2006) Innate immune responses in rainbow trout (*Oncorhynchus mykiss*, Walbaum) induced by probiotics. *Fish and Shellfish Immunology* 21, 513 - 524.
[55] Kumar, G., Menanteau-Ledouble, S., Saleh, M. and El-Matbouli, M. (2015) *Yersinia ruckeri*, the causative agent of enteric redmouth disease in fish. *Veterinary Research* 46, 1 - 10.
[56] Lucangeli, C., Morabito, S., Caprioli, A., Achene, L., Busani, L. *et al*. (2000) Molecular fingerprinting of strains of *Yersinia ruckeri* serovar O1 and *Photobacterium damsela* subsp. *piscicida* isolated in Italy. *Veterinary Microbiology* 76, 273 - 281.
[57] Mammeri, H., Poirel, L., Nazik, H. and Nordmann, P. (2006) Cloning and functional characterization of the ambler class C β-lactamase of *Yersinia ruckeri*. *FEMS Microbiology Letters* 257, 57 - 62.
[58] Michel, C., Kerouault, B. and Martin, C. (2003) Chloramphenicol and florfenicol susceptibility of fishpathogenic bacteria isolated in France: comparison of minimum inhibitory concentration, using recommended provisory standards for fish bacteria. *Journal of Applied Microbiology* 95, 1008 - 1015.
[59] Millard, C.M., Baiano, J.C.F., Chan, C., Yuen, B., Aviles, F. *et al*. (2012) Evolution of the capsular operon of *Strepto coccus iniae* in response to vaccination. *Applied and Environmental Microbiology* 78, 8219 - 8226.
[60] Newaj-Fyzul, A., Al-Harbi, A.H. and Austin, B. (2014) Review: developments in the use of probiotics for disease control in aquaculture. *Aquaculture* 431, 1 - 11.
[61] O'Leary, P.J. (1977) *Enteric redmouth bacterium of salmonids: a biochemical and serological comparison of selected isolates*. PhD Thesis, Oregon State University, Corvallis, Oregon.
[62] Ohtani, M., Villumsen, K.R., Strøm, H.K. and Raida, M.K. (2014) 3D Visualization of the initial *Yersinia ruckeri* infection route in rainbow trout (*Oncorhynchus mykiss*) by optical projection tomography. *PLoS One* 9(2): e89672.
[63] Ormsby, M.J., Caws, T., Burchmore, R., Wallis, T., Verner-Jeffreys, D.W. and Davies, R.L. (2016) *Yersinia ruckeri* isolates recovered from diseased Atlantic salmon (*Salmo salar*) in Scotland are more diverse than those from rainbow trout (*Oncorhynchus mykiss*) and represent distinct sub-populations. *Applied and Environmental Microbiology* 82, 5785 - 5794.
[64] Orozova, P., Chikova, V. and Sirakov, I. (2014) Diagnostics and antibiotic resistance of *Yersinia ruckeri* strains isolated from trout fish farms in Bulgaria. *International Journal of Development Research* 4, 2727 - 2733.
[65] Post, G. (1987) Pathogenic Yersinia ruckeri, enteric redmouth disease (yersiniosis). In:

Post, G (ed.) *Textbook of Fish Health*. THF Publications, Neptune City, New Jersey, pp. 47 - 51.
[66] Pyle, S. W. and Schill, W. B. (1985) Rapid serological analysis of bacterial lipopolysaccharides by electrotransfer to nitrocellulose. *Journal of Immunological Methods* 85, 371 - 382.
[67] Pyle, S. W., Ruppenthal, T., Cipriano, R. and Shotts, E. B. (1987) Further characterization of biochemical and serological characteristics of *Yersinia ruckeri* from different geographic areas. *Microbios Letters* 35, 87 - 93.
[68] Raida, M. K., Larsen, J. L., Nielsen, M. E. and Buchmann, K. (2003) Enhanced resistance of rainbow trout, *Oncorhynchus mykiss* (Walbaum), against *Yersinia ruckeri* challenge following oral administration of *Bacillus subtilis* and *Bacillus licheniformis* (BioPlus2B). *Journal of Fish Diseases* 26, 495 - 498.
[69] Rintamaki, P., Tellerovo Valtonen, E. and Frerichs, G. N. (1986) Occurrence of *Yersinia ruckeri* infection in farmed whitefish, *Coregonus peled* Gmelin and *Coregonus muksun* Pallas, and Atlantic salmon, *Salmo salar* L., in northern Finland. *Journal of Fish Diseases* 9, 137 - 140.
[70] Rodgers, C.J. (1992) Development of a selective differential medium for the isolation of *Yersinia ruckeri* and its application in epidemiological studies. *Journal of Fish Diseases* 15, 243 - 254.
[71] Rodgers, C.J. (2001) Resistance of *Yersinia ruckeri* to antimicrobial agents in vitro. *Aquaculture* 196, 325 - 345.
[72] Romalde, J.L. and Toranzo, A.E. (1993) Pathological activities of *Yersinia ruckeri*, the enteric redmouth (ERM) bacterium. *FEMS Microbiology Letters* 112, 291 - 299.
[73] Romalde, J. L., Conchas, R. F. and Toranzo, A. E. (1991) Evidence that *Yersinia ruckeri* possesses a high affinity iron uptake system. *FEMS Microbiology Letters* 15, 121 - 125.
[74] Romalde, J.L., Magarinos, B., Barja, J.L. and Toranzo, A.E. (1993) Antigenic and molecular characterization of *Yersinia ruckeri*: proposal for a new intraspecies classification. *Systematic and Applied Microbiology* 16, 411 - 419.
[75] Romalde, J. L., Barja, J. L., Magarinos, B. and Toranzo, A. E. (1994) Starvation-survival processes of the bacterial fish pathogen *Yersinia ruckeri*. *Systematic and Applied Microbiology* 17, 161 - 168.
[76] Romalde, J.L., Magarinos, B., Fouz, B., Bandin, I., Nunez, S. and Toranzo, A.E. (1995) Evaluation of BIONOR Mono-kits for rapid detection of bacterial fish pathogens. *Diseases of Aquatic Organisms* 21, 25 - 34.
[77] Ross, A. J. and Klontz, W. (1965) Oral immunization of rainbow trout (*Salmo gairdneri*) against an etiological agent of 'redmouth disease'. *Journal of the Fisheries Research Board of Canada* 22, 3 - 9.
[78] Ross, A. J., Rucker, R. R. and Ewing, W. H. (1966) Description of a bacterium associated with redmouth disease of rainbow trout (*Salmo gairdneri*). *Canadian Journal of Microbiology* 12, 763 - 770.
[79] Rucker, R.R. (1966) Redmouth disease of rainbow trout (*Salmo giardneri*). *Bulletin de l' Office International des Epizooties* 65, 825 - 830.
[80] Ryckaert, J., Bossier, P., D'Herde, K., Diez-Fraile, A., Sorgeloos, P. *et al*. (2010) Persistence of *Yersinia ruckeri* in trout macrophages. *Fish and Shellfish Immunology* 29, 648 - 655.
[81] Saleh, M., Soliman, H. and El-Matbouli, M. (2008) Loopmediated isothermal amplification as an emerging technology for detection of *Yersinia ruckeri* the causative agent of enteric redmouth disease in fish. *BMC Veterinary Research* 4: 31.
[82] Schill, W. B., Phelps, S. R. and Pyle, S. W. (1984) Multilocus electrophoretic

assessment of the genetic structure and diversity of *Yersinia ruckeri*. *Applied and Environmental Microbiology* 48, 975 - 979.

[83] Scott, C.J.W., Austin, B., Austin, D.A. and Morris, P.C. (2013) Non-adjuvanted flagellin elicits a non-specific protective immune response in rainbow trout (*Oncorhynchus mykiss*, Walbaum) towards bacterial infections. *Vaccine* 31, 3262 - 3267.

[84] Secades, P. and Guijarro, J. A. (1999) Purification and characterization of an extracellular protease from the fish pathogen *Yersinia ruckeri* and effect of culture conditions on production. *Applied and Environmental Microbiology* 65, 3969 - 3975.

[85] Seker, E., Karahan, M., Ispir, U., Cetinkaya, B., Saglam, N. and Sarieyyupoglu, M. (2012) Investigation of *Yersinia ruckeri* infection in rainbow trout (*Oncorhynchus mykiss*, Walbaum 1792) farms by polymerase chain reaction (PCR) and bacteriological culture. *The Journal of the Faculty of Veterinary Medicine*, *University of Kafkas* 18, 913 - 916.

[86] Shah, S.Q.A., Karatas, S., Nilsen, H., Steinum, T.M., Colquhoun, D.J. and Sørum, H. (2012) Characterization and expression of the *gyrA* gene from quinolone resistant *Yersinia ruckeri* strains isolated from Atlantic salmon (*Salmo salar* L.) in Norway. *Aquaculture* 350 - 353, 37 - 41.

[87] Shaowu, L., Di, W., Hongbai, L. and Tongyan, L. (2013) Isolation of *Yersinia ruckeri* strain H01 from farmraised Amur sturgeon *Acipenser schrencki* in China. *Journal of Aquatic Animal Health* 25, 9 - 14.

[88] Sica, M.G., Brugnoni, L.I., Marucci, P.L. and Cubitto, M.A. (2012) Characterization of probiotic properties of lactic acid bacteria isolated from an estuarine environment for application in rainbow trout (*Oncorhynchus mykiss*, Walbaum) farming. *Antonie Van Leeuwenhoek* 101, 869 - 879.

[89] Smith, A.M., Goldring, O.L. and Dear, G. (1987) The production and methods of use of polyclonal antisera to the pathogenic organisms *Aeromonas salmonicida*, *Yersinia ruckeri* and *Renibacterium salmonicida*. *Journal of Fish Biology* 31, 225 - 226.

[90] Sousa, J.A., Magariños, B., Eiras, J.C., Toranzo, A.E. and Romalde, J.L. (2001) Molecular characterization of Portuguese strains of *Yersinia ruckeri* isolated from fish culture systems. *Journal of Fish Diseases* 24, 151 - 159.

[91] Stevenson, R. M. W. and Airdrie, D. W. (1984) Isolation of *Yersinia ruckeri* bacteriophages. *Applied and Environmental Microbiology* 47, 1201 - 1205.

[92] Stevenson, R.M.W. and Daly, J.G. (1982) Biochemical and serological characteristics of Ontario isolates of *Yersinia ruckeri*. *Canadian Journal of Fisheries and Aquatic Sciences* 39, 870 - 876.

[93] Stock, I., Henrichfreise, B. and Wiedemann, B. (2002) Natural antibiotic susceptibility and biochemical profiles of *Yersinia enterocolitica*-like strains: *Y. bercovieri*, *Y. mollaretii*, *Y. aldovae* and '*Y. ruckeri*'. *Journal of Medical Microbiology* 51, 56 - 69.

[94] Strom-Bestor, M., Mustamaki, N., Heinikainen, S., Hirvela-Koski, V., Verner-Jeffreys, D. and Wiklund, T. (2010) Introduction of *Yersinia ruckeri* biotype 2 into Finnish fish farms. *Aquaculture* 308, 1 - 5.

[95] Tatner, M. F. and Horne, M. (1985) The effects of vaccine dilution, length of immersion time, and booster vaccinations on the protection levels induced by direct immersion vaccination of brown trout, *Salmo trutta*, with *Yersinia ruckeri* (ERM) vaccine. *Aquaculture* 46, 11 - 18.

[96] Temprano, A., Yugueros, J., Hernanz, C., Sunchez, M., Berzal, B. and Luengo, J. M. (2001) Rapid identification of *Yersinia ruckeri* by PCR amplification of *yruI* - *yruR* quorum sensing. *Journal of Fish Diseases* 24, 253 - 261.

[97] Temprano, A., Riaño, J., Yugueros, J., González, P., de Castro, L. *et al*. (2005)

Potential use of a *Yersinia ruckeri* O1 auxotrophic *aroA* mutant as a live attenuated vaccine. *Journal of Fish Diseases* 28, 419 - 427.

[98] Thorsen, B.K., Enger, O., Norland, S. and Hoff, K.A. (1992) Long-term starvation survival of *Yersinia ruckeri* at different salinities studied by microscopical and flow cytometric methods. *Applied and Environmental Microbiology* 58, 1624 - 1628.

[99] Tinsley, J.W., Austin, D.A., Lyndon, A.R. and Austin, B. (2011a) Novel non-motile phenotypes of *Yersinia ruckeri* suggest expansion of the current clonal complex theory. *Journal of Fish Diseases* 34, 311 - 317.

[100] Tinsley, J.W., Lyndon, A.R. and Austin, B. (2011b) Antigenic and cross-protection studies of biotype 1 and biotype 2 isolates of *Yersinia ruckeri* in rainbow trout, *Oncorhynchus mykiss* (Walbaum). *Journal of Applied Microbiology* 111, 8 - 16.

[101] Tobback, E. (2009) Early pathogenesis of *Yersinia ruckeri* infections in rainbow trout (*Oncorhynchus mykiss*, Walbaum). PhD thesis, Ghent University, Ghent, Belgium.

[102] Tobback, E., Decostere, A., Hermans, K., Haesebrouck, F. and Chiers, K. (2007) Review of *Yersinia ruckeri* infections in salmonid fish. *Journal of Fish Diseases* 30, 257 - 268.

[103] Tobback, E., Decostere, A., Hermans, K., Ryckaert, J., Duchateau, L. et al. (2009) Route of entry and tissue distribution of *Yersinia ruckeri* in experimentally infected rainbow trout *Oncorhynchus mykiss*. *Diseases of Aquatic Organisms* 84, 219 - 228.

[104] Tobback, E., Hermans, K., Decostere, A., Van Den Broeck, W., Haesebrouck, F. and Chiers, K. (2010a) Interactions of virulent and avirulent *Yersinia ruckeri* strains with isolated gill arches and intestinal explants of rainbow trout *Oncorhynchus mykiss*. *Diseases of Aquatic Organisms* 90, 175 - 179.

[105] Tobback, E., Decostere, A., Hermans, K., Van den Broeck, W., Haesebrouck, F. and Chiers, K. (2010b) *In vitro* markers for virulence in *Yersinia ruckeri*. *Journal of Fish Diseases* 33, 197 - 209.

[106] Verner-Jeffreys, D.W. and Taylor, N.J. (2015) *SARF100 — Review of Freshwater Treatments Used in the Scottish Freshwater Rainbow Trout Aquaculture Industry*. Cefas (Centre for Environment, Fisheries and Aquaculture Science, Weymouth Laboratory, Weymouth, UK) contract report C6175A commissioned and published by SARF. Scottish Aquaculture Research Forum, Pitlochry, UK. Available at: http://www.sarf.org.uk/cms-assets/documents/208213 - 793666.sarf100.pdf (accessed 23 November 2016).

[107] Vuillaume, A., Brun, R., Chene, P., Sochon, E. and Lesel, R. (1987) First isolation of *Yersinia ruckeri* from sturgeon, *Acipenser baeri* Brandt, in south west of France. *Bulletin of the European Association of Fish Patho logists* 7, 18 - 19.

[108] Waltman, W.D. and Shotts, E.B. (1984) A medium for the isolation and differentiation of *Yersinia ruckeri*. *Canadian Journal of Fisheries and Aquatic Sciences* 41, 804 - 806.

[109] Welch, T.J. and LaPatra, S. (2016) *Yersinia ruckeri* lipopolysaccharide is necessary and sufficient for eliciting a protective immune response in rainbow trout (*Oncorhynchus mykiss*, Walbaum). *Fish and Shellfish Immunology* 49, 420 - 426.

[110] Welch, T.J., Verner-Jeffreys, D.W., Dalsgaard, I., Wiklund, T., Evenhuis, J.P. *et al*. (2011) Independent emergence of *Yersinia ruckeri* biotype 2 in the United States and Europe. *Applied and Environmental Microbiology* 77, 3493 - 3499.

[111] Wheeler, R.W., Davies, R.L., Dalsgaard, I., Garcia, J., Welch, T.J. *et al*. (2009) *Yersinia ruckeri* biotype 2 isolates from mainland Europe and the UK likely represent different clonal groups. *Diseases of Aquatic Organisms* 84, 25 - 33.